Class Time

BEFORE CLASS

- Review your notes from the previous class session.

- Read the section(s) of your textbook that will be covered in class to get familiar with the material. Read these sections carefully. Skimming may result in your not understanding some of the material and your inability to do the homework. If you do not understand something in the text, reread it more thoroughly or seek assistance.

DURING CLASS

- Pay attention and try to understand every point your instructor makes.

- Take good notes.

- Ask questions if you do not understand something. It is best to ask questions during class. Chances are that someone else has the same question, but may not be comfortable asking it. If you feel that way also, then write your question in your notebook and ask your instructor after class or see the instructor during office hours.

AFTER CLASS

- Review your notes as soon as possible after class. Insert additional steps and comments to help clarify the material.

- Reread the section(s) of your textbook. After reading through an example, cover it up and try to do it on your own. Do the practice problem that is paired with the example. (The answers to the practice problems are given in the back of the textbook.)

Homework

The best way to learn math is by doing it. Homework is designed to help you learn and apply concepts and to master related skills.

TIPS FOR DOING HOMEWORK

- Do your homework the *same day* that you have class. Keeping up with the class requires you to do homework regularly rather than "cramming" right before tests.

- Review the section of the textbook that corresponds to the homework.

- Review your notes.

- If you get stuck on a problem, look for a similar example in your textbook or notes.

- Write a question mark next to any problems that you just cannot figure out. Get help from your instructor or the tutoring center or call someone from your study group.

- Check your answer after each problem. The answers to the odd-numbered problems are in the back of the textbook. If you are assigned even-numbered problems, try the odd-numbered problem first and check the answer. If your answer is correct, then you should be able to do the even-numbered problem correctly.

Form a study group. A study group provides an opportunity to discuss class material and homework problems. Find at least two other people in your class who are committed to being successful. Exchange contact information and plan to meet or work together regularly throughout the semester either in person or via e-mail or phone.

Use your book's study resources. Additional resources and support materials to help you succeed are available with this book.

INCLUDED WITH YOUR TEXT

- The resources on the **Pass the Test CD** will help you prepare for your tests. After you've taken the Posttest at the end of each chapter in the textbook, check your work by watching an instructor work through the full solutions to all of the Posttest exercises on the Pass the Test CD. You will also find vocabulary flashcards and additional study tips on this CD.

ADDITIONAL RESOURCES

The following resources are also available through your school bookstore or www.mypearsonstore.com:

- *Student's Solutions Manual*
- *Video Lectures on CD*
- *Video Lectures on DVD*
- *Worksheets for Classroom or Lab Practice*
- *MathXL® Tutorials on CD*
- *MathXL®*
- *MyMathLab®*

Full descriptions are available in the preface of the textbook.

Notebook and Note Taking

Taking good notes and keeping a neat, well-organized notebook are important factors in being successful.

YOUR NOTEBOOK

Use a loose-leaf binder divided into four sections:

1. notes
2. homework
3. graded tests and quizzes
4. handouts

TAKING NOTES

- Copy all important information written on the board. Also, write down all points that are not clear to you so that you can discuss them with your instructor, a tutor, or your study group.

- Write explanations of what you are doing in your own words next to each step of a practice problem.

- Listen carefully to what your instructor emphasizes and make note of it.

Math Study Skills

Your overall success in mastering the material that this textbook covers depends on you. The key is to be ***committed*** to doing your best in this course. This commitment means dedicating the time needed to study math and to do your homework.

In order to succeed in math, you must know how to study it. The goal is to study math so that you understand and not just memorize it. The following tips and strategies will help you develop good study habits that lead to understanding.

General Tips

Attend every class. Be on time. If you must miss class, be sure to get a copy of the notes and any handouts. (Get notes from someone in your class who takes good, neat notes.)

Manage your time. Work, family, and other commitments place a lot of demand on your time. To be successful in college, you must be able to devote time to study math every day. Writing out a weekly schedule that lists your class schedule, work schedule, and all other commitments with times that are not flexible will help you to determine when you can study.

Do not wait to get help. If you are having difficulty, get help immediately. Since the material presented in class usually builds on previous material, it is very easy to fall behind. Ask your instructor if he or she is available during office hours, or get help at the tutoring center on campus.

Instructor Contact Information
Name:
Office Hours:
Office Location:
Phone Number:
E-mail Address:

Campus Tutoring Center
Location:
Hours:

Tests

Tests are a source of anxiety for many students. Being well prepared to take a test can help ease anxiety.

BEFORE A TEST

- Review your notes and the sections of the textbook that will be covered on the test.
- Read through the Key Concepts and Skills at the end of the chapter in the textbook and your own summary from your notes.
- Do additional practice problems. Select problems from your homework to rework. The textbook also contains Mixed Practice and Review Exercises that provide opportunities to strengthen your skills
- Use the Posttest at the end of the chapter as your practice test. While taking the practice test, do not refer to your notes or the textbook for help. Keep track of how long it takes to complete the test. Check your answers. If you cannot complete the practice test within the time you are allotted for the real test, you need additional practice in the tutoring center to speed up.

DURING A TEST

- Read through the test before starting.
- If you find yourself panicking, relax, take a few slow breaths, and try to do some of the problems that seem easy.
- Do the problems you know how to do first, and then go back to the ones that are more difficult.
- Watch your time. Do not spend too much time on any one problem. If you get stuck while working on a problem, skip it and move on to the next problem.
- Check your work, if there is time. Correct any errors you find.

AFTER A TEST

- When you get your test back, look through all of the problems.
- On a separate sheet of paper, do any problems that you missed. Use your notes and textbook, if necessary.
- Get help from your instructor or a tutor if you cannot figure out how to do a problem, or set up a meeting with your study group to go over the test together. Make sure you understand your errors.
- Attach the corrections to the test and place it in your notebook.

Basic Mathematics through Applications

Fourth Edition

Geoffrey Akst   Sadie Bragg
Borough of Manhattan Community College,
City University of New York

Boston • San Francisco • New York • London • Toronto
Sydney • Tokyo • Singapore • Madrid • Mexico City
Munich • Paris • Cape Town • Hong Kong • Montreal

Editorial Director	Christine Hoag
Editor in Chief	Maureen O'Connor
Production Manager	Ron Hampton
Executive Project Manager	Kari Heen
Associate Editor	Joanne Doxey
Developmental Editor	Laura Wheel
Senior Designer	Barbara T. Atkinson
Text and Cover Design	Leslie Haimes
Composition	Pre-Press PMG
Media Producer	Ashley O'Shaughnessy
Software Development	TestGen: Mary Durnwald; MathXL: Jozef Kubit
Marketing Manager	Michelle Renda
Marketing Coordinator	Nathaniel Koven
Senior Prepress Supervisor	Caroline Fell
Manufacturing Manager	Evelyn Beaton
Senior Media Buyer	Ginny Michaud
Cover Photo	© Peter Stroumtos/Alamy

About the Cover: Well-designed sunglasses are not just cool—they also make it easy to see the world more clearly. Akst and Bragg's distinctive side-by-side example/practice exercise format encourages students to practice as they learn, while the authors' use of motivating applications connects mathematics to real life. With clear and simple explanations that make mathematics understandable, the Akst/Bragg series brings the world of developmental mathematics into focus for students.

Photo Credits: p. 1: PhotoDisc (PP); p. 11, 19, 27, 37, 42, 74, 250, 279, 284, 312, 472: Digital Vision; p. 11, 449: Jon Feingersh/zefa/CORBIS; p. 12: AP Photo/Al Behrman; p.12, 111, 509: PhotoLink; p. 13: Neal Preston/CORBIS; p. 29: epa/CORBIS; p. 37, 218, 472: NASA; p. 55, 175, 238, 257: Getty Editorial; p. 61: Photographer's Choice RF/Getty Royalty Free; p. 75: AP Photo; p. 81: NASA Headquarters; p. 82, 156, 239: National Geographic/Getty Royalty Free; p. 85: Radius Images/Punchstock Royalty Free; p. 111: Jeff Haynes/ Pool/Reuters/CORBIS; p. 156: Jaysen F. Snow/Airliners.net; p. 156: DAJ/Getty Royalty Free; p. 161: Blend Images/Getty Royalty Free; p. 162, 170, 174, 192, 255, 278, 300, 335, 346: PhotDisc; p. 173: Photo Researchers; p. 174: Indianapolis Speedway; p. 209: Peter Guttman/CORBIS; p. 219: fStop Getty Royalty Free; p. 245: Hisham F. Ibrahim; p. 250, 472: Stockbyte/Getty Royalty Free; p. 255: NASA/Johnson Space Center; p. 257: Purestock/Getty Royalty Free; p. 259: PhotoDisc Red; p. 265: Bob Krist/CORBIS; p. 278: Blend Images/Getty Royalty Free; p. 287: AP Wideworld Photos; p. 291, 295, 372, 447: CORBIS Royalty Free; p. 299, 307, 311, 315, 366, 441, 479, 481, 483: Shutterstock; p. 324: G. Baden/zefa/CORBIS; p. 325: Scot Frei/CORBIS; p. 330, 386, 460: Bettmann/CORBIS; p. 331: DAJ/Getty Royalty Free; p. 341: Rich Kane/Icon SMI/CORBIS; p. 344: Pete Saloutos/CORBIS; p. 345: Nick Koudis; p. 353: Sandro Vannini/CORBIS; p. 364: Rod Catanach, Woods Hole Oceanographic Institution; p. 377: Wu hong/epa/CORBIS; p. 378: ABC/ Courtesy: Everett Collection; p. 419: Comstock IMAGES; p. 420: Jules Frazier; p. 435: Duomo/ CORBIS; p. 440: R. Morley/PhotoLink; p. 456: Warner Brothers/courtesy Everett Collection; p. 457: Jerry Cooke/CORBIS; p. 460: Library of Congress; p. 462: Reprinted by permission of the publisher from Struik, Dirk J., *A Concise History of Mathematics*, © 1967, Dover Publications Inc., NY; p. 472: Stockbyte/Getty Royalty Free; p. 477: Smithsonian Inst; p. 478: Charles Platiau/Reuters/CORBIS; p. 480: Nik Wheeler/CORBIS; p. 510: Charles O'Rear/CORBIS

Library of Congress Cataloging-in-Publication Data

Akst, Geoffrey.

Basic mathematics through applications / Geoffrey Akst, Sadie Bragg.—4th ed.

p. cm.

Includes index.

ISBN-13: 978-0-321-50011-3 ISBN-10: 0-321-50011-3 (student ed. : alk. paper)

1. Mathematics—Textbooks. I. Bragg, Sadie II. Title.

QA39.3.A47 2008

510—dc22 2007060134

4 5 6 7 8 9 10—CRK—11 10 09

Contents

Preface

From the Authors

Our goal in writing *Fundamental Mathematics through Applications,* Fourth Edition, was to help motivate students and to establish a strong foundation for their success in a developmental mathematics program. Our text provides the appropriate coverage and review of whole numbers, fractions, decimals, ratio and proportion, and percents. Compared to other texts on the market, we have introduced algebra earlier in our text to stress the importance of algebraic concepts and skills.

For all topics covered in this text, we have carefully selected applications that we believe are relevant, interesting, and motivating. This thoroughly integrated emphasis on applications reflects our view that college students need to master basic mathematics not so much for its own sake but rather to be able to apply this understanding to their everyday lives and to the demands of subsequent college courses.

Our goal throughout the text has been to address many of the issues raised by the American Mathematical Association of Two-Year Colleges and the National Council of Teachers of Mathematics by writing a flexible, approachable, and readable text that reflects:

- an emphasis on applications that model real-world situations
- explanations that foster conceptual understanding
- exercises with connections to other disciplines
- an early introduction to algebraic concepts and skills
- appropriate use of technology
- an emphasis on estimation
- the integration of geometric visualization with concepts and applications
- exercises in student writing and groupwork that encourage interactive and collaborative learning
- the use of real data in charts, tables, and graphs

This text is part of a series that includes the following books:

- *Fundamental Mathematics through Applications,* Fourth Edition
- *Basic Mathematics through Applications,* Fourth Edition
- *Introductory Algebra through Applications,* Second Edition
- *Intermediate Algebra through Applications,* Second Edition
- *Introductory and Intermediate Algebra through Applications,* Second Edition

The following key content and key features stem from our strong belief that mathematics is logical, useful, and fun.

Geoffrey Akst and Sadie Bragg

Key Content

Applications One of the main reasons to study mathematics is its application to a wide range of disciplines, to a variety of occupations, and to everyday situations. Each chapter begins with a real-world application to show the usefulness of the topic under discussion and to motivate student interest. These opening applications vary widely from fractions and cooking to percents and surveys (see pages 85 and 287). Applications, appropriate to the section content, are highlighted in section exercise sets with an Applications heading (see pages 54, 110, and 278).

Concepts Explanations in each section foster intuition by promoting student understanding of underlying concepts. To stress these concepts, we include discovery-type exercises on reasoning and pattern recognition that encourage students to be logical in their problem-solving techniques, promote self-confidence, and allow students with varying learning styles to be successful (see pages 32, 289, and 330).

Skills Practice is necessary to reinforce and maintain skills. In addition to comprehensive chapter problem sets, chapter review exercises include mixed applications that require students to choose among and use skills covered in that chapter (see pages 215, 254, and 330).

Writing Writing both enhances and demonstrates students' understanding of concepts and skills. In addition to the user-friendly worktext format, open-ended questions throughout the text give students the opportunity to explain their answers in full sentences. Students can build on these questions by keeping individual journals (see pages 47, 103, and 319).

Estimation Students need to develop estimation skills to distinguish between reasonable and unreasonable solutions, as well as to check their solutions. The chapters on whole numbers (Chapter 1), fractions (Chapter 2), decimals (Chapter 3), and percents (Chapter 6) cover these skills (see pages 1, 85, 161, and 287).

Use of Geometry Students need to develop their abilities to visualize and compare objects. Throughout this text, students have opportunities to use geometric concepts and drawings to solve problems (see pages 36, 228, and 245). In addition, Chapter 11 is dedicated to geometry topics (see page 481).

Use of Technology Each student should be familiar with a range of computational techniques—mental, paper-and-pencil, and calculator arithmetic—depending on the problem and the student's level of mathematical preparation. This text includes optional calculator inserts (see pages 63, 192, and 309), which provide explanations for calculator techniques. These inserts also feature paired side-by-side examples and practice exercises, as well as a variety of optional calculator exercises in the section exercise sets that use the power of scientific calculators to perform arithmetic operations (see pages 195, 208, and 280). Both calculator inserts and calculator exercises are indicated by a calculator icon.

Key Features

Pretests and Posttests To promote individualized learning—particularly in a self-paced or lab environment—pretests and posttests help students gauge their level of understanding of chapter topics both at the beginning and at the end of each chapter. The pretests and posttests also allow students to target topics for which they may need to do extra work to achieve mastery. All answers to pretests and posttests are given in the answer section of the student edition. Students can watch an instructor working through the full solutions to the Posttest exercises on the Pass the Test CD.

Section Objectives At the beginning of each section, clearly stated learning objectives help students and instructors identify and organize individual competencies covered in the upcoming content.

Side-by-Side Example/Practice Format A distinctive side-by-side format pairs each numbered example with a corresponding practice exercise, encouraging students to get actively involved in the mathematical content from the start. Examples are immediately followed by solutions so that students can have a ready guide to follow as they work (see pages 17, 91, and 179).

Tips Throughout the text, students find helpful suggestions for understanding concepts, skills, or rules, and advice on avoiding common mistakes (see pages 93, 166, and 275).

Cultural Notes To show how mathematics has evolved over the centuries—in many cultures and throughout the world—each chapter features a compelling Cultural Note that investigates and illustrates the origins of mathematical concepts. Cultural notes give students further evidence that mathematics grew out of a universal need to find efficient solutions to everyday problems. Diverse topics include the evolution of digit notation, the popularization of decimals, and the role that taxation played in the development of the percent concept (see pages 176, 265, and 301).

Mindstretchers For every appropriate section in the text, related investigation, critical thinking, mathematical reasoning, pattern recognition, and writing exercises—along with corresponding group work and historical connections—are incorporated into one broad-ranged problem set called mindstretchers. Mindstretchers target different levels and types of student understanding and can be used for enrichment, homework, or extra credit (see pages 112, 196, and 313).

For Extra Help These boxes, found at the top of the first page of every section's exercise set, direct students to helpful resources that will aid in their study of the material.

Key Concepts and Skills At the end of each chapter, a comprehensive chart organized by section relates the key concepts and skills to a corresponding description and example, giving students a unique tool to help them review and translate the main points of the chapter.

Chapter Review Exercises Following the Key Concepts and Skills at the end of each chapter, a variety of relevant exercises organized by section helps students test their comprehension of the chapter content. As mentioned earlier, included in these exercises are mixed applications, which give students an opportunity to practice their reasoning skills by requiring them to choose and apply an appropriate problem-solving method from several previously presented (see pages 218, 253, and 283).

Cumulative Review Exercises At the end of Chapter 2, and for every chapter thereafter, Cumulative Review Exercises help students maintain and build on the skills learned in previous chapters.

What's New in the Fourth Edition?

NEW Math Study Skills Foldout This full-color foldout provides students with tips on organization, test preparation, time management, and more.

NEW Scientific Notation Appendix A brief appendix on scientific notation introduces the basic mathematics student to this alternative way of writing numbers.

NEW Tab Your Way to Success Guide The Tab Your Way to Success guide provides students with color-coded Post-it® tabs to mark important pages of the text that they may need to revisit for review work, test preparation, instructor help, and so forth.

NEW Mathematically Speaking Exercises Located at the beginning of nearly every section's exercise set, these new exercises have been added to help students understand and use standard mathematical vocabulary.

NEW Mixed Practice Exercises Mixed Practice exercises, located in nearly every section's exercise set, reinforce the student's knowledge of topics and problem solving skills covered in the section.

Updated Exercise Sets This revision includes over 1,000 new exercises, including the new Mathematically Speaking and Mixed Practice exercises.

NEW Pass the Test CD Included with every new copy of the book, the Pass the Test CD includes video footage of an instructor working through the complete solutions for all Posttest exercises for each chapter, vocabulary flashcards, interactive Spanish glossary, and additional tips on improving time management.

What Supplements Are Available?

For a complete list of the supplements and study aids that accompany *Basic Mathematics through Applications*, Fourth Edition, see pp. xi through xiii.

Acknowledgments

We are grateful to everyone who has helped to shape this textbook by responding to questionnaires, participating in telephone surveys and focus groups, reviewing the manuscript, and using the text in their classes. We wish to thank all of them, especially the following diary and manuscript reviewers who provided feedback for this revision: Yon Kim, *Passaic County Community College;* James Morgan, *Holyoke Community College;* Margaret Patin, *Vernon College;* Lee H. LaRue, *Paris Junior College;* Susan Santolucito, *Delgado Community College;* Marcia Swope, *Santa Fe Community College;* Carol Marinas, *Barry College;* Kate Horton, *Portland Community College;* James Cochran, *Kirkwood Community College;* Sylvia Brown, *Mountain Empire Community College;* and LeAnn L. Lotz-Todd, *Metropolitan Community College–Longview.* In addition, we would like to extend our gratitude to our accuracy checkers: Robert Holt, *Queensborough Community College, City University of New York;* Sharon Testone, *Onondaga Community College;* Sharon O'Donnell, *Chicago State University;* Janis Cimperman, *St. Cloud State University;* and Perian Herring, *Okaloosa-Walton College.*

Writing a textbook requires the contributions of many individuals. Special thanks go to Greg Tobin, our publisher at Addison-Wesley, for encouraging and supporting us throughout the entire process. We are very grateful to Laura Wheel, our developmental editor, who assisted us in more ways than one could imagine and whose unwavering support made our work more manageable. We thank Kari Heen for her patience and tact, Michelle Renda and Maureen O'Connor for keeping us abreast of market trends, Joanna Doxey for attending to the endless details connected with the project, Ron Hampton and Laura Houston for their support throughout the production process, Leslie Haimes for the text and cover design, and the entire Addison-Wesley developmental mathematics team for helping to make this text one of which we are very proud.

Geoffrey Akst
gakst@nyc.rr.com

Sadie Bragg
sbragg@bmcc.cuny.edu

For Mag Dora Chavis and Harriet Young, and to the memory of Maxine Jefferson and Anne Akst

Student Supplements

Student's Solutions Manual

By Beverly Fusfield

- Provides detailed solutions to the odd-numbered exercises in each exercise set and solutions to all chapter pretests and posttests, practice exercises, review exercises, and cumulative review exercises

ISBN-10: 0-321-50051-2 ISBN-13: 978-0-321-50051-9

New Video Lectures on CD or DVD

- Complete set of digitized videos on CD-ROM or DVD for students to use at home or on campus
- Includes a full lecture for each section of the text
- Optional captioning in English is available

CD: ISBN-10: 0-321-49871-2 and
ISBN-13: 978-0-321-49871-7

DVD: ISBN-10: 0-321-53580-4 and
ISBN-13: 978-0-321-53580-1

Instructor Supplements

Annotated Instructor's Edition

- Provides answers to all text exercises in color next to the corresponding problems
- Includes teaching tips

ISBN-10: 0-321-50058-X ISBN-13: 978-0-321-50058-8

Instructor's Solutions Manual

- Provides complete solutions to even-numbered section exercises
- Contains answers to all Mindstretcher problems

ISBN-10: 0-321-50052-0 ISBN-13: 978-0-321-50052-6

Instructor and Adjunct Support Manual

- Includes resources designed to help both new and adjunct faculty with course preparation and classroom management, including sample syllabi, tips for using supplements and technology, and useful external resources
- Offers helpful teaching tips correlated to the sections of the text

ISBN-10: 0-321-50054-7 ISBN-13: 978-0-321-50054-0

Student Supplements (*continued*)

New Pass the Test CD

Automatically included with the book, this CD-ROM contains

- Video footage of an instructor working through the complete solutions for all Posttest problems
- Vocabulary flashcards
- An interactive Spanish glossary
- A short video offering tips on time management

ISBN-10: 0-321-53965-6 ISBN-13: 978-0-321-53965-6

New Worksheets for Classroom or Lab Practice

By Mark Stevenson, *Oakland Community College*

- Provides one worksheet for each section of the text, organized by section objective
- Each worksheet lists the associated objectives from the text, provides fill-in-the-blank vocabulary practice, and exercises for each objective.

ISBN-10: 0-321-53631-2 ISBN-13: 978-0-321-53631-0

MathXL® Tutorials on CD

- Provides algorithmically-generated practice exercises that correlate at the objective level to the exercises in the textbook.
- Includes an example and a guided solution for every exercise; selected exercises also include a video clip
- Provides helpful feedback for incorrect answers and generates printed summaries of students' progress

ISBN-10: 0-321-50055-5 ISBN-13: 978-0-321-50055-7

Math Tutor Center

- Staffed by qualified mathematics instructors
- Provides tutoring on examples and odd-numbered exercises from the textbook through a registration number with a new textbook or if purchased separately
- Accessible via toll-free telephone, toll-free fax, e-mail, or the Internet

www.aw-bc/tutorcenter

Instructor Supplements (*continued*)

Printed Test Bank

By Kay Haralson, *Austin Peay State University* and Nancy Matthews, *Montgomery Central Middle School*

- Contains three free-response and one multiple-choice test forms per chapter, and two final exams

ISBN 10: 0-321-50050-4 ISBN-13: 978-0-321-50050-2

PowerPoint Lecture Slides (available online)

- Present key concepts and definitions from the text

New Active Learning Lecture Slides (available online)

- Provide several multiple-choice questions for each section of the book, allowing instructors to quickly assess mastery of material in class
- Available in PowerPoint for use with classroom response systems

TestGen® (available online)

- *New* Now includes a premade test for each chapter that has been correlated problem-by-problem to the chapter tests in the book
- Enables instructors to build, edit, print, and administer tests using a computerized bank of questions developed to cover all text objectives
- Algorithmically based, TestGen allows instructors to create multiple but equivalent versions of the same question or test with the click of a button.
- Instructors can also modify test bank questions or add new questions.
- Tests can be printed or administered online.
- Software and test bank are available for download from the Pearson Education online catalog.
- Available on a dual-platform Windows/Macintosh CD-ROM.

Student Supplements *(continued)*

InterAct Math Tutorial Website
www.interactmath.com

- Get practice and tutorial help online
- Provides algorithmically generated practice exercises that correlate directly to the textbook exercises
- Retry an exercise as many times as desired with new values each time for unlimited practice and mastery
- Every exercise is accompanied by an interactive guided solution that gives the student helpful feedback when an incorrect answer is entered
- View the steps of a worked-out sample problem similar to the one that has been worked on

Instructor Supplements *(continued)*

Adjunct Support Center

The Math Adjunct Support Center is staffed by qualified mathematics instructors with over 50 years combined experience at both the community college and university level. Assistance is provided for faculty in the following areas:

- Suggested syllabus consultation
- Tips on using materials packaged with the text
- Book-specific content assistance
- Teaching suggestions including advice on classroom strategies

For more information visit
www.aw-bc.com/tutorcenter/math-adjunct.html

Available for Students and Instructors

MathXL® MathXL® is a powerful online homework, tutorial, and assessment system that accompanies Pearson Education's textbooks in mathematics or statistics. With MathXL, instructors can create, edit, and assign online homework and tests using algorithmically-generated exercises correlated at the objective level to the textbook. Instructors can also create and assign their own online exercises and import TestGen tests for added flexibility. All student work is tracked in MathXL's online gradebook. Students can take chapter tests in MathXL and receive personalized study plans based on their test results. The study plan diagnoses weaknesses and links students directly to tutorial exercises for the objectives they need to study and retest. Students can also access supplemental video clips and animations directly from selected exercises.

MyMathLab® MyMathLab is a series of text-specific, easily customizable online courses for Pearson Education's textbooks in mathematics and statistics. Powered by CourseCompass™ (our online teaching and learning environment) and MathXL® (our online homework, tutorial, and assessment system), MyMathLab gives you the tools you need to deliver all or a portion of your course online, whether your students are in a lab setting or working from home. MyMathLab provides a rich and flexible set of course materials, featuring free-response exercises that are algorithmically generated for unlimited practice and mastery. Students can also use online tools, such as video lectures, animations, and a multimedia textbook, to independently improve their understanding and performance. Instructors can use MyMathLab's homework and test managers to select and assign online exercises correlated directly to the textbook, and they can also create and assign their own online exercises and import TestGen tests for added flexibility. MyMathLab's online gradebook—designed specifically for mathematics and statistics—automatically tracks students' homework and test results and gives the instructor control over how to calculate final grades. Instructors can also add offline (paper-and-pencil) grades to the gradebook. MyMathLab is available to qualified adopters. For more information, visit our Web site at www.mymathlab.com or contact your sales representative.

Feature Walk-Through

Chapter Openers The focus of this textbook is applications, and you will find them everywhere. Each chapter opener introduces students to the material that lies ahead through an interesting real-life application that grabs students' attention and helps them understand the relevance of topics they are learning.

CHAPTER 2

Fractions

Fractions and Cooking

Using recipes in cooking has a very long and complicated history that goes back thousands of years. For instance, archaeologists have found Mesopotamian recipes written on clay tablets dating from 1700 B.C. Until the Industrial Revolution, at which time standard measurements and precise cooking directions were introduced, recipes gave just a list of ingredients and a general description for cooking a dish.

Today, it is much easier to follow recipes and to cook satisfying dishes because standard measurements and detailed instructions are given in recipes. Many recipes give fractional amounts of ingredients. As a result, cooks need to know how to use fractions.

For example, consider the ingredients given in a recipe for Italian-style meatloaf. The recipe, which serves 8 people, calls for $1\frac{1}{2}$ pounds of ground meat, of which $\frac{1}{3}$ is pork and $\frac{2}{3}$ is beef. To follow this recipe, a cook uses multiplying fractions to determine that for $1\frac{1}{2}$ pounds of ground meat, the amount of ground pork needed is $\frac{1}{3} \times 1\frac{1}{2}$, or $\frac{1}{2}$ pound, and the amount of ground beef needed is $\frac{2}{3} \times 1\frac{1}{2}$, or 1 pound. Multiplying fractions is also used if the cook wants to increase or decrease the number of servings that the recipe yields. (*Source:* http://www.foodtimeline.org)

85

Pretests Pretests, found at the beginning of each chapter, help students gauge their understanding of the chapter ahead. Answers can be found in the back of the book.

86 Chapter 2 | Fractions

Chapter 2 PRETEST

To see if you have already mastered the topics in this chapter, take this test.

1. Find all the factors of 20.
2. Express 72 as the product of prime factors.
3. What fraction does the shaded part of the diagram represent?
4. Write $20\frac{1}{3}$ as an improper fraction.
5. Express $\frac{31}{30}$ as a mixed number.
6. Write $\frac{9}{12}$ in simplest form.
7. What is the least common multiple of 10 and 4?
8. Which is greater, $\frac{1}{8}$ or $\frac{1}{9}$?

Add.

9. $\frac{1}{2} + \frac{7}{10}$
10. $7\frac{1}{3} + 5\frac{1}{2}$

Subtract.

11. $8\frac{1}{4} - 6$
12. $12\frac{1}{2} - 7\frac{7}{8}$

Multiply.

13. $2\frac{1}{3} \times 1\frac{1}{2}$
14. $\frac{5}{8} \times 96$
15. Divide: $3\frac{1}{3} \div 5$
16. Calculate: $2 + 1\frac{1}{3} \div \frac{4}{5}$

Solve. Write your answer in simplest form.

17. In 2006, the Pittsburgh Steelers won their fifth Super Bowl championship game. If 40 Super Bowl championship games have been played, what fraction of these games has this team won? (*Source:* National Football League)
18. In a biology class, three-fourths of the students received a passing grade. If there are 24 students in the class, how many students received failing grades?
19. Find the perimeter of the traffic island shown.

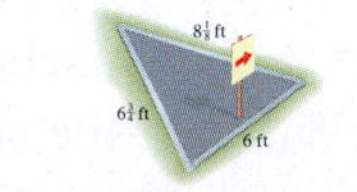

20. According to the nutrition information given, one serving of Honey Nut Cheerios® contains 22 grams of carbohydrates. If one serving is $\frac{3}{4}$ cup, what amount of carbohydrates is contained in $2\frac{1}{4}$ cups of Honey Nut Cheerios? (*Source:* General Mills)

• *Check your answers on page A-3.*

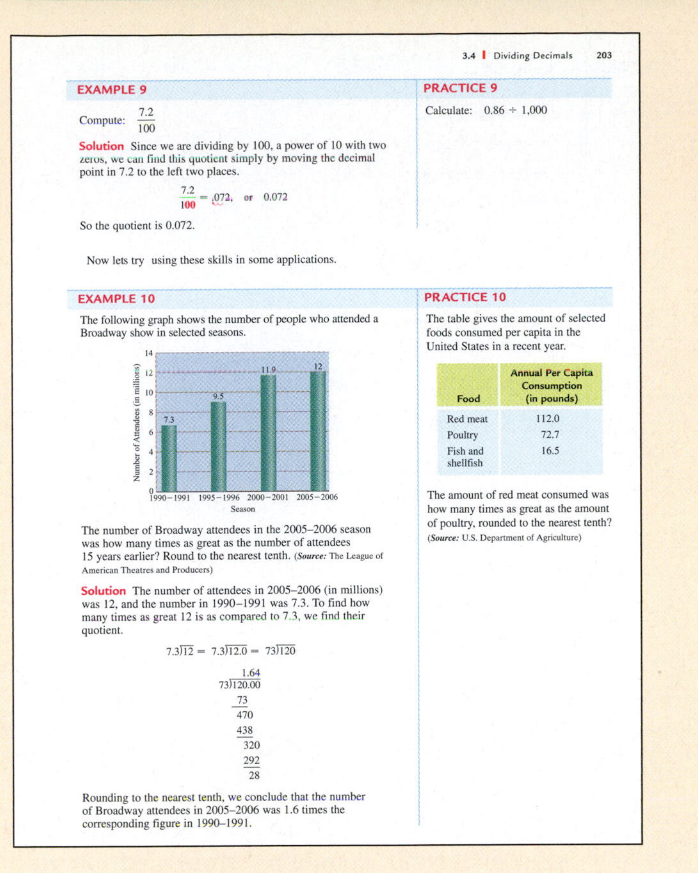

3.4 | Dividing Decimals 203

EXAMPLE 9

Compute: $\frac{7.2}{100}$

Solution Since we are dividing by 100, a power of 10 with two zeros, we can find this quotient simply by moving the decimal point in 7.2 to the left two places.

$$\frac{7.2}{100} = .072, \text{ or } 0.072$$

So the quotient is 0.072.

PRACTICE 9

Calculate: $0.86 \div 1{,}000$

Now lets try using these skills in some applications.

EXAMPLE 10

The following graph shows the number of people who attended a Broadway show in selected seasons.

The number of Broadway attendees in the 2005–2006 season was how many times as great as the number of attendees 15 years earlier? Round to the nearest tenth. (*Source:* The League of American Theatres and Producers)

Solution The number of attendees in 2005–2006 (in millions) was 12, and the number in 1990–1991 was 7.3. To find how many times as great 12 is as compared to 7.3, we find their quotient.

$$7.3\overline{)12} = 7.3\overline{)12.0} = 73\overline{)120}$$

$$\begin{array}{r} 1.64 \\ 73\overline{)120.00} \\ \underline{73} \\ 470 \\ \underline{438} \\ 320 \\ \underline{292} \\ 28 \end{array}$$

Rounding to the nearest tenth, we conclude that the number of Broadway attendees in 2005–2006 was 1.6 times the corresponding figure in 1990–1991.

PRACTICE 10

The table gives the amount of selected foods consumed per capita in the United States in a recent year.

Food	Annual Per Capita Consumption (in pounds)
Red meat	112.0
Poultry	72.7
Fish and shellfish	16.5

The amount of red meat consumed was how many times as great as the amount of poultry, rounded to the nearest tenth? (*Source:* U.S. Department of Agriculture)

Side-by-Side Format This format pairs examples and their step-by-step solutions side by side with corresponding practice exercises, encouraging active learning from the start. Students use this format for solving skill exercises, application problems, and technology exercises throughout the text.

Using the Tabs

Customize Your Textbook and Make It Work for You!

These removable and reusable tabs offer you five ways to be successful in your math course by letting you bookmark pages with helpful reminders.

- **Important** — Use these tabs to flag anything your instructor indicates is important.
- **Review This** — Mark important definitions, procedures, or key terms to review later.
- **Ask for Help** — Not sure of something? Need more instruction? Place these tabs in your textbook to address any questions with your instructor during your next class meeting or with your tutor during your next tutoring session.
- **On the Test** — If your instructor alerts you that something will be covered on a test, use these tabs to bookmark it.
- Write your own notes or create more of the preceding tabs to help you succeed in your math course.

New Tab Your Way to Success The Tab Your Way to Success guide provides students with color-coded Post-it® tabs to mark important pages of the text that they may need to revisit for review work, test preparation, instructor help, and so forth.

New Math Study Skills Foldout This insert, found at the very front of the book, provides students with tips on organization, test preparation, time management, and more.

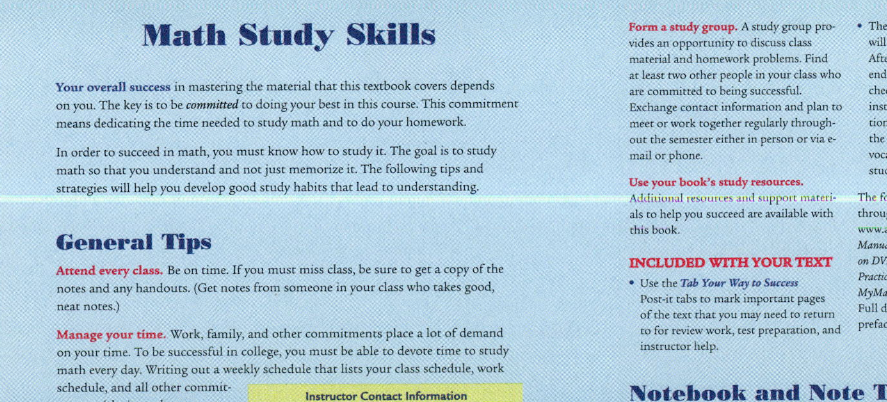

Math Study Skills

Your overall success in mastering the material that this textbook covers depends on you. The key is to be *committed* to doing your best in this course. This commitment means dedicating the time needed to study math and to do your homework.

In order to succeed in math, you must know how to study it. The goal is to study math so that you understand and not just memorize it. The following tips and strategies will help you develop good study habits that lead to understanding.

General Tips

Attend every class. Be on time. If you must miss class, be sure to get a copy of the notes and any handouts. (Get notes from someone in your class who takes good, neat notes.)

Manage your time. Work, family, and other commitments place a lot of demand on your time. To be successful in college, you must be able to devote time to study math every day. Writing out a weekly schedule that lists your class schedule, work schedule, and all other commitments with times that are not flexible will help you to determine when you can study.

Instructor Contact Information
Name:

Form a study group. A study group provides an opportunity to discuss class material and homework problems. Find at least two other people in your class who are committed to being successful. Exchange contact information and plan to meet or work together regularly throughout the semester either in person or via e-mail or phone.

Use your book's study resources. Additional resources and support materials to help you succeed are available with this book.

INCLUDED WITH YOUR TEXT

- Use the *Tab Your Way to Success* Post-it tabs to mark important pages of the text that you may need to return to for review work, test preparation, and instructor help.
- The resources on the **Pass the Test CD** will help you prepare for your tests. After you've taken the Posttest at the end of each chapter in the textbook, check your work by watching an instructor work through the full solutions to all of the Posttest exercises on the Pass the Test CD. You will also find vocabulary flashcards and additional study tips on this CD.

The following resources are also available through your school bookstore or www.aw-bc.com/math—*Student's Solutions Manual, Video Lectures on CD, Video Lectures on DVD, Worksheets for Classroom or Lab Practice, MathXL Tutorials on CD, MathXL, MyMathLab,* and the *Math Tutor Center.* Full descriptions are available in the preface of the textbook.

Notebook and Note Taking

Taking good notes and keeping a neat, well-organized notebook are important

TAKING NOTES

- Copy all important information written

Teaching Tips These tips, found only in the Annotated Instructor's Edition, help instructors with explanations, reminders of previously covered material, and tips on encouraging students to write in a journal.

Mention that another way of reading this decimal is "zero point nine."

Point out that to find the fraction equivalent to 0.21, we use the order of operations rule.

- The decimal 0.9, or .9, is another way of writing $(0 \times 1) + \left(9 \times \frac{1}{10}\right)$, or $\frac{9}{10}$.
  This decimal is read the same as the equivalent fraction: "nine tenths."
- The decimal 0.21 represents 2 tenths + 1 hundredth. This expression simplifies to the following:
  $$\left(2 \times \frac{1}{10}\right) + \left(1 \times \frac{1}{100}\right) = \frac{2}{10} + \frac{1}{100} = \frac{20}{100} + \frac{1}{100}, \text{ or } \frac{21}{100}$$
  So 0.21 is read "twenty-one hundredths."
- The decimal 0.149 stands for $\frac{149}{1{,}000}$.
  $$\left(1 \times \frac{1}{10}\right) + \left(4 \times \frac{1}{100}\right) + \left(9 \times \frac{1}{1{,}000}\right) = \frac{149}{1{,}000}$$
  So 0.149 is read "one hundred forty-nine thousandths."

Let's summarize these examples.

Point out that in each example, the number of decimal places *equals* the number of zeros in the equivalent fraction's denominator.

3 decimal places

$0.\overbrace{149} = \frac{149}{\underbrace{1{,}000}_{3 \text{ zeros}}}$

Decimal	Equivalent Fraction	Read as
0.9	$\frac{9}{10}$	Nine tenths
0.21	$\frac{21}{100}$	Twenty-one hundredths
0.149	$\frac{149}{1{,}000}$	One hundred forty-nine thousandths

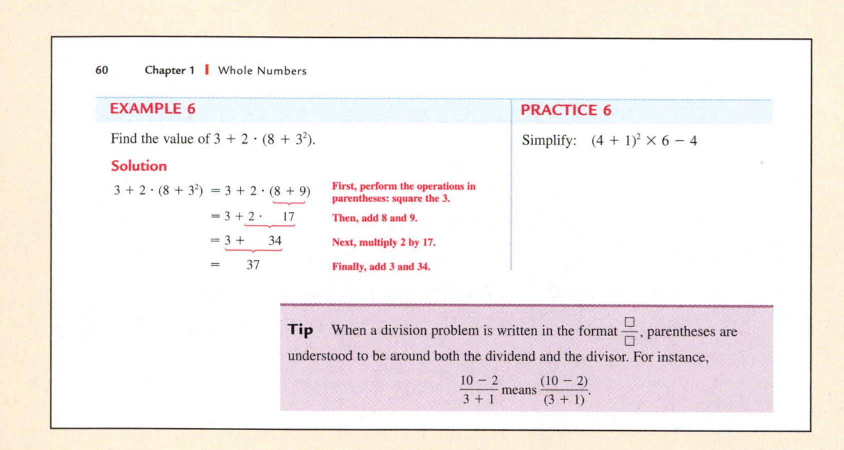
60 Chapter 1 | Whole Numbers

EXAMPLE 6

Find the value of $3 + 2 \cdot (8 + 3^2)$.

Solution

$3 + 2 \cdot (8 + 3^2) = 3 + 2 \cdot (8 + 9)$ First, perform the operations in parentheses: square the 3.

$= 3 + 2 \cdot 17$ Then, add 8 and 9.

$= 3 + 34$ Next, multiply 2 by 17.

$= 37$ Finally, add 3 and 34.

PRACTICE 6

Simplify: $(4 + 1)^2 \times 6 - 4$

Tip When a division problem is written in the format $\frac{\square}{\square}$, parentheses are understood to be around both the dividend and the divisor. For instance,

$$\frac{10 - 2}{3 + 1} \text{ means } \frac{(10 - 2)}{(3 + 1)}.$$

Student Tips These insightful tips help students avoid common errors and provide other helpful suggestions to foster understanding of the material at hand.

Geometry The authors integrate geometry throughout the text so students can see its relevance to their surroundings.

Estimation In order to help students judge the reasonableness of answers, the authors integrate the topic of estimation throughout the text.

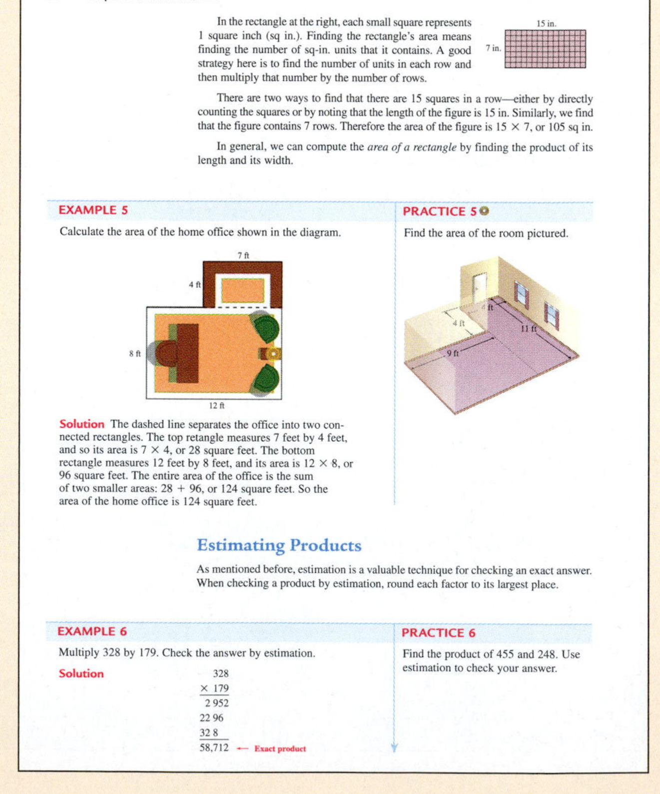
36 Chapter 1 | Whole Numbers

In the rectangle at the right, each small square represents 1 square inch (sq in.). Finding the rectangle's area means finding the number of sq-in. units that it contains. A good strategy here is to find the number of units in each row and then multiply that number by the number of rows.

There are two ways to find that there are 15 squares in a row—either by directly counting the squares or by noting that the length of the figure is 15 in. Similarly, we find that the figure contains 7 rows. Therefore the area of the figure is 15 × 7, or 105 sq in.

In general, we can compute the *area of a rectangle* by finding the product of its length and its width.

EXAMPLE 5

Calculate the area of the home office shown in the diagram.

Solution The dashed line separates the office into two connected rectangles. The top retangle measures 7 feet by 4 feet, and so its area is 7 × 4, or 28 square feet. The bottom rectangle measures 12 feet by 8 feet, and its area is 12 × 8, or 96 square feet. The entire area of the office is the sum of two smaller areas: 28 + 96, or 124 square feet. So the area of the home office is 124 square feet.

PRACTICE 5

Find the area of the room pictured.

Estimating Products

As mentioned before, estimation is a valuable technique for checking an exact answer. When checking a product by estimation, round each factor to its largest place.

EXAMPLE 6

Multiply 328 by 179. Check the answer by estimation.

Solution

```
   328
 × 179
 2 952
22 96
32 8
58,712  ← Exact product
```

PRACTICE 6

Find the product of 455 and 248. Use estimation to check your answer.

Cultural Notes Necessity is the mother of invention, and mathematics was created out of a need to solve problems in everyday life. Cultural Notes investigate the origins of mathematical concepts, discussing and illustrating the evolution of mathematics throughout the world.

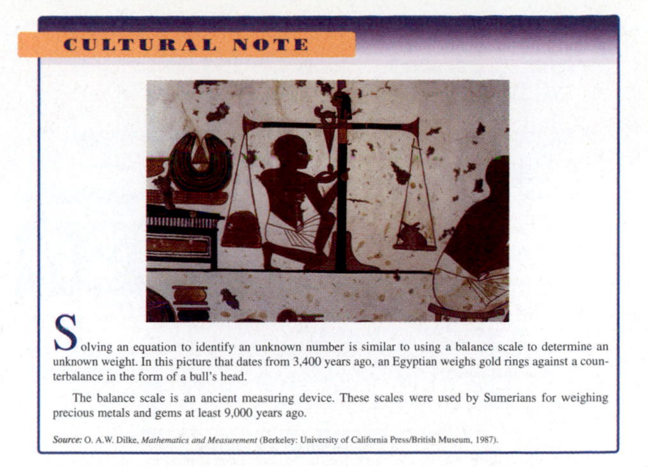

CULTURAL NOTE

Solving an equation to identify an unknown number is similar to using a balance scale to determine an unknown weight. In this picture that dates from 3,400 years ago, an Egyptian weighs gold rings against a counterbalance in the form of a bull's head.

The balance scale is an ancient measuring device. These scales were used by Sumerians for weighing precious metals and gems at least 9,000 years ago.

Source: O. A.W. Dilke, *Mathematics and Measurement* (Berkeley: University of California Press/British Museum, 1987).

Multiplying Decimals on a Calculator

Multiply decimals on a calculator by entering each decimal as you would enter a whole number, but insert a decimal point as needed. If there are too many decimal places in your answer to fit in the display, investigate how your calculator displays the answer.

EXAMPLE 12

Compute 8,278.55 × 0.875, rounding your answer to the nearest hundredth. Then check the answer by estimating.

Solution

Press	Display
8278.55 [×] 0.875 [ENTER]	8278.55 * 0.875 7243.73125

Now 7,243.73125 rounded to the nearest hundredth is 7,243.73. Checking by estimating, we get 8,000 × 0.9, or 7,200, which is close to our exact answer.

PRACTICE 12

Find the product of 2,471.66 and 0.33, rounding to the nearest tenth. Check the answer.

EXAMPLE 13

Find $(1.9)^2$

Solution

Press	Display
1.9 [^] 2 [ENTER]	1.9 ^ 2 3.61

Now let's check by estimating. Since 1.9 rounded to the nearest whole number is 2, $(1.9)^2$ should be close to 2^2, or 4, which is close to our exact answer, 3.61.

PRACTICE 13

Calculate: $(2.1)^3$

Technology Inserts In order to familiarize students with a range of computational methods—mental, paper and pencil, and calculator arithmetic—the authors include optional technology inserts that instruct students on how to use the scientific calculator to perform arithmetic operations. The side-by-side format is also used here to provide consistency across the text.

Exercise Variety The Akst/Bragg texts provide instructors with a variety of exercise types.

3.2 | Adding and Subtracting Decimals 187

MINDSTRETCHERS

Groupwork

1. Working with a partner, find the missing entries in the following magic square, in which 3.75 is the sum of every row, column, and diagonal.

0.75	1.25	
2		

Mathematical Reasoning

2. Suppose that a spider is sitting at point *A* on the rectangular web shown. If the spider wants to crawl along the web horizontally and vertically to munch on the delicious fly caught at point *B*, how long is the shortest route that the spider can take?

0.9 1 B 0.87 1.3 1.55 2.75

Writing

3. **a.** How many pairs of whole numbers are there whose sum is 7?
   b. How many pairs of decimals are there whose sum is 0.7?
   c. Explain why (a) and (b) have different answers.

Mindstretchers Found at the end of most sections, Mindstretchers are engaging activities that incorporate investigation, critical thinking, reasoning, pattern recognition, and writing exercises along with corresponding group work and historical connections in one comprehensive problem set. These problem sets target different levels and types of student understanding.

6.3 | More on Percents 317

Discount

In buying or selling merchandise, the term **discount** refers to a reduction on the merchandise's original price. The rate of discount is usually expressed as a percent of the selling price.

EXAMPLE 5

A drugstore gives senior citizens a 10% discount. If some pills normally sell for $16 a bottle, how much will a senior citizen pay?

Solution Note that, because senior citizens get a discount of 10%, they pay 100% − 10%, or 90%, of the normal price.

The question then becomes: What is 90% of $16?

We multiply. $x = 0.9 \cdot 16 = 14.4$

So a senior citizen will pay $14.40 for a bottle of the pills.

Note that another way to solve this problem is first to compute the amount of the discount (10% of $16) and then to subtract this discount from the original price. With this approach, do we get the same answer?

PRACTICE 5

Find the sale price.

FAMOUS DESIGNER JEANS REGULARLY $87 20% OFF TODAY ONLY

Have students write their responses in a journal.

Markup

A retail firm must sell goods at a higher price (the selling price) than it pays for the merchandise (the cost) to stay in business. The **markup** on an item is the difference between the selling price and the cost. Often the markup rate on merchandise is expressed as a fixed percent of the selling price.

Video Lectures on CD The icon indicates examples, practice exercises, and exercises that are covered in the Video Lectures on CD.

Writing Exercises Students will understand a concept better if they have to explain it in their own words. Journal assignments (provided as instructor's edition teaching tips) allow students to work on their mathematical communication skills, thus improving their understanding of concepts.

Calculator Exercises These optional exercises, denoted with a icon, can be found in the exercise sets, giving students the opportunity to use a calculator to solve a variety of real-life applications.

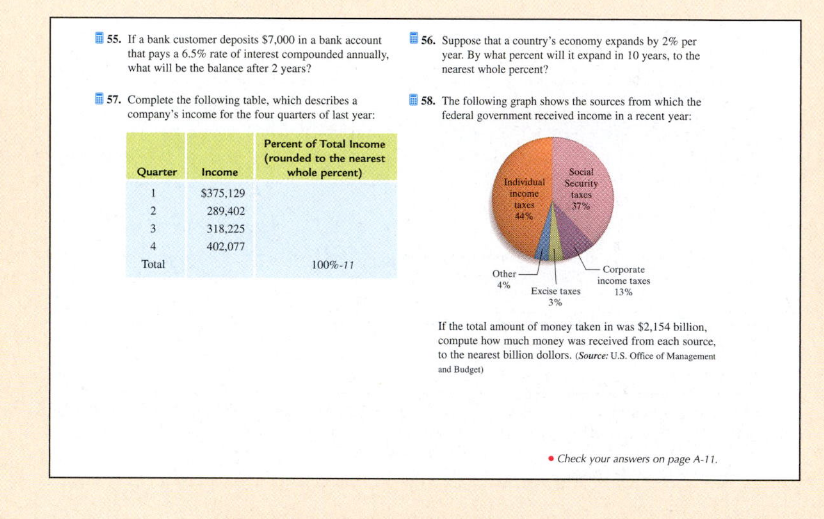

55. If a bank customer deposits $7,000 in a bank account that pays a 6.5% rate of interest compounded annually, what will be the balance after 2 years?

56. Suppose that a country's economy expands by 2% per year. By what percent will it expand in 10 years, to the nearest whole percent?

57. Complete the following table, which describes a company's income for the four quarters of last year:

Quarter	Income	Percent of Total Income (rounded to the nearest whole percent)
1	$375,129	
2	289,402	
3	318,225	
4	402,077	
Total		100%-11

58. The following graph shows the sources from which the federal government received income in a recent year:

If the total amount of money taken in was $2,154 billion, compute how much money was received from each source, to the nearest billion dollars. (*Source:* U.S. Office of Management and Budget)

• *Check your answers on page A-11.*

2.2 | Introduction to Fractions 107

2.2 Exercises FOR EXTRA HELP MyMathLab MathXL PRACTICE WATCH DOWNLOAD READ REVIEW

Mathematically Speaking

Fill in each blank with the most appropriate term or phrase from the given list.

improper fraction	proper fraction	mixed
greatest common factor	simplify	composite
convert	least common denominator	
equivalent	like fractions	

1. A fraction whose numerator is smaller than its denominator is called a(n) ______.

2. The improper fraction $\frac{5}{2}$ can be expressed as a(n) ______ number.

3. The fractions $\frac{6}{8}$ and $\frac{3}{4}$ are ______.

4. Divide the numerator and denominator of a fraction by the same whole number in order to ______ it.

For Extra Help These boxes, found at the top of the first page of every section's exercise set, direct students to helpful resources that will aid in their study of the material.

***New* Mathematically Speaking Exercises** Located at the begining of nearly every exercise section of the text, Mathematically Speaking exercises help students understand and master mathematical vocabulary.

***New* Mixed Practice Exercises** Mixed Practice exercises reinforce skills within the section and encourage review.

Mixed Practice

55. Solve and check: $\frac{\frac{3}{4}}{15} = \frac{x}{8}$

56. Solve and check: $\frac{1.6}{x} = \frac{2.4}{27}$

57. Solve and check: $\frac{3}{2} = \frac{2\frac{2}{5}}{x}$

58. Determine whether the proportion 8 is to 1 as 2 is to $\frac{1}{4}$ is true.

59. Is $\frac{4}{9} = \frac{3}{8}$ a true or false statement?

60. Solve and check: $\frac{x}{9} = \frac{5}{6}$

End-of-Chapter Material At the end of each chapter, students will find a wealth of review- and retention-oriented material to reinforce the concepts presented in current and previous chapters.

KEY CONCEPTS AND SKILLS CONCEPT SKILL

Concept/Skill	Description	Example
[4.1] Variable	A letter that represents an unknown number.	x, y, t
[4.1] Constant	A known number.	$2, \frac{1}{3}, 5.6$
[4.1] Algebraic Expression	An expression that combines variables, constants, and arithmetic operations.	$x + 3, \frac{1}{8}n$
[4.1] To Evaluate an Algebraic	• Substitute the given value for each variable.	Evaluate $8 - x$ for $x = 3.5$:

Key Concepts and Skills These give students quick reminders of the chapter's most important elements and provide a one-stop quick review of the chapter material. Each concept/skill is keyed to the section in which it was introduced, and students are given a brief description and example for each.

Chapter 4 Review Exercises

To help you review this chapter, solve these problems.

[4.1] *Translate each algebraic expression to words.*

1. $x + 1$ 2. $y + 4$ 3. $w - 1$ 4. $s - 3$

5. $\frac{c}{7}$ 6. $\frac{a}{10}$ 7. $2x$ 8. $6y$

9. $y \div 0.1$ 10. $n \div 1.6$ 11. $\frac{1}{3}x$ 12. $\frac{1}{10}w$

Chapter Review Exercises These exercises are keyed to the corresponding sections for easy student reference. Numerous mixed application problems complete each of these exercise sets, reinforcing the applicability of what students are learning.

Chapter 4 POSTTEST FOR EXTRA HELP Pass the Test Test solutions are found on the enclosed CD.

To see if you have mastered the topics in this chapter, take this test.

Write each algebraic expression in words.

1. $x + \frac{1}{2}$ 2. $\frac{a}{3}$

Translate each word phrase to an algebraic expression.

3. 10 less than a number 4. The quotient of 8 and p

Chapter Posttest Just as every chapter begins with a Pretest to test student understanding *before* attempting the material, every chapter ends with a Posttest to measure student understanding *after* completing the chapter material. Answers to these tests are provided in the back of the book. Students can watch an instructor working through the full solutions to the Posttest exercises on the Pass the Test CD.

Cumulative Review Exercises

To help you review, solve the following:

1. Subtract: $8\frac{1}{4} - 2\frac{7}{8}$
2. Find the quotient: $7.5 \div 1{,}000$
3. Decide whether 2 is a solution to the equation $w + 3 = 5$
4. Multiply: 804×29
5. Round 3.14159 to the nearest hundredth.
6. Solve and check: $n - 3.8 = 4$
7. Solve and check: $\frac{x}{2} = 16$
8. In animating a cartoon, artists had to draw 24 images to appear during 1 second of screen time. How many images did they have to draw to produce a 5-minute cartoon?

Cumulative Review Exercises Beginning at the end of Chapter 2, students have the opportunity to maintain their skills by completing the Cumulative Review Exercises. These exercises are invaluable, especially when students need to recall a previously learned concept or skill before beginning the next chapter, or when studying for midterm and final examinations.

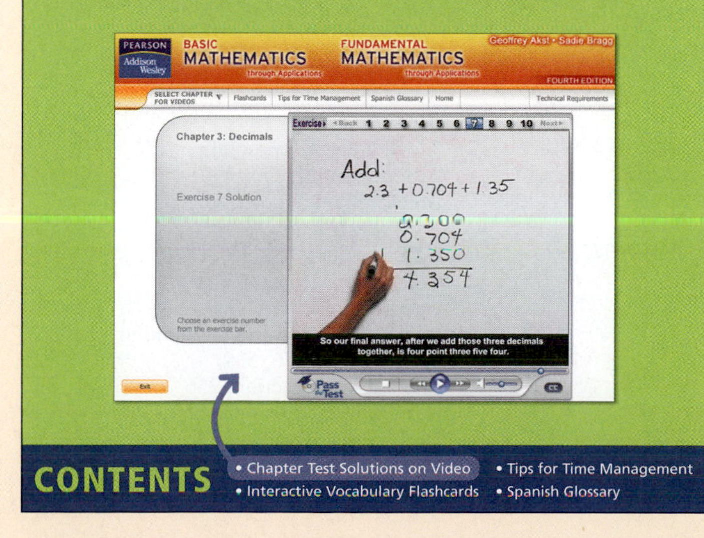

***New* Pass the Test CD** Included with every new copy of the book, the Pass the Test CD includes the following resources:

- video footage of an instructor working through the complete solutions for all Posttest exercises for each chapter
- vocabulary flashcards
- a video that offers an interactive Spanish glossary
- additional tips on improving time management

Index of Applications

Environment

Finance

Geology

Geometry

Government

Health/Life Sciences

Labor

Miscellaneous

Physics

Sports

Statistics/Demographics

Technology

Transportation

CHAPTER 1

Whole Numbers

Whole Numbers and the Census

Every ten years, the U.S. Bureau of the Census attempts to count and gather information about each man, woman, and child in the nation. The government then uses this information both to reapportion the 435 seats in the House of Representatives and to reallocate billions of dollars in federal funds.

The census also paints a picture of the nation, showing how it has changed since the last count. For example, in the decade that elapsed between the 1990 census and the 2000 census, the number of males rose from 94 million to 101 million, and the number of females grew from 101 million to 108 million. The total of family households swelled from 93 million to 103 million, and the number of students enrolled in college increased from 14 million to 15 million. (*Source:* U.S. Bureau of the Census)

Chapter 1 PRETEST

To see if you have already mastered the topics in this chapter, take this test.

1. Insert commas as needed in the number 2 0 5 0 0 7. Then write the number in words.

2. Write the number one million, two hundred thirty-five thousand in standard form.

3. What place does the digit 8 occupy in 805,674?

4. Round 8,143 to the nearest hundred.

5. Add: $38 + 903 + 7{,}285$

6. Subtract 286 from 5,000.

7. Subtract: $734 - 549$

8. Find the product of 809 and 36.

9. Find the quotient: $27\overline{)7{,}020}$

10. Divide: $13{,}558 \div 44$

11. Write $2 \cdot 2 \cdot 2$, using exponents.

12. Evaluate: 6^2

Simplify.

13. $26 - 7 \cdot 3$

14. $3 + 2^3 \cdot (8 - 3)$

Solve and check.

15. The mathematician Benjamin Banneker was born in 1731 and died in 1806. About how old was he when he died? (*Source:* *The New Encyclopedia Britannica*)

16. At a certain college, students pay \$75 for each college credit. If a student takes 9 credits, how much will it cost?

17. Tiger Woods had scores of 74, 66, 65, and 71 for his four rounds of golf at a Masters Tournament. What was his average score for a round of golf? (*Source: The Augusta Chronicle*, 2005)

18. The HP Photosmart 8050 photo printer can print a 4-inch by 6-inch photo in 27 seconds, and the 8250 model printer can print the same size photo in 14 seconds. How much longer would it take the HP 8050 printer to print 12 4-inch by 6-inch photos? (*Source:* Hewlett-Packard)

19. New London County Insurance Company (NLC) offers an installment plan for paying auto insurance premiums. For a \$540 policy, NLC requires a down payment of \$81. The balance is paid in 9 equal installments of \$54, which includes a billing fee. How much would be saved by paying for this policy without using the installment plan? (*Source:* New London County Insurance Company, 2006)

20. Which of the rooms pictured has the largest area? (feet = ft)

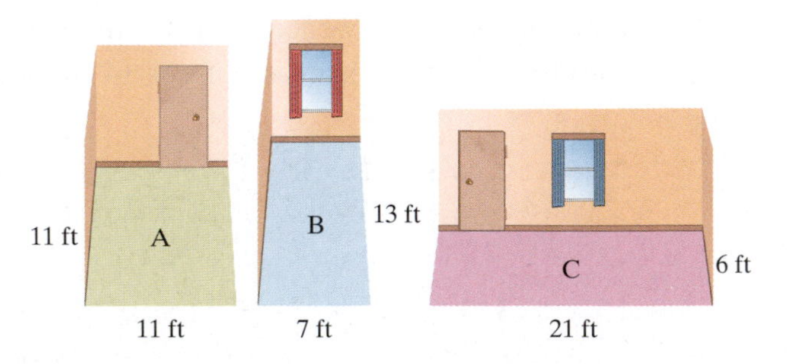

• *Check your answers on page A-1.*

1.1 Introduction to Whole Numbers

What the Whole Numbers Are and Why They Are Important

OBJECTIVES

- To read and write whole numbers
- To write whole numbers in expanded form
- To round whole numbers

We use whole numbers for counting, whether it is the number of *e*'s on this page, the number of stars in the sky, or the number of runs, hits, and errors in a baseball game.

The whole numbers are 0, 1, 2, 3, 4, 5, 6, 7, 8, 9, 10, 11, 12, 13, ... An important property of whole numbers is that there is always a next whole number. This property means that they go on without end, as the three dots above indicate.

Every whole number is either *even* or *odd*. The even whole numbers are 0, 2, 4, 6, 8, 10, 12, ... The odd whole numbers are 1, 3, 5, 7, 9, 11, 13, ...

We can represent the whole numbers on a number line. Similar to a ruler, the number line starts with 0 and extends without end to the right, as the arrow indicates.

0 1 2 3 4 5 6 7 8 ...

Reading and Writing Whole Numbers

Generally speaking, we *read* whole numbers in words, but we use the **digits** 0, 1, 2, 3, 4, 5, 6, 7, 8, and 9 to *write* them. For instance, we read the whole number *fifty-one* but write it *51*, which we call **standard form**.

Each of the digits in a whole number in standard form has a **place value**. Our **place value** system is very important because it underlies both the way we write and the way we compute with numbers.

The following chart shows the place values in whole numbers up to 12 digits long. For instance in the number 1,234,056 the digit 2 occupies the hundred thousands place. Study the place values in the chart now.

BILLIONS			MILLIONS			THOUSANDS			ONES			← Period
Hundreds	Tens	Ones	Hundreds	Tens	Ones	Hundreds	Tens	Ones	Hundreds	Tens	Ones	← Place value
					1	2	3	4	0	5	6	
		8	1	6	8	9	3	1	0	4	7	

Tip We read whole numbers from left to right, but it is easier in the place value chart to learn the names of the places *from right to left.*

When we write a large whole number in standard form, we insert *commas* to separate its digits into groups of three, called **periods**. For instance the number 8,168,931,047 has four periods: *ones, thousands, millions*, and *billions*.

EXAMPLE 1

In each number, identify the place that the digit 7 occupies.

a. 207

b. 7,654,000

c. 5,700,000,001

Solution

a. The ones place

b. The millions place

c. The hundred millions place

PRACTICE 1

What place does the digit 8 occupy in each number?

a. 278,056

b. 803,746

c. 3,080,700,059

The following rule provides a shortcut for *reading a whole number*.

To Read a Whole Number

Working from left to right,

- read the number in each period and then
- name the period in place of the comma.

For instance, 1,234,056 is read "one million, two hundred thirty-four thousand, fifty-six." Note that the ones period is not read.

EXAMPLE 2

How do you read the number 422,000,085?

Solution Beginning at the left in the millions period, we read this number as "four hundred twenty-two million, eighty-five." Note that because there are all zeros in the thousands period, we do not read "thousands."

PRACTICE 2

Write 8,000,376,052 in words.

EXAMPLE 3

The display on a calculator shows the answer 3578002105. Insert commas in this answer and then read it.

Solution The number with commas is 3,578,002,105. It is read "three billion, five hundred seventy-eight million, two thousand, one hundred five."

PRACTICE 3

A company is worth $7372050. After inserting commas, read this amount.

Until now, we have discussed how to *read* whole numbers in standard form. Now let's turn to the question of how they are *written* in standard form. We simply reverse the process just described. For instance, the number eight billion, one hundred sixty-eight million, nine hundred thirty-one thousand, forty-seven in standard form is 8,168,931,047. Here, we use the 0 as a **placeholder** in the hundreds place because there are no hundreds.

To Write a Whole Number

Working from left to right,

- write the number named in each period and
- replace each period name with a comma.

When writing large whole numbers in standard form, we must remember that the number of commas is always one less than the number of periods. For instance, the number one million, two hundred thirty-four thousand, fifty-six—1,234,056—has three periods and two commas. Similarly, the number 8,168,931,047 has four periods and three commas.

EXAMPLE 4

Write the number eight billion, seven in standard form.

Solution This number involves billions, so there are four periods—billions, millions, thousands, and ones—and three commas. Writing the number named in each period and replacing each period name with a comma, we get 8,000,000,007. Note that we write three 0's when no number is named in a period.

PRACTICE 4

Use digits and commas to write the amount ninety-five million, three dollars.

EXAMPLE 5

The treasurer of a company writes a check in the amount of four hundred thousand seven hundred dollars. Using digits, how would she write this amount on the check?

Solution This quantity is written with one comma, because its largest period is thousands. So the treasurer writes $400,700, as shown on the check below.

PRACTICE 5

A rich alumna donates three hundred seventy-five thousand dollars to her college's scholarship fund.

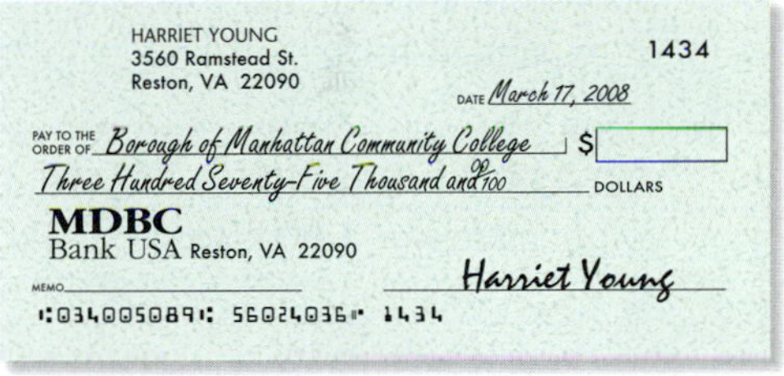

Using digits, how would she write this amount on the check?

____

When writing checks, we write the amount in both digits and words. Why do we do this?

Writing Whole Numbers in Expanded Form

We have just described how to write whole numbers in standard form. Now let's turn to how we write these numbers in **expanded form**.

Let's consider the whole number 4,325 and examine the place value of its digits.

$$4{,}325 = 4 \text{ thousands} + 3 \text{ hundreds} + 2 \text{ tens} + 5 \text{ ones}$$

This last expression is called the expanded form of the number, and it can be written as follows

$$4{,}000 + 300 + 20 + 5$$

The expanded form of a number spells out its value in terms of place value, helping us understand what the number really means. For instance, think of the numbers 92 and 29. By representing them in *expanded* form, can you explain why they differ in value even though their *standard* form consists of the same digits?

EXAMPLE 6

Write in expanded form:

a. 906

b. 3,203,000

Solution

a. The 6 is in the ones place, the 0 is in the tens place, and the 9 is in the hundreds place.

ONES		
Hundreds	Tens	Ones
9	0	6

So 906 is 9 hundreds + 0 tens + 6 ones = 900 + 0 + 6 in expanded form.

b. Using the place value chart, we see that 3,203,000 = 3 millions + 2 hundred thousands + 3 thousands = 3,000,000 + 200,000 + 3,000.

PRACTICE 6

Express in expanded form.

a. 27,013

b. 1,270,093

Rounding Whole Numbers

Most people equate mathematics with precision, but some problems require sacrificing precision for simplicity. In this case, we use the technique called **rounding** to approximate the exact answer with a number that ends in a given number of zeros. Rounded numbers have special advantages: They seem clearer to us than other numbers, and they make computation easier—especially when we are trying to compute in our heads.

Of these two headlines, which do you prefer? Why?

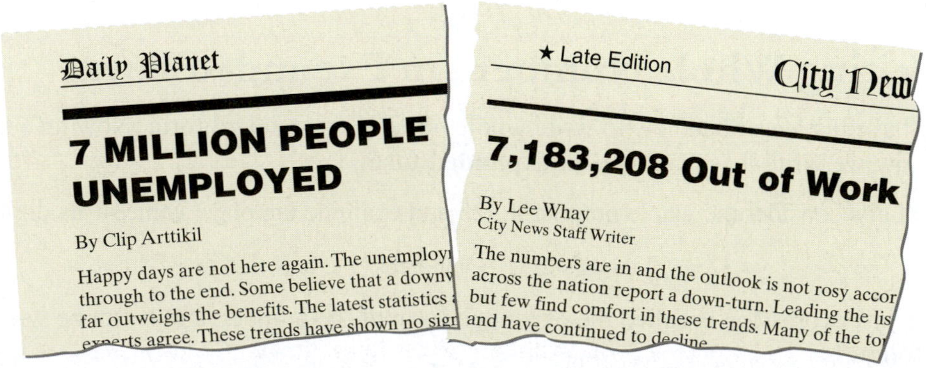

Daily Planet
7 MILLION PEOPLE UNEMPLOYED
By Clip Arttikil
Happy days are not here again. The unemploy
through to the end. Some believe that a downw
far outweighs the benefits. The latest statistics
experts agree. These trends have shown no sign
★ Late Edition
City New
7,183,208 Out of Work
By Lee Whay
City News Staff Writer
The numbers are in and the outlook is not rosy accor
across the nation report a down-turn. Leading the lis
but few find comfort in these trends. Many of the to
and have continued to decline

Study the following chart to see the connection between place value and rounding

Rounding to the nearest	Means that the rounded number ends in at least
10	One 0
100	Two 0's
1,000	Three 0's
10,000	Four 0's
100,000	Five 0's
1,000,000	Six 0's

Note in the chart that the place value tells us how many 0's the rounded number must have at the end. Having more 0's than indicated is possible. Can you think of an example?

When rounding, we use an underlined digit to indicate the place to which we are rounding.

Now let's consider the following rule for rounding whole numbers.

To Round a Whole Number

- Underline the place to which you are rounding.
- Look at the digit to the right of the underlined digit, called the **critical digit.** If this digit is 5 or more, add 1 to the underlined digit; if it is less than 5, leave the underlined digit unchanged.
- Replace all the digits to the right of the underlined digit with zeros.

EXAMPLE 7

Round 79,630 to

a. the nearest thousand

b. the nearest hundred.

Solution

a. $79{,}630 = 7\underline{9}{,}630$ ← **Underline the digit in the thousands place.**

$= 7\underline{9}{,}630$ ← **The critical digit 6 is greater than 5; add 1 to the underlined digit.**

$\approx 80{,}000$ ← **Change the digits to the right of the underlined digit to 0's.**

↑ **This symbol means "is approximately equal to."**

Note that adding 1 to the underlined digit gave us 10 and forced us to write 0 and carry 1 to the next column, changing the 7 to 8.

b. First, we underline the 6 because that digit occupies the hundreds place: $79{,}\underline{6}30$. The critical digit is **3**: $79{,}\underline{6}30$. Since 3 is less than 5, we leave the underlined digit unchanged. Then, we replace all digits to the right with 0's, getting 79,600. We write $79{,}630 \approx 79{,}600$, meaning that 79,630 when rounded to the nearest hundred is 79,600.

PRACTICE 7

Round 51,760 to

a. the nearest thousand

b. the nearest ten thousand.

For Example 7, consider this number line.

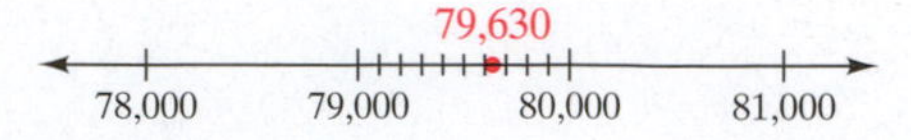

The number line shows that 79,630 lies between 79,000 and 80,000 and that it is closer to 80,000, as the rule indicates.

EXAMPLE 8

In an anatomy and physiology class, a student learned that the adult human skeleton contains 206 bones. How many bones is this to the nearest hundred bones?

Solution We first write 2̲06. The critical digit 0 is less than 5, so we do *not* add 1 to the underlined digit. However, we do change both the digits to the right of the 2 to 0's. So 2̲06 ≈ 200, and there are approximately 200 bones in the human body.

PRACTICE 8

Based on current population data, the U.S. Bureau of the Census projects that the U.S. resident population will be 419,845,000 in the year 2050. What is the projected population to the nearest million?

EXAMPLE 9

The following table lists five of the highest-grossing movies of all time, and the amount of money they took in.

Film	Year	World Total (in U.S. dollars)
Titanic	1997	1,845,034,188
Star Wars: Episode I, The Phantom Menace	1999	925,600,000
Harry Potter and the Sorcerer's Stone	2001	985,817,659
The Lord of the Rings: The Two Towers	2002	926,287,400
The Lord of the Rings: The Return of the King	2003	1,118,888,979

(**Source:** *The Top Ten of Everything 2006*)

a. Write in words the amount of money taken in by the film with the largest world total.

b. Round to the nearest ten million dollars the world total for *Harry Potter and the Sorcerer's Stone.*

Solution

a. *Titanic* has the largest world total. This total is read "one billion, eight hundred forty-five million, thirty-four thousand, one hundred eighty-eight dollars."

b. The world total for *Harry Potter and the Sorcerer's Stone* is $985,817,659. To round, we underline the digit in the ten millions place: 98̲5,817,659. Since the critical digit is 5, we add 1 to the underlined digit, and change the digits to the right to 0's. So the rounded total is $990,000,000.

PRACTICE 9

This chart gives the number of male and female faculty members in U.S. colleges during a recent year.

Faculty Members	Number
Men	382,808
Women	248,788

(**Source:** *The Chronicle of Higher Education,* August 25, 2006)

a. Write in words the number of female college faculty members.

b. What is the number of male college faculty members rounded to the nearest hundred thousand?

Mathematically Speaking

Fill in each blank with the most appropriate term or phrase from the given list.

calculated	rounded	periods	odd
even	digits	whole numbers	standard form
placeholder	place value	expanded form	

1. The _______ are 0, 1, 2, 3, 4, 5, … .

2. The numbers 0, 2, 4, 6, 8, 10, … are _______.

3. The numbers 1, 3, 5, 7, 9, … are _______.

4. The whole numbers are written with the _______ 0, 1, 2, 3, 4, 5, 6, 7, 8, and 9.

5. The number thirty-seven, when written as 37, is said to be in _______.

6. In the number 528, the _______ of the 5 is hundreds.

7. In the number 206, the 0 is used as a _______ in the tens place.

8. Commas separate the digits in a large whole number into groups of three called _______.

9. When the number 973 is written as 9 hundreds + 7 tens + 3 ones, it is said to be in _______.

10. The number 545 _______ to the nearest hundred is 500.

Underline the digit that occupies the given place.

11. 4,867 Thousands place

12. 975 Hundreds place

13. 316 Tens place

14. 41,722 Ten thousands place

15. 28,461,013 Millions place

16. 762,800 Hundred thousands place

Identify the place occupied by the underlined digit.

17. $\underline{6}91{,}400$

18. $7\underline{2}{,}109$

19. $7{,}\underline{3}80$

20. $3\underline{5}1$

21. $\underline{8}{,}450{,}000{,}000$

22. $\underline{3}5{,}832{,}775$

Insert commas as needed, and then write the number in words.

23. 4 8 7 5 0 0

24. 5 2 8 0 5 0

25. 2 3 5 0 0 0 0

26. 1 3 5 0 1 3 2

27. 9 7 5 1 3 5 0 0 0

28. 2 1 0 0 0 1 3 2

29. 2 0 0 0 0 0 0 3 5 2

30. 4 1 0 0 0 0 0 0 0 7

31. 1 0 0 0 0 0 0 0 0 0

32. 3 7 9 0 5 2 0 0 0

Write each number in standard form.

33. Ten thousand, one hundred twenty

34. Three billion, seven hundred million

35. One hundred fifty thousand, eight hundred fifty-six

36. Twenty million, five thousand

37. Six million, fifty-five

38. Two million, one hundred twenty-two

39. Fifty million, six hundred thousand, one hundred ninety-five

40. Nine hundred thousand, eight hundred eleven

41. Four hundred thousand, seventy-two

42. Nine hundred billion

Write each number in expanded form.

43. 3

44. 6,300

45. 858

46. 9,000,000

47. 2,500,004

48. 7,251,380

Round to the indicated place.

49. 671 to the nearest ten

50. 838 to the nearest hundred

51. 7,103 to the nearest hundred

52. 46,099 to the nearest thousand

53. 28,241 to the nearest ten thousand

54. 7,802,555 to the nearest million

55. 705,418 to its largest place

56. 96 to its largest place

57. 31,972 to its largest place

58. 4,913,440 to its largest place

Round each number as indicated.

59.

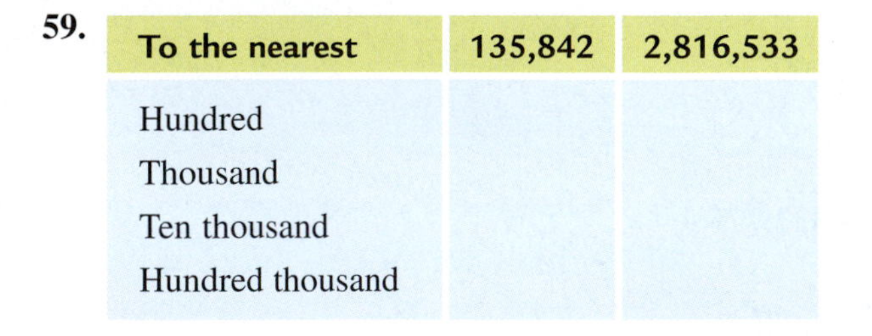

To the nearest	135,842	2,816,533
Hundred		
Thousand		
Ten thousand		
Hundred thousand		

60.

To the nearest	972,055	3,189,602
Thousand		
Ten thousand		
Hundred thousand		
Million		

Mixed Practice

Solve.

61. Write 12,051 in expanded form.

62. Identify the place occupied by the underlined digit in $\underline{2}$6,543,009.

63. Underline the digit that occupies the ten thousands place in 40,059.

64. Write five hundred forty-two thousand, sixty-seven in standard form.

65. Insert commas as needed, and then write 1 0 5 6 1 0 0 in words.

66. Round 26,255 to the nearest thousand.

Applications

Write each whole number in words.

67. Biologists have classified more than 900,000 species of insects. (*Source:* Smithsonian Institution)

68. Each pair of human lungs contains some 300,000,000 tiny air sacs.

69. The total land area of the Caribbean island of Puerto Rico is 8,959 square kilometers. (*Source: The World Fact-Book*, 2006)

70. Mercury is the closest planet to the Sun, a distance of approximately 36,000,000 miles.

71. There are 37,842 WiFi hotspots in the United States. (*Source: JiWire*, March 2006)

72. The Pyramid of Khufu in Egypt has a base of approximately 2,315,000 blocks. (*Source: The New Encyclopedia Britannica*)

Write each whole number in standard form.

73. Some one hundred billion nerve cells are part of the human brain.

74. Son of Beast, a roller coaster at Paramount's Kings Island in Ohio, has a track length of seven thousand thirty-two feet. (*Source:* American Coasters Network)

75. One of the largest giant sequoias in the United States is three thousand, two hundred eighty-eight inches tall. (*Source:* U.S. National Park Service)

76. The total land area of the United States is nine million, six hundred thirty-one thousand, four hundred eighteen square kilometers. (*Source: The World Factbook*, 2006)

77. The number of registered nurses employed in the United States is expected to grow to two million, nine hundred eight thousand by the year 2012. (*Source:* U.S. Bureau of the Census, *Statistical Abstract of the United States,* 2005)

78. George W. Bush received sixty-two million, thirty-nine thousand, seventy-three votes in the 2004 presidential election. (*Source:* U.S. National Archives and Records Administration)

Round to the indicated place.

79. The Statue of Liberty is 152 feet high. What is its height to the nearest 10 feet?

80. The Nile, with a length of 4,180 miles, is the longest river in the world. Find this length to the nearest thousand miles.

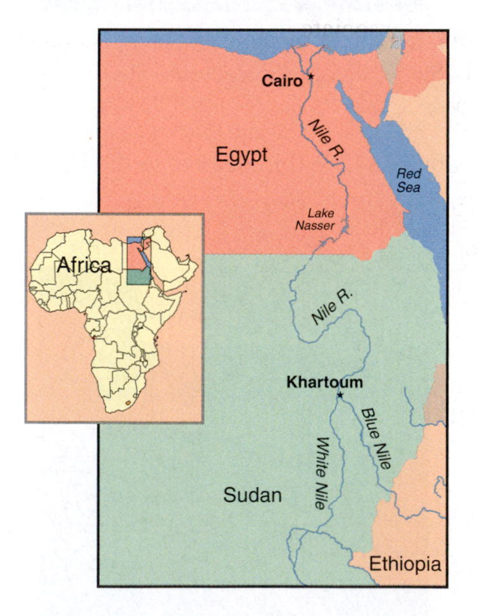

81. In 1949, Air Force Captain James Gallagher led the first team to make an around-the-world flight. The team flew 23,452 miles. What is this distance to the nearest ten thousand miles? (*Source:* Taylor and Mondey, *Milestones of Flight*)

82. The element copper changes from a liquid to a gas at the temperature 2,567 degrees Celsius (°C). Find this temperature to the nearest hundred degrees Celsius.

83. A weight of 454 grams is equivalent to 1 pound. How many grams is this to the nearest hundred?

84. The Rose Bowl stadium has a seating capacity of 92,542. Round this number to the nearest ten thousand. (*Source:* Rose Bowl Operating Company)

85. This chart displays the number of degrees awarded in the United States during a recent year.

Degree	Number Awarded
Associate	665,301
Bachelor's	1,399,542
Master's	558,940
Doctorate	48,378
Professional	83,041

(*Source: The Chronicle of Higher Education, Almanac,* 2006)

a. Write in words the number of bachelor's degrees awarded.

b. Round, to the nearest hundred thousand, the number of associate degrees awarded.

86. The following table lists the amount of gold produced during a recent year in six countries that are leading producers.

Country	Amount of Gold (in troy ounces)
South Africa	13,316,820
USA	10,108,180
Australia	10,050,308
China	7,552,199
Russia	6,462,290

(*Source:* Gold Fields Mineral Services Ltd., Gold Survey 2004)

a. Write in words the gold production for China.

b. Which country or countries produced ten million troy ounces, rounded to the nearest million troy ounces?

• *Check your answers on page A-1.*

MINDSTRETCHERS

Mathematical Reasoning

1. I am thinking of a certain whole number. My number, rounded to the nearest hundred, is 700. When it is rounded to the nearest ten, it is 750. What numbers could I be thinking of?

Writing

2. How does the number 10 play a special role in the way that we write whole numbers? Would it be possible to have the number 2 play this role? Explain.

Groupwork

3. Here are three ways of writing the number seven: 7 VII 𝍸||

Working with a partner, express each of the numbers 1, 2, … , 9 in these three ways.

1.2 Adding and Subtracting Whole Numbers

OBJECTIVES

- To add and subtract whole numbers
- To estimate the sum and difference of whole numbers
- To solve word problems involving the addition or subtraction of whole numbers

The Meaning and Properties of Addition and Subtraction

Addition is perhaps the most fundamental of all operations. One way to think about this operation is as *combining sets.* For example, suppose that we have two distinct sets of pens, with 5 pens in one set and 3 in the other. If we put the two sets together, we get a single set that has 8 pens.

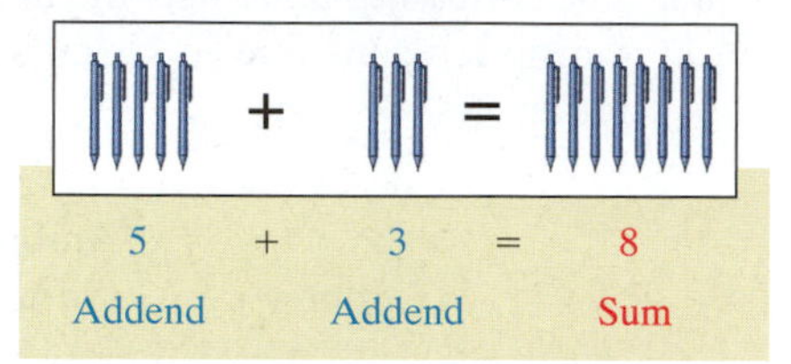

So we can say that 3 added to 5 is 8, or here, 5 pens plus 3 pens equals 8 pens. Numbers being added are called *addends*. The result is called the *sum*, or *total.*

In the above example, note that we are adding quantities of the same thing, or *like quantities.*

Another good way to think about the addition of whole numbers is as *moving to the right on a number line.* In this way, we start at the point on the line corresponding to the first number, 5. Then to add 3, we move 3 units to the right, ending on the point that corresponds to the answer, 8.

Move 3 units to the right.

0 1 2 3 4 5 6 7 8 9 10

Start End

Now let's look at subtraction. One way to look at this operation is as *taking away.* For instance, when we subtract 5 pens from 8 pens, we take 5 pens away from 8 pens, leaving 3 pens.

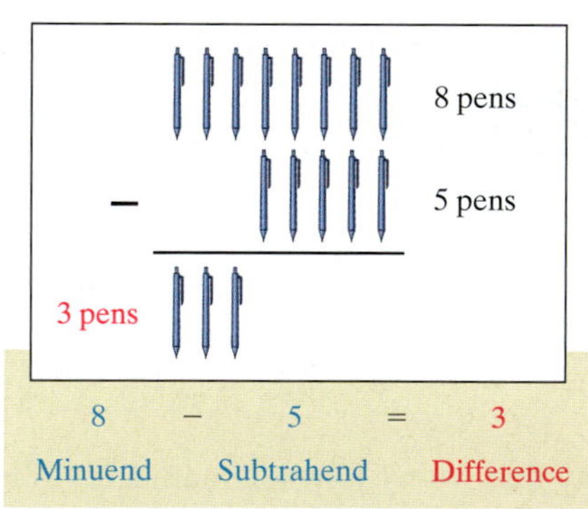

In a subtraction problem, the number from which we subtract is called the *minuend,* the number being subtracted is called the *subtrahend*, and the result is called the *difference*.

As in the preceding example, we can only subtract *like quantities*: we cannot subtract 5 pens from 8 scissors.

We can also think of subtraction as the *opposite of addition.*

$8 - 5 = 3$ because $5 + 3 = 8$

Subtraction **Related addition**

Note in this example that, if we add the 5 pens to the 3 pens, we get 8 pens.

Addition and subtraction problems can be written either horizontally or vertically.

$5 + 3 = 8 \qquad 8 - 5 = 3$

Horizontal

$$\begin{array}{r} 5 \\ +3 \\ \hline 8 \end{array} \qquad \begin{array}{r} 8 \\ -5 \\ \hline 3 \end{array}$$

Vertical

Either format gives the correct answer. But it is generally easier to figure out the sum and difference of large numbers if the problems are written vertically.

Now let's briefly consider several special properties of addition that we use frequently. Examples appear to the right of each property.

The Identity Property of Addition

The sum of a number and zero is the original number.

$3 + 0 = 3$
$0 + 5 = 5$

The Commutative Property of Addition

Changing the order in which two numbers are added does not affect their sum.

$3 + 2 = 2 + 3$
↓ 5 ↓ 5

The Associative Property of Addition

When adding three numbers, regrouping addends gives the same sum. Note that the parentheses tell us which numbers to add first.

We add inside the parentheses first

$(4 + 7) + 2 = 4 + (7 + 2)$
$11 + 2 = 4 + 9$
$13 \qquad 13$

Adding Whole Numbers

We add whole numbers by arranging the numbers vertically, keeping the digits with the same place value in the same column. Then we add the digits in each column.

Consider the sum 32 + 65. In the vertical format at the right, the sum of the digits in each column is 9 or less. The sum is 97. When the sum of the digits in a column is greater than 9, we must **regroup** and **carry**, because only a single digit can occupy a single place. Example 1 illustrates this process.

$$\begin{array}{r} 32 \\ +65 \\ \hline 97 \end{array}$$

EXAMPLE 1

Add 47 and 28.

Solution First, we write the addends in expanded form. Then, we add down the ones column.

$$\begin{array}{rcrcr} 47 & = & 4 \text{ tens} + 7 \text{ ones} & = & \overset{1 \text{ ten}}{4 \text{ tens}} + 7 \text{ ones} \\ +28 & = & 2 \text{ tens} + 8 \text{ ones} & = & 2 \text{ tens} + 8 \text{ ones} \\ \hline & & 15 \text{ ones} & & 5 \text{ ones} \end{array}$$

By regrouping, we express 15 ones as 1 ten + 5 ones. Then we carry the 1 ten to the tens place.

Next, we add down the tens column.

$$\begin{array}{l} \text{1 ten} \\ 4 \text{ tens} + 7 \text{ ones} \\ 2 \text{ tens} + 8 \text{ ones} \\ \hline 7 \text{ tens} + 5 \text{ ones} = 75 \end{array}$$

PRACTICE 1

Add: 178 + 207

The following rule tells how to add whole numbers without using expanded form.

To Add Whole Numbers

- Write the addends vertically, lining up the place values.
- Add the digits in the ones column, writing the rightmost digit of the sum on the bottom. If the sum has two digits, carry the left digit to the top of the next column on the left.
- Add the digits in the tens column, as in the preceding step.
- Repeat this process until you reach the last column on the left, writing the entire sum of that column on the bottom.

EXAMPLE 2

Add: 9,824 + 356 + 2,976

Solution We write the problem vertically, with the addends lined up on the right.

$$\begin{array}{r} {\scriptstyle 1} \\ 9,824 \\ 356 \\ +2,976 \\ \hline 6 \end{array}$$

← The sum of the ones digits is 16 ones. We write the 6 and carry the 1 to the tens column.

$$\begin{array}{r} {\scriptstyle 1\,1} \\ 9,824 \\ 356 \\ +2,976 \\ \hline 56 \end{array}$$

The sum of the tens digits is 15 tens. We write the 5 and carry the 1 to the hundreds column.

$$\begin{array}{r} {\scriptstyle 2\,1\,1} \\ 9,824 \\ 356 \\ +2,976 \\ \hline 156 \end{array}$$

The sum of the hundreds digits is 21 hundreds. We write the 1 and carry the 2 to the thousands column.

$$\begin{array}{r} {\scriptstyle 2\,1\,1} \\ 9,824 \\ 356 \\ +2,976 \\ \hline 13,156 \end{array}$$

The sum of the digits in the thousands column is 13, which we write completely—no need to carry here.

The sum is 13,156.

PRACTICE 2

Find the total: 838 + 96 + 9,502

In Example 3, let's apply the operation of addition to finding the geometric perimeter of a figure. The **perimeter** is the distance around a figure, which we can find by adding the lengths of its sides.

EXAMPLE 3

What is the perimeter of the region marked off for the construction of a swimming pool and an adjacent pool cabana?

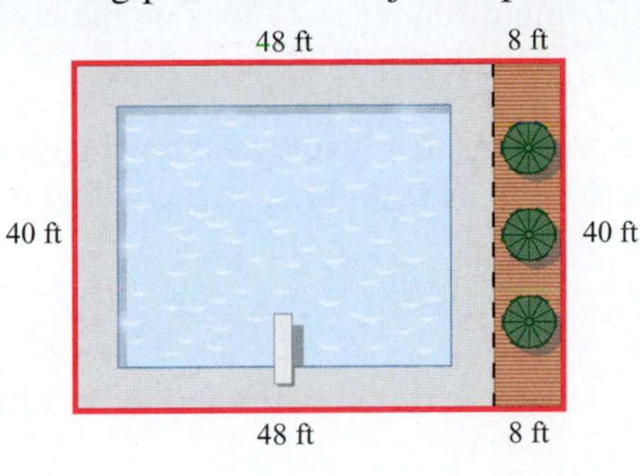

Solution This figure consists of two rectangles placed side by side. We note that the opposite sides of each rectangle are equal in length.

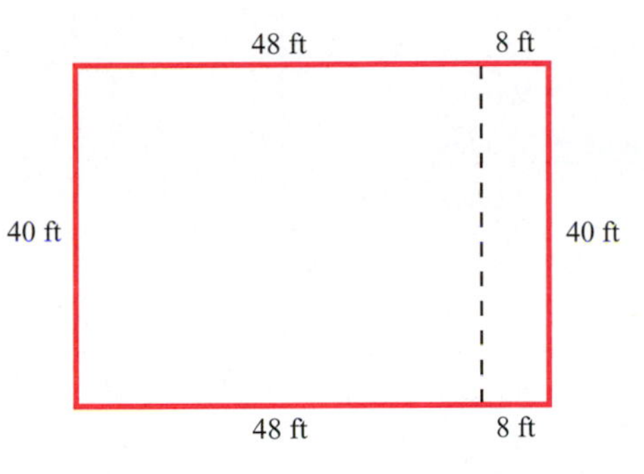

To compute the figure's perimeter, we need to add the lengths of all its sides.

$$\begin{array}{r} 40 \\ 48 \\ 8 \\ 40 \\ 8 \\ +\ 48 \\ \hline 192 \end{array}$$

The figure's perimeter is 192 feet.

PRACTICE 3

How long a fence is needed to enclose the piece of land sketched?

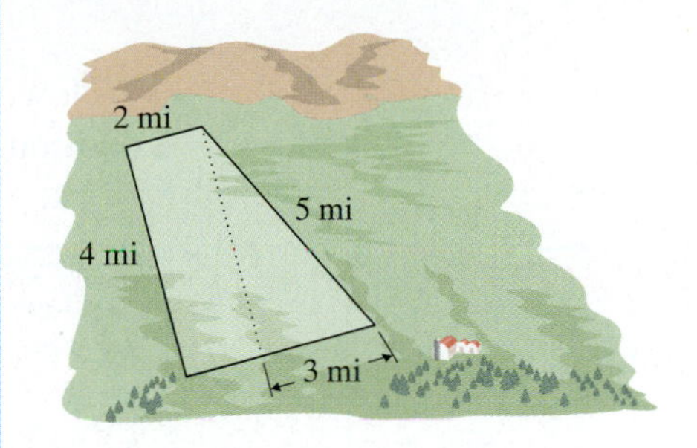

Subtracting Whole Numbers

Consider the subtraction problem $59 - 36$, written vertically at the right. We write the whole numbers underneath one another, lined up on the right, so each column contains digits with the same place value. Subtracting the digits within each column, the bottom digit from the top, the result is a difference of 23.

$$\begin{array}{r} 59 \\ -36 \\ \hline 23 \end{array}$$

Keep in mind two useful properties of subtraction.

- When we subtract a number from itself, the result is 0: $6 - 6 = 0$
- When we subtract 0 from a number, the result is the original number: $25 - 0 = 25$

Tip When writing a subtraction problem vertically, be sure that

- the minuend—the number from which we are subtracting—goes on the top and that
- the subtrahend—the number being taken away—goes on the bottom.

Now we consider subtraction problems that involve *borrowing*. In these problems a digit on the bottom is too large to subtract from the corresponding digit on top.

EXAMPLE 4

Subtract: 329 − 87

Solution We first write these numbers vertically in expanded form.

$$\begin{array}{rcr} 329 = & & 3 \text{ hundreds} + 2 \text{ tens} + 9 \text{ ones} \\ -\ 87 = & - & 8 \text{ tens} + 7 \text{ ones} \\ \hline \end{array}$$

We then subtract the digits in the ones column: 7 ones from 9 ones gives 2 ones.

$$\begin{array}{lr} & 3 \text{ hundreds} + 2 \text{ tens} + 9 \text{ ones} \\ - & 8 \text{ tens} + 7 \text{ ones} \\ \hline & 2 \text{ ones} \end{array}$$

10 tens + 2 tens = 12 tens

We next go to the tens column. We cannot take 8 tens from 2 tens. But we can *borrow* 1 hundred from the 3 hundreds, leaving 2 in the hundreds place. We *exchange* this hundred for 10 tens (1 hundred = 10 tens). Then combining the 10 tens with the 2 tens gives 12 tens.

$$\begin{array}{lr} & \overset{2}{\not{3}} \text{ hundreds} + \overset{1}{\ }2 \text{ tens} + 9 \text{ ones} \\ - & 8 \text{ tens} + 7 \text{ ones} \\ \hline & 2 \text{ ones} \end{array}$$

We next take 8 from 12 in the tens column, giving 4 tens. Finally, we bring down the 2 hundreds. The difference is 242 in standard form.

$$\begin{array}{lr} & \overset{2}{\not{3}} \text{ hundreds} + \overset{1}{\ }2 \text{ tens} + 9 \text{ ones} \\ - & 8 \text{ tens} + 7 \text{ ones} \\ \hline & 2 \text{ hundreds} + 4 \text{ tens} + 2 \text{ ones} = 242 \end{array}$$

PRACTICE 4

Subtract: 748 − 97

Although we can always rewrite whole numbers in expanded form so as to subtract them, the following rule provides a shortcut.

To Subtract Whole Numbers

- On top, write the number *from which* we are subtracting. On the bottom, write the number that is being taken *away*, lining up the place values. Subtract in each column separately.
- Start with the ones column.
  - **a.** If the digit on top is *larger* than or *equal* to the digit on the bottom, subtract and write the difference below the bottom digit.
  - **b.** If the digit on top is *smaller* than the digit on the bottom, borrow from the digit to the left on top. Then subtract and write the difference below the bottom digit.
- Repeat this process until the last column on the left is finished.

Adding and Subtracting Whole Numbers on a Calculator

Calculators are handy and powerful tools for carrying out complex computations. But it is easy to press a wrong key, so be sure to estimate the answer and compare this estimate to the displayed answer to see if it is reasonable.

EXAMPLE 11

On a calculator, compute the sum of 3,125 and 9,391.

Solution

Press	Display
3125 [+] 9391 [ENTER]	3125+9391 12516.

To check this answer, we mentally round the addends and then add.

$$\begin{aligned} 3125 &\approx 3{,}000 \\ 9391 &\approx \underline{9{,}000} \\ & 12{,}000 \end{aligned}$$

This estimate is reasonably close to our answer 12,516.

PRACTICE 11

Use a calculator to add: 39,822 + 9,710

Pressing the clear key cancels the number in the display. Press this key after completing a computation to be sure that no number remains to affect the next problem. Note that calculator models vary as to how they work, so it may be necessary to consult the manual for a particular model.

EXAMPLE 12

Calculate: 39 + 48 + 277

Solution

Press	Display
39 [+] 48 [+] 277 [ENTER]	39+48+277 364.

A reasonable estimate is the sum of 40, 50, and 300, or 390—close to our calculated answer 364.

PRACTICE 12

Find the sum on a calculator:
23,801 + 7,116 + 982

When using a calculator to subtract,

- enter the numbers in the correct order—first enter the number **from which** we are subtracting and then the number **being** subtracted; and
- do not confuse the *negative sign key* [(−)] that some calculators have with the *subtraction key* [−].

EXAMPLE 13

Subtract on a calculator: 3,000 − 973

Solution

Press	Display
3000 [−] 973 [ENTER]	3000−973 2027.

A good estimate is 3,000 − 1,000, or 2,000, which is close to 2,027.

PRACTICE 13

Use a calculator to find the difference between 5,280 feet and 2,781 feet.

Mathematically Speaking

Fill in each blank with the most appropriate term or phrase from the given list.

subtrahend	addends	left	estimates
Commutative Property of Addition	right	Identity Property of Addition	Associative Property of Addition
	difference	sum	
	minuend		

1. The operation of addition can be thought of as moving to the _______ on a number line.

2. The _______ states that the sum of a number and zero is the original number.

3. The result of addition is called the _______.

4. The _______ states that changing the order in which two numbers are added does not affect the sum.

5. The _______ states that when adding three numbers, regrouping addends gives the same sum.

6. In an addition problem, the numbers being added are called _______.

7. In a subtraction problem, the number being subtracted is called the _______.

8. The result of subtraction is called the _______.

Add and check by estimation.

9. $\begin{array}{r} 100{,}250 \\ +\ 77{,}528 \\ \hline \end{array}$

10. $\begin{array}{r} 3{,}505 \\ +\ \ \ 11 \\ \hline \end{array}$

11. $\begin{array}{r} 8{,}132 \\ +6{,}578 \\ \hline \end{array}$

12. $\begin{array}{r} 60{,}725 \\ +38{,}928 \\ \hline \end{array}$

13. $\begin{array}{r} 7{,}481 \\ 702 \\ +5{,}819 \\ \hline \end{array}$

14. $\begin{array}{r} 99{,}103 \\ 33{,}450 \\ +\ 6{,}627 \\ \hline \end{array}$

15. $\begin{array}{r} 49{,}002 \\ 1{,}999 \\ +\ 5{,}187 \\ \hline \end{array}$

16. $\begin{array}{r} 55{,}998 \\ 40{,}003 \\ +17{,}827 \\ \hline \end{array}$

17. 1,903 + 5,075

18. 7,406 + 12,381

19. 800 + 20 + 4,000

20. 40,000 + 800 + 60

21. 31 + 93 + 277 + 12

22. 418 + 47 + 365 + 95

23. 3,911 + 2,947 + 8,007

24. 5,374 + 4,055 + 20,173

25. 6,482 meters + 9,027 meters

26. 17,812 miles + 4,283 miles

27. 35 hours + 47 hours

28. 225 square feet + 896 square feet

29. $92,258 + $7,447 + $5,126

30. $55,709 + $2,822 + $30,819

31. $1,863 + $1,089 + $9,772

32. 5,009 feet + 7,993 feet

33. 8,300 tons + 22,900 tons

34. 420,057 pounds + 900,808 pounds

35.
$$\begin{array}{r} 3{,}088{,}281 \\ 5{,}658{,}137 \\ +4{,}550{,}239 \\ \hline \end{array}$$

36.
$$\begin{array}{r} 638{,}719 \\ 40{,}003 \\ +984{,}035 \\ \hline \end{array}$$

37.
$$\begin{array}{r} 2{,}008{,}490 \\ 8{,}948{,}227 \\ +11{,}956{,}174 \\ \hline \end{array}$$

38.
$$\begin{array}{r} 1{,}938{,}722 \\ 325{,}411 \\ +\ 517{,}827 \\ \hline \end{array}$$

In each addition table, fill in the empty spaces. Check that the sum in the shaded empty space is the same working both downward and across.

39.

+	400	200	1,200	300	Total
300					
800					
Total					

40.

+	4,000	300	3,000	2,000	Total
100					
900					
Total					

41.

+	389	172	1,155	324	Total
255					
799					
Total					

42.

+	3,749	279	2,880	1,998	Total
134					
896					
Total					

In each group of three sums, one is wrong. Use estimation to identify which sum is incorrect.

43. a.
$$\begin{array}{r} 814 \\ 9{,}106 \\ +2{,}811 \\ \hline 15{,}731 \end{array}$$

b.
$$\begin{array}{r} 30{,}812 \\ 47{,}045 \\ +\ 9{,}338 \\ \hline 87{,}195 \end{array}$$

c.
$$\begin{array}{r} 183{,}066 \\ 78{,}911 \\ +\ 96{,}527 \\ \hline 358{,}504 \end{array}$$

44. a.
$$\begin{array}{r} 1{,}035 \\ 5{,}210 \\ +7{,}992 \\ \hline 14{,}237 \end{array}$$

b.
$$\begin{array}{r} 5{,}801 \\ 3{,}882 \\ +12{,}644 \\ \hline 32{,}327 \end{array}$$

c.
$$\begin{array}{r} 801{,}716 \\ 78{,}001 \\ +5{,}009{,}635 \\ \hline 5{,}889{,}352 \end{array}$$

45. **a.** $\begin{array}{r} \$711{,}488 \\ 102{,}663 \\ +\ 95{,}003 \\ \hline \$809{,}154 \end{array}$ **b.** $\begin{array}{r} \$62{,}933 \\ 51{,}858 \\ +\ 49{,}612 \\ \hline \$164{,}403 \end{array}$ **c.** $\begin{array}{r} \$106{,}729 \\ 99{,}821 \\ +\ 103{,}277 \\ \hline \$309{,}827 \end{array}$

46. **a.** $\begin{array}{r} \$9{,}512{,}622 \\ 8{,}038{,}517 \\ +\ 2{,}615{,}334 \\ \hline \$20{,}166{,}473 \end{array}$ **b.** $\begin{array}{r} \$4{,}277{,}020 \\ 915{,}611 \\ +\ 3{,}688{,}402 \\ \hline \$8{,}881{,}033 \end{array}$ **c.** $\begin{array}{r} \$200{,}312 \\ 102{,}683 \\ +\ 504{,}113 \\ \hline \$707{,}108 \end{array}$

Subtract and check.

47. $\begin{array}{r} 379 \\ -162 \\ \hline \end{array}$ **48.** $\begin{array}{r} 362 \\ -110 \\ \hline \end{array}$ **49.** $\begin{array}{r} 200 \\ -110 \\ \hline \end{array}$ **50.** $\begin{array}{r} 210 \\ -100 \\ \hline \end{array}$

51. $\begin{array}{r} 401 \\ -\ 39 \\ \hline \end{array}$ **52.** $\begin{array}{r} 728 \\ -539 \\ \hline \end{array}$ **53.** $\begin{array}{r} 70{,}000 \\ -\ 1{,}759 \\ \hline \end{array}$ **54.** $\begin{array}{r} 8{,}000 \\ -1{,}691 \\ \hline \end{array}$

55. $\begin{array}{r} 5{,}062 \\ -2{,}777 \\ \hline \end{array}$ **56.** $\begin{array}{r} 3{,}005 \\ -1{,}666 \\ \hline \end{array}$ **57.** $\begin{array}{r} 72{,}000 \\ -19{,}001 \\ \hline \end{array}$ **58.** $\begin{array}{r} 2{,}001 \\ -\ \ 2 \\ \hline \end{array}$

59. $\begin{array}{r} 3{,}000 \\ -\ \ 57 \\ \hline \end{array}$ **60.** $\begin{array}{r} 52{,}947 \\ -27{,}997 \\ \hline \end{array}$ **61.** $\begin{array}{r} 261{,}406 \\ -57{,}941 \\ \hline \end{array}$ **62.** $\begin{array}{r} 729{,}888 \\ -192{,}889 \\ \hline \end{array}$

Find the difference and check.

63. 550 − 182

64. 1,448 − 962

65. 6,000 − 1,004

66. 8,602 − 907

67. 3,570 − 2,588

68. 2,182 − 899

69. 5,000 miles − 3,005 miles

70. 701 square feet − 206 square feet

71. $800 − $131

72. 622 hours − 137 hours

73. $4,812 − $1,203

74. 402 miles − 57 miles

75. 500 books − 227 books

76. $537 − $196

77. 527 meters − 318 meters

78. 1,266 tons − 597 tons

79. $\begin{array}{r} 30{,}000{,}000 \\ -27{,}999{,}000 \\ \hline \end{array}$ **80.** $\begin{array}{r} 1{,}973{,}000 \\ -\ 997{,}001 \\ \hline \end{array}$ **81.** $\begin{array}{r} 3{,}402{,}331 \\ -2{,}588{,}902 \\ \hline \end{array}$ **82.** $\begin{array}{r} 14{,}500{,}007 \\ -13{,}972{,}008 \\ \hline \end{array}$

In each group of three differences, one is wrong. Use estimation to identify which difference is incorrect.

83. **a.** $\begin{array}{r} 817{,}770 \\ -502{,}966 \\ \hline 314{,}804 \end{array}$ **b.** $\begin{array}{r} 11{,}172{,}055 \\ -\ 7{,}892{,}106 \\ \hline 3{,}279{,}949 \end{array}$ **c.** $\begin{array}{r} 71{,}384{,}612 \\ -32{,}016{,}594 \\ \hline 29{,}368{,}018 \end{array}$

84. **a.** $\begin{array}{r} 67{,}812 \\ -12{,}180 \\ \hline 55{,}632 \end{array}$ **b.** $\begin{array}{r} 3{,}997{,}401 \\ -1{,}125{,}166 \\ \hline 1{,}872{,}235 \end{array}$ **c.** $\begin{array}{r} 316{,}134 \\ -\ 89{,}164 \\ \hline 226{,}970 \end{array}$

85. a.
$$\begin{array}{r} \$381{,}882 \\ -\ 173{,}552 \\ \hline \$108{,}330 \end{array}$$

b.
$$\begin{array}{r} \$479{,}116 \\ -\ 102{,}663 \\ \hline \$376{,}453 \end{array}$$

c.
$$\begin{array}{r} \$200{,}072{,}639 \\ -\ 150{,}038{,}270 \\ \hline \$\ 50{,}034{,}369 \end{array}$$

86. a.
$$\begin{array}{r} \$3{,}810{,}662 \\ -\ \ 299{,}137 \\ \hline \$3{,}511{,}525 \end{array}$$

b.
$$\begin{array}{r} \$4{,}718{,}287 \\ -\ 1{,}002{,}875 \\ \hline \$5{,}721{,}162 \end{array}$$

c.
$$\begin{array}{r} \$381{,}975 \\ -\ 117{,}263 \\ \hline \$264{,}712 \end{array}$$

Mixed Practice

Perform the indicated operation.

87.
$$\begin{array}{r} 7{,}415 \\ -\ 350 \\ \hline \end{array}$$

88.
$$\begin{array}{r} 90{,}316 \\ 10{,}882 \\ +\ 5{,}281 \\ \hline \end{array}$$

89. 281 + 758 + 104 + 533

90. \$5,233 + \$481 + \$82

91. 8,286 − 3,100

92. 410,700 miles − 280,900 miles

Applications

Solve and check.

93. In 1900, the population of the United States was approximately 76,000,000. During the next 100 years, the population grew by about 205,000,000 people. What was the population in 2000? (*Source:* U.S. Bureau of the Census)

94. It is recommended that a person drink 64 ounces of water each day. Will a person who drank 32 ounces so far today and then drinks another 24 ounces meet the recommended daily amount?

95. Of the 6,000,0000 square miles of tropical rainforest that originally existed on earth, 3,400,000 square miles have been lost due to deforestation. How many square miles of rainforest still exist? (*Source:* The Nature Conservancy, 2006)

96. The Great Blue Norther of 1911 was the largest cold snap in U.S. history. On the day of the Norther, the temperature in Oklahoma City dropped from a record high of 83°F to a record low of 17°F in a 24-hour period. By how much did the temprature drop that day? (*Source:* National Weather Service)

97. The chart shows the 2006 Winter Olympic medal counts of selected countries.

Country	Gold	Silver	Bronze
Austria	9	7	7
Canada	7	10	7
Germany	11	12	6
Russia	8	6	8
United States	9	9	7

a. Calculate the total number of medals won by each country.

b. Which country won the most medals? (*Source:* http://www.olympic.org)

98. Consider the deposit slip shown.

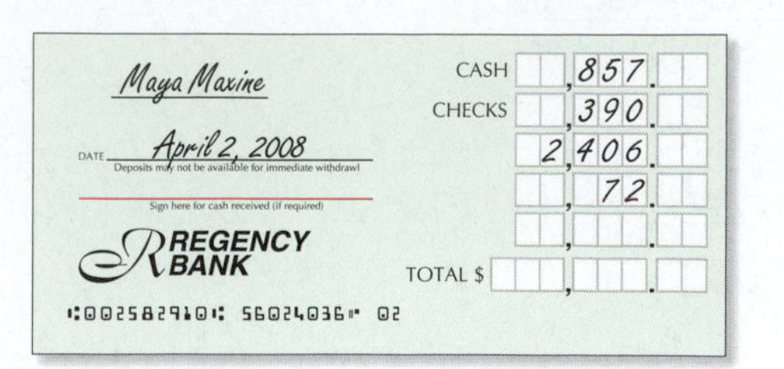

a. Estimate how much money is being deposited.

b. Fill in the exact total.

99. Blues singer Bessie Smith was born in 1894 and died in 1937. About how old was she when she died? (*Source: Encyclopedia of World Biography*)

100. The United States entered the First World War in 1917 and the Second World War in 1941. Approximately how many years apart were these two events?

101. A sign in an elevator reads: MAXIMUM CAPACITY: 1,000 POUNDS. The passengers in the elevator weigh 187 pounds, 147 pounds, 213 pounds, 162 pounds, 103 pounds, and 151 pounds. Will the elevator be overloaded?

102. A student would like to install a computer program that requires 128 megabytes (MB) of memory. The memory in his old computer is only 80 MB. If he increases the memory in his computer by 64 MB, will there be enough memory for him to run the program?

103. The thermometer at the right shows the boiling point and the freezing point of water in degrees Fahrenheit (°F). What is the difference between these two temperatures?

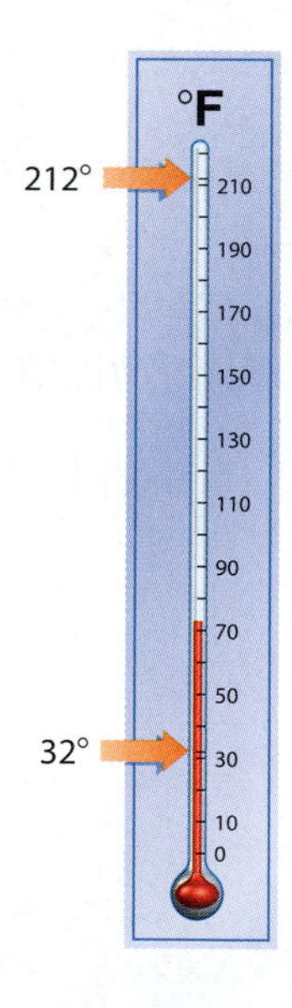

104. The following ad for a hybrid car was listed in a local newspaper. How much below the MSRP (manufacturer's suggested retail price) is the selling price?

105. Some friends cycle from town A to town B, to town C, to town D, and then back to A, as shown below. How far did they cycle in all?

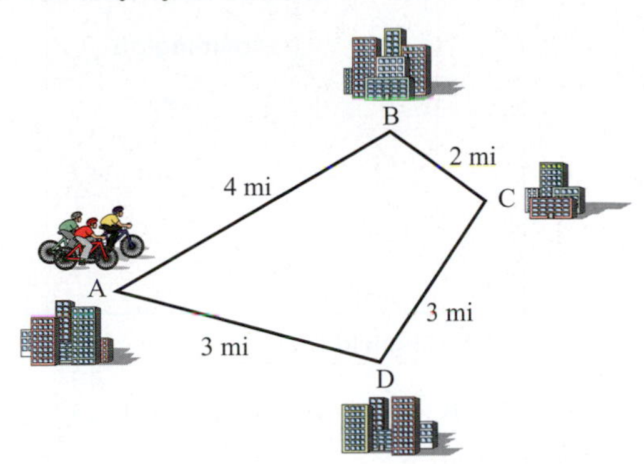

106. What is the length of the molding along the perimeter of the room pictured (yards = yd)?

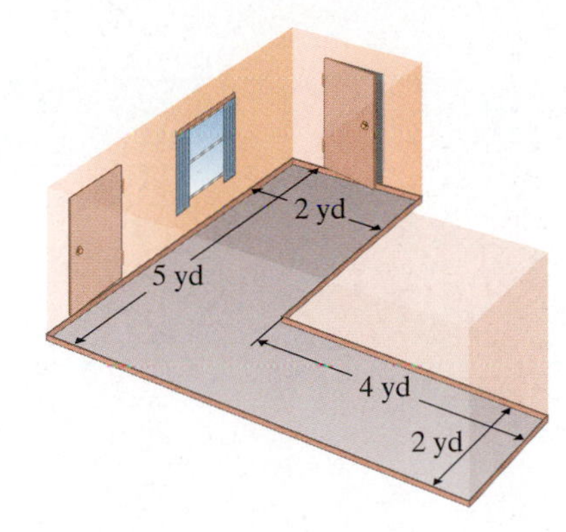

107. The total number of hours a typical person spends surfing the Web is expected to increase from 200 hours in 2005 to 236 hours in 2008. How big an increase is this? (*Source: Statistical Analysis*, Veronis Suhler Stevenson, 2006)

108. In a recent year, 4,822 merchant vessels were registered under the flag of Panama, in contrast to 412 under the U.S. flag. What is the difference between the number of merchant vessels in these fleets? (*Source:* Maritime Administration, U.S. Department of Commerce)

109. Find the total amount of money deposited in a checking account according to the following deposit slip.

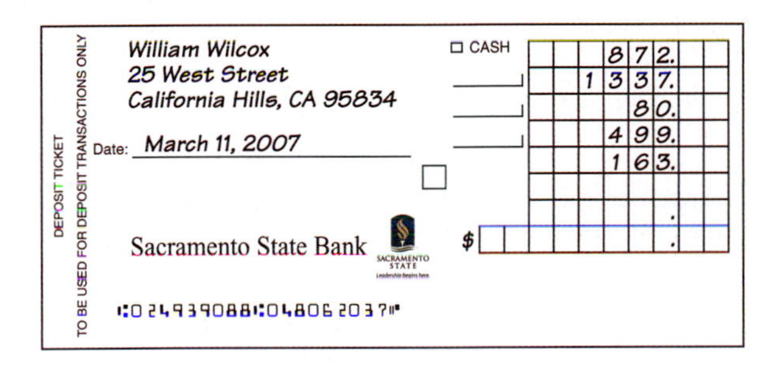

110. An oil tanker broke apart at sea. It spilled 150,000 gallons (gal) of crude oil the first day, 400,000 gal the second day, and 1,000,000 gal the third day. How much oil was spilled in all?

111. An airline limits the size of luggage that passengers can take with them. For each bag, the sum of the outside dimensions, that is, the length, width, and height, is limited to 62 inches for bags checked free of charge and 45 inches for carry-on bags. The the bag that a passanger wishes to take on a flight measures 21 inches by 10 inches by 19 inches. (*Source:* http://www.aa.com)

a. Will the passenger be allowed to carry the bag onto the plane?

b. To check the bag free of charge?

112. A particular credit card has a credit limit of $3,500. On this card, there is a balance due of $2,367.

a. Is there enough credit available to pay a tuition bill of $1,295?

b. If, instead of the tuition bill, a DVD recorder for $253 were purchased on the card, how much credit on the card would still be available?

113. The following graph gives the amount of red meat produced during a recent year in leading American states.

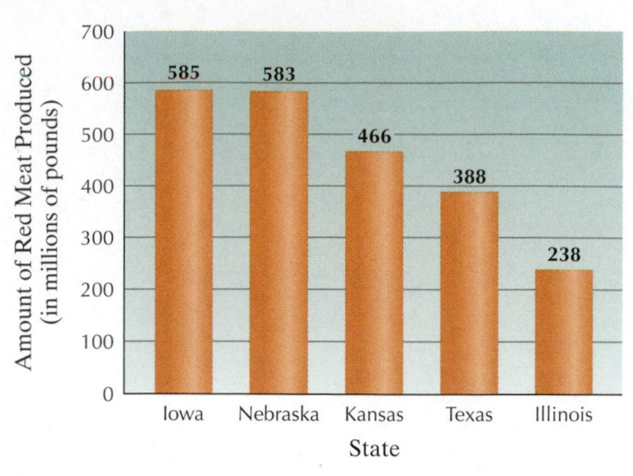

(*Source:* USDA)

a. How much more red meat did Iowa produce than Illinois?

b. Estimate the total amount of red meat produced in these five states.

c. Find the total amount of red meat produced in these five states.

114. A real estate broker based in Springfield, Illinois, covers the region shown below.

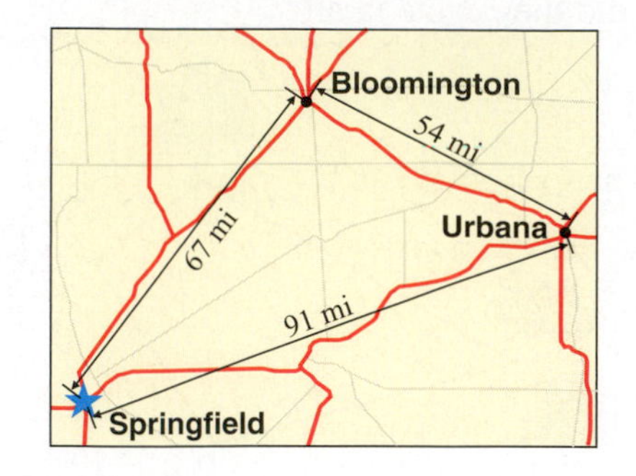

a. How much further from Springfield is Urbana than Bloomington?

b. How far does he drive going from Springfield to Bloomington by way of Urbana?

c. If the broker then drives back home to Springfield directly from Bloomington, how much shorter is the trip returning home than the earlier trip from Springfield?

Use a calculator to solve each problem, giving (a) the operation(s) carried out in the solution, (b) the exact answer, and (c) an estimate of the answer.

115. Is the total amount of the deposits shown on the following bank statement correct?

MBU Bank & Trust Co.

Your Account

Deposits	Date
$ 83	2/13
$ 59	2/14
$ 727	2/16
$ 183	2/17
$ 511	2/21

TOTAL $1,563

116. At its first eight games this season, a professional baseball team had the following paid attendance:

Game	Attendance
1	11,862
2	18,722
3	14,072
4	9,713
5	25,913
6	28,699
7	19,302
8	18,780

What was the combined attendance for these games?

• *Check your answers on page A-1.*

MINDSTRETCHERS

Writing

1. There are many different ways of putting numerical expressions into words.

   a. For example, 3 + 2 can be expressed as

   the sum of 3 and 2, 2 more than 3, or 3 increased by 2

   What are some other ways of reading this expression?

   b. For example, 5 − 2 can be expressed as

   the difference between 5 and 2, 5 take away 2, or 5 decreased by 2

   Write two other ways.

Critical Thinking

2. In a **magic square**, the sum of every row, column, and diagonal is the same number. Using the given information, complete the square at the right, which contains the whole numbers from 1 to 16. (*Hint:* The sum of every row, column, and diagonal is 34.)

16	3	2	
	10	11	
	6	7	

Groupwork

3. Two methods for borrowing in a subtraction problem are illustrated as follows. In method (a)—the method that we have already discussed—we borrow by taking 1 from the top, and in method (b) by adding 1 to the bottom.

$$\text{a.}\quad \begin{array}{r} {}^{7}\cancel{8}\,{}^{1}5\;9 \\ -3\;7\;6 \\ \hline 4\;8\;3 \end{array} \qquad\qquad \text{b.}\quad \begin{array}{r} 8\;{}^{1}5\;9 \\ -{}^{4}\cancel{3}\;7\;6 \\ \hline 4\;8\;3 \end{array}$$

Note that we get the same answer with both methods. Working with a partner, discuss the advantages of each method.

1.3 Multiplying Whole Numbers

OBJECTIVES

- To multiply whole numbers
- To estimate the product of whole numbers
- To solve word problems involving the multiplication of whole numbers

The Meaning and Properties of Multiplication

What does it mean to multiply whole numbers? A good answer to this question is *repeated addition.*

For instance, suppose that you buy 4 packages of pens and each package contains 3 pens. How many pens are there altogether?

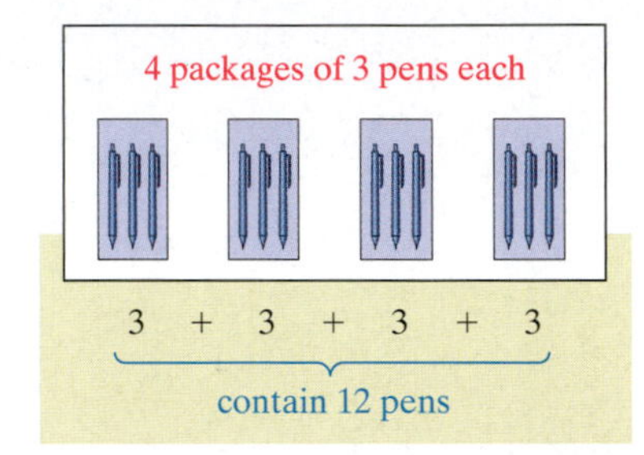

That is, $4 \times 3 = 3 + 3 + 3 + 3 = 12$. Generally, *multiplication means adding the same number repeatedly.*

We can also picture multiplication in terms of a rectangular figure, like this one, that represents 4×3.

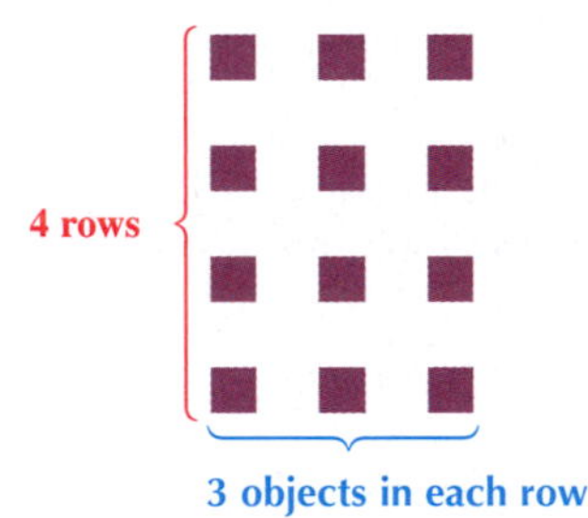

In a multiplication problem, the numbers being multiplied are called *factors*. The result is the *product*.

There are several ways to write a multiplication problem.

		Factor ↓		Factor ↓		Product ↓
×	the times sign	4	×	3	=	12
·	a multiplication dot	4	·	3	=	12
( )( )	parentheses	(4)(3)			=	12

Like addition and subtraction, multiplication problems can be written either horizontally or vertically.

$8 \times 5 = 40$
Horizontal

$$\begin{array}{r} 8 \\ \times\ 5 \\ \hline 40 \end{array}$$
Vertical

The operation of multiplication has several important properties that we use frequently.

The Identity Property of Multiplication

The product of any number and 1 is that number.

$$1 \times 12 = 12$$
$$5 \times 1 = 5$$

The Multiplication Property of 0

The product of any number and 0 is 0.

$$49 \times 0 = 0$$
$$0 \times 8 = 0$$

The Commutative Property of Multiplication

Changing the order in which two numbers are multiplied does not affect their product.

$$2 \times 9 = 9 \times 2$$
$$18 = 18$$

The Associative Property of Multiplication

When multiplying three numbers, regrouping the factors gives the same product.

We multiply inside the parentheses first.

$$(3 \times 4) \times 5 = 3 \times (4 \times 5)$$
$$12 \times 5 = 3 \times 20$$
$$60 = 60$$

The next—and last—property of multiplication also involves addition.

The Distributive Property

Multiplying a factor by the sum of two numbers gives the same result as multiplying the factor by each of the two numbers and then adding.

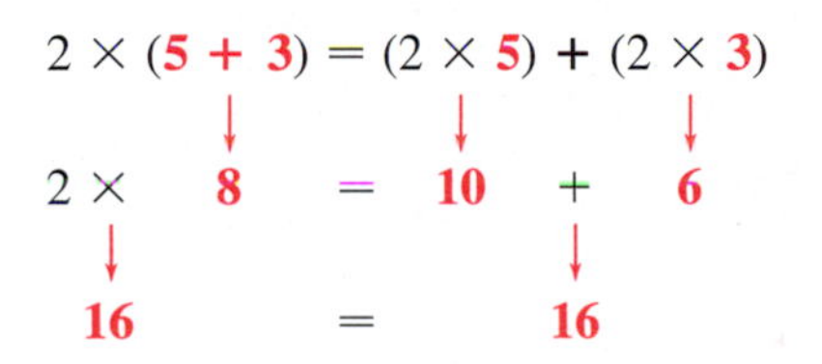

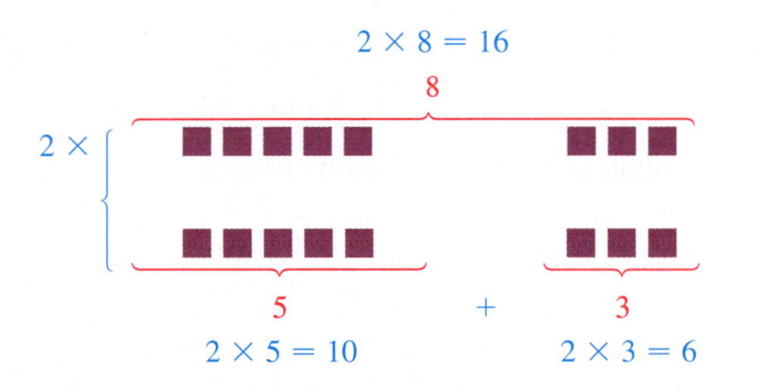

Before going on to the next section, study these properties of multiplication.

Multiplying Whole Numbers

Now let's consider problems in which we multiply any whole number by a single-digit whole number.

Note that, to multiply whole numbers with reasonable speed, you must commit to memory the products of all single-digit whole numbers.

EXAMPLE 1

Multiply: $98 \cdot 4$

Solution We recall that the dot means multiplication. We first write the problem vertically.

We recall that the 9 in 98 means 9 tens.

So the product of 98 and 4 is 392.

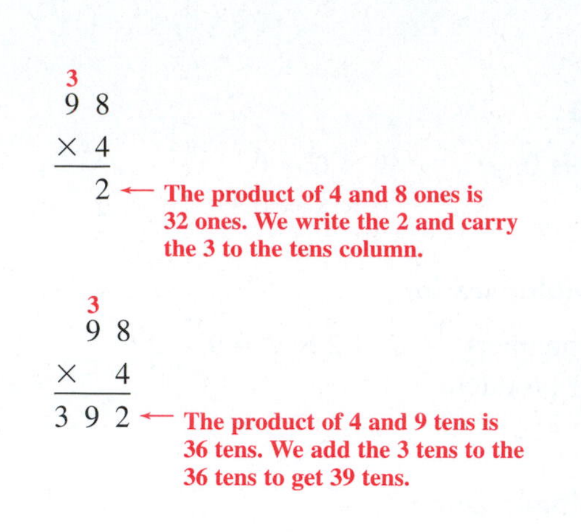

3
9 8
× 4
2
The product of 4 and 8 ones is 32 ones. We write the 2 and carry the 3 to the tens column.
3
9 8
× 4
3 9 2
The product of 4 and 9 tens is 36 tens. We add the 3 tens to the 36 tens to get 39 tens.

PRACTICE 1

Find the product of 76 and 8.

EXAMPLE 2

Calculate: (806) (7)

Solution We recall that parentheses side-by-side means to multiply. We write this problem vertically.

$$\begin{array}{r} \overset{4}{8\,0}\,6 \\ \times \quad 7 \\ \hline 5{,}6\,4\,2 \end{array}$$

Here, 7×0 tens $= 0$ tens. Add the carried 4 tens to the 0 tens to get 4 tens.

The product of 806 and 7 is 5,642.

PRACTICE 2

Find the product: (705)(6)

Now let's look at multiplying any two whole numbers.

Consider multiplying 32 by 48. We can write 32×48 as follows.

$$32 \times 48 = 32 \times (40 + 8)$$

We then use the distributive property to get the answer.

$$\begin{aligned} 32 \times (40 + 8) &= (32 \times 40) + (32 \times 8) \\ &= 1{,}280 + 256 \\ &= 1{,}536 \end{aligned}$$

Generally, we solve this problem vertically.

$$\begin{array}{r} \overset{1}{3}\,2 \\ \times \quad 4\,8 \\ \hline 2\,5\,6 \\ 1\,2\,8\,0 \\ \hline 1{,}5\,3\,6 \end{array}$$

- 256 ← Partial product (8×32)
- 1280 ← Partial product (40×32)
- 1,536 ← Add the partial products.

Shortcut

$$\begin{array}{r} 32 \\ \times\ 48 \\ \hline 256 \\ 1\,28 \\ \hline 1{,}536 \end{array}$$

- 256 ← (8×32)
- 1 28 ← (4×32)

If we use just the tens digit 4, we must write the product 128 leftward, starting at the tens column.

Example 2 suggests the following rule for multiplying whole numbers.

To Multiply Whole Numbers

- Multiply the top factor by the ones digit in the bottom factor, and write down this product.
- Multiply the top factor by the tens digit in the bottom factor, and write this product leftward, beginning with the tens column.
- Repeat this process until all the digits in the bottom factor are used.
- Add the partial products, writing down this sum.

EXAMPLE 3

Multiply: 300×50

Solution

```
   300
 ×  50
   000  ← 0 × 300 = 0
 15 00  ← 5 × 300 = 1,500
15,000
```

PRACTICE 3

Find the product of 1,200 and 400.

In Example 3, note that the number of zeros in the product equals the total number of zeros in the factors. This result suggests a shortcut for multiplying factors that end in zeros.

```
   300  ← 2 zeros
 ×  50  ← 1 zero
15,000  ← 2 + 1 = 3 zeros
```

Tip When multiplying two whole numbers that end in zeros, multiply the nonzero parts of the factors and then attach the total number of zeros to the product.

EXAMPLE 4

Simplify: $739 \cdot 305$

Solution

```
    739
 ×  305
  3 695  ← 5 × 739
  0 00   ← 0 × 739 = 0
221 7    ← 3 × 739
225,395
```

We don't have to write the row 000. Here is a shortcut.

```
    739
 ×  305
  3 695
221 70   ← This one 0 represents the product of the tens digit 0
225,395     and 739. This 0 lines up the products correctly.
```

PRACTICE 4

Find the product of 987 and 208.

Now let's apply the operation of multiplication to geometric area. Area means the number of square units that a figure contains.

In the rectangle at the right, each small square represents 1 square inch (sq in.). Finding the rectangle's area means finding the number of sq-in. units that it contains. A good strategy here is to find the number of units in each row and then multiply that number by the number of rows.

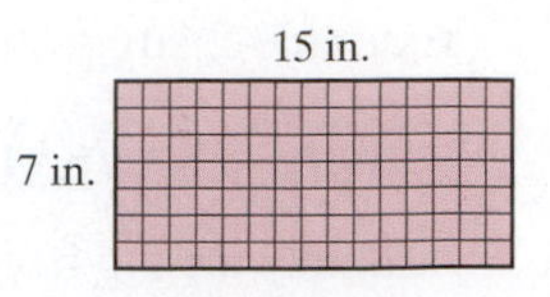

There are two ways to find that there are 15 squares in a row—either by directly counting the squares or by noting that the length of the figure is 15 in. Similarly, we find that the figure contains 7 rows. Therefore the area of the figure is 15 × 7, or 105 sq in.

In general, we can compute the *area of a rectangle* by finding the product of its length and its width.

EXAMPLE 5

Calculate the area of the home office shown in the diagram.

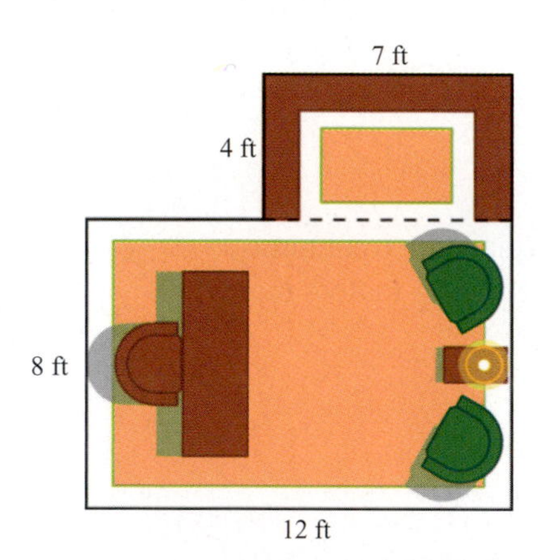

Solution The dashed line separates the office into two connected rectangles. The top retangle measures 7 feet by 4 feet, and so its area is 7 × 4, or 28 square feet. The bottom rectangle measures 12 feet by 8 feet, and its area is 12 × 8, or 96 square feet. The entire area of the office is the sum of two smaller areas: 28 + 96, or 124 square feet. So the area of the home office is 124 square feet.

PRACTICE 5

Find the area of the room pictured.

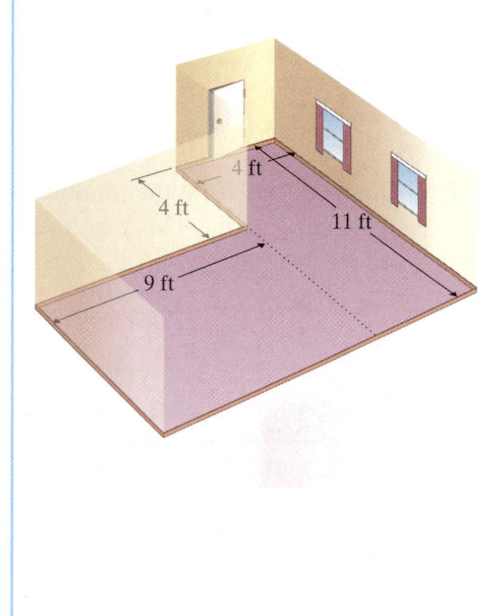

Estimating Products

As mentioned before, estimation is a valuable technique for checking an exact answer. When checking a product by estimation, round each factor to its largest place.

EXAMPLE 6

Multiply 328 by 179. Check the answer by estimation.

Solution

```
   328
 × 179
 2 952
22 96
32 8
58,712  ← Exact product
```

PRACTICE 6

Find the product of 455 and 248. Use estimation to check your answer.

Check

$$\begin{array}{rrl} 328 \approx & 300 & \leftarrow \text{The largest place is hundreds.} \\ \times\ 179 \approx & \times\ 200 & \leftarrow \text{The largest place is hundreds.} \\ \hline 58{,}712 & 60{,}000 & \leftarrow \text{Estimated product} \end{array}$$

Our exact product (58,712) and the estimated product (60,000) are fairly close.

When solving some multiplication problems, we are willing to settle for—or even prefer—an approximate answer.

EXAMPLE 7

A couple planning their wedding set aside $2,000 for floral centerpieces. Each centerpiece costs $72, and a total of 19 centerpieces are needed. By estimating, decide if the couple has set aside enough money for the centerpieces.

Solution To estimating a product, we first round each factor to its largest place value so that every digit after the first digit is 0.

$$\begin{array}{rl} 72 \approx 70 & \leftarrow \text{The largest place is tens.} \\ \times 19 \approx 20 & \leftarrow \text{The largest place is tens.} \end{array}$$

Then, we multiply the rounded factors.

$$70 \times 20 = 1{,}400$$

Since the centerpieces will cost about $1,400 and since $2,000 is greater than $1,400, we conclude that the couple has set aside enough money for the centerpieces.

PRACTICE 7

Producing flyers for your college's registration requires 25,000 sheets of paper. If the college buys 38 reams of paper and there are 500 sheets in a ream, estimate to decide if there is enough paper to produce the flyers.

Multiplying Whole Numbers on a Calculator

Now let's use a calculator to find a product. When you are using a calculator to multiply large whole numbers, the answer may be too big to fit in the display. When this occurs the answer may be displayed in scientific notation (see Appendix xx).

EXAMPLE 8

Use a calculator to multiply: $3{,}192 \times 41$

Solution

Press	Display
3192 [×] 41 [ENTER =]	3192*41 130872.

A reasonable estimate for this product is $3{,}000 \times 40$, or 120,000, which supports our answer, 130,872.

PRACTICE 8

Find the product: $2{,}811 \times 365$

EXAMPLE 9

Calculate: $61 \cdot 24 \cdot 19$

Solution

Press	Display
61 [×] 24 [×] 19 [ENTER =]	61*24*19 27816.

A good estimate is $60 \cdot 20 \cdot 20$, or 24,000—in the ballpark of 27,816.

PRACTICE 9

Multiply: $2{,}133 \cdot 18 \cdot 9$

Mathematically Speaking

Fill in each blank with the most approppriate term or phrase from the given list.

Identity Property of Multiplication	addition	Distributive Property	product
subtraction	perimeter	sum	area
Multiplication Property of 0	Associative Property		

1. The result of multiplying two factors is called their __________.

2. The __________ is illustrated by $3 \times (7 + 2) = (3 \times 7) + (3 \times 2)$.

3. The __________ states that the product of any number and 1 is that number.

4. The __________ states that the product of any number and 0 is 0.

5. The multiplication of whole numbers can be thought of as repeated __________.

6. The __________ of a figure is the number of square units that it contains.

Compute.

7. 4×100

8. $1{,}000 \times 12$

9. 710×200

10. 270×50

11. $8{,}500 \times 20$

12. 680×300

13. $10{,}000 \times 700$

14. $1{,}000 \times 8{,}000$

Multiply and check by estimation.

15. $6{,}350 \times 2$

16. $8{,}864 \times 7$

17. 209×2

18. 703×9

19. $812{,}000 \times 4$

20. $19{,}250 \times 8$

21. 882×74

22. 881×28

23. $43 \cdot 19$

24. $85 \cdot 72$

25. $709 \cdot 48$

26. $602 \cdot 34$

27. 273×11

28. 607×65

29. 301×12

30. 513×34

31. $3{,}001 \times 19$

32. $4{,}005 \times 72$

33. $5{,}072 \times 48$

34. $8{,}801 \times 25$

35. $5{,}003 \times 40$

36. $2{,}881 \times 70$

Find the product and check by estimation.

37. (372)(403)

38. (699)(101)

39. 8,500 × 17

40. 700 × 207

41. 406 × 305

42. 702 × 509

43. 46 · 8 · 9

44. 13 · 11 · 5

45. 81 × 2 × 13

46. 3 × 5 × 88

47. (10)(10)(400)

48. (20)(80)(30)

49. 57 × 81 × 5

50. 73 × 4 × 33

51. 8,972 × 365

52. 7,552 × 841

53. 18,650 × 2,949

54. 8,783 × 7,159

In each group of three products, one is wrong. Use estimation to indentify which product is incorrect.

55. a. 802 × 755 = 605,510 **b.** 39 × 4,722 = 184,158 **c.** 77 × 6,005 = 46,385

56. a. 618 × 555 = 342,990 **b.** 86,331 × 21 = 18,129,511 **c.** 380 × 772 = 293,360

57. a. 9 × 37,118 = 334,062 **b.** 82 × 961 = 7,882 **c.** 13 × 986 = 12,818

58. a. 3,002 × 9 = 2,718 **b.** 58 × 891 = 51,678 **c.** 106 × 68 = 7,208

Mixed Practice

Multiply and check by estimation.

59. 48 · 5 · 12

60. 89 × 10,000

61. 9,605 × 24

62. (809)(201)

63. 357,000 × 3

64. 301 · 34

65. (50)(60)(100)

66. 495 × 21

Applications

Solve. Then check by estimation.

67. Underwater explorers in the eastern Mediterranean Sea found the wreck of an Egyptian ship that had sunk 33 centuries earlier. How long ago in years did the ship sink? (*Hint:* 1 century = 100 years)

68. Each day, an athlete in training takes two capsules. If each capsule contains 1,600 international units (IU) of vitamin A, how much vitamin A does he take daily?

69. The walls of a human heart are made of muscles that contract about 100,000 times a day.

a. How many contractions are there in 30 days? (*Source: American Heart Association's Your Heart: An Owner's Manual*)

b. How many more contractions are there in 40 days than in 30 days?

70. It is estimated that India would have to create 10 million new jobs per year to maintain the present unemployment rate.

a. How many new jobs altogether would have to be created during a 5-year period for the present unemployment rate to remain constant?

b. If 35 million new jobs are created during this time period, how short of the goal is this? (*Source:* Narayana Murthy, http://www.knowledgeplex.org/news/147232.html)

71. The 2006 Honda Civic gets about 40 miles per gallon of gasoline. If the fuel tank holds about 13 gallons of gasoline, can a person drive from San Francisco to Los Angeles, a distance of 276 miles, without refilling the car's fuel tank? (*Source:* American Honda Motor Company, Inc., 2006)

72. The area of a football field in the Canadian Football League (CFL) is 87,750 square feet. A football field in the National Football League (NFL) is 360 feet long and 160 feet wide. Is the area of a football field in the NFL larger than the area of a football field in the CFL?

73. Find the area of the countertop shown in the diagram.

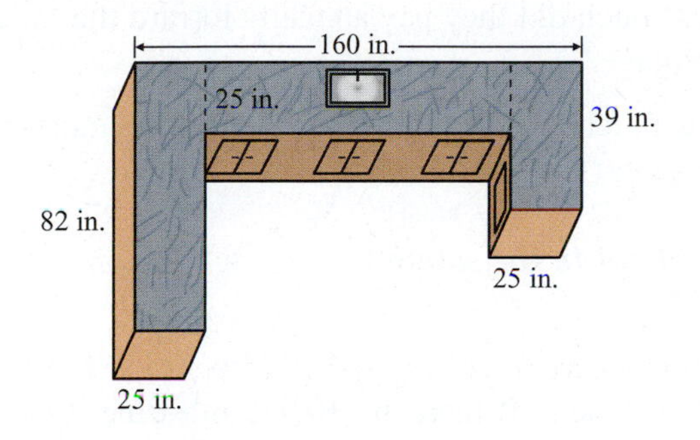

74. Calculate the area of the deck shown in the diagram.

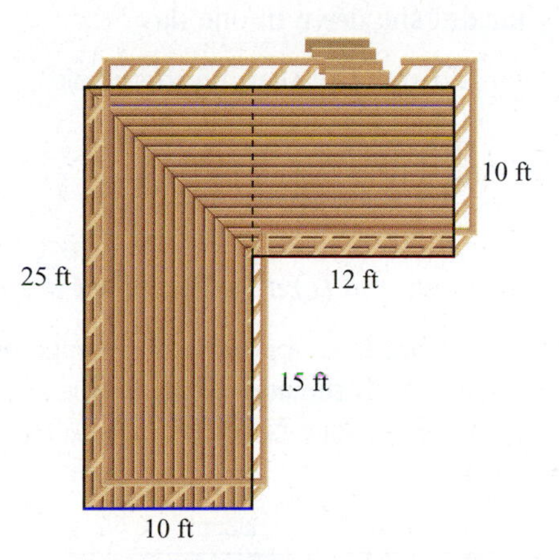

75. On the following map, 1 inch corresponds to 250 miles in the real world. How many miles actually separate towns A and B?

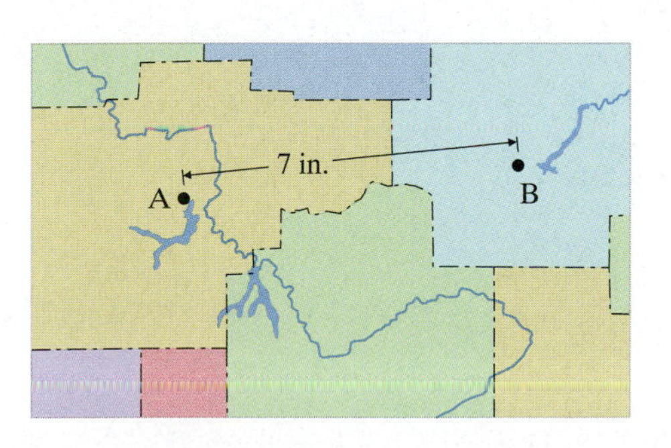

76. Angles are measured in either degrees (°) or radians (rad). A radian is about 57°. Express in degrees the measure of the angle shown.

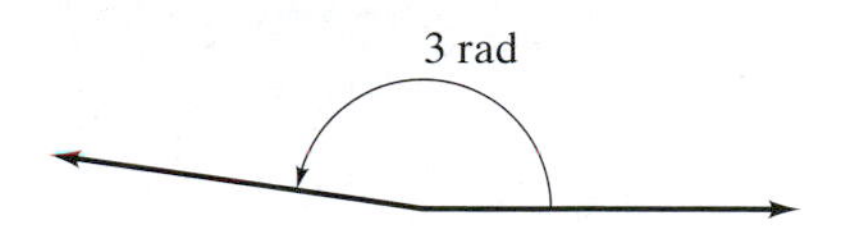

77. It costs \$130 to join a local health club and \$26 for each month of the membership. How much does a 1-year membership cost?

78. A customer bought a plasma television on the installment plan offered by the store. If the total cost of the television is \$1,599 and the customer pays \$134 each month, how much does he have left to pay off after 7 months of payments?

79. During a 5-day week, a truck driver daily drove 42 miles an hour for 7 hours.

a. How far did she drive in one day?

b. How far did she drive during the week?

80. A young couple took out a mortgage on a condo. They paid \$790 per month for 15 years.

a. How much did they pay annually toward the mortgage?

b. How much did it cost them to pay off the mortgage altogether?

Use a calculator to solve each problem, giving (a) the operation(s) carried out in the solution, (b) the exact answer, and (c) an estimate of the answer.

81. The state of Colorado is approximately rectangular in shape, as shown. If the area of Kansas is about 82,000 sq mi, which state is larger? (*Source: The Columbia Gazeteer of the World*)

388 mi

275 mi

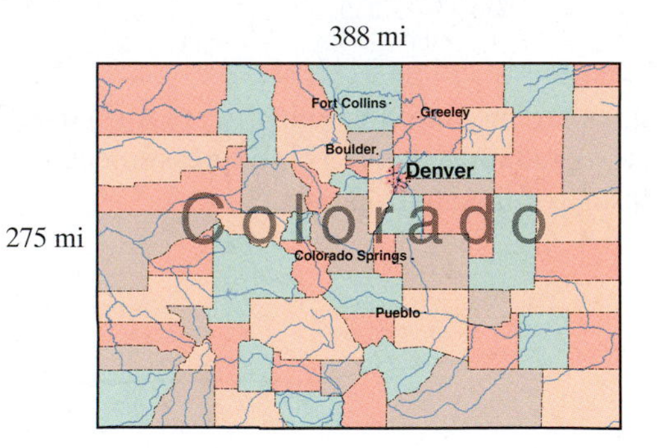

82. Tuition at a certain college is \$2,125 per year for every full-time student. If there are 10,975 full-time students at the college, how much revenue is generated from student tuition?

• *Check your answers on page A-2.*

MINDSTRETCHERS

Writing

1. Study the following diagram. Explain how it justifies the Distributive Property.

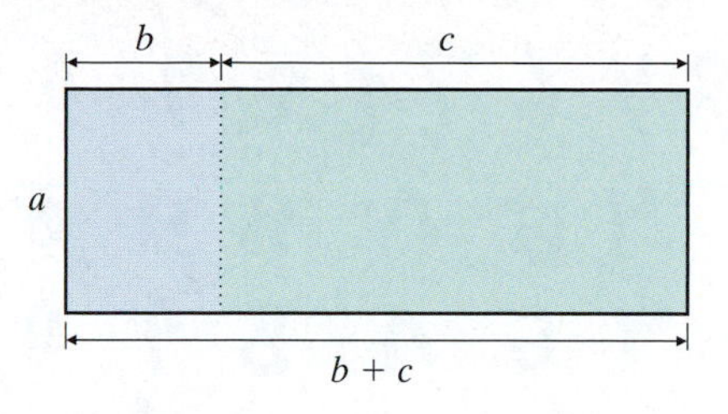

Mathematical Reasoning

2. Consider the six digits 1, 3, 5, 7, 8, and 9. Fill in the blanks with these digits, using each digit only once, so as to form the largest possible product.

$$\underline{9}\ \underline{5}\ \underline{1} \times \underline{8}\ \underline{7}\ \underline{3}$$

Historical

3. Centuries ago in India and Persia, the **lattice method** of multiplication was popular. The following example, in which we multiply 57 by 43, illustrates this method. Explain how it works.

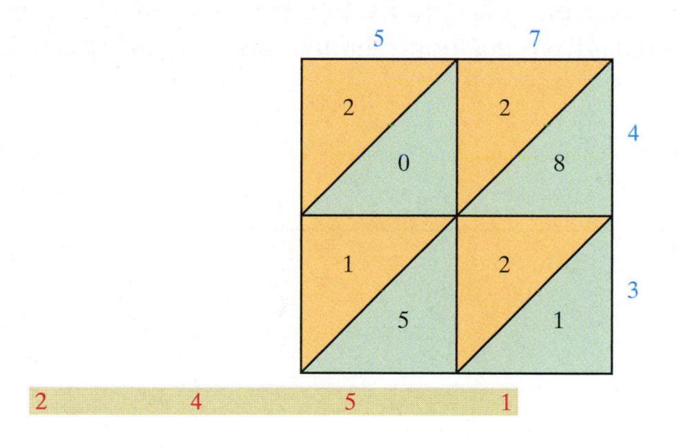

CULTURAL NOTE

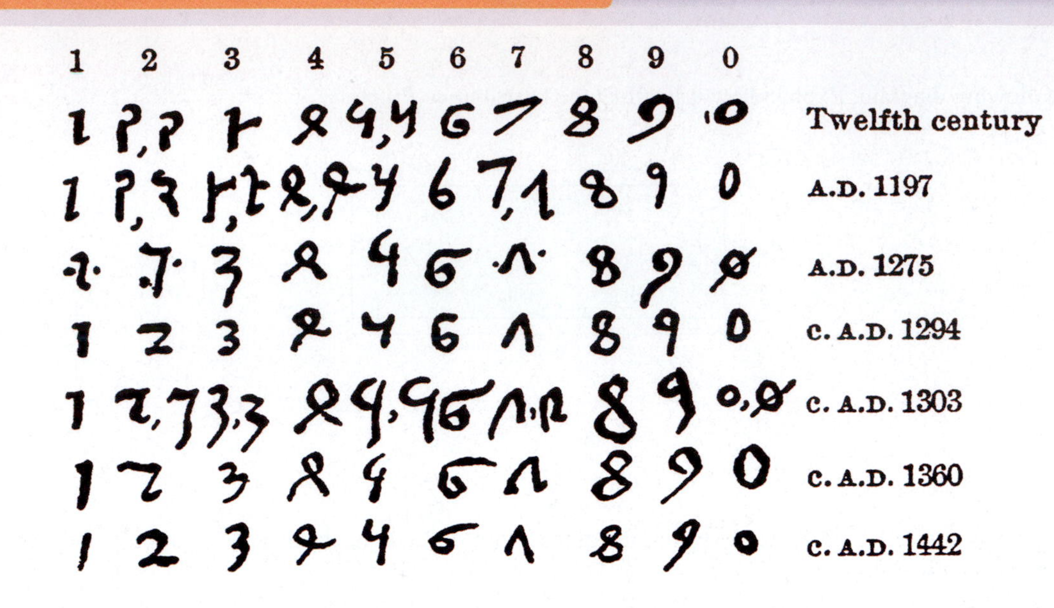

The way the 10 digits are written has evolved over time. Early Hindu symbols found in a cave in India date from more than two thousand years ago. About twelve hundred years ago, an Indian manuscript on arithmetic, which had been translated into Arabic, was carried by merchants to Europe where it was later translated into Latin.

This table shows European examples of digit notation from the twelfth to the fifteenth century, when the printing press led to today's standardized notation. Through international trade, these symbols became known throughout the world.

Source: David Eugene Smith and Jekuthiel Ginsburg, *Numbers and Numerals, A Story Book for Young and Old* (New York: Bureau of Publications, Teachers College, Columbia University, 1937)

1.4 Dividing Whole Numbers

OBJECTIVES

- To divide whole numbers
- To estimate the quotient of whole numbers
- To solve word problems involving the division of whole numbers

The Meaning and Properties of Division

What does it mean to divide? One good answer is to think of division as *breaking up a set of objects* into a given number of equal smaller sets.

For instance, suppose that we want to split a set of 15 objects, say pens, evenly among 3 boxes.

From the diagram we see that each box ends up with 5 pens. We therefore say that 15 divided by 3 is 5, which we can write as follows:

$$\begin{array}{rl} & 5 \quad \textbf{Quotient} \\ \textbf{Divisor} \quad 3 & \overline{)15} \quad \textbf{Dividend} \end{array}$$

In a division problem, the number that is being used to divide another number is called the *divisor*. The number being divided is the *dividend*. The result is the *quotient*.

We can also think of division as the *opposite* (*inverse*) of multiplication. Consider the following pair of problems that illustrate this point.

$$\underset{\textbf{Division}}{3\overline{)15}}^{\;5} \quad \text{because} \quad \underset{\textbf{Related multiplication}}{5 \times 3 = 15}$$

The following relationship connects multiplication and division.

Quotient × Divisor = Dividend

Note that this relationship allows us to check our answer to a division problem by multiplying.

There are several common ways to write a division problem.

$$3\overline{)15}^{\,5}, \quad \frac{15}{3} = 5, \quad \text{or} \quad 15 \div 3 = 5$$

Usually, we use the first of these to compute the answer. However, no matter which way we write this problem, 3 is the divisor, 15 is the dividend, and 5 is the quotient.

Tip When reading a division problem, we say that we are dividing either the divisor *into* the dividend or the dividend *by* the divisor. For instance, $3\overline{)15}$ is read either "3 divided into 15" or "15 divided by 3."

When calculating a quotient, we frequently use the following properties of division.

	Division	Related Multiplication
• Any whole number (except 0) divided by itself is 1.	$6\overline{)6}$ with quotient 1	$1 \times 6 = 6$
• Any whole number divided by 1 is the number itself.	$1\overline{)12}$ with quotient 12	$12 \times 1 = 12$
• Zero divided by any whole number (other than 0) is 0.	$8\overline{)0}$ with quotient 0	$0 \times 8 = 0$
• Division by 0 is not permitted.	$0\overline{)5}$ with quotient ?	$? \times 0 = 5$ **There is no number that when multiplied by 0 equals 5.**

Dividing Whole Numbers

Multiplication is the opposite of division. So in the simple division problem $3\overline{)15}$, we know that the answer is 5 because we have memorized that $5 \cdot 3$ is 15. But what should we do when the dividend is a larger number?

Consider the following problem: Divide 9 into 5,112 and check the answer.

- We start with the greatest place (thousands) in the dividend. We consider the dividend to be 5 thousands and think $9\overline{)5}$. Since $9 \cdot 1 = 9$ and 9 is larger than 5, there are no thousands in the quotient.

```
    0  ← Thousands
9)5,112
```

- So we go to the hundreds place in the dividend. We consider the dividend to be 51 hundreds and think $9\overline{)51}$. Since $9 \cdot \mathbf{5} = 45$, we position the **5** in the hundreds place of the quotient.

```
     5  ← Hundreds
9)5,112
 -4,500 ← 500 · 9 = 4500
    612 ← Difference
```

- Next, we move to the tens place of the difference, 612. We consider the new dividend to be 61 tens and think $9\overline{)61}$. Since $9 \cdot \mathbf{6} = 54$, we position the **6** in the tens place of the quotient.

```
     56  (6 ← Tens)
9)5,112
 -4,500
    612
   -540 ← 60 · 9 = 540
     72 ← Difference
```

- Finally, we go to the ones place of the difference, 72. We consider the new dividend to be 72 ones. So we think $9\overline{)72}$. Since $9 \cdot \mathbf{8} = 72$, we position the **8** in the ones place of the quotient.

So 568 is our answer.

```
    568  (8 ← Ones)
9)5,112
 -4,500
    612
   -540
     72
    -72 ← 8 · 9 = 72
      0 ← Difference
```

Instead of writing 0's as placeholders, we can use the following shortcut.

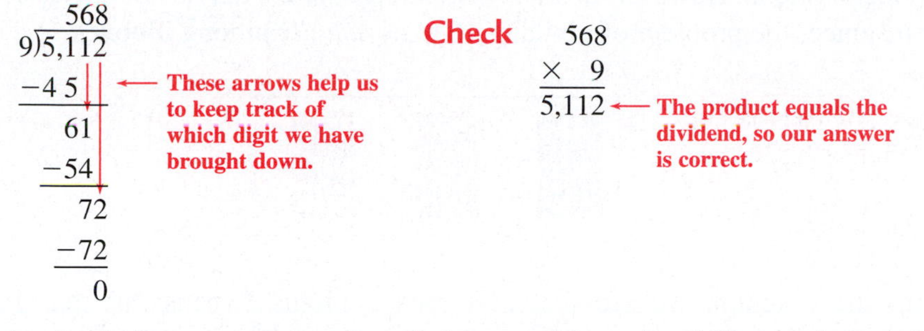

568
9)5,112
−4 5
61
−54
72
−72
0
These arrows help us to keep track of which digit we have brought down.
Check
568
× 9
5,112
The product equals the dividend, so our answer is correct.

Note that each time we subtract in a division problem, the difference is less than the divisor. Why must that be true?

EXAMPLE 1

Divide and check: 4,263 ÷ 7

Solution

```
  609        Think 7)42
7)4,263
 −4 2   ← 6 × 7 = 42. Subtract.
    06  ← Think 7)6. There are zero 7's in 6.
    −0  ← 0 × 7 = 0. Subtract.
    63  ← Think 7)6.
   −63  ← 9 × 7 = 63. Subtract.
     0
```

Check

```
  609
×   7
4,263
```

The product agrees with our dividend. Note the 0 in the quotient. Can you explain why the 0 is needed?

PRACTICE 1

Compute 9)7,263 and then check your answer.

Tip In writing your answer to a division problem, position the first digit of the quotient over the *right digit* of the number into which you are dividing (the 6 over the 2 in Example 1).

```
  609
  ↓↓↓
7)4,263
```

EXAMPLE 2

Compute $\frac{2{,}709}{9}$. Then check your answer.

Solution

```
  301
9)2,709
 −2 7
    00
    −0
    09
    −9
     0
```

Check

```
  301
×   9
2,709
```

PRACTICE 2

Carry out the following division and check your answer.

8)56,016

In Examples 1 and 2, note that the remainder is 0; that is, the divisor goes evenly into the dividend. However, in some division problems, that is not the case. Consider, for instance, the problem of dividing 16 pens *equally* among 3 boxes.

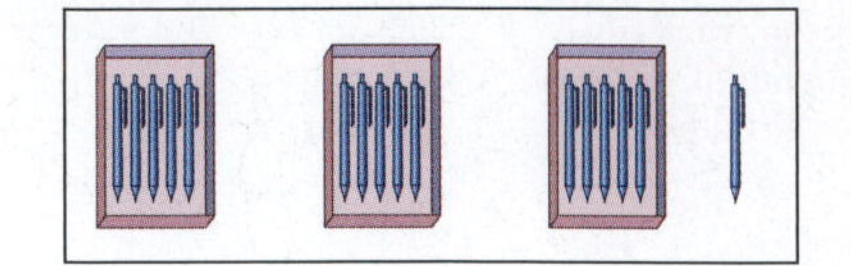

From the diagram, we see that each box contains 5 pens *but* that 1 pen—the *remainder*—is left over.

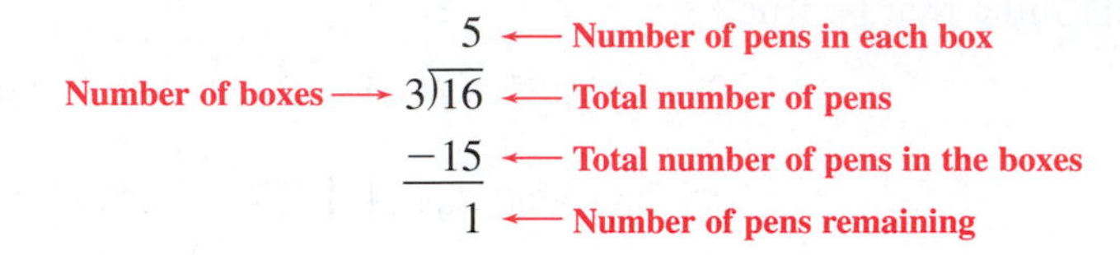

We write the answer to this problem as 5 R1 (read "5 Remainder 1"). Note that $(3 \times 5) + 1 = 16$. The following relationship is always true.

(Quotient × Divisor) + Remainder = Dividend

When a division problem results in a remainder as well as a quotient, we use this relationship for checking.

EXAMPLE 3

Find the quotient of 55,811 and 6. Then check.

Solution

```
    9,301 R5
 6)55,811
  -54
    1 8
   -1 8
     01
    -0
     11
    -6
      5
```

Our answer is therefore 9,301 R5.

(Quotient × Divisor) + Remainder × Dividend

Check $(9{,}301 \times 6) + 5 =$
$55{,}806 + 5 = 55{,}811$

Since this matches the dividend, our answer checks.

PRACTICE 3

Compute $8\overline{)42{,}329}$ and check.

Now let's consider division problems in which a divisor has more than one digit. Notice that such problems involve rounding.

EXAMPLE 4

Compute $\dfrac{2{,}574}{34}$ and check.

Solution In order to estimate the first digit of the quotient, we round 34 to 30 and 257 to 260.

$$\begin{array}{r} 8 \\ 34\overline{)2{,}574} \\ \underline{-2\,72} \end{array}$$

← Think 260 ÷ 30, or 26 ÷ 3. The quotient 8 goes over the 7 because we are dividing 34 into 257.

← 8 × 34 = 272. Try to subtract.

Because 272 is too large, we reduce our estimate in the quotient by 1 and try 7.

$$\begin{array}{r} 76 \\ 34\overline{)2{,}574} \\ \underline{-2\,38} \\ 194 \\ \underline{-204} \end{array}$$

← 7 × 34 = 238. Subtract.

← Think 190 ÷ 30 or 19 ÷ 3.

← 6 × 34 = 204. Try to subtract.

Because 204 is too large, we reduce our estimate in the quotient by 1 and try 5.

$$\begin{array}{r} 75 \text{ R24} \\ 34\overline{)2{,}574} \\ \underline{-2\,38} \\ 194 \\ \underline{-170} \\ 24 \end{array}$$

So our answer is 75 R24.

Check $(75 \times 34) + 24 = 2{,}574$

Since 2,574 is the dividend, our answer checks.

PRACTICE 4

Divide 23 into 1,818. Then check.

EXAMPLE 5

Divide $26\overline{)1{,}849}$ and then check.

Solution First we round 26 to 30 and 184 to 180. Think $180 \div 30 = 6$.

$$\begin{array}{r} 6 \\ 26\overline{)1{,}849} \\ \underline{-1\,56} \\ 28 \end{array}$$

← This difference is larger than the divisor, so we increase the 6 in the quotient by 1.

$$\begin{array}{r} 71 \\ 26\overline{)1{,}849} \\ \underline{-182} \\ 29 \\ \underline{-26} \\ 3 \end{array}$$

Our answer is therefore 71 R3.

Check $(71 \times 26) + 3 = 1{,}849$

PRACTICE 5

Compute and check: $15\overline{)1{,}420}$

Tip If the divisor has more than one digit, estimate each digit in the quotient by rounding and then dividing. If the product is too large or too small, adjust it up or down by 1 and then try again.

EXAMPLE 6

Find the quotient of 13,559 and 44. Then check.

Solution

$$\begin{array}{r} 308 \text{ R7} \\ 44\overline{)13{,}559} \\ -13\,2 \\ \hline 35 \\ -0 \\ \hline 359 \\ -352 \\ \hline 7 \end{array}$$

← This number is smaller than the divisor, so the next digit in the quotient is 0.

Check $(308 \times 44) + 7 = 13{,}559$

PRACTICE 6

Divide 16,999 by 28. Then check your answer.

EXAMPLE 7

Divide and check: $6{,}000 \div 20$

Solution We set up the problem as before.

$$\begin{array}{r} 300 \\ 20\overline{)6{,}000} \\ -60 \\ \hline 00 \\ -00 \\ \hline 00 \\ -00 \\ \hline 0 \end{array}$$

Check

$$\begin{array}{r} 300 \\ \times\ 20 \\ \hline 6{,}000 \end{array}$$

Because the divisor and dividend both end in zero, a quicker way to do Example 7 is by dropping zeros.

$20\!\!\!/\overline{)6{,}00\!\!\!/0}$ ← Drop one 0 from both the divisor and the dividend.

$$\begin{array}{r} 300 \\ 2\overline{)600} \end{array}$$ ← Then divide.

PRACTICE 7

Compute $40\overline{)8{,}000}$ and then check.

Tip Dropping the same number of zeros at the right end of both the divisor and the dividend does not change the quotient.

Estimating Quotients

As for other operations, estimating is an important skill for division. Checking a quotient by estimation is faster than checking it by multiplication, although less exact. And in some division problems, we need only an approximate answer.

How do we estimate a quotient? A good way is to round the divisor to its greatest place. The new divisor then contains only one nonzero digit and so is relatively easy to divide by mentally. Then we round the dividend to the place of our choice.

Finally, we compute the estimated quotient by calculating its first digit and then attaching the appropriate number of zeros.

EXAMPLE 8

Calculate $\frac{7,004}{34}$ and then check by estimation.

Solution

$$\begin{array}{r} 206 \\ 34\overline{)7,004} \\ -6\,8 \\ \hline 204 \\ -204 \\ \hline \end{array}$$ ← Exact quotient

Check $34\overline{)7,004}$ → $30\overline{)7,000}$ Round 34 to 30 and round 7,004 to 7,000.

Think 70 ÷ 30 or 7 ÷ 3.

$$\begin{array}{r} 200 \\ 30\overline{)7,000} \end{array}$$ ← Estimated quotient

Note that, to the right of the 2 in the estimated quotient, we added a 0 over each of the digits in the dividend. Our answer (206) is close to our estimate (200), and so our answer is reasonable.

PRACTICE 8

Compute 100,568 ÷ 104 and use estimation to check.

EXAMPLE 9

Sound travels at about 340 meters per second, whereas light travels at 299,792,458 meters per second. Estimate how many times as fast as the speed of sound is the speed of light.

Solution To estimate a quotient, we first round the divisor and the dividend to their largest place value.

$340\overline{)299,792,458}$ → $300\overline{)300,000,000}$ Round 340 to 300 and 299,792,458 to 300,000,000.

Then we divide.

$$\begin{array}{r} 1,000,000 \\ 300\overline{)300,000,000} \end{array}$$

So the speed of light is about 1,000,000 times faster than the speed of sound.

PRACTICE 9

Based on population projections, China will have a population of 1,366,205,049 in the year 2012. In that same year, Brazil will have a population of 199,083,155. Estimate how many times the population of Brazil the population of China will be in 2012. (*Source:* U.S. Bureau of the Census, International Database)

Dividing Whole Numbers on a Calculator

When using a calculator to divide, we must enter the numbers in the correct order to get the correct answer. We first enter the number *into* which we are dividing (the dividend) and then the number *by* which we are dividing (the divisor).

EXAMPLE 10

Use a calculator to divide $18\overline{)11{,}718}$.

Solution

Press	Display
11718 [÷] 18 [ENTER =]	11718 / 18 651.

A reasonable estimate is 10,000 ÷ 20, or 500, which is fairly close to 651.

PRACTICE 10

Find the following quotient with a calculator:

$$\frac{47{,}034}{78}$$

Mathematically Speaking

Fill in each blank with the most appropriate term or phrase from the given list.

subtraction	product	increased
quotient	divisor	multiplication
divided		

1. When dividing, the dividend is divided by the __________.

2. The result of dividing is called the __________.

3. The opposite operation of division is __________.

4. Any whole number __________ by 1 is equal to the number itself.

Divide and check.

5. $5\overline{)2{,}000}$

6. $5\overline{)10{,}000}$

7. $5\overline{)2{,}800}$

8. $8\overline{)12{,}504}$

9. $9\overline{)2{,}709}$

10. $2\overline{)5{,}780}$

11. $7\overline{)21{,}021}$

12. $5\overline{)27{,}450}$

13. $3\overline{)606}$

14. $2\overline{)30{,}534}$

15. $9\overline{)4{,}500}$

16. $3\overline{)4{,}512}$

Find the quotient and check.

17. $300 \div 10$

18. $400 \div 20$

19. $700 \div 50$

20. $6{,}000 \div 20$

21. $\frac{8{,}400}{200}$

22. $\frac{7{,}500}{300}$

23. $\frac{16{,}000}{40}$

24. $\frac{48{,}000}{20}$

25. $6{,}996 \div 44$

26. $9{,}660 \div 92$

27. $80{,}295 \div 15$

28. $936 \div 72$

29. $39{,}078 \div 39$

30. $49{,}497 \div 21$

31. $249{,}984 \div 36$

32. $499{,}992 \div 24$

33. $52\overline{)52{,}052}$

34. $24\overline{)48{,}072}$

35. $12\overline{)36{,}600}$

36. $36\overline{)25{,}560}$

37. $6{,}512 \div 10$

38. $8{,}922 \div 25$

39. $304 \div 27$

40. $206 \div 45$

41. $\frac{10{,}175}{87}$

42. $\frac{21{,}109}{25}$

43. $\frac{63{,}002}{90}$

44. $\frac{12{,}509}{61}$

45. $47\overline{)34,000}$ **46.** $66\overline{)99,980}$ **47.** $14\overline{)6,000}$ **48.** $32\overline{)3,007}$

49. $537\overline{)387,177}$ **50.** $265\overline{)197,160}$ **51.** $638\overline{)98,890}$ **52.** $152\overline{)34,048}$

In each group of three quotients, one is wrong. Use estimation to identify which quotient is incorrect.

53. a. $455,260 \div 65 = 704$ **b.** $11,457 \div 57 = 201$ **c.** $10,044 \div 93 = 108$

54. a. $18,473 \div 91 = 203$ **b.** $43,364 \div 74 = 586$ **c.** $14,562 \div 18 = 8,009$

55. a. $43,710 \div 93 = 47$ **b.** $71,048 \div 107 = 664$ **c.** $11,501 \div 31 = 371$

56. a. $178,267 \div 89 = 2,003$ **b.** $350,007 \div 21 = 1,667$ **c.** $37,185 \div 37 = 1,005$

Mixed Practice

Divide and check.

57. $38,095 \div 42$ **58.** $\frac{63,147}{21}$ **59.** $6\overline{)12,000}$ **60.** $4,907 \div 7$

61. $\frac{48,000}{20}$ **62.** $36\overline{)249,986}$ **63.** $\frac{3,330}{9}$ **64.** $4,090 \div 91$

Applications

Solve and check.

65. A part-time student is taking 9 credit-hours this semester at a local community college. If her tuition bill is $1,215, how much does each credit-hour cost?

66. A car used 15 gallons of gas on a 300-mile trip. How many miles per gallon (mpg) of gas did the car get?

67. The area of the Pacific Ocean is about 64 million square miles, and the area of the Atlantic Ocean is approximately 32 million square miles. The Pacific is how many times as large as the Atlantic? (***Source:*** *The New Encyclopedia Britannica*)

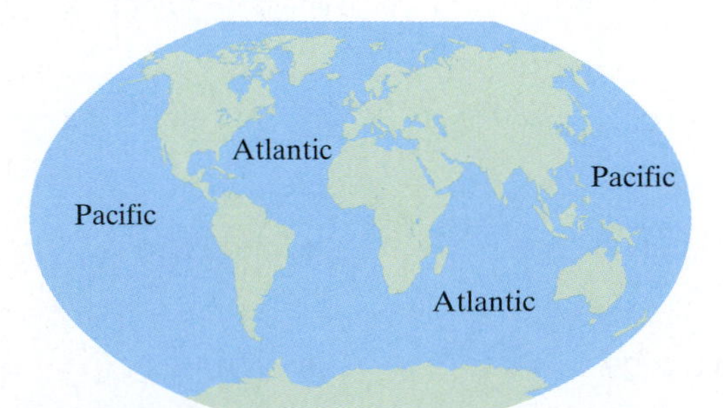

68. The diameter of Earth is about 8,000 miles, whereas the diameter of the Moon is about 2,000 miles. How many times the Moon's diameter is Earth's? (***Source:*** *The New Encyclopedia Britannica*)

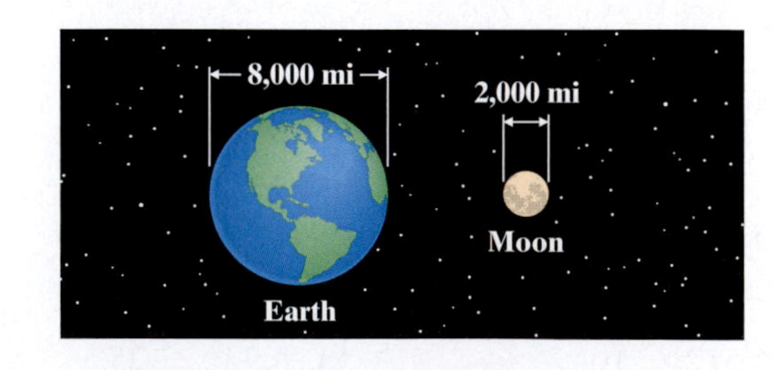

69. In the year 2030, Ohio is projected to have a population of about 12,300,000 people. If Ohio has a total land area of about 41,000 square miles, how many people per square mile will there be in 2030? (***Source:*** Ohio Department of Development)

70. A certified medical assistant has an annual salary of $26,472. What is her gross monthly income?

71. A 150-pound person can burn about 360 calories in 1 hour doing yoga. How many calories are burned in 1 minute? (*Source:* American Cancer Society)

72. Derek Jeter signed a 10-year contract for $189,000,000 with the New York Yankees in 2001. What is his pay per year from the contract? (*Source: The New York Times*)

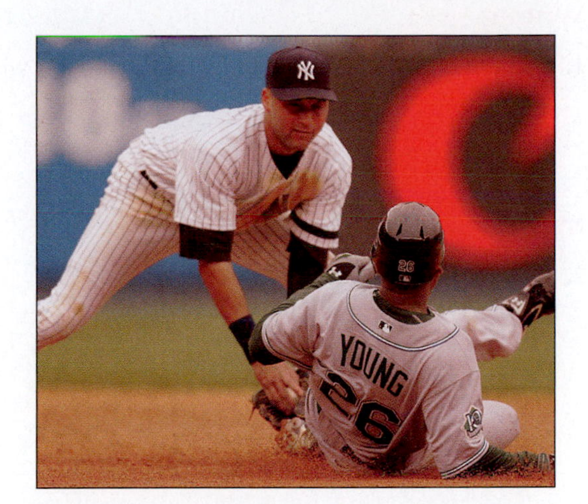

Use a calculator to solve each problem, giving (a) the operation(s) carried out in the solution, (b) the exact answer, and (c) an estimate of the answer.

73. A homeowner is remodeling a bathroom with dimensions 96 inches and 114 inches. For the floor, she has selected tiles that measure 6 inches by 6 inches.

a. How many tiles must she purchase?

b. The tiles come in boxes of 12. How many boxes of tiles must she purchase?

c. If each box of tiles costs $18, how much will she spend on the tiles for the floor?

74. The group admission rate for 15 or more people at Six Flags Great Adventure amusement park is $30 per person. A student group hosted a field trip to the park and charged $46 per ticket, covering both the cost of admission to the park and the bus transportation. (*Source:* Six Flags Great Adventure, 2006)

a. If the total amount the group collected for tickets was $1,656, how many students went on the field trip?

b. Calculate the total cost of admissions for the students on the field trip.

c. What was the cost of the bus transportation?

75. Although the areas of the United States and of China are roughly the same, the United States has a much smaller population. The population of China is 1,306,313,800, whereas that of the United States is only 295,734,100. Is the population of China more or less than 4 times that of the United States? Explain. (*Source: The World Factbook*, 2005)

76. A couple set aside $3,300 for mortgage payments. If they pay $281 per month toward the mortgage, for how many months can they make full payments?

• *Check your answers on page A-2.*

MINDSTRETCHERS

Writing

1. Use the problem $10 \div 2 = 5$ to help explain why division can be thought of as repeated subtraction.

Mathematical Reasoning

2. Consider the following pair of problems.

   a. $2\overline{)7}$ **b.** $4\overline{)13}$

   Are the answers the same? Explain.

Groupwork

3. In the following division problem, A, B, and C each stand for a different digit. Working with a partner, identify all the digits. (*Hint:* There are two answers.)

$$
\begin{array}{r}
\text{ABA} \\
\text{AB}\overline{)\text{CACAB}} \\
\underline{-\text{CAB}} \\
\text{CA} \\
\underline{-\text{B}} \\
\text{CAB} \\
\underline{-\text{CAB}}
\end{array}
$$

1.5 Exponents, Order of Operations, and Averages

OBJECTIVES

- To evaluate expressions involving exponents
- To evaluate expressions using the rule for order of operations
- To compute averages

Exponents

There are many mathematical situations in which we multiply a number by itself repeatedly. Writing such expressions in **exponential form** provides a shorthand method for representing this repeated multiplication of the same factor.

For instance, we can write $5 \cdot 5 \cdot 5 \cdot 5$ in exponential form as

$$5^4$$

(4: Exponent; 5: Base)

This expression is read "5 to the fourth *power*" or simply "5 to the fourth."

Definition

An **exponent** (or **power**) is a number that indicates how many times another number (called the **base**) is used as a factor.

We read the power 2 or the power 3 in a special way. For instance, 5^2 is usually read "5 *squared*" rather than "5 to the second power." Similarly, we usually read 5^3 as "5 *cubed*" instead of "5 to the third power."

Let's look at a number written in exponential form—namely, 2^4. To evaluate this expression, we multiply 4 factors of 2.

$$\begin{aligned} 2^4 &= \underbrace{2 \cdot 2} \cdot 2 \cdot 2 \\ &= \underbrace{4 \cdot 2} \cdot 2 \\ &= \underbrace{8 \cdot 2} \\ &= 16 \end{aligned}$$

In short, $2^4 = 16$. Do you see the difference between 2^4 and $2 \cdot 4$?

Sometimes we prefer to shorten expressions by writing them in exponential form. For instance, we can write $3 \cdot 3 \cdot 4 \cdot 4 \cdot 4$ in terms of powers of 3 and 4.

$$\underbrace{3 \cdot 3}_{\text{2 factors of 3}} \cdot \underbrace{4 \cdot 4 \cdot 4}_{\text{3 factors of 4}} = 3^2 \cdot 4^3$$

EXAMPLE 1

Rewrite

$$6 \cdot 6 \cdot 6 \cdot 10 \cdot 10 \cdot 10 \cdot 10$$

in exponential form.

Solution

$$\underbrace{6 \cdot 6 \cdot 6}_{\text{3 factors of 6}} \cdot \underbrace{10 \cdot 10 \cdot 10 \cdot 10}_{\text{4 factors of 10}} = 6^3 \cdot 10^4$$

PRACTICE 1

Write

$$5 \cdot 5 \cdot 5 \cdot 5 \cdot 5 \cdot 2 \cdot 2$$

in terms of powers.

EXAMPLE 2

Compute:

a. 1^5

b. 22^2

Solution

a. $1^5 = \underbrace{1 \cdot 1 \cdot 1 \cdot 1 \cdot 1}$
$= 1$

Note that 1 raised to any power is 1.

b. $22^2 = 22 \cdot 22$
$= 484$

After considering this example, can you explain the difference between squaring and doubling a number?

PRACTICE 2

Calculate:

a. 1^8

b. 11^3

EXAMPLE 3

Write $4^3 \cdot 5^3$ in standard form.

Solution

$$\begin{aligned} 4^3 \cdot 5^3 &= (4 \cdot 4 \cdot 4) \cdot (5 \cdot 5 \cdot 5) \\ &= 64 \cdot 125 \\ &= 8{,}000 \end{aligned}$$

From this example, do you see the difference between cubing and tripling a number?

PRACTICE 3

Express $7^2 \cdot 2^4$ in standard form.

It is especially easy to compute powers of 10.

$$10^2 = 10 \cdot 10 = 1\underbrace{00}_{\text{2 zeros}}, \qquad 10^3 = 10 \cdot 10 \cdot 10 = 1{,}\underbrace{000}_{\text{3 zeros}}$$

$$10^4 = 10 \cdot 10 \cdot 10 \cdot 10 = 1\underbrace{0{,}000}_{\text{4 zeros}}$$

and so on.

Do you see the pattern?

EXAMPLE 4

Astronomical distances are commonly expressed in terms of light-years. Our galaxy is approximately 100,000 light-years in diameter. Express this number in terms of a power of 10. (*Source: The Time Almanac 2006*)

Solution

$$1\underbrace{00{,}000}_{\text{5 zeros}} = 10^5$$

So the diameter of our galaxy is 10^5 light-years.

PRACTICE 4

In 1850, the world population was approximately 1,000,000,000. Represent this number as a power of 10. (*Source: World Almanac and Book of Facts 2006*)

Order of Operations

Some mathematical expressions involve more than one mathematical operation. For instance, consider $5 + 3 \cdot 2$. This expression seems to have two different values, depending on the order in which we perform the given operations.

Adding first	**Multiplying first**
$\underbrace{5 + 3} \cdot 2$	$5 + \underbrace{3 \cdot 2}$
$= \underbrace{8 \cdot 2}$	$= \underbrace{5 + 6}$
$= 16$	$= 11$

How are we to know which operation to carry out first? By consensus we agree to follow the rule called the **order of operations** so that everyone always gets the same value for an answer.

Order of Operations Rule

To evaluate mathematical expressions, carry out the operations *in the following order*.

1. First, perform the operations within any grouping symbols, such as parentheses ( ) or brackets [ ].
2. Then, raise any number to its power $\blacksquare^{\blacksquare}$.
3. Next, perform all multiplications and divisions as they appear from left to right.
4. Finally, do all additions and subtractions as they appear from left to right.

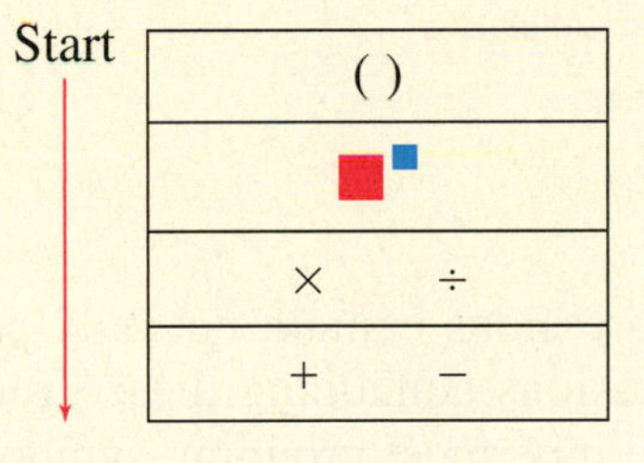

Applying this rule to the preceding example gives us the following result.

$$
\begin{aligned}
& 5 + \underbrace{3 \cdot 2} && \text{Multiply first.} \\
&= \underbrace{5 + 6} && \text{Then add.} \\
&= 11
\end{aligned}
$$

So 11 is the correct answer.

Let's consider more examples that depend on the order of operations rule.

EXAMPLE 5

Simplify: $18 - 7 \cdot 2$

Solution Applying the rule, we multiply first, and then subtract.

$$18 - \underbrace{7 \cdot 2} =$$
$$18 - 14 = 4$$

PRACTICE 5

Evaluate: $2 \cdot 8 + 4 \cdot 3$

EXAMPLE 6

Find the value of $3 + 2 \cdot (8 + 3^2)$.

Solution

$$
\begin{aligned}
3 + 2 \cdot (8 + 3^2) &= 3 + 2 \cdot \underbrace{(8 + 9)} && \text{First, perform the operations in parentheses: square the 3.} \\
&= 3 + \underbrace{2 \cdot 17} && \text{Then, add 8 and 9.} \\
&= \underbrace{3 + 34} && \text{Next, multiply 2 by 17.} \\
&= 37 && \text{Finally, add 3 and 34.}
\end{aligned}
$$

PRACTICE 6

Simplify: $(4 + 1)^2 \times 6 - 4$

Tip When a division problem is written in the format $\frac{\square}{\square}$, parentheses are understood to be around both the dividend and the divisor. For instance,

$$\frac{10 - 2}{3 + 1} \text{ means } \frac{(10 - 2)}{(3 + 1)}.$$

EXAMPLE 7

Evaluate: $6 \cdot 2^3 - \frac{21 - 11}{2}$

Solution

$$
\begin{aligned}
6 \cdot 2^3 - \frac{21 - 11}{2} &= 6 \cdot 2^3 - \frac{10}{2} && \text{First, simplify the dividend by subtracting.} \\
&= 6 \cdot 8 - \frac{10}{2} && \text{Then, cube.} \\
&= 48 - 5 && \text{Next, multiply and divide.} \\
&= 43 && \text{Finally, subtract.}
\end{aligned}
$$

PRACTICE 7

Simplify: $10 + \frac{24}{12 - 8} - 3 \times 4$

Some arithmetic expressions contain not only parentheses but also brackets. When simplifying expressions containing these grouping symbols, first perform the operations within the innermost grouping symbols and then continue to work outward.

EXAMPLE 8

Simplify: $5 + [4(10 - 3^2) - 2]$

Solution

$$
\begin{aligned}
5 + [4(10 - 3^2) - 2] &= 5 + [4(10 - 9) - 2] && \text{Perform the operation in parentheses: square the 3.} \\
&= 5 + [4 \cdot 1 - 2] && \text{Subtract 9 from 10.} \\
&= 5 + [4 - 2] && \text{Multiply.} \\
&= 5 + 2 && \text{Subtract.} \\
&= 7 && \text{Add.}
\end{aligned}
$$

PRACTICE 8

Evaluate: $[4 + 3(2^3 - 5)] + 10$

EXAMPLE 9

Young's Rule is a rule of thumb for calculating the dose of medicine recommended for a child of a given age. According to this rule, the dose of acetaminophen in milligrams (mg) for a child who is eight years old can be calculated using the expression.

$$\frac{8 \times 500}{8 + 12}$$

What is the recommended dose?

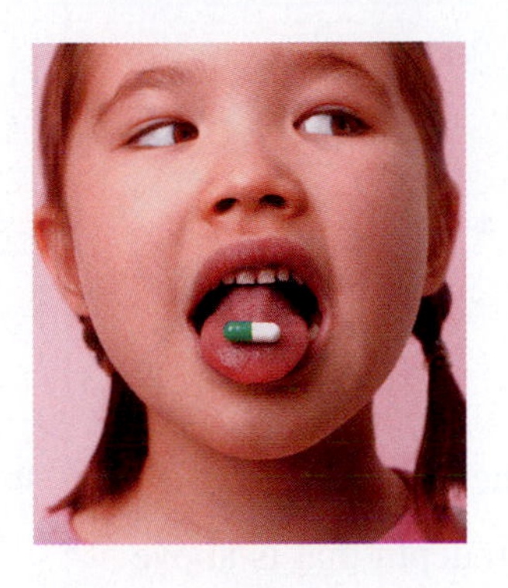

Solution

$$\frac{8 \times 500}{8 + 12} = \frac{4{,}000}{20}$$ First, simplify the dividend and the divisor.

$$= 200$$ Then, divide.

So the recommended dose is 200 milligrams.

PRACTICE 9

The minimum distance (in feet) that it takes a car to stop if it is traveling on a particular road surface at a speed of 30 miles per hour is given by the expression.

$$\frac{10 \times 30^2}{30 \times 5}$$

What is this minimum stopping distance?

Averages

We use an **average** to represent a set of numbers. Averages allow us to compare two or more sets. (For example, do the men or the women in your class spend more time studying?) Averages also allow us to compare an individual with a set. (For example, is the amount of time you spend studying above or below the class average?) The following definition shows how to compute an average.

Definition

The **average** (or **mean**) of a set of numbers is the sum of those numbers divided by however many numbers are in the set.

EXAMPLE 10

What is the average of 100, 94, and 100?

Solution The average equals the sum of these three numbers divided by 3.

$$\frac{100 + 94 + 100}{3} = \frac{294}{3} = 98$$

PRACTICE 10

Find the average of $30, $0, and $90.

EXAMPLE 11

The following map shows the five Great Lakes. The maximum depth of each of these lakes is given in the table. (*Source:* U.S. Environmental Protection Agency)

Superior
Huron
Ontario
Michigan
Erie

Lake	Maximum Depth (in meters)
Erie	64
Huron	229
Michigan	282
Ontario	244
Superior	406

a. What is the average maximum depth of the Great Lakes?

b. Which of the Great Lakes has a maximum depth that is above the average?

Solution

a. $\frac{\text{The sum of the depths}}{\text{The number of lakes}} = \frac{64 + 229 + 282 + 244 + 406}{5}$

$= \frac{1{,}225}{5}$

$= 245$

So the average maximum depth is 245 meters.

b. Lake Michigan and Lake Superior are deeper than the average.

PRACTICE 11

The table shown gives the number of fatalities due to tornadoes in the United States in each year from 2002 through 2005. (*Source:* NOAA/National Weather Service, Storm Prediction Center)

Year	Number of Fatalities
2002	55
2003	54
2004	36
2005	39

a. What was the average annual number of fatalities for these years?

b. In which years was the number of fatalities below the average?

Powers and Order of Operations on a Calculator

Let's use a calculator to carry out computations that involve either powers or the order of operations rule.

EXAMPLE 12

Calculate 23^3.

Solution

Press	Display
23 [^] 3 [ENTER =]	23^3 12167.

PRACTICE 12

Use a calculator to compute 375^2.

EXAMPLE 13

Combine: $2 + 3 \times 4$

Solution

Press	Display
2 [+] 3 [×] 4 [ENTER =]	2+3*4 14.

Note that some calculators do not follow the order of operations rule. When using this kind of calculator, enter the operations in the order specified by the order of operations rule to get the correct answer.

PRACTICE 13

On a calculator, compute $135 - 44 \div 11$.

Mathematically Speaking

Fill in each blank with the most appropriate term or phrase from the given list.

product	sum	adding	listing
subtracting	grouping	power	base

1. An exponent indicates how many times the _______ is used as a factor.

2. Parentheses and brackets are examples of _______ symbols.

3. An average of numbers on a list is found by _______ the numbers and then dividing by how many numbers there are on the list.

4. In evaluating an expression involving both a sum and a product, the _______ is evaluated first.

Complete each table by squaring the numbers given.

5.

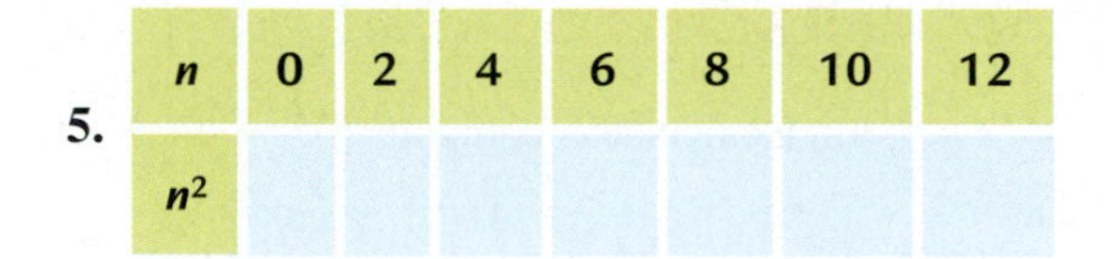

n	0	2	4	6	8	10	12
n^2							

6.

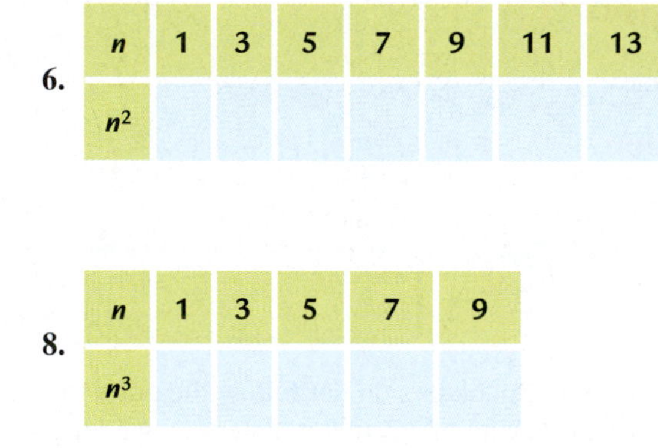

n	1	3	5	7	9	11	13
n^2							

Complete each table by cubing the numbers given.

7.

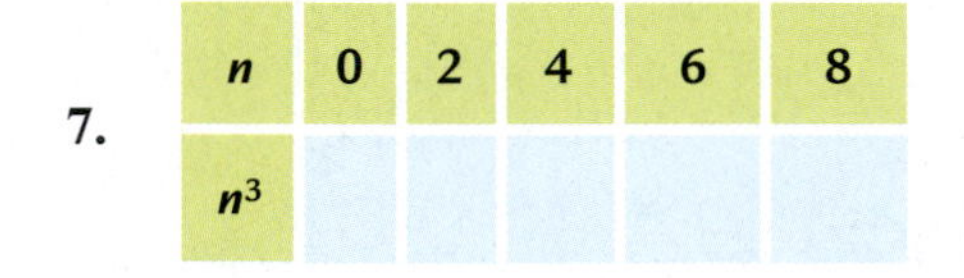

n	0	2	4	6	8
n^3					

8.

n	1	3	5	7	9
n^3					

Express each number as a power of 10.

9. $100 = 10^{\square}$

10. $1{,}000 = 10^{\square}$

11. $10{,}000 = 10^{\square}$

12. $100{,}000 = 10^{\square}$

13. $1{,}000{,}000 = 10^{\square}$

14. $10{,}000{,}000 = 10^{\square}$

Write each number in terms of powers.

15. $2 \cdot 2 \cdot 3 \cdot 3 = 2^{\square} \cdot 3^{\square}$

16. $2 \cdot 2 \cdot 5 \cdot 2 \cdot 5 = 2^{\square} \cdot 5^{\square}$

17. $5 \cdot 4 \cdot 4 \cdot 4 = 4^{\square} \cdot 5^{\square}$

18. $6 \cdot 7 \cdot 6 \cdot 7 \cdot 6 \cdot 7 = 6^{\square} \cdot 7^{\square}$

Write each number in standard form.

19. $6^2 \cdot 5^2$

20. $10^3 \cdot 9^2$

21. $2^5 \cdot 7^2$

22. $3^4 \cdot 4^3$

Evaluate.

23. $8 + 5 \cdot 2$

24. $9 + 10 \div 2$

25. $8 - 2 \times 3$

26. $12 - 6 \div 2$

27. $10 + 5^2$

28. $9 - 2^3$

29. $(9 - 2)^3$

30. $(10 + 5)^2$

31. 10×5^2

32. $12 \div 2^2$

33. $(12 \div 2)^2$

34. $(10 \times 5)^2$

35. $(24 \div 4) + 2$

36. $(15 \cdot 6) - 2$

37. $15 \cdot 6 + 2$

38. $24 \div 4 - 2$

39. $15 \cdot (6 - 2)$

40. $24 \div (4 + 2)$

41. $2^6 - 6^2$

42. $3^5 - 5^3$

43. $8 + 5 - 3 - 2 \times 2$

44. $7 - 1 + 2 + 3 \cdot 2$

45. $(10 - 1)(10 + 1)$

46. $(8 - 1)(8 + 1)$

47. $10^2 - 1$

48. $8^2 - 1$

49. $\left(\frac{8 + 2}{7 - 2}\right)^2$

50. $\left(\frac{9 - 1}{3 + 5}\right)^3$

51. $\frac{5^3 - 2^3}{3}$

52. $\frac{3^2 + 5^2}{2}$

53. $(2 + 14) \div 4(9 - 5)$

54. $3 + 10(20 - 2^3)$

55. $10 \cdot 3^2 + \frac{10 - 4}{2}$

56. $\frac{3^3 + 1^3 + 2^3}{4}$

57. $[9 + 2(3^2 - 8)] + 7$

58. $15 + [3(8 - 2^2) - 6]$

59. $32 + 9 \cdot 215 \div 5$

60. $84 \cdot 27 + 32 \cdot 27^2 \div 2$

61. $48(48 - 31)(48 - 24)(48 - 41)$

62. $137^2 - 4(36)(22)$

In each exercise, the three squares stand for the numbers 4, 6, and 8 in some order. Fill in the squares to make true statements.

63. $\square \cdot 3 + \square \cdot 5 + \square \cdot 7 = 98$

64. $\square + 10 \times \square - \frac{\square}{2} = 42$

65. $(\square)(3 + \square) - 2 \cdot \square = 44$

66. $\square \cdot 3 + \square \cdot 5 + \square \cdot 7 = 82$

67. $\square + 10 \times \square - \square \div 2 = 45$

68. $\frac{48}{\square} - \frac{\square}{2} + (3 + \square)^2 = 127$

Insert parentheses, if needed, to make the expression on the left equal to the number on the right.

69. $5 + 2 \cdot 4^2 = 112$

70. $5 + 2 \cdot 4^2 = 69$

71. $5 + 2 \cdot 4^2 = 169$

72. $5 + 2 \cdot 4^2 = 37$

73. $8 - 4 \div 2^2 = 1$

74. $8 - 4 \div 2^2 = 7$

Find the area of each shaded region.

75.

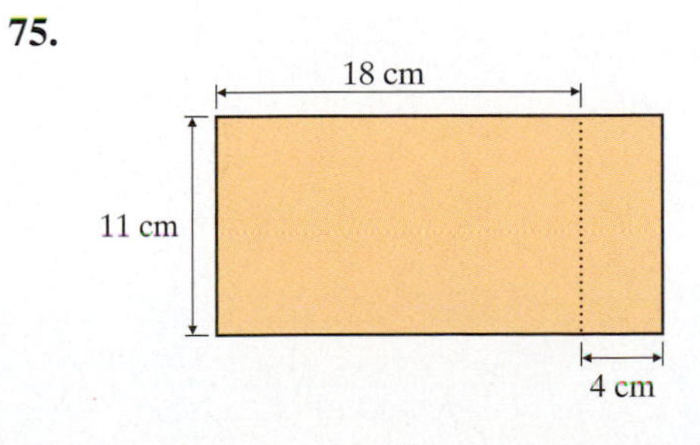

76.

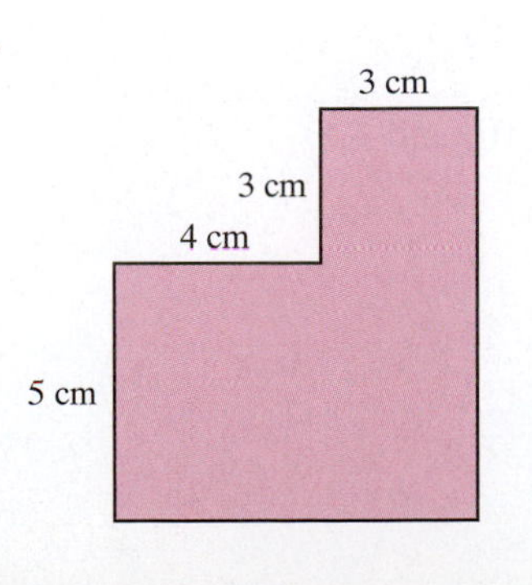

77.

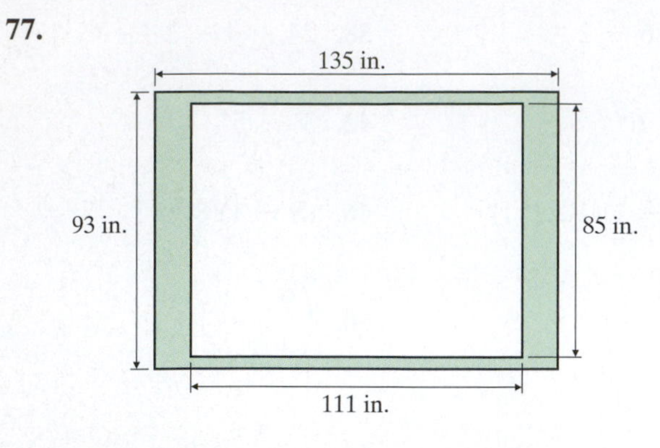

78.

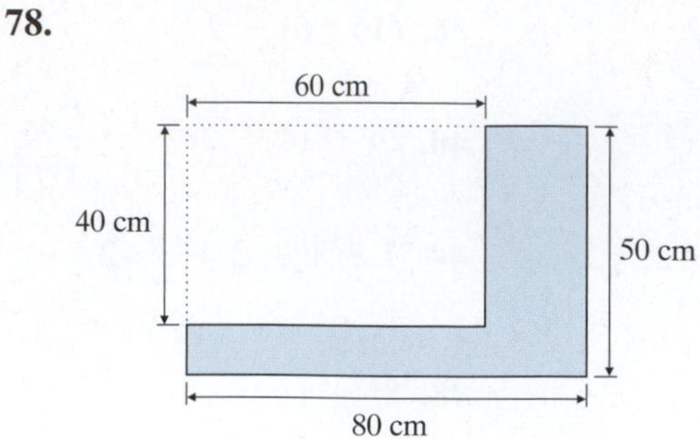

Complete each table.

79.

Input	Output
0	$21 + 3 \times \mathbf{0} =$
1	$21 + 3 \times \mathbf{1} =$
2	$21 + 3 \times \mathbf{2} =$

80.

Input	Output
0	$14 - 5 \times \mathbf{0} =$
1	$14 - 5 \times \mathbf{1} =$
2	$14 - 5 \times \mathbf{2} =$

Find the average of each set of numbers.

81. 20 and 30

82. 10 and 50

83. 30, 60, and 30

84. 17, 17, and 26

85. 10, 0, 3, and 3

86. 5, 7, 7, and 17

87. 3,527 miles, 1,788 miles, and 1,921 miles

88. 7 hours, 6 hours, 10 hours, 9 hours, and 8 hours

89. Six 10's and four 5's

90. Sixteen 5's and four 0's

Mixed Practice

Solve.

91. Express 100,000,000 as a power of 10.

92. Find the area of the shaded region.

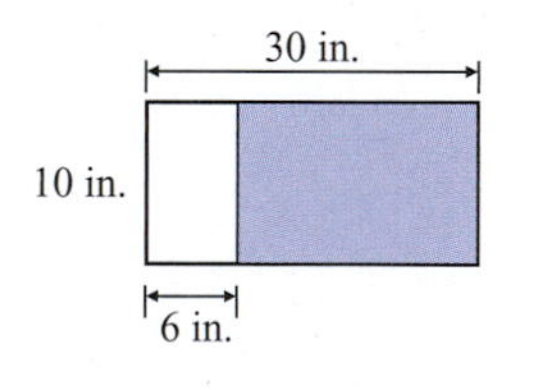

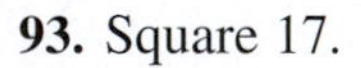

93. Square 17.

94. Rewrite in terms of powers of 2 and 7:
$2 \cdot 2 \cdot 2 \cdot 7 \cdot 7 = 2 \cdot 7$

95. Simplify: $50 - 2(10 - 3^2)$.

96. Cube 10.

97. Find the average of 10, 10, and 4.

98. Evaluate: 6×4^2.

Applications

Solve and check.

99. A 40-story office building has 25,000 square feet of space to rent. What is the average rental space on a floor?

100. The total area of the 50 states in the United States is about 3,700,000 square miles. If the state of Georgia's area is about 60,000 square miles, is its size above the average of all the states? Explain. (*Source: The New Encyclopedia Britannica*)

101. In a branch of mathematics called number theory, the numbers 3, 4, and 5 are called a *Pythagorean triple* because $3^2 + 4^2 = 5^2$ (that is, $9 + 16 = 25$). Show that 5, 12, and 13 are a Pythagorean triple.

102. If an object is dropped off a cliff, after 10 seconds it will have fallen $\frac{32 \cdot 10^2}{2}$ feet, ignoring air resistance. Express this distance in standard form, without exponents.

103. The solar wind streams off the Sun at speeds of about 1,000,000 miles per hour. Express this number as a power of 10. (*Source:* NASA)

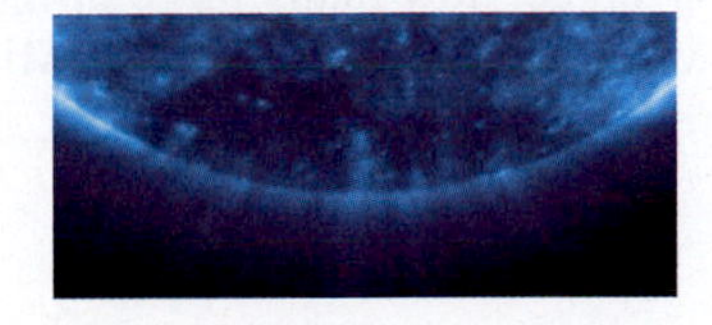

104. In 2005, about \$100,000,000,000 worth of twenty-dollar bills were in circulation. Represent this number as a power of 10. (*Source:* Financial Management Service *Treasury Bulletin*, March 2006)

105. The following table shows a lab assistant's salary in various years.

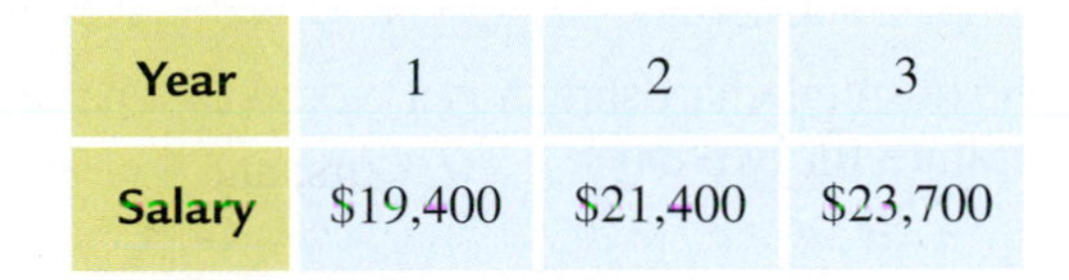

Year	1	2	3
Salary	\$19,400	\$21,400	\$23,700

a. Find the average salary for the three years.

b. How much greater was her average salary for the last two years than for all three years?

106. The following grade book shows a student's math test scores.

Test 1	Test 2	Test 3	Test 4
85	63	98	82

a. What is the average of his math scores?

b. If he were to get a 92 on the next math test, by how much would his average score increase?

107. In the last four home games, a college basketball team had scores of 68, 79, 57, and 72.

a. What was the average score for these games?

b. The average score for the last four away games was 64. On average, did the team score more at home or away? Explain.

108. A small theater company's production of *Romeo and Juliet* had 10 performances over two weekends. The attendance for each performance during the second weekend was 171, 297, 183, 347, and 232.

a. What was the average attendance for the performances during the second weekend?

b. If the average attendance at a performance during the first weekend was 272, was this average greater in the first or second weekend? Explain.

109. The number of hours the typical American spent watching broadcast television and cable and satellite television each year is given in the following table. (*Source:* Veronis Suhler Stevenson)

Year	Broadcast Television	Cable and Satellite Television
2001	833	843
2002	787	918
2003	769	975
2004	782	1,010
2005	785	1,042
2006	790	1,068

a. Find the average annual number of hours spent watching broadcast television and the average annual number of hours spent watching cable and satellite television.

b. Which average is higher? By how much?

110. The following table shows how many one-family homes were bought in the Northeast and Midwest regions of the United States each year from 2001 through 2005. (*Source:* National Association of Realtors)

Year	Northeast	Midwest
2001	710,000	1,155,000
2002	730,000	1,217,000
2003	769,000	1,322,000
2004	821,000	1,389,000
2005	839,000	1,410,000

a. Find the average annual number of one-family homes bought in each region for the 5-year period.

b. On the average, how many more of these homes were bought per year in the Midwest as compared to the Northeast?

Use a calculator to solve each problem giving (a) the operation(s) carried out in the solution, (b) the exact answer, and (c) an estimate of the answer.

111. Last week, newspaper A had an average daily circulation of 72,073. The daily circulation for newspaper B was as follows.

Day	B's Circulation
M	85,774
Tu	72,503
W	68,513
Th	74,812
F	89,002
Sa	92,331
Su	102,447

Which newspaper had a greater daily average circulation last week? By how many newspapers?

112. The hospital chart shown is a record of a patient's temperature for two days.

Time	Temp. (°F)	Time	Temp. (°F)
6 A.M.	98	6 A.M.	101
10 A.M.	100	10 A.M.	102
2 P.M.	98	2 P.M.	101
6 P.M.	100	6 P.M.	102
10 P.M.	98	10 P.M.	100
2 A.M.	100	2 A.M.	100

What was her average temperature for this period of time?

• *Check your answers on page A-2.*

MINDSTRETCHERS

Writing

1. Evaluate the expressions in parts (a) and (b).

a. $7^2 + 4^2$ ____

b. $(7 + 4)^2$ ____

c. Are the answers to parts (a) and (b) the same? ____ If not, explain why not.

Mathematical Reasoning

2. The square of any whole number (called a **perfect square**) can be represented as a geometric square, as follows:

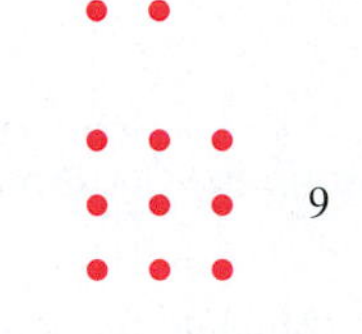

Try to represent the numbers 16, 25, 5, and 8 the same way.

16 25 5 8

Critical Thinking

3. Find the average of the whole numbers from 1 through 999.

1.6 More on Solving Word Problems

OBJECTIVE

- To solve word problems involving the addition, subtraction, multiplication, or division of whole numbers using various problem-solving strategies

What Word Problems Are and Why They Are Important

In this section, we consider some general tips to help solve word problems.

Word problems can deal with any subject—from shopping to physics and geography to business. Each problem is a brief story that describes a particular situation and ends with a question. Our job, after reading and thinking about the problem, is to answer that question by using the given information.

Although there is no magic formula for solving word problems, you should keep the following problem-solving steps in mind.

To Solve Word Problems

- Read the problem carefully.
- Choose a strategy (such as drawing a picture, breaking up the question, substituting simpler numbers, or making a table).
- Decide which basic operation(s) are relevant and then translate the words into mathematical symbols.
- Perform the operations.
- Check the solution to see if the answer is reasonable. If it is not, start again by rereading the problem.

Reading the Problem

In a math problem, each word counts. So it is important to read the problem slowly and carefully, and not to scan it as if it were a magazine or newspaper article.

When reading a problem, we need to understand the problem's key points: *What information is given* and *what question is posed*. Once these points are clear, jot them down so as to help keep them in mind.

After taking notes, decide on a plan of action that will lead to the answer. For many problems, just thinking back to the meaning of the four basic operations will be helpful.

Operation	Meaning
$+$	Combining
$-$	Taking away
$\times$	Adding repeatedly
$\div$	Splitting up

Many word problems contain *clue words* that suggest performing particular operations. If we spot a clue word in a problem, we should consider whether the operation indicated in the following table will lead us to a solution.

+	−	×	÷
• add	• subtract	• multiply	• divide
• sum	• difference	• product	• quotient
• total	• take away	• times	• over
• plus	• minus	• double	• split up
• more	• less	• twice	• fit into
• increase	• decrease	• triple	• per
• gain	• loss	• of	• goes into

However, be on guard—a clue word can be misleading. For instance, in the problem *What number increased by 2 is 6?*, we solve by subtracting, not adding.

Consider the following "translations" of these clues.

The patient's fever **increased by 5°**.	**+ 5**
The number of unemployed people **tripled**.	**× 3**
The length of the bedroom is **8 feet less** than the kitchen's.	**− 8**
The company's earnings were **split** among the **four** partners.	**÷ 4**

Choosing a Strategy

If no method of solution comes to mind after reading problem, there are a number of problem-solving strategies that may help. Here we discuss four of these strategies: drawing a picture, breaking up the question, substituting simpler numbers, and making a table.

Drawing a Picture

Sketching even a rough representation of a problem—say, a diagram or a map—can provide insight into its solution, if the sketch accurately reflects the given information.

EXAMPLE 1

In an election, everyone voted for one of three candidates. The winner received 188,000 votes, and the second-place candidate got 177,000 votes. If 380,000 people voted in the election, how many people voted for the third candidate?

Solution To help us understand the given information, let's draw a diagram to represent the situation.

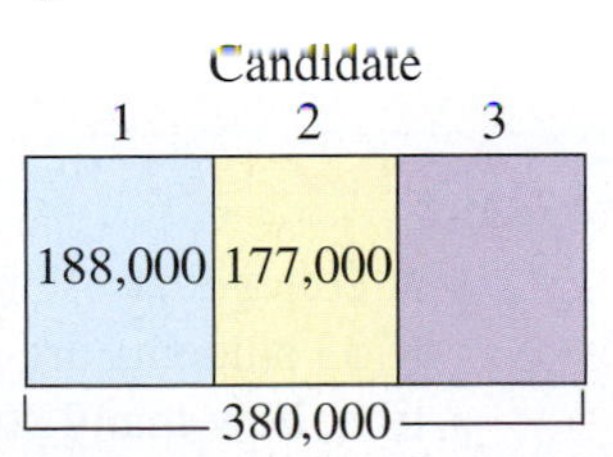

We see from this diagram that to find the answer we need to do two things.

PRACTICE 1

A company slashed its workforce by laying off 1,150 employees during one month and laying off 2,235 employees during another month. Afterward, 7,285 employees remained. How many employees worked for the company before the layoffs began?

- First, we need to add 188,000 to 177,000.

$$\begin{array}{r} 188{,}000 \\ +\ 177{,}000 \\ \hline 365{,}000 \end{array}$$

- Then, we need to subtract this sum from 380,000.

$$\begin{array}{r} 380{,}000 \\ -365{,}000 \\ \hline 15{,}000 \end{array}$$

A good way to check our answer here is by adding.

$$188{,}000 + 177{,}000 + 15{,}000 = 380{,}000$$

Our answer checks, so 15,000 people voted for the third candidate.

Breaking Up the Question

Another effective problem-solving strategy is to break up the given question into a chain of simpler questions.

EXAMPLE 2

Suppose that a student took a math test consisting of 20 questions, and that she answered 3 questions wrong. How many more questions did she get right than wrong?

Solution Say that we do not know how to answer this question directly. Try to split it into several easier questions that lead to a solution.

- How many questions did the student get *right*? $20 - 3 = 17$
- How many questions did she get *wrong*? 3
- How many *more* questions did she get right than wrong? $17 - 3 = 14$

So the student had 14 more questions right than wrong.

This answer seems reasonable, because it must be less than 17, the number of questions that the student answered correctly.

PRACTICE 2

Teddy and Franklin Roosevelt were both U.S. presidents. Teddy was born in 1858 and died in 1919. Franklin was born in 1882 and died in 1945. How much longer did Franklin live than Teddy?

(*Source:* Foner and Garraty, *The Reader's Companion to American History*)

Substituting Simpler Numbers

A word problem involving large numbers often seems difficult just because of these numbers. A good problem-solving strategy here is to consider first the identical problem but with simpler numbers. Solve the revised problem and then return to the original problem.

EXAMPLE 3

Raffle tickets cost $4 each. How many tickets must be sold for the raffle to break even if the prizes total $4,736?

Solution Suppose that we are not sure which operation to perform to solve this problem. Let's try substituting a simpler amount (say, $8) for the break-even amount of $4,736 and see if we can solve the resulting problem.

PRACTICE 3

A college has 47 sections of Math 110. If 33 students are enrolled in each section, how many students are taking Math 110?

The question would then become: How many \$4 tickets must be sold to make back \$8? Because it is a "fit-in" question, we must *divide* the \$8 by the \$4. Going back to the original problem, we see that we must divide \$4,736 by 4.

$$\$4{,}736 \div 4 = 1{,}184 \text{ tickets}$$

Is this answer reasonable? We can check either by estimating ($5{,}000 \div 4 = 1{,}250$, which is close to our answer) or by multiplying ($1{,}184 \times 4 = 4{,}736$), which also checks).

Making a Table

Finally, let's consider a strategy for solving word problems that involve many numbers. Organizing these numbers into a table often leads to a solution.

EXAMPLE 4

A borrower promises to pay back \$50 per month until a \$1,000 loan is settled. What is the remaining loan balance at the end of 5 months?

Solution We can solve by organizing the information in a table.

After Month	Remaining Balance
1	$1{,}000 - 50 = 950$
2	$950 - 50 = 900$
3	$900 - 50 = 850$
4	$850 - 50 = 800$
5	$800 - 50 = 750$

From the table, we see that the remaining balance after 5 months is \$750.

We can also solve this problem by breaking up the question into simpler questions.

- How much money did the borrower pay after 5 months? $5 \cdot 50 = \$250$
- How much money did the borrower still owe after 5 months? $1{,}000 - 250 = \$750$

Again, the remaining balance after 5 months is \$750.

PRACTICE 4

An athlete weighs 210 pounds and decides to go on a diet. If he loses 2 pounds a week while on the diet, how much will he weigh after 15 weeks?

1.6 Exercises

FOR EXTRA HELP **MyMathLab**

Choose a strategy. Solve and check.

1. In retailing, the difference between the gross sales and customer returns and allowances is called the net sales. If a store's gross sales were $2,538 and customer returns and allowances amounted to $388, what was the store's net sales?

2. The population of the United States in 1800 was 5,308,483. Ten years later, the population had grown to 7,239,881. During this period of time, did the country's population double? Justify your answer. (***Source:*** *The Time Almanac 2000*)

3. Suppose that you drive 27 miles north, 31 miles east, 45 miles west, and 14 miles east. How far are you from your starting point?

4. In your office, there are 19 reams of paper. If you need 7,280 sheets of paper for a printing job, do you have enough reams? (***Hint:*** A ream is 500 sheets of paper.)

5. A blue whale weighs about 300,000 pounds, and a great white shark weighs about 4,000 pounds. How many times the weight of a great white shark is the weight of a blue whale? (***Source:*** wikipedia.com)

6. A movie fan installed shelves for his collection of 400 DVDs. If 36 DVDs fit on each shelf, how many shelves did he need to house his entire collection?

7. A sales representative flew from Los Angeles to Miami (2,339 miles), then to New York (1,092 miles), and finally back to LA (2,451 miles). How many total miles did he fly?

8. Two major naval disasters of the twentieth century involved the sinking of British ships—the *Titanic* and the *Lusitania.* The *Titanic,* which weighed about 93,000,000 pounds, was the most luxurious liner of its time; it struck an iceberg on its maiden voyage in 1912. The *Lusitania,* which weighed about 63,000,000 pounds, was sunk by a German submarine in 1915. How much heavier was the *Titanic* than the *Lusitania?* (***Source:*** *The Oxford Companion to Ships and the Sea*)

The New York Times. EXTRA

LUSITANIA SUNK BY A SUBMARINE, PROBABLY 1,260 DEAD; TWICE TORPEDOED OFF IRISH COAST; SINKS IN 15 MINUTES; CAPT. TURNER SAVED, FROHMAN AND VANDERBILT MISSING; WASHINGTON BELIEVES THAT A GRAVE CRISIS IS AT HAND

9. Immigrants from all over the world came to the United States between 1931 and 1940 in the following numbers: 348,289 (Europe), 15,872 (Asia), 160,037 (Americas, outside the United States), 1,750 (Africa), and 2,231 (Australia/New Zealand). What was the total number of immigrants? (***Source:*** George Thomas Kurian, *Datepedia of the United States*)

10. In 2004, there were 36,652 movie screens in the United States. A year later, there were 37,740 movie screens. What was the increase in the number of movie screens from 2004 to 2005? (***Source:*** National Association of Theater Owners, 2006)

11. For each 4×6 print of a digital photo, a lab usually charges 10¢. During a promotion, the first 20 prints of each order are free. How much does the lab charge for 50 4×6 prints during the promotion?

12. Because of a noisy neighbor, a young man decided to put acoustical tiles on the living room ceiling, which measures 21 feet by 18 feet. The tiles are square, with a side length of 1 foot. If the tiles cost $3 apiece, what is the total cost to cover the ceiling?

13. After it was on the market for 4 months, the sellers of a home reduced the asking price by $14,000. After 6 months, they reduced the asking price a second and final time. If the original asking price had been $229,000 and the final asking price was $198,000, by how much did the sellers reduce the price the final time?

14. A nurse sets the drip rate for an IV medication at 25 drops each minute. How many drops does a patient receive in 2 hours?

15. A car dealer offered to lease a car for $1,500 down and $189 per month. If a customer accepted a lease contract of 2 years, how much did the customer have to pay over the lease period?

16. A doctor instructed a patient to take 100 milligrams of a medication daily for 4 weeks. The local pharmacy dispensed 120 tablets, each containing 25 milligrams of the medication. After taking the tablets for 4 weeks, how many remained?

17. The part-time tuition rates per credit-hour at a community college were $95 for in-state residents and $257 for out-of-state residents. To take 9 credit-hours, how much more than an in-state resident does an out-of-state resident pay?

18. A shirt placed on sale is marked down by $16. At the register, the customer receives an additional discount of $6. If the final sale price of the shirt was $18, what was the original price?

19. An office manager needs to order 1,000 pens from an office supply catalog. If the catalog sells pens by the gross (that is, in sets of 144) and 7 gross were ordered, how many extra pens did the manager order?

20. While shopping, a mother decides to buy three shirts costing $39 apiece and two pairs of shoes at $62 per pair. If she has $300 with her, is that enough money to pay for these items? Explain.

21. Eisenhower beat Stevenson in the 1952 and 1956 presidential elections. In 1952, Eisenhower received 442 electoral votes and Stevenson 89. In 1956, Eisenhower got 457 electoral votes and Stevenson 73. Which election was closer? By how many electoral votes? (*Source: World Almanac*)

22. A garden is rectangular in shape—26 feet in length and 14 feet in width. If fencing costs $13 a foot, how much will it cost to enclose the garden with this fencing?

Use a calculator to solve each problem, giving (a) the operation(s) carried out in the solution, (b) the exact answer, and (c) an estimate of the answer.

23. A couple agrees to pay the seller of the house of their dreams $165,000. They put down $23,448 and promise to pay the balance in 144 equal installments. How much money will each installment be?

24. Earth revolves around the Sun in 365 days, but the planet Mercury does so in only 88 days. Compared to Earth, how many more complete revolutions will Mercury make in 1,000 days?

• *Check your answers on page A-3.*

KEY CONCEPTS AND SKILLS

CONCEPT SKILL

Concept/Skill	Description	Example
[1.1] Place value	<table><tr><th colspan="3">Thousands</th><th colspan="3">Ones</th></tr><tr><td>Hundreds</td><td>Tens</td><td>Ones</td><td>Hundreds</td><td>Tens</td><td>Ones</td></tr></table>	846,120 ↑ 4 is in the ten thousands place.
[1.1] To read a whole number	Working from left to right, • read the number in each period, and then • name the period in place of the comma.	71,400 is read "seventy-one thousand, four hundred".
[1.1] To write a whole number	Working from left to right, • write the number named in each period, and • replace each period name with a comma.	"Five thousand, twelve" is written 5,012.
[1.1] To round a whole number	• Underline the place to which you are rounding. • Look at the digit to the right of the underlined digit, called the *critical digit.* If this digit is 5 or more, add 1 to the underlined digit; if it is less than 5, leave the underlined digit unchanged. • Replace all the digits to the right of the underlined digit with zeros.	$3\underline{8}6 \approx 390$ $4,\underline{8}17 \approx 4,800$
[1.2] Addend, Sum	In an addition problem, the numbers being added are called *addends.* The result is called their *sum.*	6 + 4 = 10 (6: Addend, 4: Addend, 10: Sum)
[1.2] The Identity Property of Addition	The sum of a number and zero is the original number.	$4 + 0 = 4$ $0 + 7 = 7$
[1.2] The Commutative Property of Addition	Changing the order in which two numbers are added does not affect their sum.	$7 + 8 = 8 + 7$
[1.2] The Associative Property of Addition	When adding three numbers, regrouping addends gives the same sum.	$(5 + 4) + 1 = 5 + (4 + 1)$
[1.2] To add whole numbers	• Write the addends vertically, lining up the place values. • Add the digits in the ones column, writing the rightmost digit of the sum on the bottom. If the sum has two digits, carry the left digit to the top of the next column on the left. • Add the digits in the tens column as in the preceding step. • Repeat this process until you reach the last column on the left, writing the entire sum of that column on the bottom.	$\begin{array}{r} \scriptstyle 1\ \ \ 1\ 1\ \ \\ 7,385 \\ 92,551 \\ +\ \ 2,007 \\ \hline 101,943 \end{array}$

continued

Concept/Skill	Description	Example
[1.2] **Minuend, Subtrahend, Difference**	In a subtraction problem, the number that is being subtracted from is called the *minuend.* The number that is being subtracted is called the *subtrahend.* The answer is called the *difference.*	$10 - 6 = 4$ (10 = Minuend, 6 = Subtrahend, 4 = Difference)
[1.2] **To subtract whole numbers**	• On top, write the number *from which* we are subtracting. On the bottom, write the number that is being *taken away,* lining up the place values. Subtract in each column separately. • Start with the ones column. **a.** If the digit on top is *larger* than or *equal* to the digit on the bottom, subtract and write the difference below. **b.** If the digit on top is *smaller* than the digit on the bottom, borrow from the digit to the left on top. Then subtract and write the difference below the bottom digit. • Repeat this process until the last column on the left is finished, subtracting and writing its difference below.	8 ¹4 1 7, 9 5 2 −1, 8 8 3 6, 0 6 9
[1.3] **Factor, Product**	In a multiplication problem, the numbers being multiplied are called *factors*. The result is called their *product.*	$4 \times 5 = 20$ (4, 5 = Factor; 20 = Product)
[1.3] **The Identity Property of Multiplication**	The product of any number and 1 is that number.	$1 \times 6 = 6$ $7 \times 1 = 7$
[1.3] **The Multiplication Property of 0**	The product of any number and 0 is 0.	$51 \times 0 = 0$ $0 \times 9 = 0$
[1.3] **The Commutative Property of Multiplication**	Changing the order in which two numbers are multiplied does not affect their product.	$3 \times 2 = 2 \times 3$
[1.3] **The Associative Property of Multiplication**	When multiplying three numbers, regrouping the factors gives the same product.	$(4 \times 5) \times 6$ $= 4 \times (5 \times 6)$
[1.3] **The Distributive Property**	Multiplying a factor by the sum of two numbers gives the same result as multiplying the factor by each of the two numbers and then adding.	$2 \times (4 + 3)$ $= (2 \times 4) + (2 \times 3)$
[1.3] **To multiply whole numbers**	• Multiply the top factor by the ones digit in the bottom factor and write this product. • Multiply the top factor by the tens digit in the bottom factor and write this product leftward, beginning with the tens column. • Repeat this process until all the digits in the bottom factor are used. • Add the partial products, writing this sum.	693 × 71 693 48 51 49,203

continued

CONCEPT SKILL

Concept/Skill	Description	Example
[1.4] Divisor, Dividend, Quotient	In a division problem, the number that is being used to divide another number is called the *divisor*. The number into which it is being divided is called the *dividend*. The result is called the *quotient*.	$4\overline{)12}$ with quotient 3 on top. Labels: Quotient (3), Divisor (4), Dividend (12)
[1.4] To divide whole numbers	• Divide 17 into 39, which gives 2. Multiply the 17 by 2 and subtract the result (34) from 39. Beside the difference (5), bring down the next digit (3) of the dividend. • Repeat this process, dividing the divisor (17) into 53. • At the end, there is a remainder of 2. Write it beside the quotient on top.	23 R2 $17\overline{)393}$ 34 53 51 2
[1.5] Exponent (or Power), Base	An *exponent* (or *power*) is a number that indicates how many times another number (called the *base*) is used as a factor.	Exponent $5^3 = 5 \times 5 \times 5$ Base
[1.5] Order of Operations Rule	To evaluate mathematical expressions, carry out the operations *in the following order.* **1.** First, perform the operations within any grouping symbols, such as parentheses ( ) or brackets [ ]. **2.** Then, raise any number to its power. **3.** Next, perform all multiplications and divisions as they appear from left to right. **4.** Finally, do all additions and subtractions as they appear from left to right. ( ) power × ÷ + −	$8 + 5 \cdot (3 + 1)^2 = 8 + 5 \cdot 4^2$ $= 8 + 5 \cdot 16$ $= 8 + 80$ $= 88$
[1.5] Average (or Mean)	The *average* (or *mean*) of a set of numbers is the sum of those numbers divided by however many numbers are in the set.	The average of 3, 4, 10, and 3 is 5 because $\frac{3 + 4 + 10 + 3}{4} = \frac{20}{4} = 5$
[1.6] To solve word problems	• Read the problem carefully. • Choose a strategy (such as drawing a picture, breaking up the question, substituting simpler numbers, or making a table). • Decide which basic operation(s) are relevant and then translate the words into mathematical symbols. • Perform the operations. • Check the solution to see if the answer is reasonable. If it is not, start again by rereading the problem.	

Chapter 1 Review Exercises

To help you review this chapter, solve these problems.

[1.1] *In each whole number, identify the place that the digit 3 occupies.*

1. 23

2. 30,802

3. 385,000,000

4. 30,000,000,000

Write each number in words.

5. 497

6. 2,050

7. 3,000,007

8. 85,000,000,000

Write each number in standard form.

9. Two hundred fifty-one

10. Nine thousand, two

11. Fourteen million, twenty-five

12. Three billion, three thousand

Express each number in expanded form.

13. 2,500,000

14. 42,707

Round each number to the place indicated.

15. 571 to the nearest hundred

16. 938 to the nearest thousand

17. 384,056 to the nearest ten thousand

18. 68,332 to its largest place

[1.2] *Find the sum and check.*

19. 102 + 4,251 + 5,133

20. 53,569 + 10,000 + 2,123

21. 48,758 + 37,226 + 87,559

22. 95,000 + 25,895 + 30,000

Add and check.

23. 972,558 + 87,055 + 36,488 + 861,724

24. $138,865 + $729 + $8,002 + $75,471

Find the difference and check.

25. 876 − 431

26. 56,000 − 45,984

27. 98,118 − 87,009

28. 7,100 − 1,590

29. 60,000,000 − 48,957,777

30. $5,000,000 − $2,937,148

31. From 67,502 subtract 56,496.

32. Subtract 89,724 from 92,713.

[1.3] *Find the product and check.*

33. $\begin{array}{r} 72 \\ \times\ 6 \\ \hline \end{array}$

34. $\begin{array}{r} 400 \\ \times\ 3 \\ \hline \end{array}$

35. $\begin{array}{r} 2,923 \\ \times\ 51 \\ \hline \end{array}$

36. $\begin{array}{r} 6,000 \\ \times\ 2,000 \\ \hline \end{array}$

37. $\begin{array}{r} 14,921 \\ \times\ 32 \\ \hline \end{array}$

38. $\begin{array}{r} 8,152 \\ \times\ 125 \\ \hline \end{array}$

Multiply and check.

39. $2,751 \cdot 508$

40. $(681)(498)(555)$

[1.4] *Divide and check.*

41. $\dfrac{975}{25}$

42. $21\overline{)6,450}$

43. $13\overline{)491}$

44. $7,488 \div 11$

Find the quotient and check.

45. $8\overline{)205,000}$

46. $347\overline{)332,079}$

[1.5] *Compute.*

47. 7^3

48. 1^{10}

49. $2^3 \cdot 3^2$

50. $3 \cdot 10^5$

51. $20 - 3 \times 5$

52. $(9 + 4)^2$

53. $10 - \dfrac{6 + 4}{2}$

54. $3 + (5 - 1)^2$

55. $5 + [4^2 - 3(2 + 1)]$

56. $17 + [2(3^2 - 6) - 5]$

57. $98(50 - 1)(50 - 2)(50 - 3)$

58. $\dfrac{28^3 + 29^3 + 37^3 - 10}{(7 - 1)^2}$

Rewrite each expression, using exponents.

59. $7 \cdot 7 \cdot 5 \cdot 5 = 7^{\blacksquare} \cdot 5^{\blacksquare}$

60. $5 \cdot 2 \cdot 5 \cdot 2 \cdot 5 = 2^{\blacksquare} \cdot 5^{\blacksquare}$

Find the average.

61. 34 and 44

62. 20, 0, and 1

63. 5, 8, and 5

64. 4, 6, 3, and 7

Mixed Applications

Solve and check.

65. Beetles about the size of a pinhead destroyed 2,400,000 acres of forest. Express this number in words.

66. Scientists in Utah found a dinosaur egg one hundred fifty million years old. Write this number in standard form.

67. In a part-time job, a graduate student earned $15,964 a year. How much money did she earn per week? (*Hint:* 1 year equals 52 weeks.)

68. Halley's Comet was sighted in 1682. If it reappeared 76 years later, in what year did it reappear? (***Source:*** *The Time Almanac 2000*)

69. What is the land area of Texas to the nearest hundred thousand square miles? (***Source:*** *Time Almanac 2006*)

70. Apple Computer sold 22,497,000 iPods in 2005. How many iPods is this to the nearest million? (***Source:*** Apple Computer, 2005 10-K Annual Report)

71. The Empire State Building is 1,250 feet high, and the Statue of Liberty is 152 feet in height. What is the minimum number of Statues of Liberty that would have to be stacked to be taller than the Empire State Building?

72. A millipede—a small insect with 68 body segments—has 4 legs per segment. How many legs does a millipede have?

73. Taipei 101 in Taiwan, the tallest building in the world as of 2006, is 67 meters taller than the Sears Tower in Chicago. If the Sears Tower is 442 meters tall, what is the height of Taipei 101? (***Source:*** Emporis Buildings)

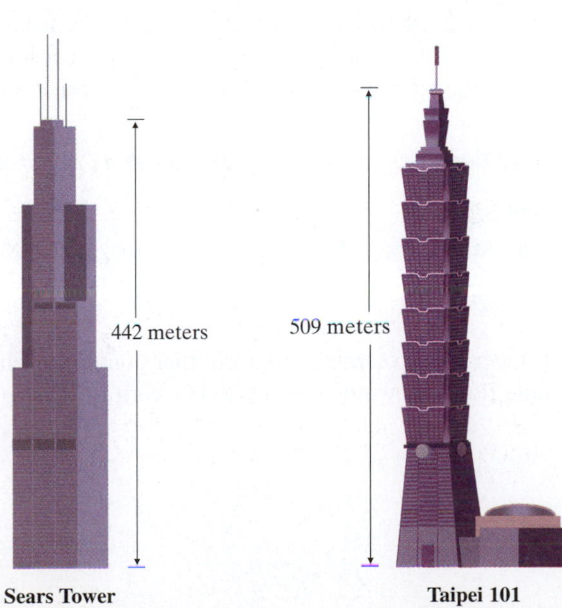

74. A landscaper needs 550 flower plants for a landscaping project. If a local garden center sells flats containing 24 plants, how many flats should the landscaper buy?

75. The following graph shows the consolidated assets of the six largest banks in the United States (in millions of U.S. dollars). Find the combined assets of Wells Fargo and Wachovia.

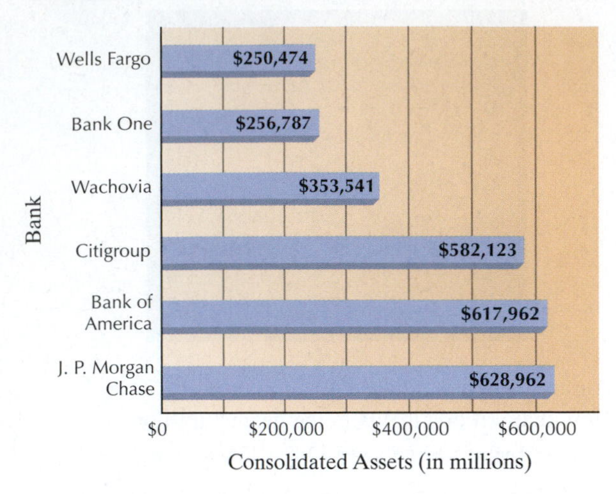

(*Source: Time Almanac 2006*)

76. Compute a company's net profit by completing the following business *skeletal profit and loss statement.*

Net sales	$430,000
− Cost of merchandise sold	− 175,000
Gross margin	$
− Operating expenses	− 135,000
Net profit	$

77. In Giza, Egypt, the pyramid of Khufu has a base that measures 230 meters by 230 meters, whereas the pyramid of Khafre has a base that measures 215 meters by 215 meters. In area, how much larger than the base of the pyramid of Khafre is that of Khufu? (*Source:* pbs.org)

78. Both a singles tennis court and a football field are rectangular in shape. A tennis court measures 78 feet by 27 feet, whereas a football field measures 360 feet by 160 feet. About how many times the area of a tennis court is that of a football field?

79. Richard Nixon ran for the U.S. presidency three times. According to the table below, which was greater—the increase from 1960 to 1972 in the number of votes he got or the increase from 1968 to 1972? (*Source: The World Almanac & Book of Facts 2000*)

Year	Number of Votes for Nixon
1960	34,108,546
1968	31,785,480
1972	47,165,234

80. On a business trip, a sales representative flew from Chicago to Los Angeles to Boston. The chart below shows the air distances in miles between these cities.

Air Distance	Chicago	Los Angeles	Boston
Chicago	—	1,745	1,042
Los Angeles	1,745	—	2,596
Boston	1,042	2,596	—

If the sales rep earned a frequent flier point for each mile flown, how many points did he earn?

81. The Tour de France is a 20-stage bicycle race held in France annually. The chart shows the distances for the first 10 stages of the 2006 Tour de France. (*Source:* Le Tour de France)

Stage	Distance (in kilometers)
1	183
2	223
3	216
4	215
5	219
6	184
7	52
8	177
9	170
10	193

a. What was the total distance covered in the first 10 stages?

b. The entire race covered a distance of 3,632 kilometers. How many kilometers were covered in the last 10 stages?

82. The number of students enrolled annually in public colleges in the United States from 2000 through 2005 is given in the table. (*Source:* National Center for Education Statistics)

Year	Enrollment
2000	11,753,000
2001	12,233,000
2002	12,752,000
2003	12,952,000
2004	13,092,000
2005	13,283,000

a. What was the average annual enrollment in the years 2000 through 2005?

b. The enrollments in 2006 and 2007 were 13,518,800 and 13,752,000, respectively. How would the average annual enrollment change if the enrollments for 2006 and 2007 were included?

83. Find the area of the figure.

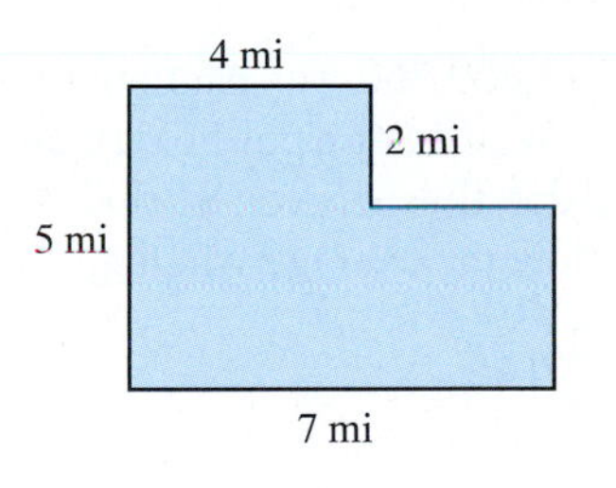

84. Find the perimeter of the figure.

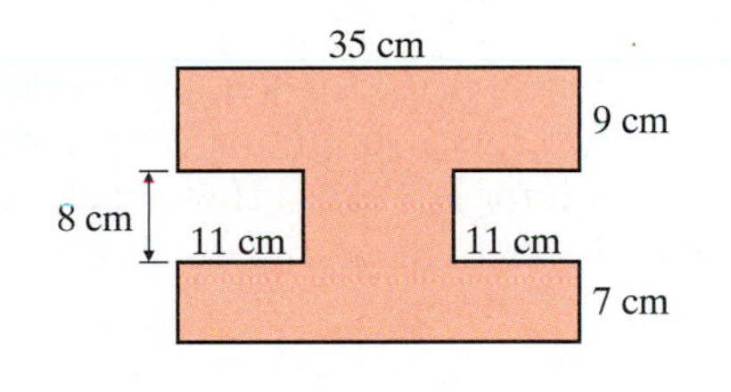

• *Check your answers on page A-3.*

Chapter 1 POSTTEST

FOR EXTRA HELP

Test solutions are found on the enclosed CD.

To see if you have already mastered the topics in this chapter, take this test.

1. Write two hundred twenty-five thousand, sixty-seven in standard form.

2. Underline the digit that occupies the ten thousands place in 1,768,405.

3. Write 1,205,007 in words.

4. Round 196,593 to the nearest hundred thousand.

5. Find the sum of 398 and 1,496.

6. Subtract 398 from 1,005.

7. Subtract: $2{,}000 - 1{,}853$

8. Multiply: 328×907

9. Compute: $\dfrac{23{,}923}{47}$

10. Find the quotient: $59\overline{)36{,}717}$

11. Evaluate: 5^4

12. Write $5 \cdot 5 \cdot 4 \cdot 4 \cdot 4$ using exponents.

Simplify.

13. $4 \cdot 9 + 3 \cdot 4^2$

14. $29 - 3^3 \cdot (10 - 9)$

Solve and check.

15. The two largest continents in the world are Asia and Africa. To the nearest hundred thousand square miles, Asia's area is 17,400,000 and Africa's is 11,700,000. How much larger is Asia than Africa? (***Source:*** *The Time Almanac 2006*)

16. In the year 2005, the state of Kansas had about 64,000 farms with an average size of 732 acres. How many acres of land in Kansas were devoted to farming? (***Source:*** U.S. Department of Agriculture, *Farms and Land in Farms,* 2006)

17. A part-time student had \$1,679 in his checking account. He wrote a \$625 check for tuition, a \$546 check for rent, and a \$39 check for groceries. How much money remained in the account after these checks cleared?

18. A total of \$27,609,360 was paid out to the top twenty places at the World Series of Poker main event in 2005. The ninth-place finisher won \$1,000,000. Was this amount above or below the average winnings? Explain. (***Source:*** World Series of Poker)

19. A homeowner wishes to carpet the hallway shown below. If the cost of carpeting is about \$10 per square foot, approximately how much will the carpeting cost?

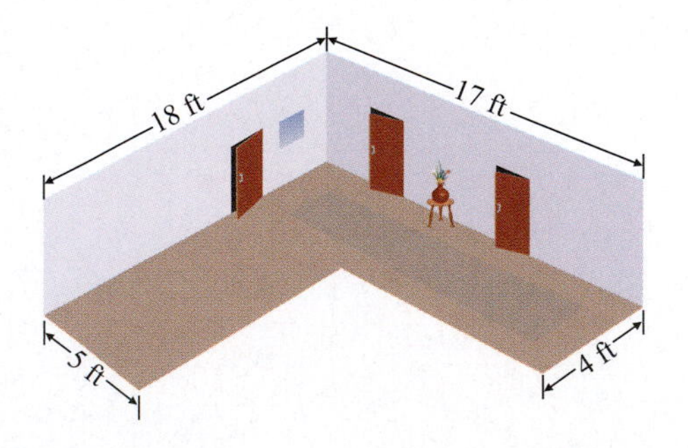

20. For breakfast, you have an 8-ounce (oz) serving of yogurt, a cup of black coffee, and 2 cups of pineapple juice. Based on the following table, how much more vitamin C do you need to reach the recommended 60 milligrams (mg)?

Food	Quantity	Vitamin C Content
Pineapple juice	1 cup	23 mg
Yogurt	8 oz	2 mg
Black coffee	1 cup	0 mg

• *Check your answers on page A-3.*

CHAPTER 2

Fractions

Fractions and Cooking

Using recipes in cooking has a very long and complicated history that goes back thousands of years. For instance, archaeologists have found Mesopotamian recipes written on clay tablets dating from 1700 B.C. Until the Industrial Revolution, at which time standard measurements and precise cooking directions were introduced, recipes gave just a list of ingredients and a general description for cooking a dish.

Today, it is much easier to follow recipes and to cook satisfying dishes because standard measurements and detailed instructions are given in recipes. Many recipes give fractional amounts of ingredients. As a result, cooks need to know how to use fractions.

For example, consider the ingredients given in a recipe for Italian-style meatloaf. The recipe, which serves 8 people, calls for $1\frac{1}{2}$ pounds of ground meat, of which $\frac{1}{3}$ is pork and $\frac{2}{3}$ is beef. To follow this recipe, a cook uses multiplying fractions to determine that for $1\frac{1}{2}$ pounds of ground meat, the amount of ground pork needed is $\frac{1}{3} \times 1\frac{1}{2}$, or $\frac{1}{2}$ pound, and the amount of ground beef needed is $\frac{2}{3} \times 1\frac{1}{2}$, or 1 pound. Multiplying fractions is also used if the cook wants to increase or decrease the number of servings that the recipe yields. (*Source:* http://www.foodtimeline.org)

Chapter 2 PRETEST

To see if you have already mastered the topics in this chapter, take this test.

1. Find all the factors of 20.

2. Express 72 as the product of prime factors.

3. What fraction does the shaded part of the diagram represent?

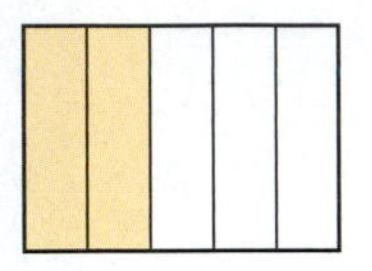

4. Write $20\frac{1}{3}$ as an improper fraction.

5. Express $\frac{31}{30}$ as a mixed number.

6. Write $\frac{9}{12}$ in simplest form.

7. What is the least common multiple of 10 and 4?

8. Which is greater, $\frac{1}{8}$ or $\frac{1}{9}$?

Add.

9. $\frac{1}{2} + \frac{7}{10}$

10. $7\frac{1}{3} + 5\frac{1}{2}$

Subtract.

11. $8\frac{1}{4} - 6$

12. $12\frac{1}{2} - 7\frac{7}{8}$

Multiply.

13. $2\frac{1}{3} \times 1\frac{1}{2}$

14. $\frac{5}{8} \times 96$

15. Divide: $3\frac{1}{3} \div 5$

16. Calculate: $2 + 1\frac{1}{3} \div \frac{4}{5}$

Solve. Write your answer in simplest form.

17. In 2006, the Pittsburgh Steelers won their fifth Super Bowl championship game. If 40 Super Bowl championship games have been played, what fraction of these games has this team won? (*Source:* National Football League)

18. In a biology class, three-fourths of the students received a passing grade. If there are 24 students in the class, how many students received failing grades?

19. Find the perimeter of the traffic island shown.

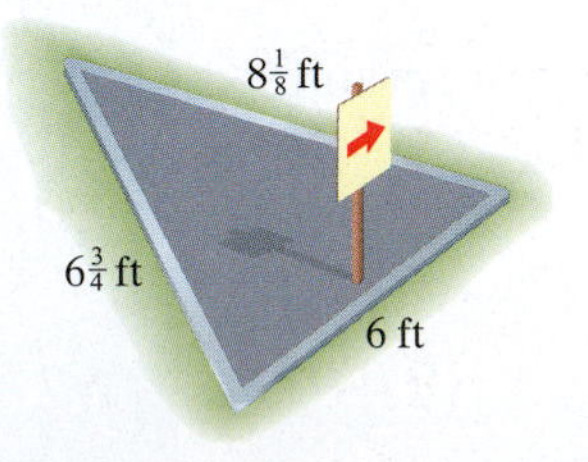

20. According to the nutrition information given, one serving of Honey Nut Cheerios® contains 22 grams of carbohydrates. If one serving is $\frac{3}{4}$ cup, what amount of carbohydrates is contained in $2\frac{1}{4}$ cups of Honey Nut Cheerios? (*Source:* General Mills)

• *Check your answers on page A-3.*

2.1 Factors and Prime Numbers

OBJECTIVES

- To identify prime and composite numbers
- To find the factors and prime factorization of a whole number
- To find the least common multiple of two or more numbers
- To solve word problems using factoring or the LCM

What Factors Mean and Why They Are Important

Recall that in a multiplication problem, the whole numbers that we are multiplying are called **factors** of the product. For instance, 2 is said to be a factor of 8 because $2 \cdot 4 = 8$. Likewise, 4 is a factor of 8.

Another way of expressing the same idea is in terms of division: We say that 8 is *divisible* by 2, meaning that there is a remainder 0 when we divide 8 by 2.

$$\frac{8}{2} = 4 \text{ R0}$$

Note that 1, 2, 4, and 8 are all factors of 8.

Although we factor whole numbers, a major application of factoring involves working with fractions, as we demonstrate in the next section.

Finding Factors

To identify the factors of a whole number, we divide the whole number by the numbers 1, 2, 3, 4, 5, 6, and so on, looking for remainders of 0.

EXAMPLE 1

Find all the factors of 6.

Solution

Starting with 1, we divide each whole number into 6.

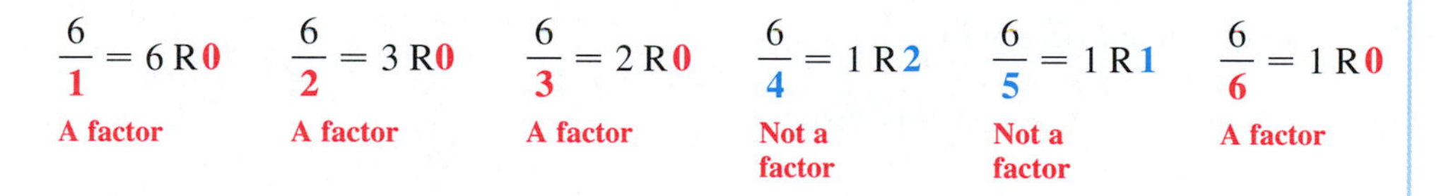

In finding the factors of 6, we do not need to divide 6 by the numbers 7 or greater. The reason is that no number larger than 6 could divide evenly into 6, that is, divide into 6 with no remainder.

So the factors of 6 are 1, 2, 3, and 6. Note that

- 1 is a factor of 6 and
- 6 is a factor of 6.

PRACTICE 1

What are the factors of 7?

Tip For any whole number, both *the number itself* and *1* are always factors. Therefore, all whole numbers (except 1) have at least two factors.

When checking to see if one number is a factor of another, it is generally faster to use the following **divisibility tests** than to divide.

The number is divisible by	if
2	the ones digit is 0, 2, 4, 6, or 8, that is, if the number is even.
3	the sum of the digits is divisible by 3.
4	the number named by the last two digits is divisible by 4.
5	the ones digit is either 0 or 5.
6	the number is even and the sum of the digits is divisible by 3.
9	the sum of the digits is divisible by 9.
10	the ones digit is 0.

EXAMPLE 2

What are the factors of 45?

Solution Let's see if 45 is divisible by 1, 2, 3, and so on, using the divisibility tests wherever they apply.

Is 45 divisible by	Answer
1?	Yes, because 1 is a factor of any number; $\frac{45}{1} = 45$, so 45 is also a factor.
2?	No, because the ones digit is not even.
3?	Yes, because the sum of the digits, $4 + 5 = 9$, is divisible by 3; $\frac{45}{3} = 15$, so 15 is also a factor.
4?	No, because 4 will not divide into 45 evenly.
5?	Yes, because the ones digit is 5; $\frac{45}{5} = 9$, so 9 is also a factor.
6?	No, because 45 is not even.
7?	No, because $45 \div 7$ has remainder 3.
8?	No, because $45 \div 8$ has remainder 5.
9?	We already know that 9 is a factor.
10?	No, because the ones digit is not 0.

The factors of 45 are, therefore, 1, 3, 5, 9, 15, and 45.

Note that we really didn't have to check to see if 9 was a factor—we learned that it was when we checked for divisibility by 5. Also, because the factors were beginning to repeat with 9, there was no need to check numbers greater than 9.

PRACTICE 2

Find all the factors of 75.

EXAMPLE 3

Identify all the factors of 60.

Solution Let's check to see if 60 is divisible by 1, 2, 3, 4, and so on.

Is 60 divisible by	Answer
1?	Yes, because 1 is a factor of all numbers; $\frac{60}{1} = 60$, so 60 is also a factor.
2?	Yes, because the ones digit is even; $\frac{60}{2} = 30$, so 30 is also a factor.
3?	Yes, because the sum of the digits, $6 + 0 = 6$, is divisible by 3; $\frac{60}{3} = 20$, so 20 is also a factor.
4?	Yes, because 4 will divide into 60 evenly; $\frac{60}{4} = 15$, so 15 is also a factor.
5?	Yes, because the ones digit is 0; $\frac{60}{5} = 12$, so 12 is also a factor.
6?	Yes, because the number is even, and the sum of the digits is divisible by 3; $\frac{60}{6} = 10$, so 10 is also a factor.
7?	No, because $60 \div 7$ has remainder 4.
8?	No, because $60 \div 8$ has remainder 4.
9?	No, because the sum of the digits, $6 + 0 = 6$, is not divisible by 9.
10?	We already know that 10 is a factor.

The factors of 60 are, therefore, 1, 2, 3, 4, 5, 6, 10, 12, 15, 20, 30, and 60.

PRACTICE 3

What are the factors of 90?

EXAMPLE 4

A presidential election takes place in the United States every year that is a multiple of 4. Was there a presidential election in 1866? Explain.

Solution The question is: Does 4 divide into 1866 evenly? Using the divisibility test for 4, we check whether 66 is a multiple of 4.

$$\frac{66}{4} = 16 \text{ R}2$$

Because $\frac{66}{4}$ has remainder 2, 4 is not a factor of 1866. So there was no presidential election in 1866.

PRACTICE 4

The doctor instructs a patient to take a pill every 3 hours. If the patient took a pill at 8:00 this morning, should she take one tomorrow at the same time? Explain.

Identifying Prime and Composite Numbers

Now let's discuss the difference between prime numbers and composite numbers.

Definitions

A **prime number** is a whole number that has exactly two different factors: itself and 1.

A **composite number** is a whole number that has more than two factors.

Note that the numbers 0 and 1 are neither prime nor composite. But every whole number greater than 1 is either prime or composite, depending on its factors.

For instance, 5 is prime because its only factors are 1 and 5. But 8 is composite because it has more than two factors (it has four factors: 1, 2, 4, and 8).

Let's practice distinguishing between primes and composites.

EXAMPLE 5

Indicate whether each number is prime or composite.

a. 2 **b.** 78 **c.** 51 **d.** 19 **e.** 31

Solution

a. The only factors of 2 are 1 and 2. Therefore, 2 is prime.

b. Because 78 is even, it is divisible by 2. Having 2 as an "extra" factor—in addition to 1 and 78—means that 78 is composite. Do you see why all even numbers, except for 2, are composite?

c. Using the divisibility test for 3, we see that 51 is divisible by 3, because the sum of the digits 5 and 1, or 6, is divisible by 3. Because 51 has more than two factors, it is composite.

d. The only factors of 19 are itself and 1. Therefore, 19 is prime.

e. Because 31 has no factors other than itself and 1, it is prime.

PRACTICE 5

Decide whether each number is prime or composite.

a. 3 **b.** 57 **c.** 29

d. 34 **e.** 17

Finding the Prime Factorization of a Number

Every composite number can be written as the product of prime factors. This product is called its **prime factorization.** For instance, the prime factorization of 12 is $2 \cdot 2 \cdot 3$.

Definition

The **prime factorization** of a whole number is the number written as the product of its prime factors.

Being able to find the prime factorization of a number is an important skill to have for working with fractions, as we show later in this chapter. A good way to find the prime factorization of a number is by making a **factor tree**, as illustrated in Example 6.

EXAMPLE 6

Write the prime factorization of 72.

Solution We start building a factor tree for 72 by dividing 72 by the smallest prime, 2. Because 72 is $2 \cdot 36$, we write both 2 and 36 underneath the 72. Then we circle the 2 because it is prime.

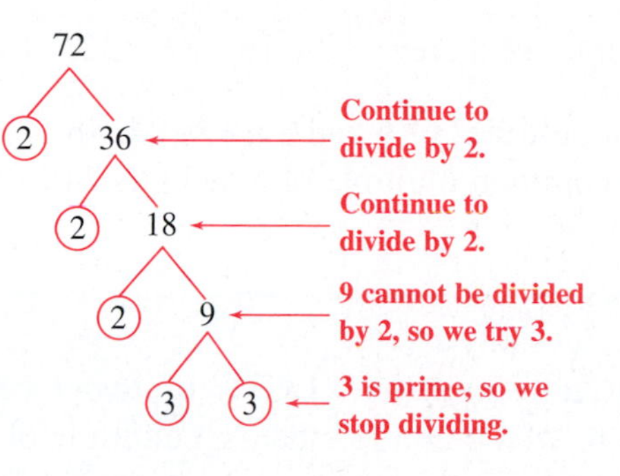

Next we divide 36 by 2, writing both 2 and 18, and circling 2 because it is prime. Below the 18, we write 2 and 9, again circling the 2. Because 9 is not divisible by 2, we divide it by the next smallest prime, 3. We continue this process until all the factors in the bottom row are prime. The prime factorization of 72 is the product of the circled factors.

$$72 = 2 \times 2 \times 2 \times 3 \times 3$$

We can also write this prime factorization as $2^3 \times 3^2$.

PRACTICE 6

Write the prime factorization of 56, using exponents.

EXAMPLE 7

Express 80 as the product of prime factors.

Solution The factor tree method for 80 is as shown.

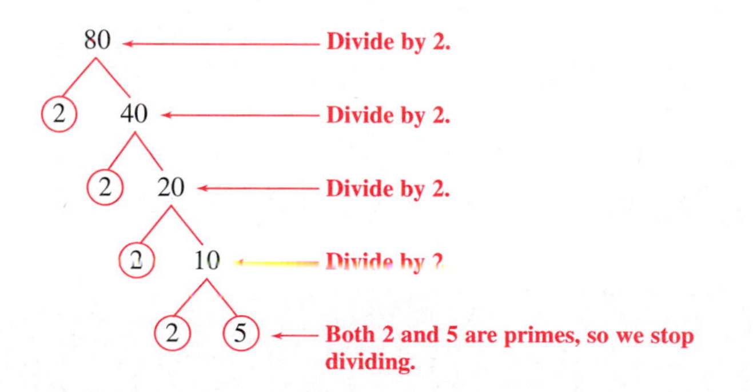

The prime factorization of 80 is $2 \times 2 \times 2 \times 2 \times 5$, or $2^4 \times 5$.

PRACTICE 7

What is the prime factorization of 75?

Finding the Least Common Multiples

The *multiples* of a number are the products of that number and the whole numbers. For instance, some multiples of 5 are the following:

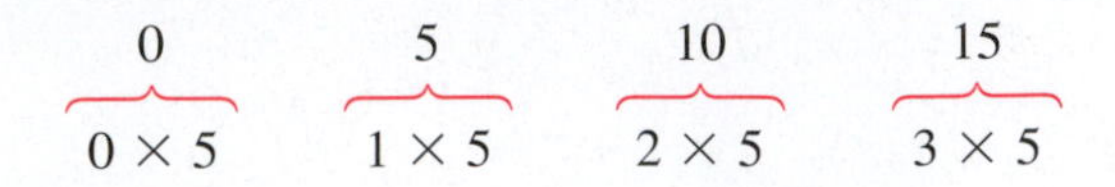

A number that is a multiple of two or more numbers is called a *common multiple* of these numbers. To find the common multiples of 6 and 8, we first list the multiples of 6 and the multiples of 8 separately.

- The multiples of 6 are 0, 6, 12, 18, 24, 30, 36, 42, 48, 54, 60, ...
- The multiples of 8 are 0, 8, 16, 24, 32, 40, 48, 56, 64, ...

So the common multiples of 6 and 8 are 0, 24, 48, ... Of the nonzero common multiples, the *least* common multiple of 6 and 8 is 24.

Definition

The **least common multiple (LCM)** of two or more whole numbers is the smallest nonzero whole number that is a multiple of each number.

A shortcut for finding the LCM—often faster than listing multiples—involves prime factorization.

To Compute the Least Common Multiple (LCM)

- Find the prime factorization of each number.
- Identify the prime factors that appear in each factorization.
- Multiply these prime factors, using each factor the greatest number of times that it occurs in any of the factorizations.

EXAMPLE 8

Find the LCM of 8 and 12.

Solution We first find the prime factorization of each number.

$$8 = 2 \times 2 \times 2 = 2^3 \qquad 12 = 2 \times 2 \times 3 = 2^2 \times 3$$

The factor 2 appears *three times* in the factorization of 8 and *twice* in the factorization of 12, so it must be included three times in forming the least common multiple. Also, the factor 3 appears once in the prime factorization of 12.

The highest power of 2

$$\text{LCM} = 2^3 \times 3 = 8 \times 3 = 24$$

As always, it is a good idea to check that our answer makes sense. We do so by verifying that 8 and 12 really are factors of 24.

PRACTICE 8

What is the LCM of 9 and 6?

EXAMPLE 9

Find the LCM of 5 and 9.

Solution First, we write each number as the product of primes.

$$5 = 5 \qquad 9 = 3 \times 3 = 3^2$$

To find the LCM, we multiply the highest power of each prime.

$$\text{LCM} = 5 \times 3^2 = 5 \times 9 = 45$$

So the LCM of 5 and 9 is 45. Note that 45 is also the product of 5 and 9.

PRACTICE 9

Find the LCM for 3 and 22.

Tip If two or more numbers have no common factor (other than 1), the LCM is their product. If one number is a multiple of another number, then their LCM is the larger of the two numbers.

Now let's find the LCM of three numbers.

EXAMPLE 10

Find the LCM of 3, 5, and 6.

Solution First, we find the prime factorizations of these three numbers.

$$3 = 3 \qquad 5 = 5 \qquad 6 = 2 \times 3$$

The LCM is therefore the product $2 \times 3 \times 5$, which is 30. Note that 30 is a multiple of 3, 5, and 6, which supports our answer.

PRACTICE 10

Find the LCM of 2, 3, and 4.

EXAMPLE 11

A gym that is open every day of the week offers aerobics classes every third day and yoga classes every fourth day. A student took both classes this morning. In how many days will the gym offer both classes on the same day?

Solution To answer this question, we ask: What is the LCM of 3 and 4? As usual, we begin by finding prime factorizations.

$$3 = 3 \qquad 4 = 2 \times 2 = 2^2$$

To find the LCM, we multiply 3 by 2^2.

$$\text{LCM} = 2^2 \times 3 = 12$$

So both classes will be offered again on the same day in 12 days.

PRACTICE 11

Suppose that a Senate seat and a House of Representatives seat were both filled this year. If the Senate seat is filled every 6 years and the House seat every 2 years, in how many years will both seats be up for election?

Mathematically Speaking

Fill in each blank with the most appropriate term or phrase from the given list.

division	least common multiple	composite	factor tree
divisibility	prime	remainders	common multiple
prime factorization	factors	multiples	

1. 1, 2, 3, 5, 6, 10, 15, and 30 are _______ of 30.

2. A(n) _______ number is a whole number that has more than two factors.

3. A(n) _______ number has exactly two different factors: itself and 1.

4. The _______ of two or more numbers is the smallest nonzero number that is a multiple of each number.

5. A number written as the product of its prime factors is called its _______.

6. The _______ test for 10 is to check if the ones digit is 0.

List all the factors of each number.

7. 21 **8.** 10 **9.** 17 **10.** 9

11. 12 **12.** 15 **13.** 31 **14.** 47

15. 36 **16.** 35 **17.** 29 **18.** 73

19. 100 **20.** 98 **21.** 28 **22.** 48

Indicate whether each number is prime or composite. If it is composite, identify a factor other than the number itself and 1.

23. 13 **24.** 7 **25.** 16 **26.** 24 **27.** 49

28. 75 **29.** 11 **30.** 31 **31.** 81 **32.** 45

Write the prime factorization of each number.

33. 8 **34.** 10 **35.** 49 **36.** 14

37. 24 **38.** 18 **39.** 50 **40.** 40

41. 77 **42.** 63 **43.** 51 **44.** 57

45. 25 **46.** 49 **47.** 32 **48.** 64

49. 21 **50.** 22 **51.** 104 **52.** 105

53. 121 **54.** 169 **55.** 142 **56.** 62

57. 100
58. 200
59. 125
60. 90
61. 135
62. 400

Find the LCM in each case.

63. 3 and 15
64. 9 and 12
65. 8 and 10
66. 4 and 6
67. 9 and 30
68. 20 and 21
69. 10 and 11
70. 15 and 60
71. 18 and 24
72. 30 and 150
73. 40 and 180
74. 100 and 90
75. 12, 5, and 50
76. 2, 8, and 10
77. 4, 7, and 12
78. 2, 3, and 5
79. 3, 5, and 7
80. 6, 8, and 12
81. 5, 15, and 20
82. 8, 24, and 56

Mixed Practice

Solve.

83. Write the prime factorization of 75.

84. Is 63 prime or composite? If it is composite, identify a factor other than the number itself and 1.

85. List all the factors of 72.

86. Find the LCM of 5, 10, and 12.

Applications

Solve.

87. The federal government conducts a census every year that is a multiple of 10. Explain whether there was a census in

a. 1995.

b. 1990.

88. Because of production considerations, the number of pages in a book that you are writing must be a multiple of 4. Can the book be

a. 196 pages long?

b. 198 pages long?

89. In 2006, the men's World Cup soccer tournament was held in Munich, Germany. If the tournament is held every 4 years, will there be a tournament in 2036? (*Source:* FIFA World Cup; Soccer Hall of Fame)

90. A car manufacturer recommends changing the oil every 3,000 miles. Would an oil change be recommended at 21,000 miles? Explain.

91. There are 9 players on a baseball team and 11 players on a soccer team. What is the smallest number of students in a college that can be split evenly into either baseball or soccer teams?

92. The Fields Medal, the highest scientific award for mathematicians, is awarded every 4 years. The Dantzig Prize, an achievement award in the field of mathematical programming, is awarded every 3 years. If both were given in 2006, in what year will both be given again? (*Source:* Eric W. Weisstein et al., "Fields Medals," Math World; A Wolfram Web Resource; Mathematical Programming Society)

93. Two friends work in a hospital. One gets a day off every 5 days, and the other every 6 days. If they were both off today, in how many days will they again both be off?

94. A family must budget for life insurance premiums every 6 months, car insurance premiums every 3 months, and payments for a home security system every 4 months. If all these bills were due this month, in how many months will they again all fall due?

Historical

1. The eighteenth-century mathematician Christian Goldbach made several famous conjectures (guesses) about prime numbers. One of these conjectures states: Every odd number greater than 7 can be expressed as the sum of three odd prime numbers. For instance, 11 can be expressed as 3 + 3 + 5. Write the following odd numbers as the sum of three odd primes.

   a. $57 = \blacksquare + \blacksquare + \blacksquare$ **b.** $81 = \blacksquare + \blacksquare + \blacksquare$

Mathematical Reasoning

2. What is the smallest whole number divisible by every whole number from 1 to 10?

Critical Thinking

3. Choose a three-digit number, say, 715. Find three prime numbers so that, when 715 is multiplied by the product of the three prime numbers, the product of all four numbers is 715,715.

2.2 Introduction to Fractions

OBJECTIVES

- To read and write fractions and mixed numbers
- To write improper fractions as mixed numbers and mixed numbers as improper fractions
- To find equivalent fractions and to write fractions in simplest form
- To compare fractions
- To solve word problems with fractions

What Fractions Are and Why They Are Important

A fraction can mean *part of a whole*. Just as a whole number answers the question How many?, a fraction answers the question What part of? Every day we use fractions in this sense. For example, we can speak of *two-thirds* of a class (meaning two of every three students) or *three-fourths* of a dollar (indicating that we have split a dollar into four equal parts and have taken three of these parts).

A fraction can also mean *the quotient of two whole numbers*. In this sense, the fraction $\frac{3}{4}$ tells us what we get when we divide the whole number 3 by the whole number 4.

Definition

A **fraction** is any number that can be written in the form $\frac{a}{b}$, where a and b are whole numbers and b is nonzero.

From this definition, $\frac{1}{2}, \frac{3}{9}, \frac{6}{5}, \frac{8}{2}$, and $\frac{0}{1}$ are all fractions.

When written as $\frac{a}{b}$, a fraction has three components: $\frac{\text{Numerator}}{\text{Denominator}}$ ← Fraction line

- The **denominator** (on the bottom) stands for the number of parts into which the whole is divided.
- The **numerator** (on top) tells us how many parts of the whole the fraction contains.
- The **fraction line** separates the numerator from the denominator and stands for "out of" or "divided by."

Alternatively, a fraction can be represented as either a decimal or a percent. We discuss decimals and percents in Chapters 3 and 6.

Fraction Diagrams and Proper Fractions

Diagrams help us work with fractions. The fraction three-fourths, or $\frac{3}{4}$, is represented by the shaded part in each of the following diagrams:

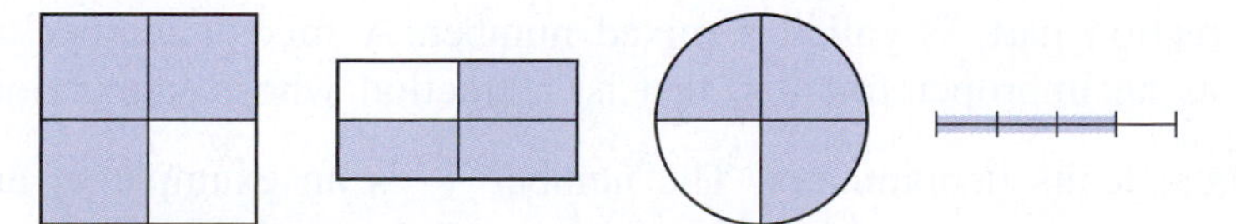

Note that in each diagram the whole has been divided into 4 *equal* parts, with 3 of the parts shaded.

The number $\frac{3}{4}$ is an example of a **proper fraction** because its numerator is smaller than its denominator. Let's consider some other examples of proper fractions.

EXAMPLE 1

In the diagram, what does the shaded portion represent?

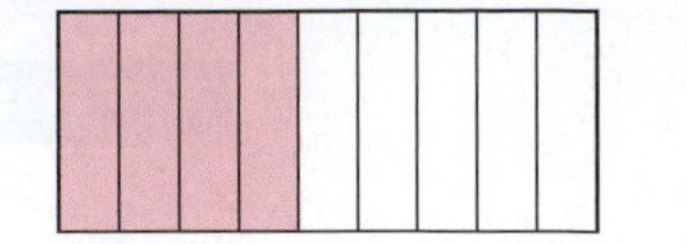

Solution In this diagram, the whole is divided into 9 equal parts, so the denominator of the fraction shown is 9. Four of these parts are shaded, so the numerator is 4. The diagram represents the fraction $\frac{4}{9}$.

PRACTICE 1

The diagram illustrates what fraction?

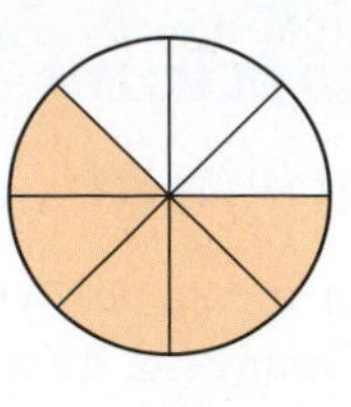

EXAMPLE 2

A college accepted 147 out of 341 applicants for admission into the nursing program. What fraction of the applicants were accepted into this program?

Solution Since there was a total of 341 applicants, the denominator of our fraction is 341. Because 147 of the applicants were accepted, 147 is the numerator. So the college accepted $\frac{147}{341}$ of the applicants into the nursing program.

PRACTICE 2

During a 30-minute television program, 7 minutes were devoted to commercials. What fraction of the time was for commercials?

EXAMPLE 3

The U.S. Senate approved a foreign-aid spending bill by a vote of 83 to 17. What fraction of the senators voted against the bill?

Solution First, we find the total number of senators. Because 83 senators voted for the bill and 17 voted against it, the total number of senators is $83 + 17$, or 100. So $\frac{17}{100}$ of the senators voted against the bill.

PRACTICE 3

Through the 2005 baseball season, the American League had won 60 World Series championships, whereas the National League had won 41. What fraction of the World Series championships were *not* won by the American League? (*Source:* http://mlb.com)

Mixed Numbers and Improper Fractions

On many jobs, if you work overtime, the rate of pay increases to one-and-a-half times the regular rate. A number such as $1\frac{1}{2}$, with a whole number part and a proper fraction part, is called a mixed number. A mixed number can also be expressed as an improper fraction, that is, a fraction whose numerator is greater than or equal to its denominator. The number $\frac{3}{2}$ is an example of an improper fraction.

Diagrams help us understand that mixed numbers and improper fractions are different forms of the same numbers, as Example 4 illustrates.

EXAMPLE 4

Draw diagrams to show that $2\frac{1}{3} = \frac{7}{3}$.

Solution First, represent the mixed number and the improper fraction in diagrams.

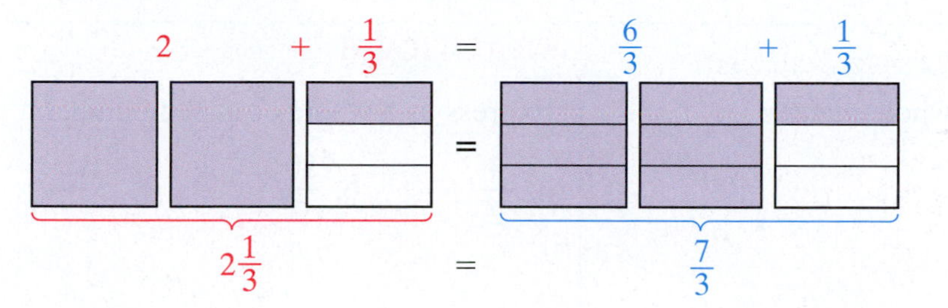

Both diagrams represent $2 + \frac{1}{3}$, so the numbers $2\frac{1}{3}$ and $\frac{7}{3}$ must be equal.

PRACTICE 4

By means of diagrams, explain why $1\frac{2}{3} = \frac{5}{3}$.

In Example 4 each unit (or square) corresponds to 1 whole, which is also three-thirds. That is why the total number of *thirds* in $2\frac{1}{3}$ is $(2 \times 3) + 1$, or 7. The number of *wholes* in $\frac{7}{3}$ is 2 wholes, with $\frac{1}{3}$ of a whole left over. We can generalize these observations into two rules.

To Change a Mixed Number to an Improper Fraction

- Multiply the denominator of the fraction by the whole-number part of the mixed number.
- Add the numerator of the fraction to this product.
- Write this sum over the denominator to form the improper fraction.

EXAMPLE 5

Write each of the following mixed numbers as an improper fraction.

a. $3\frac{2}{9}$ **b.** $12\frac{1}{4}$

Solution

a. $3\frac{2}{9} = \frac{(9 \times 3) + 2}{9}$ Multiply the denominator 9 by the whole number 3, adding the numerator 2. Place over the denominator.

$= \frac{27 + 2}{9} = \frac{29}{9}$ Simplify the numerator.

b. $12\frac{1}{4} = \frac{(4 \times 12) + 1}{4}$

$= \frac{48 + 1}{4} = \frac{49}{4}$

PRACTICE 5

Express each mixed number as an improper fraction.

a. $5\frac{1}{3}$ **b.** $20\frac{2}{5}$

To Change an Improper Fraction to a Mixed Number

- Divide the numerator by the denominator.
- If there is a remainder, write it over the denominator.

EXAMPLE 6

Write each improper fraction as a mixed or whole number.

a. $\frac{11}{2}$ **b.** $\frac{20}{20}$ **c.** $\frac{42}{5}$

Solution

a. $\frac{11}{2} = 2\overline{)11}$ with quotient 5 R1 — Divide the numerator by the denominator.

$\frac{11}{2} = 5\frac{1}{2}$ — Write the remainder over the denominator.

In other words, 5 R1 means that in $\frac{11}{2}$ there are 5 wholes with $\frac{1}{2}$ of a whole left over.

b. $\frac{20}{20} = 1$

c. $\frac{42}{5} = 8\frac{2}{5}$

PRACTICE 6

Express as a whole or mixed number.

a. $\frac{4}{2}$ **b.** $\frac{50}{9}$ **c.** $\frac{8}{3}$

Changing an improper fraction to a mixed number is important when we are dividing whole numbers: It allows us to express any remainder as a fraction. Previously, we would have said that the problems $2\overline{)7}$ and $4\overline{)13}$ both have the answer 3 R1. But by interpreting these problems as improper fractions, we see that their answers are different.

$$\frac{7}{2} = 3\frac{1}{2} \quad \text{but} \quad \frac{13}{4} = 3\frac{1}{4}$$

When a number is expressed as a mixed number, we know its size more readily than when it is expressed as an improper fraction. For instance, consider the mixed number $11\frac{7}{8}$. We immediately see that it is larger than 11 and smaller than 12 (that is, between 11 and 12). We could not reach this conclusion so easily if we were to examine only $\frac{95}{8}$, its improper form. However, there are situations—when we multiply or divide fractions—in which the use of improper fractions is preferable.

Equivalent Fractions

Some fractions that at first glance appear to be different from one another are really the same.

For instance, suppose that we cut a pizza into 8 equal slices, and then eat 4 of the slices. The shaded portion of the diagram at the right represents the amount eaten. Can you explain why in this diagram the fractions $\frac{4}{8}$ and $\frac{1}{2}$ describe the same part of the whole pizza? We say that these fractions are **equivalent**.

Any fraction has infinitely many equivalent fractions. To see why, let's consider the fraction $\frac{1}{3}$. We can draw different diagrams representing one-third of a whole.

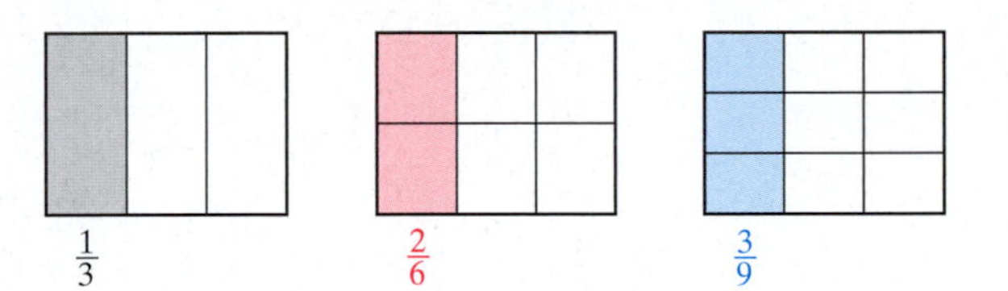

All the shaded portions of the diagrams are identical, so $\frac{1}{3} = \frac{2}{6} = \frac{3}{9}$.

A faster way to generate fractions equivalent to $\frac{1}{3}$ is to multiply both its numerator and denominator by the *same* whole number. Any whole number except 0 will do.

$$\frac{1}{3} = \frac{1 \cdot 2}{3 \cdot 2} = \frac{2}{6}$$

$$\frac{1}{3} = \frac{1 \cdot 3}{3 \cdot 3} = \frac{3}{9}$$

$$\frac{1}{3} = \frac{1 \cdot 4}{3 \cdot 4} = \frac{4}{12}$$

$$\frac{1}{3} = \frac{1 \cdot 5}{3 \cdot 5} = \frac{5}{15}$$

So $\frac{1}{3} = \frac{2}{6} = \frac{3}{9} = \frac{4}{12} = \frac{5}{15} = \ldots$.

Can you explain how you would generate fractions equivalent to $\frac{3}{5}$?

To Find an Equivalent Fraction

Multiply the numerator and denominator of $\frac{a}{b}$ by the same whole number n,

$$\frac{a}{b} = \frac{a \cdot n}{b \cdot n},$$

where both b and n are nonzero.

An important property of equivalent fractions is that their **cross products** are always equal.

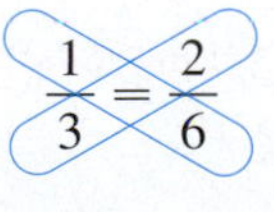

In this case, $1 \cdot 6 = 3 \cdot 2 = 6$

EXAMPLE 7

Find two fractions equivalent to $\frac{1}{7}$.

Solution Let's multiply the numerator and denominator by 2 and then by 6.

$$\frac{1}{7} = \frac{1 \cdot 2}{7 \cdot 2} = \frac{2}{14} \quad \text{and} \quad \frac{1}{7} = \frac{1 \cdot 6}{7 \cdot 6} = \frac{6}{42}$$

We use cross products to check.

$$\frac{1}{7} \stackrel{?}{=} \frac{2}{14}$$

$$1 \cdot 14 \stackrel{?}{=} 7 \cdot 2$$

$$14 \stackrel{\checkmark}{=} 14$$

So $\frac{1}{7}$ and $\frac{2}{14}$ are equivalent.

$$\frac{1}{7} \stackrel{?}{=} \frac{6}{42}$$

$$1 \cdot 42 \stackrel{?}{=} 7 \cdot 6$$

$$42 \stackrel{\checkmark}{=} 42$$

So $\frac{1}{7}$ and $\frac{6}{42}$ are equivalent.

PRACTICE 7

Identify three fractions equivalent to $\frac{2}{5}$.

EXAMPLE 8

Write $\frac{3}{7}$ as an equivalent fraction whose denominator is 35.

Solution $\frac{3}{7} = \frac{3 \cdot 5}{7 \cdot 5} = \frac{15}{35}$ **Multiply the numerator and denominator by 5.**

Therefore, $\frac{15}{35}$ is equivalent to $\frac{3}{7}$. To check, we find the cross products: Both $3 \cdot 35$ and $7 \cdot 15$ equal 105.

PRACTICE 8

Express $\frac{5}{8}$ as a fraction whose denominator is 72.

Writing a Fraction in Simplest Form

In the preceding section, we showed that $\frac{4}{8}$ and $\frac{1}{2}$ are equivalent fractions. Note that we could have written $\frac{4}{8}$ in its simplest form by dividing its numerator and denominator by 4.

$$\frac{4}{8} = \frac{4 \div 4}{8 \div 4} = \frac{1}{2}$$ **1 and 2 have no common factor except 1.**

A fraction is said to be in **simplest form** (or **reduced to lowest terms**) when the only common factor of its numerator and its denominator is 1.

To Simplify (Reduce) a Fraction

Divide the numerator and denominator of $\frac{a}{b}$ by the same whole number n,

$$\frac{a}{b} = \frac{a \div n}{b \div n},$$

where both b and n are nonzero.

EXAMPLE 9

Express $\frac{3}{15}$ in simplest form.

Solution To reduce this fraction, we can divide both its numerator and its denominator by 3.

$$\frac{3}{15} = \frac{3 \div 3}{15 \div 3} = \frac{1}{5}$$

To be sure that we have not made an error, let's check whether the cross products are equal: $3 \cdot 5 = 15$ and $1 \cdot 15 = 15$.

PRACTICE 9

Reduce $\frac{14}{21}$ to lowest terms.

In Example 9, we simplified the fraction $\frac{3}{15}$. How would you simplify the mixed number $2\frac{3}{15}$?

To reduce a fraction to lowest terms, we divide the numerator and denominator by all the factors that they have in common. To find these common factors, it is often helpful to express both the numerator and denominator as the product of prime factors. We can then divide out or *cancel* all common factors.

EXAMPLE 10

Write $\frac{42}{28}$ in lowest terms.

Solution

$\frac{42}{28} = \frac{2 \cdot 3 \cdot 7}{2 \cdot 2 \cdot 7}$ **Express the numerator and denominator as the product of primes.**

$= \frac{\overset{1}{\cancel{2}} \cdot 3 \cdot \overset{1}{\cancel{7}}}{\underset{1}{\cancel{2}} \cdot 2 \cdot \underset{1}{\cancel{7}}}$ **Divide out the common factors, noting that 1 remains.**

$= \frac{3}{2}$ **Multiply the remaining factors.**

PRACTICE 10

Simplify $\frac{42}{18}$.

EXAMPLE 11

Suppose that a couple's annual income is $75,000. If they pay $9,000 for rent and $3,000 for food per year, rent and food account for what fraction of their income? Simplify the answer.

Solution First, we must find the total part of the income that is paid for rent and food per year.

$$\underbrace{\$9{,}000}_{\text{Rent}} + \underbrace{\$3{,}000}_{\text{Food}} = \underbrace{\$12{,}000}_{\text{Total part}}$$

The total part is $12,000 and the whole is $75,000, so the fraction is $\frac{12{,}000}{75{,}000}$. We can simplify this fraction in the following way:

$$\frac{12{,}000}{75{,}000} = \frac{12{,}0\not{0}\not{0}}{75{,}0\not{0}\not{0}} = \frac{12}{75}$$ Note that canceling a 0 is the same as dividing by 10.

$$= \frac{3 \cdot 4}{3 \cdot 25} = \frac{\overset{1}{\not{3}} \cdot 4}{\underset{1}{\not{3}} \cdot 25} = \frac{4}{25}$$

Therefore, $\frac{4}{25}$ of the couple's income goes for rent and food.

PRACTICE 11

An acre is a unit of area approximately equal to 4,800 square yards. A developer is selling parcels of land of 50 yards by 30 yards. What fraction of an acre is each parcel?

Comparing Fractions

Some situations require us to *compare* fractions, that is, to rank them in order of size.

For instance, suppose that $\frac{5}{8}$ of one airline's flights arrive on time, in contrast to $\frac{3}{5}$ of another airline's flights. To decide which airline has a better record for on time arrivals, we need to compare the fractions.

Or to take another example, suppose that the drinking water in your home, according to a lab report, has 2 parts per million (ppm) of lead. Is the water safe to drink if the federal limit on lead in drinking water is 15 parts per billion (ppb)? Again, we need to compare fractions.

One way to handle such problems is to draw diagrams corresponding to the fractions in question. The larger fraction corresponds to the larger shaded region.

For instance, the diagrams to the right show that $\frac{3}{4}$ is greater than $\frac{1}{4}$. The symbol $>$ stands for "greater than."

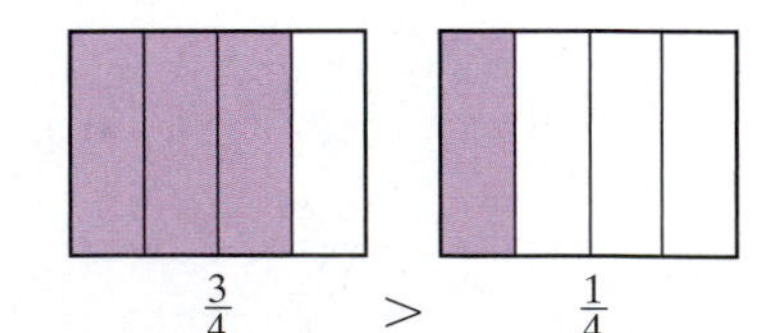

$\frac{3}{4} > \frac{1}{4}$

Both $\frac{3}{4}$ and $\frac{1}{4}$ have the same denominator, so we can rank them simply by comparing their numerators.

$$\frac{3}{4} > \frac{1}{4} \quad \text{because} \quad 3 > 1$$

For **like fractions**, the fraction with the larger numerator is the larger fraction.

Definitions

Like fractions are fractions with the same denominator.

Unlike fractions are fractions with different denominators.

To Compare Fractions

- If the fractions are like, compare their numerators.
- If the fractions are unlike, write them as equivalent fractions with the same denominator and then compare their numerators.

EXAMPLE 12

Compare $\frac{7}{15}$ and $\frac{4}{9}$.

Solution These fractions are unlike because they have different denominators. Therefore, we need to express them as equivalent fractions having the same denominator. But what should that denominator be?

One common denominator that we can use is the *product of the denominators:* $15 \cdot 9 = 135$.

$\frac{7}{15} = \frac{7 \cdot 9}{15 \cdot 9} = \frac{63}{135}$ **$135 = 15 \cdot 9$, so the new numerator is $7 \cdot 9$ or 63.**

$\frac{4}{9} = \frac{4 \cdot 15}{9 \cdot 15} = \frac{60}{135}$ **$135 = 9 \cdot 15$, so the new numerator is $4 \cdot 15$ or 60.**

Next, we compare the numerators of the like fractions that we just found.

Because $63 > 60$, $\frac{63}{135} > \frac{60}{135}$. Therefore, $\frac{7}{15} > \frac{4}{9}$.

Another common denominator that we can use is the least common multiple of the denominators.

$$15 = 3 \times 5 \qquad 9 = 3 \times 3 = 3^2$$

The LCM is $3^2 \times 5 = 9 \times 5 = 45$. We then compute the equivalent fractions.

$\frac{7}{15} = \frac{7 \cdot 3}{15 \cdot 3} = \frac{21}{45}$ **Multiply the numerator and denominator by 3.**

$\frac{4}{9} = \frac{4 \cdot 5}{9 \cdot 5} = \frac{20}{45}$ **Multiply the numerator and denominator by 5.**

Because $\frac{21}{45} > \frac{20}{45}$, we know that $\frac{7}{15} > \frac{4}{9}$.

PRACTICE 12

Which is larger, $\frac{13}{24}$ or $\frac{11}{16}$?

Note that in Example 12 we computed the LCM of the two denominators. This type of computation is used frequently in working with fractions.

Definition

For two or more fractions, their **least common denominator** (LCD) is the least common multiple of their denominators.

In Example 13, pay particular attention to how we use the LCD.

EXAMPLE 13

Order from smallest to largest: $\frac{3}{4}, \frac{7}{10}$, and $\frac{29}{40}$

Solution Because these fractions are unlike, we need to find equivalent fractions with a common denominator. Let's use their LCD as that denominator.

$$4 = 2 \times 2 = 2^2$$
$$10 = 2 \times 5$$
$$40 = 2 \times 2 \times 2 \times 5 = 2^3 \times 5$$

The LCD $= 2^3 \times 5 = 8 \times 5 = 40$. Check: 4 and 10 are both factors of 40.

We write each fraction with a denominator of 40.

$$\frac{3}{4} = \frac{3 \cdot 10}{4 \cdot 10} = \frac{30}{40} \qquad \frac{7}{10} = \frac{7 \cdot 4}{10 \cdot 4} = \frac{28}{40} \qquad \frac{29}{40} = \frac{29}{40}$$

Then we order the fractions from smallest to largest. (The symbol $<$ stands for "less than.")

$$\frac{28}{40} < \frac{29}{40} < \frac{30}{40} \quad \text{or} \quad \frac{7}{10} < \frac{29}{40} < \frac{3}{4}$$

PRACTICE 13

Arrange $\frac{9}{10}, \frac{23}{30}$, and $\frac{8}{15}$ from smallest to largest.

EXAMPLE 14

About $\frac{7}{10}$ of Earth's surface is covered by water and $\frac{1}{20}$ is covered by desert. Does water or desert cover more of Earth?

Solution We need to compare $\frac{7}{10}$ with $\frac{1}{20}$. The LCD is 20.

$$\frac{7}{10} = \frac{14}{20}$$
$$\frac{1}{20} = \frac{1}{20}$$

Since $\frac{14}{20} > \frac{1}{20}, \frac{7}{10} > \frac{1}{20}$. Therefore, water covers more of Earth than desert does.

PRACTICE 14

In 2005, about $\frac{1}{22}$ of the commercial radio stations in the United States had a top-40 format, and $\frac{2}{11}$ had a country format. In that year, were there more top-40 stations or country stations? (*Source: The M Street Radio Directory*)

Mathematically Speaking

Fill in each blank with the most appropriate term or phrase from the given list.

improper fraction	proper fraction	mixed
greatest common factor	simplify	composite
convert	least common denominator	
equivalent	like fractions	

1. A fraction whose numerator is smaller than its denominator is called a(n) _______.

2. The improper fraction $\frac{5}{2}$ can be expressed as a(n) _______ number.

3. The fractions $\frac{6}{8}$ and $\frac{3}{4}$ are _______.

4. Divide the numerator and denominator of a fraction by the same whole number in order to _______ it.

5. Fractions with the same denominator are said to be _______.

6. The _______ of two or more fractions is the least common multiple of their denominators.

Identify a fraction or mixed number represented by the shaded portion of each figure.

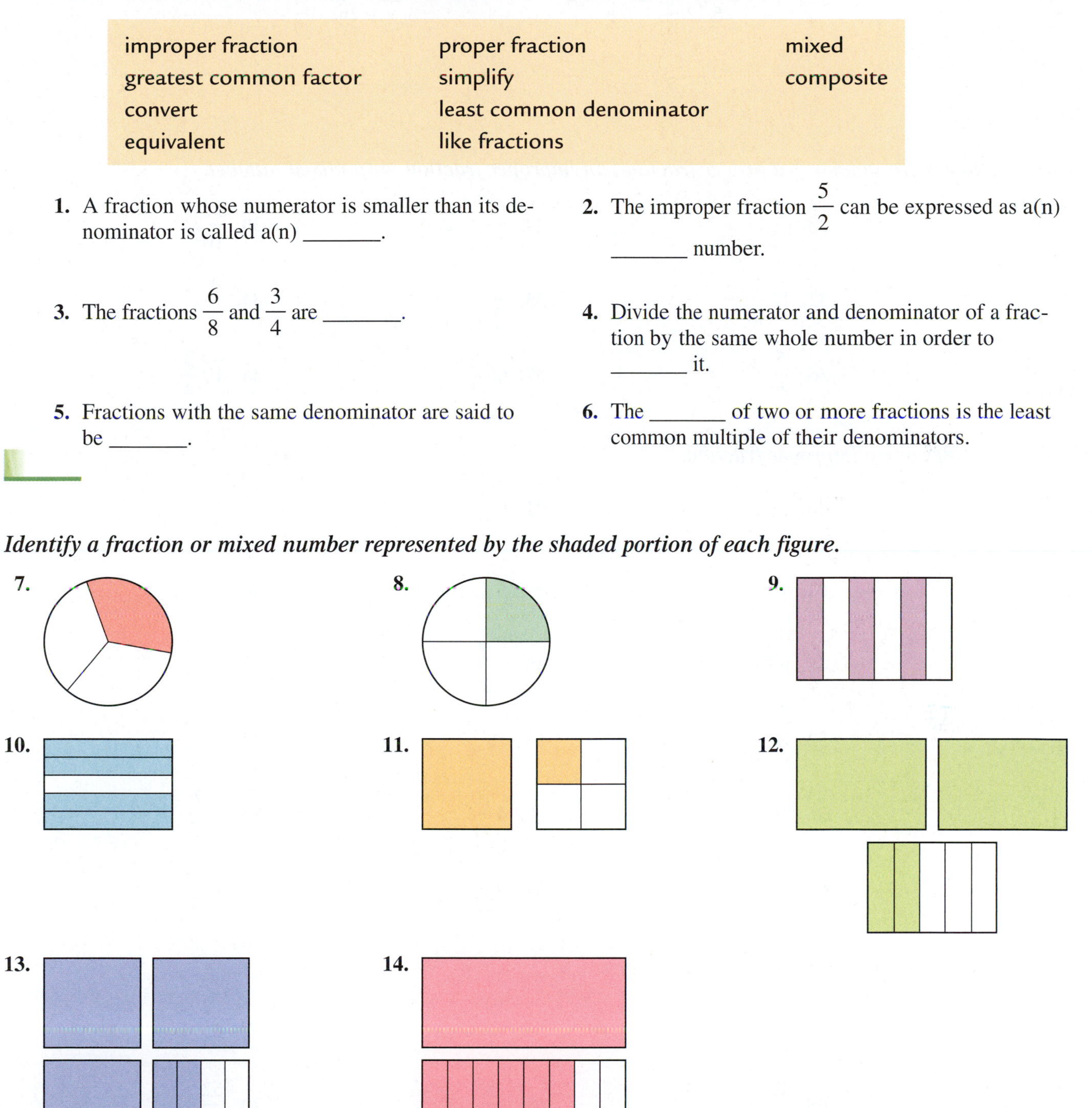

Draw a diagram to represent each fraction or mixed number.

15. $\frac{5}{7}$

16. $\frac{6}{11}$

17. $\frac{2}{9}$

18. $\frac{4}{10}$ **19.** $\frac{6}{6}$ **20.** $\frac{11}{11}$

21. $\frac{6}{5}$ **22.** $\frac{8}{3}$ **23.** $2\frac{1}{2}$

24. $4\frac{1}{5}$ **25.** $2\frac{1}{3}$ **26.** $3\frac{4}{9}$

Indicate whether each number is a proper fraction, an improper fraction, or a mixed number.

27. $\frac{2}{5}$ **28.** $\frac{7}{12}$ **29.** $\frac{10}{9}$ **30.** $\frac{11}{10}$

31. $16\frac{2}{3}$ **32.** $12\frac{1}{2}$ **33.** $\frac{5}{5}$ **34.** $\frac{4}{4}$

35. $\frac{4}{9}$ **36.** $\frac{5}{6}$ **37.** $66\frac{2}{3}$ **38.** $10\frac{3}{4}$

Write each number as an improper fraction.

39. $2\frac{3}{5}$ **40.** $1\frac{1}{3}$ **41.** $6\frac{1}{9}$ **42.** $10\frac{2}{3}$

43. $11\frac{2}{5}$ **44.** $12\frac{3}{4}$ **45.** 5 **46.** 8

47. $7\frac{3}{8}$ **48.** $6\frac{5}{6}$ **49.** $9\frac{7}{9}$ **50.** $10\frac{1}{2}$

51. $13\frac{1}{2}$ **52.** $20\frac{1}{8}$ **53.** $19\frac{3}{5}$ **54.** $11\frac{5}{7}$

55. 14 **56.** 10 **57.** $4\frac{10}{11}$ **58.** $2\frac{7}{13}$

59. $8\frac{3}{14}$ **60.** $4\frac{1}{6}$ **61.** $8\frac{2}{25}$ **62.** $14\frac{1}{10}$

Express each fraction as a mixed or whole number.

63. $\frac{4}{3}$ **64.** $\frac{6}{5}$ **65.** $\frac{10}{9}$ **66.** $\frac{12}{5}$

67. $\frac{9}{3}$ **68.** $\frac{12}{12}$ **69.** $\frac{15}{15}$ **70.** $\frac{62}{3}$

71. $\frac{99}{5}$ **72.** $\frac{31}{2}$ **73.** $\frac{82}{9}$ **74.** $\frac{100}{100}$

75. $\frac{45}{45}$ **76.** $\frac{40}{3}$ **77.** $\frac{74}{9}$ **78.** $\frac{41}{8}$

79. $\frac{27}{2}$ **80.** $\frac{58}{11}$ **81.** $\frac{100}{9}$ **82.** $\frac{19}{1}$

83. $\frac{27}{1}$ **84.** $\frac{72}{9}$ **85.** $\frac{56}{7}$ **86.** $\frac{38}{3}$

Find two fractions equivalent to each fraction.

87. $\frac{1}{8}$ **88.** $\frac{3}{10}$ **89.** $\frac{2}{11}$ **90.** $\frac{1}{10}$

91. $\frac{3}{4}$ **92.** $\frac{5}{6}$ **93.** $\frac{1}{9}$ **94.** $\frac{3}{5}$

Write an equivalent fraction with the given denominator.

95. $\frac{3}{4} = \frac{}{12}$ **96.** $\frac{2}{9} = \frac{}{18}$ **97.** $\frac{5}{8} = \frac{}{24}$ **98.** $\frac{7}{10} = \frac{}{20}$

99. $4 = \frac{}{10}$ **100.** $5 = \frac{}{15}$ **101.** $\frac{3}{5} = \frac{}{60}$ **102.** $\frac{4}{9} = \frac{}{63}$

103. $\frac{5}{8} = \frac{}{64}$ **104.** $\frac{3}{10} = \frac{}{40}$ **105.** $3 = \frac{}{18}$ **106.** $2 = \frac{}{21}$

107. $\frac{4}{9} = \frac{}{81}$ **108.** $\frac{7}{8} = \frac{}{24}$ **109.** $\frac{6}{7} = \frac{}{49}$ **110.** $\frac{5}{6} = \frac{}{48}$

111. $\frac{2}{17} = \frac{}{51}$ **112.** $\frac{1}{3} = \frac{}{90}$ **113.** $\frac{7}{12} = \frac{}{84}$ **114.** $\frac{1}{4} = \frac{}{100}$

115. $\frac{2}{3} = \frac{}{48}$ **116.** $\frac{7}{8} = \frac{}{56}$ **117.** $\frac{3}{10} = \frac{}{100}$ **118.** $\frac{5}{6} = \frac{}{144}$

Simplify, if possible.

119. $\frac{6}{9}$ **120.** $\frac{9}{12}$ **121.** $\frac{10}{10}$ **122.** $\frac{21}{21}$

123. $\frac{5}{15}$ **124.** $\frac{4}{24}$ **125.** $\frac{9}{20}$ **126.** $\frac{25}{49}$

127. $\frac{25}{100}$ **128.** $\frac{75}{100}$ **129.** $\frac{125}{1{,}000}$ **130.** $\frac{875}{1{,}000}$

131. $\frac{20}{16}$ **132.** $\frac{15}{9}$ **133.** $\frac{66}{32}$ **134.** $\frac{30}{18}$

135. $\frac{18}{32}$ **136.** $\frac{36}{45}$ **137.** $\frac{7}{24}$ **138.** $\frac{19}{51}$

139. $\frac{27}{9}$ **140.** $\frac{36}{144}$ **141.** $\frac{12}{84}$ **142.** $\frac{21}{36}$

143. $3\frac{38}{57}$ **144.** $11\frac{51}{102}$ **145.** $2\frac{100}{100}$ **146.** $1\frac{144}{144}$

Between each pair of numbers, insert the appropriate sign: <, =, or >.

147. $\frac{7}{20}$ $\frac{11}{20}$

148. $\frac{5}{10}$ $\frac{3}{10}$

149. $\frac{1}{8}$ $\frac{1}{9}$

150. $\frac{5}{6}$ $\frac{7}{8}$

151. $\frac{2}{3}$ $\frac{6}{9}$

152. $\frac{9}{12}$ $\frac{3}{4}$

153. $2\frac{1}{3}$ $2\frac{9}{15}$

154. $2\frac{3}{7}$ $1\frac{1}{2}$

Arrange in increasing order.

155. $\frac{1}{2}, \frac{1}{3}, \frac{1}{4}$

156. $\frac{3}{2}, \frac{3}{3}, \frac{3}{4}$

157. $\frac{2}{3}, \frac{7}{12}, \frac{5}{6}$

158. $\frac{3}{4}, \frac{5}{6}, \frac{7}{8}$

159. $\frac{3}{5}, \frac{2}{3}, \frac{8}{9}$

160. $\frac{5}{8}, \frac{1}{2}, \frac{4}{11}$

Mixed Practice

Solve.

161. Choose the number whose value is between the other two: $\frac{7}{10}, \frac{8}{9}, \frac{5}{6}$.

162. Express $\frac{32}{6}$ as a mixed number.

163. Find two fractions equivalent to $\frac{2}{9}$.

164. Draw a diagram to represent $\frac{9}{10}$.

165. Write an equivalent fraction for $\frac{4}{5}$ with denominator 15.

166. Write $2\frac{3}{8}$ as an improper fraction.

Applications

Solve. Write your answer in simplest form.

167. During the last 5 days, a student spent 11 hours studying mathematics at home. On the average, how much time is this per day?

168. A recipe for pasta with garlic and oil calls for 6 garlic cloves, peeled and chopped. If the recipe serves 4, how many garlic cloves on the average are in each serving?

169. As of 2005, the Nobel Prize was awarded to 33 women and 725 men. (*Source:* nobelprize.org)

a. What fraction of the Nobel prize winners were women?

b. What fraction were men?

170. In 2010, it is projected that there will be 3,256,000 teachers in public elementary and secondary schools and 424,000 teachers in private elementary and secondary schools. (*Source:* Natonal Center for Education Statistics, *Digest of Education Statistics 2005*)

a. What fraction of these teachers will be in public schools?

b. What fraction will be in private schools?

171. Of the 206 bones in the human skeleton, 106 are in the hands and feet. What fraction of these bones are not in the hands and feet? (*Source:* Henry Gray, *Anatomy of the Human Body*)

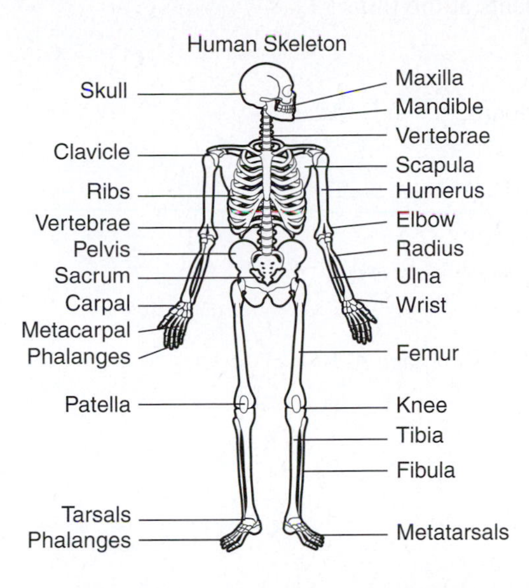

172. Perhaps because of an ankle injury, Rasheed Wallace, the Detroit Pistons power forward, made just 3 of 10 shots in game 1 of the Eastern Conference finals. Of the shots that Wallace took, what fraction did he not make? (*Source:* Yahoo! Sports 2006)

173. The gutter on your roof overflows whenever more than $\frac{1}{4}$ inch of rain falls. Yesterday, $\frac{23}{100}$ inch of rain fell. Did the gutter overflow? Explain.

174. In a course on probability and statistics, a student learns that when rolling a pair of dice, the probability of getting a 5 is $\frac{1}{9}$, and the probability of getting a 6 is $\frac{5}{36}$. Does getting a 5 or a 6 have a greater probability? Explain.

175. According to projections, $\frac{2}{5}$ of the total energy consumption in the United States in the year 2020 will come from petroleum products, $\frac{3}{40}$ will come from nuclear power, and $\frac{7}{30}$ will come from natural gas. (*Source: Annual Energy Outlook 2006*)

a. The greatest consumption of energy will come from which energy resource?

b. Will more nuclear power or natural gas be consumed in 2020?

176. When fog hit the New York City area, visibility was reduced to one-sixteenth mile at Kennedy Airport, one-eighth mile at LaGuardia Airport, and one-half mile at Newark Airport.

a. Which of the three airports had the best visibility?

b. Which of the three airports had the worst visibility?

177. In a recent year, the weights (in pounds) of the six centers that played for the Sabres, the hockey team from Buffalo, were 178, 186, 180, 217, 194, and 187. What was the average weight of these centers? (*Source:* http://sabers.com).

178. The following chart gives the age of the first six American presidents at the time of their inauguration.

President	Washington	J. Adams	Jefferson	Madison	Monroe	J. Q. Adams
Age (in years)	57	61	57	57	58	57

What was their average age at inauguration? (*Source: Significant American Presidents of the United States*)

• *Check your answers on page A-3.*

MINDSTRETCHERS

Mathematical Reasoning

1. Identify the fraction that the shaded portion of the figure to the right represents.

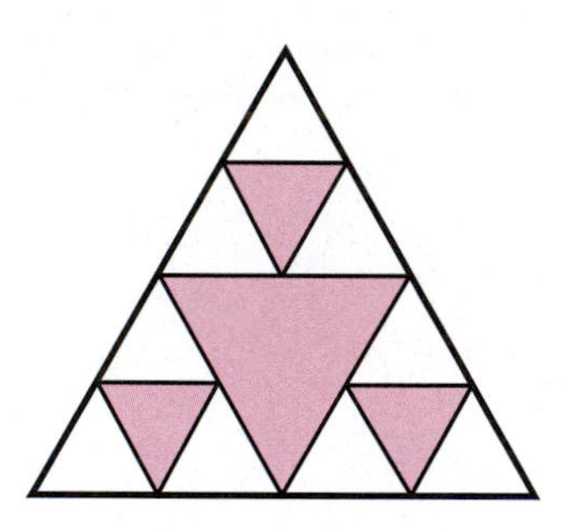

Groupwork

2. Working with a partner, determine how many fractions there are between the numbers 1 and 2.

Critical Thinking

3. Consider the three equivalent fractions shown. Note that the numerators and denominators are made up of the digits 1, 2, 3, 4, 5, 6, 7, 8, and 9—each appearing once.

$$\frac{3}{6} = \frac{7}{14} = \frac{29}{58}$$

a. Verify that these fractions are equivalent by making sure that their cross products are equal.

b. Complete the following fractions to form another trio of equivalent fractions that use the same nine digits only once.

$$\frac{2}{4} = \quad =$$

2.3 Adding and Subtracting Fractions

OBJECTIVES

- To add and subtract fractions and mixed numbers
- To estimate sums and differences involving mixed numbers
- To solve word problems involving the addition or subtraction of fractions or mixed numbers

In Section 2.2 we examined what fractions mean, how they are written, and how they are compared. In the rest of this chapter, we discuss computations involving fractions, beginning with sums and differences.

Adding and Subtracting Like Fractions

Let's first discuss how to add and subtract like fractions. Suppose that an employee spends $\frac{1}{7}$ of his weekly salary for food and $\frac{2}{7}$ for rent. What part of his salary does he spend for food and rent combined? A diagram can help us understand what is involved. First we shade one-seventh of the diagram, then another two-sevenths. We see in the diagram that the total shaded area is three-sevenths, $\frac{1}{7}+\frac{2}{7}=\frac{3}{7}$. Note that we added the original numerators to get the numerator of the answer but that *the denominator stayed the same*.

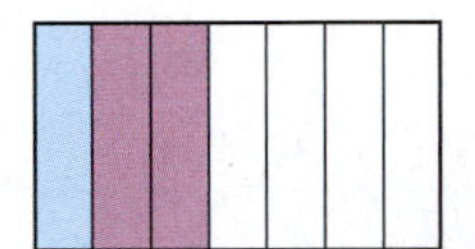

The diagram at the right illustrates the subtraction of like fractions, namely, $\frac{3}{7}-\frac{1}{7}$. If we shade three-sevenths of the diagram and then remove the shading in one-seventh, two-sevenths remain shaded. Therefore, $\frac{3}{7}-\frac{1}{7}=\frac{2}{7}$. Note that we could have gotten this answer simply by subtracting numerators without changing the denominator.

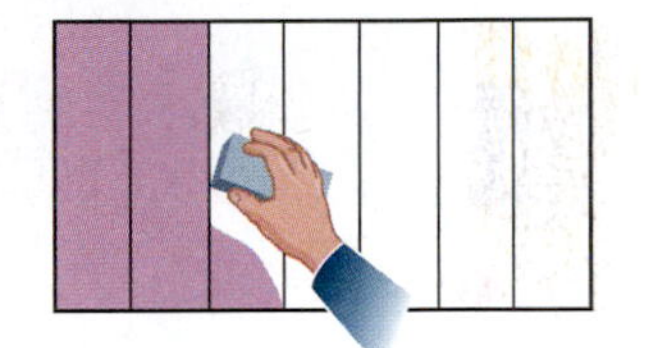

The following rule summarizes how to add or subtract fractions, *provided that they have the same denominator*.

To Add (or Subtract) Like Fractions

- Add (or subtract) the numerators.
- Use the given denominator.
- Write the answer in simplest form.

EXAMPLE 1

Add: $\frac{7}{12}+\frac{2}{12}$

Solution Applying the rule, we get $\frac{7}{12}+\frac{2}{12}=\frac{7+2}{12}=\frac{9}{12}$

Add the numerators. Keep the same denominator.

$=\frac{3\cdot 3}{4\cdot 3}=\frac{3\cdot \cancel{3}^{1}}{4\cdot \cancel{3}_{1}}=\frac{3}{4}$.

Simplest form

PRACTICE 1

Find the sum of $\frac{7}{15}$ and $\frac{3}{15}$.

Tip Be careful *not* to add the denominators when adding fractions.

EXAMPLE 2

Find the sum of $\frac{12}{16}$, $\frac{3}{16}$, and $\frac{9}{16}$.

Solution

Answer as a mixed number

$$\frac{12}{16} + \frac{3}{16} + \frac{9}{16} = \frac{24}{16} = \frac{3}{2}, \text{ or } 1\frac{1}{2}$$

So the sum of $\frac{12}{16}$, $\frac{3}{16}$, and $\frac{9}{16}$ is $1\frac{1}{2}$.

PRACTICE 2

Add: $\frac{13}{40}$, $\frac{11}{40}$ and $\frac{23}{40}$

EXAMPLE 3

Find the difference between $\frac{11}{7}$ and $\frac{3}{7}$.

Solution

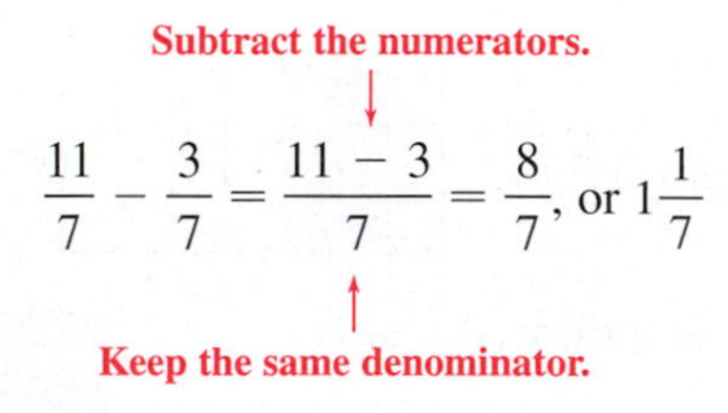

$$\frac{11}{7} - \frac{3}{7} = \frac{11 - 3}{7} = \frac{8}{7}, \text{ or } 1\frac{1}{7}$$

PRACTICE 3

Subtract: $\frac{19}{20} - \frac{11}{20}$

EXAMPLE 4

In the following diagram,

a. how far is it from the college to the library via city hall?

b. which route from the college to the library is shorter—via city hall or via the hospital? By how much?

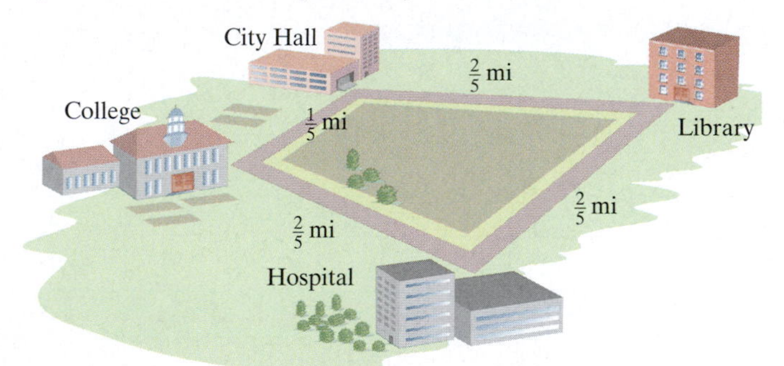

Solution

a. Examining the diagram, we see that

- the distance from the college to city hall is $\frac{1}{5}$ mile, and
- the distance from city hall to the library is $\frac{2}{5}$ mile.

To find the distance from the college to the library via city hall, we add.

$$\frac{1}{5} + \frac{2}{5} = \frac{3}{5}$$

So this distance is $\frac{3}{5}$ mile.

PRACTICE 4

A doctor prescribed $\frac{9}{20}$ gram of pain medication for a patient to take every 4 hours.

a. If the dosage were increased by $\frac{3}{20}$ gram, what would the new dosage be?

b. If the original dosage were decreased by $\frac{1}{20}$ gram, find the new dosage.

b. To find the distance from the college to the library via the hospital, we again add.

$$\frac{2}{5} + \frac{2}{5} = \frac{4}{5}$$

So this distance is $\frac{4}{5}$ mile. Since $\frac{3}{5} < \frac{4}{5}$, the route from the college to the library via city hall is shorter than the route via the hospital. Now we find the difference.

$$\frac{4}{5} - \frac{3}{5} = \frac{1}{5}$$

Therefore, the route via city hall is $\frac{1}{5}$ mile shorter than the route via the hospital.

Adding and Subtracting Unlike Fractions

Adding (or subtracting) **unlike fractions** is more complicated than adding (or subtracting) like fractions. An extra step is required: changing the unlike fractions to equivalent like fractions. For instance, suppose that we want to add $\frac{1}{10}$ and $\frac{2}{15}$. Even though we can use any common denominator for these fractions, let's use their *least* common denominator to find equivalent fractions.

$$10 = 2 \cdot 5$$
$$15 = 3 \cdot 5$$
$$\text{LCD} = 2 \cdot 3 \cdot 5 = 30$$

Let's rewrite the fractions vertically as equivalent fractions with the denominator 30.

$$\frac{1}{10} = \frac{1 \cdot 3}{10 \cdot 3} = \frac{3}{30}$$
$$+\frac{2}{15} = \frac{2 \cdot 2}{15 \cdot 2} = +\frac{4}{30}$$

Now we add the equivalent like fractions.

$$\frac{3}{40}$$
$$+\frac{4}{30}$$
$$\overline{\frac{7}{30}}$$

So $\frac{1}{10} + \frac{2}{15} = \frac{7}{30}$.

We can also add and subtract unlike fractions horizontally.

$$\frac{1}{10} + \frac{2}{15} = \frac{3}{30} + \frac{4}{30} = \frac{3+4}{30} = \frac{7}{30}$$

To Add (or Subtract) Unlike Fractions

- Write the fractions as equivalent fractions with the same denominator, usually the LCD.
- Add (or subtract) the numerators, keeping the same denominator.
- Write the answer in simplest form.

EXAMPLE 5

Add: $\frac{5}{12} + \frac{5}{16}$

Solution First, we find the LCD, which is 48. After finding equivalent fractions, we add the numerators, keeping the same denominator.

$$\begin{array}{rcccr} \frac{5}{12} & = & \frac{5 \cdot 4}{12 \cdot 4} & = & \frac{20}{48} \\ +\frac{5}{16} & = & \frac{5 \cdot 3}{16 \cdot 3} & = & +\frac{15}{48} \\ \hline & & & & \frac{35}{48} \end{array}$$

← Already in lowest terms

PRACTICE 5

Add: $\frac{11}{12} + \frac{3}{4}$

EXAMPLE 6

Subtract $\frac{1}{12}$ from $\frac{1}{3}$.

Solution Because 3 is a factor of 12, the LCD is 12. Again, let's set up the problem vertically.

$$\begin{array}{rcr} \frac{1}{3} & = & \frac{4}{12} \\ -\frac{1}{12} & = & -\frac{1}{12} \\ \hline & & \frac{3}{12} = \frac{1}{4} \end{array}$$

Subtract the numerators, keeping the same denominator.

Reduce $\frac{3}{12}$ to lowest terms.

PRACTICE 6

Calculate: $\frac{4}{5} - \frac{1}{2}$

EXAMPLE 7

Combine: $\frac{1}{3} + \frac{1}{6} - \frac{3}{10}$

Solution First, we find the LCD of all three fractions. The LCD is 30.

$$\frac{1}{3} = \frac{10}{30}, \quad \frac{1}{6} = \frac{5}{30}, \quad \text{and} \quad \frac{3}{10} = \frac{9}{30}.$$

So $\frac{1}{3} + \frac{1}{6} - \frac{3}{10} = \frac{10}{30} + \frac{5}{30} - \frac{9}{30} = \frac{10 + 5 - 9}{30} = \frac{6}{30} = \frac{1}{5}$.

PRACTICE 7

Combine: $\frac{1}{3} - \frac{2}{9} + \frac{7}{8}$

EXAMPLE 8

Find the perimeter of the piece of stained glass.

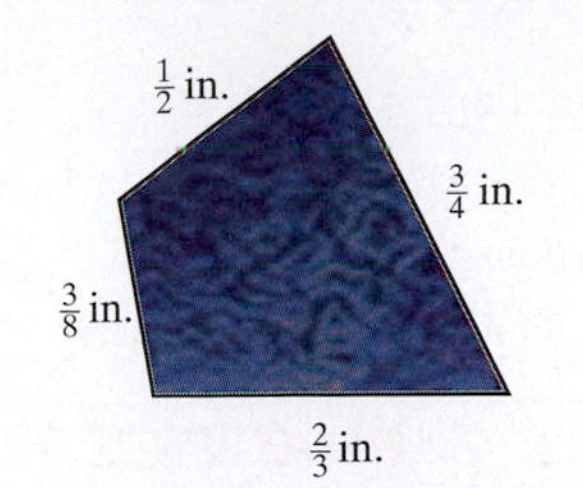

Solution Recall that the perimeter of a figure is the sum of the lengths of its sides.

$$\text{Perimeter} = \frac{3}{8} + \frac{1}{2} + \frac{3}{4} + \frac{2}{3} \qquad \begin{aligned} \frac{3}{8} &= \frac{9}{24} \leftarrow \text{LCD} \\ \frac{1}{2} &= \frac{12}{24} \\ \frac{3}{4} &= \frac{18}{24} \\ +\frac{2}{3} &= +\frac{16}{24} \\ \hline &\quad \frac{55}{24}, \text{ or } 2\frac{7}{24} \end{aligned}$$

The perimeter of the piece of stained glass is $2\frac{7}{24}$ inches.

PRACTICE 8

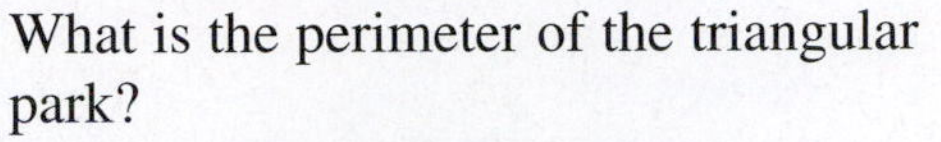

What is the perimeter of the triangular park?

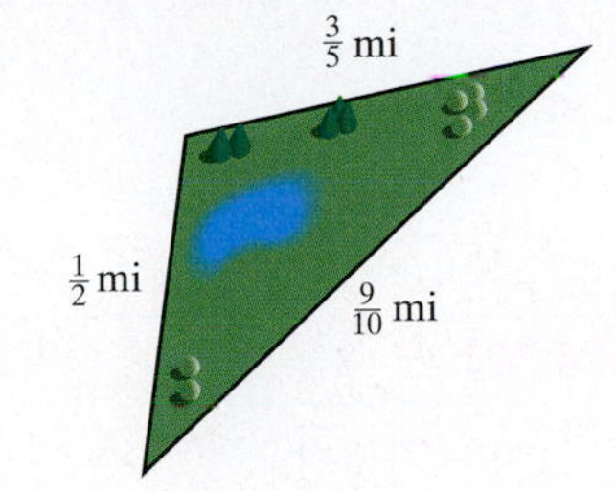

Adding Mixed Numbers

Now let's consider how to add **mixed numbers**, starting with those that have the same denominator.

Suppose, for instance, that we want to add $1\frac{1}{5}$ and $2\frac{1}{5}$. Let's draw a diagram to represent this sum.

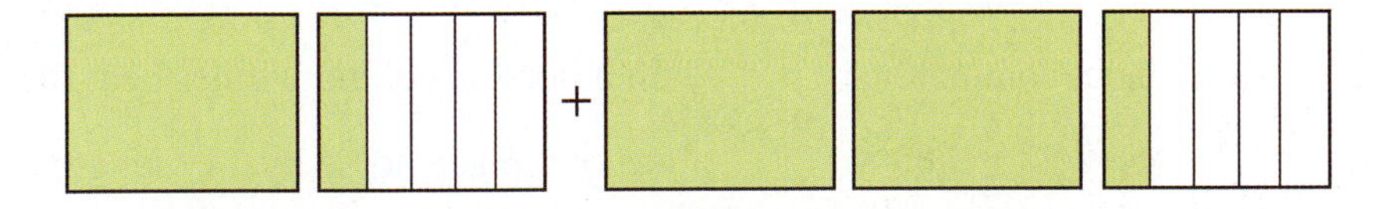

We can rearrange the elements of the diagram by combining the whole numbers and the fractions separately.

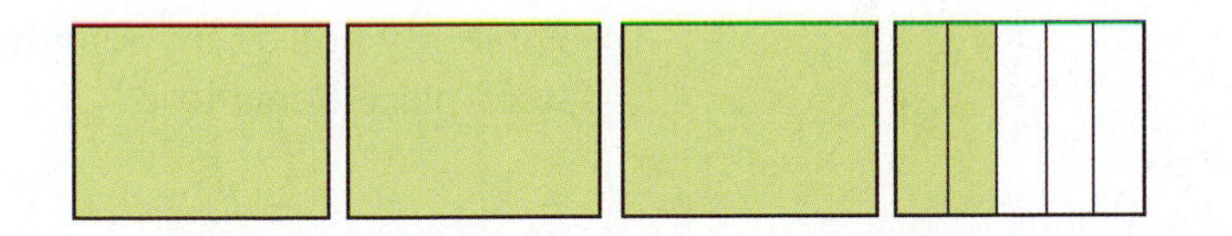

This diagram shows that the sum is $3\frac{2}{5}$.

Note that we can also write and solve this problem vertically.

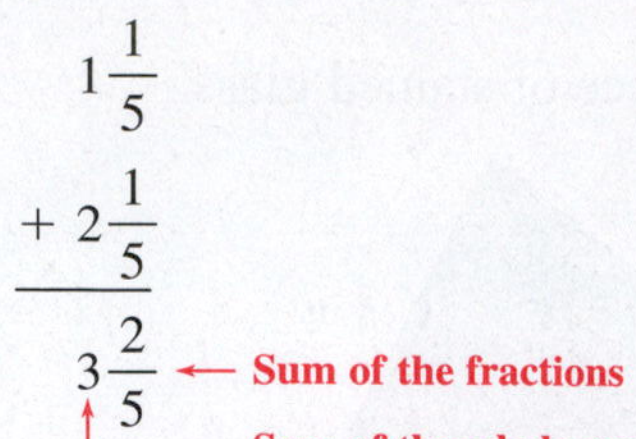

EXAMPLE 9

Add: $8\frac{5}{9} + 10\frac{1}{9}$

Solution

$$\begin{array}{r} 8\frac{5}{9} \\ +\ 10\frac{1}{9} \\ \hline 18\frac{6}{9} = 18\frac{2}{3} \end{array}$$

PRACTICE 9

Add: $25\frac{3}{10} + 9\frac{1}{10}$

EXAMPLE 10

Find the sum of $3\frac{3}{5}$, $2\frac{4}{5}$, and 6.

Solution Add the fractions and then add the whole numbers.

$$\begin{array}{r} 3\frac{3}{5} \\ 2\frac{4}{5} \\ +\ 6 \\ \hline 11\frac{7}{5} = 12\frac{2}{5} \end{array}$$

Since $\frac{7}{5} = 1\frac{2}{5}$, we get $11\frac{7}{5} = 11 + 1\frac{2}{5} = 12\frac{2}{5}$.

So the sum is $12\frac{2}{5}$.

PRACTICE 10

Find the sum of $2\frac{5}{16}$, $1\frac{3}{16}$, and 4.

EXAMPLE 11

Two movies are shown back-to-back on TV without commercial interruption. The first runs $1\frac{3}{4}$ hours, and the second $2\frac{1}{4}$ hours. How long will it take to watch both movies?

Solution

We need to add $1\frac{3}{4}$ and $2\frac{1}{4}$.

$$\begin{array}{r} 1\frac{3}{4} \\ +2\frac{1}{4} \\ \hline 3\frac{4}{4} = 3 + 1 = 4 \end{array}$$

Therefore, it will take 4 hours to watch the two movies.

PRACTICE 11

In a horse race, the winner beat the second-place horse by $1\frac{1}{2}$ lengths, and the second-place horse finished $2\frac{1}{2}$ lengths ahead of the third-place horse. By how many lengths did the third-place horse lose?

We have previously shown that when we add fractions with different denominators, we must first change the unlike fractions to equivalent like fractions. The same applies to adding mixed numbers that have different denominators.

To Add Mixed Numbers

- Write the fractions as equivalent fractions with the same denominator, usually the LCD.
- Add the fractions.
- Add the whole numbers.
- Write the answer in simplest form.

EXAMPLE 12

Find the sum of $3\frac{1}{5}$ and $7\frac{2}{3}$.

Solution The LCD is 15. Add the fractions and then add the whole numbers.

$$\begin{array}{rcr} 3\frac{1}{5} & = & 3\frac{3}{15} \\ +\,7\frac{2}{3} & = & +\,7\frac{10}{15} \\ \hline & & 10\frac{13}{15} \end{array}$$

The sum of $3\frac{1}{5}$ and $7\frac{2}{3}$ is $10\frac{13}{15}$.

PRACTICE 12

Add $4\frac{1}{8}$ to $3\frac{1}{2}$.

EXAMPLE 13

Find the sum of $1\frac{2}{3}$, $8\frac{1}{4}$, and $3\frac{4}{5}$.

Solution Set up the problem vertically and use the LCD, which is 60. Add the fractions and then add the whole numbers.

$$\begin{array}{rcr} 1\frac{2}{3} & = & 1\frac{40}{60} \\ 8\frac{1}{4} & = & 8\frac{15}{60} \\ +\,3\frac{4}{5} & = & +\,3\frac{48}{60} \\ \hline \end{array}$$

$$12\frac{103}{60} = 12 + 1\frac{43}{60} = 13\frac{43}{60}$$

PRACTICE 13

What is the sum of $5\frac{5}{8}$, $3\frac{1}{6}$, and $2\frac{5}{12}$?

Subtracting Mixed Numbers

Now let's discuss how to subtract mixed numbers, beginning with those that have the same denominator.

For instance, suppose that we want to subtract $2\frac{1}{5}$ from $3\frac{2}{5}$. We draw a diagram to represent $3\frac{2}{5}$.

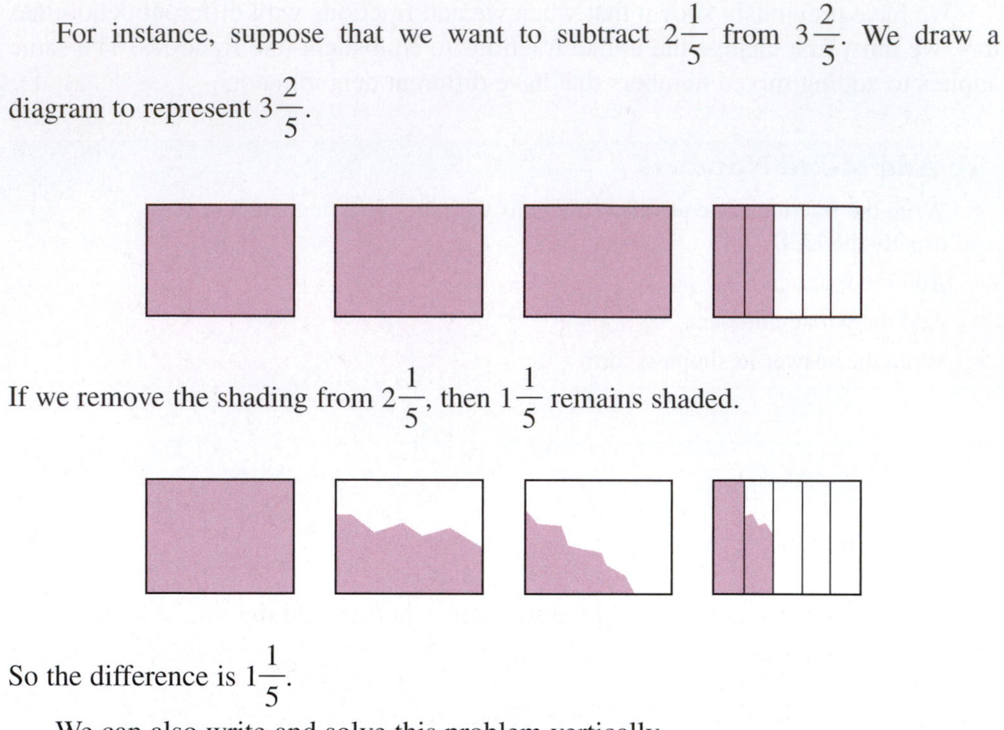

If we remove the shading from $2\frac{1}{5}$, then $1\frac{1}{5}$ remains shaded.

So the difference is $1\frac{1}{5}$.

We can also write and solve this problem vertically.

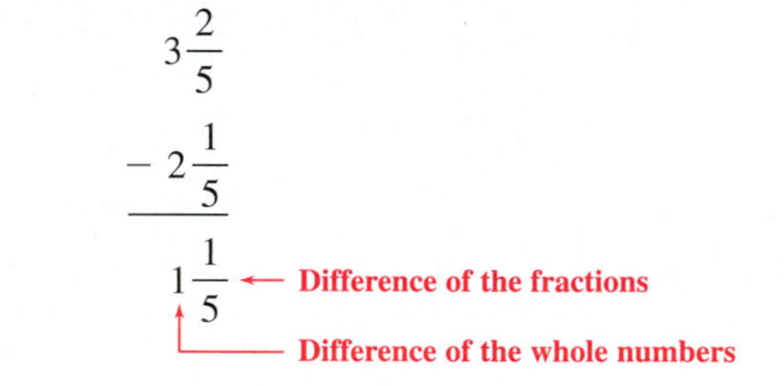

$$\begin{array}{r} 3\frac{2}{5} \\ -\ 2\frac{1}{5} \\ \hline 1\frac{1}{5} \end{array}$$

EXAMPLE 14

Subtract: $4\frac{5}{6} - 2\frac{1}{6}$

Solution We set up the problem vertically. Subtract the fractions and then subtract the whole numbers.

$$\begin{array}{r} 4\frac{5}{6} \\ -\ 2\frac{1}{6} \\ \hline 2\frac{4}{6} \end{array} = 2\frac{2}{3}$$

Therefore, the difference is $2\frac{2}{3}$.

PRACTICE 14

Subtract $5\frac{3}{10}$ from $9\frac{7}{10}$.

EXAMPLE 15

A construction job was scheduled to last $5\frac{3}{4}$ days, but was finished in $4\frac{1}{4}$ days. How many days ahead of schedule was the job?

Solution

This question asks us to subtract $4\frac{1}{4}$ from $5\frac{3}{4}$.

$$\begin{array}{r} 5\frac{3}{4} \\ -4\frac{1}{4} \\ \hline 1\frac{2}{4} = 1\frac{1}{2} \end{array}$$

So the job was $1\frac{1}{2}$ days ahead of schedule.

PRACTICE 15

A photograph is displayed in a frame. What is the difference between the height of the frame and the height of the photo?

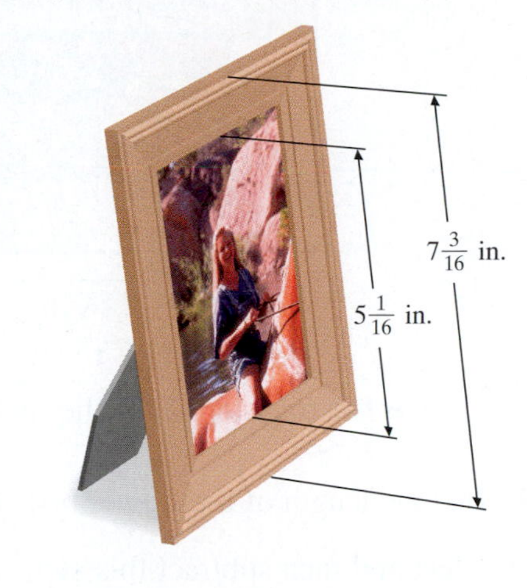

Subtracting mixed numbers that have different denominators is similar to adding mixed numbers.

EXAMPLE 16

Subtract $2\frac{7}{100}$ from $5\frac{9}{10}$.

Solution As usual, we use the LCD (which is 100) to find equivalent fractions. Then we subtract the equivalent mixed numbers with the same denominator. Again, let's set up the problem vertically. Subtract the fractions and then subtract the whole numbers.

$$\begin{array}{rcr} 5\frac{9}{10} & = & 5\frac{90}{100} \\ -\ 2\frac{7}{100} & = & -\ 2\frac{7}{100} \\ \hline & & 3\frac{83}{100} \end{array}$$

The answer is $3\frac{83}{100}$.

PRACTICE 16

Calculate: $8\frac{2}{3} - 4\frac{1}{12}$

EXAMPLE 17

Find the length of the flower bed.

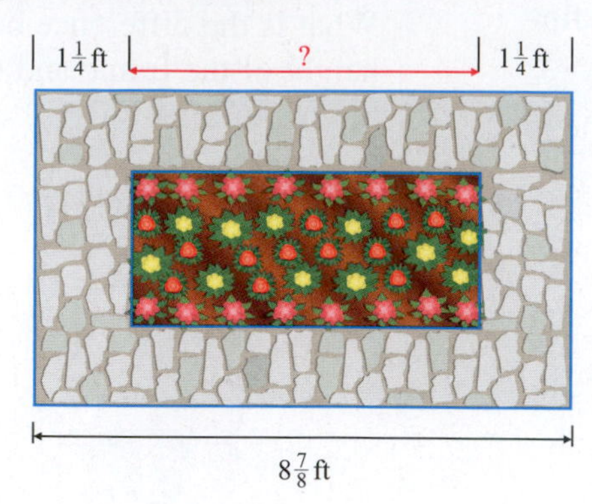

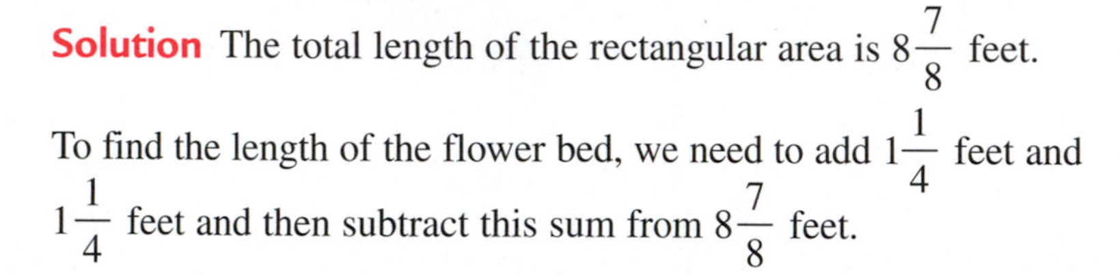

Solution The total length of the rectangular area is $8\frac{7}{8}$ feet.

To find the length of the flower bed, we need to add $1\frac{1}{4}$ feet and $1\frac{1}{4}$ feet and then subtract this sum from $8\frac{7}{8}$ feet.

$$\begin{array}{r} 1\frac{1}{4} \\ +\ 1\frac{1}{4} \\ \hline 2\frac{2}{4} = 2\frac{1}{2} \end{array} \qquad \begin{array}{rcr} 8\frac{7}{8} & = & 8\frac{7}{8} \\ -\ 2\frac{1}{2} & = & -\ 2\frac{4}{8} \\ \hline & & 6\frac{3}{8} \end{array}$$

So the length of the flower bed is $6\frac{3}{8}$ feet. We can check this answer by adding $1\frac{1}{4}$, $6\frac{3}{8}$, and $1\frac{1}{4}$, getting $8\frac{7}{8}$.

PRACTICE 17

The figure below is called a **trapezoid**. Suppose that this trapezoid's perimeter is $20\frac{1}{2}$ miles. How long is the left side?

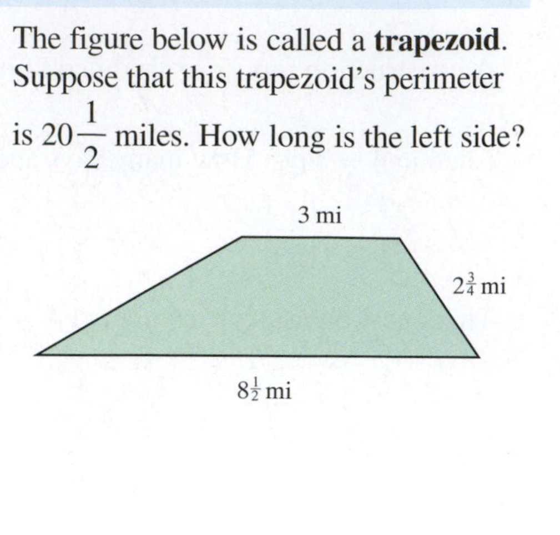

Recall from our discussion of subtracting whole numbers that, in problems in which a digit in the subtrahend is larger than the corresponding digit in the minuend, we need to borrow.

$$\begin{array}{r} \overset{2}{\cancel{3}}\ \overset{1}{2}\ 9 \\ -\ 8\ 7 \\ \hline 2\ 4\ 2 \end{array}$$

A similar situation can arise when we are subtracting mixed numbers. If the fraction on the bottom is larger than the fraction on top, we *rename* (or *borrow from*) the whole number on top.

EXAMPLE 18

Subtract: $6 - 1\frac{1}{3}$

Solution Let's rewrite the problem vertically.

$$\begin{array}{r} 6 \\ -1\frac{1}{3} \\ \hline \end{array}$$ There is no fraction on top from which to subtract $\frac{1}{3}$.

$$\begin{array}{r} 5\frac{3}{3} \\ -1\frac{1}{3} \\ \hline \end{array}$$ Rename 6 as 5 + 1, or $5 + \frac{3}{3}$, or $5\frac{3}{3}$.

$$\begin{array}{r} 5\frac{3}{3} \\ -1\frac{1}{3} \\ \hline 4\frac{2}{3} \end{array}$$ Now subtract.

So $6 - 1\frac{1}{3} = 4\frac{2}{3}$.

As in any subtraction problem, we can check our answer by addition.

$$4\frac{2}{3} + 1\frac{1}{3} = 5\frac{3}{3} = 6$$

PRACTICE 18

Subtract: $9 - 7\frac{5}{7}$

In Example 18, the answer is $4\frac{2}{3}$. Would we get the same answer if we compute $6\frac{1}{3} - 1$?

We have already discussed subtracting mixed numbers without renaming (borrowing) as well as subtracting a mixed number from a whole number. Now let's consider the general rule for subtracting mixed numbers.

To Subtract Mixed Numbers

- Write the fractions as equivalent fractions with the same denominator, usually the LCD.
- Rename (or borrow from) the whole number on top if the fraction on the bottom is larger than the fraction on top.
- Subtract the fractions.
- Subtract the whole numbers.
- Write the answer in simplest form.

EXAMPLE 19

Compute: $13\frac{2}{9} - 7\frac{8}{9}$

Solution First, we write the problem vertically.

$$\begin{array}{r} 13\frac{2}{9} \\ -7\frac{8}{9} \\ \hline \end{array}$$

Because $\frac{8}{9}$ is larger than $\frac{2}{9}$, we need to rename $13\frac{2}{9}$.

$$13\frac{2}{9} = 12 + 1 + \frac{2}{9} = 12 + \frac{9}{9} + \frac{2}{9} = 12\frac{11}{9}$$

$$\begin{array}{r} 12\frac{11}{9} \\ -7\frac{8}{9} \\ \hline \end{array}$$

Finally, we subtract and then write the answer in simplest form.

$$\begin{array}{rcr} 13\frac{2}{9} & = & 12\frac{11}{9} \\ -7\frac{8}{9} & = & -7\frac{8}{9} \\ \hline & & 5\frac{3}{9}, \text{ or } 5\frac{1}{3} \end{array}$$

PRACTICE 19

Find the difference between $15\frac{1}{12}$ and $9\frac{11}{12}$.

EXAMPLE 20

Find the difference between $10\frac{1}{4}$ and $1\frac{5}{12}$.

Solution First, we write the equivalent fractions, using the LCD.

$$\begin{array}{rcr} 10\frac{1}{4} & = & 10\frac{3}{12} \\ -1\frac{5}{12} & = & -1\frac{5}{12} \\ \hline \end{array}$$

Then, we subtract.

$$\begin{array}{rcr} 10\frac{3}{12} & = & 9\frac{15}{12} \\ -1\frac{5}{12} & = & -1\frac{5}{12} \\ \hline & & 8\frac{10}{12} = 8\frac{5}{6} \end{array}$$

We rename $10\frac{3}{12}$: $10\frac{3}{12} = 9 + \frac{12}{12} + \frac{3}{12} = 9\frac{15}{12}$

PRACTICE 20

Find the difference between $16\frac{3}{5}$ and $3\frac{1}{10}$.

EXAMPLE 21

In Oregon's Columbia River Gorge, a hiker walks along the Eagle Creek Trail, headed for Punchbowl Falls $2\frac{1}{10}$ miles away. After reaching Metlako Falls, does he have more or less than $\frac{1}{2}$ mile left to go? (*Source:* USDA Forest Service)

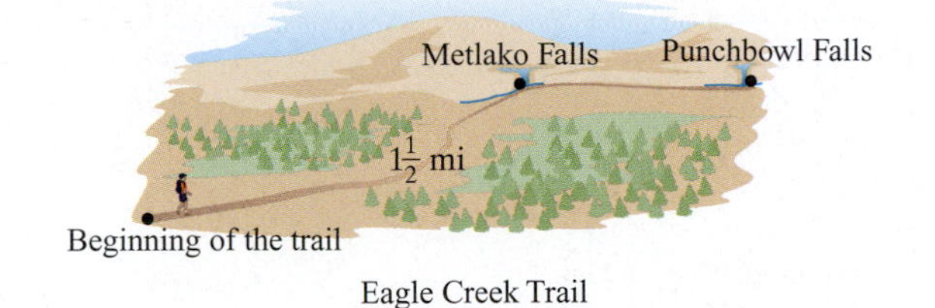

Solution First, we must find the difference between the length of the trail from its begining to Punchbowl Falls, namely, $2\frac{1}{10}$ miles, and the distance already hiked, $1\frac{1}{2}$ miles.

$$\begin{array}{rcrcr} 2\frac{1}{10} & = & 2\frac{1}{10} & = & 1\frac{11}{10} \\ -1\frac{1}{2} & = & -1\frac{5}{10} & = & -1\frac{5}{10} \\ \hline & & & & \frac{6}{10} = \frac{3}{5} \end{array}$$

So the distance left to hike is $\frac{3}{5}$ mile. Finally, we compare $\frac{3}{5}$ mile and $\frac{1}{2}$ mile.

$$\frac{3}{5} = \frac{6}{10} \qquad \frac{1}{2} = \frac{5}{10}$$

Because $6 > 5$, $\frac{6}{10} > \frac{5}{10}$. Therefore, $\frac{3}{5} > \frac{1}{2}$, and the hiker has more than $\frac{1}{2}$ mile left to go from Metlako Falls to Punchbowl Falls.

PRACTICE 21

A homeowner purchased a roll of wallpaper that unrolls to $30\frac{1}{2}$ yards long and used $26\frac{7}{8}$ yards from the roll to paper a room. Is there enough paper left on the roll for a job that requires 4 yards of paper?

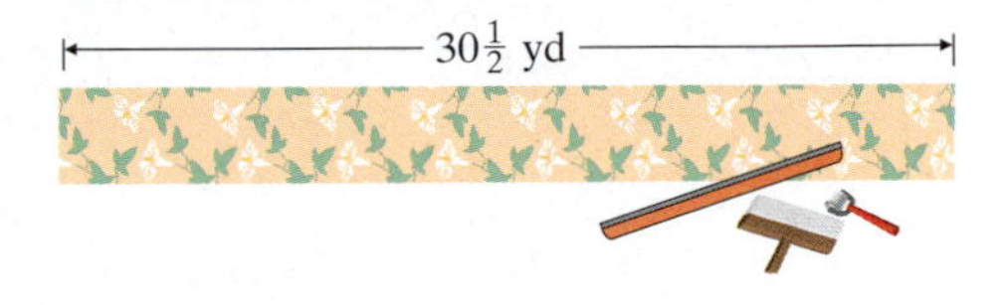

Another Method of Adding and Subtracting Mixed Numbers

Recall that any mixed number can be rewritten as an improper fraction. So when adding or subtracting mixed numbers, we can first express them as improper fractions. In a subtraction problem, this method has an advantage over the method previously discussed; namely, we never have to borrow. However, expressing mixed numbers as improper fractions may have the disadvantage of involving unnecessarily large numbers, as the following examples show.

EXAMPLE 22

Add: $14\frac{1}{6} + 8\frac{2}{3}$

Solution We begin by writing each mixed number as an improper fraction.

$$14\frac{1}{6} + 8\frac{2}{3} = \frac{85}{6} + \frac{26}{3}$$ Express $14\frac{1}{6}$ and $8\frac{2}{3}$ as improper fractions.

$$= \frac{85}{6} + \frac{52}{6}$$ Write the fractions to be added as equivalent fractions.

$$= \frac{137}{6}$$ Add the like fractions.

$$= 22\frac{5}{6}$$ Express the improper fraction as a mixed number.

PRACTICE 22

Find the sum: $7\frac{4}{5} + 2\frac{3}{4}$

Can you show that this method gives the same sum that we would have gotten if we had not expressed the mixed numbers as improper fractions? Explain.

EXAMPLE 23

Find the difference: $8\frac{5}{6} - 4\frac{9}{10}$

Solution $8\frac{5}{6} - 4\frac{9}{10} = \frac{53}{6} - \frac{49}{10}$ Write as improper fractions.

$$= \frac{265}{30} - \frac{147}{30}$$ Write as equivalent fractions.

$$= \frac{118}{30}$$ Subtract the like fractions.

$$= 3\frac{28}{30}$$ Express as a mixed number.

$$= 3\frac{14}{15}$$ Simplify.

Check that this answer is the same as we would have gotten without changing the mixed numbers to improper fractions.

PRACTICE 23

Subtract: $13\frac{1}{4} - 11\frac{7}{8}$

Estimating Sums and Differences of Mixed Numbers

When adding or subtracting mixed numbers, we can check by *estimating,* determining whether our estimate and our answer are close. Note that when we round mixed numbers, we round to the nearest whole number.

Checking a Sum by Estimating

$$\begin{array}{rcrl} 1\frac{1}{5} & \rightarrow & 1 & \text{Because } \frac{1}{5} < \frac{1}{2}\text{, round } \textit{down} \text{ to the whole number 1.} \\ +2\frac{3}{5} & \rightarrow & +3 & \text{Because } \frac{3}{5} > \frac{1}{2}\text{, round } \textit{up} \text{ to the whole number 3.} \\ \hline 3\frac{4}{5} & & 4 & \text{Our answer, } 3\frac{4}{5}\text{, is close to 4, the sum of the rounded addends (1 and 3).} \end{array}$$

Checking a Difference by Estimating

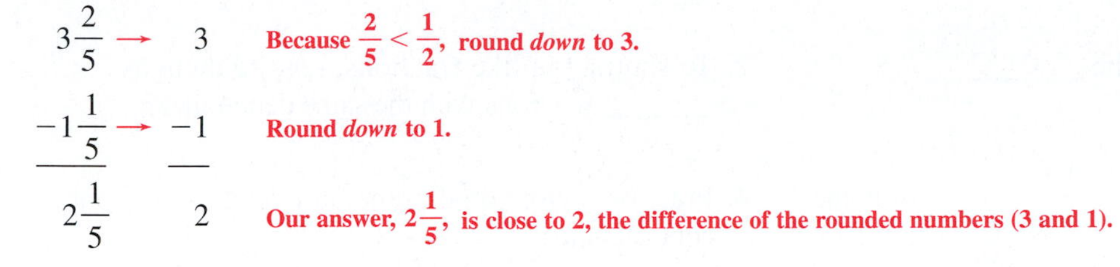

$3\frac{2}{5} \rightarrow 3$ Because $\frac{2}{5} < \frac{1}{2}$, round *down* to 3.
$-1\frac{1}{5} \rightarrow -1$ Round *down* to 1.
$2\frac{1}{5}$ 2 Our answer, $2\frac{1}{5}$, is close to 2, the difference of the rounded numbers (3 and 1).

EXAMPLE 24

Combine and check: $5\frac{1}{3} - \left(2\frac{4}{5} + 1\frac{1}{10}\right)$

Solution Following the order of operations rule, we begin by adding the two mixed numbers in parentheses.

$$\begin{array}{rcr} 2\frac{4}{5} & = & 2\frac{8}{10} \\ +1\frac{1}{10} & = & +1\frac{1}{10} \\ \hline & & 3\frac{9}{10} \end{array}$$

Next we subtract this sum from $5\frac{1}{3}$.

$$\begin{array}{rcrcr} 5\frac{1}{3} & = & 5\frac{10}{30} & = & 4\frac{40}{30} \\ -3\frac{9}{10} & = & -3\frac{27}{30} & = & -3\frac{27}{30} \\ \hline & & & & 1\frac{13}{30} \end{array}$$

So $5\frac{1}{3} - \left(2\frac{4}{5} + 1\frac{1}{10}\right) = 1\frac{13}{30}$.

Now let's check this answer by estimating:

$$\begin{array}{ccccc} 5\frac{1}{3} & - & \Big(2\frac{4}{5} & + & 1\frac{1}{10}\Big) \\ \downarrow & & \downarrow & & \downarrow \\ 5 & - & (3 & + & 1) = 5 - 4 = 1 \end{array}$$

The estimate, 1, is sufficiently close to $1\frac{13}{30}$ to confirm our answer.

PRACTICE 24

Calculate and check:

$8\frac{1}{4} - \left(3\frac{2}{5} - 1\frac{9}{10}\right)$

2.3 Exercises

Mathematically Speaking

Fill in each blank with the most appropriate term or phrase from the given list.

denominators	borrow	equivalent
subtract	numerators	improper

1. To add like fractions, add the _______.

2. To subtract unlike fractions, rewrite them as _______ fractions with the same denominator.

3. When subtracting $2\frac{4}{5}$ from $7\frac{1}{5}$, _______ from the 7 on the top.

4. Fractions with equal numerators and _______ are equivalent to 1.

Add and simplify.

5. $\frac{5}{8} + \frac{5}{8}$

6. $\frac{7}{10} + \frac{9}{10}$

7. $\frac{11}{12} + \frac{7}{12}$

8. $\frac{71}{100} + \frac{79}{100}$

9. $\frac{1}{5} + \frac{1}{5} + \frac{2}{5}$

10. $\frac{1}{7} + \frac{3}{7} + \frac{2}{7}$

11. $\frac{3}{20} + \frac{1}{20} + \frac{8}{20}$

12. $\frac{1}{10} + \frac{3}{10} + \frac{1}{10}$

13. $\frac{2}{3} + \frac{1}{2}$

14. $\frac{1}{4} + \frac{2}{5}$

15. $\frac{1}{2} + \frac{3}{8}$

16. $\frac{1}{6} + \frac{2}{3}$

17. $\frac{7}{10} + \frac{7}{100}$

18. $\frac{5}{6} + \frac{1}{12}$

19. $\frac{4}{5} + \frac{1}{8}$

20. $\frac{3}{4} + \frac{3}{7}$

21. $\frac{4}{9} + \frac{5}{6}$

22. $\frac{9}{10} + \frac{4}{5}$

23. $\frac{87}{100} + \frac{3}{10}$

24. $\frac{7}{20} + \frac{3}{4}$

25. $\frac{1}{3} + \frac{1}{4} + \frac{1}{6}$

26. $\frac{1}{5} + \frac{1}{6} + \frac{1}{3}$

27. $\frac{3}{8} + \frac{1}{10} + \frac{3}{16}$

28. $\frac{3}{10} + \frac{1}{3} + \frac{1}{9}$

29. $\frac{2}{9} + \frac{5}{8} + \frac{1}{4}$

30. $\frac{1}{2} + \frac{1}{3} + \frac{1}{4}$

31. $\frac{7}{8} + \frac{1}{5} + \frac{1}{4}$

32. $\frac{1}{10} + \frac{2}{5} + \frac{5}{6}$

Add and simplify. Then check by estimating.

33. $1 + 2\frac{1}{3}$

34. $4\frac{1}{5} + 2$

35. $8\frac{1}{10} + 7\frac{3}{10}$

36. $6\frac{1}{12} + 4\frac{1}{12}$

37. $7\frac{3}{10} + 6\frac{9}{10}$

38. $8\frac{2}{3} + 6\frac{2}{3}$

39. $5\frac{1}{6} + 9\frac{5}{6}$

40. $2\frac{3}{10} + 7\frac{9}{10}$

41. $5\frac{1}{4} + 5\frac{1}{6}$

42. $17\frac{3}{8} + 20\frac{1}{5}$

43. $3\frac{1}{3} + \frac{2}{5}$

44. $4\frac{7}{10} + \frac{7}{20}$

45. $8\frac{1}{5} + 5\frac{2}{3}$

46. $4\frac{1}{9} + 20\frac{7}{10}$

47. $\frac{2}{3} + 6\frac{1}{8}$

48. $\frac{1}{6} + 3\frac{2}{5}$

49. $9\frac{2}{3} + 10\frac{7}{12}$

50. $20\frac{3}{5} + 4\frac{1}{2}$

51. $6\frac{1}{10} + 3\frac{93}{100}$

52. $4\frac{8}{9} + 5\frac{1}{3}$

53. $4\frac{1}{2} + 6\frac{7}{8}$

54. $10\frac{5}{6} + 8\frac{1}{4}$

55. $30\frac{21}{100} + 5\frac{17}{20}$

56. $8\frac{3}{10} + 2\frac{321}{1{,}000}$

57. $80\frac{1}{3} + \frac{3}{4} + 10\frac{1}{2}$

58. $\frac{1}{3} + 25\frac{7}{24} + 100\frac{1}{2}$

59. $2\frac{1}{3} + 2 + 2\frac{1}{6}$

60. $4\frac{1}{8} + 4\frac{3}{16} + \frac{5}{4}$

61. $6\frac{7}{8} + 2\frac{3}{4} + 1\frac{1}{5}$

62. $1\frac{2}{3} + 5\frac{5}{6} + 3\frac{1}{4}$

63. $2\frac{1}{2} + 5\frac{1}{4} + 3\frac{5}{8}$

64. $4\frac{2}{3} + 2\frac{11}{36} + 1\frac{1}{2}$

Subtract and simplify.

65. $\frac{4}{5} - \frac{3}{5}$

66. $\frac{7}{9} - \frac{5}{9}$

67. $\frac{7}{10} - \frac{3}{10}$

68. $\frac{11}{12} - \frac{5}{12}$

69. $\frac{23}{100} - \frac{7}{100}$

70. $\frac{3}{2} - \frac{1}{2}$

71. $\frac{3}{4} - \frac{1}{4}$

72. $\frac{7}{9} - \frac{4}{9}$

73. $\frac{12}{5} - \frac{2}{5}$

74. $\frac{1}{8} - \frac{1}{8}$

75. $\frac{3}{4} - \frac{2}{3}$

76. $\frac{2}{5} - \frac{1}{6}$

77. $\frac{4}{9} - \frac{1}{6}$

78. $\frac{9}{10} - \frac{3}{100}$

79. $\frac{4}{5} - \frac{3}{4}$

80. $\frac{5}{6} - \frac{1}{8}$

81. $\frac{4}{7} - \frac{1}{2}$

82. $\frac{2}{5} - \frac{2}{9}$

83. $\frac{4}{9} - \frac{3}{8}$

84. $\frac{11}{12} - \frac{1}{3}$

85. $\frac{6}{8} - \frac{1}{2}$

86. $\frac{5}{6} - \frac{2}{3}$

Subtract and simplify. Then check either by adding or by estimating.

87. $5\frac{3}{7} - 1\frac{1}{7}$

88. $6\frac{2}{3} - 1\frac{1}{3}$

89. $3\frac{7}{8} - 2\frac{1}{8}$

90. $10\frac{5}{6} - 2\frac{5}{6}$

91. $20\frac{1}{2} - \frac{1}{2}$

92. $7\frac{3}{4} - \frac{1}{4}$

93. $8\frac{1}{10} - 4$

94. $2\frac{1}{3} - 2$

95. $6 - 2\frac{2}{3}$

96. $4 - 1\frac{1}{5}$

97. $8 - 4\frac{7}{10}$

98. $2 - 1\frac{1}{2}$

99. $10 - 3\frac{2}{3}$

100. $5 - 4\frac{9}{10}$

101. $6 - \frac{1}{2}$

102. $9 - \frac{3}{4}$

103. $7\frac{1}{4} - 2\frac{3}{4}$

104. $5\frac{1}{10} - 2\frac{3}{10}$

105. $6\frac{1}{8} - 2\frac{7}{8}$

106. $3\frac{1}{5} - 1\frac{4}{5}$

107. $12\frac{2}{5} - \frac{3}{5}$

108. $3\frac{7}{10} - \frac{9}{10}$

109. $8\frac{1}{3} - 1\frac{2}{3}$

110. $2\frac{1}{5} - \frac{4}{5}$

111. $13\frac{1}{2} - 5\frac{2}{3}$

112. $7\frac{1}{10} - 2\frac{1}{7}$

113. $9\frac{3}{8} - 5\frac{5}{6}$

114. $2\frac{1}{10} - 1\frac{27}{100}$

115. $20\frac{2}{9} - 4\frac{5}{6}$

116. $9\frac{13}{100} - 6\frac{7}{10}$

117. $3\frac{4}{5} - \frac{5}{6}$

118. $1\frac{2}{8} - \frac{2}{6}$

119. $1\frac{3}{4} - 1\frac{1}{2}$

120. $2\frac{1}{2} - 1\frac{3}{4}$

121. $10\frac{1}{12} - 4\frac{2}{3}$

122. $7\frac{1}{4} - 1\frac{5}{16}$

123. $22\frac{7}{8} - 8\frac{9}{10}$

124. $9\frac{1}{10} - 3\frac{1}{2}$

125. $3\frac{1}{8} - 2\frac{3}{4}$

126. $3\frac{1}{4} - 2\frac{5}{16}$

Combine and simplify.

127. $\frac{5}{8} + \frac{9}{10} - \frac{1}{4}$

128. $\frac{2}{3} - \frac{1}{5} + \frac{1}{2}$

129. $12\frac{1}{6} + 5\frac{9}{10} - 1\frac{3}{10}$

130. $7\frac{1}{3} - 2\frac{4}{5} - 1\frac{1}{3}$

131. $15\frac{1}{2} - 3\frac{4}{5} - 6\frac{1}{2}$

132. $4\frac{1}{10} + 2\frac{9}{10} - 3\frac{3}{4}$

133. $20\frac{1}{10} - \left(\frac{1}{20} + 1\frac{1}{2}\right)$

134. $19\frac{1}{6} - \left(8\frac{9}{10} - \frac{1}{5}\right)$

Mixed Practice

Perform the indicated operations and simplify.

135. Subtract $1\frac{7}{8}$ from 6.

136. Add: $6\frac{1}{10} + 3\frac{7}{15}$

137. Calculate: $12\frac{2}{3} - \left(8\frac{5}{6} - 4\frac{1}{2}\right)$

138. Find the sum of $\frac{3}{8}$, $\frac{1}{2}$, and $\frac{1}{3}$.

139. Find the difference between $4\frac{3}{5}$ and $1\frac{2}{3}$.

140. Subtract: $\frac{9}{10} - \frac{1}{4}$

Applications

Solve. Write the answer in simplest form.

141. A $\frac{7}{8}$-inch nail was hammered through a $\frac{3}{4}$-inch door. How far did it extend from the door?

142. A building occupies $\frac{1}{4}$ acre on a $\frac{7}{8}$-acre plot of land. What is the area of the land not occupied by the building?

143. The Kentucky Derby, Belmont Stakes, and the Preakness Stakes are three prestigious horse races that comprise the Triple Crown. (*Source:* http://infoplease.com)

a. Horses run $1\frac{3}{16}$ mile in the Preakness Stakes. If the Preakness Stakes is $\frac{5}{16}$ mile shorter than the Belmont Stakes, how far do horses run in the Belmont Stakes?

b. Horses run $1\frac{1}{4}$ miles in the Kentucky Derby. How much farther do horses run in the Belmont Stakes than in the Kentucky Derby?

144. In the year 2010, the total amount of electricity consumed anywhere in the world is projected to be approximately 16 billion kilowatt-hours. Of this amount, $\frac{1}{4}$ is expected to be consumed in the United States, $\frac{1}{9}$ in China, and $\frac{1}{20}$ in Japan. (*Source: International Energy Outlook 2004*)

a. According to these projections, the combined electrical consumption of China and Japan will be what fraction of the world consumption?

b. As a fraction of world consumption, how much greater is the U.S. consumption than the combined consumption in China and Japan?

145. The first game of a baseball doubleheader lasted $2\frac{1}{4}$ hours. The second game began after a $\frac{1}{4}$-hour break and lasted $2\frac{1}{2}$ hours. How long did the doubleheader take to play?

146. Three student candidates competed in a student government election. The winner got $\frac{5}{8}$ of the votes, and the second-place candidate got $\frac{1}{4}$ of the votes. If the rest of the votes went to the third candidate, what fraction of the votes did that student get?

147. In the hallway pictured, how much greater is the length than the width?

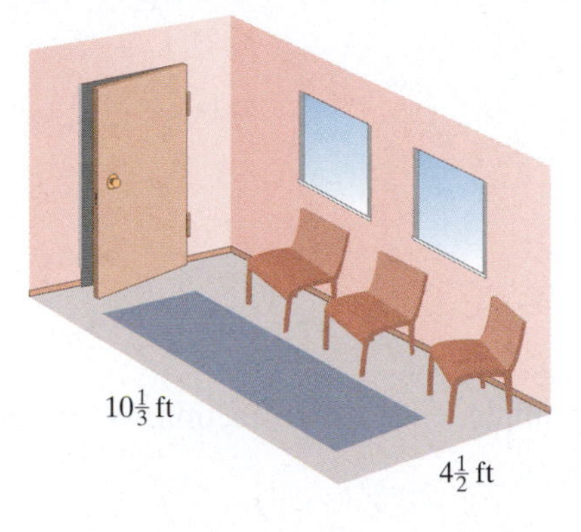

148. Find the perimeter of the figure shown.

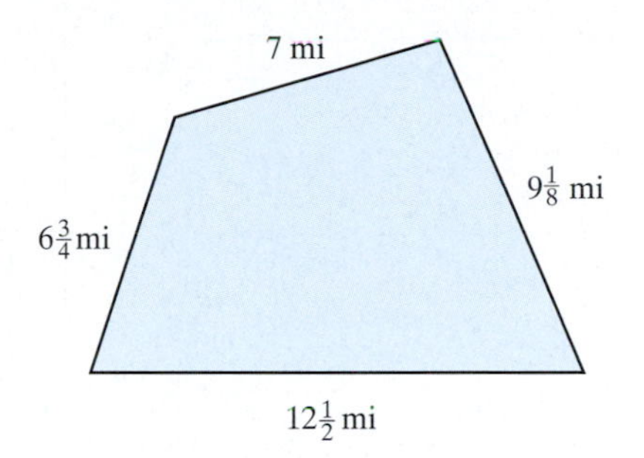

149. In testing a new drug, doctors found that $\frac{1}{2}$ of the patients given the drug improved, $\frac{2}{5}$ showed no change in their condition, and the remainder got worse. What fraction got worse?

150. According to a growth chart for young girls, their average weight is $38\frac{3}{4}$ pounds at age 4, $47\frac{1}{2}$ pounds at age 6, and $60\frac{3}{4}$ pounds at age 8. On the average, do girls gain more weight from age 4 to age 6 or from age 6 to age 8? (*Source:* http://www.babybag.com)

151. Suppose that four packages are placed on a scale, as shown. If the scale balances, how heavy is the small package on the right?

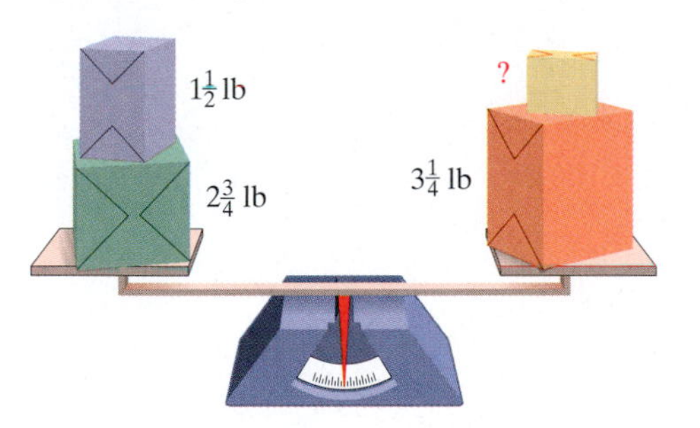

152. If the scale pictured balances, how heavy is the small package on the left?

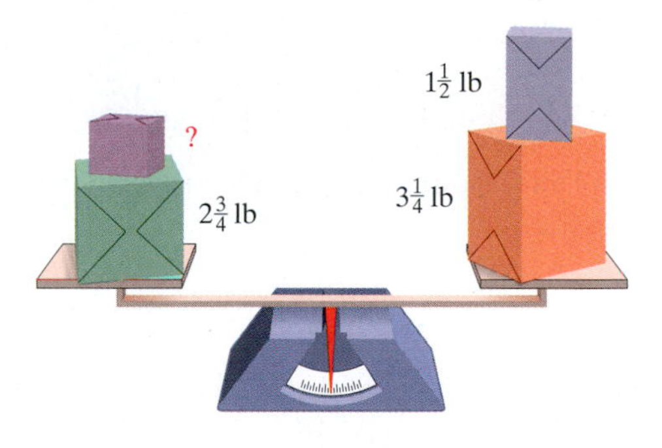

• *Check your answers on page A-4.*

MINDSTRETCHERS

Groupwork

1. Working with a partner, complete the following magic square in which each row, column, and diagonal adds up to the same number.

$1\frac{1}{4}$		
	1	
$\frac{11}{12}$		$\frac{3}{4}$

Mathematical Reasoning

2. A fraction with 1 as the numerator is called a **unit fraction**. For example, $\frac{1}{7}$ is a unit fraction. Write $\frac{3}{7}$ as the sum of three unit fractions, using no unit fraction more than once.

$$\frac{3}{7} = \frac{1}{\square} + \frac{1}{\square} + \frac{1}{\square}$$

Writing

3. Consider the following two ways of subtracting $2\frac{4}{5}$ from $4\frac{1}{5}$.

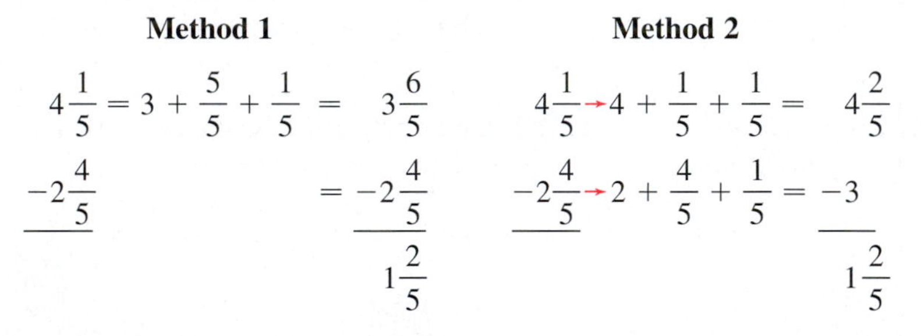

a. Explain the difference between the two methods.

b. Explain which method you prefer.

c. Explain why you prefer that method.

2.4 Multiplying and Dividing Fractions

OBJECTIVES

- To multiply and divide fractions and mixed numbers
- To estimate products and quotients involving mixed numbers
- To solve word problems involving the multiplication or division of fractions or mixed numbers

This section begins with a discussion of multiplying fractions. We then move on to multiplying mixed numbers and conclude with dividing fractions and mixed numbers.

Multiplying Fractions

Many situations require us to multiply fractions. For instance, suppose that a mixture in a chemistry class calls for $\frac{4}{5}$ gram of sodium chloride. If we make only $\frac{2}{3}$ of that mixture, we need

$$\frac{2}{3} \text{ of } \frac{4}{5}$$
$$\downarrow$$
$$\frac{2}{3} \times \frac{4}{5}$$

that is, $\frac{2}{3} \times \frac{4}{5}$ gram of sodium chloride.

To illustrate how to find this product, we diagram these two fractions.

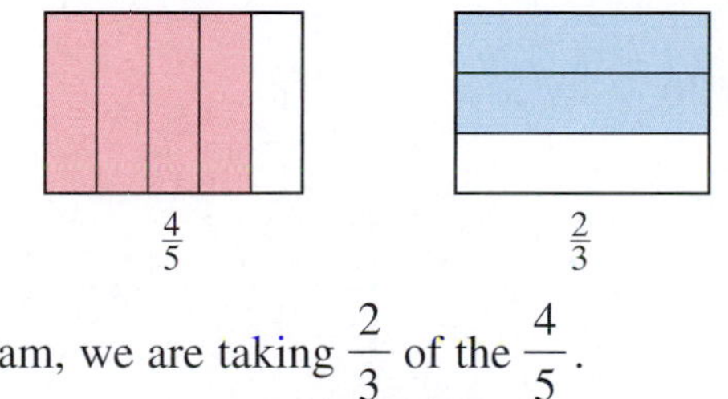

In the following diagram, we are taking $\frac{2}{3}$ of the $\frac{4}{5}$.

Note that we divided the whole into 15 parts and that our product, containing 8 of the 15 small squares, represents the double-shaded region. The answer is, therefore, $\frac{8}{15}$ of the original whole, which we can compute as follows.

$$\frac{2}{3} \times \frac{4}{5} = \frac{8}{15}$$

The numerator and denominator of the answer are the products of the original numerators and denominators.

To Multiply Fractions

- Multiply the numerators.
- Multiply the denominators.
- Write the answer in simplest form.

EXAMPLE 1

Multiply: $\frac{7}{8} \cdot \frac{9}{10}$

Solution

Multiply the numerators.

$$\frac{7}{8} \cdot \frac{9}{10} = \frac{7 \cdot 9}{8 \cdot 10} = \frac{63}{80}$$

Multiply the denominators.

PRACTICE 1

Find the product of $\frac{3}{4}$ and $\frac{5}{7}$.

EXAMPLE 2

Calculate: $\left(\frac{4}{5}\right)^2$

Solution

$$\left(\frac{4}{5}\right)^2 = \frac{4}{5} \cdot \frac{4}{5} = \frac{4 \cdot 4}{5 \cdot 5} = \frac{16}{25}$$

PRACTICE 2

Square $\frac{9}{10}$.

EXAMPLE 3

What is $\frac{3}{8}$ of 10?

Solution Finding $\frac{3}{8}$ of 10 means multiplying $\frac{3}{8}$ by 10.

$$\frac{3}{8} \times 10 = \frac{3}{8} \times \frac{10}{1} = \frac{3 \times 10}{8 \times 1} = \frac{30}{8} = \frac{15}{4}, \text{ or } 3\frac{3}{4}$$

PRACTICE 3

What is $\frac{2}{3}$ of 30?

In Example 3, we multiplied the two fractions first and then simplified the answer. It is preferable, however, to reverse these steps: Simplify first and then multiply. By first simplifying, sometimes referred to as *canceling*, we divide *any* numerator and *any* denominator by a common factor. Canceling before multiplying allows us to work with smaller numbers and still gives us the same answer.

EXAMPLE 4

Find the product of $\frac{4}{9}$ and $\frac{5}{8}$.

Solution

$$\frac{4}{9} \times \frac{5}{8} = \frac{\overset{1}{\cancel{4}}}{9} \times \frac{5}{\underset{2}{\cancel{8}}}$$

Divide the numerator 4 and the denominator 8 by 4.

$$= \frac{1 \times 5}{9 \times 2}$$

Multiply the resulting fractions.

$$= \frac{5}{18}$$

PRACTICE 4

Multiply: $\frac{7}{10} \cdot \frac{5}{11}$

EXAMPLE 5

Multiply: $\frac{9}{8} \times \frac{6}{5} \times \frac{7}{9}$

Solution We simplify and then multiply.

$$\frac{9}{8} \times \frac{6}{5} \times \frac{7}{9} = \frac{\overset{1}{\cancel{9}}}{\underset{4}{\cancel{8}}} \times \frac{\overset{3}{\cancel{6}}}{5} \times \frac{7}{\underset{1}{\cancel{9}}}$$

Divide the numerator 9 and the denominator 9 by 9. Divide the numerator 6 and the denominator 8 by 2.

$$= \frac{21}{20}, \text{ or } 1\frac{1}{20}$$

PRACTICE 5

Multiply: $\frac{7}{27} \cdot \frac{9}{4} \cdot \frac{8}{21}$

EXAMPLE 6

At a college, $\frac{3}{5}$ of the students take a math course. Of these students, $\frac{1}{6}$ take elementary algebra. What fraction of the students in the college take elementary algebra?

Solution We must find $\frac{1}{6}$ of $\frac{3}{5}$.

$$\frac{1}{6} \times \frac{3}{5} = \frac{1}{\underset{2}{\cancel{6}}} \times \frac{\overset{1}{\cancel{3}}}{5} = \frac{1 \times 1}{2 \times 5} = \frac{1}{10}$$

One-tenth of the students in the college take elementary algebra.

PRACTICE 6

A flight from New York to Los Angeles took 7 hours. With the help of the jet stream, the return trip took $\frac{3}{4}$ the time. How long did the trip from Los Angeles to New York take?

EXAMPLE 7

Of the 639 employees at a company, $\frac{4}{9}$ responded to a voluntary survey distributed by the human resources department. How many employees did not respond to the survey?

Solution Apply the strategy of breaking the problem into two parts.

- First, find $\frac{4}{9}$ of 639.
- Then, subtract the result from 639.

In short, we can solve this problem by computing $639 - \left(\frac{4}{9} \times 639\right)$.

$$639 - \left(\frac{4}{9} \times 639\right) = 639 - \left(\frac{4}{\underset{1}{\cancel{9}}} \times \frac{\overset{71}{\cancel{639}}}{1}\right) = 639 - 284 = 355$$

So 355 employees did not respond to the survey.

PRACTICE 7

The state sales tax on a car in Wisconsin is $\frac{1}{20}$ of the price of the car. What is the total amount a consumer would pay for a $19,780 car?

Multiplying Mixed Numbers

Some situations require us to multiply mixed numbers. For instance, suppose that your regular hourly wage is $\$7\frac{1}{2}$ and that you make time-and-a-half for working overtime. To find your overtime hourly wage, you need to multiply $1\frac{1}{2}$ by $7\frac{1}{2}$. The key here is to first rewrite each mixed number as an improper fraction.

$$1\frac{1}{2} \times 7\frac{1}{2} = \frac{3}{2} \times \frac{15}{2} = \frac{45}{4}, \text{ or } 11\frac{1}{4}$$

So you make $\$11\frac{1}{4}$ per hour overtime.

To Multiply Mixed Numbers

- Write the mixed numbers as improper fractions.
- Multiply the fractions.
- Write the answer in simplest form.

EXAMPLE 8

Multiply $2\frac{1}{5}$ by $1\frac{1}{4}$.

Solution $2\frac{1}{5} \times 1\frac{1}{4} = \frac{11}{5} \times \frac{5}{4}$ — Write each mixed number as an improper fraction.

$$= \frac{11 \times \overset{1}{\cancel{5}}}{\underset{1}{\cancel{5}} \times 4}$$ — Simplify and multiply.

$$= \frac{11}{4}, \text{ or } 2\frac{3}{4}$$

PRACTICE 8

Find the product of $3\frac{3}{4}$ and $2\frac{1}{10}$.

EXAMPLE 9

Multiply: $\left(4\frac{3}{8}\right)\left(4\right)\left(2\frac{2}{5}\right)$

Solution

$$\left(4\frac{3}{8}\right)\left(4\right)\left(2\frac{2}{5}\right) = \left(\frac{35}{8}\right)\left(\frac{4}{1}\right)\left(\frac{12}{5}\right)$$

$$= \left(\frac{\overset{7}{\cancel{35}}}{\underset{\underset{1}{2}}{\cancel{8}}}\right)\left(\frac{\overset{1}{\cancel{4}}}{1}\right)\left(\frac{\overset{6}{\cancel{12}}}{\underset{1}{\cancel{5}}}\right) = 42$$

Note in this problem that, although there are several ways to simplify, the answer always comes out the same.

PRACTICE 9

Multiply: $\left(1\frac{3}{4}\right)\left(5\frac{1}{3}\right)\left(3\right)$

EXAMPLE 10

A lawn surrounding a garden is to be installed, as depicted in the following drawing.

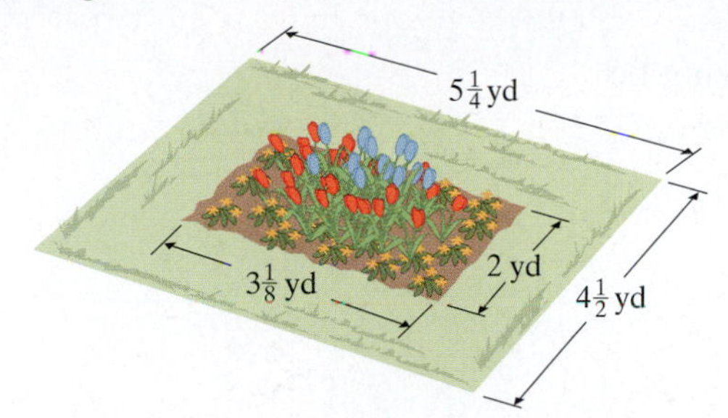

How many square yards of turf will we need to cover the lawn?

Solution Let's break this problem into three steps. First, we find the area of the rectangle with dimensions $5\frac{1}{4}$ yards and $4\frac{1}{2}$ yards. Then, we find the area of the small rectangle whose length and width are $3\frac{1}{8}$ yards and 2 yards, respectively.

$$5\frac{1}{4} \times 4\frac{1}{2} = \frac{21}{4} \times \frac{9}{2}$$

$$= \frac{189}{8}, \text{ or } 23\frac{5}{8}$$ The area of the large rectangle is $23\frac{5}{8}$ square yards.

$$3\frac{1}{8} \times 2 = \frac{25}{8} \times \frac{2}{1}$$

$$= \frac{25}{4}, \text{ or } 6\frac{1}{4}$$ The area of the small rectangle is $6\frac{1}{4}$ square yards.

Finally, we subtract the area of the small rectangle from the area of the large rectangle.

$$\begin{array}{rcr} 23\frac{5}{8} & = & 23\frac{5}{8} \\ -6\frac{1}{4} & = & -6\frac{2}{8} \\ \hline & & 17\frac{3}{8} \end{array}$$

The area of the lawn is, therefore, $17\frac{3}{8}$ square yards. So we will need $17\frac{3}{8}$ square yards of turf for the lawn.

PRACTICE 10

How much greater is the area of a sheet of legal-size paper than a sheet of letter-size paper?

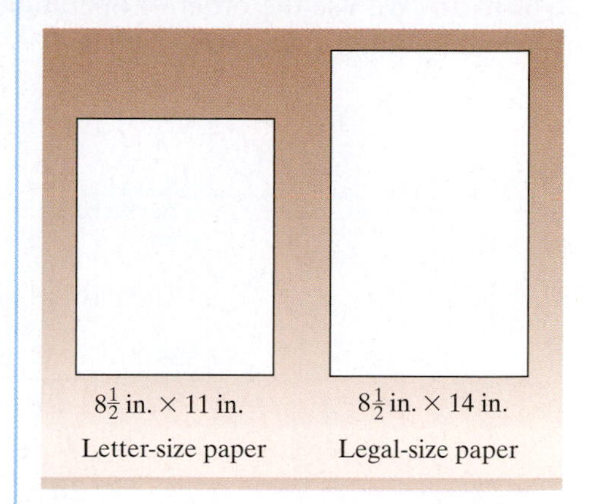

$8\frac{1}{2}$ in. × 11 in.
Letter-size paper

$8\frac{1}{2}$ in. × 14 in.
Legal-size paper

EXAMPLE 11

Simplify: $16\frac{1}{4} - 2 \cdot 4\frac{3}{5}$

Solution We use the order of operations rule, multiplying before subtracting.

$$\begin{aligned} 16\frac{1}{4} - 2 \cdot 4\frac{3}{5} &= 16\frac{1}{4} - \frac{2}{1} \cdot \frac{23}{5} \\ &= 16\frac{1}{4} - \frac{46}{5} \\ &= 16\frac{1}{4} - 9\frac{1}{5} \\ &= 16\frac{5}{20} - 9\frac{4}{20} \\ &= 7\frac{1}{20} \end{aligned}$$

PRACTICE 11

Calculate: $6 + \left(3\frac{1}{2}\right)^2$

Dividing Fractions

We now turn to quotients, beginning with dividing a fraction by a whole number. Suppose, for instance, that you want to share $\frac{1}{3}$ of a pizza with a friend, that is, to divide the $\frac{1}{3}$ into two equal parts. What part of the whole pizza will each of you receive?

This diagram shows $\frac{1}{3}$ of a pizza.

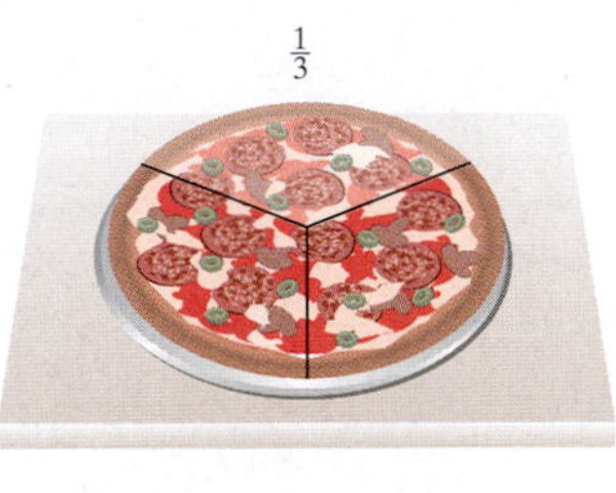

If we split each third into two equal parts, each part is $\frac{1}{6}$ of the pizza.

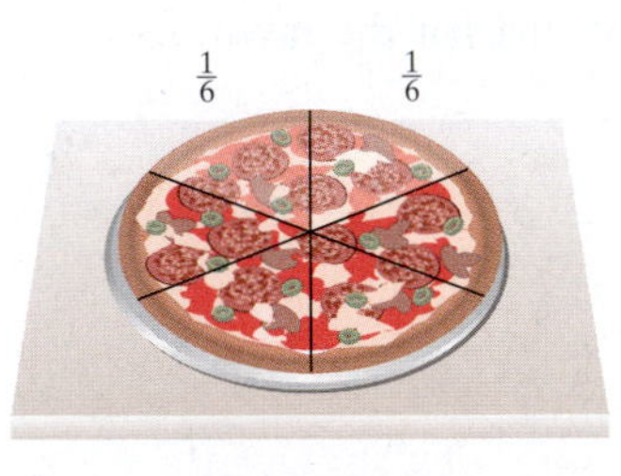

You and your friend will each get $\frac{1}{6}$ of the whole pizza, which you can compute as follows.

$$\frac{1}{3} \div 2 = \frac{1}{6}$$

Note that dividing a number by 2 is the same as taking $\frac{1}{2}$ of it. This equivalence suggests the procedure for dividing fractions shown next.

$$\frac{1}{3} \div 2 = \frac{1}{3} \div \frac{2}{1} = \frac{1}{3} \times \frac{1}{2} = \frac{1 \times 1}{3 \times 2} = \frac{1}{6}$$

Divisor

$\frac{2}{1}$ and $\frac{1}{2}$ are reciprocals.

This procedure involves *inverting*, or finding the *reciprocal* of the divisor. The reciprocal is found by switching the numerator and denominator.

To Divide Fractions

- Change the divisor to its reciprocal, and multiply the resulting fractions.
- Write the answer in simplest form.

EXAMPLE 12

Divide: $\frac{4}{5} \div \frac{3}{10}$

Solution $\frac{4}{5} \div \frac{3}{10} = \frac{4}{5} \times \frac{10}{3}$ — Change the divisor to its reciprocal and multiply.

$= \frac{4}{\cancel{5}_1} \times \frac{\cancel{10}^2}{3}$ — Divide the numerator 10 and the denominator 5 by 5.

$= \frac{4 \times 2}{1 \times 3}$ — Multiply the fractions.

$= \frac{8}{3}$ — Simplify.

$= 2\frac{2}{3}$

As in any division problem, we can check our answer by multiplying it by the divisor. $\frac{\cancel{8}^4}{\cancel{3}_1} \times \frac{\cancel{3}^1}{\cancel{10}_5} = \frac{4}{5}$

Because $\frac{4}{5}$ is the dividend, we have confirmed our answer.

PRACTICE 12

Divide: $\frac{3}{4} \div \frac{1}{8}$

Tip In a division problem, the fraction to the right of the division sign is the divisor. Always invert the divisor (the second fraction) and not the dividend (the first fraction).

EXAMPLE 13

What is $\frac{4}{7}$ divided by 20?

Solution $\frac{4}{7} \div 20 = \frac{4}{7} \times \frac{1}{20}$ Invert $\frac{20}{1}$ and multiply.

$= \frac{\overset{1}{\cancel{4}}}{7} \times \frac{1}{\underset{5}{\cancel{20}}}$ Divide the numerator 4 and the denominator 20 by 4.

$= \frac{1 \times 1}{7 \times 5} = \frac{1}{35}$

PRACTICE 13

Compute the following quotient:

$5 \div \frac{5}{8}$

EXAMPLE 14

To stop the developing process, photographers use a chemical called stop bath. Suppose that a photographer needs $\frac{1}{4}$ bottle of stop bath for each roll of film. If the photographer has $\frac{2}{3}$ bottle of stop bath left, can he develop three rolls of film?

Solution We want to find out how many $\frac{1}{4}$'s there are in $\frac{2}{3}$, that is, to compute $\frac{2}{3} \div \frac{1}{4}$.

$$\frac{2}{3} \div \frac{1}{4} = \frac{2}{3} \times \frac{4}{1} = \frac{8}{3} \text{ or } 2\frac{2}{3}$$

Find the reciprocal of the divisor, $\frac{1}{4}$, and then multiply.

So the photographer cannot develop three rolls of film.

PRACTICE 14

A house is built on ground that is sinking $\frac{3}{4}$ inch per year. How many years will it take the house to sink 2 inches?

Dividing Mixed Numbers

Dividing mixed numbers is similar to dividing fractions, except that there is an additional step.

To Divide Mixed Numbers

- Write the mixed numbers as improper fractions.
- Divide the fractions.
- Write the answer in simplest form.

EXAMPLE 15

Find: $9 \div 2\frac{7}{10}$.

Solution $9 \div 2\frac{7}{10} = \frac{9}{1} \div \frac{27}{10}$ **Write the whole number and the mixed number as improper fractions.**

$= \frac{\overset{1}{\cancel{9}}}{1} \times \frac{10}{\underset{3}{\cancel{27}}}$ **Invert and multiply.**

$= \frac{10}{3}$, or $3\frac{1}{3}$

PRACTICE 15

Divide: $6 \div 3\frac{3}{4}$

EXAMPLE 16

What is $2\frac{1}{2} \div 4\frac{1}{2}$?

Solution $2\frac{1}{2} \div 4\frac{1}{2} = \frac{5}{2} \div \frac{9}{2} = \frac{5}{\underset{1}{\cancel{2}}} \times \frac{\overset{1}{\cancel{2}}}{9} = \frac{5}{9}$

Invert and multiply.

PRACTICE 16

Divide $2\frac{3}{8}$ by $5\frac{3}{7}$.

EXAMPLE 17

There are $6\frac{3}{4}$ yards of silk in a roll. If it takes $\frac{3}{4}$ yards to make one designer tie, how many ties can be made from the roll?

Solution The question is: How many $\frac{3}{4}$'s fit into $6\frac{3}{4}$? It tells us that we must divide.

$6\frac{3}{4} \div \frac{3}{4} = \frac{\overset{9}{\cancel{27}}}{\underset{1}{\cancel{4}}} \times \frac{\overset{1}{\cancel{4}}}{\underset{1}{\cancel{3}}} = 9$

So nine ties can be made from the roll of silk.

PRACTICE 17

According to a newspaper advertisement for a "diet shake," a man lost 33 pounds in $5\frac{1}{2}$ months. How much weight did he lose per month?

Estimating Products and Quotients of Mixed Numbers

As with adding or subtracting mixed numbers, it is important to check our answers when multiplying or dividing. We can check a product or a quotient of mixed numbers by estimating the answer and then confirming that our estimate and answer are reasonably close.

Checking a Product by Estimating

$$2\frac{1}{5} \times 7\frac{2}{3} = \frac{11}{5} \times \frac{23}{3} = \frac{253}{15}, \text{ or } 16\frac{13}{15}$$

$$\downarrow \quad \downarrow$$

$$2 \times 8 = 16$$

Our answer, $16\frac{13}{15}$, is close to 16, the product of the rounded factors.

Because $16\frac{13}{15}$ is near 16, $16\frac{13}{15}$ is a reasonable answer.

Checking a Quotient by Estimating

$$6\frac{1}{4} \div 2\frac{7}{10} = \frac{25}{4} \div \frac{27}{10} = \frac{25}{\cancel{4}_{2}} \times \frac{\cancel{10}^{5}}{27} = \frac{125}{54}, \text{ or } 2\frac{17}{54}$$

$$\downarrow \quad \downarrow$$

$$6 \div 3 = 2$$

Our answer, $2\frac{17}{54}$, is close to 2, the quotient of the rounded dividend and divisor.

Because 2 is near $2\frac{17}{54}$, $2\frac{17}{54}$ is a reasonable answer.

EXAMPLE 18

Simplify and check: $3\frac{3}{4} \times 5\frac{1}{3} \div 2\frac{7}{9}$

Solution Following the order of operations rule, we work from left to right, multiplying the first two mixed numbers.

$$3\frac{3}{4} \times 5\frac{1}{3} = \frac{\cancel{15}^{5}}{\cancel{4}_{1}} \times \frac{\cancel{16}^{4}}{\cancel{3}_{1}} = 20$$

Then we divide 20 by $2\frac{7}{9}$ to get the answer.

$$20 \div 2\frac{7}{9} = \frac{20}{1} \div \frac{25}{9} = \frac{\cancel{20}^{4}}{1} \times \frac{9}{\cancel{25}_{5}} = \frac{36}{5}, \text{ or } 7\frac{1}{5}$$

Now let's check by estimating.

$$3\frac{3}{4} \times 5\frac{1}{3} \div 2\frac{7}{9}$$

$$\downarrow \quad \downarrow \quad \downarrow$$

$$4 \times 5 \div 3 = 20 \div 3 \approx 7$$

The answer, $7\frac{1}{5}$, and the estimate, 7, are reasonably close, confirming the answer.

PRACTICE 18

Compute and check:

$$5\frac{3}{5} \div 2\frac{1}{10} \times 2\frac{1}{4}$$

EXAMPLE 19

Calculate and check: $12 \div 1\frac{2}{3} + 5 \cdot 2\frac{9}{10}$

Solution According to the order of operations rule, we divide and multiply before adding.

$$\begin{aligned}12 \div 1\frac{2}{3} + 5 \cdot 2\frac{9}{10} &= \frac{12}{1} \div \frac{5}{3} + \frac{5}{1} \cdot \frac{29}{10}\\ &= \frac{12}{1} \cdot \frac{3}{5} + \frac{5}{1} \cdot \frac{29}{10}\\ &= \frac{12}{1} \cdot \frac{3}{5} + \frac{\overset{1}{\cancel{5}}}{1} \cdot \frac{29}{\underset{2}{\cancel{10}}}\\ &= \frac{36}{5} + \frac{29}{2}\\ &= 7\frac{1}{5} + 14\frac{1}{2}\\ &= 7\frac{2}{10} + 14\frac{5}{10}\\ &= 21\frac{7}{10}\end{aligned}$$

Now we estimate the answer in order to check.

$$12 \div 1\frac{2}{3} + 5 \cdot 2\frac{9}{10}$$

$$\downarrow \quad \downarrow \quad \downarrow \quad \downarrow$$

$$12 \div 2 + 5 \cdot 3 \approx 21$$

The estimate and the answer are close, confirming the answer.

PRACTICE 19

Compute and check:

$14\frac{1}{3} \div 2 - 6 \div 2\frac{1}{4}$

Mathematically Speaking

Fill in each blank with the most approppriate term or phrase from the given list.

reverse	proper fraction	multiply
divide	simplify	reciprocal
		improper fraction

1. To find the product of the fractions $\frac{1}{7}$ and $\frac{5}{8}$, _______ 1 and 5, and 7 and 8.

2. To multiply mixed numbers, change each mixed number to its equivalent _______.

3. The fraction $\frac{2}{3}$ is said to be the _______ of the fraction $\frac{3}{2}$.

4. To _______ fractions, change the divisor to its reciprocal, and multiply the resulting fractions.

Multiply.

5. $\frac{1}{3} \times \frac{2}{5}$

6. $\frac{7}{8} \times \frac{1}{2}$

7. $\left(\frac{5}{8}\right)\left(\frac{2}{3}\right)$

8. $\left(\frac{3}{10}\right)\left(\frac{1}{4}\right)$

9. $\left(\frac{3}{4}\right)^2$

10. $\left(\frac{1}{8}\right)^2$

11. $\frac{4}{5} \times \frac{2}{5}$

12. $\frac{1}{2} \times \frac{3}{2}$

13. $\frac{7}{8} \times \frac{5}{4}$

14. $\frac{20}{3} \times \frac{2}{7}$

15. $\frac{5}{2} \cdot \frac{9}{8}$

16. $\frac{11}{10} \cdot \frac{9}{5}$

17. $\left(\frac{2}{5}\right)\left(\frac{5}{9}\right)$

18. $\left(\frac{4}{5}\right)\left(\frac{1}{4}\right)$

19. $\frac{7}{9} \times \frac{3}{4}$

20. $\frac{4}{5} \times \frac{1}{2}$

21. $\left(\frac{1}{8}\right)\left(\frac{6}{10}\right)$

22. $\left(\frac{4}{6}\right)\left(\frac{3}{8}\right)$

23. $\frac{10}{9} \times \frac{93}{100}$

24. $\frac{12}{5} \times \frac{15}{4}$

25. $\frac{2}{3} \times 20$

26. $\frac{5}{6} \times 5$

27. $\left(\frac{10}{3}\right)(4)$

28. $\frac{5}{3} \times 7$

29. $\frac{2}{3} \times 24$

30. $\frac{3}{4} \times 12$

31. $\frac{2}{3} \cdot 6$

32. $100 \cdot \frac{2}{5}$

33. $18 \cdot \frac{2}{9}$

34. $20 \cdot \frac{4}{5}$

35. $\frac{7}{8} \times 10$

36. $\frac{5}{8} \times 12$

37. $\left(\frac{7}{8}\right)\left(1\frac{1}{2}\right)$

38. $\left(4\frac{1}{3}\right)\left(\frac{1}{5}\right)$

39. $\frac{1}{4} \cdot 8\frac{1}{2}$

40. $\frac{1}{3} \cdot 2\frac{1}{5}$

41. $\left(\frac{5}{6}\right)\left(1\frac{1}{9}\right)$

42. $\left(\frac{9}{10}\right)\left(2\frac{1}{7}\right)$

43. $\frac{1}{2} \times 5\frac{1}{3}$

44. $4\frac{1}{2} \times \frac{2}{3}$

45. $\frac{4}{5} \cdot 1\frac{1}{4}$

46. $\frac{3}{8} \cdot 5\frac{1}{3}$

47. $\left(\frac{3}{16}\right)\left(4\frac{2}{3}\right)$

48. $\left(\frac{7}{9}\right)\left(2\frac{1}{4}\right)$

49. $1\frac{1}{7} \times 1\frac{1}{5}$

50. $2\frac{1}{3} \times 1\frac{1}{2}$

51. $\left(2\frac{1}{10}\right)^2$

52. $\left(1\frac{1}{2}\right)^2$

53. $3\frac{9}{10} \cdot 2$

54. $5 \cdot 1\frac{1}{2}$

55. $100 \times 3\frac{3}{4}$

56. $1\frac{5}{6} \times 20$

57. $1\frac{1}{2} \times 5\frac{1}{3}$

58. $5\frac{1}{4} \times 1\frac{1}{9}$

59. $\left(2\frac{1}{2}\right)\left(1\frac{1}{5}\right)$

60. $\left(1\frac{3}{10}\right)\left(2\frac{4}{9}\right)$

61. $12\frac{1}{2} \cdot 3\frac{1}{3}$

62. $5\frac{1}{10} \cdot 1\frac{2}{3}$

63. $66\frac{2}{3} \times 1\frac{7}{10}$

64. $37\frac{1}{2} \times 1\frac{3}{5}$

65. $1\frac{5}{9} \times \frac{3}{8} \times 2$

66. $\frac{1}{8} \times 2\frac{1}{4} \times 6$

67. $\left(\frac{1}{2}\right)^2\left(2\frac{1}{3}\right)$

68. $\left(1\frac{1}{4}\right)^2\left(\frac{1}{5}\right)$

69. $\frac{4}{5} \times \frac{7}{8} \times 1\frac{1}{10}$

70. $8\frac{1}{3} \times \frac{3}{10} \times \frac{5}{6}$

71. $\left(1\frac{1}{2}\right)^3$

72. $\left(2\frac{1}{2}\right)^3$

Divide.

73. $\frac{3}{5} \div \frac{2}{3}$

74. $\frac{2}{3} \div \frac{3}{5}$

75. $\frac{4}{5} \div \frac{7}{8}$

76. $\frac{7}{8} \div \frac{4}{5}$

77. $\frac{1}{2} \div \frac{1}{7}$

78. $\frac{1}{7} \div \frac{1}{2}$

79. $\frac{5}{9} \div \frac{1}{8}$

80. $\frac{1}{8} \div \frac{5}{9}$

81. $\frac{4}{5} \div \frac{8}{15}$

82. $\frac{3}{10} \div \frac{6}{5}$

83. $\frac{7}{8} \div \frac{3}{8}$

84. $\frac{10}{3} \div \frac{5}{6}$

85. $\frac{9}{10} \div \frac{3}{4}$

86. $\frac{5}{6} \div \frac{1}{3}$

87. $\frac{1}{10} \div \frac{2}{5}$

88. $\frac{3}{4} \div \frac{6}{5}$

89. $\frac{2}{3} \div 7$

90. $\frac{7}{10} \div 10$

91. $\frac{2}{3} \div 6$

92. $\frac{1}{20} \div 2$

93. $8 \div \frac{1}{5}$

94. $8 \div \frac{2}{9}$

95. $7 \div \frac{3}{7}$

96. $10 \div \frac{2}{5}$

97. $4 \div \frac{3}{10}$

98. $10 \div \frac{2}{3}$

99. $1 \div \frac{1}{7}$

100. $3 \div \frac{1}{8}$

101. $2\frac{5}{6} \div \frac{3}{7}$

102. $5\frac{1}{9} \div \frac{2}{3}$

103. $1\frac{1}{3} \div \frac{4}{5}$

104. $7\frac{1}{10} \div \frac{1}{2}$

105. $8\frac{5}{6} \div \frac{9}{10}$

106. $6\frac{1}{2} \div \frac{1}{2}$

107. $20\frac{1}{10} \div \frac{1}{5}$

108. $15\frac{2}{3} \div \frac{5}{6}$

109. $\frac{1}{6} \div 2\frac{1}{7}$

110. $\frac{2}{7} \div 1\frac{1}{3}$

111. $\frac{1}{2} \div 2\frac{3}{5}$

112. $\frac{3}{4} \div 3\frac{1}{9}$

113. $4 \div 1\frac{1}{4}$

114. $7 \div 1\frac{9}{10}$

115. $2\frac{1}{10} \div 20$

116. $5\frac{6}{7} \div 14$

117. $2\frac{1}{2} \div 3\frac{1}{7}$

118. $3\frac{1}{7} \div 2\frac{1}{2}$

119. $8\frac{1}{10} \div 5\frac{3}{4}$

120. $1\frac{7}{10} \div 5\frac{1}{8}$

121. $2\frac{1}{3} \div 4\frac{1}{2}$

122. $8\frac{1}{6} \div 2\frac{1}{2}$

123. $6\frac{3}{8} \div 2\frac{5}{6}$

124. $1\frac{2}{3} \div 1\frac{2}{5}$

Simplify.

125. $\frac{1}{2} + \frac{2}{3} \times 1\frac{1}{3}$

126. $\frac{9}{10} + \frac{4}{5} \cdot 8$

127. $5 - \frac{1}{3} \times \frac{2}{5}$

128. $3 \div \frac{2}{5} - 2\frac{1}{3}$

129. $2\frac{3}{4} \times \frac{1}{8} + \frac{1}{5}$

130. $\frac{3}{8} \cdot \frac{1}{2} - \frac{1}{10}$

131. $4 - \frac{2}{9} \div \frac{3}{4}$

132. $6 \div 5 \times \frac{1}{4}$

133. $3\frac{1}{2} \times 6 \div 5$

134. $4 \cdot \frac{2}{3} - 1\frac{1}{8}$

135. $10 \times \frac{1}{8} \times 2\frac{1}{2}$

136. $\frac{1}{3} \div \frac{1}{6} \times \frac{2}{3}$

137. $8 \div 1\frac{1}{5} + 3 \cdot 1\frac{1}{2}$

138. $3\frac{1}{8} \div 5 + 4 \div 2\frac{1}{2}$

139. $\left(1\frac{1}{2} \div \frac{1}{3}\right)^2 + \left(1 - \frac{1}{4}\right)^2$

140. $\left(3\frac{1}{2}\right)^2 + 2\left(1\frac{1}{2} - 1\frac{1}{3}\right)$

Mixed Practice

141. Divide $6\frac{1}{8}$ by $2\frac{3}{4}$.

142. Compute: $14 - 3 \div \left(\frac{4}{5}\right)^2$

143. Find the product of $\frac{3}{5}$ and $\frac{7}{8}$.

144. Find the quotient of $\frac{9}{10}$ and $\frac{2}{5}$.

145. Multiply $\frac{2}{3}$ by 12.

146. Calculate: $\left(4\frac{1}{2}\right)\left(6\frac{2}{3}\right)$

Applications

Solve. Write the answer in simplest form.

147. In a local town, $\frac{5}{6}$ of the voting-age population is registered to vote. If $\frac{7}{10}$ of the registered voters voted in the election for mayor, what fraction of the voting-age population voted?

148. Last year, $\frac{1}{8}$ of the emergency room visits at a hospital were injury related. Of these, $\frac{2}{5}$ were due to motor vehicle accidents. What fraction of the emergency room visits were due to motor vehicle accidents?

149. The house that a couple wants to buy is selling for \$240,000. They need to put $\frac{1}{20}$ of the selling price down, and take out a mortgage for the rest. How much money do they need to put down?

150. There is a rule of thumb that no one should spend more than $\frac{1}{4}$ of their income on rent. If someone makes \$24,000 a year, what is the most he or she should spend per month on rent according to this rule?

151. Students in an astronomy course learn that a first-magnitude star is $2\frac{1}{2}$ times as bright as a second-magnitude star, which in turn is $2\frac{1}{2}$ times as bright as a third-magnitude star. How many times as bright as a third-magnitude star is a first-magnitude star?

152. Which of these rooms has the larger area?

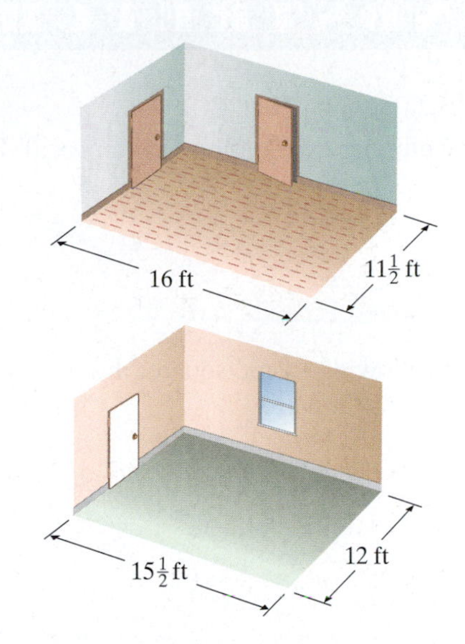

153. Find the cost of buying carpeting at $\$7\frac{1}{2}$ per square-foot for the hallway shown.

154. Some people believe that gasohol is superior to gasoline as an automotive fuel. Gasohol is a mixture of gasoline $\left(\frac{9}{10}\right)$ and ethyl alcohol $\left(\frac{1}{10}\right)$. How much more gasoline than ethyl alcohol is there in $10\frac{1}{2}$ gallons of gasohol?

155. Because of evaporation, a pond loses $\frac{1}{4}$ of its remaining water each month of summer. If it is full at the beginning of summer, what fraction of the original amount will the pond contain after three summer months?

156. A scientist is investigating the effects of cold on human skin. In one of the scientist's experiments, the temperature starts at 70°F and drops by $\frac{1}{10}^{\circ}$ every 2 minutes. What is the temperature after 6 minutes?

157. A trip to a nearby island takes $3\frac{1}{2}$ hours by boat and $\frac{1}{2}$ hour by airplane. How many times as fast as the boat is the plane?

158. Each dose of aspirin weighs $\frac{3}{4}$ grain. If a nurse has 9 grains of aspirin on hand, how many doses can he administer?

159. A store sells two types of candles. The scented candle is 8 inches tall and burns $\frac{1}{2}$ inch per hour, whereas the unscented candle is 10 inches tall and burns $\frac{1}{3}$ inch per hour.

a. In an hour, which candle will burn more?

b. Which candle will last longer?

160. A college-wide fund-raising campaign collected \$3 million in $1\frac{1}{2}$ years for student scholarships.

a. What was the average amount collected per year?

b. By how much would this average increase if an additional \$1 million were collected?

• *Check your answers on page A-4.*

MINDSTRETCHERS

Writing

1. Every number except 0 has a reciprocal. Explain why 0 does not have a reciprocal.

Groupwork

2. In the following magic square, the *product* of every row, column, and diagonal is 1. Working with a partner, complete the square.

$\frac{2}{3}$		$1\frac{1}{2}$
$\frac{1}{2}$		

Patterns

3. Find the product: $1\frac{1}{2} \cdot 1\frac{1}{3} \cdot 1\frac{1}{4} \cdot \cdots \cdot 1\frac{1}{99} \cdot 1\frac{1}{100}$

KEY CONCEPTS AND SKILLS

CONCEPT SKILL

Concept/Skill	Description	Example
[2.1] **Prime number**	A whole number that has exactly two different factors: itself and 1.	2, 3, 5
[2.1] **Composite number**	A whole number that has more than two factors.	4, 8, 9
[2.1] **Prime factorization of a whole number**	The number written as the product of its prime factors.	$30 = 2 \cdot 3 \cdot 5$
[2.1] **Least common multiple (LCM) of two or more whole numbers**	The smallest nonzero whole number that is a multiple of each number.	The LCM of 30 and 45 is 90.
[2.1] **To compute the least common multiple (LCM)**	• Find the prime factorization of each number. • Identify the prime factors that appear in each factorization. • Multiply these prime factors, using each factor the greatest number of times that it occurs in any of the factorizations.	$20 = 2 \cdot 2 \cdot 5$ $= 2^2 \cdot 5$ $30 = 2 \cdot 3 \cdot 5$ The LCM of 20 and 30 is $2^2 \cdot 3 \cdot 5$, or 60.
[2.2] **Fraction**	Any number that can be written in the form $\frac{a}{b}$, where a and b are whole numbers and b is nonzero.	$\frac{3}{11}, \frac{9}{5}$
[2.2] **Proper fraction**	A fraction whose numerator is smaller than its denominator.	$\frac{2}{7}, \frac{1}{2}$
[2.2] **Mixed number**	A number with a whole-number part and a proper fraction part.	$5\frac{1}{3}, 4\frac{5}{6}$
[2.2] **Improper fraction**	A fraction whose numerator is greater than or equal to its denominator.	$\frac{9}{4}, \frac{5}{5}$
[2.2] **To change a mixed number to an improper fraction**	• Multiply the denominator of the fraction by the whole-number part of the mixed number. • Add the numerator of the fraction to this product. • Write this sum over the denominator to form the improper fraction.	$4\frac{2}{3} = \frac{3 \times 4 + 2}{3}$ $= \frac{14}{3}$
[2.2] **To change an improper fraction to a mixed number**	• Divide the numerator by the denominator. • If there is a remainder, write it over the denominator.	$\frac{14}{3} = 4\frac{2}{3}$
[2.2] **To find an equivalent fraction**	Multiply the numerator and denominator of $\frac{a}{b}$ by the same whole number; that is, $\frac{a}{b} = \frac{a \cdot n}{b \cdot n}$, where both b and n are nonzero.	$\frac{3}{4} = \frac{3 \cdot 2}{4 \cdot 2} = \frac{6}{8}$
[2.2] **To simplify (reduce) a fraction**	Divide the numerator and denominator of $\frac{a}{b}$ by the same whole number n; that is, $\frac{a}{b} = \frac{a \div n}{b \div n}$, where both b and n are nonzero.	$\frac{6}{8} = \frac{6 \div 2}{8 \div 2} = \frac{3}{4}$

continued

CONCEPT SKILL

Concept/Skill	Description	Example
[2.2] Like fractions	Fractions with the same denominator.	$\frac{2}{5}, \frac{3}{5}$
[2.2] Unlike fractions	Fractions with different denominators.	$\frac{3}{5}, \frac{3}{10}$
[2.2] To compare fractions	• If the fractions are like, compare their numerators. • If the fractions are unlike, write them as equivalent fractions with the same denominator and then compare their numerators.	$\frac{6}{8}, \frac{7}{8}$ $6 < 7$, so $\frac{6}{8} < \frac{7}{8}$ $\frac{2}{3}, \frac{12}{15}$ or $\frac{10}{15}, \frac{12}{15}$ $12 > 10$, so $\frac{12}{15} > \frac{2}{3}$
[2.2] Least common denominator (LCD) of two or more fractions	The least common multiple of their denominators.	The LCD of $\frac{11}{30}$ and $\frac{7}{45}$ is 90.
[2.3] To add (or subtract) like fractions	• Add (or subtract) the numerators. • Use the given denominator. • Write the answer in simplest form.	$\frac{1}{8} + \frac{1}{8} = \frac{2}{8} = \frac{1}{4}$ $\frac{3}{8} - \frac{1}{8} = \frac{2}{8} = \frac{1}{4}$
[2.3] To add (or subtract) unlike fractions	• Write the fractions as equivalent fractions with the same denominator, usually the LCD. • Add (or subtract) the numerators, keeping the same denominator. • Write the answer in simplest form.	$\frac{2}{3} + \frac{1}{2} = \frac{4}{6} + \frac{3}{6}$ $= \frac{7}{6}$, or $1\frac{1}{6}$ $\frac{5}{12} - \frac{1}{6}$ $= \frac{5}{12} - \frac{2}{12} = \frac{3}{12}$, or $\frac{1}{4}$
[2.3] To add mixed numbers	• Write the fractions as equivalent fractions with the same denominator, usually the LCD. • Add the fractions. • Add the whole numbers. • Write the answer in simplest form.	$4\frac{1}{2} = 4\frac{3}{6}$ $+6\frac{2}{3} = +6\frac{4}{6}$ $10\frac{7}{6} = 11\frac{1}{6}$
[2.3] To subtract mixed numbers	• Write the fractions as equivalent fractions with the same denominator, usually the LCD. • Rename (or borrow from) the whole number on top if the fraction on the bottom is larger than the fraction on top. • Subtract the fractions. • Subtract the whole numbers. • Write the answer in simplest form.	$4\frac{1}{5} = 3\frac{6}{5}$ $-1\frac{2}{5} = -1\frac{2}{5}$ $2\frac{4}{5}$
[2.4] To multiply fractions	• Multiply the numerators. • Multiply the denominators. • Write the answer in simplest form.	$\frac{1}{2} \cdot \frac{3}{5} = \frac{3}{10}$

continued

Concept/Skill	Description	Example
[2.4] **To multiply mixed numbers**	• Write the mixed numbers as improper fractions. • Multiply the fractions. • Write the answer in simplest form.	$2\frac{1}{2} \cdot 1\frac{2}{3} = \frac{5}{2} \cdot \frac{5}{3}$ $= \frac{25}{6}$, or $4\frac{1}{6}$
[2.4] **Reciprocal of** $\frac{a}{b}$	The fraction $\frac{b}{a}$ formed by switching the numerator and denominator.	The reciprocal of $\frac{4}{3}$ is $\frac{3}{4}$.
[2.4] **To divide fractions**	• Change the divisor to its reciprocal, and multiply the resulting fractions. • Write the answer in simplest form.	$\frac{2}{5} \div \frac{3}{7} = \frac{2}{5} \cdot \frac{7}{3}$ $= \frac{14}{15}$
[2.4] **To divide mixed numbers**	• Write the mixed numbers as improper fractions. • Divide the fractions. • Write the answer in simplest form.	$2\frac{1}{2} \div 1\frac{1}{3} =$ $\frac{5}{2} \div \frac{4}{3} =$ $\frac{5}{2} \cdot \frac{3}{4} = \frac{15}{8}$, or $1\frac{7}{8}$

CULTURAL NOTE

I/IIII Dieſſe figur iſt vñ bedeüt ain fiertel von ainem ganʒen/alſo mag man auch ain fünfftail/ayn ſechſtail/ain ſybentail oder zwai ſechſtail 2c. vnd alle ander brüch beſchreiben/Als I/V | I/VI | I/VII | II/VI 2c.

VI/VIII Diß ſein Sechs achtail/das ſein ſechstail der acht ain ganʒ machen.

IX/XI Diß Figur bezaigt ann newn ayilfftail das ſeyn IX tail/der XI ain ganʒ machen.

XX/XXXI Diß Figur bezaichet/zwenʒigt ainundreyſigt tail /das ſein zwenʒigt tail der ainsundreiſſigt ain ganʒ machen.

IIC/IIIIC.LX Diß ſein zwaihundert tail/der Fierhundert vnd ſechʒigt ain ganʒ machen.

Source: David Eugene Smith and Jekuthiel Ginsburg, *Numbers and Numerals, a Story Book for Young and Old* (New York: Bureau of Publications, Teachers College, Columbia University, 1937).

In societies throughout the world and across the centuries, people have written fractions in strikingly different ways. In ancient Greece, for example, the fraction $\frac{1}{4}$ was written Δ'' where Δ (read "delta") is the fourth letter of the Greek alphabet.

At one time, people wrote the numerator and denominator of fractions in Roman numerals, as shown at the left in a page from a sixteenth-century German book. In today's notation, the last fraction is $\frac{200}{460}$.

Chapter 2 Review Exercises

To help you review this chapter, solve these problems.

[2.1] *Find all the factors of each number.*

1. 150 **2.** 180 **3.** 57 **4.** 70

Indicate whether each number is prime or composite.

5. 23 **6.** 33 **7.** 87 **8.** 67

Write the prime factorization of each number, using exponents.

9. 36 **10.** 75 **11.** 99 **12.** 54

Find the LCM.

13. 6 and 14 **14.** 5 and 10 **15.** 18, 24, and 36 **16.** 10, 15, and 20

[2.2] *Identify the fraction or mixed number represented by the shaded portion of each figure.*

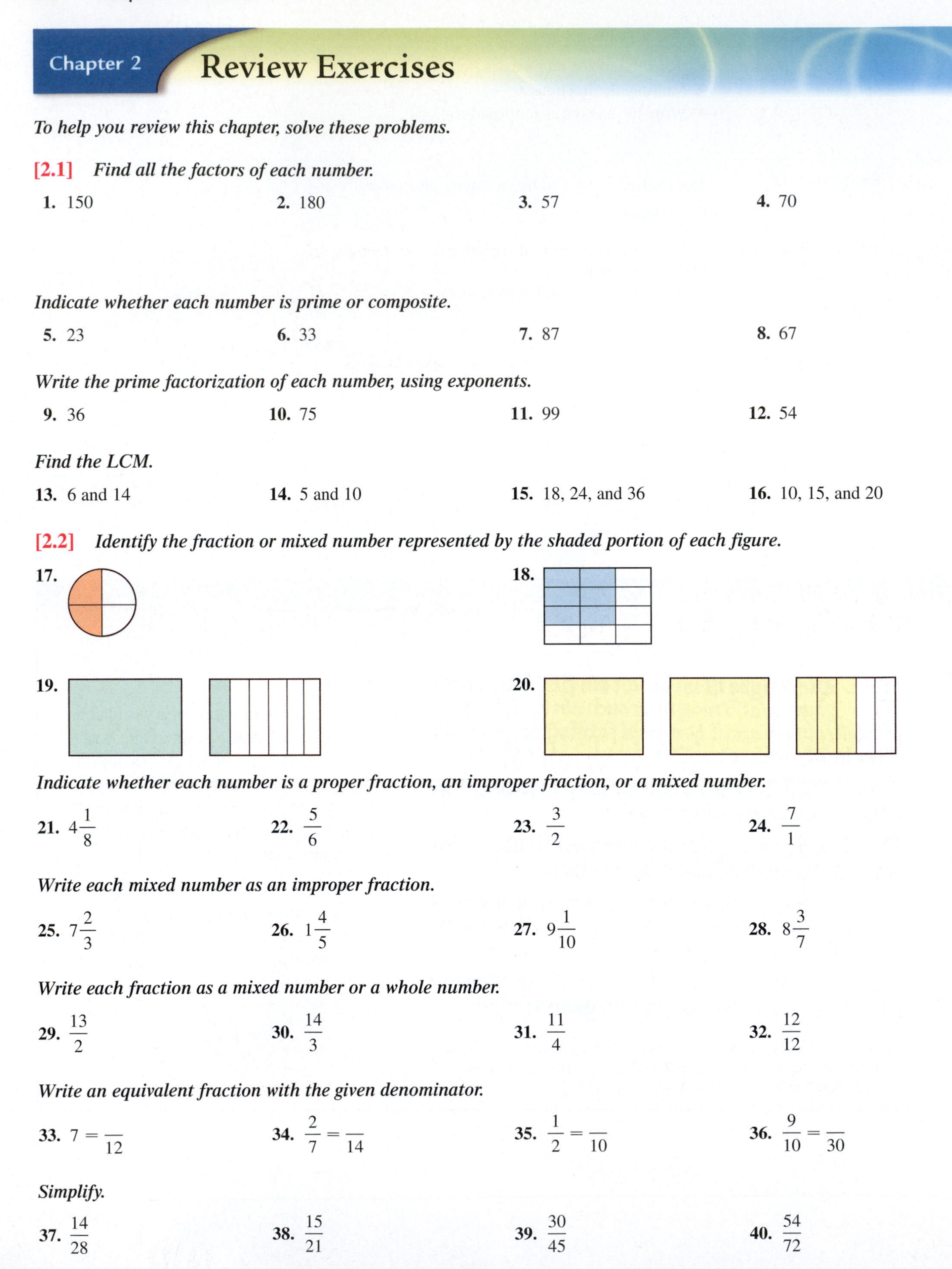

17. **18.** **19.** **20.**

Indicate whether each number is a proper fraction, an improper fraction, or a mixed number.

21. $4\frac{1}{8}$ **22.** $\frac{5}{6}$ **23.** $\frac{3}{2}$ **24.** $\frac{7}{1}$

Write each mixed number as an improper fraction.

25. $7\frac{2}{3}$ **26.** $1\frac{4}{5}$ **27.** $9\frac{1}{10}$ **28.** $8\frac{3}{7}$

Write each fraction as a mixed number or a whole number.

29. $\frac{13}{2}$ **30.** $\frac{14}{3}$ **31.** $\frac{11}{4}$ **32.** $\frac{12}{12}$

Write an equivalent fraction with the given denominator.

33. $7 = \frac{}{12}$ **34.** $\frac{2}{7} = \frac{}{14}$ **35.** $\frac{1}{2} = \frac{}{10}$ **36.** $\frac{9}{10} = \frac{}{30}$

Simplify.

37. $\frac{14}{28}$ **38.** $\frac{15}{21}$ **39.** $\frac{30}{45}$ **40.** $\frac{54}{72}$

41. $5\frac{2}{4}$ **42.** $8\frac{10}{15}$ **43.** $6\frac{12}{42}$ **44.** $8\frac{45}{63}$

Insert the appropriate sign: <, =, or >.

45. $\frac{5}{8} \quad \frac{3}{8}$ **46.** $\frac{5}{6} \quad \frac{1}{6}$ **47.** $\frac{2}{3} \quad \frac{4}{5}$ **48.** $\frac{9}{10} \quad \frac{7}{8}$

49. $\frac{3}{4} \quad \frac{5}{8}$ **50.** $\frac{7}{10} \quad \frac{5}{9}$ **51.** $3\frac{1}{5} \quad 1\frac{9}{10}$ **52.** $5\frac{1}{8} \quad 5\frac{1}{9}$

Arrange in increasing order.

53. $\frac{2}{7}, \frac{3}{8}, \frac{1}{2}$ **54.** $\frac{1}{5}, \frac{1}{3}, \frac{2}{15}$ **55.** $\frac{4}{5}, \frac{9}{10}, \frac{3}{4}$ **56.** $\frac{7}{8}, \frac{7}{9}, \frac{13}{18}$

[2.3] *Add and simplify.*

57. $\frac{2}{5} + \frac{4}{5}$ **58.** $\frac{7}{20} + \frac{8}{20}$ **59.** $\frac{5}{8} + \frac{7}{8} + \frac{3}{8}$ **60.** $\frac{3}{10} + \frac{1}{10} + \frac{2}{10}$

61. $\frac{1}{3} + \frac{2}{5}$ **62.** $\frac{7}{8} + \frac{5}{6}$ **63.** $\frac{9}{10} + \frac{1}{2} + \frac{2}{5}$ **64.** $\frac{3}{8} + \frac{4}{5} + \frac{3}{4}$

65. $2 + 3\frac{7}{8}$ **66.** $6\frac{1}{4} + 3\frac{1}{4}$ **67.** $8\frac{7}{10} + 1\frac{9}{10}$ **68.** $5\frac{5}{6} + 2\frac{1}{6}$

69. $2\frac{1}{3} + 4\frac{1}{3} + 5\frac{2}{3}$ **70.** $1\frac{3}{10} + \frac{9}{10} + 2\frac{1}{10}$ **71.** $5\frac{2}{5} + \frac{3}{10}$ **72.** $9\frac{1}{6} + 8\frac{3}{8}$

73. $10\frac{2}{3} + 12\frac{3}{4}$ **74.** $20\frac{1}{2} + 25\frac{7}{8}$ **75.** $10\frac{3}{5} + 7\frac{9}{10} + 2\frac{1}{4}$ **76.** $20\frac{7}{8} + 30\frac{5}{6} + 4\frac{1}{3}$

Subtract and simplify.

77. $\frac{3}{8} - \frac{1}{8}$ **78.** $\frac{7}{9} - \frac{1}{9}$ **79.** $\frac{5}{3} - \frac{2}{3}$ **80.** $\frac{4}{6} - \frac{4}{6}$

81. $\frac{3}{10} - \frac{1}{20}$ **82.** $\frac{1}{2} - \frac{1}{8}$ **83.** $\frac{3}{5} - \frac{1}{4}$ **84.** $\frac{1}{3} - \frac{1}{10}$

85. $12\frac{1}{2} - 5$ **86.** $4\frac{3}{10} - 2$ **87.** $8\frac{7}{8} - 5\frac{1}{8}$ **88.** $20\frac{3}{4} - 2\frac{1}{4}$

89. $12 - 5\frac{1}{2}$ **90.** $4 - 2\frac{3}{10}$ **91.** $7 - 4\frac{1}{3}$ **92.** $1 - \frac{4}{5}$

93. $6\frac{1}{10} - 4\frac{3}{10}$ **94.** $2\frac{5}{8} - 1\frac{7}{8}$ **95.** $5\frac{1}{4} - 2\frac{3}{4}$ **96.** $7\frac{1}{6} - 3\frac{5}{6}$

97. $3\frac{1}{10} - 2\frac{4}{5}$ **98.** $7\frac{1}{2} - 4\frac{5}{8}$ **99.** $5\frac{1}{12} - 4\frac{1}{2}$ **100.** $6\frac{2}{9} - 2\frac{1}{3}$

101. $\frac{1}{3} + \frac{5}{6} - \frac{1}{2}$ **102.** $7\frac{9}{10} - 1\frac{1}{5} + 2\frac{3}{4}$

[2.4] *Multiply and simplify.*

103. $\frac{3}{4} \times \frac{1}{4}$ **104.** $\frac{1}{2} \times \frac{7}{8}$ **105.** $\left(\frac{5}{6}\right)\left(\frac{3}{4}\right)$ **106.** $\left(\frac{2}{3}\right)\left(\frac{1}{4}\right)$

107. $\frac{2}{3} \cdot 8$ **108.** $\frac{1}{10} \cdot 7$ **109.** $\left(\frac{1}{5}\right)^3$ **110.** $\left(\frac{2}{3}\right)^3$

111. $\frac{1}{2} \times \frac{2}{3} \times \frac{3}{4}$ **112.** $\frac{7}{8} \times \frac{2}{5} \times \frac{1}{6}$ **113.** $\frac{4}{5} \times 1\frac{1}{5}$ **114.** $\frac{2}{3} \times 2\frac{1}{3}$

115. $5\frac{1}{3} \cdot \frac{1}{2}$ **116.** $\frac{1}{10} \cdot 6\frac{2}{3}$ **117.** $1\frac{1}{3} \cdot 4\frac{1}{2}$ **118.** $3\frac{1}{4} \cdot 5\frac{2}{3}$

119. $6\frac{3}{4} \times 1\frac{1}{4}$ **120.** $8\frac{1}{2} \times 2\frac{1}{2}$ **121.** $\frac{7}{8} \times 1\frac{1}{5} \times \frac{3}{7}$ **122.** $1\frac{3}{8} \times \frac{10}{11} \times 1\frac{1}{4}$

123. $\left(3\frac{1}{3}\right)^3$ **124.** $\left(1\frac{1}{2}\right)^3$ **125.** $\frac{5}{8} + \frac{1}{2} \cdot 5$ **126.** $1\frac{9}{10} - \left(\frac{2}{3}\right)^2$

127. $4\left(\frac{2}{5}\right) + 3\left(\frac{1}{6}\right)$ **128.** $6\left(1\frac{1}{2} - \frac{3}{10}\right)$

Find the reciprocal.

129. $\frac{2}{3}$ **130.** $1\frac{1}{2}$ **131.** 8 **132.** $\frac{1}{4}$

Divide and simplify.

133. $\frac{7}{8} \div 5$ **134.** $\frac{5}{9} \div 9$ **135.** $\frac{2}{3} \div 5$

136. $\frac{1}{100} \div 2$ **137.** $\frac{1}{2} \div \frac{2}{3}$ **138.** $\frac{2}{3} \div \frac{1}{2}$

139. $6 \div \frac{1}{5}$ **140.** $7 \div \frac{4}{5}$ **141.** $\frac{7}{8} \div \frac{3}{4}$ **142.** $\frac{9}{10} \div \frac{1}{2}$

143. $\frac{3}{5} \div \frac{3}{10}$ **144.** $\frac{2}{3} \div \frac{1}{6}$ **145.** $3\frac{1}{2} \div 2$ **146.** $2 \div 3\frac{1}{2}$

147. $6\frac{1}{3} \div 4$ **148.** $4 \div 6\frac{1}{3}$ **149.** $8\frac{1}{4} \div 1\frac{1}{2}$ **150.** $3\frac{2}{5} \div 1\frac{1}{3}$

151. $4\frac{1}{2} \div 2\frac{1}{4}$ **152.** $7\frac{1}{5} \div 2\frac{2}{5}$ **153.** $\left(5 - \frac{2}{3}\right) \div \frac{4}{9}$ **154.** $6\frac{1}{2} \div \left(\frac{1}{2} + 4\frac{1}{2}\right)$

155. $7 \div 2\frac{1}{4} + 5 \div \left(1\frac{1}{2}\right)^2$ **156.** $\left(1\frac{2}{3}\right)^2 \times 2 + 9 \div 4\frac{1}{2}$

Mixed Applications

Solve. Write the answer in simplest form.

157. The Summer Olympic Games are held during each year divisible by 4. Were the Olympic Games held in 1990?

158. What is the smallest amount of money that you can pay in both all quarters and all dimes?

159. Eight of the 32 human teeth are incisors. What fraction of human teeth are incisors? (*Source:* Ilsa Goldsmith, *Human Anatomy for Children*)

160. The planets in the solar system (including the "dwarf planet" Pluto) consist of Earth, two planets closer to the Sun than Earth, and six planets farther from the Sun than Earth. What fraction of the planets in the solar system are closer than Earth to the Sun? (*Source:* Patrick Moore, *Astronomy for the Beginner*)

161. A Filmworks camera has a shutter speed of $\frac{1}{8,000}$ second and a Lensmax camera has a shutter speed of $\frac{1}{6,000}$ second. Which shutter is faster? (*Hint:* The faster shutter has the smaller shutter speed.)

162. Of the approximately 12,000 women that started the 2006 New York City Marathon, only about 300 did not finish the race. What fraction of the women did finish the race? (*Source:* www.nycmarathon.org)

163. An insurance company reimbursed a patient \$275 on a dental bill of \$700. Did the patient get more or less than $\frac{1}{3}$ of the money paid back? Explain.

164. A union goes on strike if at least $\frac{2}{3}$ of the workers voting support the strike call. If 23 of the 32 voting workers support a strike, should a strike be declared? Explain.

165. A grand jury has 23 jurors. Sixteen jurors are needed for a quorum, and a vote of 12 jurors is needed to indict.

a. What fraction of the full jury is needed to indict?

b. Suppose that 16 jurors are present. What fraction of those present is needed to indict?

166. In a tennis match, Lisa Gregory went to the net 12 times, winning the point 7 times. By contrast, Monica Yates won the point 4 of the 6 times that she went to the net.

a. Which player went to the net more often?

b. Which player had a better rate of winning points at the net?

167. In a math course, $\frac{3}{5}$ of a student's grade is based on four in-class exams, and $\frac{3}{20}$ of the grade is based on homework. What fraction of a student's grade is based on in-class exams and homework?

168. A metal alloy is made by combining $\frac{1}{4}$ ounce of copper with $\frac{2}{3}$ ounce of tin. Find the alloy's total weight.

169. The weight of a diamond is measured in carats. What is the difference in weight between a $\frac{1}{2}$-carat and a $\frac{3}{4}$-carat diamond?

170. During a sale, the price of a sweater was marked $\frac{1}{4}$ off the original price of \$45. Using a coupon, a customer received an additional $\frac{1}{5}$ off the sale price. What fraction of the original price was the final sale price?

171. In a math class, $\frac{3}{8}$ of the students are chemistry majors and $\frac{2}{3}$ of those students are women. If there are 48 students in the math class, how many women are chemistry majors?

172. Three-eighths of the undergraduate students at a two-year college receive financial aid. If the college has 4,296 undergraduate students, how many undergraduate students do not receive financial aid?

173. An investor bought $1,000 worth of a technology stock. At the beginning of last year, it had increased in value by $\frac{2}{5}$. During the year, the value of the stock declinedby $\frac{1}{4}$. What was the value of the stock at the end of last year?

174. A sea otter eats about $\frac{1}{5}$ of its body weight each day. How much will a 35-pound otter eat in a day? (***Source:*** Karl W. Kenyon, *The Sea Otter in the Eastern Pacific Ocean*)

175. The regular price of roses is $27 a dozen. What is the cost of the roses on sale?

Roses $\frac{1}{3}$ off!

176. In Roseville, 40 of every 1,000 people who want to work are unemployed, in contrast to 8 of every 100 people in Georgetown. How many times as great as the unemployment rate in Roseville is the unemployment rate in Georgetown?

177. A brother and sister want to buy as many goldfish as possible for their new fish tank. A rule of thumb is that the total length of fish, in inches, should be less than the capacity of the tank in gallons. If they have a 10-gallon tank and goldfish average $\frac{1}{2}$ inch in length, how many fish should they buy?

178. A commuter is driving to the city of Denver 15 miles away. If he has already driven $3\frac{1}{4}$ miles, how far is he from Denver?

179. The wingspan of a Boeing 717 jet is $93\frac{1}{4}$ feet, whereas the wingspan of a Boeing 767 jet is $156\frac{1}{12}$ feet. How much longer is the wingspan of a Boeing 767 jet?

180. The Acela Express train from Boston to New York took $3\frac{5}{12}$ hours, whereas the Regional Service train took $4\frac{1}{4}$ hours. How much faster than the Regional Service train was the Acela Express train? (***Source:*** Amtrak).

181. An airplane is flying $1\frac{1}{2}$ times the speed of sound. If sound travels at about 1,000 feet per second, at what speed is the plane flying?

182. When you stand upright, the pressure per square inch on your hip joint is about $2\frac{1}{2}$ times your body weight. If you weigh 200 lb, what is that pressure?

183. A cubic foot of water weighs approximately $62\frac{1}{2}$ pounds. If a basin contains $4\frac{1}{2}$ cubic feet of water, how much does the water weigh?

184. During a housing boom, the market value of a home was $2\frac{1}{2}$ times its original value. If its market value is $115,000, what was its original value?

185. It took the space shuttle Endeavor $1\frac{1}{2}$ hours to orbit Earth. How many orbits did the Endeavor make in 12 hours? (*Source:* NASA)

186. What is the area of the highway billboard sign shown below?

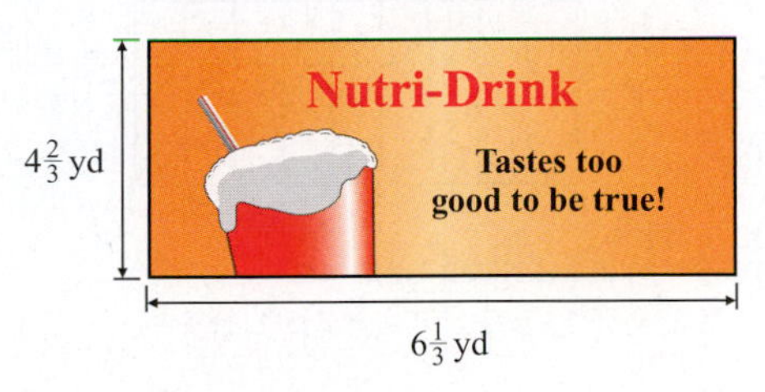

187. The following chart is a record of the amount of time (in hours) that two employees spent working the past weekend. Complete the chart.

Employee	Saturday	Sunday	Total
L. Chavis	$7\frac{1}{2}$	$4\frac{1}{4}$	
R. Young	$5\frac{3}{4}$	$6\frac{1}{2}$	
Total			

188. Complete the following chart.

Worker	Hours per Day	Days Worked	Total Hours	Wage per Hour	Gross Pay
Maya	5	3		\$7	
Noel	$7\frac{1}{4}$	4		\$10	
Alisa	$4\frac{1}{2}$	$5\frac{1}{2}$		\$9	

189. According to a newspaper advertisement, a man on a diet lost 60 pounds in $5\frac{1}{2}$ months. On the average, how much weight did he lose per month?

190. According to the nutrition label on a box of cereal, one serving is $1\frac{1}{4}$ cups. If the box contains 18 servings of cereal, how many cups of cereal does it contain?

• *Check your answers on page A-4.*

Chapter 2 POSTTEST

FOR EXTRA HELP

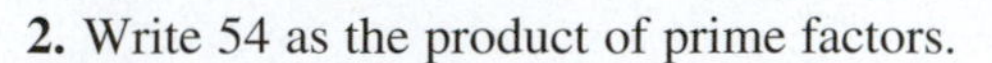

Test solutions are found on the enclosed CD.

To see if you have mastered the topics in this chapter, take this test.

1. List all the factors of 63.

2. Write 54 as the product of prime factors.

3. What fraction of the diagram is shaded?

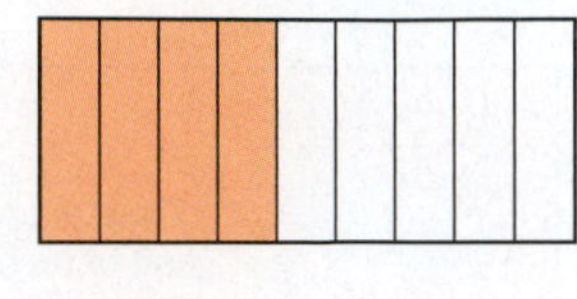

4. Write 12 as an improper fraction.

5. Express $\frac{41}{4}$ as a mixed number.

6. Write $\frac{875}{1{,}000}$ in simplest form

7. Which is smaller, $\frac{2}{3}$ or $\frac{5}{10}$?

8. What is the LCD for $\frac{3}{8}$ and $\frac{1}{12}$?

Add.

9. $\frac{2}{3} + \frac{1}{8} + \frac{3}{4}$

10. $6\frac{7}{8} + 1\frac{3}{10}$

Subtract.

11. $6 - 1\frac{5}{7}$

12. $10\frac{1}{6} - 4\frac{2}{5}$

Multiply.

13. $\left(\frac{1}{9}\right)^2$

14. $2\frac{2}{3} \times 4\frac{1}{2}$

15. Divide: $2\frac{1}{3} \div 3$

16. Calculate: $14\frac{1}{2} - 5 \cdot 1\frac{1}{3}$

Solve. Write your answer in simplest form.

17. Seven of the 42 men who have served as president of the United States were born in Ohio. What fraction of these men were *not* born in Ohio? (***Source:*** *The World Almanac 2006*)

18. In an Ironman triathlon, an athlete completed the 112-mile bike ride in $5\frac{5}{6}$ hours. What was the average number of miles she bicycled each hour?

19. Find the area of the rectangular floor pictured.

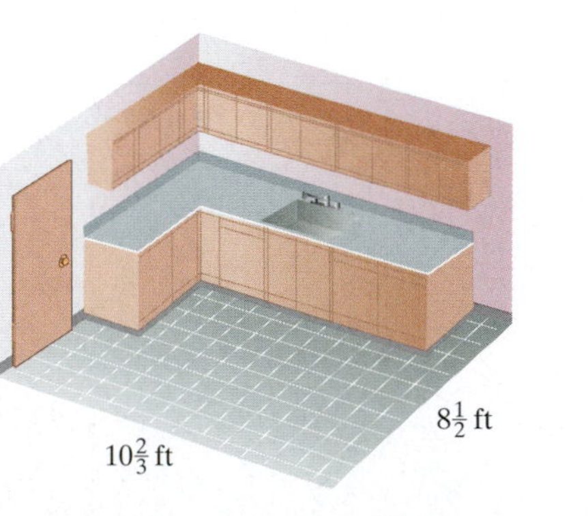

20. Find the distance across the hubcap of the tire shown in the diagram.

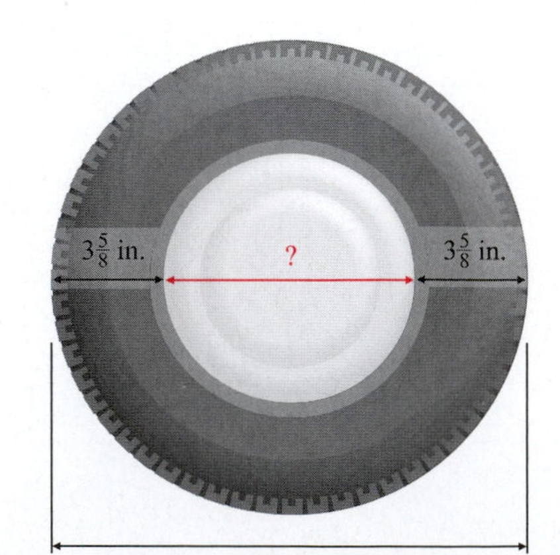

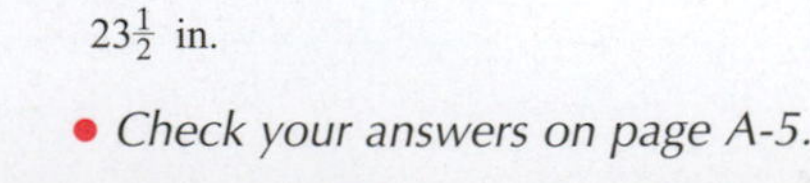
• *Check your answers on page A-5.*

Cumulative Review Exercises

To help you review, solve the following:

1. Write in words: 5,000,315

2. Multiply: $5{,}814 \times 100$

3. Find the quotient: $89\overline{)80{,}812}$

4. Write $\frac{75}{100}$ in simplest form.

5. Subtract: $8 - 1\frac{3}{5}$

6. Find the product: $1\frac{1}{2} \cdot 4\frac{2}{3}$

7. Which is larger, $\frac{1}{4}$ or $\frac{3}{8}$?

8. A jury decided on punishments in an oil spill case. The jury ordered the captain of the oil barge to pay $5,000 in punitive damages and the oil company to pay $5 billion. The amount that the company had to pay is how many times the amount that the captain had to pay?

9. The counter on a photocopier keeps track of the number of copies made. How many copies did you make if the counter showed 23,459 copies before you started copying and 24,008 after you finished?

10. A homeowner wants to refinish his basement over three weekends. He completes $\frac{1}{4}$ of the job the first weekend and $\frac{5}{12}$ of the job the second weekend. What fraction of the job remains to be completed?

• *Check your answers on page A-5.*

CHAPTER 3

Decimals

Decimals and Blood Tests

Blood tests reveal a great deal about a person's health—whether to reduce the cholesterol level to lower the risk of heart disease, or raise the red blood cell count to prevent anemia. And blood tests identify a variety of diseases, for example, AIDS and mononucleosis.

Blood analyses are typically carried out in clinical laboratories. Technicians in these labs operate giant machines that perform thousands of blood tests per hour.

What these blood tests, known as "chemistries," actually do is to analyze blood for a variety of substances, such as creatinine and calcium.

In any blood test, doctors look for abnormal levels of the substance being measured. For instance, the normal range on the creatinine test is typically from 0.7 to 1.5 milligrams (mg) per unit of blood. A high level may mean kidney disease; a low level, muscular dystrophy.

The normal range on the calcium test may be 9.0 to 10.5 mg per unit of blood. A result outside this range is a clue for any of several diseases.

(***Source:*** Dixie Farley, "Top 10 Laboratory Tests: Blood Will Tell," *FDA Consumer*, Vol. 23)

Chapter 3 PRETEST

To see if you have already mastered the topics in this chapter, take this test.

1. In the number 27.081, what place does the 8 occupy?

2. Write in words: 4.012

3. Round 3.079 to the nearest tenth.

4. Which is largest: 0.00212, 0.0029, or 0.000888?

Perform the indicated operations.

5. $7.02 + 3.5 + 11$

6. $2.37 + 5.0038$

7. $13.79 - 2.1$

8. $9 - 2.7 + 3.51$

9. $8.3 \times 1{,}000$

10. 8.01×2.3

11. $(0.12)^2$

12. $5 + 3 \times 0.7$

13. $6.05 \div 1{,}000$

14. $\frac{9.81}{0.3}$

Express as a decimal.

15. $\frac{7}{8}$

16. $2\frac{5}{6}$, rounded to the nearest hundredth

Solve.

17. In a science course, a student learns that an acid is stronger if it has a lower pH value. Which is stronger, an acid with a pH value of 3.7 or an acid with a pH value of 2.95?

18. In 2005, Microsoft Corporation had quarterly revenues of \$9.189 billion, \$10.818 billion, \$9.62 billion, and \$10.161 billion. What was Microsoft's total revenue that year? (***Source:*** Microsoft Corporation, 2005 Annual Report)

19. A serving of iceberg lettuce contains 3.6 milligrams (mg) of vitamin C, whereas romaine lettuce contains 11.9 mg of vitamin C. Romaine lettuce is how many times as rich in vitamin C as iceberg lettuce? Round the answer to the nearest whole number. (***Source:*** *The Concise Encyclopedia of Foods and Nutrition*)

20. Suppose that a long-distance telephone call costs \$0.85 for the first 3 minutes and \$0.17 for each additional minute. What is the cost of a 20-minute call?

• *Check your answers on page A-5.*

3.1 Introduction to Decimals

OBJECTIVES

- To read and write decimals
- To find the fraction equivalent to a decimal
- To compare decimals
- To round decimals
- To solve word problems involving decimals

What Decimals Are and Why They Are Important

Decimal notation is in common use. When we say that the price of a book is $32.75, that the length of a table is 1.8 meters, or that the answer displayed on a calculator is 5.007, we are using decimals.

A number written as a **decimal** has

- a whole-number part, which *precedes* the decimal point, and
- a fractional part, which *follows* the decimal point.

Whole-number part ↓ Fractional part ↓

4.51

↑ Decimal point

A decimal without a decimal point shown is understood to have one at the right end and is the same as a whole number. For instance, 3 and 3. are the same number.

The fractional part of any decimal has as its denominator a power of 10, such as 10, 100, or 1,000. The use of the word *decimal* reminds us of the importance of the number 10 in this notation, just as decade means 10 years or December meant the 10th month of the year (which it was for the early Romans).

In many problems, we can choose to work with either decimals or fractions. Therefore, we need to know how to work with both if we are to use the easier approach to solve a particular problem.

Decimal Places

Each digit in a decimal has a place value. The place value system for decimals is an extension of the place value system for whole numbers.

The places to the right of the decimal point are called **decimal places**. For instance, the number 64.149 is said to have three decimal places.

Recall that, for a whole number, place values are powers of 10: 1, 10, 100, ... By contrast, each place value for the fractional part of a decimal is the reciprocal of a power of 10: $\frac{1}{10}, \frac{1}{100}, \frac{1}{1{,}000}, \ldots$.

The first decimal place after the decimal point is the tenths place. Working to the right, the next decimal places are the hundredths place, the thousandths place, and so on.

The following table shows the place values in the decimals 0.54 and 0.30716.

Ones	.	Tenths	Hundredths	Thousandths	Ten-thousandths	Hundred-thousandths
0	.	5	4			
0	.	3	0	7	1	6

The next table shows the place values for the decimals 7,204.5 and 513.285.

Thousands	Hundreds	Tens	One	.	Tenths	Hundredths	Thousandths
7	2	0	4	.	5		
	5	1	3	.	2	8	5

EXAMPLE 1

In each number, identify the place that the digit 3 occupies.

a. 0.134 **b.** 92.388 **c.** 0.600437

Solution

a. The hundredths place

b. The tenths place

c. The hundred-thousandths place

PRACTICE 1

What place does the digit 1 occupy in each number?

a. 566.184

b. 43.57219

c. 0.921

Changing Decimals to Fractions

Knowing the place value system is the key to understanding what decimals mean, how to read them, and how to write them.

- The decimal 0.9, or .9, is another way of writing $(0 \times 1) + \left(9 \times \frac{1}{10}\right)$, or $\frac{9}{10}$.
  This decimal is read the same as the equivalent fraction: "nine tenths."
- The decimal 0.21 represents 2 tenths + 1 hundredth. This expression simplifies to the following:

$$\left(2 \times \frac{1}{10}\right) + \left(1 \times \frac{1}{100}\right) = \frac{2}{10} + \frac{1}{100} = \frac{20}{100} + \frac{1}{100}, \text{ or } \frac{21}{100}$$

  So 0.21 is read "twenty-one hundredths."

- The decimal 0.149 stands for $\frac{149}{1,000}$.

$$\left(1 \times \frac{1}{10}\right) + \left(4 \times \frac{1}{100}\right) + \left(9 \times \frac{1}{1,000}\right) = \frac{149}{1,000}$$

  So 0.149 is read "one hundred forty-nine thousandths."

Let's summarize these examples.

Decimal	Equivalent Fraction	Read as
0.9	$\frac{9}{10}$	Nine tenths
0.21	$\frac{21}{100}$	Twenty-one hundredths
0.149	$\frac{149}{1,000}$	One hundred forty-nine thousandths

Note that in each of these decimals, the fractional part is the same as the numerator of the equivalent fraction: $0.149 = \frac{149}{1,000}$.

We can use the following rule to rewrite any decimal as a fraction or a mixed number.

To Change a Decimal to the Equivalent Fraction or Mixed Number

- Copy the nonzero whole-number part of the decimal and drop the decimal point.
- Place the fractional part of the decimal in the numerator of the equivalent fraction.
- Make the denominator of the equivalent fraction 1 followed by as many zeros as the decimal has decimal places.
- Simplify the resulting fraction, if possible.

EXAMPLE 2

Express 0.75 in fractional form and simplify.

Solution The whole-number part of the decimal is 0. We drop the decimal point. The fractional part (75) of the decimal becomes the numerator of the equivalent fraction. Since the decimal has two decimal places, we make the denominator of the equivalent fraction 1 followed by two zeros (100). So we can write 0.75 as $\frac{75}{100}$, which simplifies to $\frac{3}{4}$.

PRACTICE 2

Write 0.875 as a fraction reduced to lowest terms.

EXAMPLE 3

Express 1.87 as a mixed number.

Solution This decimal is equivalent to a mixed number whose whole-number part is 1. The fractional part (87) of the decimal is the numerator of the equivalent fraction. The decimal has two decimal places, so the fraction's denominator has two zeros (that is, it is 100).

$$1.87 = 1\frac{87}{100}$$

Do you see that the answer can also be written as $\frac{187}{100}$?

PRACTICE 3

The decimal 2.03 is equivalent to what mixed number?

EXAMPLE 4

Find the equivalent fraction of each decimal.

a. 3.2 **b.** 3.200

Solution

a. 3.2 represents $3\frac{2}{10}$, or $3\frac{1}{5}$.

b. 3.200 equals $3\frac{200}{1,000}$, or $3\frac{1}{5}$.

PRACTICE 4

Express each decimal in fractional form.

a. 5.6 **b.** 5.6000

Tip Adding zeros in the rightmost decimal places does not change a decimal's value. However, generally decimals can be written without these extra zeros.

EXAMPLE 5

Write each decimal as a mixed number.

a. 1.309 **b.** 1.39

Solution

a. $1.309 = 1\frac{309}{1,000}$

b. $1.39 = 1\frac{39}{100}$

PRACTICE 5

What mixed number is equivalent to each decimal?

a. 7.003 **b.** 4.1

Knowing how to change a decimal to its equivalent fraction also helps us read the decimal.

EXAMPLE 6

Express each decimal in words.

a. 0.319 **b.** 2.71 **c.** 0.08

Solution

a. $0.319 = \frac{319}{1,000}$ We read the decimal as "three hundred nineteen thousandths."

b. $2.71 = 2\frac{71}{100}$ The decimal point is read as "and." We read the decimal as "two and seventy-one hundredths."

c. $0.08 = \frac{8}{100}$ We read the decimal as "eight hundredths."
Note that we *do not simplify* the equivalent fraction when reading the decimal.

PRACTICE 6

Express each decimal in words.

a. 0.61

b. 4.923

c. 7.05

EXAMPLE 7

Write each number in decimal notation.

a. Seven tenths **b.** Five and thirty-two thousandths

Solution

a. Since 7 is in the tenths place, the decimal is written as 0.7.

b.

The whole number preceding *and* is in the ones place. → 5

The last digit of 32 is in the thousandths place. → 2

5.0 3 2

We replace *and* with the decimal point. → .

We need a 0 to hold the tenths place. → 0

The answer is 5.032.

PRACTICE 7

Write each number in decimal notation.

a. Forty-three thousandths

b. Ten and twenty-six hundredths

EXAMPLE 8

For hay fever, an allergy sufferer takes a decongestant pill that has a tablet strength of three hundredths of a gram. Write the equivalent decimal.

Solution "Three hundredths" is written 0.03, with the digit 3 in the hundredths place.

PRACTICE 8

The number pi (usually written π) is approximately three and fourteen hundredths. Write this approximation as a decimal.

Comparing Decimals

Suppose that we want to compare two decimals—say, 0.6 and 0.7. The key is to rethink the problem in terms of fractions.

$$0.6 = \frac{6}{10} \qquad 0.7 = \frac{7}{10}$$

Because $\frac{7}{10} > \frac{6}{10}$, $0.7 > 0.6$.

The following procedure provides another way to compare decimals that is faster than converting the decimals to fractions.

To Compare Decimals

- Rewrite the numbers vertically, lining up the decimal points.
- Working from left to right, compare the digits that have the same place value. The decimal which has the largest digit with this place value is the largest decimal.

EXAMPLE 9

Which is larger, 0.729 or 0.75?

Solution First let's line up the decimal point.

0.729
0.75

We see that both decimals have a 0 in the ones place. We next compare the digits in the tenths place and see that, again, they are the same. Looking to the right in the hundredths place, we see that $5 > 2$. Therefore, $0.75 > 0.729$. Note that the decimal with more digits is not necessarily the larger decimal.

PRACTICE 9

Which is smaller, 0.83 or 0.8297?

EXAMPLE 10

Rank from smallest to largest: 2.17, 2.1, and 0.99

Solution First, we position the decimals so that the decimal points are aligned.

2.17
2.1
0.99

PRACTICE 10

Rewrite in decreasing order: 3.5, 3.51, and 3.496

Working from left to right, we see that in the ones place, the first two decimals have a 2 and the third decimal has a 0, so the third decimal is the smallest of the three. To decide which of the first two decimals is smaller, we compare the digits in the tenths place. Since both of these decimals have a 1 in the tenths place, we proceed to the hundredths place. A 0 is understood to the right of the 1 in 2.1, so we compare 0 and 7.

$$\begin{array}{l} 2.17 \\ 2.10 \\ 0.99 \end{array}$$

Since $0 < 7$, we conclude that $2.10 < 2.17$. Therefore, the three decimals from smallest to largest are 0.99, 2.1, and 2.17.

EXAMPLE 11

Plastic garbage bags come in three thicknesses (or gauges): 0.003 inch, 0.0025 inch, and 0.002 inch. The three gauges are called lightweight, regular weight, and heavyweight. Which is the lightweight gauge?

Solution To find the smallest of the decimals, we first line up the decimal points.

$$\begin{array}{l} 0.003 \\ 0.0025 \\ 0.002 \end{array}$$

Working from left to right, we see that the three decimals have the same digits until the thousandths place, where $3 > 2$. Therefore, 0.003 must be the heavyweight gauge. To compare 0.0025 and 0.002, we look at the ten-thousandths place. The 5 is greater than the 0 that is understood to be there. So 0.0025 inch must be the regular-weight gauge, and 0.002 the lightweight gauge.

PRACTICE 11

The higher the energy efficiency rating (EER) of an air conditioner, the more efficiently it uses electricity. Which of the following air conditioners is least efficient with EER ratings shown to be 8.2, 9, and 8.1?

(*Source: Consumer Guide*)

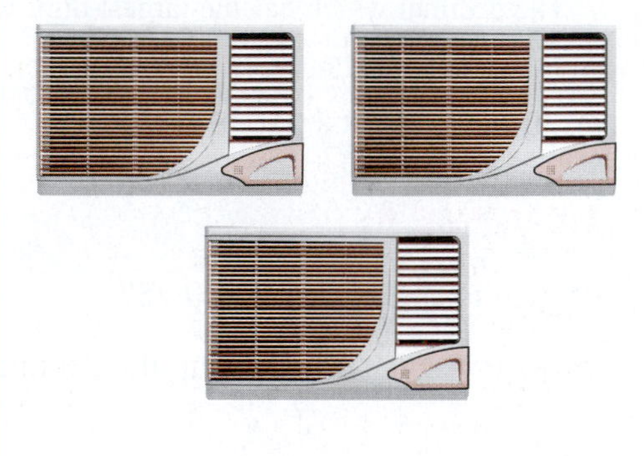

Rounding Decimals

As with whole numbers, we can round decimals to a given place value. For instance, suppose that we want to round the decimal 1.38 to the nearest tenth. The decimal 1.38 lies between 1.3 and 1.4, so one of these two numbers will be our answer—but which? To decide, let's take a look at a number line.

1.31 1.33 1.35 1.37 1.39
1.3 1.32 1.34 1.36 1.38 1.4

Do you see from this diagram that 1.38 is closer to 1.4 than to 1.3?

$$1.38 \approx 1.4$$

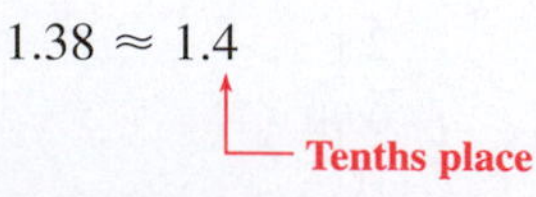

Rounding a decimal to the nearest tenth means that the last digit lies in the tenths place.

The following table shows the relationship between the place to which we are rounding and the number of decimal places in our answer.

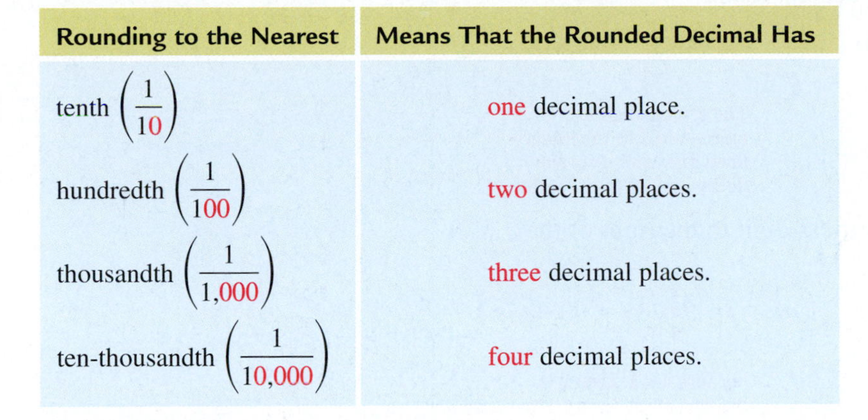

Rounding to the Nearest	Means That the Rounded Decimal Has
tenth $\left(\frac{1}{10}\right)$	one decimal place.
hundredth $\left(\frac{1}{100}\right)$	two decimal places.
thousandth $\left(\frac{1}{1{,}000}\right)$	three decimal places.
ten-thousandth $\left(\frac{1}{10{,}000}\right)$	four decimal places.

Note that the number of decimal places is the same as the number of zeros in the corresponding denominator.

The following rule can be used to round decimals.

To Round a Decimal to a Given Decimal Place

- Underline the digit in the place to which the number is being rounded.
- Look at the digit to the right of the underlined digit—the critical digit. If this digit is 5 or more, add 1 to the underlined digit; if it is less than 5, leave the underlined digit unchanged.
- Drop all decimal places to the right of the underlined digit.

Let's apply this rule to the problem that we just considered—namely, rounding 1.38 to the nearest tenth.

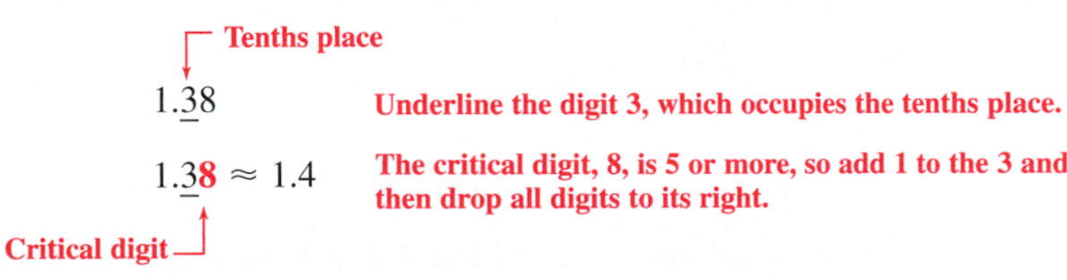

The following examples illustrate this method of rounding.

EXAMPLE 12

Round 94.735 to

a. the nearest tenth.

b. two decimal places.

c. the nearest thousandth.

d. the nearest ten.

e. the nearest whole number.

PRACTICE 12

Round 748.0772 to

a. the nearest tenth.

b. the nearest hundredth.

c. three decimal places.

d. the nearest whole number.

e. the nearest hundred.

Solution

a. First, we underline the digit 7 in the tenths place: 94.735. The critical digit, 3, is less than 5, so we do not add 1 to the underlined digit. Dropping all digits to the right of the 7, we get 94.7. Note that our answer has only one decimal place because we are rounding to the nearest tenth.

b. We need to round 94.735 to two decimal places (to the nearest hundredth).

$94.735 \approx 94.74$

The critical digit is 5 or more. Add 1 to the underlined digit and drop the decimal place to the right.

c. $94.735 \approx 94.735$ because the critical digit to the right of the 5 is understood to be 0.

d. We are rounding 94.735 to the nearest 10 (*not tenth*), which is a whole-number place.

$94.735 \approx 90$

Because 4 < 5, keep 9 in the tens place, insert 0 in the ones place and drop all decimal places.

e. Rounding to the nearest whole number means rounding to the nearest 1.

$94.735 \approx 95$

Because 7 > 5, change the 4 to 5 and drop all decimal places.

EXAMPLE 13

Round 3.982 to the nearest tenth.

Solution First, we underline the digit 9 in the tenths place and identify the critical digit: 3.982. The critical digit, 8, is more than 5, so we add 1 to the 9, get 10, and write down the 0. We add the carried 1 to 3 getting 4, and drop the 8 and 2.

$3.982 \approx 4.0$

Drop

The answer is 4.0. Note that we do not drop the 0 in the tenths place of the answer to indicate that we have rounded to that place.

PRACTICE 13

Round 7.2962 to two decimal places.

EXAMPLE 14

On the commodities market, prices are often quoted in terms of thousandths or even ten-thousandths of a dollar. Suppose that the commodities market price of a pound of coffee is \$2.0883. What is this price to the nearest cent?

Solution A cent is one-hundredth of a dollar. Therefore, we need to round 2.0883 to the nearest hundredth.

$2.0883 \approx 2.09$

The price to the nearest cent is \$2.09.

PRACTICE 14

Mount Waialeale on the Hawaiian island of Kauai is one of the world's wettest places, with an average annual rainfall of 11.43 meters. What is the amount of rainfall to the nearest tenth of a meter?

Mathematically Speaking

Fill in each blank with the most appropriate term or phrase from the given list.

less	greater	increasing	left
decreasing	ten	hundredths	multiple
thousandths	power	right	tenth

1. A decimal place is a place to the _______ of the decimal point.

2. The fractional part of a decimal has as its denominator a _______ of 10.

3. The decimal 0.17 is equivalent to the fraction seventeen _______.

4. The decimal 209.95 rounded to the nearest _______ is 210.0.

5. The decimal 0.371 is _______ than the decimal 0.3499.

6. The decimals 0.48, 0.4, and 0.371 are written in _______ order.

Underline the digit that occupies the given place.

7. 2.78 Tenths place

8. 9.01 Hundredths place

9. 2.00175 Ten-thousandths place

10. 6.835 Tenths place

11. 358.02 Tens place

12. 823.001 Thousandths place

13. 0.772 Hundredths place

14. 135.83 Hundreds place

Identify the place occupied by the underlined digit.

15. $25.\underline{7}1$

16. $3.00\underline{2}$

17. $8.1\underline{8}3$

18. $\underline{4}9.771$

19. $1{,}077.04\underline{2}$

20. $2.8371\underline{0}7$

21. $\$25\underline{3}.72$

22. $\$7{,}571.3\underline{9}$

For each decimal, find the equivalent fraction or mixed number, reduced to lowest terms.

23. 0.6

24. 0.8

25. 0.39

26. 0.27

27. 1.5

28. 9.8

29. 8.000

30. 6.700

31. 5.012

32. 20.304

Write each decimal in words.

33. 0.53

34. 0.72

35. 0.305

36. 0.849

37. 0.6

38. 0.3

39. 5.72

40. 3.89

41. 24.002

42. 370.081

Write each number in decimal notation.

43. Eight tenths

44. Eleven hundredths

45. One and forty-one thousandths

46. Five and sixty-three hundredths

47. Sixty and one hundredth

48. Eighteen and four thousandths

49. Four and one hundred seven thousandths

50. Ninety-two and seven hundredths

51. Three and two tenths meters

52. Ninety-eight and six tenths degrees

Between each pair of numbers, insert the appropriate sign, $<$, $=$, or $>$, to make a true statement.

53. 3.21   2.5

54. 8.66   4.952

55. 0.71   0.8

56. 1.2   1.38

57. 9.123   9.11

58. 0.5   0.52

59. 4   4.000

60. 7.60   7.6

61. 8.125 feet   8.2 feet

62. 2.45 pounds   2.5 pounds

Rearrange each group of numbers from smallest to largest.

63. 7.1, 7, 7.07

64. 0.002, 0.2, 0.02

65. 5.001, 4.9, 5.2

66. 3.85, 3.911, 2

67. 9.6 miles, 9.1 miles, 9.38 miles

68. 2.7 seconds, 2.15 seconds, 2 seconds

Round as indicated.

69. 17.36 to the nearest tenth

70. 8.009 to two decimal places

71. 3.5905 to the nearest thousandth

72. 3.5902 to the nearest thousandth

73. 37.08 to one decimal place

74. 3.08 to the nearest whole number

75. 0.396 to the nearest hundredth

76. 0.978 to the nearest tenth

77. 7.0571 to two decimal places

78. 3.038 to one decimal place

79. 8.7 miles to the nearest mile

80. $35.75 to the nearest dollar

Round to the indicated place.

81.

To the Nearest	8.0714	0.9916
Tenth		
Hundredth		
Ten		

82.

To the Nearest	0.8166	72.3591
Tenth		
Hundredth		
Ten		

Mixed Practice

Solve.

83. In the decimal 0.024, underline the digit in the tenths place.

84. What is the equivalent fraction of 3.8?

85. Round 870.062 to the nearest hundredth.

86. Write four and thirty-one thousandths in decimal notation.

87. Write in increasing order: 2.14 meters, 2.4 meters, and 2.04 meters.

88. Write 0.05 in words.

Applications

The following statements involve decimals. Write all decimals in words.

89. It takes the Earth 23.934 hours to rotate once about its axis. (*Source:* NASA)

90. A chemistry text gives 55.847 as the atomic weight of iron.

91. Over two years, the average score on a college admissions exam increased from 18.7 to 18.8.

92. Male Rufous hummingbirds weigh an average of 0.113 ounces and have a wingspan of 4.25 inches. (*Source:* Lanny Chambers, *Facts about Hummingbirds*)

93. The following table shows the number of people per square mile living on various continents in a recent year.

Continent	Number of People per Square Mile
Asia	301.3
Africa	55.9
Europe	268.2
North America	46.6
South America	43.6

94. The coefficient of friction is a measure of the amount of friction produced when one surface rubs against another. The following table gives these coefficients for various surfaces.

Materials	Coefficient of Friction
Wood on wood	0.3
Steel on steel	0.15
Steel on wood	0.5
A rubber tire on a dry concrete road	0.7
A rubber tire on a wet concrete road	0.5

(*Source:* *CRC Handbook of Chemistry and Physics*)

95. Bacteria are single-celled organisms that typically measure from 0.00001 inch to 0.00008 inch across.

96. In one month, the consumer confidence index rose from 71.9 points to 80.2 points.

Write each number in decimal notation.

97. The area of a plot of land is one and two tenths acres.

98. The lead in many mechanical pencils is seven tenths millimeter thick.

99. At the first Indianapolis 500 auto race in 1911, the winning speed was seventy-four and fifty-nine hundredths miles per hour. (***Source:*** Jack Fox, *The Indianapolis 500*)

100. According to the owner's manual, the voltage produced by a camcorder battery is nine and six tenths volts (V).

101. At sea level, the air pressure on each square inch of surface area is fourteen and seven tenths pounds.

102. A doctor prescribed a dosage of one hundred twenty-five thousandths milligram of Prolixin.

103. In 1796, there was a U.S. coin in circulation, the half cent, worth five thousandths of a dollar.

104. In preparing an injection, a nurse measured out one and eight tenths milliliters of sterile water.

105. The electrical usage in a tenant's apartment last month amounted to three hundred fifty-two and one tenth kilowatt hours (kWh).

106. In one day, the Dow Jones Industrial Average fell by three and sixty-three hundredths points.

Solve.

107. A jury awarded a plaintiff \$1.85 million. Was this award more or less than the \$2.1 million that the plaintiff had demanded?

108. The more powerful an earthquake is, the higher its magnitude is on the Richter scale. Great earthquakes, such as the 1906 San Francisco earthquake, have magnitudes of 8.0 or higher. Is an earthquake with magnitude 7.8 considered to be a great earthquake? (***Source:*** *The New Encyclopedia Britannica*)

109. Last winter, a homeowner's average daily heating bill was for 8.75 units of electricity. This winter, it was for 8.5 units. During which winter was the average higher?

110. Suppose that in order to qualify for the dean's list at a community college, a student's grade point average (GPA) must be 3.5 or above. Did a student with a GPA of 3.475 make the dean's list?

111. A person with reasonably good vision can see objects as small as 0.0004 inch long. Can such a person see a mite—a tiny bug that is 0.003 inch long?

112. At the 2006 Winter Olympics, the two top skaters in the 500-meter women's speed skating event finished in 76.57 seconds and 76.78 seconds. Which time was better? (***Source:*** www.nbcolympics.com)

113. As part of her annual checkup, a patient had a blood test. The normal range for a particular substance is 1.1 to 2.3. If she scored 0.95, was her blood in the normal range?

114. Last year, an electronics factory released 1.5 million pounds of toxic gas into the air. During the same time, a food factory and a chemical factory released 1.4 and 1.48 million pounds of toxic emissions, respectively. Which of the three factories was the worst polluter?

Round to the indicated place.

115. A bank pays interest on all its accounts to the nearest cent. If the interest on an account is $57.0285, how much interest does the bank pay?

116. A city's sales tax rate, expressed as a decimal, is 0.0825. What is this rate to the nearest hundredth?

117. According to the organizers of a lottery, the probability of winning the lottery is 0.0008. Round this probability to three decimal places.

118. One day last week, a particular foreign currency was worth 0.7574 U.S. dollars ($US). How much is this currency worth to the nearest tenth of a dollar?

119. According to a recent survey, the cost of medical care is 1.77 times what it was a decade ago. Express this decimal to the nearest tenth.

120. The length of the Panama Canal is 50.7 miles. Round this length to the nearest mile. (***Source:*** *The New Encyclopedia Britannica*)

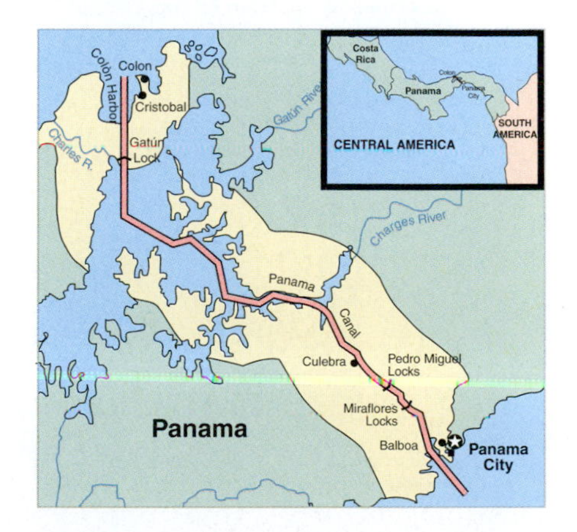

• *Check your answers on page A-5.*

MINDSTRETCHERS

Critical Thinking

1. For each question, either give the answer or explain why there is none.

a. Find the *smallest* decimal that when rounded to the nearest tenth is 7.5.

b. Find the *largest* decimal that when rounded to the nearest tenth is 7.5.

Writing

2. The next whole number after 7 is 8. What is the next decimal after 0.7? Explain.

Groupwork

3. Working with a partner, list fifteen numbers between 2.5 and 2.6.

CULTURAL NOTE

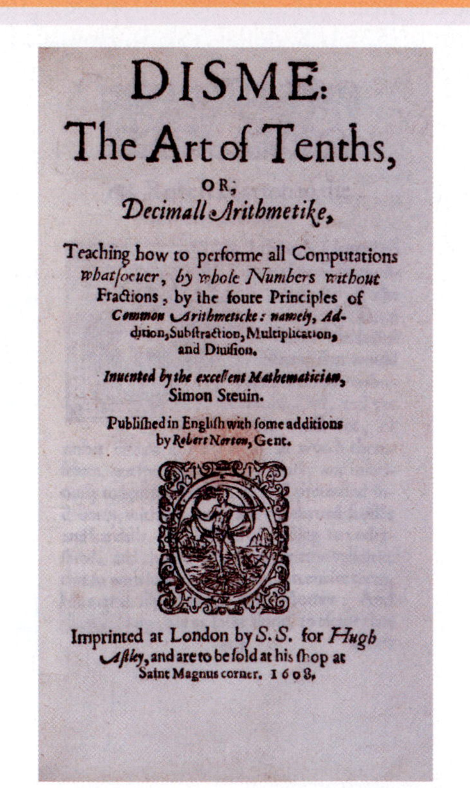

DISME:
The Art of Tenths,
OR,
Decimall Arithmetike,
Teaching how to performe all Computations *whatſoeuer, by whole Numbers without* Fractions, by the foure Principles of *Common Arithmeticke: namely,* Addition, Subſtraction, Multiplication, and Diuiſion.
Inuented by the excellent Mathematician, Simon Steuin.
Publiſhed in Engliſh with ſome additions by *Robert Norton*, Gent.
Imprinted at London by *S. S.* for *Hugh Aſtley*, and are to be ſold at his ſhop at Saint Magnus corner. 1608.

In 1585, Simon Stevin, a Dutch engineer, published a book entitled *The Art of Tenths* (*La Disme* in French) in which he presented a thorough account of decimals. Stevin sought to teach everyone "with an ease unheard of, all computations necessary between men by integers without fractions."

Stevin did not invent decimals; their history dates back thousands of years to ancient China, medieval Arabia, and Renaissance Europe. However, Stevin's writings popularized decimals and also supported the notion of decimal coinage—as in American currency, where there are 10 dimes to the dollar.

Source: Morris Kline, *Mathematics, a Cultural Approach* (Reading, Massachusetts: Addison-Wesley Publishing Company, 1962), p. 614.

3.2 Adding and Subtracting Decimals

OBJECTIVES

- To add and subtract decimals
- To estimate the sum and difference of decimals
- To solve word problems involving the addition or subtraction of decimals

In Section 3.1 we discussed the meaning of decimals and how to compare and round them. Now we turn our attention to computing with decimals, starting with addition and subtraction.

Adding Decimals

Adding decimals is similar to adding whole numbers: We add the digits in each place value position, carrying when necessary. Suppose that we want to find the sum of two decimals: 1.2 + 3.5. First, we rewrite the problem vertically, lining up the decimal points in the addends. Then we add as usual, inserting the decimal point below the other decimal points.

$$\begin{array}{r} 1.2 \\ +3.5 \\ \hline 4.7 \end{array} \quad \text{This addition is equivalent to} \quad \begin{array}{r} 1\frac{2}{10} \\ +3\frac{5}{10} \\ \hline 4\frac{7}{10} \end{array}$$

Note that, when we added the mixed numbers corresponding to the decimals, we got $4\frac{7}{10}$, which is equivalent to 4.7. This example suggests the following rule.

To Add Decimals

- Rewrite the numbers vertically, lining up the decimal points.
- Add.
- Insert a decimal point in the sum below the other decimal points.

EXAMPLE 1

Find the sum: 2.7 + 80.13 + 5.036

Solution

$$\begin{array}{r} 2.7 \\ 80.13 \\ +5.036 \\ \hline 87.866 \end{array}$$

Rewrite the addends with decimal points lined up vertically.

Add.

Insert the decimal point in the sum.

PRACTICE 1

Add: 5.12 + 4.967 + 0.3

EXAMPLE 2

Compute: 2.367 + 5 + 0.143

Solution Recall that 5 and 5. are equivalent.

$$\begin{array}{r} 2.367 \\ 5. \\ +\ 0.143 \\ \hline 7.510 \end{array} = 7.51$$

Line up the decimal points and add.

Insert the decimal point in the sum.

We can drop the extra 0 at the right end.

PRACTICE 2

What is the sum of 7.31, 8, and 23.99?

EXAMPLE 3

A runner's time was 0.06 second longer than the world record of 21.71 seconds. What was the runner's time?

Solution We need to compute the sum of the two numbers. The runner's time was 21.77 seconds.

$$\begin{array}{r} 0.06 \\ +\ 21.71 \\ \hline 21.77 \end{array}$$

PRACTICE 3

A child has the flu. This morning, his body temperature was 99.4°F. What was his temperature after it went up by 2.7°?

Subtracting Decimals

Now let's discuss subtracting decimals. As with addition, subtracting decimals is similar to subtracting whole numbers. To compute the difference between 12.83 and 4.2, we rewrite the problem vertically, lining up the decimal points. Then we subtract as usual, inserting a decimal point below the other decimal points.

$$\begin{array}{r} 12.83 \\ -4.2 \\ \hline 8.63 \end{array} \quad \text{is equivalent to} \quad \begin{array}{rcr} 12\frac{83}{100} & = & 12\frac{83}{100} \\ -4\frac{2}{10} & = & -4\frac{20}{100} \\ \hline & & 8\frac{63}{100}, \text{ or } 8.63 \end{array}$$

Again, note that when we subtracted the equivalent mixed numbers, we got the same difference.

As in any subtraction problem, we can check this answer by adding the subtrahend (4.2) to the difference (8.63), confirming that we get the original minuend (12.83). This example suggests the following rule.

$$\begin{array}{r} 8.63 \\ +4.2 \\ \hline 12.83 \end{array}$$

To Subtract Decimals

- Rewrite the numbers vertically, lining up the decimal points.
- Subtract, inserting extra zeros in the minuend if necessary for borrowing.
- Insert a decimal point in the difference, below the other decimal points.

EXAMPLE 4

Subtract and check: 5.038 − 2.11

Solution

$$\begin{array}{r} 5.038 \\ -2.11 \\ \hline 2.928 \end{array}$$

Rewrite the problem with decimal points lined up vertically.

Subtract. Borrow when necessary.

Insert the decimal point in the answer.

Check To verify that our difference is correct, we check by addition.

$$\begin{array}{r} 2.928 \\ +2.11 \\ \hline 5.038 \end{array}$$

PRACTICE 4

Find the difference and check:
71.3825 − 25.17

EXAMPLE 5

65 is how much larger than 2.04?

Solution Recall that 65 and 65. are equivalent.

Insert zeros needed for borrowing.

$$\begin{array}{r} 65.00 \\ -2.04 \\ \hline 62.96 \end{array}$$

Line up the decimal points.

Subtract.

Insert the decimal point in the answer.

Check

$$\begin{array}{r} 62.96 \\ +\ 2.04 \\ \hline 65.00 = 65 \end{array}$$

PRACTICE 5

How much greater is \$735 than \$249.57?

EXAMPLE 6

A McDonald's hamburger contains 0.53 grams of sodium, whereas a McDonald's cheeseburger contains 0.74 grams. How much more sodium does the cheeseburger contain? (***Source:*** McDonald's)

Solution We need to find the difference between 0.74 and 0.53.

$$\begin{array}{r} 0.74 \\ -0.53 \\ \hline 0.21 \end{array}$$

So a McDonald's cheeseburger contains 0.21 grams more sodium.

PRACTICE 6

A swimmer is competing in the 28.5 mile swim around the island of Manhattan. After swimming 15 miles, how much farther does she have to go?

EXAMPLE 7

Suppose that a part-time employee's salary is \$350 a week, less deductions. The following table shows these deductions.

Deduction	Amount
Federal, state, and city taxes	\$100.80
Social Security	13.50
Union dues	8.88

What is the employee's take-home pay?

Solution Let's use the strategy of breaking the question into two simpler questions.

- *How much money is deducted per week?* The weekly deductions (\$100.80, \$13.50, and \$8.88) add up to \$123.18.
- *How much of the salary is left after subtracting the total deductions?* The difference between \$350 and \$123.18 is \$226.82, which is the employee's take-home pay.

PRACTICE 7

A sales rep, working in Ohio, wants to drive from Circleville to Columbus. How much shorter is it to drive directly to Columbus instead of going by way of Lancaster?

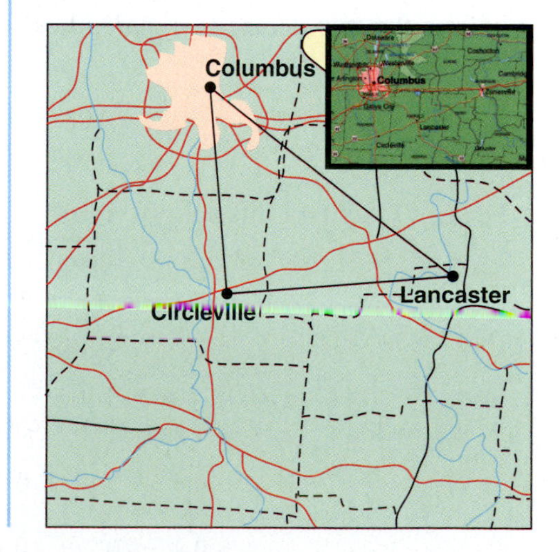

Estimating Sums and Differences

Being able to estimate in your head the sum or difference between two decimals is a useful skill, if only for checking an exact answer. To estimate, simply round the numbers to be added or subtracted and then carry out the operation on the rounded numbers.

EXAMPLE 8

Compute the sum $0.17 + 0.4 + 0.083$. Use estimation to check.

Solution First, we add. Then, to check, we round the addends—say, to the nearest tenth—and add the rounded numbers.

$$\begin{array}{rcr} 0.17 & \approx & 0.2 \\ 0.4 & \approx & 0.4 \\ +0.083 & \approx & +0.1 \\ \hline 0.653 & & 0.7 \end{array}$$

Exact sum → 0.653; 0.7 ← Estimated sum

Our exact sum is close to our estimated sum, and in fact, rounds to it. So we can have confidence that our answer, 0.653, is correct.

PRACTICE 8

Add 0.093, 0.008, and 0.762. Then check by estimating.

EXAMPLE 9

Subtract $0.713 - 0.082$. Then check by estimating.

Solution First we find the exact answer and then round the given numbers to get an estimate.

$$\begin{array}{rcr} 0.713 & \approx & 0.7 \\ -0.082 & \approx & -0.1 \\ \hline 0.631 & & 0.6 \end{array}$$

Exact difference → 0.631; 0.6 ← Estimated difference

Our exact answer, 0.631, is close to 0.6, so we can feel confident that it is correct.

PRACTICE 9

Compute: $0.17 - 0.091$. Use estimation to check.

EXAMPLE 10

Combine and check: $0.4 - (0.17 + 0.082)$

Solution Following the order of operations rule, we begin by adding the two decimals in parentheses.

$$\begin{array}{r} 0.17 \\ +0.082 \\ \hline 0.252 \end{array}$$

Next, we subtract this sum from 0.4.

$$\begin{array}{r} 0.400 \\ -0.252 \\ \hline 0.148 \end{array}$$

So $0.4 - (0.17 + 0.082) = 0.148$.

Now let's check this answer by estimating:

$$\begin{array}{l} 0.4 - (0.17 + 0.082) \\ \downarrow \qquad \downarrow \qquad \downarrow \\ 0.4 - (0.2 + 0.1) = 0.4 - 0.3 = 0.1 \end{array}$$

The estimate, 0.1, is sufficiently close to 0.148 to confirm our answer.

PRACTICE 10

Calculate and check:
$0.813 - (0.29 - 0.0514)$

EXAMPLE 11

A movie budgeted at \$7.25 million ended up costing \$1.655 million more. Estimate the final cost of the movie.

Solution Let's round each number to the nearest million dollars.

$$
\begin{aligned}
1.655 &\approx 2 \text{ million} \\
7.25 &\approx \underline{+7 \text{ million}} \\
& 9 \text{ million}
\end{aligned}
$$

Adding the rounded numbers, we see that the movie cost approximately \$9 million.

PRACTICE 11

From the deposit ticket shown below, estimate the total amount deposited.

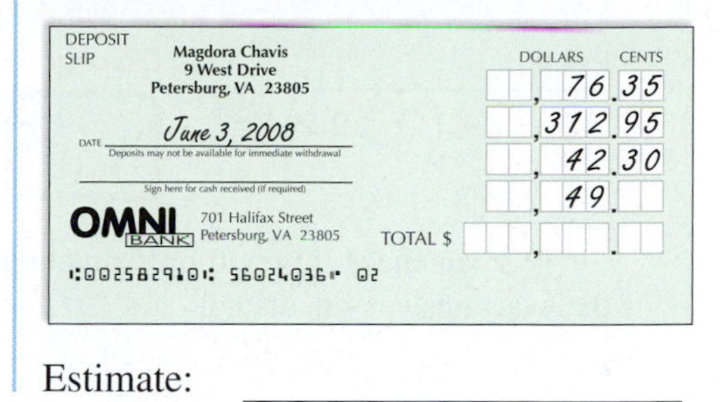
DEPOSIT SLIP
Magdora Chavis
9 West Drive
Petersburg, VA 23805
DATE June 3, 2008
Deposits may not be available for immediate withdrawal
Sign here for cash received (if required)
OMNI BANK 701 Halifax Street Petersburg, VA 23805

DOLLARS	CENTS
76	35
312	95
42	30
49	
TOTAL \$	

Estimate: ________________

EXAMPLE 12

When the underwater tunnel connecting the United Kingdom and France was built, French and British construction workers dug from their respective countries. They met at the point shown on the map.

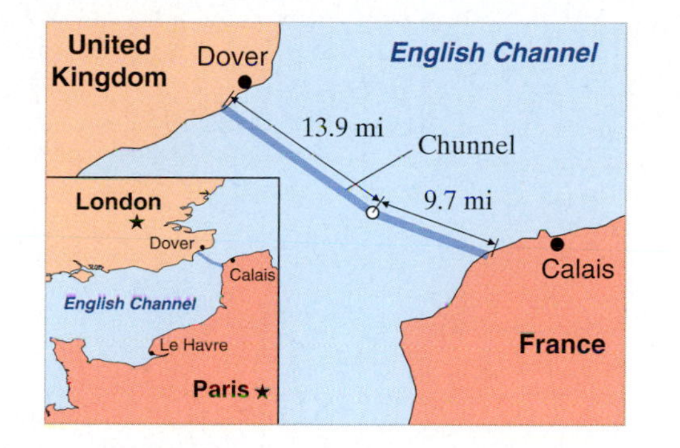

Estimate how much farther the British workers had dug than the French workers. (***Source:*** *The New York Times*)

Solution We can round 13.9 to 14 and 9.7 to 10. The difference between 14 and 10 is 4, so the British workers dug about 4 miles farther than the French workers.

PRACTICE 12

An art collector bought a painting for \$2.3 million. A year later, she sold the painting for \$4.1 million. Estimate her profit on the sale.

Adding and Subtracting Decimals on a Calculator

When adding or subtracting decimals, press the [.] key to enter the decimal point. If a sum or difference ends with a 0 in the rightmost decimal place, does your calculator drop the 0? If a sum or difference has no whole-number part, does your calculator insert a 0?

EXAMPLE 13

Compute: 2.7 + 4.1 + 9.2

Solution

Press	Display
2.7 [+] 4.1 [+] 9.2 [ENTER =]	2.7 + 4.1 + 9.2 16.

We can check this sum by estimating: 3 + 4 + 9 = 16, which is the exact answer calculated.

PRACTICE 13

Find the sum: 3.82 + 9.17 + 66.24

EXAMPLE 14

Find the difference: 83.71 − 83.70002

Solution

We can check this difference by adding: 0.00998 + 83.70002 = 83.71.

PRACTICE 14

Compute: 5.00003 − 5.00001

Mathematically Speaking

Fill in each blank with the most appropriate term or phrase from the given list.

sum	decimal points	difference
any number	rightmost digits	zeros

1. When adding decimals, rewrite the numbers vertically, lining up the _______.

2. Inserting _______ at the right end of a decimal does not change its value.

3. To estimate the _______ of 0.31 and 0.108, add 0.3 and 0.1.

4. To estimate the _______ between 0.31 and 0.108, subtract 0.1 from 0.3.

Find the sum. Check by estimating.

5. $3.89 + 5.44$ **6.** $2.17 + 4.29$ **7.** $0.6 + 0.3$ **8.** $12.7 + 3.9$

9. $6.03 + 2.1$ **10.** $0.4 + 3.96$ **11.** $13.05 + 8.4$ **12.** $3.922 + 5.1$

13. $2.67 + 5$ **14.** $8 + 4.99$ **15.** \$74 + \$3.21 **16.** \$8.77 + \$6

17. $0.49023 + 0.5997$ **18.** $1.002 + 0.20013$ **19.** $8.01 + 6.7 + 9.45$ **20.** $9.73 + 5.99 + 3.688$

21. $34.7 + 5.84 + 3 + 0.882$

22. $75.285 + 2 + 3.871 + 0.5$

23. 7 millimeters + 3.5 millimeters + 9.82 millimeters

24. 10.35 inches + 32 inches + 54.9 inches

25. 4.7 kilograms + 2.98 kilograms + 9.002 kilograms

26. 0.85 second + 1.72 seconds + 3.009 seconds

27. $3.861 + 2.89 + 3.775 + 9.00813 + 3.77182$

28. \$8.99 + \$3.99 + \$17.83 + \$15 + \$201.75

Find the difference. Check either by estimating or by adding.

29. $0.8 - 0.1$ **30.** $12.98 - 5.73$ **31.** $20.72 - 3.92$ **32.** $0.68 - 0.59$

33. $23.81 - 5.4$ **34.** $17.49 - 10.2$ **35.** $80.2 - 4.57$ **36.** $9.71 - 3.225$

37. $25.99 - 3.666$ **38.** $80.2 - 3.51$ **39.** $0.27 - 0.1$ **40.** $4.92 - 1.01$

41. $1.032 - 0.9178$ **42.** $0.01 - 0.0001$ **43.** $13.2 - 7$ **44.** $9.662 - 4$

45. $20 - 4.63$ **46.** $8 - 2.55$ **47.** $10 - 4.1$ **48.** $13 - 7.2$

49. $8 - 1.79$ **50.** $20 - 4.63$ **51.** 3.2 pounds − 1.35 pounds

52. 23.5 seconds − 2.8 seconds **53.** 103.7°F − 98.8°F **54.** 32.5 grams − 19.27 grams

Compute.

55. 35.2 − 2.86 + 9.07 − 1.658

56. 10 − 2.38 − 4.92 + 6.02

57. 30 milligrams − 0.5 milligram − 1.6 milligrams

58. \$20.93 + \$1.07 − \$19.58

59. 5.21 − (1.03 + 0.975)

60. 6.953 − (4.09 − 0.008)

61. 41.075 − 2.87104 − 17.005

62. 0.00661 + 1.997 − 0.05321

In each group of three computations, one answer is wrong. Use estimation to identify which answer is incorrect.

63. a. $\begin{array}{r} 0.059 \\ 0.00234 \\ +0.036 \\ \hline 0.09734 \end{array}$ **b.** $\begin{array}{r} 0.1903 \\ 0.074 \\ +0.2051 \\ \hline 0.4694 \end{array}$ **c.** $\begin{array}{r} 0.00441 \\ 0.06882 \\ +0.0103 \\ \hline 0.8353 \end{array}$

64. a. $\begin{array}{r} \$32.71 \\ 43.09 \\ +\ 8.27 \\ \hline \$74.07 \end{array}$ **b.** $\begin{array}{r} \$19.37 \\ 2. \\ +\ 7.22 \\ \hline \$28.59 \end{array}$ **c.** $\begin{array}{r} \$139.26 \\ 82.87 \\ +\ 3.01 \\ \hline \$225.14 \end{array}$

65. a. $\begin{array}{r} 0.35 \\ -0.1007 \\ \hline 0.2493 \end{array}$ **b.** $\begin{array}{r} 0.072 \\ -0.0056 \\ \hline 0.664 \end{array}$ **c.** $\begin{array}{r} 0.03 \\ -0.008 \\ \hline 0.022 \end{array}$

66. a. $\begin{array}{r} 8.551 \\ -2.9995 \\ \hline 5.5515 \end{array}$ **b.** $\begin{array}{r} 78.328 \\ -\ 5.5 \\ \hline 7.2828 \end{array}$ **c.** $\begin{array}{r} 65 \\ -\ 2.778 \\ \hline 62.222 \end{array}$

Mixed Practice

Solve.

67. Calculate: 4.78 + 13 − 10.009

68. Find the difference between 90.1 and 12.58.

69. Add: 0.5 pound + 3 pounds + 4.25 pounds

70. Subtract: \$20 − \$6.95

71. Compute: 8 − 2.4 + 6.0013

72. What is the sum of 1.265, 7, and 0.14?

Applications

Solve and check.

73. A paperback book that normally sells for \$13 is now on sale for \$11.97. What is the discount in dollars and cents?

74. During a drought, the mayor of a city attempted to reduce daily water consumption to 3.1 million gallons. If daily water consumption fell to 1.948 million gallons above that goal, estimate the city's consumption.

75. A skeleton was found at an archaeological dig. Radiocarbon dating—a technique used for estimating age—indicated that the skeleton was 56 centuries old, plus or minus 0.8 centuries. According to this estimate, what is the greatest possible age of the skeleton?

76. A college launched a campaign to collect \$3 million to build a new technology complex. If \$1.316 million has been collected so far, how much more money, to the nearest million dollars, is needed?

77. As an investment, a couple bought an apartment house for \$2.3 million. Two years later, they sold the apartment house for \$4 million. What was their profit?

78. A woman sued her business partner and was awarded \$1.5 million. On appeal, however, the award was reduced to \$0.75 million. By how much was the award reduced?

79. In setting up a page in word processing, the margins of a page are usually expressed in decimal parts of an inch. The page shown is 8.5 inches wide. How long is each typed line on the page?

8.5 in.

0.83 in. 0.83 in.

80. In the boat pictured below, what is the distance between the bottom of the rudder and the bottom of the lake?

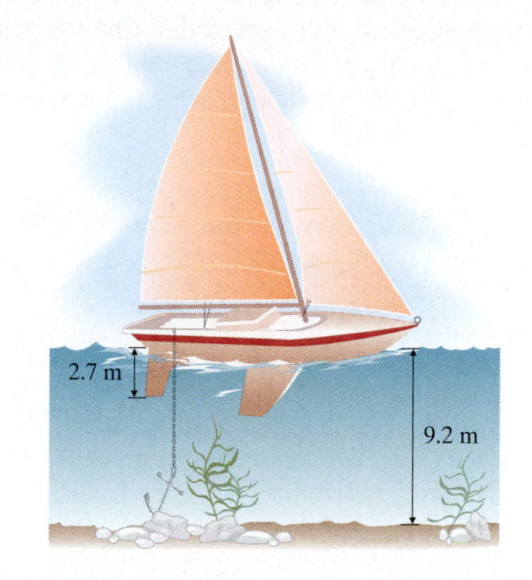

81. A radio disc jockey wants to choose among compact disc tracks that last 3.5, 2.8, 2.9, 2.6, and 1.6 minutes. Can he select tracks so as to get between 9.8 and 10 minutes of music? Explain.

82. A shopper plans to buy three items that cost \$4.99, \$7.99, and \$2.99 each. If she has \$15 with her, will she have enough money to pay for all three items? Explain.

83. When gymnasts compete, they receive scores in four separate events: vault (VT), uneven bar (UB), balance bar (BB), and floor exercises (FX). The total of these four event scores is called the all-around score (AA). The following chart shows the results for two Indiana high school gymnasts at a 2006 competition:

Gymnast	VT	UB	BB	FX	AA
Madeline Whiteman	9.2	9.275	8.6	8.05	
Jordyn Stengel	9	9	8.65	8.45	

(***Source:*** http://www.sigsgym.org/Results/2005-2006/State_Championships_5.asp)

a. Calculate the all-around scores for these two competitors.

b. Which of the competitors had the higher all-around score?

84. The graph shown gives the number of households that watched TV programs in a recent week according to the Nielsen Top 20 ratings. (***Source:*** http://tv.yahoo.com/nielsen/)

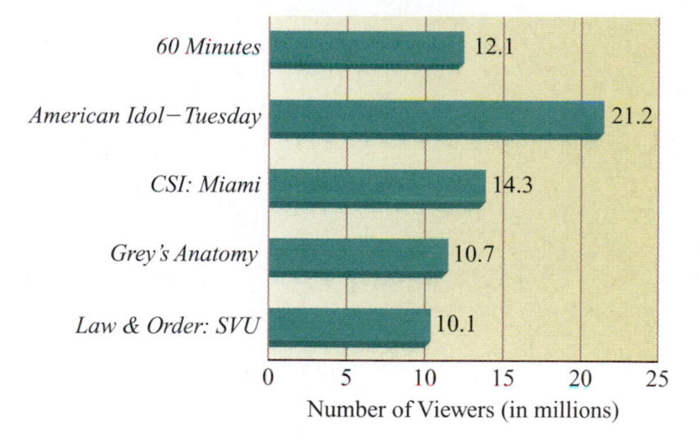

a. How many more households watched *American Idol—Tuesday* than *CSI: Miami*?

b. How many households altogether watched the five programs?

Use a calculator to solve each problem, giving (a) the operation(s) carried out in the solution, (b) the exact answer, and (c) an estimate of the answer.

85. To prevent anemia, a doctor advises his patient to take at least 18 milligrams of iron each day. The following table shows the amount of iron in the food that the patient ate yesterday. Did she get enough iron? If not, how much more does she need?

Food	Iron (mg)
Tomato juice	1.1
Oat flakes	2.4
Milk	0.1
Peanut butter sandwich	1.8
Carrot	0.5
Dried apricots	1
Oatmeal raisin cookies	1.4
Fish	1.1
Baked potato	1.1
Green peas	1.5
Romaine lettuce	0.8
Chocolate pudding	0.4

86. When filling a prescription, buying a generic drug rather than a brand-name drug can often save money. The following table shows the prices of various brand-name and generic drugs.

Drug	Brand-Name Price	Generic Price
Metformin	\$36.89	\$27.19
Allopurinal	\$47	\$18.55
Imipramine	\$43.90	\$10.15
Propranolol	\$27.60	\$10.55
Forosemide	\$13.99	\$10.29

How much money will a shopper save if he buys all five generic drugs rather than the brand-name drugs?

• *Check your answers on page A-6.*

MINDSTRETCHERS

Groupwork

1. Working with a partner, find the missing entries in the following magic square, in which 3.75 is the sum of every row, column, and diagonal.

0.75	1.25	
2		

Mathematical Reasoning

2. Suppose that a spider is sitting at point *A* on the rectangular web shown. If the spider wants to crawl along the web horizontally and vertically to munch on the delicious fly caught at point *B*, how long is the shortest route that the spider can take?

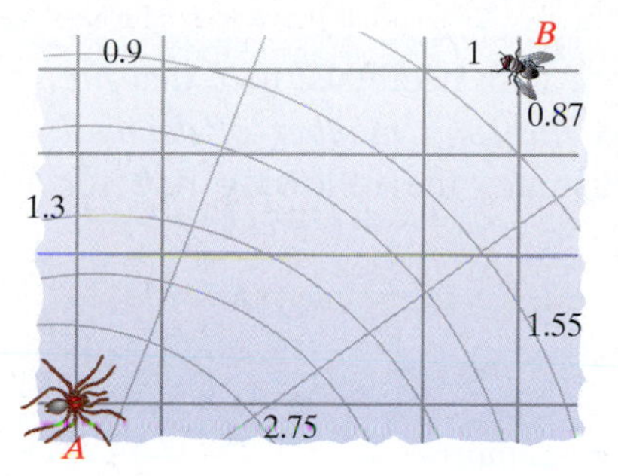

Writing

3. **a.** How many pairs of whole numbers are there whose sum is 7?

   b. How many pairs of decimals are there whose sum is 0.7?

   c. Explain why (a) and (b) have different answers.

3.3 Multiplying Decimals

OBJECTIVES

- To multiply decimals
- To estimate the product of decimals
- To solve word problems involving the multiplication of decimals

In this section, we discuss how to multiply two or more decimals, finding both the exact product and an estimated product.

Multiplying Decimals

To find the product of two decimals—say, 1.02 and 0.3—we multiply the same way we multiply whole numbers. But with decimals we need to know where the decimal point goes in the product. To find out, let's change each decimal to its fractional equivalent.

$$\left.\begin{array}{r} 1.02 \\ \times\ 0.3 \end{array}\right\} \text{ is equivalent to } \longrightarrow \quad 1\frac{2}{100} \times \frac{3}{10}$$

$$306 \quad \text{Where should we place the decimal point?}$$

$$= \frac{102}{100} \times \frac{3}{10} = \frac{306}{1{,}000}, \quad \text{or} \quad 0.306$$

The product is in thousandths, so it has three decimal places.

Looking at the multiplication problem with decimals, note that *the product has as many decimal places as the total number of decimal places in the factors*. This example illustrates the following rule for multiplying decimals.

$$\begin{array}{r} 1.02 \\ 0.3 \\ \hline 0.306 \end{array}$$

To Multiply Decimals

- Multiply the factors as if they were whole numbers.
- Find the total number of decimal places in the factors.
- Count that many places from the right end of the product, and insert a decimal point.

EXAMPLE 1

Multiply: 6.1 × 3.7

Solution First, multiply 61 by 37, ignoring the decimal points.

$$\begin{array}{r} 6.1 \\ \times\ 3.7 \\ \hline 427 \\ 183 \\ \hline 2257 \end{array}$$

Then count the total number of decimal places in the factors.

$$\begin{array}{rl} 6.1 & \longleftarrow \text{One decimal place} \\ \times\ 3.7 & \longleftarrow \text{One decimal place} \\ \hline 427 & \\ 183 & \\ \hline 22.57 & \longleftarrow \text{Two decimal places in the product} \end{array}$$

Insert the decimal point two places from the right end. So 22.57 is the product.

PRACTICE 1

Find the product: 2.81 × 3.5

EXAMPLE 2

Find the product of 0.75 and 4.

Solution Let's multiply 0.75 by 4, ignoring the decimal point.

$$\begin{array}{r} 0.75 \\ \times \quad 4 \\ \hline 300 \end{array}$$

Count the total number of decimal places.

$$\begin{array}{rl} 0.75 & \leftarrow \text{Two decimal places.} \\ \times \quad 4 & \leftarrow \text{Zero decimal places (4 is a whole number)} \\ \hline 3.00 & \leftarrow \text{Two decimal places in the product} \end{array}$$

So the product is 3.00, which simplifies to 3.

PRACTICE 2

Multiply: 0.28×5

EXAMPLE 3

Multiply 0.03 and 0.25, rounding the answer to the nearest thousandth.

Solution

$$\begin{array}{r} 0.25 \\ \times \; 0.03 \\ \hline 075 \\ 000 \\ \hline 0.0075 \end{array}$$

Rounding to the nearest thousandth, we get 0.008.

PRACTICE 3

What is the product of 0.44 and 0.03, rounded to the nearest hundredth?

EXAMPLE 4

Multiply: $(1.1)\,(3.5)\,(0.8)$

Solution To find the product of three factors, we can first multiply the two left factors and then multiply this product by the third factor:

$$\underbrace{(1.1)\,(3.5)}\,(0.8) =$$
$$(3.85)\,(0.8) = 3.08$$

So 3.08 is the final product.

PRACTICE 4

Evaluate: $(0.2)\,(0.3)\,(0.4)$

EXAMPLE 5

Simplify: $3 + (1.2)^2$

Solution Recall that, according to the order of operations rule, we first must find $(1.2)^2$ and then add 3.

$$3 + (1.2)^2 =$$
$$3 + 1.44 = 4.44$$

PRACTICE 5

Evaluate: $10 - (0.3)^2$

EXAMPLE 6

Multiply: 8.274×100

Solution

$$\begin{array}{r} 8.274 \leftarrow \text{Three decimal places} \\ \times 100 \leftarrow \text{Zero decimal places} \\ \hline 827.400 \leftarrow \text{Three decimal places in the product} \end{array}$$

So the product is 827.400, or 827.4 after we drop the extra zeros.

Note that the second factor (100) is a power of 10 ending in **two** zeros and that the product is identical to the first factor except that the decimal point is moved to the right **two** places.

PRACTICE 6

Compute: $0.325 \times 1{,}000$

As Example 6 illustrates, a shortcut for multiplying a decimal by a power of 10 is to *move the decimal point to the right the same number of places as the power of 10 has zeros.* Let's apply this shortcut in the next example.

EXAMPLE 7

Find the product: $1{,}000 \times 2.89$

Solution We see that 1,000 is a power of 10 and has three zeros. So to multiply 1,000 by 2.89, we simply move the decimal point in 2.89 to the right three places.

Add a 0 to move three places. $2.890 = 2890. = 2{,}890$

So the product is 2,890, with the 0 serving as a placeholder.

PRACTICE 7

Multiply 32.7 by 10,000.

EXAMPLE 8

The popularity of a television show is measured in ratings, where each rating point represents 900,000 homes in which the show is watched. After examining the table at the right, answer each question.

Show	Rating
1	17.3
2	14.25

a. In how many homes was show 1 watched?

b. In how many more homes was show 1 watched than show 2?

Solution

a. To find the number of homes in which show 1 was watched, we multiply its rating, 17.3, by 900,000, which gives us 15,570,000.

b. To compare the popularity of show 1 and show 2, we compute the number of homes in which show 2 was viewed: $14.25 \times 900{,}000 = 12{,}825{,}000$. The number of show 1 homes exceeds the number of show 2 homes by $15{,}570{,}000 - 12{,}825{,}000$, or 2,745,000 homes.

PRACTICE 8

A chemistry student learns that a molecule is made up of atoms. For instance, the water molecule, H_2O, consists of two atoms of hydrogen, H, and one atom of oxygen, O. Each of these atoms has an atomic weight.

Atom	Atomic Weight
H	1.008
O	15.999

a. After examining the chart above, compute the weight of the water molecule.

b. Round this weight to the nearest whole number.

Estimating Products

A good way to estimate the product of decimals is to round each factor so that it has only one nonzero digit. Then multiply the rounded factors.

For instance, suppose that we want to estimate the product of the decimals 19.0382 and 0.061.

$$\begin{array}{rcr} 19.0382 & \approx & 20 \\ \times 0.061 & \approx & \times 0.06 \\ \hline & & 01.20 = 1.2 \end{array}$$

Before we dropped the extra 0, the estimated product, 1.20, has two decimal places—the total number of decimal places in the factors.

EXAMPLE 9

Multiply 0.703 by 0.087 and check the answer by estimating.

Solution First, we multiply the factors to find the exact product. Then, we round each factor and multiply them.

$$\begin{array}{rrcrl} & 0.703 & \approx & 0.7 & \leftarrow \text{Rounded to have one nonzero digit} \\ & \times\ 0.087 & \approx & \times 0.09 & \leftarrow \text{Rounded to have one nonzero digit} \\ \hline \text{Exact product} \rightarrow & 0.061161 & & 0.063 & \leftarrow \text{Estimated product} \end{array}$$

We see that the exact product and the estimated product are fairly close, as expected.

PRACTICE 9

Find the product of 0.0037×0.092, estimating to check.

EXAMPLE 10

Calculate and check: (4.061) (0.72) + (0.91) (0.258)

Solution Following the order of operations rule, we begin by finding the two products.

$$(4.061)(0.72) = 2.92392 \qquad (0.91)(0.258) = 0.23478$$

Then we add these two products.

$$2.92392 + 0.23478 = 3.1587$$

So (4.061) (0.72) + (0.91) (0.258) = 3.1587.

Now, let's check this answer by estimating.

$$\begin{array}{c} (4.061)(0.72) + (0.91)(0.258) \\ \downarrow \quad \downarrow \quad \downarrow \quad \downarrow \\ (4)\ (0.7) + (0.9)\ (0.3) = 2.8 + 0.27 = 3.07 \approx 3 \end{array}$$

The estimate, 3, is sufficiently close to 3.1587 to confirm our answer.

PRACTICE 10

Compute and check:
(0.488) (9.1) − (3.5) (0.227)

EXAMPLE 11

The sound waves of an elephant call can travel through both the ground and the air. Through the air, the waves may travel 6.63 miles. If they travel 1.5 times as far through the ground, what is the estimated ground distance? (*Source:* http://www.abc.net.au/science/k2/moments/s434107.htm)

Solution We know that the waves may travel 6.63 miles through the air and 1.5 times as far through the ground. To find the estimated ground distance, we compute this product.

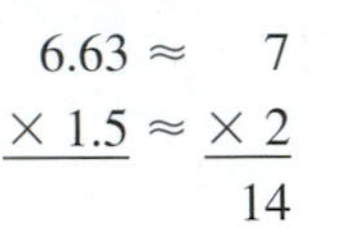

$$\begin{array}{r} 6.63 \approx 7 \\ \underline{\times\, 1.5} \approx \underline{\times\, 2} \\ 14 \end{array}$$

So the estimated ground distance of the sound waves of an elephant call is about 14 miles.

PRACTICE 11

The Earth travels through space at a speed of 18.6 miles per second. Estimate how far the Earth travels in 60 seconds. (*Source:* The Diagram Group, *Comparisons*)

Multiplying Decimals on a Calculator

Multiply decimals on a calculator by entering each decimal as you would enter a whole number, but insert a decimal point as needed. If there are too many decimal places in your answer to fit in the display, investigate how your calculator displays the answer.

EXAMPLE 12

Compute 8,278.55 × 0.875, rounding your answer to the nearest hundredth. Then check the answer by estimating.

Solution

Now 7,243.73125 rounded to the nearest hundredth is 7,243.73. Checking by estimating, we get 8,000 × 0.9, or 7,200, which is close to our exact answer.

PRACTICE 12

Find the product of 2,471.66 and 0.33, rounding to the nearest tenth. Check the answer.

EXAMPLE 13

Find $(1.9)^2$

Solution

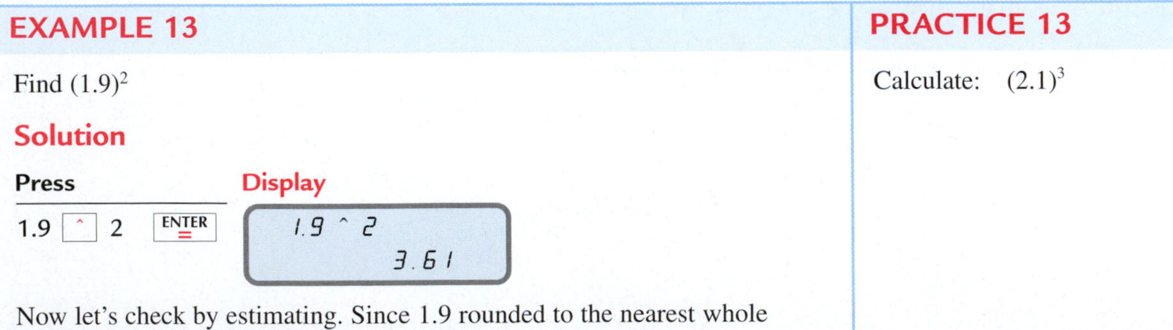

Now let's check by estimating. Since 1.9 rounded to the nearest whole number is 2, $(1.9)^2$ should be close to 2^2, or 4, which is close to our exact answer, 3.61.

PRACTICE 13

Calculate: $(2.1)^3$

Mathematically Speaking

Fill in each blank with the most appropriate term or phrase from the given list.

add	three	factors	five
first factor	four	multiplication	
square	two	division	

1. The operation understood in the expression (3.4) (8.9) is _______.

2. When multiplying decimals, the number of decimal places in the product is equal to the total number of decimal places in the _______.

3. To multiply a decimal by 100, move the decimal point _______ places to the right.

4. The product of 0.27 and 8.18 has _______ decimal places.

5. To compute the expression $(8.5)^2 + 2.1$, first _______.

6. To multiply a decimal by 1,000, move the decimal point _______ places to the right.

Insert a decimal point in each product. Check by estimating.

7. $2.356 \times 1.27 = 299212$

8. $97.26 \times 5.3 = 515478$

9. $3{,}144 \times 0.065 = 204360$

10. $837 \times 0.15 = 12555$

11. $71.2 \times 35 = 24920$

12. $0.002 \times 37 = 0074$

13. $0.0019 \times 0.051 = 969$

14. $0.0089 \times 0.0021 = 1869$

15. $2.87 \times 1{,}000 = 287000$

16. $492.31 \times 10 = 492310$

17. $\$4.25 \times 0.173 = \$73525$

18. 11.2 feet $\times$ 0.75 = 8400 feet

Find the product. Check by estimating.

19. 0.6×0.9

20. 0.8×0.7

21. 0.5×0.8

22. 0.6×0.8

23. 0.1×0.2

24. 0.09×0.5

25. 0.04×0.07

26. 0.03×0.01

27. 2.55×0.3

28. 80.7×0.6

29. 0.96×2.1

30. 0.043×0.02

31. 38.01×0.2

32. 1.22×0.09

33. 125×0.004

34. 0.003×0.7

35. 3.8×1.54

36. 9.51×0.7

37. 13.74×11

38. $1{,}245 \times 2.5$

39. 12.459×0.3

40. 72.558×0.2

41. $(0.675)(2.66)$

42. $(4.003)(0.59)$

43. 83.127×100

44. 4.9×10

45. $0.0023 \times 10{,}000$

46. $0.0135 \times 1{,}000$

47. (1.5) (0.6) (0.1)

48. (12) (3.5) (0.2)

49. (0.03) (1.4) (25)

50. (2.6) (0.5) (0.9)

51. $(0.001)^3$

52. $(0.1)^4$

53. 17 feet × 2.5

54. 5 hours × 0.75

55. 3.5 miles × 0.4

56. 9.1 meters × 1,000

57. $\begin{array}{r} 43.87 \\ \times\ 0.075 \\ \hline \end{array}$

58. $\begin{array}{r} 18{,}275.33 \\ \times\ \quad 0.39 \\ \hline \end{array}$

59. $\begin{array}{r} 99{,}125 \\ \times\ \quad 2.75 \\ \hline \end{array}$

60. $\begin{array}{r} 3.512 \\ \times\ 1.47 \\ \hline \end{array}$

Simplify.

61. 0.7×10^2

62. 0.6×10^4

63. $30 - 2.5 \times 1.7$

64. $8 + 4.1 \times 2$

65. $1 + (0.3)^2$

66. $6 - (1.2)^2$

67. $0.8(1.3 + 2.9) - 0.5$

68. $4 - 2.1(3.5 - 1.8)$

Complete each table.

69.

Input	Output
1	3.8 × **1** − 0.2 =
2	3.8 × **2** − 0.2 =
3	3.8 × **3** − 0.2 =
4	3.8 × **4** − 0.2 =

70.

Input	Output
1	7.5 × **1** + 0.4 =
2	7.5 × **2** + 0.4 =
3	7.5 × **3** + 0.4 =
4	7.5 × **4** + 0.4 =

Each product is rounded to the nearest hundredth. In each group of three products, one is wrong. Use estimation to explain which product is incorrect.

71. a. 51.6 × 0.813 = 419.51 **b.** 2.93 × 7.283 = 21.34 **c.** $(5.004)^2 = 25.04$

72. a. 0.004 × 3.18 = 0.01 **b.** 2.99 × 0.287 = 0.86 **c.** $(1.985)^3 = 10.82$

73. a. 4.913 × 2.18 = 10.71 **b.** 0.023 × 0.71 = 0.16 **c.** (8.92)(1.0027) = 8.94

74. a. \$138.28 × 0.075 = \$10.37 **b.** 0.19 × \$487.21 = \$92.57 **c.** 0.77 × \$6,005.79 = \$462.45

Mixed Practice

Solve.

75. Simplify: $9 - (0.5)^2$

76. Compute: $2.1 + 5 \times 0.6$

77. Multiply 0.75 and 0.09, rounding the answer to the nearest thousandth.

78. Multiply: (2.3)(4.5)(0.6)

79. Find the product of 0.56 and 8.

80. Find the product: 3.01 × 1,000

Applications

Solve. Check by estimating.

81. Sound travels at approximately 1,000 feet per second (fps). If a jet is flying at Mach 2.9 (that is, 2.9 times the speed of sound), what is its speed?

82. If insurance premiums of \$323.50 are paid yearly for 10 years for a life insurance policy, how much did the policy holder pay altogether in premiums?

83. The planet in the solar system closest to the Sun is Mercury. The average distance between these two bodies is 57.9 million kilometers. Express this distance in standard form. (***Source:*** Jeffrey Bennett et al., *The Cosmic Perspective*)

84. According to the first American census in 1790, the population of the United States was approximately 3.9 million. Write this number in standard form. (***Source:*** *The Statistical History of the United States*)

85. The area of this circle is approximately $3.14 \times (2.5)^2$ square feet. Find this area to the nearest tenth.

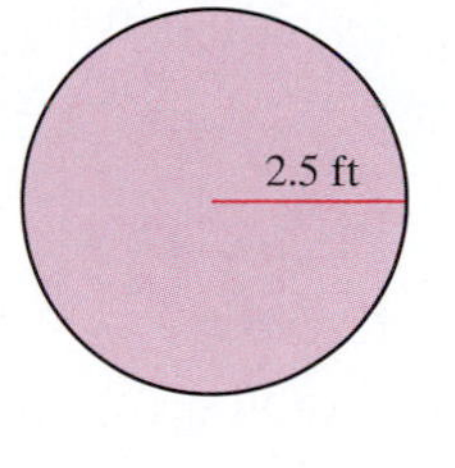

86. Find the area (in square meters) of the room pictured.

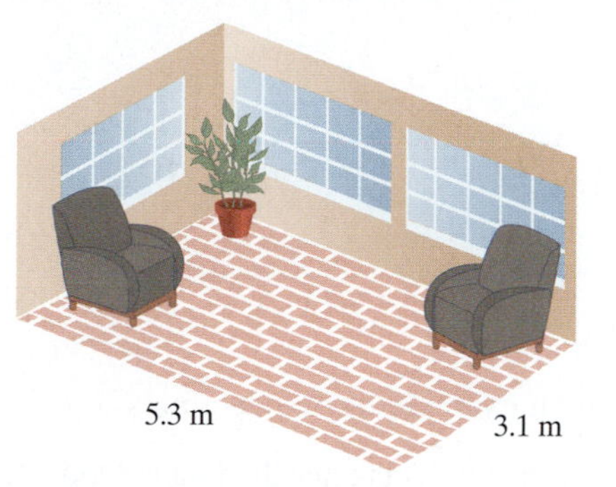

87. Over a 5-day period, a nurse administered 10 tablets to a patient. If each tablet contained 0.125 milligram of the drug Digoxin, how much Digoxin did the nurse administer?

88. Water weighs approximately 62.5 pounds per cubic foot (lb/ft^3). If a bathtub contains about 30 ft^3 of water, how much does the water in the bathtub weigh?

89. A tennis player weighing 180 pounds burns 10.9 calories per minute while playing singles tennis. How many calories would he burn in 2 hours? (***Source:*** http://www.caloriesperhour.com)

90. A plumber is paid \$37.50 per hour for the first 40 hours worked. She gets time and a half, \$56.25, for any time over her 40-hour week. If she works 49 hours in a week, how much is her pay?

91. The sales receipt for a shopper's purchases is as follows.

Purchase	Quantity	Unit Price	Price
Belt	1	\$11.99	\$___.__
Shirt	3	\$16.95	\$___.__
Total Price			\$___.__

a. Complete the table.

b. If the shopper pays for these purchases with four \$20 bills, how much change should he get?

92. On an electric bill, *usage* is the difference between the meter's *current reading* and the *previous reading* in kilowatt hours (kWh). The *amount due* is the product of the usage and the *rate per kWh.* Find the two missing quantities in the table, rounding to the nearest hundredth.

Previous Reading	750.07 kWh
Current Reading	1,115.14 kWh
Usage	_____ kWh
Rate per kWh	\$0.10
Amount Due	\$_____

Use a calculator to solve each problem, giving (a) the operation(s) carried out in the solution, (b) the exact answer, and (c) an estimate.

93. Scientists have discovered a relationship between the length of a person's bones and the person's overall height. For instance, an adult male's height (in inches) can be predicted from the length of his femur bone by using the formula $(1.9 \times \text{femur}) + 32.0$. With this formula, estimate the height of the German giant Constantine, whose femur measured 29.9 inches. (***Source:*** *Guinness World Records 2007*)

94. In order to buy a \$125,000 house, a couple puts down \$25,000 and takes out a mortgage on the balance. To pay off the mortgage, they pay \$877.57 per month for the following 360 months. How much more will they end up paying for the house than the original price of \$125,000?

• *Check your answers on page A-6.*

MINDSTRETCHERS

Patterns

1. When $(0.001)^{100}$ is multiplied out, how many decimal places will it have?

Mathematical Reasoning

2. Give an example of two decimals
   a. whose sum is greater than their product, and
   b. whose product is greater than their sum.

Groupwork

3. In the product to the right, each letter stands for a different digit. Working with a partner, identify all the digits.

$$\begin{array}{r} A.B \\ \times\ B.A \\ \hline C\ D \end{array}$$

3.4 Dividing Decimals

OBJECTIVES

- To find the decimal equivalent to a fraction
- To divide decimals
- To estimate the quotient of decimals
- To solve word problems involving the division of decimals

In this section, we first consider changing a fraction to its decimal equivalent, which involves both division and decimals. Then we move on to our main concern—the division of decimals.

Changing a Fraction to the Equivalent Decimal

Earlier in this chapter, we discussed how to change a decimal to its equivalent fraction. Now let's consider the opposite problem—how to change a fraction to its equivalent decimal.

When the denominator of a fraction is already a power of 10, the problem is simple. For example, the decimal equivalent of $\frac{43}{100}$ is 0.43.

But what about the more difficult problem where the denominator is *not* a power of 10? A good strategy is to find an equivalent fraction that does have a power of 10 as its denominator. Consider, for instance, the fraction $\frac{3}{4}$. Since 4 is a factor of 100, which is a power of 10, we can easily find an equivalent fraction having a denominator of 100.

$$\frac{3}{4} = \frac{3 \cdot \mathbf{25}}{4 \cdot \mathbf{25}} = \frac{75}{100} = 0.75$$

So 0.75 is the decimal equivalent of $\frac{3}{4}$.

There is a faster way to show that $\frac{3}{4}$ is the same as 0.75, without having to find an equivalent fraction. Because $\frac{3}{4}$ can mean $3 \div 4$, we divide the numerator (3) by the denominator (4). Note that if we continue to divide to the hundredths place, there is no remainder.

$$\begin{array}{r} 0.75 \\ 4\overline{)3.00} \\ \underline{0} \\ 3\,0 \\ \underline{2\,8} \\ 20 \\ \underline{20} \end{array}$$

Insert the decimal point directly above the decimal point in the dividend.

The decimal point is after the 3. Insert enough 0's to continue dividing as far as necessary.

So this division also tells us that $\frac{3}{4}$ equals 0.75.

To Change a Fraction to the Equivalent Decimal

- Divide the denominator of the fraction into the numerator, inserting to its right both a decimal point and enough zeros to get an answer either without a remainder or rounded to a given decimal place.
- Place a decimal point in the quotient directly above the decimal point in the dividend.

EXAMPLE 1

Express $\frac{1}{2}$ as a decimal.

Solution To find the decimal equivalent, we divide the fraction's numerator by its denominator.

$$\begin{array}{r} 0.5 \\ 2\overline{)1.0} \\ \underline{0} \\ 1\,0 \\ \underline{1\,0} \end{array}$$

Add a decimal point and a 0 to the right of the 1.

So 0.5 is the decimal equivalent of $\frac{1}{2}$.

Check We verify that the fractional equivalent of 0.5 is $\frac{1}{2}$.

$$0.5 = \frac{5}{10} = \frac{1}{2}$$

The answer checks.

PRACTICE 1

Write the fraction $\frac{3}{8}$ as a decimal.

EXAMPLE 2

Convert $2\frac{3}{5}$ to a decimal.

Solution Let's first change this mixed number to an improper fraction: $2\frac{3}{5} = \frac{13}{5}$. We can then change this improper fraction to a decimal by dividing its numerator by its denominator.

$$\begin{array}{r} 2.6 \\ 5\overline{)13.0} \\ \underline{10} \\ 3\,0 \\ \underline{3\,0} \end{array}$$

So 2.6 is the decimal form of $2\frac{3}{5}$.

Check We convert this answer back from a decimal to its mixed number form.

$$2.6 = 2\frac{6}{10} = 2\frac{3}{5}$$ The answer checks.

PRACTICE 2

Write $7\frac{5}{8}$ as a decimal.

When converting some fractions to decimal notation, we keep getting a remainder as we divide. In this case, we round the answer to a given decimal place.

EXAMPLE 3

Convert $4\frac{8}{9}$ to a decimal, rounded to the nearest hundredth.

Solution First, we change this mixed number to an improper fraction. Then, we convert it to a decimal.

$$4\frac{8}{9} = \frac{44}{9} = 9\overline{)44.000} = 4.888$$

$$\begin{array}{r} 4.888 \\ 9\overline{)44.000} \\ \underline{36} \\ 8\,0 \\ \underline{7\,2} \\ 80 \\ \underline{72} \\ 80 \\ \underline{72} \end{array}$$

← **In order to round to the nearest hundredth, we must continue to divide to the thousandths place. So we insert three 0's.**

Finally, we round to the nearest hundredth: $4.8\underline{8}8 \approx 4.89$

PRACTICE 3

Express $83\frac{1}{3}$ as a decimal, rounded to the nearest tenth.

In Example 3, note that if instead of rounding we had continued to divide we would have gotten as our answer the *repeating decimal* 4.88888 . . . (also written $4.\overline{8}$). Can you think of any other fraction that is equivalent to a repeating decimal?

Let's now consider some word problems in which we need to convert fractions to decimals.

EXAMPLE 4

A share of stock sells for $5\frac{7}{8}$ dollars. Express this amount in dollars and cents, to the nearest cent.

Solution To solve, we must convert the mixed number $5\frac{7}{8}$ to a decimal.

$$5\frac{7}{8} = \frac{47}{8} = 8\overline{)47.000}$$

$$\begin{array}{r} 5.8\underline{7}5 \approx 5.88 \\ 8\overline{)47.000} \\ \underline{40} \\ 7\,0 \\ \underline{6\,4} \\ 60 \\ \underline{56} \\ 40 \\ \underline{40} \end{array}$$

So a share of this stock sells for $5.88, to the nearest cent.

PRACTICE 4

The gas nitrogen makes up about $\frac{39}{50}$ of the air in the atmosphere. Express this fraction as a decimal, rounded to the nearest tenth. (***Source:*** *World of Scientific Discovery*)

Dividing Decimals

Before we turn our attention to dividing one decimal by another, let's consider simpler problems in which we are dividing a decimal by a whole number. An example of such a problem is $0.6 \div 2$. We can write this expression as the fraction $\frac{0.6}{2}$, which can be rewritten as the quotient of two whole numbers by multiplying the numerator and denominator by 10.

$$\frac{0.6}{2} = \frac{0.6 \times 10}{2 \times 10} = \frac{6}{20}$$

We then convert this fraction to the equivalent decimal, as we have previously discussed.

$$\begin{array}{r} 0.3 \\ 20\overline{)\,6.0} \\ \underline{0} \\ 6\,0 \\ \underline{6\,0} \end{array}$$

So $\dfrac{0.6}{2} = 0.3$

Note that we get the same quotient if we divide the number in the original problem as follows:

$$\begin{array}{r} 0.3 \leftarrow \textbf{Quotient} \\ \textbf{Divisor} \rightarrow 2\overline{)0.6} \leftarrow \textbf{Dividend} \\ \underline{0} \\ 6 \\ \underline{6} \end{array}$$

It is important to write the decimal point in the quotient directly above the decimal point in the dividend.

Next, let's consider a division problem where we are dividing one decimal by another: $0.006 \div 0.02$. Writing this expression as a fraction, we get $\dfrac{0.006}{0.02}$. Since we have already discussed how to divide a decimal by a whole number, the goal here is to find a fraction equivalent to $\dfrac{0.006}{0.02}$ where the denominator is a whole number. Multiplying the numerator and denominator by 100 will do just that.

$$0.006 \div 0.02 = \frac{0.006}{0.02} = \frac{0.006 \times 100}{0.02 \times 100} = \frac{0.006}{0.02} = \frac{0.6}{2}$$

We know from the previous problem that $\dfrac{0.6}{2} = 0.3$. Since $\dfrac{0.006}{0.02} = \dfrac{0.6}{2}$, we conclude $\dfrac{0.006}{0.02} = 0.3$.

A shortcut to multiply by 100 in both the divisor and the dividend is to move the decimal point two places to the right.

So $0.02\overline{)0.006}$ (quotient 0.3) is equivalent to $2\overline{)0.6}$ (quotient 0.3)

As in any division problem, we can check our answer by confirming that the product of the quotient and the *original divisor* equals the *original dividend.*

Division Problem	**Check**
$0.6 \div 2 = 0.3$	$0.3 \times 2 = 0.6$
$0.006 \div 0.02 = 0.3$	$0.3 \times 0.02 = 0.006$

These examples suggest the following rule.

To Divide Decimals

- Move the decimal point in the divisor to the right end of the number.
- Move the decimal point in the dividend the same number of places to the right as in the divisor.
- Insert a decimal point in the quotient directly above the decimal point in the dividend.
- Divide the new dividend by the new divisor, inserting zeros at the right end of the dividend as necessary.

EXAMPLE 5

What is 0.035 divided by 0.25?

Solution Move the decimal point to the right end, making the divisor a whole number.

$$0.25\overline{)0.035}$$

Move the decimal point in the dividend the same number of places.

Now, we divide 3.5 by 25, which gives us 0.14.

$$\begin{array}{r} 0.14 \\ 25\overline{)3.50} \\ \underline{2\,5} \\ 1\,00 \\ \underline{1\,00} \end{array}$$

Check We see that the product of the quotient and the original divisor is equal to the original dividend.

$$\begin{array}{r} 0.14 \\ \underline{\times\,0.25} \\ 0\,7\,0 \\ \underline{0\,2\,8} \\ 0.03\,5\,0 \end{array} = 0.035$$

PRACTICE 5

Divide and check: 2.706 ÷ 0.15

EXAMPLE 6

Find the quotient: 6 ÷ 0.0012. Check the answer.

Solution The decimal point is moved four places to the right.

$$0.0012\overline{)6.0000}$$

To move the decimal point four places to the right, we must add four 0's as placeholders.

$$\begin{array}{r} 5{,}000 \\ 12\overline{)60{,}000} \end{array}$$

Check

$$\begin{array}{r} 5{,}000 \\ \underline{\times\;\;0.0012} \\ 0006.0000 \end{array} = 6$$

The answer checks.

PRACTICE 6

Divide $\frac{8.2}{0.004}$ and then check.

EXAMPLE 7

Divide and round to the nearest hundredth: $0.7\overline{)40.2}$
Then check.

Solution $0.7\overline{)40.2}$

$$
\begin{array}{r}
57.428 \\
7\overline{)402.000} \\
\underline{35} \\
52 \\
\underline{49} \\
3\,0 \\
\underline{2\,8} \\
20 \\
\underline{14} \\
60 \\
\underline{56} \\
4
\end{array}
\approx 57.43 \text{ to the nearest hundredth}
$$

Check

$$
\begin{array}{r}
57.43 \\
\times\ 0.7 \\
\hline
40.201
\end{array}
\approx 40.2
$$

Because we rounded our answer, the check gives us a product only approximately equal to the original dividend.

PRACTICE 7

Find the quotient of 8.07 and 0.11, rounded to the nearest tenth.

EXAMPLE 8

Compute and check: $8.319 \div 1{,}000$

Solution

$$
\begin{array}{r}
0.008319 \\
1{,}000\overline{)8.319000} \\
\underline{8\,000} \\
3190 \\
\underline{3000} \\
1900 \\
\underline{1000} \\
9000 \\
\underline{9000}
\end{array}
$$

Check

$$
\begin{array}{r}
0.008319 \\
\times \qquad 1{,}000 \\
\hline
0008.319000
\end{array}
= 8.319
$$

Note that the divisor (1,000) is a power of 10 ending in three zeros, and that the quotient is identical to the dividend except that the decimal point is moved to the left three places.

$$\frac{8.319}{1{,}000} = 0.008319$$

PRACTICE 8

Divide: $100\overline{)3.41}$

As Example 8 illustrates, a shortcut for dividing a decimal by a power of 10 is to *move the decimal point to the left the same number of places as the power of 10 has zeros.* Can you explain the difference between the shortcuts for multiplying and for dividing by a power of 10?

EXAMPLE 9

Compute: $\dfrac{7.2}{100}$

Solution Since we are dividing by 100, a power of 10 with two zeros, we can find this quotient simply by moving the decimal point in 7.2 to the left two places.

$$\frac{7.2}{100} = .072, \quad \text{or} \quad 0.072$$

So the quotient is 0.072.

PRACTICE 9

Calculate: $0.86 \div 1{,}000$

Now let's try using these skills in some applications.

EXAMPLE 10

The following graph shows the number of people who attended a Broadway show in selected seasons.

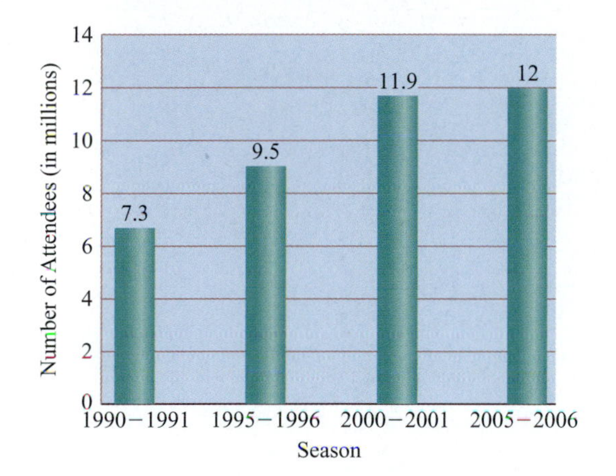

The number of Broadway attendees in the 2005–2006 season was how many times as great as the number of attendees 15 years earlier? Round to the nearest tenth. (***Source:*** The League of American Theatres and Producers)

Solution The number of attendees in 2005–2006 (in millions) was 12, and the number in 1990–1991 was 7.3. To find how many times as great 12 is as compared to 7.3, we find their quotient.

$$7.3\overline{)12} = 7.3\overline{)12.0} = 73\overline{)120}$$

$$\begin{array}{r} 1.64 \\ 73\overline{)120.00} \\ \underline{73} \\ 470 \\ \underline{438} \\ 320 \\ \underline{292} \\ 28 \end{array}$$

Rounding to the nearest tenth, we conclude that the number of Broadway attendees in 2005–2006 was 1.6 times the corresponding figure in 1990–1991.

PRACTICE 10

The table gives the amount of selected foods consumed per capita in the United States in a recent year.

Food	Annual Per Capita Consumption (in pounds)
Red meat	112.0
Poultry	72.7
Fish and shellfish	16.5

The amount of red meat consumed was how many times as great as the amount of poultry, rounded to the nearest tenth? (***Source:*** U.S. Department of Agriculture)

Estimating Quotients

As we have shown, one way to check the quotient of two decimals is by multiplying. Another way is by estimating.

To check a decimal quotient by estimating, we can round each decimal to one nonzero digit and then mentally divide the rounded numbers. But we must be careful to position the decimal point correctly in our estimate.

EXAMPLE 11

Divide and check by estimating: $3.36 \div 0.021$

Solution $0.021\overline{)3.360}$

We compute the exact answer.

$$\begin{array}{r} 160 \\ 21\overline{)3{,}360} \\ \underline{2\,1} \\ 1\,26 \\ \underline{1\,26} \\ 00 \\ \underline{00} \end{array}$$

So 160 is our quotient.

Now let's check by estimating. Because $3.36 \approx 3$ and $0.021 \approx 0.02$, $3.36 \div 0.021 \approx 3 \div 0.02$. We mentally divide to get the estimate.

$$\begin{array}{r} 150 \\ 0.02\overline{)3.00} \end{array}$$

Our estimate, 150, is reasonably close to our exact answer, 160, and so confirms it.

PRACTICE 11

Compute and check by estimating:
$8.229 \div 0.39$

EXAMPLE 12

Calculate and check: $(9.13) \div (0.2) + (4.6)^2$

Solution Following the order of operations rule, we begin by finding the square and then the quotient.

$$(4.6)^2 = 21.16$$
$$(9.13) \div (0.2) = 45.65$$

Then we add these two answers.

$$21.16 + 45.65 = 66.81$$

So $(9.13) \div (0.2) + (4.6)^2 = 66.81$.

Now, let's check this answer by estimating.

$$(9.13) \div (0.2) + (4.6)^2$$
$$\underbrace{9 \div 0.2}_{45} + 25 \approx 45 + 25 \approx 70$$

The estimate, 70, is sufficiently close to 66.81 to confirm our answer.

PRACTICE 12

Compute and check:
$13.07 + (8.4 \div 0.5)^2$

EXAMPLE 13

The water in a filled aquarium weighs 638.25 pounds. If 1 cubic foot of water weighs 62.5 pounds, estimate how many cubic feet of water there are in the aquarium.

Solution We know that the water in the aquarium weighs 638.25 pounds. Since 1 cubic foot of water weighs 62.5 pounds, we can estimate the number of cubic feet of water in the aquarium by computing the quotient $638.25 \div 62.5$, which is approximately $600 \div 60$, or 10. So a reasonable estimate for the amount of water in the aquarium is 10 cubic feet.

PRACTICE 13

The following graph shows the number of farms, in a recent year, in five states.

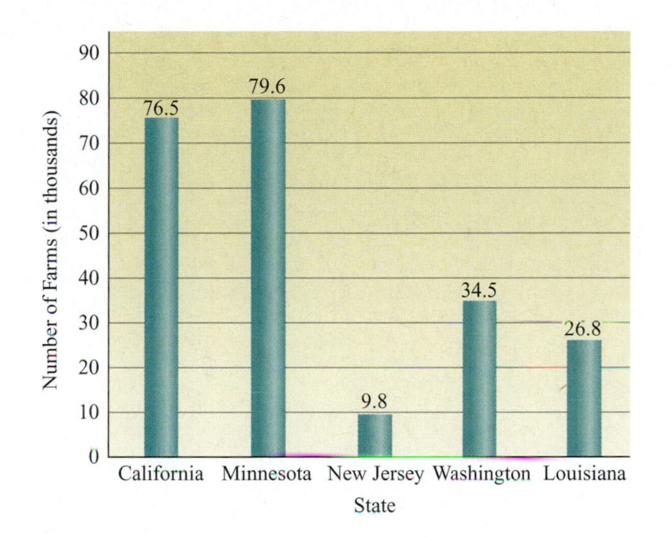

The number of farms in Minnesota is about how many times as great as the number in New Jersey? (***Source:*** U.S. Department of Agriculture)

Dividing on a Calculator

When dividing decimals on a calculator, be careful to enter the dividend first and then the divisor. Note that when the dividend is larger than the divisor, the quotient is greater than 1, and when the dividend is smaller than the divisor, the quotient is less than 1.

EXAMPLE 14

Calculate $8.6 \div 1.6$ and round to the nearest tenth.

Solution

Press	Display
8.6 [÷] 1.6 [ENTER =]	8.6/1.6 5.375

The answer, when rounded to the nearest tenth, is 5.4. As expected, the answer is greater than 1, because $8.6 > 1.6$.

PRACTICE 14

Compute the quotient $8.6\overline{)1.6}$ and round to the nearest tenth.

EXAMPLE 15

Divide $0.3\overline{)0.07}$, rounding to the nearest hundredth.

Solution

Press	Display
0.07 [÷] 0.3 [ENTER/=]	0.07/0.3 0.2333333333

The answer, when rounded to the nearest hundredth, is 0.23. As expected, the answer is less than 1, because $0.07 < 0.3$.

PRACTICE 15

Find the quotient, rounding to the nearest hundredth: $0.3 \div 0.07$

3.4 Exercises

Mathematically Speaking

Fill in each blank with the most appropriate term or phrase from the given list.

quotient	three	divisor	decimal
dividend	right	terminating	four
fraction	product	left	repeating

1. To change a fraction to the equivalent _______, divide the numerator of the fraction by its denominator.

2. An example of a(n) _______ decimal is 0.3333 . . . .

3. When dividing decimals, move the decimal point in the divisor to the _______ end.

4. To divide a decimal by 1,000, move the decimal point _______ places to the left.

5. To estimate the _______ of 0.813 and 0.187, divide 0.8 by 0.2.

6. When dividing a decimal by a whole number, the decimal point in the quotient is placed above the decimal point in the _______.

Change to the equivalent decimal. Then check.

7. $\frac{1}{2}$ **8.** $\frac{3}{5}$ **9.** $\frac{1}{4}$ **10.** $\frac{1}{8}$

11. $\frac{37}{10}$ **12.** $\frac{517}{100}$ **13.** $1\frac{5}{8}$ **14.** $10\frac{3}{4}$

15. $2\frac{7}{8}$ **16.** $8\frac{2}{5}$ **17.** $21\frac{3}{100}$ **18.** $60\frac{17}{100}$

Divide and check.

19. $4\overline{)17}$ **20.** $2\overline{)35}$ **21.** $5\overline{)21}$ **22.** $6\overline{)33}$

23. $8\overline{)11}$ **24.** $6\overline{)9}$ **25.** $18\overline{)153}$ **26.** $14\overline{)217}$

Change to the equivalent decimal. Round to the nearest hundredth.

27. $\frac{2}{3}$ **28.** $\frac{5}{6}$ **29.** $\frac{7}{9}$ **30.** $\frac{1}{3}$

31. $3\frac{1}{9}$ **32.** $2\frac{4}{7}$ **33.** $5\frac{1}{16}$ **34.** $10\frac{11}{32}$

Divide. Express any remainder as a decimal rounded to the nearest thousandth.

35. $7\overline{)23}$ **36.** $9\overline{)41}$ **37.** $11\overline{)3}$ **38.** $13\overline{)2}$

39. $7\overline{)46}$ **40.** $6\overline{)82}$ **41.** $13\overline{)911}$ **42.** $12\overline{)208}$

Insert the decimal point in the appropriate place.

43. $\begin{array}{r} 5\,8\,8\,2 \\ 0.7\overline{)4\,1.1\,7\,4} \end{array}$

44. $\begin{array}{r} 5\,7 \\ 3\overline{)0.0\,1\,7\,1} \end{array}$

45. $\begin{array}{r} 6\,6\,3 \\ 0.5\,8\overline{)0.0\,3\,8\,4\,5\,4} \end{array}$

46. $\begin{array}{r} 6\,9 \\ 3.9\overline{)2\,6.9\,1} \end{array}$

Divide. Check, either by multiplying or by estimating.

47. $8\overline{)23.1}$

48. $2\overline{)0.0035}$

49. $7\overline{)2.002}$

50. $6\overline{)24.042}$

51. $\frac{17.2}{4}$

52. $\frac{0.75}{5}$

53. $\frac{0.003}{2}$

54. $\frac{1.04}{8}$

55. $8.65 \div 5$

56. $0.42 \div 3$

57. $11.5 \div 4$

58. $7.3 \div 2$

59. $0.2\overline{)0.8}$

60. $0.3\overline{)0.6}$

61. $0.05\overline{)3.52}$

62. $0.04\overline{)1.92}$

63. $\frac{47}{0.5}$

64. $\frac{86}{0.2}$

65. $\frac{5}{0.4}$

66. $\frac{9}{0.6}$

67. $0.03 \div 0.1$

68. $1.2 \div 0.01$

69. $0.38 \div 1.9$

70. $0.075 \div 0.25$

71. $95.2 \div 100$

72. $81.6 \div 10$

73. $0.082 \div 100$

74. $9.03 \div 1{,}000$

Divide, rounding to the nearest hundredth. Check, either by multiplying or by estimating.

75. $0.8\overline{)307.1}$

76. $0.4\overline{)81.9}$

77. $0.9\overline{)0.0057}$

78. $0.2\overline{)0.057}$

79. $\frac{3.69}{0.4}$

80. $\frac{3.995}{0.7}$

81. $\frac{87}{0.009}$

82. $\frac{23}{0.06}$

83. $41 \div 0.021$

84. $9.13 \div 0.007$

85. $35.77 \div 0.11$

86. $0.291 \div 0.17$

87. $49.071 \div 0.728$

88. $18.3 \div 7.96$

89. $3 \div 0.0721$

90. $100 \div 3.89$

Perform the indicated operations.

91. $\frac{10.71}{5} \cdot \frac{0.4}{5}$

92. $\frac{2.04}{3} + 1$

93. $\frac{51.3}{10} - 5$

94. $\frac{26.77 - 10.1}{0.4}$

95. $\frac{13.05}{7.27 - 7.02}$

96. $\frac{81.51}{3} - \frac{25.2}{9}$

97. $\frac{8.1 \times 0.2}{0.4}$

98. $(82.9 - 3.6) \div (0.21 - 0.01)$

Complete each table.

99.

Input	Output
1	$\mathbf{1} \div 5 - 0.2 =$
2	$\mathbf{2} \div 5 - 0.2 =$
3	$\mathbf{3} \div 5 - 0.2 =$
4	$\mathbf{4} \div 5 - 0.2 =$

100.

Input	Output
1	$\mathbf{1} \div 4 + 0.4 =$
2	$\mathbf{2} \div 4 + 0.4 =$
3	$\mathbf{3} \div 4 + 0.4 =$
4	$\mathbf{4} \div 4 + 0.4 =$

Each of the following quotients is rounded to the nearest hundredth. In each group of three quotients, one is wrong. Use estimation to identify which quotient is incorrect.

101. a. $5.7 \div 89 \approx 0.06$ **b.** $0.77 \div 0.0019 \approx 405.26$ **c.** $31.5 \div 0.61 \approx 516.39$

102. **a.** $\frac{9.83}{4.88} \approx 0.20$ **b.** $\frac{2.771}{0.452} \approx 6.13$ **c.** $\frac{389.224}{1.79} \approx 217.44$

103. **a.** $61.27 \div 0.057 \approx 1{,}074.91$ **b.** $0.614 \div 2.883 \approx 2.13$ **c.** $0.0035 \div 0.00481 \approx 0.73$

104. **a.** $\$365 \div \$4.89 \approx 7.46$ **b.** $\$17{,}358.27 \div \$365 \approx 47.56$ **c.** $\$3{,}000 \div \$2.54 \approx 1{,}181.10$

Mixed Practice

Solve.

105. Express $\frac{4}{5}$ as a decimal.

106. Divide $1.6\overline{)8.5}$ and round to the nearest tenth.

107. Change $1\frac{1}{6}$ to a decimal, rounded to the nearest hundredth.

108. Simplify: $81.5 - \frac{32}{0.4}$

109. What is 0.063 divided by 0.14?

110. Find the quotient: $9 \div 0.0072$

Applications

Solve and check.

111. A stalactite is an icicle-shaped mineral deposit that hangs from the roof of a cave. If it took a thousand years for a stalactite to grow to a length of 3.7 inches, how much did it grow per year?

112. In a strong earthquake, the damage to 100 houses was estimated at \$12.7 million. What was the average damage per house?

113. So far this season, the women's softball team has won 21 games and lost 14 games. The men's softball team has won 22 games and lost 18.

a. The women's team has won what fraction of the games that it played, expressed as a decimal?

b. The men's team has won what fraction of its games, expressed as a decimal?

c. Which team has a better record? Explain.

114. Yesterday, 0.08 inches of rain fell. Today, $\frac{1}{4}$ inch of rain fell.

a. How much rain fell today, expressed as a decimal?

b. Which day did more rain fall? Explain.

115. The table shown gives the best gasoline mileage for a road test of three SUVs.

SUVs	Distance Driven (in miles)	Gasoline Used (in gallons)	Miles per Gallon
A	17.4	1.2	
B	8.4	0.6	
C	23.4	1.2	

a. For each SUV, compute how many miles it gets per gallon.

b. Which SUV gives the best mileage?

116. The following table shows the number of assists that three basketball players handed out over the same three-year career period.

Player	No. of Games	No. of Assists	Average
Vince Carter	229	1,013	
Richard Hamilton	234	957	
Dwayne Wade	213	1,298	

a. Compute the average number of assists per game for each player, expressed as a decimal rounded to the nearest tenth.

b. Decide which player has the highest average.

117. If Rite Aid stock sells for \$4.20 a share, how many shares can be bought for \$8,400? (***Source:*** finance.yahoo.com, 2006)

118. A light microscope can distinguish two points separated by 0.0005 millimeters, whereas an electron microscope can distinguish two points separated by 0.0000005 millimeters. The electron microscope is how many times as powerful as the light microscope?

119. Typically, the heaviest organ in the body is the skin, weighing about 9 pounds. By contrast, the heart weighs approximately 0.7 pound. About how many times the weight of the heart is that of the skin? (***Source:*** *World of Scientific Discovery*)

120. At a community college, each student enrolled pays a \$19.50 student fee per semester. In a given semester, if the college collected \$39,000 in student fees, how many students were enrolled?

121. A shopper buys four organic chickens. The chickens weigh 3.2 pounds, 3.5 pounds, 2.9 pounds, and 3.6 pounds. How much less than the average weight of the four chickens was the weight of the lightest one?

122. A dieter joins a weight-loss club. Over a 5-month period, she loses 8 pounds, 7.8 pounds, 4 pounds, 1.5 pounds, and 0.8 pound. What was her average monthly weight loss, to the nearest tenth of a pound?

Use a calculator to solve the the following problems, giving (a) the operation(s) carried out in your solution, (b) the exact answer, and (c) an estimate of the answer.

123. Babe Ruth got 2,873 hits in 8,398 times at bat, resulting in a batting average of $\frac{2,873}{8,398}$, or approximately .342. Another great player, Ty Cobb, got 4,189 hits out of 11,434 times at bat. What was his batting average, expressed as a decimal rounded to the nearest thousandth? (Note that batting averages don't have a zero to the left of the decimal point because they can never be greater than 1.) (***Source:*** http://www.baseball-reference.com)

124. Light travels at a speed of 186,000 miles per second. If Earth is about 93,000,000 miles from the Sun, how many seconds, to the nearest tenth of a second, does it take for light to reach Earth from the Sun?

• *Check your answers on page A-6.*

MINDSTRETCHERS

Patterns

1. In the *repeating decimal* 0.142847142847142847… , identify the 994th digit to the right of the decimal point.

Groupwork

2. In the following magic square, 3.375 is the *product* of the numbers in every row, column, and diagonal. Working with a partner, fill in the missing numbers.

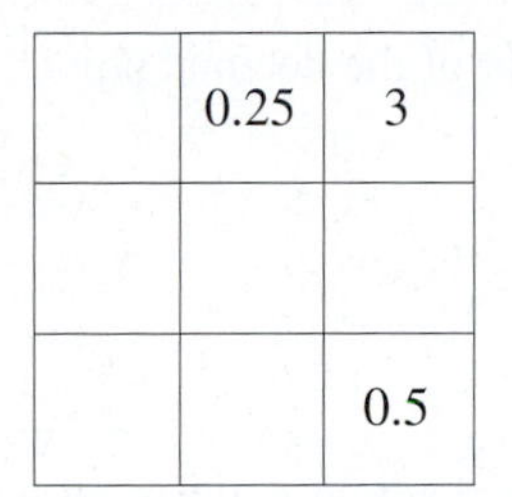

	0.25	3
		0.5

Writing

3. a. $0.5 \div 0.8 = ?$

b. $0.8 \div 0.5 = ?$

c. Find the product of your answers in parts (a) and (b). Explain how you could have predicted this product.

KEY CONCEPTS AND SKILLS

CONCEPT SKILL

Concept/Skill	Description	Example
[3.1] **Decimal**	A number written with two parts: a whole number, which precedes the decimal point, and a fractional part, which follows the decimal point.	Whole-number part → 3.721 ← Fractional part; ↑ Decimal point
[3.1] **Decimal place**	A place to the right of the decimal point.	Decimal places: 8.<u>035</u> (Tenths, Hundredths, Thousandths)
[3.1] **To change a decimal to the equivalent fraction or mixed number**	• Copy the nonzero whole-number part of the decimal and drop the decimal point. • Place the fractional part of the decimal in the numerator of the equivalent fraction. • Make the denominator of the equivalent fraction 1 followed by as many zeros as the decimal has decimal places. • Simplify the resulting fraction, if possible.	The decimal 3.25 is equivalent to the mixed number $3\frac{25}{100}$ or $3\frac{1}{4}$.
[3.1] **To compare decimals**	• Rewrite the numbers vertically, lining up the decimal points. • Working from left to right, compare the digits that have the same place value. The decimal which has the largest digit with this place value is the largest decimal.	1.073 1.06999 In the ones place and the tenths place, the digits are the same. But in the hundredths place, $7 > 6$, so $1.073 > 1.06999$.
[3.1] **To round a decimal to a given decimal place**	• Underline the digit in the place to which the number is being rounded. • Look at the digit to the right of the underlined digit—the critical digit. If this digit is 5 or more, add 1 to the underlined digit; if it is less than 5, leave the underlined digit unchanged. • Drop all decimal places to the right of the underlined digit.	$23.9\underline{3}81 \approx 23.94$ ↑ Critical digit
[3.2] **To add decimals**	• Rewrite the numbers vertically, lining up the decimal points. • Add. • Insert a decimal point in the sum below the other decimal points.	0.035 0.08 +0.00813 0.12313
[3.2] **To subtract decimals**	• Rewrite the numbers vertically, lining up the decimal points. • Subtract, inserting extra zeros in the minuend if necessary for borrowing. • Insert a decimal point in the difference below the other decimal points.	0.90370 −0.17052 0.73318

continued

CONCEPT SKILL

Concept/Skill	Description	Example
[3.3] **To multiply decimals**	• Multiply the factors as if they were whole numbers. • Find the total number of decimal places in the factors. • Count that many places from the right end of the product, and insert a decimal point.	21.07 ← Two decimal places × 0.18 ← Two decimal places 3.7926 ← Four decimal places
[3.4] **To change a fraction to the equivalent decimal**	• Divide the denominator of the fraction into the numerator, inserting to its right both a decimal point and enough zeros to get an answer either without a remainder or rounded to a given decimal place. • Place a decimal point in the quotient directly above the decimal point in the dividend.	$\frac{7}{8} = 8\overline{)7.000}$ with quotient 0.875
[3.4] **To divide decimals**	• Move the decimal point in the divisor to the right end of the number. • Move the decimal point in the dividend the same number of places to the right as in the divisor. • Insert a decimal point in the quotient directly above the decimal point in the dividend. • Divide the new dividend by the new divisor, inserting zeros at the right end of the dividend as necessary.	$3.5\overline{)71.05} =$ 20.3 $35\overline{)710.5}$ 70 10 0 105 105

Chapter 3 Review Exercises

To help you review this chapter, solve these problems.

[3.1] *Name the place that each underlined digit occupies.*

1. $8.3\underline{5}9$ **2.** $13.\underline{0}05$ **3.** $8{,}024.\underline{5}$ **4.** $0.000\underline{3}$

Express each number as a fraction, mixed number, or whole number.

5. 0.35 **6.** 8.2 **7.** 4.007 **8.** 10.000

Write each decimal in words.

9. 0.72 **10.** 5.6

11. 3.0009 **12.** 510.036

Write each decimal in decimal notation.

13. Seven thousandths **14.** Two and one tenth

15. Nine hundredths **16.** Seven and forty-one thousandths

Between each pair of numbers, insert the appropriate sign, <, =, or >, to make a true statement.

17. 0.037 0.04 **18.** 2.031 2.0301 **19.** 5.12 4.71932 **20.** 2 1.8

Rearrange each group of numbers from largest to smallest.

21. 0.72, 0.8, 1.002 **22.** 0.003, 0.00057, 0.004

Round as indicated.

23. 7.31 to the nearest tenth **24.** 0.0387 to the nearest thousandth

25. 4.3868 to two decimal places **26.** $899.09 to the nearest dollar

[3.2] *Perform the indicated operations. Check by estimating.*

27. $8.2 + 3.91$ **28.** $50 + 2.7 + 0.05$ **29.** $\$8 + \$3.25 + \$12.88$ **30.** 8.4 m $+ 3.6$ m

31. $30.7 - 1.92$ **32.** $93 - 5.248$ **33.** $2.5 - (0.72 - 0.054)$ **34.** $54.17 - (8 - 2.731)$

35. $5.398 + 8.72 + 92.035 + 0.7723 - 3.714 - 5.008$ **36.** $\$87{,}259.39 + \$2{,}098.35 + \$1{,}387.92 + \$203.14$

[3.3] *Find the product. Check by estimating.*

37. 7.28×0.4 **38.** $(288)(3.5)$ **39.** 0.005×0.002 **40.** $(3.7)^2$

41. $2.71 \cdot 1{,}000$ **42.** 0.0034×10 **43.** $8 - (1.5)^2$ **44.** $3(2.4) + 7(0.9)$

45. $18{,}772.35 \times 0.0836$ **46.** $(74.862)(5.901)$

[3.4] *Change to the equivalent decimal.*

47. $\frac{5}{8}$ **48.** $90\frac{1}{5}$ **49.** $4\frac{1}{16}$ **50.** $\frac{45}{1{,}000}$

Express each fraction as a decimal. Round to the nearest hundredth.

51. $\frac{1}{6}$ **52.** $\frac{2}{7}$ **53.** $8\frac{1}{3}$ **54.** $11\frac{2}{9}$

Divide and check.

55. $2\overline{)1.3}$ **56.** $\frac{4.8}{3}$ **57.** $0.7 \div 4$ **58.** $\frac{2.77}{10}$

Divide. Round to the nearest tenth.

59. $4.67 \div 0.9$ **60.** $\frac{2.35}{0.73}$ **61.** $\frac{7.11}{0.3}$ **62.** $0.06\overline{)981.5}$

63. $18.74 \div 9.7$ **64.** $220 \div 0.61$ **65.** $81.37\overline{)247.062}$ **66.** $247.062\overline{)81.37}$

Simplify.

67. $\frac{(1.3)^2 - 1.1}{0.5}$ **68.** $\frac{2.5 - (0.4)^2}{0.02}$ **69.** $\frac{13.75}{9.6 - 9.2}$ **70.** $(2.5)(3.5) \div 6.25$

Mixed Applications

Solve.

71. The venom of a certain South American frog is so poisonous that 0.0000004 ounce of the venom can kill a person. How is this decimal read?

72. On a certain day, the closing price of one share of Apple Computer stock was $59.58. If the closing price was $1.72 higher than the opening price, what was the opening price? (***Source:*** http://www.nasdaq.com)

73. Recently, a champion swimmer swam 50 meters in 25.2 seconds and then swam 100 meters in 29.29 seconds longer. How long did she take to swim the 150 meters?

74. Find the missing length.

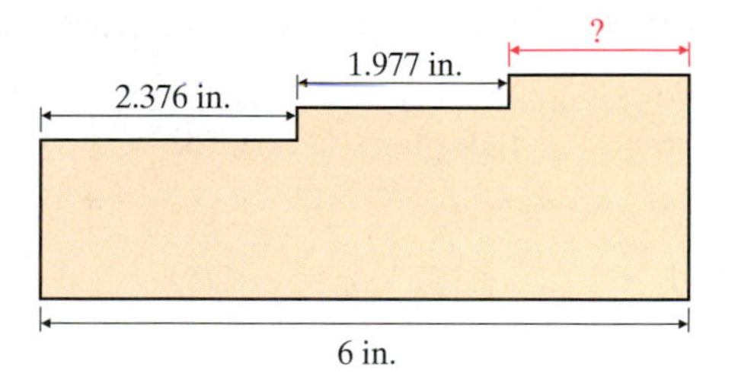

75. The fastest speed ever recorded for a spider was 1.17 miles per hour. Crawling at this speed for $\frac{1}{2}$ hour, could a spider reach a wall $\frac{3}{4}$ mile away? (***Source:*** *World Almanac and Book of Facts*)

76. In the United States Congress, there are 100 senators and 435 representatives. How many times as many representatives as senators are there?

77. A supermarket sells a 4-pound package of ground meat for $5.20 and a 5-pound package of ground meat for $6.20. What is the difference between the two prices per pound?

78. Most compact discs are sold in plastic boxes. Suppose that in a CD collection there are 29 boxes 0.4 inch thick and 3 boxes 0.94 inch thick. Estimate how many inches of shelf space is needed to house this collection.

79. In a chemistry lab, a student weighs a compound three times, getting 7.15 grams, 7.18 grams, and 7.23 grams. What is the average of these weights, to the nearest hundredth of a gram?

80. A team of geologists scaled a mountain. At the base of the mountain, the temperature had been 11°C. The temperature fell 0.75 degrees for every 100 meters the team climbed. After they climbed 1,000 meters, what was the temperature?

81. The following form was adapted from the *U.S. Individual Income Tax Return.* Find the total income in line 23.

7	Wages, salaries, tips, etc.	7	28,774.71
8	Taxable interest income	8	
9	Dividend income	9	232.55
10	Taxable refunds, credits, or offsets of state and local income taxes	10	349.77
11	Alimony received	11	
12	Business income or (loss)	12	
13	Capital gain or (loss)	13	511.74
14	Capital gain distributions not reported on line 13	14	
15	Other gains or (losses)	15	5,052.71
16	Total IRA distributions: taxable amount	16	
17	Total pensions and annuities: taxable amount	17	
18	Rents, royalties, partnerships, estates, trusts, etc.	18	1,240.97
19	Farm income or (loss)	19	
20	Unemployment compensation	20	
21	Social Security benefits: taxable amount	21	
22	Other income	22	
23	Add the amounts shown in the far right column	23	

82. The following table shows the quarterly revenues, in billions of dollars, for Google and Yahoo! for 2005. (***Source:*** http://Google.com and http://Yahoo.com)

Quarter	Google	Yahoo!
1st	1.257	1.174
2nd	1.384	1.253
3rd	1.578	1.330
4th	1.919	1.501

How much more were Google's earnings than Yahoo's for the year?

• *Check your answers on page A-7.*

Chapter 3 POSTTEST

FOR EXTRA HELP

Test solutions are found on the enclosed CD.

To see if you have mastered the topics in this chapter, take this test.

1. In the number 0.79623, which digit occupies the thousandths place?

2. Write in words: 5.102

3. Round 320.1548 to the nearest hundredth.

4. Which is smallest, 0.04, 0.0009, or 0.00028?

5. Express 3.04 as a mixed number.

6. Write as a decimal: four thousandths

Perform the indicated operations.

7. $2.3 + 0.704 + 1.35$

8. $\$5.27 + \$9 - \$8.61$

9. 2.09×10

10. 5.2×1.1

11. $(0.1)^3$

12. $\frac{3.52}{2} + \frac{4.8}{3}$

13. $2.9 \div 1{,}000$

14. $\frac{9.81}{0.3}$

Express as a decimal.

15. $\frac{3}{8}$

16. $4\frac{1}{6}$, rounded to the nearest hundredth.

Solve.

17. The element hydrogen is so light that 1 cubic foot (ft^3) of hydrogen weighs only 0.005611 pound. Round this weight to the nearest hundredth of a pound.

18. Historically, a mile was the distance that a Roman soldier covered when he took 2,000 steps. If a mile is 5,280 feet, how many feet, to the nearest tenth of a foot, was a Roman's step?

19. The Triple Crown consists of three horse races—the Kentucky Derby (1.25 miles), the Preakness Stakes (1.1875 miles), and the Belmont Stakes (1.5 miles). Which race is longest? (***Source:*** *World Almanac*)

20. To compute the annual property tax on a house in Keller, Texas, multiply the house's assessed value by 10.02807207 and then round to the nearest cent. What is the tax on a house if its assessed value is $100,000? (***Source:*** http://www.cityofkeller.com)

• *Check your answers on page A-7.*

Cumulative Review Exercises

To help you review, solve the following:

1. Round 591,622 to the nearest million.

2. Estimate: $7\frac{9}{10} \times 4\frac{1}{13}$

3. Which is larger, $1\frac{1}{2}$ or $1\frac{3}{8}$?

4. Multiply: (409)(67)

5. Subtract: $5 - 2\frac{1}{3}$

6. Calculate: $\frac{2}{5} + \frac{1}{3} \cdot \frac{1}{2}$

7. Divide: $29.89 \div 0.049$

8. A dating service advertises that it has been introducing thousands of singles for 20 years, resulting in 6,500 successful marriages. On the average, how many marriages is this per year?

9. A satellite orbiting the Earth travels at 16,000 miles per hour. An orbit takes 1.6 hours. How far will the satellite travel once around, to the nearest thousand miles?

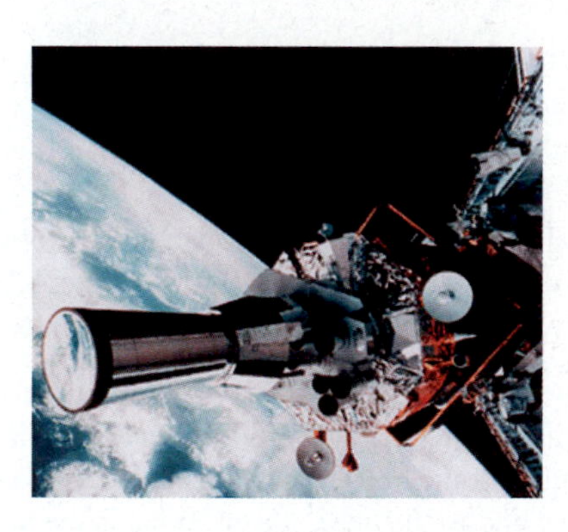

10. An electric company charges $0.09693 per kilowatt hour. If a restaurant used 2,000 kilowatt hours of electricity in a certain week, what was its weekly cost?

• *Check your answers on page A-7.*

CHAPTER 4

Basic Algebra: Solving Simple Equations

Algebra and Physics

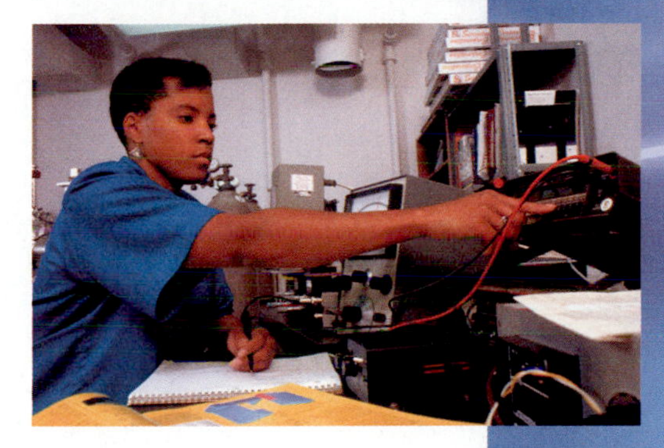

Physicists use algebra to describe the relationship between physical quantities. For instance, consider the situation in which two children are riding a seesaw. Physicists have observed that when the seesaw is balanced, the product of each child's weight and distance from the pivot are equal.

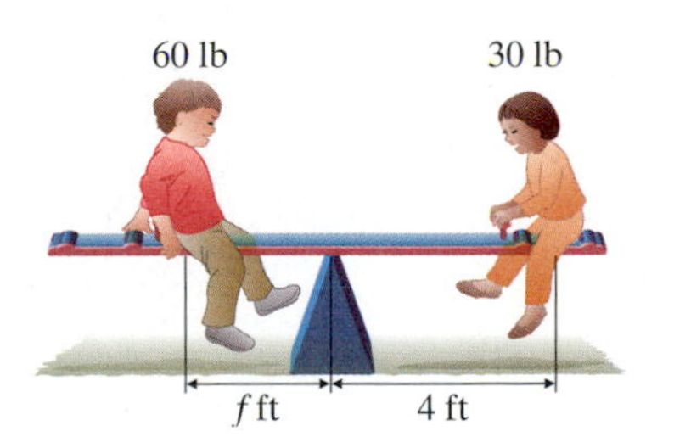

In the example pictured, if the 30-pound child is 4 feet from the pivot, we can conclude that

$$60 \cdot f = 30 \cdot 4,$$

where f represents the distance of the 60-pound child from the pivot.

Only one value of f will make the seesaw balance. To find that value, algebra can be used to solve the equation.

(***Source:*** W. Thomas Griffith, *The Physics of Everyday Phenomena,* Wm. C. Brown, 1992)

Chapter 4 PRETEST

To see if you have already mastered the topics in this chapter, take this test.

Write each algebraic expression in words.

1. $t - 4$

2. $\frac{y}{3}$

Translate each phrase to an algebraic expression.

3. 8 more than m

4. Twice n

Evaluate each algebraic expression.

5. $\frac{x}{4}$, for $x = 16$

6. $5 - y$, for $y = 3\frac{1}{2}$

Translate each sentence to an equation.

7. The sum of x and 3 equals 5.

8. The product of 4 and y is 12.

Solve and check.

9. $x + 4 = 10$

10. $t - 1 = 9$

11. $2n = 26$

12. $\frac{a}{4} = 3$

13. $8 = m + 1.9$

14. $15 = 0.5n$

15. $7m = 3\frac{1}{2}$

16. $\frac{n}{10} = 1.5$

Write an equation. Solve and check.

17. The planet Jupiter has 36 more moons than the planet Uranus. If Jupiter has 63 moons, how many does Uranus have? (***Source:*** NASA)

18. Tickets for all movies shown before 5:00 P.M. at a local movie theater qualify for the bargain matinee price, which is \$2.75 less than the regular ticket price. If the bargain price is \$6.75, what is the regular price?

19. In Michigan, about two-fifths of the area is covered with water. This portion of the state represents about 39,900 square miles. What is the area of Michigan? (***Source:*** *The New York Times Almanac, 2006*)

20. An 8-ounce cup of regular tea has about 40 milligrams of caffeine, which is 10 times the amount of caffeine in a cup of decaffeinated tea. How much caffeine is in a cup of decaffeinated tea?

• *Check your answers on page A-7.*

4.1 Introduction to Basic Algebra

What Algebra Is and Why It Is Important

In this chapter, we discuss some of the basic ideas in algebra. These ideas will be important throughout the rest of this book.

In algebra, we use letters to represent unknown numbers. The expression $2 + 3$ is arithmetic, whereas the expression $x + y$ is algebraic, since x and y represent numbers whose values are not known. With *algebraic expressions*, such as $x + y$, we can make general statements about numbers and also find the value of unknown numbers.

We can think of algebra as a *language*: The idea of translating ordinary words to algebraic notation and vice versa is the key. Often, just writing a problem algebraically makes the problem much easier to solve. We present ample proof of this point repeatedly in the chapters that follow.

We begin our discussion of algebra by focusing on what algebraic expressions mean and how to translate and evaluate them.

OBJECTIVES

- To translate phrases to algebraic expressions and vice versa
- To evaluate an algebraic expression for a given value of the variable
- To solve word problems involving algebraic expressions

Translating Phrases to Algebraic Expressions and Vice Versa

To apply mathematics to a real-world situation, we often need to be able to express that situation algebraically. Consider the following example of this kind of translation.

Suppose that you are enrolled in a college course that meets 50 minutes a day for 3 days a week. The course therefore meets for $50 \cdot 3$, or 150 minutes, a week.

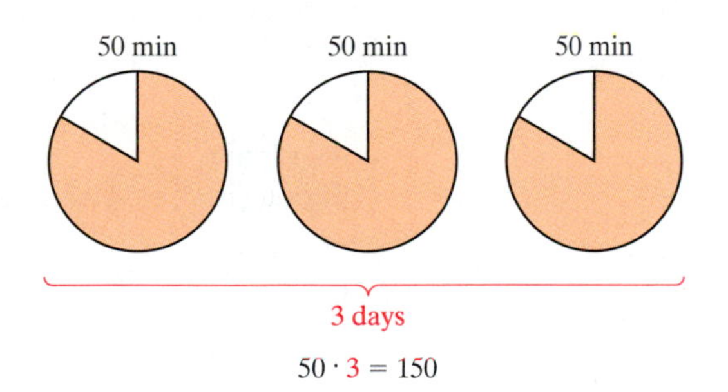

Now suppose that in a semester the 50-minute class meets d days but that we do not know what number the letter d represents. How many hours per semester does the class meet? Do you see that we can express the answer as $50d$, that is, 50 times d min?

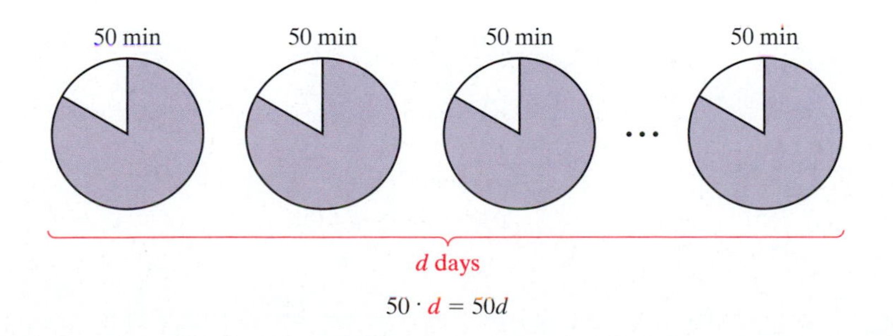

In algebra, a *variable* is a letter, or other symbol, used to represent an unknown number. In the algebraic expression $50d$, for instance, d is a variable and 50 is a *constant*. Note that in writing an algebraic expression, we usually omit any multiplication symbol: $50d$ means $50 \cdot d$.

Definitions

A **variable** is a letter that represents an unknown number.

A **constant** is a known number.

An **algebraic expression** is an expression that combines variables, constants, and arithmetic operations.

There are many translations of an algebraic expression to words, as the following table indicates.

$x + 4$ translates to	$n - 3$ translates to	$\frac{3}{4} \cdot y$ or $\frac{3}{4}y$ translates to	$z \div 5$ or $\frac{z}{5}$ translates to
• x plus 4 • x increased by 4 • the sum of x and 4 • 4 more than x • 4 added to x	• n minus 3 • n decreased by 3 • the difference between n and 3 • 3 less than n • 3 subtracted from n	• $\frac{3}{4}$ times y • the product of $\frac{3}{4}$ and y • $\frac{3}{4}$ of y	• z divided by 5 • the quotient of z and 5 • z over 5

EXAMPLE 1

Translate each algebraic expression in the table to words.

Solution

Algebraic Expression	Translation
a. $\frac{p}{3}$	p divided by 3
b. $x - 4$	4 less than x
c. $5f$	5 times f
d. $2 + y$	the sum of 2 and y
e. $\frac{2}{3}a$	$\frac{2}{3}$ of a

PRACTICE 1

Translate each algebraic expression to words.

Algebraic Expression	Translation
a. $\frac{1}{2}p$	
b. $5 - x$	
c. $y \div 4$	
d. $n + 3$	
e. $\frac{3}{5}b$	

EXAMPLE 2

Translate each word phrase in the table to an algebraic expression.

Solution

	Word Phrase	Translation
a.	16 more than m	$m + 16$
b.	the product of 5 and b	$5b$
c.	the quotient of 6 and z	$6 \div z$
d.	a decreased by 4	$a - 4$
e.	$\frac{3}{8}$ of t	$\frac{3}{8}t$

PRACTICE 2

Express each word phrase as an algebraic expression.

	Word Phrase	Translation
a.	x plus 9	
b.	10 times y	
c.	the difference between n and 7	
d.	p divided by 5	
e.	$\frac{2}{5}$ of v	

As we have seen, any letter or symbol can be used to represent a variable. For example, *five less than a number* can be translated to $n - 5$, where n represents the number.

Let's consider the following example.

EXAMPLE 3

Express each phrase as an algebraic expression.

Solution

	Word Phrase	Translation
a.	2 less than a number	$n - 2$, where n represents the number
b.	an amount divided by 10	$\frac{a}{10}$, where a represents the amount
c.	$\frac{3}{8}$ of a price	$\frac{3}{8}p$, where p represents the price

PRACTICE 3

Translate each word phrase to an algebraic expression.

	Word Phrase	Translation
a.	a quantity increased by 12	
b.	the quotient of 9 and an account balance	
c.	a cost multiplied by $\frac{2}{7}$	

Now let's look at word problems that involve translations.

EXAMPLE 4

Suppose that p partners share equally in the profits of a business. What is each partner's share if the profit was \$2,000?

Solution Each partner should get the quotient of 2,000 and p, which can be written algebraically as $\frac{2{,}000}{p}$ dollars.

PRACTICE 4

Next weekend, a student wants to study for his four classes. If he has h hours to study in all and he wants to devote the same amount of time to each class, how much time will he study per class?

EXAMPLE 5

At registration, n out of 100 classes are closed. How many classes are not closed?

Solution Since n classes are closed, the remainder of the 100 classes are not closed. So we can represent the number of classes that are not closed by the algebraic expression $100 - n$.

PRACTICE 5

Of s shrubs in front of a building, 3 survived the winter. How many shrubs died over the winter?

Evaluating Algebraic Expressions

In this section, we look at how to evaluate algebraic expressions. Let's begin with a simple example.

Suppose that the balance in a savings account is \$200. If d dollars is then deposited, the balance will be $(200 + d)$ dollars.

To evaluate the expression $200 + d$ for a particular value of d, we replace d with that number. If \$50 is deposited, we replace d by 50:

$$200 + d = 200 + \mathbf{50} = 250$$

So the new balance will be \$250.

The following rule is helpful for evaluating expressions.

To Evaluate an Algebraic Expression

- Substitute the given value for each variable.
- Carry out the computation.

Now let's consider some more examples.

EXAMPLE 6

Evaluate each algebraic expression.

Solution

	Algebraic Expression	Value
a.	$n + 8$, if $n = 15$	$\mathbf{15} + 8 = 23$
b.	$9 - z$, if $z = 7.89$	$9 - \mathbf{7.89} = 1.11$
c.	$\frac{2}{3}r$, if $r = 18$	$\frac{2}{3} \cdot \mathbf{18} = 12$
d.	$y \div 4$, if $y = 3.6$	$\mathbf{3.6} \div 4 = 0.9$

PRACTICE 6

Find the value of each algebraic expression.

	Algebraic Expression	Value
a.	$\frac{s}{4}$, if $s = 100$	
b.	$0.2y$, if $y = 1.9$	
c.	$x - 4.2$, if $x = 9$	
d.	$25 + z$, if $z = 1.6$	

The following examples illustrate how to write and evaluate expressions to solve word problems.

EXAMPLE 7

Power consumption for a period of time is measured in watt-hours, where a watt-hour means 1 watt of power for 1 hour. How many watt-hours of energy will a 60-watt bulb consume in h hours? In 3 hours?

Solution The expression that represents the number of watt-hours used in h hours is $60h$. So for $h = 3$, the number of watt-hours is $60 \cdot \mathbf{3}$, or 180. Therefore, 180 watt-hours of energy will be consumed in 3 hours.

PRACTICE 7

To avoid paying private mortgage insurance, home buyers are required to make a down payment that is at least one-fifth of the purchase price of the home. What is the required down payment for a home with a purchase price of p dollars? With a purchase price of \$349,000?

EXAMPLE 8

Suppose that there are 180 days in the local school year. How many days was a student present at school if she was absent d days? 9 days?

Solution If d represents the number of days that the student was absent, the expression $180 - d$ represents the number of days that she was present. If she was absent 9 days, we substitute 9 for d in the expression:

$$180 - d = 180 - 9 = 171$$

So the student was present 171 days.

PRACTICE 8

At a coffee shop, a lunch bill came to \$15.45 plus the tip. What is the total amount of the lunch, including a tip of t dollars? A tip of \$3?

CULTURAL NOTE

Solving an equation to identify an unknown number is similar to using a balance scale to determine an unknown weight. In this picture that dates from 3,400 years ago, an Egyptian weighs gold rings against a counterbalance in the form of a bull's head.

The balance scale is an ancient measuring device. These scales were used by Sumerians for weighing precious metals and gems at least 9,000 years ago.

Source: O. A.W. Dilke, *Mathematics and Measurement* (Berkeley: University of California Press/British Museum, 1987).

Mathematically Speaking

Fill in each blank with the most appropriate term or phrase from the given list.

arithmetic	constant	evaluate
translate	variable	algebraic

1. A(n) _______ is a letter that represents an unknown number.

2. A(n) _______ is a known number.

3. A(n) _______ expression combines variables, constants, and arithmetic operations.

4. To _______ an algebraic expression, replace each variable with the given number, and carry out the computation.

Translate each algebraic expression to two different word phrases.

5. $t + 9$

6. $8 + r$

7. $c - 12$

8. $x - 5$

9. $c \div 3$

10. $\frac{z}{7}$

11. $10s$

12. $11t$

13. $y - 10$

14. $w - 1$

15. $7a$

16. $4x$

17. $x \div 6$

18. $\frac{y}{5}$

19. $x - \frac{1}{2}$

20. $x - \frac{1}{3}$

21. $\frac{1}{4}w$

22. $\frac{4}{5}y$

23. $2 - x$

24. $8 - y$

25. $1 + x$

26. $n + 7$

27. $3p$

28. $2x$

29. $n - 1.1$

30. $x - 6.5$

31. $y \div 0.9$

32. $\frac{n}{2.4}$

Translate each word phrase to an algebraic expression.

33. x plus 10

34. d increased by 12

35. 1 less than n

36. b decreased by 9

37. the sum of y and 5

38. 11 more than x

39. t divided by 6

40. r over 2

41. the product of 10 and y

42. 5 times p

43. w minus 5

44. the difference between n and 5

45. n increased by $\frac{4}{5}$

46. the sum of x and 0.4

47. the quotient of z and 3

48. n divided by 10

49. $\frac{2}{7}$ of x

50. the product of $\frac{2}{3}$ and y

51. 6 subtracted from k

52. 8 less than z

53. 12 more than a number

54. the sum of a number and 18

55. the difference between a number and 5.1

56. $\frac{3}{4}$ less than a number

Evaluate each algebraic expression.

57. $y + 7$, if $y = 19$

58. $3 + n$, if $n = 2.9$

59. $7 - x$, if $x = 4.5$

60. $y - 19$, if $y = 25$

61. $\frac{3}{4}p$, if $p = 20$

62. $\frac{4}{5}n$, if $n = 1\frac{1}{4}$

63. $x \div 2$, if $x = 2\frac{1}{3}$

64. $\frac{n}{3}$, if $n = 7.5$

65. $p - 7.9$, if $p = 9$

66. $20.1 + y$, if $y = 7$

67. $x \div \frac{5}{6}$, if $x = \frac{1}{6}$

68. $\frac{1}{3}y$, if $y = \frac{1}{2}$

Complete each table.

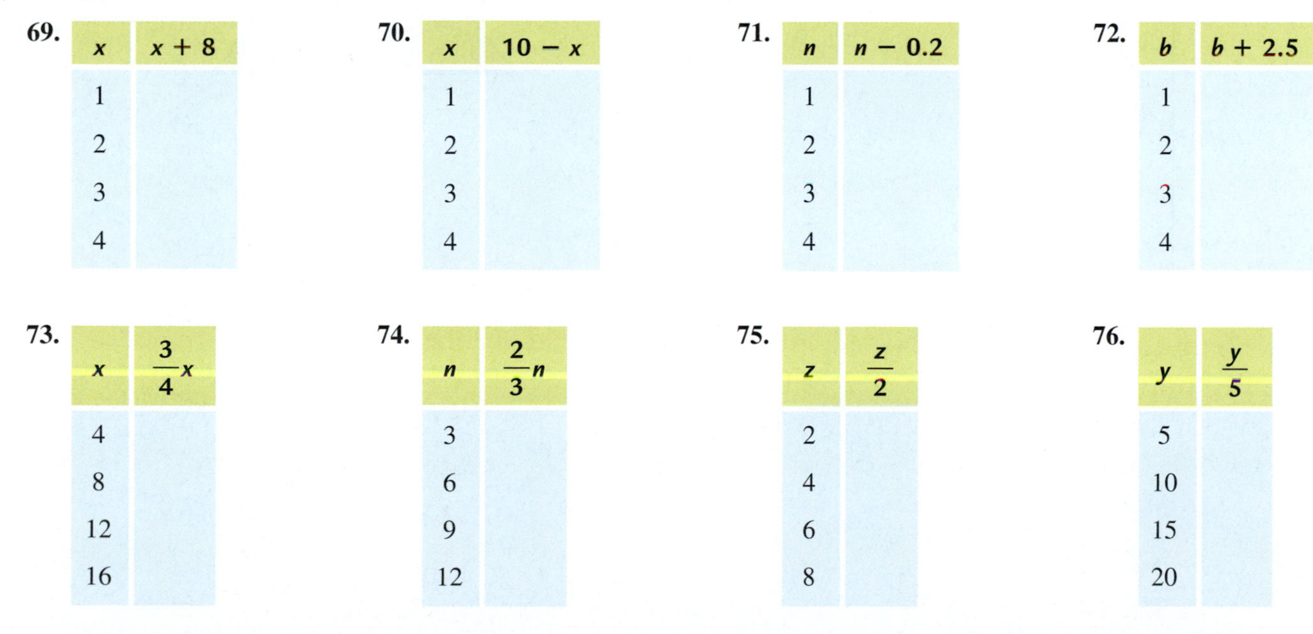

69.

x	$x + 8$
1	
2	
3	
4	

70.

x	$10 - x$
1	
2	
3	
4	

71.

n	$n - 0.2$
1	
2	
3	
4	

72.

b	$b + 2.5$
1	
2	
3	
4	

73.

x	$\frac{3}{4}x$
4	
8	
12	
16	

74.

n	$\frac{2}{3}n$
3	
6	
9	
12	

75.

z	$\frac{z}{2}$
2	
4	
6	
8	

76.

y	$\frac{y}{5}$
5	
10	
15	
20	

Mixed Practice

Solve.

77. Translate the phrase "7 less than x" to an algebraic expression.

78. Evaluate the algebraic expression $0.5t$, if $t = 8$.

79. Translate the algebraic expression $\frac{n}{2}$ to two different phrases.

80. Evaluate the algebraic expression $\frac{1}{4}y$, if $y = \frac{2}{3}$.

81. Translate the phrase "the product of 3.5 and t" to an algebraic expression.

82. Evaluate the algebraic expression $x + 1$, if $x = 4$.

83. Translate the algebraic expression $x + 6$ to two different phrases.

84. Evaluate the algebraic expression $n - 20$, if $n = 30$.

Applications

Solve.

85. A patient receives m milligrams of medication per dose. Her doctor orders her medication to be decreased by 25 milligrams. How much medication will she then receive per dose?

86. When you take out a mortgage, each monthly payment has two parts. One part goes toward the principal and the other toward the interest. If the principal payment is \$344.86 and the interest payment is i, write an algebraic expression for the total payment.

87. For the triangle shown, write an expression for the sum of the measures of the three angles.

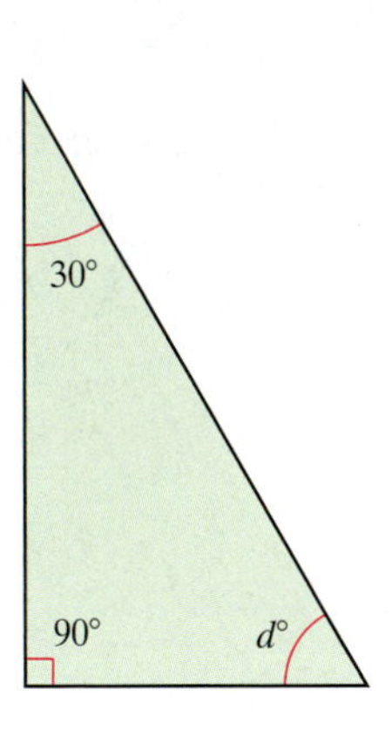

88. Write an expression for the sum of the lengths of the sides in the trapezoid shown.

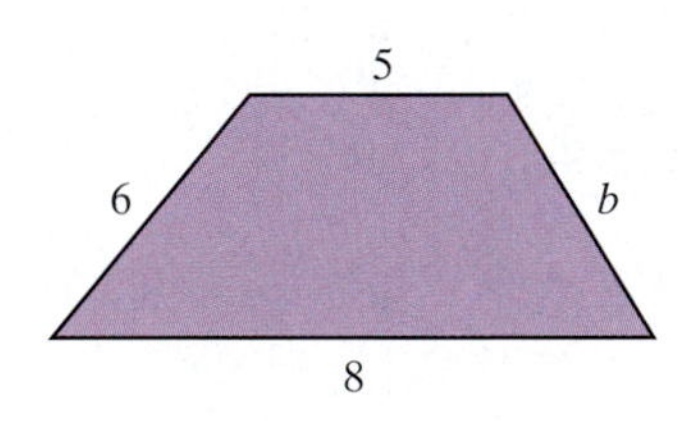

89. If a long-distance trucker drives at a speed of r miles per hour for t hours, she will travel a distance of $r \cdot t$ miles. How far will she travel at a speed of 55 miles per hour in 4 hours?

90. If a basketball player makes b baskets in a attempts, his field goal average is defined to be $\frac{b}{a}$. Find the field goal average of a player who made 12 baskets in 25 attempts.

91. A bank charges customers a fee of \$1.50 for each withdrawal made at its ATMs.

a. Write an expression for the total fee charged to a customer for w of these withdrawals

b. Find the total fee if the customer makes 9 withdrawals.

92. A computer network technician charges \$80 per hour for labor.

a. Write an expression for the cost of h hours of work.

b. Find the cost of a networking job that takes $2\frac{1}{2}$ hours.

• *Check your answers on page A-7.*

MINDSTRETCHERS

Mathematical Reasoning

1. Consider the expression $x + x$.

a. Why does this expression mean the same as the expression $2x$?

b. What does the expression $\underbrace{x + x + x + \cdots + x}_{n \text{ times}}$ mean in terms of multiplication?

Groupwork

2. Working with a partner, consider the areas of the following rectangles. For some values of x, the rectangle on the left has a larger area; for other values of x, the rectangle on the right is larger.

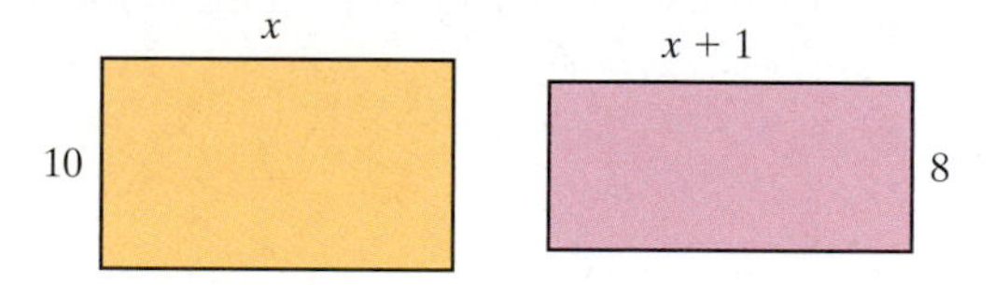

a. Find a value of x for which the rectangle on the left has the larger area.

b. Find a value of x for which the area of the rectangle on the right is larger.

Writing

3. Algebra is universal; that is, it is used in all countries of the world regardless of the language spoken. If you know how to speak a language other than English, translate each of the following algebraic expressions to that language.

a. $7x$ **b.** $x - 2$ **c.** $3 + x$ **d.** $\frac{x}{3}$

4.2 Solving Addition and Subtraction Equations

OBJECTIVES

- To translate sentences to equations involving addition or subtraction
- To solve addition and subtraction equations
- To solve word problems involving equations with addition or subtraction

What an Equation Is

An equation contains two expressions separated by an equal sign.

$$\underbrace{x + 3}_{\text{Left side}} \overset{\text{Equal sign}}{=} \underbrace{5}_{\text{Right side}}$$

Definition

An **equation** is a mathematical statement that two expressions are equal.

For example,

$$1 + 2 = 3$$
$$x - 5 = 6$$
$$2 + 7 + 3 = 12$$
$$3x = 9$$

are all equations.

Equations are used to solve a wide range of problems. A key step in solving a problem is to translate the sentences that describe the problem to an equation that models the problem. In this section, we focus on equations that involve either addition or subtraction. In the next section, we consider equations involving multiplication or division.

Translating Sentences to Equations

In translating sentences to equations, certain words and phrases mean the same as the equal sign:

- equals
- is
- is equal to
- is the same as
- yields
- results in

Let's look at some examples of translating sentences to equations that involve addition or subtraction and vice versa.

EXAMPLE 1

Translate each sentence in the table to an equation.

Solution

	Sentence	Equation
a.	The sum of y and 3 is equal to $7\frac{1}{2}$.	$y + 3 = 7\frac{1}{2}$
b.	The difference between x and 9 is the same as 14.	$x - 9 = 14$
c.	Increasing a number by 1.5 yields 3.	$n + 1.5 = 3$
d.	6 less than a number is 10.	$n - 6 = 10$

PRACTICE 1

Write an equation for each word phrase or sentence.

	Sentence	Equation
a.	n decreased by 5.1 is 9.	
b.	y plus 2 is equal to 12.	
c.	The difference between a number and 4 is the same as 11.	
d.	5 more than a number is $7\frac{3}{4}$.	

EXAMPLE 2

In a savings account, the previous balance, P, plus a deposit of \$7.50 equals the new balance of \$43.25. Write an equation that represents this situation.

Solution

The previous balance plus the deposit equals the new balance.

$$P \quad + \quad 7.50 \quad = \quad 43.25$$

So the equation is $P + 7.50 = 43.25$.

PRACTICE 2

The sale price of a jacket is \$49.95. This amount is \$6 less than the regular price p. Write an equation that represents this situation.

Equations Involving Addition and Subtraction

Suppose that you are told that five *more than some number* is equal to seven. You can find that number by solving the addition equation $x + 5 = 7$. To solve an equation means to find a number that, when substituted for the variable x, makes the equation a true statement. Such a number is called a *solution* of the equation.

To solve the equation $x + 5 = 7$, we can think of a balance scale like the one shown.

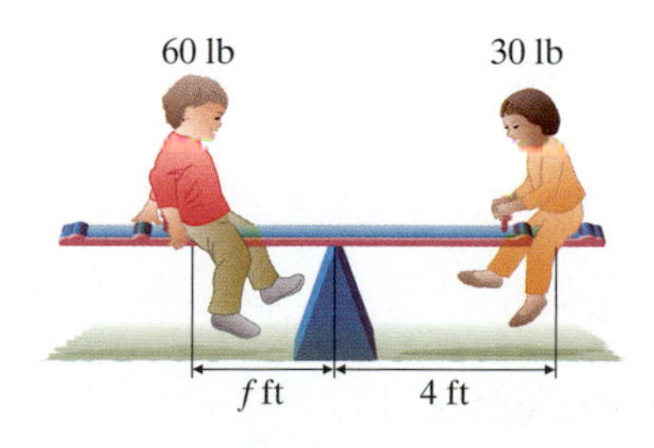

For the balance to remain level, whatever we do to one side, we must also do to the other side. In this case, if we subtract 5 g from each side of the balance, we can conclude that the unknown weight, x, must be 2 g, as shown on the next page. So 2 is the solution of the equation $x + 5 = 7$.

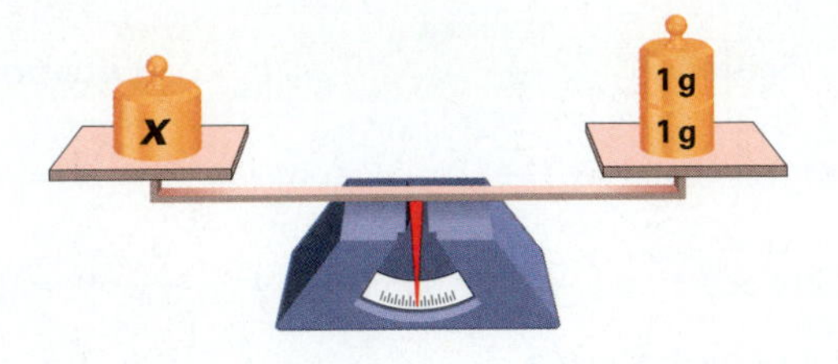

Similarly in the *subtraction equation* $x - 5 = 7$, if we add 5 to each side of the equation, we find that x equals 12.

In solving these and other equations, the key is to **isolate the variable**, that is, to get the variable alone on one side of the equation.

These examples suggest the following rule.

To Solve Addition or Subtraction Equations

- For an addition equation, *subtract* the same number from each side of the equation in order to isolate the variable on one side.
- For a subtraction equation, *add* the same number to each side of the equation in order to isolate the variable on one side.
- In either case, check the solution by substituting the value of the unknown in the original equation to verify that the resulting equation is true.

Because addition and subtraction are **opposite operations**, one operation "undoes" the other. The following examples illustrate how to perform an opposite operation to each side of an equation when you are solving for the unknown.

EXAMPLE 3

Solve and check: $y + 9 = 17$

Solution

$$y + 9 = 17$$

$$\underbrace{y + 9 - 9}_{} = \underbrace{17 - 9}_{}$$ **Subtract 9 from each side of the equation.**

$$\underbrace{y + 0}_{} = 8$$

$$y = 8$$ **Any number added to 0 is the number.**

Check

$$y + 9 = 17$$

$$8 + 9 \stackrel{?}{=} 17$$ **Substitute 8 for y in the original equation.**

$$17 \stackrel{\checkmark}{=} 17$$ **The equation is true, so 8 is the solution to the equation.**

Note that, because 9 was added to y in the original equation, we solved by subtracting 9 from both sides of the equation in order to isolate the variable.

PRACTICE 3

Solve and check: $x + 5 = 14$

EXAMPLE 4

Solve and check: $n - 2.5 = 0.7$

Solution

$$n - 2.5 = 0.7$$

$$\underbrace{n - 2.5 + 2.5}_{n - 0} = \underbrace{0.7 + 2.5}_{3.2}$$

Add 2.5 to each side of the equation.

$$\underbrace{n - 0}_{n} = 3.2$$

$$n = 3.2$$

Check

$$n - 2.5 = 0.7$$

$$3.2 - 2.5 \stackrel{?}{=} 0.7$$

Substitute 3.2 for n in the original equation.

$$0.7 \stackrel{\checkmark}{=} 0.7$$

Can you explain why checking an answer is important?

PRACTICE 4

Solve and check: $t - 0.9 = 1.8$

EXAMPLE 5

Solve and check: $x + \frac{1}{3} = 3\frac{1}{2}$

Solution

$$x + \frac{1}{3} = 3\frac{1}{2}$$

$$x + \frac{1}{3} - \frac{1}{3} = 3\frac{1}{2} - \frac{1}{3}$$

Subtract $\frac{1}{3}$ from each side of the equation.

$$x = 3\frac{1}{6}$$

Check

$$x + \frac{1}{3} = 3\frac{1}{2}$$

$$3\frac{1}{6} + \frac{1}{3} \stackrel{?}{=} 3\frac{1}{2}$$

Substitute $3\frac{1}{6}$ for x in the original equation.

$$3\frac{1}{2} \stackrel{\checkmark}{=} 3\frac{1}{2}$$

PRACTICE 5

Solve and check: $m + \frac{1}{4} = 5\frac{1}{2}$

Equations are often useful **mathematical models** of real-world situations, as the following examples show. To derive these models, we need to be able to translate word sentences to algebraic equations, which we then solve.

EXAMPLE 6

Write each sentence as an equation. Then solve and check.

Solution

	Sentence	Equation	Check
a.	15 is equal to y increased by 9.	$15 = 9 + y$ $15 - 9 = 9 - 9 + y$ $6 = y$, or $y = 6$	$15 = 9 + y$ $15 \stackrel{?}{=} 9 + 6$ $15 \stackrel{\checkmark}{=} 15$
b.	10 is equal to m decreased by 8.	$10 = m - 8$ $10 + 8 = m - 8 + 8$ $18 = m$, or $m = 18$	$10 = m - 8$ $10 \stackrel{?}{=} 18 - 8$ $10 \stackrel{\checkmark}{=} 10$

Note that we isolated the variable on the right side of the equation instead of the left side. The result is the same.

PRACTICE 6

Translate each sentence to an algebraic equation. Then solve and check.

a. 11 is 4 less than m.

b. The sum of 12 and n equals 21.

EXAMPLE 7

Suppose that a chemistry experiment requires students to find the mass of the water in a flask. If the mass of the flask with water is 21.49 grams and the mass of the empty flask is 9.56 grams, write an equation to find the mass of the water. Then solve and check.

Solution Recall that some problems can be solved by drawing a diagram. Let's use that strategy here.

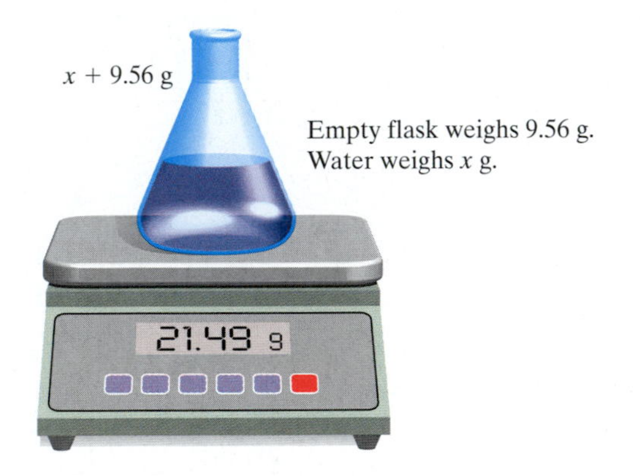

The diagram suggests the equation $21.49 = x + 9.56$, where x represents the mass of the water. Solving this equation, we get

$$21.49 = x + 9.56$$
$$21.49 - 9.56 = x + 9.56 - 9.56$$
$$11.93 = x, \text{ or } x = 11.93$$

The weight of the water is 11.93 grams.

Check
$$21.49 = x + 9.56$$
$$21.49 \stackrel{?}{=} 11.93 + 9.56$$
$$21.49 \stackrel{\checkmark}{=} 21.49$$

PRACTICE 7

An online discount book retailer charges a shipping fee of \$3.99. The total cost of a book, including the shipping fee, was \$27.18. Write an equation to determine the cost of the book without the shipping fee. Then solve and check.

EXAMPLE 8

Harvard College (in Cambridge, Massachusetts) and the College of William and Mary (in Williamsburg, Virginia) are the two oldest institutions of higher learning in the United States. Harvard, founded in 1636, is 57 years older than William and Mary. When was William and Mary founded? (*Source: The Top Ten of Everything, 2006*)

Solution Let x represent the year in which William and Mary was founded. We know that 57 years earlier than the year x is 1636, the year in which Harvard was founded. This gives us the equation

$$x - 57 = 1636$$

Now we solve for the unknown:

$$x - 57 + 57 = 1636 + 57$$
$$x = 1693$$

So William and Mary was founded in 1693.

PRACTICE 8

The two U.S. states with the largest area are Alaska and Texas. The land area of Texas, 262,000 square miles, is approximately 308,000 square miles smaller than that of Alaska. Write an equation to determine Alaska's land area. Then solve and check. (*Source: Time Almanac, 2006*)

4.2 Exercises

Mathematically Speaking

Fill in each blank with the most appropriate term or phrase from the given list.

constant	subtract	equation
translates	simplifies	variable
add	sentence	

1. A(n) _______ is a mathematical statement that two expressions are equal.

2. A solution of an equation is a number that when substituted for the _______ makes the equation a true statement.

3. In the equation $x + 2 = 5$, _______ from each side of the equation in order to isolate the variable.

4. The equation $x - 1 = 6$ _______ to the sentence "The difference between x and 1 is 6."

Translate each sentence to an equation.

5. z minus 9 is 25.

6. x decreased by 7 yields 29.

7. The sum of 7 and x is 25.

8. m plus 19 equals 34.

9. t decreased by 3.1 equals 4.

10. r minus 5 is equal to 6.4.

11. $\frac{3}{2}$ increased by a number yields $\frac{9}{2}$.

12. The sum of a number and $2\frac{1}{3}$ is 8.

13. $3\frac{1}{2}$ less than a number is equal to 7.

14. The difference between a number and $1\frac{1}{2}$ is the same as $7\frac{1}{4}$.

By answering yes or no, indicate whether the value of x shown is a solution of the given equation.

15.

Equation	Value of x	Solution?
a. $x + 1 = 9$	8	
b. $x - 3 = 4$	5	
c. $x + 0.2 = 5$	4.8	
d. $x - \frac{1}{2} = 1$	$\frac{1}{2}$	

16.

Equation	Value of x	Solution?
a. $x - 39 = 5$	44	
b. $x - 2 = 6$	4	
c. $x + 2.8 = 4$	1.2	
d. $x - \frac{2}{3} = 1$	$1\frac{2}{3}$	

Identify the operation to perform on each side of the equation to isolate the variable.

17. $x + 4 = 6$

18. $x - 6 = 9$

19. $x - 11 = 4$

20. $x + 10 = 17$

21. $x - 7 = 24$

22. $x + 21 = 25$

23. $3 = x + 2$

24. $10 = x - 3$

Solve and check.

25. $a - 7 = 24$

26. $x - 9 = 13$

27. $y + 19 = 21$

28. $z + 23 = 31$

29. $x - 2 = 10$

30. $t - 4 = 19$

31. $n + 9 = 13$

32. $d + 12 = 12$

33. $5 + m = 7$

34. $17 + d = 20$

35. $39 = y - 51$

36. $44 = c - 3$

37. $z + 2.4 = 5.3$

38. $t + 2.3 = 6.7$

39. $n - 8 = 0.9$

40. $c - 0.7 = 6$

41. $y + 8.1 = 9$

42. $a + 0.7 = 2$

43. $x + \frac{1}{3} = 9$

44. $z + \frac{2}{5} = 11$

45. $m - 1\frac{1}{3} = 4$

46. $s - 4\frac{1}{2} = 8$

47. $x + 3\frac{1}{4} = 7$

48. $t + 1\frac{1}{2} = 5$

49. $c - 14\frac{1}{5} = 33$

50. $a - 9\frac{7}{10} = 27\frac{2}{3}$

51. $x - 3.4 = 9.6$

52. $m - 12.5 = 13.7$

53. $5 = y - 1\frac{1}{4}$

54. $3 = t - 1\frac{2}{3}$

55. $5\frac{3}{4} = a + 2\frac{1}{3}$

56. $4\frac{1}{3} = n + 3\frac{1}{2}$

57. $2.3 = x - 5.9$

58. $4.1 = d - 6.9$

59. $y - 7.01 = 12.9$

60. $x - 3.2 = 5.23$

61. $x + 3.443 = 8$

62. $x + 0.035 = 2.004$

63. $2.986 = y - 7.265$

64. $3.184 = y - 1.273$

Translate each sentence to an equation. Solve and check.

65. 3 more than n is 11.

66. The sum of x and 15 is the same as 33.

67. 6 less than y equals 7.

68. The difference between t and 4 yields 1.

69. If 10 is added to n, the sum is 19.

70. 25 added to a number m gives a result of 53.

71. x increased by 3.6 is equal to 9.

72. n plus $3\frac{1}{2}$ equals 7.

73. A number minus $4\frac{1}{3}$ is the same as $2\frac{2}{3}$.

74. A number decreased by 1.6 is 5.9.

Choose the equation that best describes the situation.

75. After 6 months of dieting and exercising, an athlete lost $8\frac{1}{2}$ pounds. If she now weighs 135 pounds, what was her original weight?

a. $w + 8\frac{1}{2} = 135$

b. $w - 126\frac{1}{2} = 8\frac{1}{2}$

c. $w - 8\frac{1}{2} = 135$

d. $w + 135 = 143\frac{1}{2}$

76. A teenager has d dollars. After buying an Xbox 360 for $59.99, he has $6.01 left. How many dollars did he have at first?

a. $d + 59.99 = 66$

b. $d - 59.99 = 6.01$

c. $d - 59.99 = 66$

d. $d + 6.01 = 59.99$

77. A gigabyte (GB) is a unit used to measure computer memory. The hard drive of a computer has 10.18 GB of used space, with the rest free space. If the total capacity of the hard drive is 69.52 GB, find the amount of free space on the hard drive.

a. $x + 10.18 = 69.52$ **b.** $x - 10.18 = 69.52$

c. $x + 69.52 = 10.18$ **d.** $x - 69.52 = 10.18$

78. At a certain college, tuition costs a student $2,000 a semester. If the student received $1,250 in financial aid, how much more money does he need for the semester's tuition?

a. $x + 1{,}250 = 2{,}000$ **b.** $x + 2{,}000 = 3{,}250$

c. $x - 2{,}000 = 1{,}250$ **d.** $x - 1{,}250 = 2{,}000$

Mixed Practice

Solve and check.

79. $10 = a - 4.5$

80. $x + \frac{1}{2} = 6$

Solve.

81. The life expectancy in the United States of a female born in the year 1990 was 78.8 years. A decade later, it was 0.9 years greater. Choose the equation to find the life expectancy of a female born in the year 2000. (*Source:* The National Center for Health Statistics)

a. $x + 78.8 = 0.9$ **b.** $x - 0.9 = 78.8$

c. $x + 0.9 = 78.8$ **d.** $x - 10 = 0.9$

82. Identify the operation to perform on each side of the equation $y - 1 = 5$ to isolate the variable.

83. Is 3 a solution to the equation $10 - x = 7$?

84. Is 6 a solution to the equation $x + 4.5 = 7.5$?

85. Identify the operation to perform on each side of the equation $n + 2 = 10$ to isolate the variable.

86. Translate the sentence "x decreased by 4 is 10" to an equation.

87. The hygienist at a dentist's office cleaned a patient's teeth. The total bill came to $125, which was partially covered by dental insurance. If the patient paid $60 out of pocket toward the bill, choose the equation to find how much of the bill was covered by insurance.

a. $x + 60 = 125$ **b.** $x - 60 = 125$

c. $x + 125 = 60$ **d.** $x - 125 = 60$

88. Translate the sentence "The sum of 4.2 and n is 8" to an equation.

Applications

Write an equation. Solve and check.

89. An article on Broadway shows reported that this week the box office receipts for a particular show were $621,000. If that amount was $13,000 less than last week's, how much money did the show take in last week?

90. The first algebra textbook was written by the Arab mathematician Muhammad ibn Musa al-Khwarazmi. The title of that book, which gave rise to the word *algebra*, was *Aljabr wa'lmuqabalah*, meaning "the art of bringing together unknowns to match a known quantity." If the book appeared in the year 825, how many years ago was this? (*Source:* R.V. Young, *Notable Mathematicians*)

91. In the triangle shown, angles A and B are complementary, that is, the sum of their measures is 90°. Find x, the number of degrees in angle B.

92. In the following diagram, angles ABD and CBD are supplementary, that is, the sum of their measures is 180°. If the measure of angle ABD is 109°, find y.

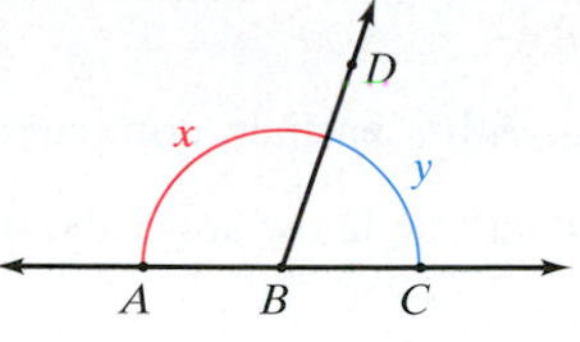

93. A local community college increased the cost of a credit hour by $12 this year. If the cost of a credit hour this year is $96, what was the cost last year?

94. Mount Kilimanjaro, the highest elevation on the continent of Africa, is 299 meters lower than Mount McKinley, the highest elevation on the continent of North America. If Mount Kilimanjaro is 5,895 meters high, how high is Mount McKinley? (**Source:** *The World Factbook, 2006*)

95. On a state freeway, the minimum speed limit is 45 miles per hour. This is 20 miles per hour lower than the maximum speed limit. What is the maximum speed limit?

96. The melting point of silver is 1,763 degrees Fahrenheit. This is 185 degrees less than the melting point of gold. What is the melting point of gold? (**Source:** *The New York Times Almanac, 2006*)

Use a calculator to solve the following problems, giving (a) the equation, (b) the exact answer, and (c) an estimate of the answer.

97. In a recent year, the U.S. charity that received the greatest private support ($1,324,089,000) was the Salvation Army. The charity that received the second greatest private support ($794,000,000) was the American Cancer Society. How much more money did the Salvation Army receive? (**Source:** *The Chronicle of Philanthropy, 2004*)

98. During a recession, an automobile company laid off 18,578 employees, reducing its workforce to 46,894. Write an equation that describes the number of employees the company had before the recession. Then solve and check.

• *Check your answers on page A-8.*

MINDSTRETCHERS

Groupwork

1. Working with a partner, compare the equations $x - 4 = 6$ and $x - a = b$.

   a. Use what you know about the first equation to solve the second equation for x.

   b. What are the similarities and the differences between the two equations?

Writing

2. Equations often serve as models for solving word problems. Write two different word problems corresponding to each of the following equations.

   a. $x + 4 = 9$

   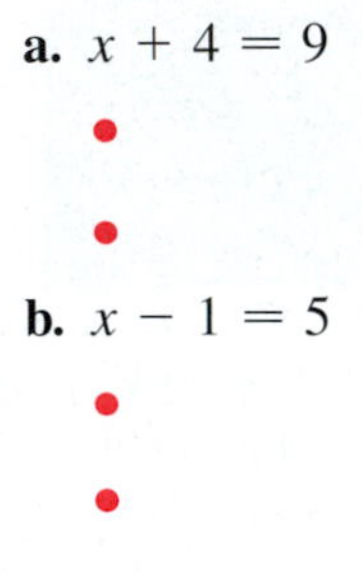

   •

   •

   b. $x - 1 = 5$

   •

   •

Critical Thinking

3. In the magic square at the right, the sum of each row, column, and diagonal is the same. Find that sum and write and solve equations to get the values of f, g, h, r, and t.

f	6	11
g	10	h
r	14	t

4.3 Solving Multiplication and Division Equations

Translating Sentences to Equations

OBJECTIVES

- To translate sentences to equations involving multiplication or division
- To solve multiplication and division equations
- To solve word problems involving equations with multiplication or division

In order to translate sentences involving multiplication or division to equations, we must recall the key words that indicate when to multiply and when to divide.

EXAMPLE 1

Translate each sentence in the table to an equation.

Solution

Sentence	Equation
a. The product of 3 and x is equal to 0.6.	$3x = 0.6$
b. The quotient of y and 4 is 15.	$\frac{y}{4} = 15$
c. Two-thirds of a number is 9.	$\frac{2}{3}n = 9$
d. One-half is equal to some number over 6.	$\frac{1}{2} = \frac{n}{6}$

PRACTICE 1

Write an equation for each sentence.

Sentence	Equation
a. Twice x is the same as 14.	
b. The quotient of a and 6 is 1.5.	
c. Some number divided by 0.3 is equal to 1.	
d. Ten is equal to one-half of some number.	

EXAMPLE 2

A house that sold for $125,000 is twice its assessed value, x. Write an equation to represent this situation.

Solution The selling price of the house is twice its assessed value, x.

$$\underbrace{\text{The selling price of the house}}_{125{,}000} \ \underset{=}{\text{is}} \ \underbrace{\text{twice its assessed value, } x.}_{2x}$$

So the equation is $125{,}000 = 2x$.

PRACTICE 2

The area of a rectangle is equal to the product of its length (3 feet) and its width (w). The rectangle's area is 15 square feet. Represent this relationship in an equation.

Equations Involving Multiplication and Division

As with addition equations, we can also solve *multiplication equations* by thinking of a balance scale like the one shown below at the left.

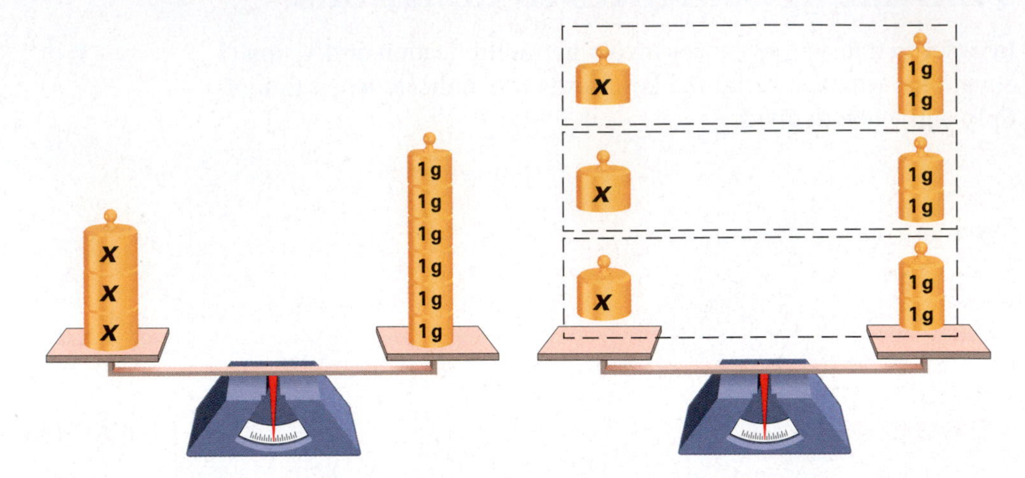

For example, consider the sentence "Three times some number x equals six," which translates to the multiplication equation $3x = 6$. We want to find the number for the variable x that, when substituted, makes this equation a true statement. To keep the balance level, whatever we do to one side we must do to the other side. In this case, dividing each side of the balance by 3 shows that in each group the unknown, x, must equal 2, as shown above at the right.

Similarly, in the division equation $\frac{x}{4} = 3$, we can multiply each side of the equation by 4 and then conclude that x equals 12.

These examples suggest the following rule.

To Solve Multiplication or Division Equations

- For a multiplication equation, divide by the same number on each side of the equation in order to isolate the variable on one side.
- For a division equation, multiply by the same number on each side of the equation in order to isolate the variable on one side.
- In either case, check the solution by substituting the value of the unknown in the original equation to verify that the resulting equation is true.

Because multiplication and division are opposite operations, one "undoes" the other. The following examples show how to perform the opposite operation on each side of an equation to solve for the unknown.

EXAMPLE 3

Solve and check: $5x = 20$

Solution $5x = 20$

$\frac{5x}{5} = \frac{20}{5}$ Divide each side of the equation by 5: $\frac{5x}{5} = 1x$, or x.

$x = 4$

Check $5x = 20$

$5(4) \stackrel{?}{=} 20$ Substitute 4 for x in the original equation.

$20 \stackrel{\checkmark}{=} 20$ The equation is true, so 4 is the solution to the original equation.

In Example 3, can you explain why $1x = x$?

PRACTICE 3

Solve and check: $6x = 30$

EXAMPLE 4

Solve and check: $5 = \frac{y}{2}$

Solution $5 = \frac{y}{2}$

$2 \cdot 5 = 2 \cdot \frac{y}{2}$ Multiply each side of the equation by 2: $2 \cdot \frac{y}{2} = \frac{2}{1} \cdot \frac{y}{2} = 1y$, or y.

$10 = y$, or $y = 10$

Check $5 = \frac{y}{2}$

$5 \stackrel{?}{=} \frac{10}{2}$ Substitute 10 for y in the original equation.

$5 \stackrel{\checkmark}{=} 5$

PRACTICE 4

Solve and check: $1 = \frac{a}{6}$

EXAMPLE 5

Solve and check: $0.2n = 4$

Solution $0.2n = 4$

$\frac{0.2n}{0.2} = \frac{4}{0.2}$ Divide each side by 0.2: $0.2\overline{)4.0}$ gives $20.$, or 20.

$n = 20$

Check $0.2n = 4$

$0.2(20) \stackrel{?}{=} 4$ Substitute 20 for n in the original equation.

$4.0 \stackrel{?}{=} 4$

$4 \stackrel{\checkmark}{=} 4$

PRACTICE 5

Solve and check: $1.5x = 6$

EXAMPLE 6

Solve and check: $\frac{m}{0.5} = 1.3$

Solution

$$\frac{m}{0.5} = 1.3$$

$$(0.5)\frac{m}{0.5} = (0.5)(1.3) \quad \text{Multiply each side by 0.5.}$$

$$m = 0.65$$

Check

$$\frac{m}{0.5} = 1.3$$

$$\frac{0.65}{0.5} \stackrel{?}{=} 1.3 \quad \text{Substitute 0.65 for } m \text{ in the original equation.}$$

$$1.3 \stackrel{\checkmark}{=} 1.3$$

PRACTICE 6

Solve and check: $\frac{a}{2.4} = 1.2$

EXAMPLE 7

Solve and check: $\frac{2}{3}n = 6$

Solution

$$\frac{2}{3}n = 6$$

$$\frac{2}{3}n \div \frac{2}{3} = 6 \div \frac{2}{3} \quad \text{Divide each side by } \frac{2}{3}.$$

$$\left(\frac{2}{3}n\right)\left(\frac{3}{2}\right) = 6\left(\frac{3}{2}\right)$$

$$\left(\frac{2}{3}\right)\left(\frac{3}{2}\right)n = 6\left(\frac{3}{2}\right)$$

$$n = 9$$

Check

$$\frac{2}{3}n = 6$$

$$\frac{2}{3}(9) \stackrel{?}{=} 6 \quad \text{Substitute 9 for } n \text{ in the original equation.}$$

$$\frac{2}{3}\left(\frac{9}{1}\right) \stackrel{?}{=} 6$$

$$6 \stackrel{\checkmark}{=} 6$$

PRACTICE 7

Solve and check: $\frac{3}{4}x = 12$

As in the case of addition and subtraction equations, multiplication and division equations can be useful mathematical models of real-world situations. To derive these models, we translate word sentences to algebraic equations and solve.

EXAMPLE 8

Write each sentence as an algebraic equation. Then solve and check.

Solution

Sentence	Equation	Check
a. Thirty-five is equal to the product of 5 and x.	$35 = 5x$ $\frac{35}{5} = \frac{5x}{5}$ $7 = x$, or $x = 7$	$35 = 5x$ $35 \stackrel{?}{=} 5(7)$ $35 \stackrel{\checkmark}{=} 35$
b. One equals p divided by 3.	$1 = \frac{p}{3}$ $3 \cdot 1 = 3 \cdot \frac{p}{3}$ $3 = p$, or $p = 3$	$1 = \frac{p}{3}$ $1 \stackrel{?}{=} \frac{3}{3}$ $1 \stackrel{\checkmark}{=} 1$

PRACTICE 8

Translate each sentence to an equation. Then solve and check.

Sentence	Equation	Check
a. Twelve is equal to the quotient of z and 6.		
b. Sixteen equals twice x.		

EXAMPLE 9

A baseball player runs 360 feet when hitting a home run.

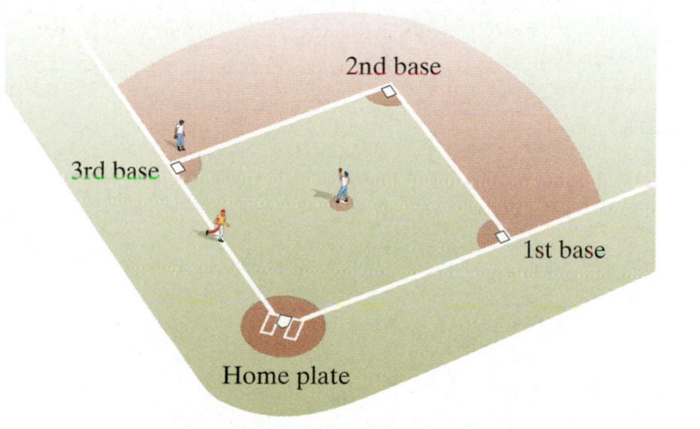

If the distances between successive bases on a baseball diamond are equal, how far is it from third base to home plate? Write an equation. Then solve and check.

Solution Let x equal the distance between successive bases. Since these distances are equal, $4x$ represents the distance around the bases.

But we know that the distance around the bases also equals 360 feet. So $4x = 360$. We solve this equation for x.

$$4x = 360$$

$$\frac{4x}{4} = \frac{360}{4} \quad \text{Divide each side by 4.}$$

$$x = 90$$

The distance from third base to home plate is 90 feet.

PRACTICE 9

The Pentagon is the headquarters of the U.S. Department of Defense.

The distance around the Pentagon is about 1.6 kilometers. If each side of the Pentagon is the same length, write an equation to find that length. Then solve and check.
(*Source:* Gene Gurney, *The Pentagon*)

EXAMPLE 10

Botswana and Andorra have the lowest and highest life expectancy, respectively, of any countries in the world. Botswana's life expectancy is only about 34 years, which is approximately $\frac{2}{5}$ of Andorra's. What is the life expectancy in Andorra? (*Source:* U.S. Bureau of the Census, International Database)

Solution Let a equal Andorra's life expectancy. Botswana's life expectancy, 34 years, is equal to $\frac{2}{5}$ of a, so we write the following equation:

$$34 = \frac{2}{5}a$$

We can solve this equation by dividing both sides by $\frac{2}{5}$.

$$34 \div \frac{2}{5} = \frac{2}{5}a \div \frac{2}{5}$$

$$(34)\left(\frac{5}{2}\right) = \left(\frac{2}{5}a\right)\left(\frac{5}{2}\right)$$

$$(34)\left(\frac{5}{2}\right) = \left(\frac{2}{5}\right)\left(\frac{5}{2}\right)a$$

$$85 = a, \text{ or } a = 85$$

So the life expectancy of Andorra is approximately 85 years.

PRACTICE 10

Six months after buying a used car, a couple sold it, taking a loss equal to $\frac{3}{8}$ of the car's original price. If their loss was \$525, what was the original price? Write an equation. Then solve and check.

Mathematically Speaking

Fill in each blank with the most appropriate term or phrase from the given list.

divide	expression	equation
addition	division	checked
substituting	solved	evaluating
		multiply

1. In the equation $2x = 6$, _______ each side of the equation by 2 in order to isolate the variable.

2. In the equation $\frac{x}{5} = 3$, _______ each side of the equation by 5 in order to isolate the variable.

3. Check whether a number is a solution of an equation by _______ the number for the variable in the equation.

4. An equation is _______ by finding its solution.

5. The equal sign separates the two sides of a(n) _______.

6. Multiplication and _______ are opposite operations.

Translate each sentence to an equation.

7. $\frac{3}{4}$ of a number y is 12.

8. The product of $\frac{2}{3}$ and x is 20.

9. A number x divided by 7 is equal to $\frac{7}{2}$.

10. The quotient of z and 1.5 is 10.

11. $\frac{1}{3}$ of x is 2.

12. 2 times m is equal to 11.

13. The quotient of a number and 3 is equal to $\frac{1}{3}$.

14. A quantity divided by 100 is 0.36.

15. The product of 9 and an amount is the same as 27.

16. $\frac{4}{5}$ of a price is equal to 24.

By answering yes or no, indicate whether the value of x shown is a solution of the given equation.

17.

Equation	Value of x	Solution?
a. $7x = 21$	3	
b. $3x = 12$	36	
c. $\frac{x}{4} = 8$	2	
d. $\frac{x}{0.2} = 4$	8	

18.

Equation	Value of x	Solution?
a. $\frac{x}{3} = 10$	30	
b. $2.5x = 5$	2	
c. $2x = \frac{1}{3}$	$\frac{1}{6}$	
d. $\frac{x}{0.4} = 3$	12	

Identify the operation to perform on each side of the equation to isolate the variable.

19. $3x = 15$ **20.** $6y = 18$ **21.** $\frac{x}{2} = 9$ **22.** $\frac{y}{6} = 1$

23. $\frac{3}{4}a = 21$ **24.** $\frac{2}{3}m = 14$ **25.** $1.5b = 15$ **26.** $2.6x = 52$

Solve and check.

27. $5x = 30$ **28.** $8y = 8$ **29.** $\frac{x}{2} = 9$ **30.** $\frac{n}{9} = 3$

31. $36 = 9n$ **32.** $125 = 5x$ **33.** $\frac{x}{7} = 13$ **34.** $\frac{w}{10} = 21$

35. $1.7y = 6.8$ **36.** $0.5a = 7.5$ **37.** $2.1b = 42$ **38.** $1.5x = 45$

39. $\frac{m}{15} = 10.5$ **40.** $\frac{p}{10} = 12.1$ **41.** $\frac{t}{0.4} = 1$ **42.** $\frac{n}{0.5} = 6$

43. $\frac{2}{3}x = 1$ **44.** $\frac{1}{8}n = 3$ **45.** $\frac{1}{4}x = 9$ **46.** $\frac{3}{7}t = 15$

47. $17t = 51$ **48.** $100x = 400$ **49.** $10y = 4$ **50.** $100n = 50$

51. $7 = \frac{n}{100}$ **52.** $40 = \frac{p}{10}$ **53.** $2.5 = \frac{x}{5}$ **54.** $4.6 = \frac{z}{2}$

55. $2 = 4x$ **56.** $3 = 5x$ **57.** $\frac{14}{3} = \frac{7}{9}m$ **58.** $\frac{4}{9} = \frac{2}{3}a$

Solve. Round to the nearest tenth. Check.

59. $3.14x = 21.3834$ **60.** $2.54x = 78.25$ **61.** $\frac{x}{1.414} = 3.5$ **62.** $\frac{x}{1.732} = 1.732$

Translate each sentence to an equation. Solve and check.

63. The product of 8 and n is 56.

64. The product of 12 and m is 3.

65. $\frac{3}{4}$ of a number y is equal to 18.

66. $\frac{1}{3}$ of a number x is 16.

67. A number x divided by 5 is 11.

68. A number y divided by 100 is 10.

69. Twice x is equal to 36.

70. 3 times m is 90.

71. $\frac{1}{2}$ of an amount is 4.

72. $\frac{5}{7}$ of a number is 10.

73. A number divided by 5 is equal to $1\frac{3}{5}$.

74. An amount divided by 14 is equal to $1\frac{1}{2}$.

75. The quotient of a number and 2.5 is 10.

76. A quantity divided by 15 equals 3.6.

Choose the equation that best describes each situation.

77. Suppose that a teenager spends \$20, which is $\frac{1}{4}$ of his total savings, m. How much money did he have in the beginning?

a. $m - \frac{1}{4} = 20$ **b.** $4m = 20$

c. $m + \frac{1}{4} = 20$ **d.** $\frac{1}{4}m = 20$

78. Find the weight of a child if $\frac{1}{3}$ of her weight is 9 pounds.

a. $3x = 9$ **b.** $\frac{1}{3}x = 9$

c. $x + 3 = 9$ **d.** $x + \frac{1}{3} = 9$

79. A high school student plans to buy an MP3 player 8 weeks from now. If the MP3 player costs \$140, how much money must the student save each week in order to buy it?

a. $8c = 140$ **b.** $c + 8 = 140$

c. $\frac{c}{8} = 140$ **d.** $c - 8 = 140$

80. The student government at a college sold tickets to a play. From the ticket sales, they collected \$300, which was twice the cost of the play. How much did the play cost?

a. $\frac{n}{2} = 300$ **b.** $n - 2 = 300$

c. $2n = 300$ **d.** $n + 2 = 300$

Mixed Practice

Solve and check.

81. $11 = 2x$

82. $\frac{x}{6} = 9$

Solve.

83. The cost of dinner at a restaurant was split evenly among three friends. If each friend paid \$25.75, choose the equation to find the amount on the check.

a. $x + 3 = 25.75$ **b.** $3x = 25.75$

c. $x - 3 = 25.75$ **d.** $\frac{x}{3} = 25.75$

84. Identify the operation to perform on each side of the equation $\frac{n}{2} = 3$ to isolate the variable.

85. Translate the sentence "The quotient of y and 3 is 6" to an equation.

86. Is 2 a solution of the equation $\frac{n}{3} = 6$?

87. Translate the sentence "Twice x is 5" to an equation.

88. The marriage rate in China is approximately 4 times that of the United States. ("Marriage rate" is the number of marriages per 1,000 annually in the latest year for which data are available.) The marriage rate in China is 35.9. Choose the equation to find the approximate marriage rate in the United States. (**Source:** *The Top Ten of Everything, 2006*)

a. $35.9 = \frac{x}{4}$ **b.** $35.9 = 4x$

c. $35.9 = x + 4$ **d.** $35.9 = x - 4$

89. Is 25 a solution of the equation $0.4x = 10$?

90. Identify the operation to perform on each side of the equation $4x = 7$ to isolate the variable.

Application

Write an equation. Solve and check.

91. In the city block shown below, the perimeter is 60 units. Find the length of one side of the square city block.

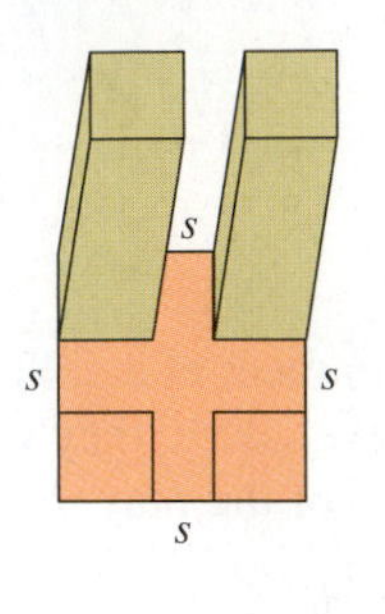

92. The area of the basketball court shown below is 4,700 square feet. Find the width of the court.

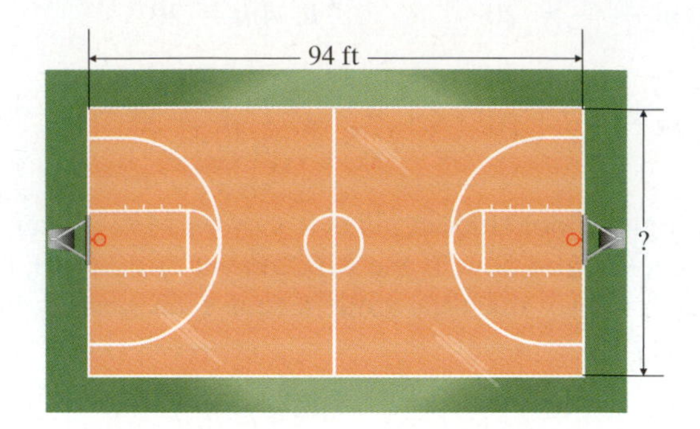

93. In an Ironman 70.3 triathlon, athletes must complete a 56-mile bike ride. This is one-half the distance of the bike ride in a regular Ironman triathlon. What distance must an athlete bike in a regular Ironman triathlon? (***Source:*** World Triathlon Corporation)

94. According to the nutrition label, one packet of regular instant oatmeal has five-eighths the calories of one packet of maple and brown sugar instant oatmeal. If the regular oatmeal has 100 calories, how many calories does the maple and brown sugar oatmeal have?

95. An online DVD movie-rental service charges customers a monthly fee for unlimited DVD rentals. If a customer paid a total of \$119.88 for one year of rental service, what was the monthly rental fee?

96. One plan offered by a long-distance phone service provider charges \$0.07 per minute for long-distance phone calls. A customer using this plan was charged \$22.26 for long-distance calls this month. How many minutes of long-distance calls did she make this month?

97. A lab technician prepared an alcohol-and-water solution that contained 60 milliliters of alcohol. This was two-fifths of the total amount of the solution.

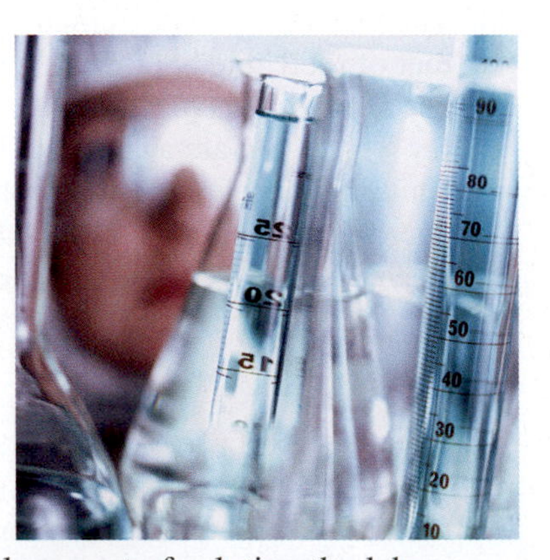

a. What was the total amount of solution the lab technician prepared?

b. How much water was in the solution?

98. A sales representative invested \$5,500 of his sales bonus in the stock market. This represents one-third of his total sales bonus.

a. What was his total sales bonus?

b. How much of his sales bonus was not invested in the stock market?

Use a calculator to solve the following problems, giving (a) the equation, (b) the exact answer, and (c) an estimate of the answer.

99. The population density of a country is the quotient of the country's population and its land area (in square miles). According to the last census, the population density of the United States was approximately 79.6 persons per square mile. Use this approximation to determine the U.S. population at the time of the census, if the land area of the United States was 3,537,441 square miles. (*Source:* U.S. Bureau of the Census)

100. In a recent year, the top two U.S. airlines in terms of passenger traffic were American Airlines and United Airlines. American flew 119,987,000,000 passenger-miles, which was approximately 1.18 times the number of passenger-miles that United flew. According to this approximation, how many passenger-miles did United fly? (*Source:* International Civil Aviation Organization)

MINDSTRETCHERS

Writing

1. Write two different word problems that are applications of each equation.

a. $4x = 20$

•

•

b. $\frac{x}{2} = 5$

•

•

Groupwork

2. The equations $\frac{r}{7} = 2$ and $\frac{7}{r} = 2$ are similar in form. Working with a partner, answer the following questions.

a. How would you solve the first equation for r?

b. How can you use what you know about the first equation to solve the second equation for r?

c. What are the similarities and differences between the two equations?

Critical Thinking

3. In a magic square with four rows and four columns, the sum of the entries in each row, column, and diagonal is the same. If the entries are the consecutive whole numbers 1 through 16, what is the sum of the numbers in each diagonal?

KEY CONCEPTS AND SKILLS

CONCEPT SKILL

Concept/Skill	Description	Example
[4.1] Variable	A letter that represents an unknown number.	x, y, t
[4.1] Constant	A known number.	$2, \frac{1}{3}, 5.6$
[4.1] Algebraic Expression	An expression that combines variables, constants, and arithmetic operations.	$x + 3, \frac{1}{8}n$
[4.1] To Evaluate an Algebraic Expression	• Substitute the given value for each variable. • Carry out the computation.	Evaluate $8 - x$ for $x = 3.5$: $8 - x = 8 - 3.5$, or 4.5
[4.2] Equation	A mathematical statement that two expressions are equal.	$2 + 4 = 6, x + 5 = 7$
[4.2] To Solve Addition or Subtraction Equations	• For an addition equation, subtract the same number from each side of the equation in order to isolate the variable on one side. • For a subtraction equation, add the same number to each side of the equation in order to isolate the variable on one side. • In either case check the solution by substituting the value of the unknown in the original equation to verify that the resulting equation is true.	$y + 9 = 15$ $y + 9 - 9 = 15 - 9$ $y = 6$ Check $y + 9 = 15$ $6 + 9 \stackrel{?}{=} 15$ $15 \stackrel{\checkmark}{=} 15$ $w - 6\frac{1}{2} = 8$ $w - 6\frac{1}{2} + 6\frac{1}{2} = 8 + 6\frac{1}{2}$ $w = 14\frac{1}{2}$ Check $w - 6\frac{1}{2} = 8$ $14\frac{1}{2} - 6\frac{1}{2} \stackrel{?}{=} 8$ $8 \stackrel{\checkmark}{=} 8$
[4.3] To Solve Multiplication or Division Equations	• For a multiplication equation, divide by the same number on each side of the equation in order to isolate the variable on one side. • For a division equation, multiply by the same number on each side of the equation in order to isolate the variable on one side. • In either case check the solution by substituting the value of the unknown in the original equation to verify that the resulting equation is true.	$1.3r = 26$ $\frac{1.3r}{1.3} = \frac{26}{1.3}$ $r = 20$ Check $1.3r = 26$ $1.3(20) \stackrel{?}{=} 26$ $26 \stackrel{\checkmark}{=} 26$ $\frac{x}{7} = 8$ $7 \cdot \frac{x}{7} = 7 \cdot 8$ $x = 56$ Check $\frac{x}{7} = 8$ $\frac{56}{7} \stackrel{?}{=} 8$ $8 \stackrel{\checkmark}{=} 8$

Chapter 4 Review Exercises

To help you review this chapter, solve these problems.

[4.1] *Translate each algebraic expression to words.*

1. $x + 1$ **2.** $y + 4$ **3.** $w - 1$ **4.** $s - 3$

5. $\frac{c}{7}$ **6.** $\frac{a}{10}$ **7.** $2x$ **8.** $6y$

9. $y \div 0.1$ **10.** $n \div 1.6$ **11.** $\frac{1}{3}x$ **12.** $\frac{1}{10}w$

Translate each word phrase to an algebraic expression.

13. Nine more than m

14. The sum of b and $\frac{1}{2}$

15. y decreased by 1.4

16. Three less than z

17. The quotient of 3 and x

18. n divided by 2.5

19. The product of an amount and 3

20. Twelve times some number

Evaluate each algebraic expression.

21. $b + 8$, for $b = 4$ **22.** $d + 12$, for $d = 7$ **23.** $a - 5$, for $a = 5$ **24.** $c - 9$, for $c = 15$

25. $1.5x$, for $x = 0.2$ **26.** $1.3t$, for $t = 5$ **27.** $\frac{1}{2}n$, for $n = 3$ **28.** $\frac{1}{6}a$, for $a = 2\frac{1}{2}$

29. $w - 9.6$, for $w = 10$ **30.** $v - 3\frac{1}{2}$, for $v = 8$ **31.** $\frac{m}{1.5}$, for $m = 2.4$ **32.** $\frac{x}{0.2}$, for $x = 1.8$

[4.2] *Solve and check.*

33. $x + 11 = 20$ **34.** $y + 15 = 24$ **35.** $n - 19 = 7$ **36.** $b - 12 = 8$

37. $a + 2.5 = 6$ **38.** $c + 1.6 = 9.1$ **39.** $x - 1.8 = 9.2$ **40.** $y - 1.4 = 0.6$

41. $w + 1\frac{1}{2} = 3$ **42.** $s + \frac{2}{3} = 1$ **43.** $c - 1\frac{1}{4} = 5\frac{1}{2}$ **44.** $p - 6 = 5\frac{2}{3}$

45. $7 = m + 2$ **46.** $10 = n + 10$ **47.** $39 = c - 39$ **48.** $72 = y - 18$

49. $38 + n = 49$ **50.** $37 + x = 62$ **51.** $4.0875 + x = 35.136$ **52.** $24.625 = m - 1.9975$

[4.2–4.3] *Translate each sentence to an equation.*

53. n decreased by 19 is 35.

54. 37 less than an amount equals 234.

55. 9 increased by a number is equal to $15\frac{1}{2}$.

56. 26 more than s is $30\frac{1}{3}$.

57. Twice y is 16.

58. The product of t and 25 is 175.

59. 34 is equal to n divided by 19.

60. 17 is the quotient of z and 13.

61. $\frac{1}{3}$ of a number equals 27.

62. $\frac{2}{5}$ of a number equals 4.

By answering yes or no, indicate whether the value of x shown is a solution to the given equation.

63.

Equation	Value of x	Solution?
a. $0.3x = 6$	2	
b. $x - \frac{1}{2} = 1\frac{2}{3}$	$2\frac{1}{6}$	
c. $\frac{x}{0.5} = 7$	3.5	
d. $x + 0.1 = 3$	3.1	

64.

Equation	Value of x	Solution?
a. $0.2x = 6$	30	
b. $x + \frac{1}{2} = 1\frac{2}{3}$	$\frac{5}{6}$	
c. $\frac{x}{0.2} = 4.1$	8.2	
d. $x + 0.5 = 7.4$	6.9	

[4.3] *Solve and check.*

65. $2x = 10$

66. $8t = 16$

67. $\frac{a}{7} = 15$

68. $\frac{n}{6} = 9$

69. $9y = 81$

70. $10r = 100$

71. $\frac{w}{10} = 9$

72. $\frac{x}{100} = 1$

73. $1.5y = 30$

74. $1.2a = 144$

75. $\frac{1}{8}n = 4$

76. $\frac{1}{2}b = 16$

77. $\frac{m}{1.5} = 2.1$

78. $\frac{z}{0.3} = 1.9$

79. $100x = 40$

80. $10t = 5$

81. $0.3 = \frac{m}{4}$

82. $1.4 = \frac{b}{7}$

83. $0.866x = 10.825$

84. $\frac{x}{0.707} = 2.1$

Mixed Applications

Write an algebraic expression for each problem. Then evaluate the expression for the given amount.

85. The temperature increases 2 degrees an hour. By how many degrees will the temperature increase in h hours? In 3 hours?

86. During the fall term, a math tutor works 20 hours per week. What is the tutor's hourly wage if she earns d dollars per week? \$191 per week?

87. The local supermarket sells a certain fruit for 89¢ per pound. How much will p pounds cost? 3 pounds?

88. After having borrowed \$3,000 from a bank, a customer must pay the amount borrowed plus a finance charge. How much will he pay the bank if the finance charge is d dollars? \$225?

Write an equation. Then solve and check.

89. After depositing \$238 in a checking account, the balance will be \$517. What was the balance before the deposit?

90. Hurricane Gilbert was one of the strongest storms to hit the Western Hemisphere in the twentieth century. A newspaper reported that the hurricane left 500,000 people, or about one-fourth of the population of Jamaica, homeless. Approximately how many people lived in Jamaica? (***Source:*** J. B. Elsner and A. B. Kara, *Hurricanes of the North Atlantic*)

91. Drinking bottled water is more popular in some countries than in others. In a recent year, the per capita consumption of bottled water for Italians was about 177 liters, or approximately $2\frac{1}{2}$ times as much as it was for Americans. Using this approximation, find the per capita consumption for Americans, to the nearest liter. (*Source:* Euromonitor)

92. A bowler's final score is the sum of her handicap and scratch score (actual score). If a bowler has a final score of 225 and a handicap of 50, what was her scratch score?

93. On the Moon, a person weighs about one-sixth of his or her weight on Earth. What is the weight on Earth of an astronaut who weighs 30 pounds on the Moon?

94. The Nile River is about 1.8 times as long as the Missouri River. If the Nile is about 6,696 kilometers long, approximately how long is the Missouri? (*Source:* Eliot Elisofon, *The Nile*)

95. The normal body temperature is 98.6°F. An ill patient had a temperature of 101°F. This temperature is how many degrees above normal?

96. This year, a community college received 8,957 applications for admission, which amounts to 256 fewer than were received last year. How many applications did the community college receive last year?

• *Check your answers on page A-8.*

Chapter 4 POSTTEST

FOR EXTRA HELP  Test solutions are found on the enclosed CD.

To see if you have mastered the topics in this chapter, take this test.

Write each algebraic expression in words.

1. $x + \frac{1}{2}$

2. $\frac{a}{3}$

Translate each word phrase to an algebraic expression.

3. 10 less than a number

4. The quotient of 8 and p

Evaluate each algebraic expression.

5. $a - 1.5$, for $a = 1.5$

6. $\frac{b}{9}$, for $b = 2\frac{1}{4}$

Translate each sentence to an equation.

7. The difference between x and 6 is $4\frac{1}{2}$.

8. The quotient of y and 8 is 3.2.

Solve and check.

9. $x + 10 = 10$

10. $y - 6 = 6$

11. $81 = 3n$

12. $82 = \frac{a}{9}$

13. $m - 1.8 = 6$

14. $1.5n = 75$

15. $10x = 5\frac{1}{2}$

16. $\frac{n}{100} = 7.6$

Write an equation. Then solve and check.

17. A recipe for seafood stew requires $2\frac{1}{4}$ pounds of fish. After buying $1\frac{3}{4}$ pounds of bluefish, a chef decides to fill out the recipe with codfish. How many pounds of codfish should he buy?

18. A newspaper reported that this year, 30,000 elephants—$\frac{1}{3}$ of all the elephants in a certain country—had been hunted down and killed for their ivory tusks. How many elephants were there at the beginning of the year?

19. The population of the world in the year 2000 is expected to be about two-thirds of the projected world population in 2050. If the population in 2000 was about 6 billion people, what is the projected world population 50 years later? (***Source:*** U.S. Bureau of the Census, International Database, 2006)

20. In chemistry, an endothermic reaction is one that absorbs heat. As a result of an endothermic reaction, the temperature of a solution dropped by 19.8 degrees Celsius to 7.6 degrees Celsius. What was the temperature of the solution before the reaction took place? (***Source:*** Timberlake, *Chemistry: An Introduction to General, Organic, and Biological Chemistry*)

Cumulative Review Exercises

To help you review, solve the following:

1. Subtract: $8\frac{1}{4} - 2\frac{7}{8}$

2. Find the quotient: $7.5 \div 1{,}000$

3. Decide whether 2 is a solution to the equation $w + 3 = 5$

4. Multiply: 804×29

5. Round 3.14159 to the nearest hundredth.

6. Solve and check: $n - 3.8 = 4$

7. Solve and check: $\frac{x}{2} = 16$

8. In animating a cartoon, artists had to draw 24 images to appear during 1 second of screen time. How many images did they have to draw to produce a 5-minute cartoon?

9. Farmers depend on bees to pollinate many crop plants, such as apples and cherries. In the American Midwest, the acreage of crops is large as compared with the number of bees, so farmers are especially concerned if the number of beehives declines. When the number of beehives in the state of Illinois dropped from 101,000 to 46,000, how big a drop was this? (***Source:*** http://www.ag.uiuc.edu)

10. Dental insurance reimbursed a patient \$200 on a bill of \$700. Did the patient get less or more than $\frac{1}{3}$ of his money back? Explain.

• *Check your answers on page A-8.*

CHAPTER 5

Ratio and Proportion

Ratio and Proportion and Pharmacology

Many of the medicines that pharmacists dispense come in solutions. An example is aminophylline, a medicine that people with asthma take to ease their breathing.

To prepare a solution of aminophylline, pharmacists dissolve 250 milligrams of aminophylline for every 10 cubic centimeters (cc or cm^3) of sterile water.

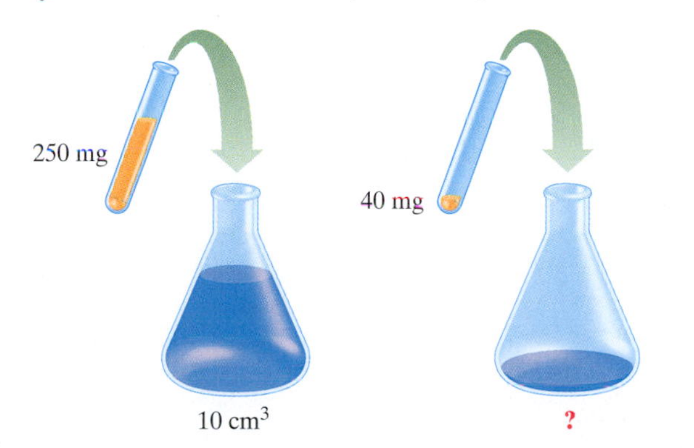

In filling a prescription for, say, 40 milligrams of aminophylline, we must determine the amount of sterile water needed. Pharmacists use the concepts of ratio and proportion to establish how much of the sterile water to dispense. (*Source:* U.S. Army Medical Department Center and School, *Pharmacology Math for the Practical Nurse*)

Chapter 5 PRETEST

To see if you have already mastered the topics in this chapter, take this test.

Write each ratio or rate in simplest form.

1. 6 to 8

2. 40 to 100

3. \$30 to \$18

4. 19 feet to 51 feet

5. 48 gallons of water in 15 minutes

6. 10 milligrams every 6 hours

Find the unit rate.

7. 12 dental assistants for every 6 dentists

8. 35 calculators for 35 students

Determine the unit price.

9. \$690 for 3 boxes of ceramic tiles

10. 12 bottles of lemon iced tea for \$6.00

Determine whether each proportion is true or false.

11. $\frac{2}{3} = \frac{16}{24}$

12. $\frac{32}{20} = \frac{8}{3}$

Solve and check.

13. $\frac{6}{8} = \frac{x}{12}$

14. $\frac{21}{x} = \frac{2}{3}$

15. $\frac{\frac{1}{2}}{4} = \frac{2}{x}$

16. $\frac{x}{6} = \frac{8}{0.3}$

Solve.

17. A contractor combines 80 pounds of sand with 100 pounds of gravel. In this mixture, what is the ratio of sand to gravel?

18. A machine at a potato chip factory can peel 12,000 pounds of potatoes in 60 minutes. At this rate, how many pounds of potatoes can it peel per minute?

19. In the first quarter of a year, a company paid \$66,000 for security. On the basis of this expense, project how much money the company will pay for security in one year.

20. The scale on a map is 3 inches to 31 miles. If two cities are 8.4 inches apart on the map, what is the actual distance, to the nearest mile, between the two cities?

• *Check your answers on page A-8.*

5.1 Introduction to Ratios

OBJECTIVES

- To write ratios of like quantities in simplest form
- To write ratios of unlike quantities in simplest form
- To solve word problems involving ratios

What Ratios Are and Why They Are Important

We frequently need to compare quantities. Sports, medicine, and business are just a few areas where we use **ratios** to make comparisons. Consider the ratios in the following examples.

- The volleyball team won 4 games for every 3 they lost.
- A physician assistant prepared a 1-to-25 boric acid solution.
- The stock's price-to-earnings ratio was 13 to 1.

Can you think of other examples of ratios in your daily life?

The preceding examples illustrate the following definition of a ratio.

Definition

A **ratio** is a comparison of two quantities expressed as a quotient.

There are, in general, three basic ways to write a ratio. For instance, we can write the ratio 1 to 25 as

$$1 \text{ to } 25 \qquad 1:25 \qquad \frac{1}{25}$$

No matter which notation we use for this ratio, it is read "1 to 25."

Simplifying Ratios

Because a ratio can be written as a fraction, we can say that, as with any fraction, a ratio is in simplest form (or reduced to lowest terms) when 1 is the only common factor of the numerator and denominator.

Let's consider some examples of writing ratios in simplest form.

EXAMPLE 1

Write the ratio 10 to 5 in simplest form.

Solution The ratio 10 to 5 expressed as a fraction is $\frac{10}{5}$.

$$\frac{10}{5} = \frac{10 \div 5}{5 \div 5} = \frac{2}{1}$$

So the ratio 10 to 5 is the same as the ratio 2 to 1. Note that the ratio 2 to 1 means that the first number is twice as large as the second number.

PRACTICE 1

Write the ratio 8:12 in simplest form.

Frequently, we deal with quantities that have units, such as months or feet. When both quantities in a ratio have the same unit, they are called **like quantities**. In a ratio of like quantities, the units drop out.

EXAMPLE 2

Express the ratio 5 months to 3 months in simplest form.

Solution The ratio 5 months to 3 months expressed as a fraction is $\frac{5 \text{ months}}{3 \text{ months}}$. Simplifying, we get $\frac{5}{3}$, which is already in lowest terms. Note that with ratios we do not rewrite improper fractions as mixed numbers because our answer must be a comparison of *two* numbers.

PRACTICE 2

Express in simplest form the ratio 9 feet to 5 feet.

EXAMPLE 3

A young couple put $58,000 down and financed $232,000 when buying a new home. What is the ratio of the amount put down to the purchase price of the home?

Solution This is a two-step problem. First we must find the purchase price of the home.

$$\$58{,}000 + \$232{,}000 = \$290{,}000$$

Then we write the ratio of the amount put down to the purchase price of the home.

$$\frac{\text{Amount put down}}{\text{Purchase price}} = \frac{\$58{,}000}{\$290{,}000} = \frac{1}{5}$$

The ratio is 1 to 5, which means the couple put down $1 of every $5 of the purchase price.

PRACTICE 3

A sales representative invests $1,500 of his $6,000 bonus in a high-risk fund and the remainder of the money in a low-risk fund. What is the ratio of the amount he placed in the high-risk fund to that in the low-risk fund?

Now let's compare **unlike quantities**, that is, quantities that have different units or are different kinds of measurement. Such a comparison is called a **rate**.

Definition

A **rate** is a ratio of unlike quantities.

For instance, suppose that your rate of pay is $52 for each 8 hours of work. Simplifying this rate, we get

$$\frac{\$52}{8 \text{ hours}} = \frac{\$13}{2 \text{ hours}}$$

So you are paid $13 for every 2 hours that you worked. Note that the units are expressed as part of the answer.

EXAMPLE 4

Simplify each rate.

a. 350 miles to 18 gallons of gas

b. 18 trees to produce 2,000 pounds of paper

Solution

a. 350 miles to 18 gallons $= \dfrac{350 \text{ miles}}{18 \text{ gallons}} = \dfrac{175 \text{ miles}}{9 \text{ gallons}}$

b. 18 trees to 2,000 pounds $= \dfrac{18 \text{ trees}}{2{,}000 \text{ pounds}} = \dfrac{9 \text{ trees}}{1{,}000 \text{ pounds}}$

PRACTICE 4

Express each rate in simplest form.

a. 150 milliliters of medication infused every 60 minutes

b. 18 pounds lost in 12 weeks

Examples 1, 2, 3, and 4 illustrate the following rule for simplifying a ratio or rate.

To Simplify a Ratio or Rate

- Write the ratio or rate as a fraction.
- Express the fraction in simplest form.
- If the quantities are alike, drop the units. If the quantities are unlike, keep the units.

Frequently, we want to find a particular kind of rate called a *unit rate.* In the rate $\dfrac{\$13}{2 \text{ hours}}$, for instance, it would be useful to know what is earned for each hour (that is, the hourly wage). We need to rewrite $\dfrac{\$13}{2 \text{ hours}}$ so that the denominator is 1 hour.

$$\frac{\$13}{2 \text{ hours}} = \frac{\$13 \div 2}{2 \text{ hours} \div 2} = \frac{\$6.50}{1 \text{ hour}} = \$6.50 \text{ per hour, or } \$6.50/\text{hr}$$

Note that "per" means "divided by."

Here, we divided the numbers in both the numerator and denominator by the number in the denominator.

Definition

A **unit rate** is a rate in which the number in the denominator is 1.

EXAMPLE 5

Write as a unit rate.

a. 275 miles in 5 hours

b. $3,453 for 6 weeks

Solution First, we write each rate as a fraction. Then we divide numbers in the numerator and denominator by the number in the denominator, getting 1 in the denominator.

PRACTICE 5

Express as a unit rate.

a. a fall of 192 feet in 4 seconds

b. 15 hits in 40 times at bat

a. 275 miles in 5 hours $= \dfrac{275 \text{ miles}}{5 \text{ hours}} = \dfrac{275 \text{ miles} \div 5}{5 \text{ hours} \div 5} = \dfrac{55 \text{ miles}}{1 \text{ hour}}$, or 55 mph

b. \$3,453 for 6 weeks $= \dfrac{\$3{,}453}{6 \text{ weeks}} = \dfrac{\$3{,}453 \div 6}{6 \text{ weeks} \div 6} = \dfrac{\$575.50}{1 \text{ week}}$, or \$575.50 per week

EXAMPLE 6

In the United States, there are approximately 1,000 public two-year colleges, with a total enrollment of about 6,000,000 students. What is the enrollment per college? (*Source: The Chronicle of Higher Education,* as reported in *Time Almanac 2006*)

Solution $\dfrac{6{,}000{,}000 \text{ students}}{1{,}000 \text{ colleges}} = \dfrac{6{,}000 \text{ students}}{1 \text{ college}}$

So the enrollment is 6,000 students per public two-year college.

PRACTICE 6

A hummingbird beats its wings 2,500 times in 5 minutes. What is the number of times it beats its wings per minute? (*Source:* The National Zoo)

In order to get the better buy, we sometimes compare prices by computing the price of a single item. This **unit price** is a type of unit rate.

Definition

A **unit price** is the price of one item, or one unit.

To find a unit price, we write the ratio of the total price of the units to the number of units and then simplify.

$$\text{Unit price} = \frac{\text{Total price}}{\text{Number of units}}$$

Let's consider some examples of unit pricing.

EXAMPLE 7

Find the unit price.

a. \$300 for 12 months of membership

b. 6 credits for \$234

c. 10-ounce box of wheat flakes for \$2.76

Solution

a. $\dfrac{\$300}{12 \text{ months}} = \$25/\text{month}$

b. $\dfrac{\$234}{6 \text{ credits}} = \$39/\text{credit}$

c. $\dfrac{\$2.76}{10 \text{ ounces}} = \$0.276/\text{ounce}$

$\approx \$0.28/\text{ounce}$ rounded to the nearest cent

PRACTICE 7

Determine the unit price.

a. 4 supersaver flights for \$696

b. \$22 for 8 hours of parking

c. \$19.80 for 20 song downloads

EXAMPLE 8

For the following two bottles of aspirin, which is the better buy?

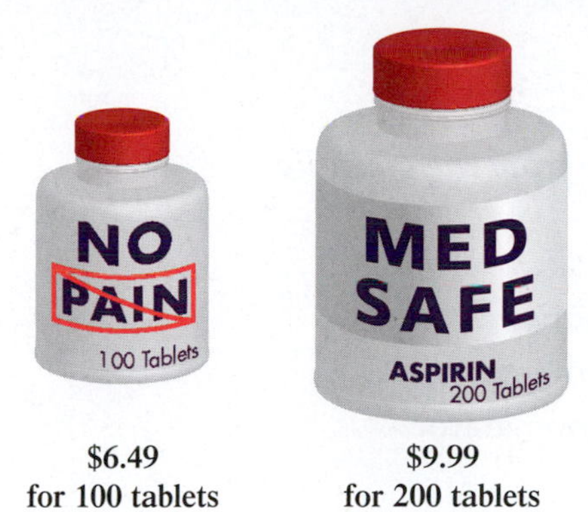

\$6.49 for 100 tablets **\$9.99 for 200 tablets**

Solution First, we find the unit price for each bottle of aspirin.

$$\text{Unit price} = \frac{\text{Total price}}{\text{Number of units}} = \frac{\$6.49}{100} = \$0.0649 \approx \$0.06 \text{ per tablet}$$

$$\text{Unit price} = \frac{\text{Total price}}{\text{Number of units}} = \frac{\$9.99}{200} = \$0.04995 \approx \$0.05 \text{ per tablet}$$

Since \$0.05 < \$0.06, the better buy is the 200-tablet bottle of aspirin.

PRACTICE 8

Which can of coffee has the lower unit price?

39-oz can for \$10.39 **13-oz can for \$3.69**

CULTURAL NOTE

The shape of a grand piano is dictated by the length of its strings. When a stretched string vibrates, it produces a particular pitch, say C. A second string of comparable tension will produce another pitch, which depends on the ratio of the string lengths. For instance, if the ratio of the second string to the first string is 18 to 16, then plucking the second string will produce the pitch B.

Around 500 B.C., the followers of the mathematician Pythagoras learned to adjust string lengths in various ratios so as to produce an entire scale. Thus the concept of ratio is central to the construction of pianos, violins, and many other musical instruments.

Sources:

John R. Pierce, *The Science of Musical Sound* (New York: Scientific American Library, 1983)

David Bergamini, *Mathematics* (New York: Time-Life Books, 1971)

Mathematically Speaking

Fill in each blank with the most appropriate term or phrase from the given list.

weight of a unit	numerator	unlike
like	difference	quotient
simplest form	fractional form	number of units
		denominator

1. A ratio is a comparison of two quantities expressed as a(n) _______.

2. A rate is a ratio of _______ quantities.

3. A ratio is said to be in _______ when 1 is the only common factor of the numerator and denominator.

4. Quantities that have the same units are called _______ quantities.

5. A unit rate is a rate in which the number in the _______ is 1.

6. To find the unit price, divide the total price of the units by the _______.

Write each ratio in simplest form.

7. 6 to 9

8. 9 to 12

9. 10 to 15

10. 21 to 27

11. 55 to 35

12. 8 to 10

13. 12 to 8

14. 25 to $1\frac{1}{4}$

15. 2.5 to 10

16. 1.25 to 100

17. 60 minutes to 45 minutes

18. \$40 to \$25

19. 10 feet to 10 feet

20. 75 tons to 75 tons

21. 30¢ to 18¢

22. 66 years to 32 years

23. 7 miles per hour to 24 miles per hour

24. 21 gallons to 20 gallons

25. 1,000 acres to 50 acres

26. 2,000 miles to 25 miles

27. 8 grams to 7 grams

28. 19 ounces to 51 ounces

29. 24 seconds to 30 seconds

30. 28 milliliters to 42 milliliters

Write each rate in simplest form.

31. 25 telephone calls in 10 days

32. 42 gallons in 4 minutes

33. 288 calories burned in 40 minutes

34. 190 e-mails in 25 days

35. 2 million hits on a website in 6 months

36. 50 million troy ounces of gold produced in 12 months

37. 68 baskets in 120 attempts

38. 18 boxes of cookies for \$45

39. 296 points in 16 games

40. 12 knockouts in 16 fights

41. 500 square feet of carpeting for $1,645

42. 300 full-time students to 200 part-time students

43. 48 males for every 9 females

44. 3 case workers for every 80 clients

45. 40 Democrats for every 35 Republicans

46. $12,500 in 6 months

47. 2 pounds of zucchini for 16 servings

48. 57 hours of work in 9 days

49. 1,535 flights in 15 days

50. 25 pounds of plaster for 2,500 square feet of wall

51. 3 pounds of grass seeds for 600 square feet of lawn

52. 684 parts manufactured in 24 hours

Determine the unit rate.

53. 3,375 revolutions in 15 minutes

54. 3,000 houses to 1,500 acres of land

55. 120 gallons of heating oil for 15 days

56. 48 yards in 8 carries

57. 3 tanks of gas to cut 10 acres of lawn

58. 192 meters in 6 seconds

59. 8 yards of material for 5 dresses

60. 648 heartbeats in 9 minutes

61. 20 hours of homework in 10 days

62. $200 for 8 hours of work

63. A run of 5 kilometers in 20 minutes

64. 56 calories in 4 ounces of orange juice

65. 140 fat calories in 2 tablespoons of peanut butter

66. 60 children for every 5 adults

Find the unit price.

67. 12 bars of soap for $5.40

68. 4 credit hours for $200

69. 6 rolls of film that cost $17.70

70. 2 notebooks that cost $6.90

71. 3 plants for $200

72. $240,000 for a 30-second primetime television commercial spot

73. 5 nights in a hotel for $495

74. 60 minutes of Internet access for $3

Complete each table. Determine which is the better buy.

75. Security envelopes

Number of Units	Total Price	Unit Price
125	$6.69	
500	$15.49	

76. Huggies® diapers

Number of Units	Total Price	Unit Price
56	$8.46	
112	$17.47	

77. Centrum® multivitamins tablets

Number of Units	Total Price	Unit Price
180	$12.99	
250	$17.49	

78. Honey jars

Number of Units (Ounces)	Total Price	Unit Price
16	$4.00	
32	$7.50	

Fill in the table. Which is the best buy?

79. Memorex® DVD-R discs

Number of Units	Total Price	Unit Price
25	$14.99	
50	$26.55	
100	$54.99	

80. Duracell® AA batteries

Number of Units	Total Price	Unit Price
4	$5.99	
8	$6.99	
16	$10.99	

Mixed Practice

Solve.

81. To the nearest cent, find the unit price of an 18-ounce jar of creamy peanut butter that costs $2.89.

82. Complete the table. Then find the best buy.

Dove® white soap bars

Number of Units	Total Price	Unit Price
2	$3.19	
4	$5.21	
8	$10.49	

83. Simplify the rate: 4 tutors for every 30 students.

84. Write as a unit rate: 50 lots to 0.2 square mile.

85. Write the ratio 20 to 4 in simplest form.

86. Express $\dfrac{30 \text{ centimeters}}{45 \text{ centimeters}}$ in simplest form.

Applications

Solve. Simplify if possible.

87. The number line shown is marked off in equal units. Find the ratio of the length of the distance x to the distance y.

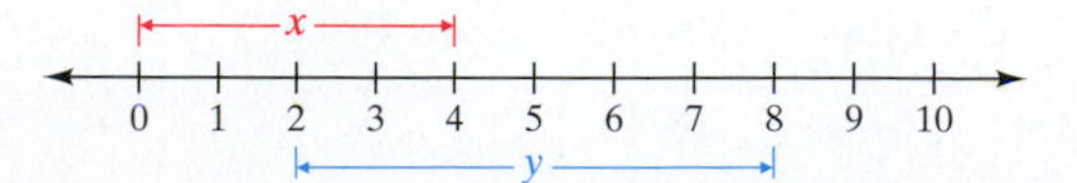

88. In the following rectangle, what is the ratio of the width to the length?

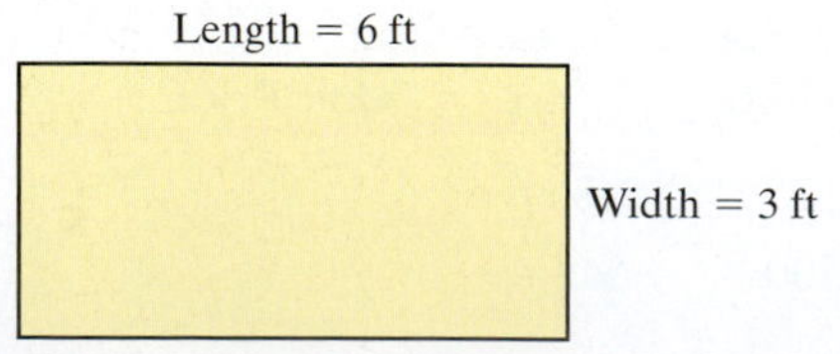

89. In 10 ounces of cashew nuts, there are 1,700 calories. How many calories are there per ounce?

90. For a building valued at $200,000 the property tax is $4,000. Find the ratio of the tax to the building's value.

91. On average, a person blinks 100 times in 4 minutes. How many times does a person blink in 1 minute? (*Source: Neurology*, May 1984, 677–8)

92. A bathtub contains 20 gallons of water. If the tub empties in 4 minutes, what is the rate of flow of the water per minute?

93. In a student government election, 1,000 students cast a vote for the incumbent, 900 voted for the opponent, and 100 cast a protest vote. What was the ratio of the incumbent's vote to the total number of votes?

94. At a college, 4,500 of the 7,500 students are female. What is the ratio of females to males at the college?

95. Russia has a population of approximately 143,000,000 people and a land area of about 17,000,000 square kilometers. The ratio of the number of people to the area is called the population density. What is the population density of Russia, rounded to the nearest tenth? (*Source: The World Factbook*)

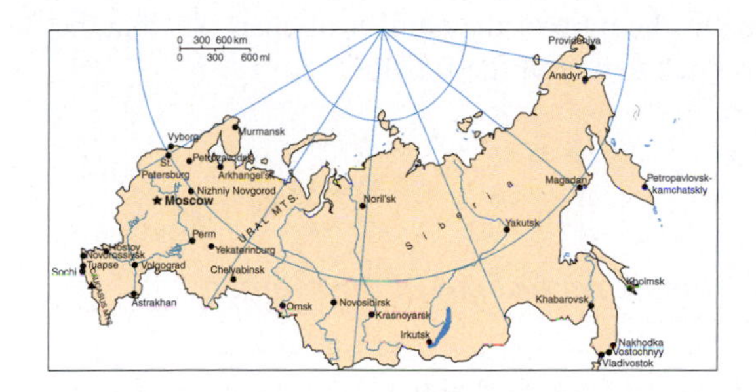

96. Eighteen thousand people can ride El Toro, a roller coaster at Six Flags Great Adventure in New Jersey, in 12 hours. How many people per hour can ride El Toro? (*Source:* Six Flags Great Adventure)

97. In a recent vote, the U.S. House of Representatives voted on House Resolution 3132 as shown.

	Yea	Nay	Present Nonvoting
Republican	195	29	7
Democrat	175	23	4
Independent	1		
Totals	371	52	11

Is the ratio of the representatives who voted against the resolution (Nay) to those who voted for it (Yea) higher or lower for Democrats than for Republicans? (*Source:* http://chocola.house.gov)

98. The table below shows the breakdown of the number of patients in two hospital units at a local city hospital. Is the ratio of nurses to patients in the intensive care unit higher or lower than the ratio of nurses to patients in the medical unit?

	Intensive Care Unit	Medical Unit
Patients	25	65
Nurses	8	11

99. The following table deals with five of the longest-reigning monarchs in history.

Monarch	Country	Reign	Length of Reign (in years)
King Louis XIV	France	1643–1715	72
King John II	Liechtenstein	1858–1929	71
Emperor Franz-Josef	Austria-Hungary	1848–1916	67
Queen Victoria	United Kingdom	1837–1901	63
Emperor Hirohito	Japan	1926–1989	62

(*Source: The Top 10 of Everything 2006*)

a. What is the ratio of Emperor Hirohito's length of reign to that of Emperor Franz-Josef?

b. What is the ratio of Queen Victoria's length of reign to that of King Louis XIV?

100. The following bar graph deals with music groups that were popular and the number of chart hits that they had in the United States.

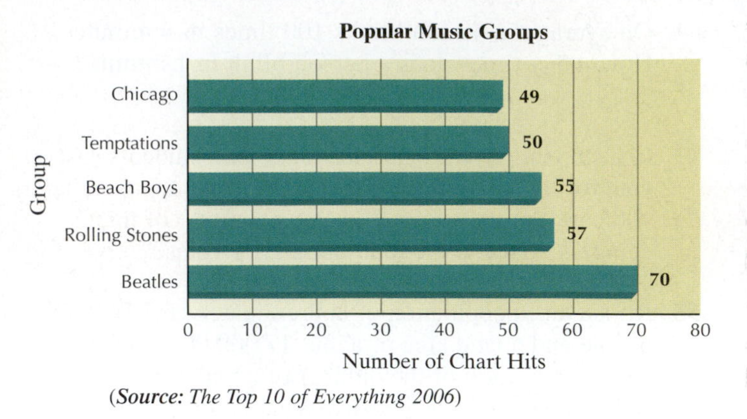

(*Source: The Top 10 of Everything 2006*)

a. Find the ratio of the number of chart hits for the Temptations as compared to the Rolling Stones.

b. What is the ratio of the number of chart hits that the Beach Boys had to that of the Beatles?

Using a calculator, solve the following problems, giving (a) the operation(s) carried out in the solution, (b) the exact answer, and (c) an estimate of the answer.

101. In the insurance industry, a **loss ratio** is the ratio of total losses paid out by an insurance company to total premiums collected for a given time period.

$$\text{Loss ratio} = \frac{\text{Losses paid}}{\text{Premiums collected}}$$

In 2 months, a certain insurance company paid losses of \$6,400,000 and collected premiums of \$12,472,000. What is the loss ratio?

102. Analysts for a brokerage firm prepare research reports on companies with stocks traded in various stock markets. One statistic that an analyst uses is the **price-to-earnings (P.E.) ratio**.

$$\text{P.E. ratio} = \frac{\text{Market price per share}}{\text{Earnings per share}}$$

Find the P.E. ratio for a stock that had a per-share market price of \$70.75 and earnings of \$5.37 per share.

• *Check your answers on page A-9.*

MINDSTRETCHERS

History

1. For a **golden rectangle**, the ratio of its length to its width is approximately 1.618 to 1 (the **golden ratio**).

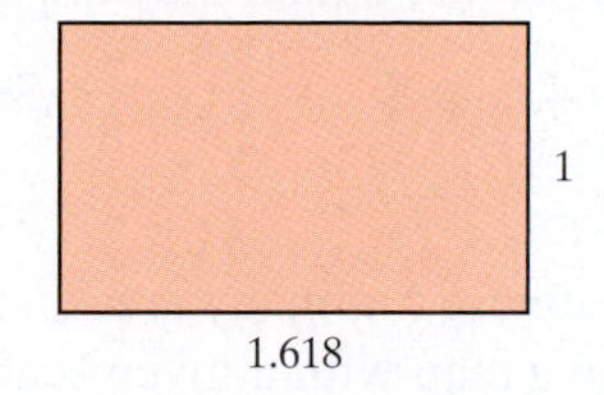

To the ancient Egyptians and Greeks, a golden rectangle was considered to be the ratio most pleasing to the eye. Show that index cards in either of the two standard sizes (3×5 and 5×8) are close approximations to the golden rectangle.

Investigation

2. The distance around a circle is called its **circumference** (C). The distance across the circle through its center is called its **diameter** (d).

   a. Use a string and ruler to measure C and d for both circles shown.

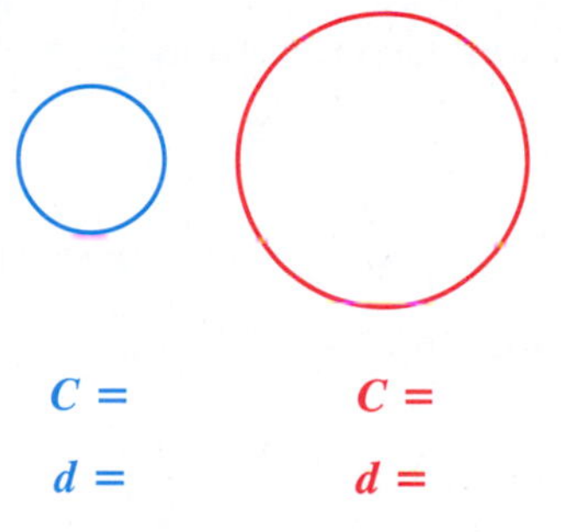

   b. Compute the ratio of C to d for each circle. Are the ratios approximately equal?

$\frac{C}{d} =$ $\qquad$ $\frac{C}{d} =$

Writing

3. Sometimes we use *differences* rather than *quotients* to compare two quantities. Give an example of each kind of comparison and any advantages and disadvantages of each approach.

5.2 Solving Proportions

OBJECTIVES

- To write and solve proportions
- To solve word problems involving proportions

Writing Proportions

When two ratios—for instance, 1 to 2 and 4 to 8—are equal, they are said to be *in proportion*. We can write "1 is to 2 as 4 is to 8" as $\frac{1}{2} = \frac{4}{8}$. Such an equation is called a **proportion**.

Proportions are common in daily life and are used in many areas, such as finding the distance between two cities from a map with a given scale.

Definition

A **proportion** is a statement that two ratios are equal.

One way to see if a proportion is true is to determine whether the *cross products* of the ratios are equal. For example, we see that the proportion

$$\frac{1}{2} = \frac{4}{8}$$

is true, because $2 \cdot 4 = 1 \cdot 8$, or $8 = 8$. However, the proportion $\frac{3}{5} = \frac{9}{10}$ is not true, since $5 \cdot 9 \neq 3 \cdot 10$.

EXAMPLE 1

Determine whether the proportion 4 is to 3 as 16 is to 12 is true.

Solution First, we write the ratios in fractional form: $\frac{4}{3} = \frac{16}{12}$.

$$\frac{4}{3} = \frac{16}{12}$$

$$3 \cdot 16 \stackrel{?}{=} 4 \cdot 12 \quad \text{Set the cross products equal.}$$

$$48 \stackrel{\checkmark}{=} 48$$

So the proportion 4 is to 3 as 16 is to 12 is true.

PRACTICE 1

Are the ratios 10 to 4 and 15 to 6 in proportion?

EXAMPLE 2

Is $\frac{15}{9} = \frac{8}{5}$ a true proportion?

Solution

$$\frac{15}{9} \stackrel{?}{=} \frac{8}{5}$$

$$9 \cdot 8 \stackrel{?}{=} 15 \cdot 5 \quad \text{Set the cross products equal.}$$

$$72 \neq 75$$

The cross products are not equal. So the proportion is not true.

PRACTICE 2

Determine whether $\frac{15}{6} = \frac{8}{3}$ is a true proportion.

EXAMPLE 3

A college claims that the student-to-faculty ratio is 13 to 1. If there are 96 faculty for 1,248 students, is the college's claim true?

Solution The college claims a student-to-faculty ratio of $\frac{13}{1}$, and the actual ratio of students to faculty is $\frac{1,248}{96}$. We want to know if these two ratios are equal.

$$\text{Students} \rightarrow \frac{13}{1} \stackrel{?}{=} \frac{1,248}{96} \leftarrow \text{Students}$$
Faculty → 1, 96 ← Faculty

$$1 \cdot 1,248 \stackrel{?}{=} 13 \cdot 96 \quad \text{Set the cross products equal.}$$

$$1,248 \stackrel{\checkmark}{=} 1,248$$

Since the cross products are equal, the college's claim is true.

PRACTICE 3

A company has a policy making the compensation of its CEO proportional to the dividends that are paid to shareholders. If the dividends increase from \$72 to \$80 and the CEO's compensation is increased from \$360,000 to \$420,000, was the company's policy followed?

Solving Proportions

Suppose that you make \$840 for working 4 weeks in a book shop. At this rate of pay, how much money will you make in 10 weeks? To solve this problem, we can write a proportion in which the rates compare the amount of pay to the time worked. We want to find the amount of pay corresponding to 10 weeks, which we call x.

$$\begin{matrix}\text{Pay} \rightarrow \\ \text{Time} \rightarrow\end{matrix} \frac{840}{4} = \frac{x}{10} \begin{matrix}\leftarrow \text{Pay} \\ \leftarrow \text{Time}\end{matrix}$$

After setting the cross products equal, we find the missing value.

$$\frac{840}{4} = \frac{x}{10}$$
$$4 \cdot x = 840 \cdot 10$$
$$4x = 8,400$$
$$\frac{4x}{4} = \frac{8,400}{4} \quad \text{Divide each side of the equation by 4.}$$
$$x = 2,100$$

So you will make \$2,100 in 10 weeks.

We can check our solution by substituting 2,100 for x in the original proportion.

$$\frac{840}{4} = \frac{x}{10}$$
$$\frac{840}{4} \stackrel{?}{=} \frac{2,100}{10}$$
$$4 \cdot 2,100 \stackrel{?}{=} 840 \cdot 10 \quad \text{Set the cross products equal.}$$
$$8,400 \stackrel{\checkmark}{=} 8,400$$

Our solution checks.

To Solve a Proportion

- Find the cross products, and set them equal.
- Solve the resulting equation.
- Check the solution by substituting the value of the unknown in the original equation to be sure that the resulting proportion is true.

EXAMPLE 4

Solve and check: $\frac{2}{3} = \frac{x}{15}$

Solution

$\frac{2}{3} = \frac{x}{15}$

$3 \cdot x = 2 \cdot 15$ **Set the cross products equal.**

$3x = 30$

$\frac{3x}{3} = \frac{30}{3}$ **Divide each side by 3.**

$x = 10$

Check

$\frac{2}{3} = \frac{x}{15}$

$\frac{2}{3} \stackrel{?}{=} \frac{10}{15}$ **Substitute 10 for x.**

$2 \cdot 15 \stackrel{?}{=} 3 \cdot 10$ **Set the cross products equal.**

$30 \stackrel{\checkmark}{=} 30$

PRACTICE 4

Solve and check: $\frac{x}{6} = \frac{12}{9}$

EXAMPLE 5

Solve and check: $\frac{\frac{1}{4}}{12} = \frac{x}{96}$

Solution

$\frac{\frac{1}{4}}{12} = \frac{x}{96}$

$12 \cdot x = \frac{1}{4} \cdot 96$ **Set the cross products equal.**

$12x = 24$

$\frac{12x}{12} = \frac{24}{12}$ **Divide each side by 12.**

$x = 2$

So $x = 2$.

Check

$\frac{\frac{1}{4}}{12} = \frac{x}{96}$

$\frac{\frac{1}{4}}{12} \stackrel{?}{=} \frac{2}{96}$ **Substitute 2 for x.**

$\frac{1}{4} \cdot (96) \stackrel{?}{=} 12(2)$ **Set the cross products equal.**

$24 \stackrel{\checkmark}{=} 24$

PRACTICE 5

Solve and check: $\frac{\frac{1}{2}}{2} = \frac{3}{x}$

EXAMPLE 6

Forty pounds of sodium hydroxide are needed to neutralize 49 pounds of sulfuric acid. At this rate, how many pounds of sodium hydroxide are needed to neutralize 98 pounds of sulfuric acid? (*Source:* Peter Atkins and Loretta Jones, *Chemistry*)

PRACTICE 6

Saffron is a powder made from crocus flowers and is used in the manufacture of perfume. Some 8,000 crocus flowers are required to make 2 ounces of saffron. How many flowers are needed to make 16 ounces of saffron? (*Source: The World Book Encyclopedia*)

Solution Let n represent the number of pounds of sodium hydroxide needed. We set up a proportion to compare the amount of sodium hydroxide to the amount of sulfuric acid.

$$\frac{40}{49} = \frac{n}{98}$$ Sodium hydroxide → 40, Sulfuric acid → 49; Sodium hydroxide → n, Sulfuric acid → 98

$$49n = 40 \cdot 98$$ Set the cross products equal.

$$49n = 3{,}920$$

$$\frac{49n}{49} = \frac{3{,}920}{49}$$ Divide each side by 49.

$$n = 80$$

Check

$$\frac{40}{49} = \frac{n}{98}$$

$$\frac{40}{49} \stackrel{?}{=} \frac{80}{98}$$ Substitute 80 for n.

$$49 \cdot 80 \stackrel{?}{=} 40 \cdot 98$$ Set the cross products equal.

$$3{,}920 \stackrel{\checkmark}{=} 3{,}920$$

So 80 pounds of sodium hydroxide are needed to neutralize 98 pounds of sulfuric acid.

Tip A good way to set up a proportion is to write quantities of the same kind in the numerators and their corresponding quantities of the other kind in the denominators.

EXAMPLE 7

The scale of the following map of Nevada indicates that $\frac{1}{2}$ inch represents 100 miles. If the two cities highlighted on the map are 1.6 inches apart, what is the actual distance between them?

PRACTICE 7

A jet files 135 miles in $\frac{1}{4}$ hour. At this rate, how far can it fly in 1.5 hours?

Solution We know that $\frac{1}{2}$ inch corresponds to 100 miles. Let's set up a proportion that compares inches to miles, letting m represent the unknown number of miles.

$$\frac{\frac{1}{2}\text{ inch}}{100\text{ miles}} = \frac{1.6\text{ inches}}{m\text{ miles}}$$

$$\frac{\frac{1}{2}}{100} = \frac{1.6}{m}$$

$$\frac{1}{2}m = (100)(1.6) \quad \text{Set the cross products equal.}$$

$$\frac{1}{2}m = 160$$

$$\frac{1}{2}m \div \frac{1}{2} = 160 \div \frac{1}{2} \quad \text{Divide each side by } \frac{1}{2}.$$

$$\frac{1}{2}m \times \frac{2}{1} = 160 \times \frac{2}{1}$$

$$m = 320$$

Check

$$\frac{\frac{1}{2}}{100} = \frac{1.6}{m}$$

$$\frac{\frac{1}{2}}{100} \stackrel{?}{=} \frac{1.6}{320}$$

$$100(1.6) \stackrel{?}{=} \frac{1}{2} \cdot (320)$$

$$160 \stackrel{\checkmark}{=} 160$$

So the cities are 320 miles apart.

EXAMPLE 8

In the following diagram, the heights and shadow lengths of the two objects shown are in proportion. Find the height of the tree, h.

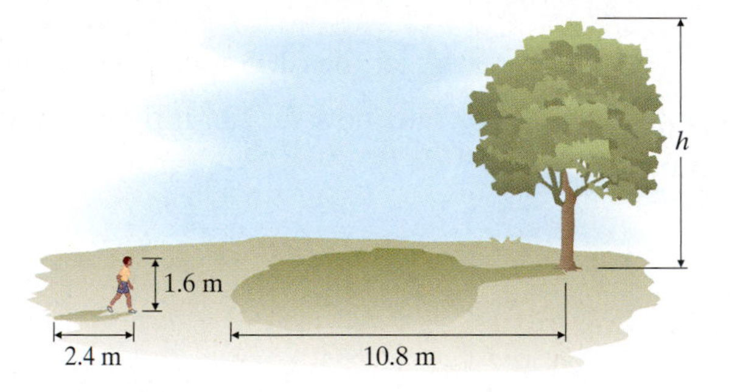

Solution The heights and shadow lengths are in proportion, so we write the following:

$$\frac{\text{Height} \rightarrow h\text{ meters}}{\text{Shadow} \rightarrow 10.8\text{ meters}} = \frac{1.6\text{ meters} \leftarrow \text{Height}}{2.4\text{ meters} \leftarrow \text{Shadow}}$$

$$\frac{h}{10.8} = \frac{1.6}{2.4}$$

$$2.4h = (10.8)(1.6)$$

$$\frac{2.4h}{2.4} = \frac{17.28}{2.4}$$

$$h = 7.2$$

Check

$$\frac{h}{10.8} = \frac{1.6}{2.4}$$

$$\frac{7.2}{10.8} \stackrel{?}{=} \frac{1.6}{2.4}$$

$$(10.8)(1.6) \stackrel{?}{=} (7.2)(2.4)$$

$$17.28 \stackrel{\checkmark}{=} 17.28$$

So the height of the tree is 7.2 meters.

PRACTICE 8

The wingspans of the Boeing 777 (pictured) and 767 passenger jets are approximately in the ratio 5 to 4. Find the wingspan of the Boeing 767 to the nearest 10 feet. (*Source:* http://boeing.com/commercial/777family/pf/pf-pf_exterior_general.html)

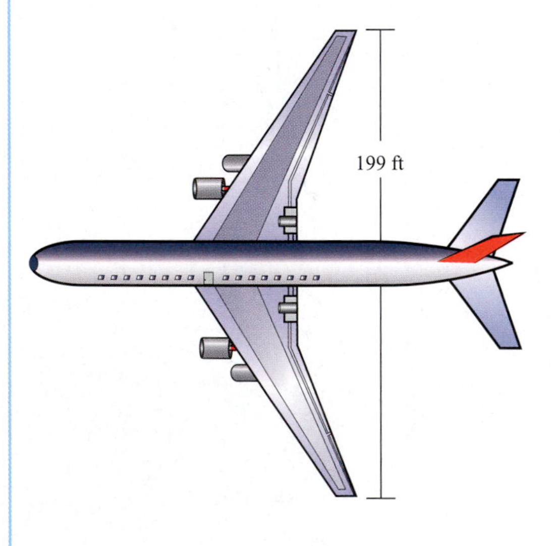

Mathematically Speaking

Fill in each blank with the most appropriate term or phrase from the given list.

equation	check	like
products	solve	cross products
as	proportion	

1. A(n) _______ is a statement that two ratios are equal.

2. To determine if a proportion is true, check whether the _______ of the ratios are equal.

3. The proportion $\frac{4}{5} = \frac{8}{10}$ can be read "4 is to 5 _______ 8 is to 10."

4. To _______ the proportion $\frac{x}{2} = \frac{4}{6}$, find the value of x that makes the proportion true.

Indicate whether each statement is true or false.

5. Thirty is to 9 as 40 is to 12.

6. Nine is to 12 as 12 is to 16.

7. Two is to 3 as 7 is to 16.

8. Three is to 8 as 10 is to 27.

9. One and one-tenth is to 0.3 as 44 is to 12.

10. One and one-half is to 2 as 0.6 is to 0.8.

11. $\frac{3}{6} = \frac{2}{5}$

12. $\frac{4}{7} = \frac{5}{8}$

13. $\frac{12}{28} = \frac{18}{42}$

14. $\frac{28}{24} = \frac{7}{6}$

15. $\frac{6}{1} = \frac{3}{\frac{1}{2}}$

16. $\frac{5}{30} = \frac{\frac{1}{3}}{2}$

Solve and check.

17. $\frac{4}{8} = \frac{10}{x}$

18. $\frac{2}{3} = \frac{x}{42}$

19. $\frac{x}{19} = \frac{10}{5}$

20. $\frac{1}{6} = \frac{x}{78}$

21. $\frac{5}{x} = \frac{15}{12}$

22. $\frac{15}{x} = \frac{6}{10}$

23. $\frac{4}{1} = \frac{52}{x}$

24. $\frac{1}{17} = \frac{x}{51}$

25. $\frac{7}{4} = \frac{14}{x}$

26. $\frac{x}{6} = \frac{15}{18}$

27. $\frac{x}{8} = \frac{3}{6}$

28. $\frac{7}{5} = \frac{35}{x}$

29. $\frac{6}{21} = \frac{x}{70}$

30. $\frac{4}{x} = \frac{92}{23}$

31. $\frac{x}{12} = \frac{25}{20}$

32. $\frac{20}{25} = \frac{x}{45}$

33. $\frac{28}{x} = \frac{36}{27}$

34. $\frac{27}{63} = \frac{24}{x}$

35. $\frac{x}{10} = \frac{4}{3}$

36. $\frac{5}{6} = \frac{2}{x}$

37. $\frac{4}{x} = \frac{\frac{2}{5}}{10}$

38. $\frac{\frac{3}{4}}{6} = \frac{3}{x}$

39. $\frac{x}{27} = \frac{1.6}{24}$

40. $\frac{24}{28} = \frac{1.8}{x}$

41. $\frac{10.5}{x} = \frac{5}{10}$

42. $\frac{32}{7.2} = \frac{x}{9}$

43. $\frac{7}{0.9} = \frac{x}{36}$

44. $\frac{18}{x} = \frac{4.8}{56}$

45. $\frac{600}{x} = \frac{3}{1\frac{1}{2}}$

46. $\frac{2\frac{1}{3}}{5} = \frac{x}{12}$

47. $\frac{15}{2} = \frac{x}{2\frac{2}{3}}$

48. $\frac{x}{11} = \frac{6}{5\frac{1}{2}}$

49. $\dfrac{\frac{1}{2}}{\frac{1}{5}} = \dfrac{x}{4}$

50. $\dfrac{2}{\frac{4}{5}} = \dfrac{\frac{2}{3}}{x}$

51. $\dfrac{\frac{1}{3}}{x} = \dfrac{2}{1.2}$

52. $\dfrac{2.5}{x} = \dfrac{\frac{1}{4}}{50}$

53. $\dfrac{x}{0.16} = \dfrac{0.15}{4.8}$

54. $\dfrac{1.5}{1.25} = \dfrac{x}{0.5}$

Mixed Practice

55. Solve and check: $\dfrac{\frac{3}{4}}{15} = \dfrac{x}{8}$

56. Solve and check: $\dfrac{1.6}{x} = \dfrac{2.4}{27}$

57. Solve and check: $\dfrac{3}{2} = \dfrac{2\frac{2}{5}}{x}$

58. Determine whether the proportion 8 is to 1 as 2 is to $\dfrac{1}{4}$ is true.

59. Is $\dfrac{4}{9} = \dfrac{3}{8}$ a true or false statement?

60. Solve and check: $\dfrac{x}{9} = \dfrac{5}{6}$

Applications

Solve and check.

61. An average adult's heart beats 8 times every 6 seconds, whereas a newborn baby's heart beats 7 times every 3 seconds. Determine whether these rates are the same. (*Source: Mosby's Medical, Nursing, and Allied Health Dictionary*)

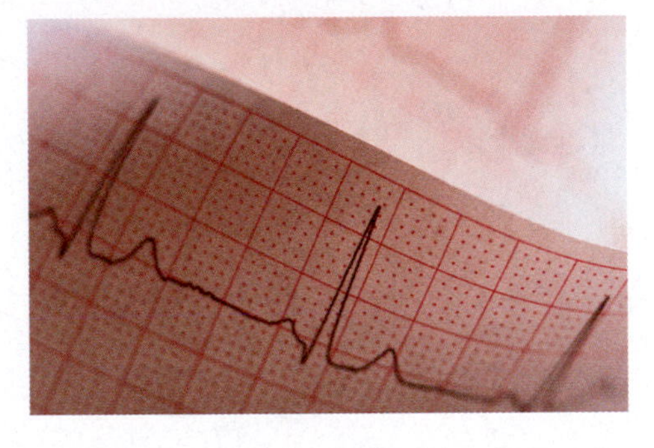

62. A full-time student at a community college pays tuition of \$1,296 for 12 credits, and a part-time student pays \$1,008 for 9 credits. Are the tuition rates the same?

63. A dripping faucet wastes about 15 gallons of water daily. About how much water is wasted in 3 hours? (*Hint:* 1 day = 24 hours)

64. An intravenous fluid is infused at a rate of 2.5 milliliters per minute. How many milliliters are infused per hour?

65. The recommended daily allowance of protein for adults is 0.8 grams for every 2.2 pounds of body weight. If you weigh 150 pounds, how many grams of protein to the nearest tenth should you consume each day? (*Source: The Nutrition Desk Reference*)

66. A homeowner is preparing a solution of insecticide and water to spray her house plants. The directions on the insecticide bottle instruct her to mix 1 part of insecticide with 50 parts of water. How much water must she mix with 2 tablespoons of insecticide?

67. In water molecules, for every 2 hydrogen atoms there is 1 oxygen atom. How many hydrogen atoms combine with 50 oxygen atoms to form water molecules?

68. The scale on a map is $\frac{1}{4}$ centimeter to 50 kilometers. Find the actual distance between two towns represented by 10 centimeters on the map.

69. The following rectangular photo is to be enlarged so that the width of the enlargement is 25 inches. If the dimensions of the photo are to remain in proportion, what should the length of the enlargement be?

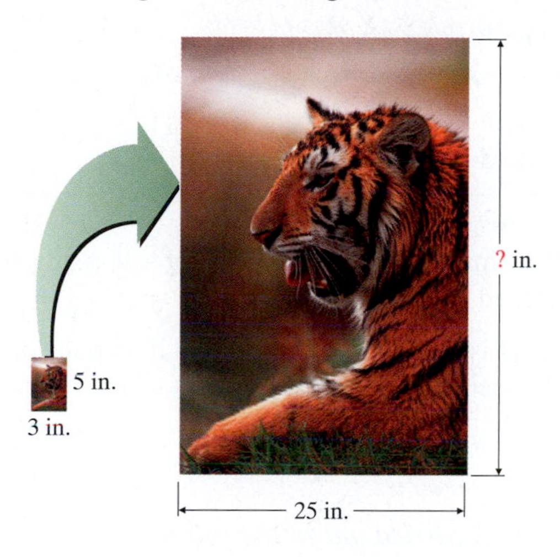

70. Architects now use computers to render their designs. If the actual length of the kitchen shown on the computer-generated floor plan is 10 feet, what is the actual length of the dining area?

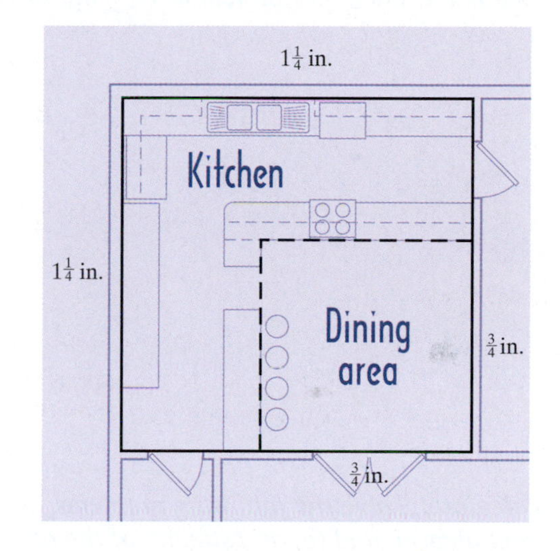

71. A popular scale for building model railroads is the N scale, where model trains are $\frac{1}{160}$ the size of actual trains. At this scale, what is the model size of a boxcar that is actually 40 feet long?

72. Thirty gallons of oil flow through a pipe in 4 hours. At this rate, how long will it take 280 gallons to flow through this same pipe?

73. The ratio of your federal income tax to state income tax is 10 to 3. How much is your state income tax if your federal income tax is $2,000?

74. A computer can download a 1,558-kilobyte file in 38 seconds. At this rate, how long will it take to download a 2,009-kilobyte file?

75. To determine the number of fish in a lake, researchers tagged 150 of them. In a later sample, they found that 6 of 480 fish were tagged. About how many fish were in the lake?

76. On a particular day, 115 Japanese yen were worth the same as 1 U.S. dollar. If a shirt cost 2,300 yen, what was its value in U.S. dollars?

77. A 5-speed bicycle has a chain linking the pedal sprocket and the gears on the rear wheel. The ratio of pedal turns to rear-wheel turns in first gear is 9 to 14. How many times in first gear does the rear wheel turn if the pedals turn 180 times?

78. The tallest land animal is the giraffe. How tall is a giraffe that casts a shadow 320 centimeters long, if a man nearby who is 180 centimeters tall casts a shadow 100 centimeters long? (*Source: Encyclopedia of Mammals*)

79. A tablet of medication consists of two substances in the ratio of 9 to 5. If the tablet contains 140 milligrams of medication, how much of each substance is in the tablet?

80. A certain metal is 5 parts tin and 2 parts lead. How many kilograms of each are there in 28 kilograms of the metal?

81. The nutrition label from a box of General Mills Total cereal indicates that a $\frac{3}{4}$-cup serving contains 23 grams of carbohydrates and 2 grams of protein.

a. How many grams of carbohydrates are there in 3 cups of cereal?

b. What is the amount of protein in $1\frac{1}{2}$ cups of cereal?

82. The following recipe is for raspberry muffins.

Raspberry Muffins	*Serves 12*
1 spray of cooking spray	
1 1/2 cup whole wheat self-rising flour	
4 Tbsp reduced-calorie margarine, softened	
2 oz ready-to-eat crisp rice cereal, divided (about 2 cups)	
1 1/2 cups raspberries, divided	
2/3 cup unpacked brown sugar, divided	
2 large eggs, lightly beaten	
2/3 cup buttermilk	

(*Source:* http://weightwatchers.com)

a. How many cups of whole wheat self-rising flour are needed for a serving of 18?

b. What is the number of servings if $4\frac{1}{2}$ cups of flour are used, with other ingredients increased proportionately?

Using a calculator, solve the following problems, giving (a) the operation(s) carried out in the solution, (b) the exact answer, and (c) an estimate of the answer.

83. A senator reported that 640 metric tons of spent nuclear fuel had produced 660,000 gallons of nuclear waste. At this rate, how much nuclear waste would be produced by 810 metric tons of fuel?

84. A car uses 0.16 gallon of gas to travel through a tunnel 3.6 miles long. At this rate, how many gallons of gas are needed to travel 2,885 miles across country?

• *Check your answers on page A-9.*

MINDSTRETCHERS

Mathematical Reasoning

1. Pictorial comparisons (called *analogies*) are used on many standardized tests. Fill in the blank.

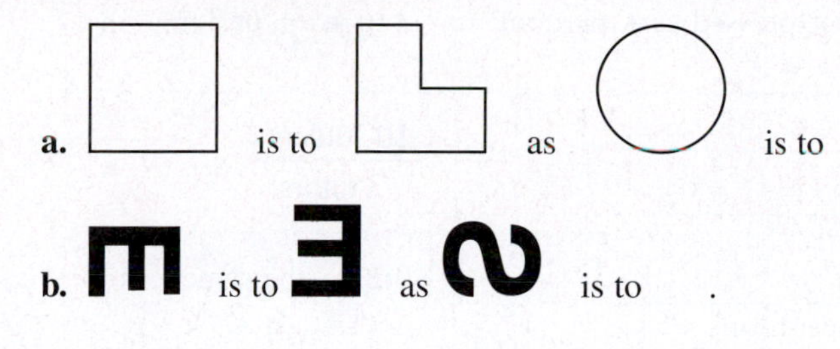

Groupwork

2. Work with a partner on the following.

   a. Complete the following table.

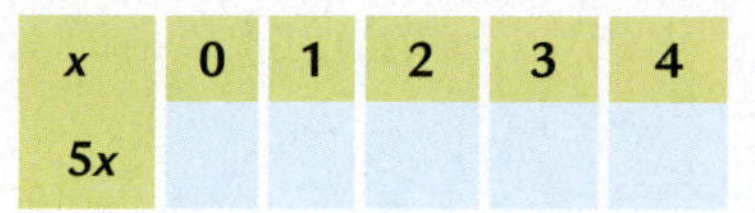

x	0	1	2	3	4
$5x$					

   b. Write as many true proportions as you can, based on the values in the table.

$\frac{\_}{\_} = \frac{\_}{\_} \quad \frac{\_}{\_} = \frac{\_}{\_} \quad \frac{\_}{\_} = \frac{\_}{\_}$

$\frac{\_}{\_} = \frac{\_}{\_} \quad \frac{\_}{\_} = \frac{\_}{\_} \quad \frac{\_}{\_} = \frac{\_}{\_}$

Technology

3. On the Web, there are many currency calculators that convert a given amount of a first currency into the equivalent amount of a second currency. Locate one such calculator. Use your knowledge of proportions to confirm that the currency calculator is working correctly.

KEY CONCEPTS AND SKILLS

CONCEPT SKILL

Concept/Skill	Description	Example
[5.1] Ratio	A comparison of two quantities expressed as a quotient.	3 to 4, $\frac{3}{4}$, or 3:4
[5.1] Rate	A ratio of unlike quantities.	$\frac{10 \text{ students}}{3 \text{ tutors}}$
[5.1] To simplify a ratio	• Write the ratio as a fraction. • Express the fraction in simplest form. • If the quantities are alike, drop the units. If the quantities are unlike, keep the units.	9:27 is the same as 1:3, because $\frac{9}{27} = \frac{1}{3}$ 21 hours to 56 hours $= \frac{21 \text{ hours}}{56 \text{ hours}} = \frac{21}{56} = \frac{3}{8}$ 175 miles per 7 gallons $= \frac{175 \text{ miles}}{7 \text{ gallons}} = \frac{25 \text{ miles}}{1 \text{ gallon}}$, or 25 mpg
[5.1] Unit rate	A rate in which the number in the denominator is 1.	$\frac{180 \text{ calories}}{1 \text{ ounce}}$, or 180 calories per ounce, or 180 cal/oz
[5.1] Unit price	The price of one item, or one unit.	\$0.69 per can, or \$0.69/can
[5.2] Proportion	A statement that two ratios are equal.	$\frac{5}{8} = \frac{15}{24}$
[5.2] To solve a proportion	• Find the cross products, and set them equal. • Solve the resulting equation. • Check the solution by substituting the value of the unknown in the original equation to verify that the resulting proportion is true.	$\frac{6}{9} = \frac{2}{x}$ $6x = 18$ $x = 3$ **Check** $\frac{6}{9} = \frac{2}{x}$ $\frac{6}{9} \stackrel{?}{=} \frac{2}{3}$ $6 \cdot 3 \stackrel{?}{=} 9 \cdot 2$ $18 \stackrel{\checkmark}{=} 18$

Chapter 5

Review Exercises

To help you review this chapter, solve these problems.

[5.1] *Write each ratio or rate in simplest form.*

1. 10 to 15

2. 28 to 56

3. 3 to 4

4. 50 to 16

5. 10,400 votes to 6,500 votes

6. 9 cups to 12 cups

7. 88 feet in 10 seconds

8. 45 applicants for 10 positions

Write each ratio as a unit rate.

9. 4 pounds of grass seed to plant in 1,600 square feet of lawn

10. 75 billion telephone calls in 150 days

11. 48 yards in 6 downs

12. 3,200 square feet covered by 8 gallons of paint

13. 21,000,000 vehicles produced in 2 years

14. 532,000 commuters traveled in 7 days

Find the unit price for each item.

15. $475 for 4 nights

16. $19.45 for 5 DVD movie rentals

17. $80,000 for 64 computer stations

18. $9,364 for 100 shares of stock

Fill in each table. Which is the better buy?

19. *The New Yorker* magazine issues

Number of Units	Total Price	Unit Price
47	$11.95	
92	$29.90	

20. Custom laser checks

Number of Units	Total Price	Unit Price
300	$59.99	
525	$74.99	

Complete each table. Determine the best buy.

21. GNC green tea extract capsules

Number of Units	Total Price	Unit Price
90	$7.19	
180	$7.43	
360	$17.91	

22. Johnson's baby oil bottles

Number of Units (Fluid Ounces)	Total Price	Unit Price
4	$1.89	
14	$3.59	
20	$4.69	

[5.2] *Indicate whether each proportion is true or false.*

23. $\frac{15}{25} = \frac{3}{5}$

24. $\frac{3}{1} = \frac{1}{3}$

25. $\frac{50}{45} = \frac{10}{8}$

26. $\frac{15}{6} = \frac{5}{2}$

Solve and check.

27. $\frac{1}{2} = \frac{x}{12}$

28. $\frac{9}{12} = \frac{x}{4}$

29. $\frac{12}{x} = \frac{3}{8}$

30. $\frac{x}{72} = \frac{5}{12}$

31. $\frac{1.6}{7.2} = \frac{x}{9}$

32. $\frac{x}{12} = \frac{1.2}{1.8}$

33. $\frac{5}{\frac{1}{2}} = \frac{7}{x}$

34. $\frac{3}{5} = \frac{x}{\frac{2}{3}}$

35. $\frac{2\frac{1}{4}}{x} = \frac{1}{30}$

36. $\frac{3}{1\frac{3}{5}} = \frac{x}{24}$

37. $\frac{\frac{5}{6}}{x} = \frac{2}{1.8}$

38. $\frac{\frac{2}{3}}{4} = \frac{x}{0.9}$

39. $\frac{0.36}{4.2} = \frac{2.4}{x}$

40. $\frac{x}{0.21} = \frac{0.12}{0.18}$

Mixed Applications

Solve and check.

41. An airplane has 12 first-class seats and 180 seats in coach. What is the ratio of first-class seats to coach seats?

42. A computer store sells $23,000 worth of desktop computers and $45,000 worth of laptop computers in a given month. What is the ratio of desktop to laptop computer sales?

43. If a personal care attendant earns $540 for a 6-day workweek, how much does she earn per day?

44. A glacier in Alaska moves about 2 inches in 16 months. How far does the glacier move per month?

45. In a recent year, approximately 200,000,000 of the 300,000,000 people in the United States were Internet users. What is the ratio of Internet users to the total population? (***Source:*** Internet World Stats, 2006)

46. A city's public libraries spend about $9.50 in operating expenses for every book they circulate. If their operating expenses amount to $475,000, how many books circulate?

47. In a college's day-care center, the required staff-to-child ratio is 2 to 5. If there are 60 children and 12 staff in the day-care center, is the center in compliance with the requirement?

48. Despite the director's protests, the 1924 silent film *Greed* was edited down from about 42 reels of film to 10 reels. If the original version was about 9 hours long, about how long was the edited version? (***Source:*** *The Film Encyclopedia*)

49. A sports car engine has an 8-to-1 compression ratio. Before compression, the fuel mixture in a cylinder takes up 440 cubic centimeters of space. How much space does the fuel mixture occupy when fully compressed?

50. On an architectural drawing of a planned community, a measurement of 25 feet is represented by 0.5 inches. If two houses are actually 62.5 feet apart, what is the distance between them on the drawing?

51. The density of a substance is the ratio of its mass to its volume. To the nearest hundredth, find the density of gasoline if a volume of 317.45 cubic centimeters has a mass of 216.21 grams.

52. The state of New Jersey has an area of 7,417 square miles and a population of 8,717,925. Of all the U.S. states, it has the highest population density—more than a dozen times that of the nation. Compute New Jersey's population density, the ratio of the number of people to the area, rounded to the nearest tenth. (***Source:*** U.S. Bureau of the Census)

• *Check your answers on page A-9.*

Chapter 5 POSTTEST

FOR EXTRA HELP

Test solutions are found on the enclosed CD.

To see if you have mastered the topics in this chapter, take this test.

Write each ratio or rate in simplest form.

1. 8 to 12

2. 15 to 42

3. 55 ounces to 31 ounces

4. 180 miles to 15 miles

5. 65 revolutions in 60 seconds

6. 3 centimeters for every 75 kilometers

Find the unit rate.

7. 340 miles in 5 hours

8. 200-meter dash in 25 seconds

Determine the unit price.

9. $4,080 for 30 days

10. 25 greeting cards for $20

Determine whether each proportion is true or false.

11. $\frac{8}{21} \stackrel{?}{=} \frac{16}{40}$

12. $\frac{7}{3} \stackrel{?}{=} \frac{63}{27}$

Solve and check.

13. $\frac{15}{x} = \frac{6}{10}$

14. $\frac{102}{17} = \frac{36}{x}$

15. $\frac{0.9}{36} = \frac{0.7}{x}$

16. $\frac{\frac{1}{3}}{4} = \frac{x}{12}$

Solve.

17. To advertise his business, an owner can purchase 3 million e-mail addresses for $120 or 5 million e-mail addresses for $175. Which is the better buy?

18. A house was originally worth $95,000 but increased in value to $110,000 after 5 years. What is the ratio of the increase to the original value?

19. A man $6\frac{1}{4}$ feet tall casts a 5-foot shadow. A nearby tree casts a 20-foot shadow. If the heights and shadow lengths of the man and tree are proportional, how tall is the tree?

20. A nurse takes his patient's pulse. What is the patient's pulse per minute if it beats 12 times in 15 seconds?

• *Check your answers on page A-10.*

Cumulative Review Exercises

To help you review, solve the following:

1. Find the difference: $3\frac{1}{10} - 2\frac{7}{10}$

2. Multiply: $8.2 \times 1{,}000$

3. Solve and check: $x + 6.5 = 9$

4. Simplify the ratio: 2.5 to 10

5. Find the unit price: 3 yards for $12

6. Estimate: $12\frac{1}{7} \div 3\frac{9}{10}$

7. Solve and check: $\frac{\frac{1}{2}}{4} = \frac{x}{6}$

8. What is the area of the singles tennis court shaded in the diagram?

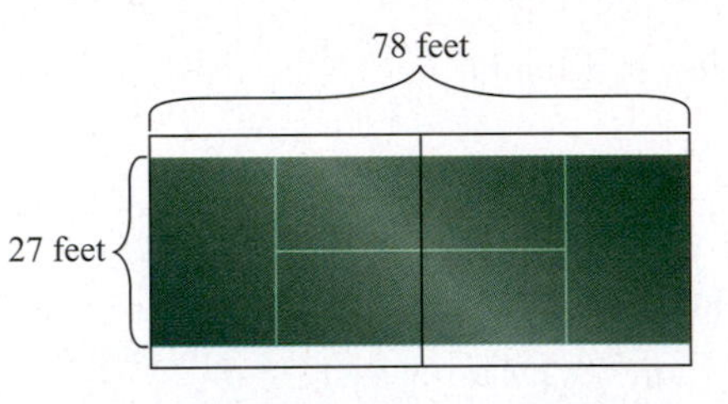

9. A rule of thumb for growing lily bulbs is to plant them 3 times as deep as they are wide. How deep should a gardener plant a lily bulb that is 2.5 inches wide?

10. At a legal firm, a part-time employee works 10 hours a week and makes $120. At this rate of pay, how much would the employee make for working 15 hours a week?

• *Check your answers on page A-10.*

CHAPTER 6

Percents

Percents and Surveys

In a recent primary election, three candidates, including Deval Patrick, were running for one political party's nomination for governor of Massachusetts. Several weeks before the primary election, a newspaper reported the results of a poll of 501 likely primary voters: 31% of those surveyed supported Patrick, with 30% in favor of a second candidate and 27% a third candidate. But every poll and survey has a *margin of error*. Allowing for this margin of error, the poll really indicated that the percent of voters in the primary supporting Patrick would probably be 26.6% to 35.4%. Because of this wide margin of error, the newspaper didn't use the headline "Patrick leads!". Instead, it correctly reported that the three candidates were in a virtual dead heat.

(**Source:** *The Boston Globe*)

Chapter 6 PRETEST

To see if you have already mastered the topics in this chapter, take this test.

Rewrite.

1. 5% as a fraction

2. $37\frac{1}{2}\%$ as a fraction

3. 250% as a decimal

4. 3% as a decimal

5. 0.007 as a percent

6. 8 as a percent

7. $\frac{2}{3}$ as a percent, rounded to the nearest whole percent

8. $1\frac{1}{10}$ as a percent

Solve.

9. What is 75% of 50 feet?

10. Find 110% of 50.

11. Estimate 84% of $61.77.

12. 2% of what number is 5?

13. What percent of 10 is 4?

14. What percent of 4 is 10?

15. In a municipal savings account, a city employee earned 4% interest on $350. How much money did the employee earn in interest?

16. The number of students enrolled at a community college rose from 2,475 last year to 2,673 this year. What was the percent increase in the college's enrollment?

17. In the depths of the Great Depression, 24% of the U.S. civilian labor force was unemployed. Write this percent as a simplified fraction. (*Source:* U.S. Bureau of the Census)

18. In a chemistry lab, a student dissolved 10 milliliters of acid in 30 milliliters of water. What percent of the solution was acid?

19. For parties of 8 or more, a restaurant automatically adds an 18% tip to the restaurant check. What tip would be added to a dinner check for a party of 10 if the total bill was $339.50?

20. A patient's health insurance covered 80% of the cost of her operation. She paid the remainder, which came to $2,000. Find the total cost of the operation.

• *Check your answers on page A-10.*

6.1 Introduction to Percents

What Percents Are and Why They Are Important

OBJECTIVES

- To read and write percents
- To find the fraction and the decimal equivalent of a given percent
- To find the percent equivalent of a given fraction or decimal
- To solve word problems involving percent conversions

Percent means divided by 100. So 50% (read "fifty percent") means 50 divided by 100 (or 50 out of 100).

A percent can also be thought of as a ratio or a fraction with denominator 100. For example, we can look at 50% either as the ratio of 50 parts to 100 parts or as the fraction $\frac{50}{100}$, or $\frac{1}{2}$. We can also think of 50% as 0.50, or 0.5, since a fraction can be written as a decimal.

In the diagram at the right, 50 of the 100 squares are shaded. This shaded portion represents 50%.

We can use diagrams to represent other percents.

In the diagram to the left, $\frac{1}{2}$% is equivalent to the shaded portion,

$$\frac{\frac{1}{2}}{100}, \text{ or } \frac{1}{200}$$

The entire diagram at the right is shaded, so 100% means $\frac{100}{100}$, or 1.

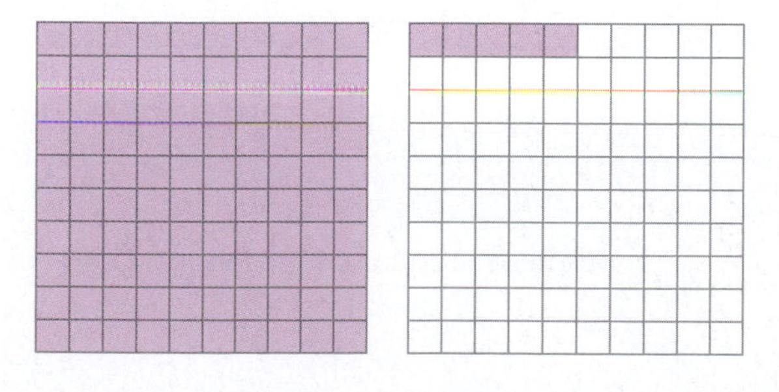

We can express 105% as $\frac{105}{100}$, or $1\frac{1}{20}$, as shown by the shaded portions of the diagrams.

Percents are commonly used, as the following statements taken from a single page of a newspaper illustrate.

- About 10% of the city's budget goes to sanitation.
- Blanket Sale—30% to 40% off!
- The number of victims of the epidemic increased by 125% in just 6 months.

A key reason for using percents so frequently is that they are easy to compare. For instance, we can tell right away that a discount of 30% is larger than a discount of 22%, simply by comparing the whole numbers 30 and 22.

To see how percents relate to fractions and decimals, let's consider finding equivalent fractions, decimals, and percents. In Chapter 3, we discussed two of the six types of conversions:

- changing a decimal to a fraction, and
- changing a fraction to a decimal.

Here, we consider the remaining four types of conversion:

- changing a percent to a fraction,
- changing a percent to a decimal,
- changing a decimal to a percent, and
- changing a fraction to a percent.

Note that each type of conversion changes the way the number is written—but not the number itself.

Changing a Percent to a Fraction

Suppose that we want to rewrite a percent—say, 30%—as a fraction. Because percent means divided by 100, we simply drop the % sign, place 30 over 100, and simplify.

$$30\% = \frac{30}{100} = \frac{3}{10}$$

Therefore, the fraction $\frac{3}{10}$ is just another way of writing the percent 30%. This result suggests the following rule.

To Change a Percent to the Equivalent Fraction

- Drop the % sign from the given percent and place the number over 100.
- Simplify the resulting fraction, if possible.

EXAMPLE 1

Write 7% as a fraction.

Solution To change this percent to a fraction, we drop the percent sign and write the 7 over 100. The fraction is already reduced to lowest term.

$$7\% = \frac{7}{100}$$

PRACTICE 1

Find the fractional equivalent of 21%.

EXAMPLE 2

Express 150% as a fraction.

Solution $150\% = \frac{150}{100} = \frac{3}{2}$, or $1\frac{1}{2}$

Note that the answer is larger than 1 because the original percent was more than 100%.

PRACTICE 2

What is the fractional equivalent of 225%?

EXAMPLE 3

Express $33\frac{1}{3}\%$ as a fraction.

Solution To find the equivalent fraction, we first drop the % sign and then put the number over 100.

$$\frac{33\frac{1}{3}}{100} = 33\tfrac{1}{3} \div 100 = 33\tfrac{1}{3} \div \frac{100}{1} = \frac{100}{3} \div \frac{100}{1} = \frac{\overset{1}{\cancel{100}}}{3} \times \frac{1}{\underset{1}{\cancel{100}}} = \frac{1}{3}$$

So $33\frac{1}{3}\%$ expressed as a fraction is $\frac{1}{3}$.

PRACTICE 3

Change $12\frac{1}{2}\%$ to a fraction.

EXAMPLE 4

The Ring of Fire contains 75% of the volcanoes on Earth. Express this percent as a fraction. (*Source:* *National Geographic*)

Solution $75\% = \frac{75}{100} = \frac{3}{4}$

So $\frac{3}{4}$ of the volcanoes on Earth are located in the Ring of Fire.

PRACTICE 4

About 86% of California's coastline is eroding. Express this percent as a fraction. (*Source:* Surfrider Foundation)

Changing a Percent to a Decimal

Now let's consider rewriting a percent as a decimal. For instance, take 75%. We begin by writing this percent as a fraction.

$$75\% = \frac{75}{100}$$

Converting this fraction to a decimal, we divide:

$$100\overline{)75.00}\quad 0.75$$

Note that we could have gotten this answer simply by moving the decimal point two places to the left and dropping the % sign.

$$75\% = 75.\% = .75, \text{ or } 0.75$$

This example suggests the following rule.

To Change a Percent to the Equivalent Decimal

- Drop the % sign from the given percent and divide the number by 100.

EXAMPLE 5

Change 42% to a decimal.

Solution Drop the % sign from 42% and divide by 100. Recall that to divide a decimal by 100, we can simply move the decimal point two places to the left.

$$42\% = 42.\% = .42$$

So the decimal equivalent of 42% is 0.42.

PRACTICE 5

Express 31% as a decimal.

Tip A shortcut for changing a percent to its equivalent decimal is dropping the percent sign and moving the decimal point *two places* to the *left*.

EXAMPLE 6

Find the decimal equivalent of 1%.

Solution The unwritten decimal point lies to the right of the 1. Moving the decimal point two places to the left and dropping the % sign, we get:

$$1\% = 01.\% = .01, \text{ or } 0.01$$

Note that we inserted a 0 as a placeholder, because there was only a single digit to the left of the 1.

PRACTICE 6

What decimal is equivalent to 5%?

EXAMPLE 7

Convert 37.5% to a decimal.

Solution 37.5% = .375, or 0.375

In this problem, the given number is a percent even though it involves a decimal point.

PRACTICE 7

Rewrite 48.2% as a decimal.

EXAMPLE 8

Change $12\frac{1}{2}\%$ to a decimal.

Solution To find the decimal equivalent of $12\frac{1}{2}\%$, we begin by converting the fraction $\frac{1}{2}$ in the mixed number to its decimal equivalent 0.5. Then we move the decimal point to the left two places, dropping the % sign.

$$\frac{1}{2} = 2\overline{)1.0} \quad 0.5, \text{ or } .5$$

$$12\frac{1}{2}\% = 12.5\% = .125, \text{ or } 0.125$$

PRACTICE 8

Express the following percent as a decimal: $62\frac{1}{4}\%$.

EXAMPLE 9

In 1945 at the end of World War II, the public debt of the United States was 602% of what it had been in 1940. Write this percent as a decimal.

Solution 602% = 602.%, or 6.02

The 1945 U.S. debt was 6.02 times what it had been 5 years earlier.

PRACTICE 9

In 2014, the enrollment in American schools and colleges is expected to be 112% of the enrollment in 2005. Express this percent as a decimal.

(*Source:* National Center for Education Statistics)

Changing a Decimal to a Percent

Suppose that we want to change 0.75 to a percent. Because 100% is the same as 1, we can multiply this number by 100% without changing its value.

$$0.75 \times 100\% = 75\%$$

Note that we could have gotten this answer simply by moving the decimal point to the right two places and adding the % sign.

$$0.75 = 075.\% = 75\%$$

Note that we dropped the decimal point in the answer because it was to the right of the units digit.

To Change a Decimal to the Equivalent Percent

- Multiply the number by 100 and insert a % sign.

EXAMPLE 10

Write 0.425 as a percent.

Solution We multiply 0.425 by 100 and add a % sign.

$$0.425 = 0.425 \times 100\% = 42.5\%, \text{ or } 42\frac{1}{2}\%$$

PRACTICE 10

What percent is equivalent to the decimal 0.025?

Tip A shortcut for changing a decimal to its equivalent percent is inserting a % sign and moving the decimal point *two places* to the *right*.

EXAMPLE 11

Convert 0.03 to a percent.

Solution $0.03 = 003.\% = 3\%$

PRACTICE 11

Change 0.09 to a percent.

EXAMPLE 12

Express 0.1 as a percent.

Solution In the given number, only a single digit is to the right of the decimal point. So to move the decimal point two places to the right, we need to insert a 0 as a placeholder.

$$0.1 = 0.10 = 10.\% = 10\%$$

PRACTICE 12

What percent is equivalent to 0.7?

EXAMPLE 13

What percent is equivalent to 2?

Solution Recall that a whole number such as 2 has a decimal point understood to its right. We move the decimal point two places to the right.

$$2 = 2. = 2.\mathbf{00} = 200.\% = 200\%$$

So the answer is 200%, which makes sense: 200% is double 100%, just as 2 is double 1.

PRACTICE 13

Rewrite 3 as a percent.

EXAMPLE 14

Express 0.2483 as a percent, rounded to the nearest whole percent.

Solution First, we obtain the exact percent equivalent.

$$0.2483 = 24.83\%$$

To round this number to the nearest whole percent, we underline the digit 4. Then we check the critical digit immediately to its right. This digit is 8, so we round up.

$$2\underline{4}.83\% \approx 25.\% = 25\%$$

PRACTICE 14

Convert 0.714 to a percent, rounded to the nearest whole percent.

EXAMPLE 15

Red blood cells make up about 0.4 of the total blood volume in the human body, whereas 55% of the total blood volume is plasma. Which makes up more of the blood volume—red blood cells or plasma? Explain. (***Source:*** Mayo Clinic; Merck)

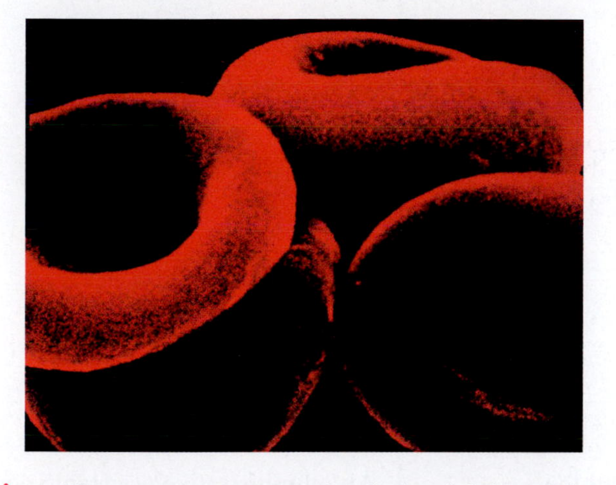

Solution We want to compare the decimal 0.4 and the percent 55%. One way is to change the decimal to a percent.

$$0.4 = 0.40 = 40.\% = 40\%$$

Since 40% is less than 55%, we conclude that plasma makes up more of the blood volume.

PRACTICE 15

Air is a mixture of many gases. For example, 0.78 of air is nitrogen, and 0.93% is argon. Is there more nitrogen or argon in air? Explain.

Changing a Fraction to a Percent

Now let's change a fraction to a percent. Consider, for instance, the fraction $\frac{1}{5}$.

To convert this fraction to a percent, multiply $\frac{1}{5}$ by 100%, which is equal to 1.

$$\frac{1}{5} = \frac{1}{5} \times 100\% = \frac{1}{\cancel{5}_1} \times \frac{\cancel{100}^{20}}{1}\% = 20\%$$

To Change a Fraction to the Equivalent Percent

- Multiply the fraction by 100 and insert a % sign.

EXAMPLE 16

Rewrite $\frac{7}{20}$ as a percent.

Solution To change the given fraction to a percent, we multiply by 100 and insert a % sign.

$$\frac{7}{20} = \frac{7}{20} \times 100\% = \frac{7}{\cancel{20}_1} \times \frac{\cancel{100}^{5}}{1}\% = 35\%$$

PRACTICE 16

Convert $\frac{4}{25}$ to a percent.

EXAMPLE 17

Which is larger: 130% or $1\frac{3}{8}$?

Solution To compare, let's express $1\frac{3}{8}$ as a percent.

$$1\frac{3}{8} = 1\frac{3}{8} \times 100\% = \frac{11}{8} \times \frac{100}{1}\%$$

$$= \frac{11}{\cancel{8}_{2}} \times \frac{\cancel{100}^{25}}{1}\% = \frac{275}{2}\% = 137\frac{1}{2}\%$$

Because $137\frac{1}{2}\%$ is larger than 130%, so is $1\frac{3}{8}$.

PRACTICE 17

True or false: $\frac{2}{3} > 60\%$. Justify your answer.

EXAMPLE 18

A student got 28 of 30 questions correct on a test. If all the questions were equal in value, what was the student's grade, rounded to the nearest whole percent?

Solution The student answered $\frac{28}{30}$ of the questions right. To find the student's grade, we change this fraction to a percent.

$$\frac{28}{30} = \frac{28}{30} \times 100\% = \frac{28}{\cancel{30}_{3}} \times \frac{\cancel{100}^{10}}{1}\%$$

$$= \frac{280}{3}\% = 93\frac{1}{3}\% = 93.3\ldots\% \approx 93\%$$

Note that the critical digit is 3, so we round down. The rounded grade was therefore 93%.

PRACTICE 18

An administrative assistant spends \$490 out of her monthly salary of \$1,834 on rent. What percent of her monthly salary is spent on rent, rounded to the nearest whole percent?

Before going any further, study the table of common fraction, decimal, and percent equivalents in the Appendix at the back of this book. These numbers come up frequently and are useful reference points for estimating the answer to percent word problems, as we demonstrate in Section 6.2.

Mathematically Speaking

Fill in each blank with the most appropriate term or phrase from the given list.

right	fraction	percent	divide
decimal	left	whole number	multiply

1. A(n) _______ is a ratio or fraction with denominator 100.

2. To change a percent to the equivalent _______, drop the % sign from the given percent, and place the number over 100.

3. To change a percent to the equivalent decimal, move the decimal point two places to the _______ and drop the % sign.

4. To change a fraction to the equivalent percent, _______ the fraction by 100 and insert a % sign.

Change each percent to a fraction or mixed number. Simplify.

5. 8% **6.** 3% **7.** 250% **8.** 110%

9. 33% **10.** 41% **11.** 18% **12.** 6%

13. 14% **14.** 45% **15.** 65% **16.** 92%

17. $\frac{3}{4}\%$ **18.** $\frac{1}{10}\%$ **19.** $\frac{3}{10}\%$ **20.** $\frac{1}{5}\%$

21. $7\frac{1}{2}\%$ **22.** $2\frac{1}{2}\%$ **23.** $14\frac{2}{7}\%$ **24.** $28\frac{4}{7}\%$

Convert each percent to a decimal.

25. 6% **26.** 9% **27.** 72% **28.** 25%

29. 0.1% **30.** 0.2% **31.** 102% **32.** 113%

33. 42.5% **34.** 10.5% **35.** 500% **36.** 400%

37. $106\frac{9}{10}\%$ **38.** $201\frac{1}{10}\%$ **39.** $3\frac{1}{2}\%$ **40.** $2\frac{4}{5}\%$

41. $\frac{9}{10}\%$ **42.** $\frac{7}{10}\%$ **43.** $\frac{3}{4}\%$ **44.** $\frac{1}{4}\%$

Express each decimal as a percent.

45. 0.31 **46.** 0.05 **47.** 0.17 **48.** 0.18

49. 0.3 **50.** 0.4 **51.** 0.04 **52.** 0.875

53. 0.125 **54.** 0.27 **55.** 1.29 **56.** 1.07

57. 2.9 **58.** 3.5 **59.** 2.87 **60.** 12.91

61. 1.016 **62.** 1.003 **63.** 9 **64.** 7

Change each fraction to a percent.

65. $\frac{3}{10}$ **66.** $\frac{1}{2}$ **67.** $\frac{1}{10}$ **68.** $\frac{3}{20}$

69. $\frac{4}{25}$ **70.** $\frac{6}{25}$ **71.** $\frac{9}{10}$ **72.** $\frac{7}{10}$

73. $\frac{3}{50}$ **74.** $\frac{1}{50}$ **75.** $\frac{5}{9}$ **76.** $\frac{2}{9}$

77. $\frac{1}{9}$ **78.** $\frac{4}{7}$ **79.** 6 **80.** 8

81. $1\frac{1}{2}$ **82.** $2\frac{3}{5}$ **83.** $2\frac{1}{6}$ **84.** $1\frac{1}{3}$

Replace ■ *with < or >.*

85. $2\frac{1}{4}$ ■ 240% **86.** $\frac{5}{6}$ ■ 80% **87.** $\frac{1}{2}\%$ ■ 50% **88.** $\frac{1}{40}$ ■ $\frac{1}{4}\%$

Express as a percent, rounded to the nearest whole percent.

89. $\frac{4}{9}$ **90.** $\frac{3}{7}$ **91.** 2.2469 **92.** 1.1633

Complete each table.

93.

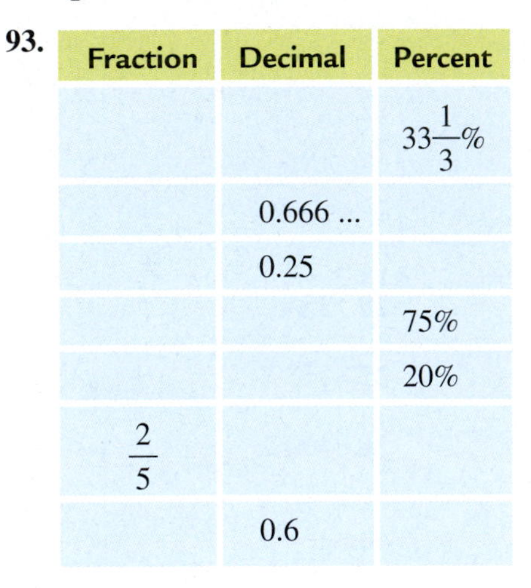

Fraction	Decimal	Percent
		$33\frac{1}{3}\%$
	0.666 ...	
	0.25	
		75%
		20%
$\frac{2}{5}$		
	0.6	

94.

Fraction	Decimal	Percent
	0.8	
$\frac{1}{6}$		
$\frac{5}{6}$		
		12.5%
	0.375	
		$62\frac{1}{2}\%$
	0.875	

Mixed Practice

Solve.

95. Change 104% to a mixed number.

96. What percent is equivalent to $\frac{2}{5}$?

97. Express $3\frac{1}{6}$ as a percent.

98. Express $62\frac{1}{2}\%$ as a fraction.

99. Convert 27.5% to a decimal.

100. Find the decimal equivalent to $\frac{3}{8}\%$

101. What percent is equivalent to 3.1?

102. Change 0.003 to a percent.

103. Which is smaller, $2\frac{5}{9}$ or 254%?

104. Express 1.2753 to the nearest whole percent.

Applications

Solve.

105. It is estimated that by the year 2010, 79% of all e-mail sent worldwide each day will be spam e-mail. Express this percent as a decimal. (*Source:* Radicati Group)

106. According to a recent study, 65% of children have had an imaginary companion by age 7. Express this percent as a fraction. (*Source:* University of Washington, http://uwnews.org)

107. The following graph shows the percent of people in various age groups who say that they get at least some of their news from cell phones, PDAs, or podcasts.

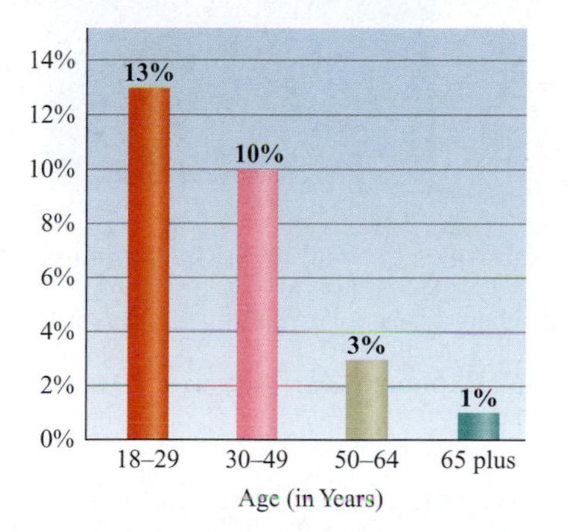

What fraction of people in the 30–49 age group do *not* get their news in this way? (*Source: USA Today*, August 14, 2006)

108. The following graph shows the distribution of investments for a retiree. Express as a decimal the percent of investments that are in equities.

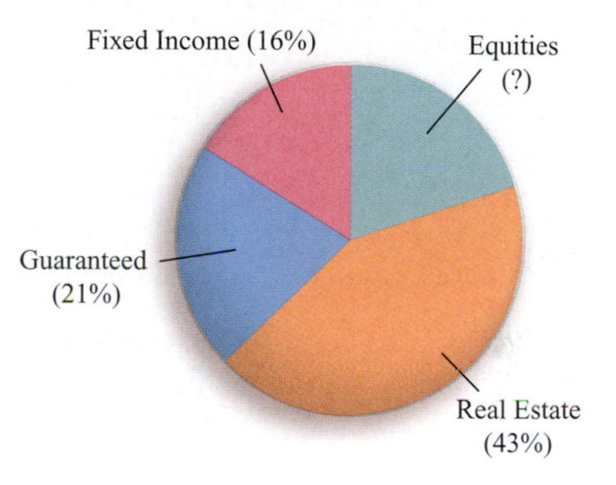

109. According to the nutrition label, one large egg contains 6 grams of protein. This is 10% of the daily value (DV) for protein. Express this percent as a fraction.

110. A bank offers a Visa credit card with a fixed annual percentage rate (APR) of 16.99%. Express the APR as a decimal.

111. New York City has the largest population of any U.S. city. But, Juneau, Alaska, has the greatest area—9.0 times that of New York City. Write this decimal as a percent. (*Source:* U.S. Bureau of the Census)

112. A medical school accepted $\frac{2}{5}$ of its applicants. What percent of the applicants did the school accept?

113. When the recession ended, the factory's output grew by 135%. Write this percent as a simplified mixed number.

114. According to a survey, 78% of the arguments that couples have are about money. Express this percent as a decimal.

115. In Nevada, the federal government controls about 84.5% of the land. Convert this percent to a decimal. (*Source:* Republican Study Committee)

116. After an oil spill, 15% of the wildlife survived. Express this percent as a fraction.

117. By age 75, about $\frac{1}{3}$ of women and about 40% of men have chronic hearing loss. Is this condition more common among men or among women? Explain.

118. The state sales tax rate in North Dakota is 5%, and in South Dakota it is $\frac{1}{25}$. Which state has a lower sales tax rate? Explain. (*Source:* Federation of Tax Administrators, 2006)

119. A quality control inspector found 2 defective machine parts out of 500 manufactured.

a. What percent of the machine parts manufactured were defective?

b. What percent of the machine parts manufactured were not defective?

120. In a survey of several hundred children, 3 out of every 25 children indicated that they wanted to become professional athletes when they grow up.

a. What percent of the children wanted to become professional athletes?

b. What percent of the children did not want to become professional athletes? (*Source: National Geographic Kids*)

Use a calculator to solve the following problems, giving (a) the operation(s) you used, (b) the exact answer, and (c) an estimate of the answer.

121. In the 2004 U.S. presidential election, 122,294,978 voters turned out. At the time, the voting-age population numbered 221,256,931. What percent of the voting-age population turned out? (*Source: Time Almanac 2006*)

122. The first Social Security retirement benefits were paid in 1940 to Ida May Fuller of Vermont. She had paid in a total of $24.85 and got back $20,897 before her death in 1975. Express the ratio of what she got back to what she put in as a percent. (*Source:* James Trager, *The People's Chronology*)

• *Check your answers on page A-10.*

MINDSTRETCHERS

Mathematical Reasoning

1. By mistake, you move the decimal point to the right instead of to the left when changing a percent to a decimal. Your answer is how many times as large as the correct answer?

Writing

2. A study of the salt content of seawater showed that the average salt content varies from 33‰ to 37‰, where the symbol ‰ (read "per mil") means "for every thousand." Explain why you think the scientist who wrote this study did not use the % symbol.

Critical Thinking

3. What percent of the region shown is shaded in?

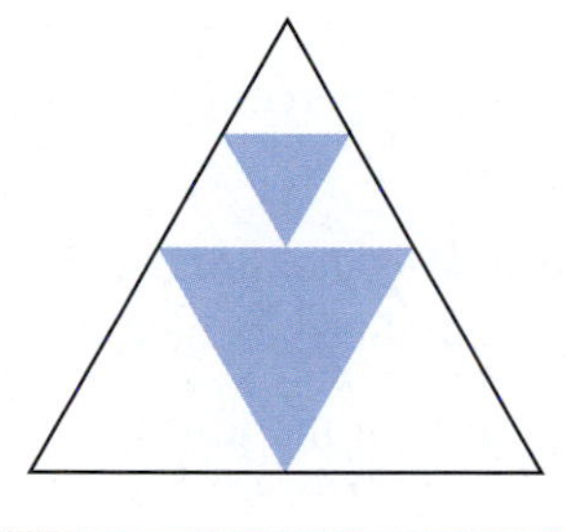

CULTURAL NOTE

Sources: Frank J. Swetz, *Capitalism and Arithmetic: The New Math of the 15th Century,* Open Court, 1987; Carolyn Webber and Aaron Wildavsky, *A History of Taxation and Expenditure in the Western World,* Simon and Schuster, 1986.

Throughout history, the concepts of percent and taxation have been interrelated. At the peak of the Roman Empire, the Emperor Augustus instituted an inheritance tax of 5% to provide retirement funds for the military. Another emperor, Julius Caesar, imposed a 1% sales tax on the population. And in Roman Asia, tax collectors exacted a tithe of 10% on crops. If landowners could not pay, the collectors offered to loan them funds at interest rates that ranged from 12% up to 48%.

Roman taxation served as a model for modern countries when these countries developed their own systems of taxation many centuries later.

6.2 Solving Percent Problems

OBJECTIVES

- To identify the amount, the base, and the percent in a percent problem
- To find the amount, the base, or the percent in a percent problem
- To estimate the amount in a percent problem
- To solve word problems involving a percent

The Three Basic Types of Percent Problems

Frequently, we think of a percent not in isolation but rather in connection with another number. In other words, we take *a percent of a number*.

Consider, for example, the problem of taking 50% of 8. This problem is equivalent to finding $\frac{1}{2}$ of 8, which gives us 4.

Note that this percent problem, like all others, involves three numbers.

$$\underset{\text{Percent}}{\underset{\uparrow}{50\%}} \text{ of } \underset{\text{Base}}{\underset{\uparrow}{8}} \text{ is } \underset{\text{Amount}}{\underset{\uparrow}{4}}.$$

- The 50% is called the **percent** (or the **rate**). The percent always contains the % sign.
- The 8 is called the **base**. The base of a percent—the number that we are taking the percent of—always follows the word *of* in the statement of the problem.
- The remaining number 4 is called the **amount** (or the **part**).

Percent problems involve finding one of the three numbers. For example, if we omit the 4 in "50% of 8 is 4," we ask the question: What is 50% of 8? Omitting the 8, we ask: 50% of what number is 4? And omitting the 50%, we ask: What percent of 8 is 4?

There are several ways to solve these three basic percent questions. In this section, we discuss two ways—the translation method and the proportion method.

The Translation Method

In the translation method, a percent problem has the form

The percent of the base is the amount.

The percent problem gives only two of the three quantities. To find the missing quantity using the translation method, we translate to a simple equation that we then solve.

This method depends on translating the words in the given problem to the appropriate mathematical symbols.

Word(s)	Math Symbol
What, what number, what percent	x (or some other letter)
is	$=$
of	$\times$ or $\cdot$
percent, %	Percent value expressed as a decimal or fraction

Let's translate several percent problems to equations.

EXAMPLE 1

Translate each question to an equation, using the translation method.

a. What is 10% of 2? **b.** 20% of what number is 5?

c. What percent of 8 is 4?

Solution

a. In this problem, we are looking for the amount.

What is 10% of 2?
↓ ↓ ↓ ↓ ↓
$x = 0.1 \cdot 2$

b. Here, we are looking for the base, that is, the number after the word *of*.

20% of what number is 5?
↓ ↓ ↓ ↓ ↓
$0.2 \cdot x = 5$

c. This problem asks "what percent?" So we are looking for the percent.

What percent of 8 is 4?
↓ ↓ ↓ ↓ ↓
$x \cdot 8 = 4$

PRACTICE 1

Use the translation method to set up an equation.

a. What is 70% of 80?

b. 50% of what number is 10?

c. What percent of 40 is 20?

Finding an Amount

Now let's apply the translation method to solve the type of percent problem in which we are given both the percent and the base and are looking for the amount.

EXAMPLE 2

What is 25% of 8?

Solution First, we translate the question to an equation.

What is 25% of 8?
↓ ↓ ↓ ↓ ↓
$x = \frac{1}{4} \cdot 8$

Then we solve this equation:

$$x = \frac{1}{4} \cdot 8 = \frac{1}{\cancel{4}_1} \cdot \frac{\cancel{8}^2}{1} = \frac{2}{1}$$
$$= 2$$

So 2 is 25% of 8. Would we have gotten the same answer if we had translated 25% to 0.25?

PRACTICE 2

What is 20% of 40?

EXAMPLE 3

Find 200% of 30.

Solution We can reword the problem as a question.

What is 200% of 30?
↓ ↓ ↓ ↓ ↓
$x = 2 \cdot 30$

Solving this equation, we get $x = 60$. So 200% of 30 is 60.

PRACTICE 3

150% of 8 is what number?

Tip When the percent is less than 100%, the amount is *less* than the base. When the percent is more than 100%, the amount is *more* than the base.

In a percent problem, we frequently want to *estimate* the amount. Sometimes an approximate answer is good enough, and other times we use the estimate to check an exact answer we have already computed.

EXAMPLE 4

Approximately how much is 67% of 14.8?

Solution Here is one way to estimate the answer. We note that 67% is close to $66\frac{2}{3}\%$, which is equivalent to the fraction $\frac{2}{3}$. Also, we see that 14.8 rounds to 15. So the answer to the given question is close to the answer to the following question.

$$\begin{array}{ccccc} \text{What} & \text{is} & \frac{2}{3} & \text{of} & 15? \\ \downarrow & \downarrow & \downarrow & \downarrow & \downarrow \\ x & = & \frac{2}{3} & \cdot & 15 \end{array}$$

We multiply mentally. $x = \frac{2}{\cancel{3}_1} \cdot \frac{\cancel{15}^5}{1} = \frac{10}{1} = 10$

So 67% of 14.8 is approximately 10. (The exact answer is 9.916, which is reasonably close to our estimate.)

PRACTICE 4

Estimate 49.3% of 401.6.

EXAMPLE 5

A marketing account manager has $3\frac{1}{2}\%$ of her monthly salary put into a 401(k) plan. How much did she put into the 401(k) plan if her monthly salary is $3,200?

Solution We are looking for the monthly amount placed into the 401(k) plan, which is $3\frac{1}{2}\%$ of $3,200.

$$\begin{array}{ccccc} \text{What} & \text{is} & 3\frac{1}{2}\% & \text{of} & \$3{,}200? \\ \downarrow & \downarrow & \downarrow & \downarrow & \downarrow \\ x & = & (0.035) & \cdot & (3{,}200) \\ & = 112 & & & \end{array}$$

So she has $112 per month put into the 401(k) plan. Note that this amount has the same unit (dollars) as the base.

PRACTICE 5

Of the 600 workers at a factory, $8\frac{1}{2}\%$ belong to a union. How many workers are in the union?

Finding a Base

Now let's consider some examples of using the translation method to find the base when we know the percent and the amount.

EXAMPLE 6

4% of what number is 8?

Solution We begin by writing the appropriate equation.

4%	of	what number	is	8?
↓	↓	↓	↓	↓
0.04	·	x	=	8

Next, we solve this equation.

$$0.04x = 8$$

$$\frac{0.04}{0.04}x = \frac{8}{0.04} \quad \text{Divide each side by 0.04.}$$

$$x = \frac{8}{0.04} = 200 \qquad 0.04\overline{)8.00} = 4\overline{)800.}\ \ 200.$$

So 4% of 200 is 8.

PRACTICE 6

6 is 12% of what number?

EXAMPLE 7

108 is 120% of what number?

Solution We consider the following question:

120%	of	what number	is	108?
↓	↓	↓	↓	↓
1.2	·	x	=	108

Solving, we get:

$$1.2x = 108$$

$$\frac{1.2}{1.2}x = \frac{108}{1.2}$$

$$x = 90$$

So 120% of 90 is 108.

PRACTICE 7

250% of what number is 18?

EXAMPLE 8

A college awarded financial aid to 3,843 students, which was 45% of the total number of students enrolled at the college. What was the student enrollment at the college?

Solution We must answer the following question:

45%	of	what number	is	3,843?
↓	↓	↓	↓	↓
0.45	·	x	=	3,843

Next we solve the equation.

$$0.45x = 3{,}843$$

$$\frac{0.45x}{0.45} = \frac{3{,}843}{0.45}$$

$$x = 8{,}540$$

So 8,540 students were enrolled at the college.

PRACTICE 8

There was a glut of office space in a city, with 400,000 square feet, or 16% of the total office space, vacant. How much office space did the city have?

Finding a Percent

Finally, let's look at the third type of percent problem, where we are given the base and the amount and are looking for the percent.

EXAMPLE 9

What percent of 80 is 60?

Solution We begin by writing the appropriate equation.

What percent of 80 is 60?

$$x \cdot 80 = 60$$

$$80x = 60 \quad \text{Write the equation in standard form.}$$

$$\frac{80}{80}x = \frac{60}{80} \quad \text{Divide each side by 80.}$$

$$x = \frac{\overset{3}{\cancel{60}}}{\underset{4}{\cancel{80}}} = \frac{3}{4} \quad \text{Simplify.}$$

Since we are looking for a percent, we change $\frac{3}{4}$ to a percent. So 75% of 80 is 60.

$$x = \frac{3}{4} = \frac{3}{\underset{1}{\cancel{4}}} \cdot \frac{\overset{25}{\cancel{100}}}{1}\% = 75\%$$

PRACTICE 9

What percent of 6 is 5?

EXAMPLE 10

What percent of 60 is 80?

Solution We begin by writing the appropriate equation, as shown to the right.

What percent of 60 is 80?

$$x \cdot 60 = 80$$

$$60x = 80$$

$$\frac{60}{60}x = \frac{80}{60}$$

$$x = \frac{\overset{4}{\cancel{80}}}{\underset{3}{\cancel{60}}} = \frac{4}{3}$$

Finally, we want to change $\frac{4}{3}$ to a percent.

$$x = \frac{4}{3} = \frac{4}{3} \cdot \frac{100}{1}\% = \frac{400}{3}\% = 133\frac{1}{3}\%$$

So $133\frac{1}{3}\%$ of 60 is 80.

PRACTICE 10

What percent of 8 is 9?

EXAMPLE 11

A young couple buys a house for \$125,000, making a down payment of \$25,000 and paying the difference over time with a mortgage. What percent of the cost of the house was the down payment?

Solution We write the question as shown to the right.

What percent of \$125,000 is \$25,000?

$$x \cdot 125{,}000 = 25{,}000$$

$$125{,}000x = 25{,}000$$

$$\frac{\cancel{125{,}000}}{\cancel{125{,}000}}x = \frac{25{,}000}{125{,}000}$$

$$x = \frac{25}{125} = \frac{1}{5}$$

Next, we change $\frac{1}{5}$ to a percent. $x = \frac{1}{5} = \frac{1}{\cancel{5}_1} \cdot \frac{\cancel{100}^{20}}{1}\% = 20\%$

So the down payment was 20% of the total cost of the house.

PRACTICE 11

Of the 400 acres on a farm, 120 were used to grow corn. What percent of the total acreage was used to grow corn?

The Proportion Method

So far, we have used the translation method to solve percent problems. Now let's consider an alternative approach, the proportion method.

Using the proportion method, we view a percent relationship in the following way.

$$\frac{\textbf{Amount}}{\textbf{Base}} = \frac{\textbf{Percent}}{\textbf{100}}$$

If we are given two of the three quantities, we set up this proportion and then solve it to find the third quantity.

EXAMPLE 12

What is 60% of 35?

Solution The base (the number after the word *of*) is 35. The percent (the number followed by the % sign) is 60. The amount is unknown. We set up the proportion, substitute into it, and solve.

$$\frac{\text{Amount}}{\text{Base}} = \frac{\text{Percent}}{100}$$

$$\frac{x}{35} = \frac{60}{100}$$

$$100x = 60 \cdot 35 \quad \textbf{Set cross products equal.}$$

$$\frac{\cancel{100}}{\cancel{100}}x = \frac{2{,}100}{100} \quad \textbf{Divide each side by 100.}$$

$$x = 21$$

So 60% of 35 is 21.

PRACTICE 12

Find 108% of 250.

EXAMPLE 13

15% of what number is 21?

Solution Here, the number after the word *of* is missing, so we are looking for the base. The amount is 21, and the percent is 15. We set up the proportion, substitute into it, and solve.

$$\frac{\text{Amount}}{\text{Base}} = \frac{\text{Percent}}{100}$$

$$\frac{21}{x} = \frac{15}{100}$$

$$15x = 2{,}100 \quad \text{Set cross products equal.}$$

$$\frac{\cancel{15}}{\cancel{15}}x = \frac{2{,}100}{15} \quad \text{Divide each side by 15.}$$

$$x = 140$$

So 15% of 140 is 21.

PRACTICE 13

2% of what number is 21.6?

EXAMPLE 14

What percent of \$45 is \$30?

Solution We know that the base is 45 and that the amount is 30 and are looking for the percent.

$$\frac{30}{45} = \frac{x}{100}$$

$$45x = 3{,}000$$

$$\frac{\cancel{45}}{\cancel{45}}x = \frac{3{,}000}{45}$$

$$x = 66\frac{2}{3}$$

So we conclude that $66\frac{2}{3}$% of \$45 is \$30.

PRACTICE 14

What percent of 63 is 21?

EXAMPLE 15

A car depreciated, that is, dropped in value, by 20% during its first year. By how much did the value of the car decline if it cost \$30,500 new?

Solution The question here is: What is 20% of \$30,500? So the percent is 20, the base is \$30,500 and we are looking for the amount. We set up the proportion and solve.

$$\frac{x}{30{,}500} = \frac{20}{100}$$

$$100x = 610{,}000$$

$$\frac{\cancel{100}}{\cancel{100}}x = \frac{610{,}000}{100}$$

$$x = 6{,}100$$

So the value of the car depreciated by \$6,100.

PRACTICE 15

A credit card company requires a minimum payment of 4% of the balance. What is the minimum payment if the credit card balance is \$2,450?

EXAMPLE 16

Each day, an adult takes tablets containing 24 milligrams of zinc. If this amount is 160% of the recommended daily allowance, how many milligrams are recommended? (*Source: Podiatry Today*)

Solution Here, we are looking for the base. The question is: 160% of what amount is 24 milligrams? We set up the proportion and solve.

$$\frac{24}{x} = \frac{160}{100}$$
$$160x = 2{,}400$$
$$\frac{\cancel{160}}{\cancel{160}}x = \frac{2{,}400}{160}$$
$$x = 15$$

Therefore, the recommended daily allowance of zinc is 15 milligrams. Note that this base is less than the amount (24 milligrams). Why must that be true?

PRACTICE 16

A Nobel Prize winner had to pay the Internal Revenue Service $129,200—or 38% of his prize—in taxes. How much was his Nobel Prize worth?

EXAMPLE 17

A college accepted 1,620 of the 4,500 applicants for admission. What was the acceptance rate, expressed as a percent?

Solution The question is: What percent of 4,500 is 1,620?

$$\frac{1{,}620}{4{,}500} = \frac{x}{100}$$
$$4{,}500x = 162{,}000$$
$$\frac{\cancel{4{,}500}}{\cancel{4{,}500}}x = \frac{162{,}0\cancel{00}}{4{,}5\cancel{00}}$$
$$x = 36$$

So the college's acceptance rate was 36%.

PRACTICE 17

A bookkeeper's annual salary was raised from $38,000 to $39,900. What percent of her original annual salary is her new annual salary?

Percents on a Calculator

Most calculators have a percent key ([%]), sometimes used with the 2nd function ([2nd]). However, the percent key functions differently on different models. Check to see if the following approach works on your machine. If it does not, experiment to find an approach that does.

EXAMPLE 18

Use a calculator to find 50% of 8.

Solution

Press	Display
50 [2nd] [%] [×] 8 [ENTER]	50% * 8 4.

PRACTICE 18

What is 8.25% of $72.37, to the nearest cent?

Mathematically Speaking

Fill in each blank with the most appropriate term or phrase from the given list.

amount	of	base
is	what	percent

1. The _______ is the number that we are taking the percent of.

2. The _______ is the result of taking the percent of the base.

3. The _______ of the base is the amount.

4. In the translation method of solving a percent problem, _______ is replaced by a multiplication symbol.

Find the amount. Check by estimating.

5. What is 75% of 8?

6. Find 50% of 48.

7. Compute 100% of 23.

8. What is 200% of 6?

9. Find 41% of 7.

10. Calculate 6% of 9.

11. What is 35% of $400?

12. 40% of 10 miles is what?

13. What is 3.1% of 20?

14. Find 0.5% of 7.

15. Compute $\frac{1}{2}$% of 20.

16. $\frac{1}{10}$% of 35 is what number?

17. What is $12\frac{1}{2}$% of 32?

18. Compute $66\frac{2}{3}$% of 33.

19. What is $7\frac{1}{8}$% of $257.13, rounded to the nearest cent?

20. Calculate 8.9% of 7,325 miles, rounded to the nearest mile.

Find the base.

21. 25% of what number is 8?

22. 30% of what number is 120?

23. $12 is 10% of how much money?

24. 1% of what salary is $195?

25. 5 is 200% of what number?

26. 70% of what amount is 14?

27. 2% of what amount of money is $5?

28. 8 meters is 20% of what length?

29. 15 is $33\frac{1}{3}$% of what number?

30. $8\frac{1}{2}$% of what number is 85?

31. 3.5 is 200% of what number?

32. 150% of what number is 8.1?

33. 0.5% of what number is 23?

34. 0.75% of what is 24?

35. 6.5% of how much money is \$3,200, rounded to the nearest cent?

36. 4,718 is $2\frac{1}{8}$% of what number?

Find the percent.

37. 50 is what percent of 100?

38. What percent of 13 is 13?

39. What percent of 8 is 6?

40. What percent of 50 is 20?

41. What percent of 12 is 10?

42. 5 is what percent of 15?

43. 2 miles is what percent of 8 miles?

44. \$16 is what percent of \$20?

45. \$30 is what percent of \$20?

46. 10 is what percent of 8?

47. 9 feet is what percent of 8 feet?

48. 35¢ is what percent of 21¢?

49. 2.5 is what percent of 4?

50. 0.1 is what percent of 8?

51. What percent of 251,749 is 76,801, rounded to the nearest percent?

52. 8,422 is what percent of 11,630, to the nearest percent?

Mixed Practice

Solve.

53. Compute $37\frac{1}{2}$% of 160

54. Calculate 0.01% of 55, rounded to the nearest hundredth.

55. What percent of 15 is 10?

56. What percent of 20 is 30?

57. 20% of what length is 35 miles?

58. 2.5% of what is 32?

59. 3 feet is what percent of 60 feet?

60. Find 7.2% of \$300.

61. 4% of what amount of money is \$20?

62. \$24 is what percent of \$300?

63. What is 40% of 25?

64. $\frac{3}{4}$% of what number is 60?

Applications

Solve.

65. During a tournament, a golfer made par on 12 of 18 holes. On what percent of the holes on the course did she make par?

66. In a dormitory, 40% of the rooms are especially equipped for disabled students. How many rooms are so equipped if the dorm has 80 rooms?

67. Flexible-fuel vehicles run on E85, an alternative fuel that is a blend of ethanol and gasoline containing 85% ethanol. How much ethanol is in 12 gallons of E85?

68. A property management company sold 80% of the condominium units in a new building with 90 units. How many units were sold?

69. Payroll deductions comprise 40% of the gross income of a student working part-time. If his deductions total $240, what is his gross income?

70. In 1862, the U.S. Congress enacted the nation's first income tax, at the rate of 3%. How much in income tax would you have paid if you made $2,500? (*Source:* U.S. Bureau of the Census)

71. A 224-pound man joined a weight-loss program and lost 56 pounds. What percent of his original weight did he lose?

72. According to the report on a country's economic conditions, 1.5 million people, or 8% of the workforce, were unemployed. How large was the workforce?

73. In a restaurant, 60% of the tables are in the no-smoking section. If the restaurant has 90 tables, how many tables are in the no-smoking section?

74. Investors buy a studio apartment for $150,000. Of this amount, they are able to put down $30,000. The down payment is what percent of the purchase price?

75. A lab technician mixed 36 milliliters of alcohol with 84 milliliters of water to make a solution. What percent of the solution was alcohol?

76. A shopper lives in a town where the sales tax is 5%. Across the river, the tax is 4%. If it costs her $6 to make the round trip across the river, should she cross the river to buy a $250 television set?

77. A student answered 90% of the questions on a math exam correctly. If she answered 36 questions correctly, how many questions were on the exam?

78. In the first quarter of last year, a steel mill produced 300 tons of steel. If this was 20% of the year's output, find that output.

79. The following graph shows the breakdown of the projected U.S population by gender in the year 2020. If the population is expected to be 340 million people, how many more women than men will there be in 2020? (*Source:* U.S. Bureau of the Census)

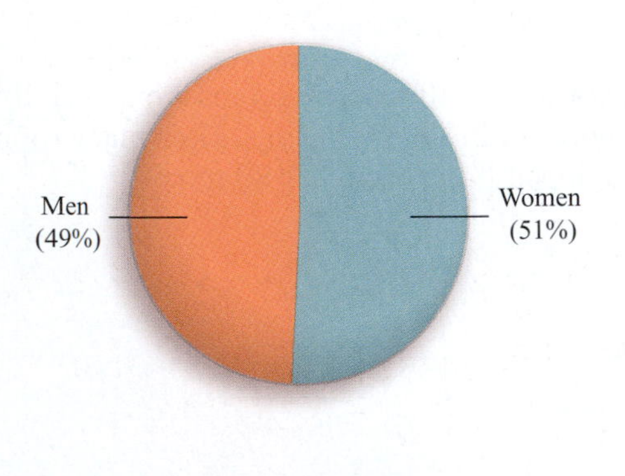

80. The percent of various kinds of degrees projected to be conferred in the year 2010 is shown in the following graph. If a total of 2,860,000 degrees are conferred, how many more Bachelor's degrees than Associate's degrees will be conferred in that year? (*Source:* National Center for Educational Statistics)

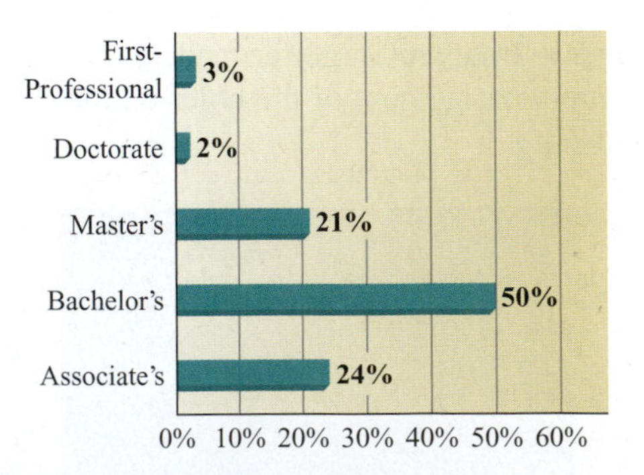

81. According to the latest census, the state of Oregon has an urban population of about 3 million people and a rural population of approximately 1 million people.

a. What is the combined population of Oregon?

b. What percent of the combined population is the urban population?
(*Source:* U.S. 2006 Census)

82. A homeowner builds a family room addition on his 1,650-square-foot house, increasing the area of the house by 495 square feet.

a. Calculate the total area of the house with the addition.

b. What percent of the original area is the total area?

83. A company's profits amounted to 10% of its sales. If the profits were $3 million, compute the company's sales.

84. The most commonly studied language in American colleges is Spanish, with some 750,000 enrollments. By contrast, the number of enrollments in French is approximately 200,000. The number of French enrollments is what percent of the number of Spanish enrollments, to the nearest 10%? (*Source: Time Almanac 2006*)

85. In a company, 85% of the employees are female. If 765 males work for the company, what is the total number of employees?

86. A quarterback completed 15 passes or 20% of his attempted passes. How many of his attempted passes did he *not* complete?

87. A math lab coordinator is willing to spend up to 25% of her income on housing. What is the most she can spend if her annual income is $36,000?

88. An office supply warehouse shipped 648 cases of copy paper. If this represents 72% of the total inventory, how many cases of paper did the warehouse have in its inventory?

• *Check your answers on page A-10.*

MINDSTRETCHERS

Writing

1. Do you prefer solving percent problems using the translation method or the proportion method? In a few sentences, explain why.

Critical Thinking

2. At a college, 20% of the women commute, in contrast to 30% of the men. Yet more women than men commute. Explain how this result is possible.

Technology

3. On the Web, go to the U.S. Bureau of the Census home page (http://www.census.gov). Pose a percent problem of interest to you involving data from the site, and solve the problem.

6.3 More on Percents

Finding a Percent Increase or Decrease

OBJECTIVES

- To solve percent increase and decrease problems
- To solve percent problems involving taxes, commissions, markups, and discounts
- To solve simple and compound interest problems
- To solve word problems of these types

Next, let's consider a type of "what percent" problem that deals with a *changing quantity*. If the quantity is increasing, we speak of a *percent increase*; if it is decreasing, of a *percent decrease*.

Here is an example: Last year, a family paid \$2,000 in health insurance, and this year, their health insurance bill was \$2,500. By what percent did this expense increase?

Note that this problem states the value of a quantity at two points in time. We are asked to find the percent increase between these two values.

To solve, we first compute the difference between the values.

$$2{,}500 - 2{,}000 = 500$$

Change in value

The question posed is expressed as follows:

$$\begin{array}{ccccc} \text{What percent} & \text{of} & 2{,}000 & \text{is} & 500? \\ \downarrow & \downarrow & \downarrow & \downarrow & \downarrow \\ x & \cdot & 2{,}000 & = & 500 \end{array}$$

It is important to note that the *base* here—as in all percent change problems—is the original value of the quantity.

Next, we solve the equation.

$$2{,}000x = 500$$

$$\frac{\cancel{2{,}000}}{\cancel{2{,}000}}x = \frac{500}{2{,}000}$$

$$x = \frac{1}{4} = 0.25, \text{ or } 25\%$$

So we conclude that the family's health insurance expense *increased* by 25%.

To Find a Percent Increase or Decrease

- Compute the difference between the two given values.
- Compute what percent this difference is of the *original value*.

EXAMPLE 1

The cost of a marriage license had been \$10. Later it rose to \$15. What percent increase was this?

PRACTICE 1

To accommodate a flood of tourists, businesses in town boosted the number of hotel beds from 25 to 100. What percent increase is this?

Solution The earlier value of the cost of the license was \$10, and the later value was \$15. The *change in value* is, therefore, \$15 − \$10, or \$5. So the question is as follows:

$$\text{What percent of } \$10 \text{ is } \$5?$$

$$x \cdot 10 = 5$$

$$10x = 5$$

$$\frac{\cancel{10}}{\cancel{10}}x = \frac{\overset{1}{\cancel{5}}}{\underset{2}{\cancel{10}}}$$

$$x = \frac{1}{2}$$

Next, we change $\frac{1}{2}$ to a percent. $\frac{1}{2} = 0.5$, or 50%

So the cost of the license increased by 50%.

EXAMPLE 2

Suppose that an animal species is considered to be endangered if its population drops by more than 60%. If a species' population fell from 40 to 18, should we consider the animal endangered?

Solution The population dropped from 40 to 18, that is, by 22. The question is how the percent decrease compares with 60%. We compute.

$$\text{What percent of } 40 \text{ is } 22?$$

$$x \cdot 40 = 22$$

$$40x = 22$$

$$x = \frac{22}{40}, \text{ or } \frac{11}{20}$$

We convert this fraction to a percent. $\frac{11}{20} = 0.55$, or 55%

Since the population decreased by less than 60%, the species is not considered to be endangered.

PRACTICE 2

Major financial crashes took place on both Tuesday, October 29, 1929, and Monday, October 19, 1987. On the earlier date, the stock index dropped from 300 to 230. On the latter date, it dropped from 2,250 to 1,750. As a percent, did the stock index drop more in 1929 or in 1987? (**Source:** *The Wall Street Journal*)

Business Applications of Percent

The idea of percent is fundamental to business and finance. Percent applications are part of our lives whenever we buy or sell merchandise, pay taxes, and borrow or invest money.

Taxes

Governments levy taxes to pay for a variety of services, from supporting schools to paving roads. There are many kinds of taxes, including sales, income, property, and import taxes.

In general, the amount of a tax that we pay is a percent of a related value. For instance, sales tax is usually computed as a percent of the price of merchandise sold. Thus, in a town where the sales tax rate is 7%, we could compute the tax on any item sold by computing 7% of the price of that item.

Similarly, property tax is commonly computed by taking a given percent (the tax rate) of the property's assessed value. And an import tax is calculated by taking a specified percent of the market value of the imported item.

EXAMPLE 3

The sales tax on a \$950 digital camcorder is \$71.25. What is the sales tax rate, expressed as a percent?

Solution We must consider the following question:

$$\begin{array}{ccccc} 71.25 & \text{is} & \text{what percent} & \text{of} & 950? \\ \downarrow & \downarrow & \downarrow & \downarrow & \downarrow \\ 71.25 & = & x & \cdot & 950 \end{array}$$

$$950x = 71.25$$

$$\frac{950x}{950} = \frac{71.25}{950}$$

$$x = 0.075, \text{ or } 7.5\%$$

So the rate of the sales tax is 7.5%, or $7\frac{1}{2}\%$.

PRACTICE 3

When registering a new car, the owner paid a 3% excise tax on the purchase price of \$18,500. How much excise tax did he pay?

Commission

To encourage salespeople to make more sales, many of them, instead of receiving a fixed salary, are paid on **commission**. Working on commission means that the amount of money that they earn is a specified percent—say, 10%—of the total sales for which they are responsible.

On some jobs, salespeople make a flat fee in addition to a commission based on sales. On other jobs, a salesperson may earn a higher rate of commission on sales over an amount previously agreed upon, as an extra incentive.

EXAMPLE 4

A real estate agent sold a condo in San Diego for \$222,000. On this amount, she received a commission of 6%.

a. Find the amount of the commission.

b. How much money will the seller make from the sale after paying the agent's fee?

Solution

a. The commission is 6% of \$222,000.

$$\begin{array}{ccccc} \text{What} & \text{is} & 6\% & \text{of} & \$222{,}000? \\ \downarrow & \downarrow & \downarrow & \downarrow & \downarrow \\ x & = & 0.06 & \cdot & 222{,}000 = 13{,}320 \end{array}$$

So the commission amounted to \$13,320.

b. The seller made \$222,000 – \$13,320, or \$208,680.

PRACTICE 4

A sales associate at a furniture store is paid a base monthly salary of \$1,500. In addition, she earns a 9% commission on her monthly sales. If her total sales this month is \$12,500, calculate

a. her commission, and

b. her total monthly income.

Discount

In buying or selling merchandise, the term **discount** refers to a reduction on the merchandise's original price. The rate of discount is usually expressed as a percent of the selling price.

EXAMPLE 5

A drugstore gives senior citizens a 10% discount. If some pills normally sell for \$16 a bottle, how much will a senior citizen pay?

Solution Note that, because senior citizens get a discount of 10%, they pay 100% − 10%, or 90%, of the normal price.

The question then becomes: What is 90% of \$16?

We multiply. $x = 0.9 \cdot 16 = 14.4$

So a senior citizen will pay \$14.40 for a bottle of the pills.

Note that another way to solve this problem is first to compute the amount of the discount (10% of \$16) and then to subtract this discount from the original price. With this approach, do we get the same answer?

PRACTICE 5

Find the sale price.

Markup

A retail firm must sell goods at a higher price (the selling price) than it pays for the merchandise (the cost) to stay in business. The **markup** on an item is the difference between the selling price and the cost. Often the markup rate on merchandise is expressed as a fixed percent of the selling price.

EXAMPLE 6

An online bookstore sells a best seller for \$35 at a markup rate of 55% based on the selling price. How much money is the markup on the best seller?

Solution We write the question shown at the right.

What is 55% of \$35?

$x = 0.55 \cdot 35 = 19.25$

So the markup on the best seller is \$19.25.

PRACTICE 6

A department store buyer purchases wallets at \$480 per dozen and sells them for \$80 each. What percent markup, based on the selling price, is the store making?

Simple Interest

Anyone who has been late in paying a credit card bill or who has deposited money in a savings account knows about **interest**. When we loan or deposit money, we make interest. When we borrow money, we pay interest.

Interest depends on the amount of money borrowed (the **principal**), the annual rate of interest (usually expressed as a percent), and the length of time the money is borrowed (usually expressed in years). We can compute the amount of interest by multiplying the principal by the rate of interest and the number of years. This type of interest is called *simple interest* to distinguish it from *compound interest* (which we discuss later).

EXAMPLE 7

How much simple interest is earned in 1 year on a principal of \$900 at an annual interest rate of 6.5%?

Solution To compute the interest, we multiply the principal by the rate of interest and the number of years.

Principal ↓ Rate of Interest ↓ Number of Years ↓

$$\text{Interest} = 900 \times 0.065 \times 1$$
$$= 58.5$$

So \$58.50 in interest is earned.

PRACTICE 7

What is the simple interest on an investment of \$20,000 for 1 year at an annual interest rate of 7.25%?

EXAMPLE 8

A customer deposited \$825 in a savings account that each year pays 5% in simple interest, which is credited to his account. What is the account balance after 2 years?

Solution To solve this problem, let's break it into two questions:

- How much interest did the customer make after 2 years?
- What is the sum of the original deposit and that interest?

First, let's find the interest. To do this, we multiply the principal by the rate of interest and the number of years.

Principal ↓ Rate of Interest ↓ Number of Years ↓

$$\text{Interest} = (825)\ (0.05)\ (2)$$
$$= 82.50$$

The customer made \$82.50 in interest.

Now, let's find the account balance by adding the amount of the original deposit to the interest made.

Original Deposit ↓ Interest ↓

$$\text{Account Balance} = 825 + 82.50$$
$$= 907.50$$

So the account balance after 2 years is \$907.50

PRACTICE 8

A bank account pays 6% simple interest on \$1,600 for 2 years. Compute the account balance after 2 years.

Compound Interest

As we have seen, simple interest is paid on the principal. Most banks, however, pay their customers *compound interest*, which is paid on both the principal and the previous interest generated.

For instance, suppose that a bank customer has \$1,000 deposited in a savings account that pays 5% interest compounded annually. There were no withdrawals or other deposits. Let's compute the balance in the account at the end of the third year.

The following table shows the account balance after the customer has left the money in the account for 3 years. After 1 year, the account will contain \$1,050 (that is, 100% of the original \$1,000 added to 5% of \$1,000, giving us 105% of \$1,000).

Year	Balance at the End of the Year
0	\$1,000
1	\$1,000 + 0.05 × \$1,000 = \$1,050.00
2	\$1,050 + 0.05 × \$1,050 = \$1,102.50
3	\$1,102.50 + 0.05 × \$1,102.50 = \$1,157.63

The balance in the account after the third year is \$1,157.63, rounded to the nearest cent.

In the following table, for each year we multiply the account balance by 1.05 to compute the balance at the end at the next year.

Year	Balance at the End of the Year
0	\$1,000
1	1.05 × \$1,000 = \$1,050.00
2	$(1.05)^2$ × \$1,000 = \$1,102.50
3	$(1.05)^3$ × \$1,000 = \$1,157.63

So the balance at the end of the third year is $(1.05)^3 \times \$1{,}000$, or \$1,157.63, in agreement with our previous computation. What would the balance be at the end of the fourth year? What is the relationship between the number of years the money has been invested and the power of 1.05?

In computing the preceding answer, we needed to raise the number 1.05 to a power. Before scientific calculators became available, compound interest problems were commonly solved by use of a compound interest table that contained information such as the following:

Number of Years	4%	5%	6%	7%
1	1.04000	1.05000	1.06000	1.07000
2	1.08160	1.10250	1.12360	1.14490
3	1.12486	1.15763	1.19102	1.22504

When using such a table to calculate a balance, we simply multiply the principal by the number in the table corresponding to the rate of interest and the number of years for which the principal is invested. For instance, after 3 years a principal of \$1,000 compounded at 5% per year results in a balance of 1.15763 × 1,000, or \$1,157.63, as we previously noted.

Today, problems of this type are generally solved on a calculator.

EXAMPLE 9

A couple deposited $7,000 in a bank account and did not make any withdrawals or deposits in the account for 3 years. The interest is compounded annually at a rate of 3.5%. What will be the amount in their account at the end of this period?

Solution Each year, the amount in the account is 100% + 3.5%, or 1.035 times the previous year's balance. So at the end of 3 years, the number of dollars in the account is calculated as follows:

Principal	First Year	Second Year	Third Year
↓	↓	↓	↓

$$7{,}000 \times 1.035 \times 1.035 \times 1.035$$

It makes sense to use a calculator to carry out this computation. One way to key in this computation on a calculator is as follows.

Press	Display
7000 [×] 1.035 [^] 3 [ENTER]	7000 * 1.035 ^ 3 7761.025125

So at the end of 3 years, they have $7,761.03 in the account, rounded to the nearest cent.

PRACTICE 9

Find the balance after 4 years on a principal amount of $2,000 invested at a rate of 6% compounded annually.

Mathematically Speaking

Fill in each blank with the most appropriate term or phrase from the given list.

discount	on salary	on commission
markup	simple	compound
final	original	

1. When computing a percent increase or decrease, the _______ value is used as the base of the percent.

2. Sellers who are paid a fixed percent of the sales for which they are responsible are said to work _______.

3. A reduction on the price of merchandise is called a(n) _______.

4. When interest is paid on both the principal and the previous interest generated, it is called _______ interest.

Find the percent increase or decrease.

5.

Original Value	New Value	Percent Increase or Decrease
\$10	\$12	
\$10	\$8	
\$6	\$18	
\$35	\$70	
\$14	\$21	
\$10	\$1	
\$8	\$6.50	
\$6	\$5.25	

6.

Original Value	New Value	Percent Increase or Decrease
\$5	\$6	
\$12	\$10	
\$4	\$9	
\$25	\$45	
\$10	\$36	
\$100	\$20	
4 ft	3 ft	
8 lb	4.5 lb	

Compute the sales tax. Round to the nearest cent.

7.

Selling Price	Rate of Sales Tax	Sales Tax
\$30.00	5%	
\$24.88	3%	
\$51.00	$7\frac{1}{2}\%$	
\$196.23	4.5%	

8.

Selling Price	Rate of Sales Tax	Sales Tax
\$40.00	6%	
\$16.98	4%	
\$85.00	$5\frac{1}{2}\%$	
\$286.38	5%	

Compute the commission. Round to the nearest cent.

9.

Sales	Rate of Commission	Commission
$700	10%	
$450	2%	
$870	$4\frac{1}{2}\%$	
$922	7.5%	

10.

Sales	Rate of Commission	Commission
$400	1%	
$670	3%	
$610	$6\frac{1}{2}\%$	
$2,500	8.25%	

Compute the discount and sale price. Round to the nearest cent.

11.

Original Price	Rate of Discount	Discount	Sale Price
$700.00	25%		
$18.00	10%		
$43.50	20%		
$16.99	5%		

12.

Original Price	Rate of Discount	Discount	Sale Price
$200.00	30%		
$21.00	50%		
$88.88	10%		
$72.50	40%		

Compute the markup and the cost. The rate of markup is based on selling price. Round to the nearest cent.

13.

Selling Price	Rate of Markup	Markup	Cost
$10.00	50%		
$23.00	70%		
$18.40	10%		
$13.55	60%		

14.

Selling Price	Rate of Markup	Markup	Cost
$20.00	40%		
$81.00	25%		
$74.20	30%		
$300.00	8.5%		

Calculate the simple interest and the final balance. Round to the nearest cent.

15.

Principal	Interest Rate	Time (in years)	Interest	Final Balance
$300	4%	2		
$600	7%	2		
$500	8%	2		
$375	10%	4		
$1,000	3.5%	3		
$70,000	6.25%	30		

16.

Principal	Interest Rate	Time (in years)	Interest	Final Balance
$100	6%	5		
$800	4%	5		
$500	3%	10		
$800	6%	10		
$250	1.5%	2		
$300,000	4.25%	20		

Calculate the final balance after compounding the interest. Round to the nearest cent.

17.

Principal	Interest Rate	Time (in years)	Final Balance
$500	4%	2	
$6,200	3%	5	
$300	5%	8	
$20,000	4%	2	
$145	3.8%	3	
$810	2.9%	10	

18.

Principal	Interest Rate	Time (in years)	Final Balance
$300	6%	1	
$2,900	5%	4	
$800	3%	5	
$10,000	3%	4	
$250	4.1%	2	
$200	3.3%	5	

Mixed Practice

Complete each table. Round to the nearest whole percent.

19.

Original Value	New Value	Percent Decrease
$5	$4.50	

20.

Original Value	New Value	Percent Increase
$220	$300	

Complete each table. Round to the nearest cent.

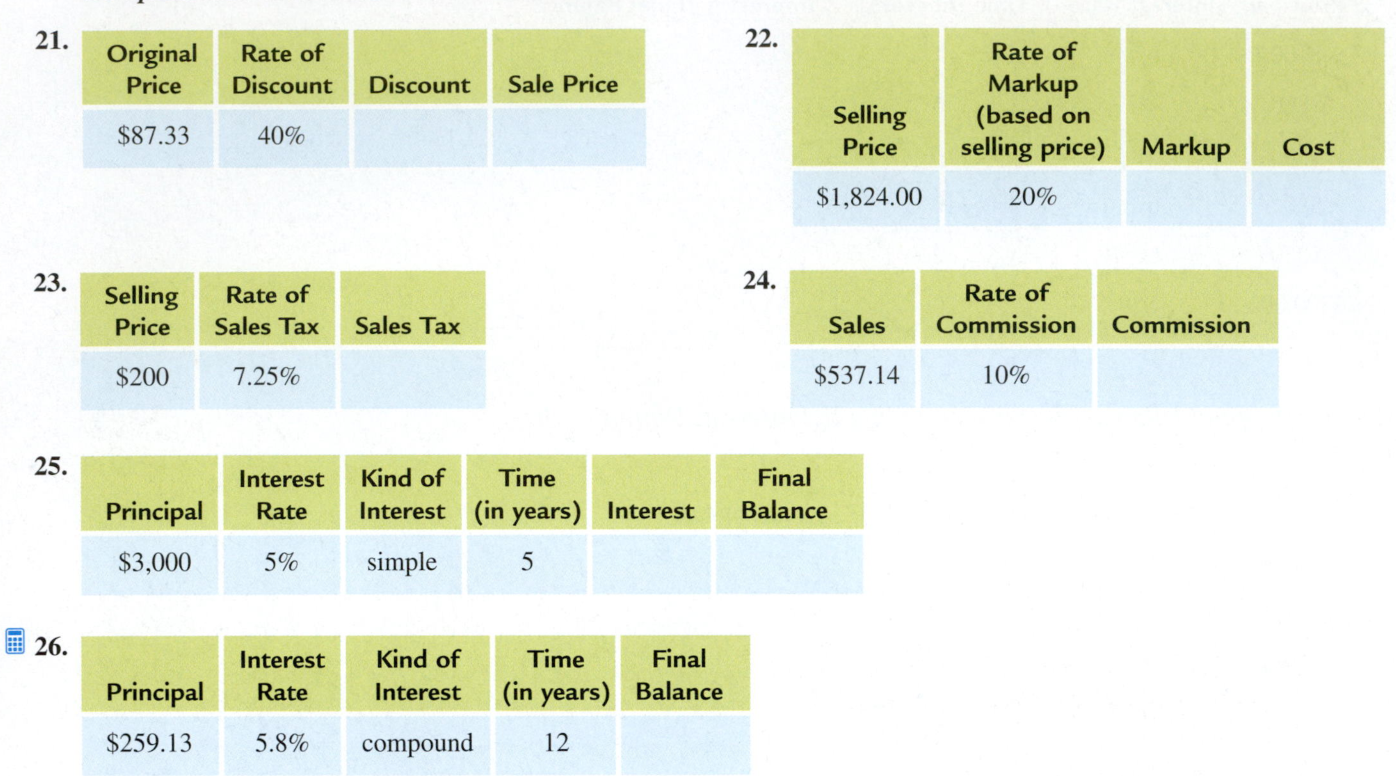

21.

Original Price	Rate of Discount	Discount	Sale Price
$87.33	40%		

22.

Selling Price	Rate of Markup (based on selling price)	Markup	Cost
$1,824.00	20%		

23.

Selling Price	Rate of Sales Tax	Sales Tax
$200	7.25%	

24.

Sales	Rate of Commission	Commission
$537.14	10%	

25.

Principal	Interest Rate	Kind of Interest	Time (in years)	Interest	Final Balance
$3,000	5%	simple	5		

26.

Principal	Interest Rate	Kind of Interest	Time (in years)	Final Balance
$259.13	5.8%	compound	12	

Mixed Applications

Solve.

27. An upscale department-store chain reported that total sales this year were $2.3 billion—up from $1.8 billion last year. Find the percent increase in sales, to the nearest whole percent.

28. Last year, a local team won 20 games. This year, it won 15 games. What was the percent decrease of games won?

29. In 9 years, the number of elderly nursing home residents rose from 200,000 to 1.3 million. By what percent did the number of residents increase?

30. Due to a decrease in demand, a manufacturing plant decreased its production from 2,400 to 1,800 units per day. What was the percent decrease in the number of units produced per day?

31. The first commercial telephone exchange was set up in New Haven, Connecticut, in 1878. Between 1880 and 1890, the number of telephones in the United States increased from 50 thousand to 200 thousand, in round numbers. What percent increase was this? (***Source:*** U.S. Bureau of the Census)

32. A patient's medication was decreased from 250 milligrams to 200 milligrams per dose. What was the percent decrease in the dosage?

33. A customer paid 5% sales tax on a notebook computer that sold for $1,699. Calculate the amount of sales tax that she paid.

34. Last year, a town assessed the value of a residential property at $272,000. If the homeowner paid property tax of $3,264, what was the property tax rate?

35. A customer bought a cell phone for $150. The total selling price of the phone, including sales tax, was $159.75. What was the sales tax rate?

36. In a town, the sales tax rate is 7%. It costs $5 to travel to a nearby town and back where the sales tax is only 5.8%. Is it worthwhile to make this trip to purchase an item that sells in both towns for $800?

37. A pharmaceutical sales representative earns a 12% monthly commission on all her sales above $5,000. Find her commission if her sales this month totaled $27,500.

38. A sales assistant earns a flat salary of $150 plus a 10% commission on sales of $3,000. What were his total earnings?

39. On a restaurant table, a customer leaves $1.35 as a tip. Assuming that the customer left a 15% tip, how much was the bill before the tip?

40. A salesperson receives a 5% commission on the first $2,000 in sales and a 7% commission on sales above $2,000. How much commission does she earn on sales of $3,500?

41. An electronics store sells an MP3 player for $59.95. If each MP3 player was marked up 40% based on the selling price, how much was the markup?

42. An antique store bought a table for $90 and resold it for $120. What is the markup percent, based on the selling price?

43. What is the markup percent, based on the selling price, on an item when the markup is $8 and the selling price is $15?

44. A store manager marked up the cost of merchandise by $5. If the cost had been $3, what percent of the selling price was the markup?

45. A store sells a television that lists for $399 at a 35% discount rate. What is the sale price?

46. An appliance store has a sale on all its appliances. A washing machine that originally sold for $800 is on sale for $680. What is the discount rate?

47. A bank customer borrowed $3,000 for 1 year at 5% simple interest to buy a computer. How much interest did the customer pay?

48. How much simple interest is earned on $600 at an 8% annual interest rate for 2 years?

49. A couple deposited $5,000 in a bank. How much interest will they have earned after 1 year if the interest rate is 5%?

50. A student borrowed $2,000 from a friend, agreeing to pay her 4% simple interest. If he promised to repay her the entire amount at the end of 3 years, how much money must he pay her?

51. A home goods store had a "20% off" sale on all its merchandise. A customer bought a down comforter that originally cost $180.

a. What was the sale price of the comforter?

b. Calculate the total amount the customer paid after 6% sales tax was added to the purchase.

52. During a sale, a shoe store marked down the price of a pair of sneakers that originally cost $80 by 40%.

a. What was the sale price of the sneakers?

b. After two weeks, the store marked down the sale price by another 60%. What percent off the original price was the sale price after the second discount was applied?

53. An investor put $3,000 in an account that pays 6% interest, compounded annually. Find the amount in the account after 2 years.

54. A bank pays 5.5% interest, compounded annually, on a 2-year certificate of deposit (CD) that initially costs $500. What is the value of the CD at the end of the 2 years, rounded to the nearest cent?

55. A city had a population of 4,000. If the city's population increased by 10% per year, what was the population 4 years later?

56. An art dealer bought a painting for $10,000. If the value of the painting increased by 50% per year, what was its value 4 years later?

• *Check your answers on page A-10.*

MINDSTRETCHERS

Writing

1. Explain the difference between simple interest and compound interest.

Technology

2. Using a spreadsheet, construct a three-column table showing the original price, the 10% discount, and the selling price for items with an original price of any whole number of dollars between $1 and $100.

Mathematical Reasoning

3. If a quantity increases by a given percent and then decreases by the same percent, will the final value be the same as the original value? Explain.

KEY CONCEPTS AND SKILLS

CONCEPT SKILL

Concept/Skill	Description	Example
[6.1] **Percent**	A ratio or fraction with denominator 100. It is written with the % sign, which means divided by 100.	$7\% = \frac{7}{100}$ (↑ Percent)
[6.1] **To change a percent to the equivalent fraction**	• Drop the % sign from the given percent and place the number over 100. • Simplify the resulting fraction, if possible.	$25\% = \frac{25}{100} = \frac{1}{4}$
[6.1] **To change a percent to the equivalent decimal**	• Drop the % sign from the given percent and divide the number by 100.	$23.5\% = .235$, or 0.235
[6.1] **To change a decimal to the equivalent percent**	• Multiply the number by 100 and insert a % sign.	$0.125 = 12.5\%$
[6.1] **To change a fraction to the equivalent percent**	• Multiply the fraction by 100 and insert a % sign.	$\frac{1}{5} = \frac{1}{5} \times 100\% = \frac{1}{\cancel{5}_1} \times \frac{\cancel{100}^{20}}{1}\%$ $= 20\%$
[6.2] **Base**	The number that we are taking the percent of. It always follows the word *of* in the statement of a percent problem.	50% of 8 is 4. (↑ Base: 8)
[6.2] **Amount**	The result of taking the percent of the base.	50% of 8 is 4. (↑ Amount: 4)
[6.2] **To solve a percent problem using the translation method**	• Translate as follows: What number, what percent → x is → = of → × or · % → decimal or fraction • Set up the equation. **The percent of the base is the amount.** • Solve.	What is 50% of 8? $x = 0.5 \cdot 8$ $x = 4$ 30% of what number is 6? $0.3 \cdot x = 6$ $\frac{\cancel{0.3}x}{\cancel{0.3}} = \frac{6}{0.3}$ $x = \frac{6}{0.3} = 20$ What percent of 8 is 2? $x \cdot 8 = 2$ $x = \frac{2}{8} = \frac{1}{4} = 25\%$

continued

CONCEPT SKILL

Concept/Skill	Description	Example
[6.2] **To solve a percent problem using the proportion method**	• Identify the amount, the base, and the percent, if known. • Set up and substitute into the proportion. $\frac{\text{Amount}}{\text{Base}} = \frac{\text{Percent}}{100}$ • Solve for the unknown quantity.	50% of 8 is what number? $\frac{x}{8} = \frac{50}{100}$ $100x = 400$ $x = 4$ 30% of what number is 6? $\frac{6}{x} = \frac{30}{100}$ $30x = 600$ $x = 20$ What percent of 8 is 2? $\frac{2}{8} = \frac{x}{100}$ $8x = 200$ $x = 25$ So the answer is 25%.
[6.3] **To find a percent increase or decrease**	• Compute the difference between the two given values. • Determine what percent this difference is of the *original value.*	Find the percent increase for a quantity that changes from 4 to 5. Difference: $5 - 4 = 1$ What percent of 4 is 1? ↓ ↓ ↓ ↓ ↓ $x \cdot 4 = 1$ $x = \frac{1}{4} = 0.25$, or 25%

Chapter 6

Review Exercises

To help you review this chapter, solve these problems.

[6.1] *Complete the following tables.*

1.

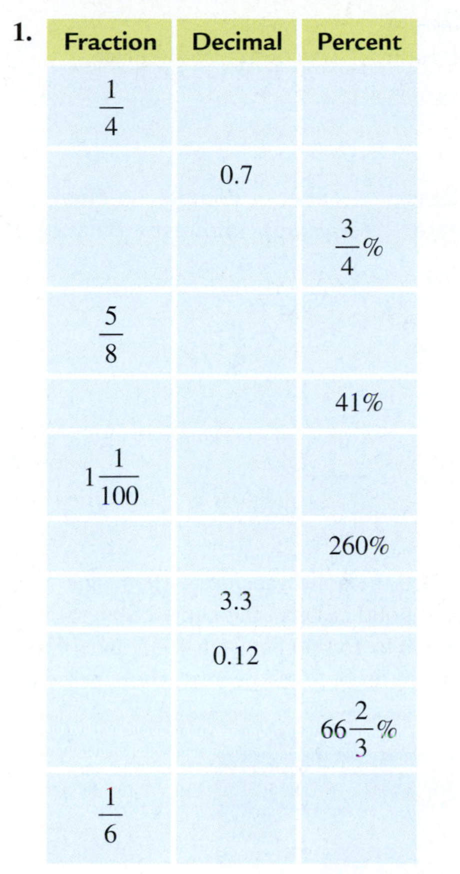

Fraction	Decimal	Percent
$\frac{1}{4}$		
	0.7	
		$\frac{3}{4}\%$
$\frac{5}{8}$		
		41%
$1\frac{1}{100}$		
		260%
	3.3	
	0.12	
		$66\frac{2}{3}\%$
$\frac{1}{6}$		

2.

Fraction	Decimal	Percent
$\frac{3}{8}$		
	0.49	
		0.1%
		150%
	0.875	
		$83\frac{1}{3}\%$
$2\frac{3}{4}$		
	1.2	
	0.75	
		10%
$\frac{1}{3}$		

[6.2] *Solve.*

3. What is 40% of 30?

4. What percent of 5 is 6?

5. 2 feet is what percent of 4 feet?

6. 30% of what number is 6?

7. What percent of 8 is 3.5?

8. Find 55% of 10.

9. $12 is 200% of what amount of money?

10. 2 is what percent of 10?

11. What is 1.2% of 25?

12. Find 115% of 400.

13. 35% of $200 is what?

14. $\frac{1}{2}\%$ of what number is 5?

15. 15 is what percent of 0.75?

16. 4.5 is what percent of 18?

17. Calculate $33\frac{1}{3}\%$ of $600.

18. What percent of $9 is $4?

19. Estimate 59% of $19.99.

20. 2.5% of how much money is $40?

21. What percent of $7.99 is $1.35, to the nearest whole percent?

22. 3.5 is $8\frac{1}{4}\%$ of what number, to the nearest hundredth?

[6.3] *Complete the following tables.*

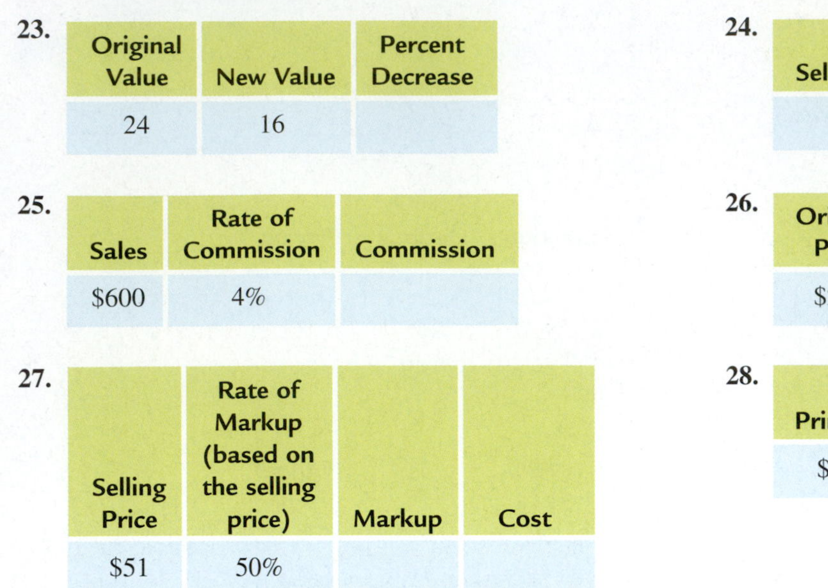

23.

Original Value	New Value	Percent Decrease
24	16	

25.

Sales	Rate of Commission	Commission
$600	4%	

27.

Selling Price	Rate of Markup (based on the selling price)	Markup	Cost
$51	50%		

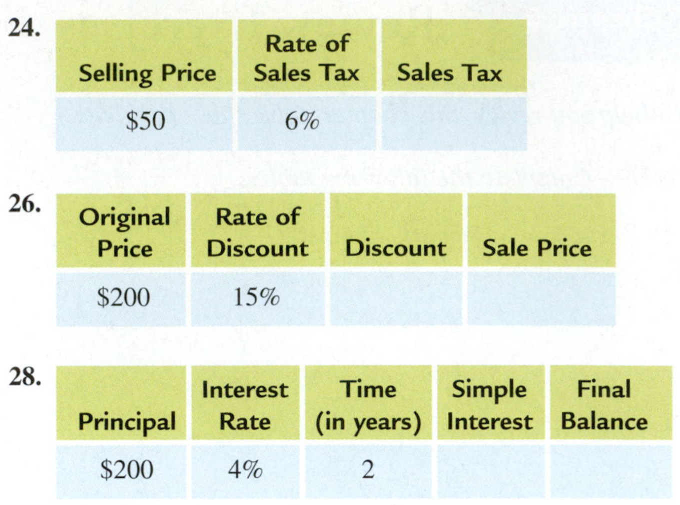

24.

Selling Price	Rate of Sales Tax	Sales Tax
$50	6%	

26.

Original Price	Rate of Discount	Discount	Sale Price
$200	15%		

28.

Principal	Interest Rate	Time (in years)	Simple Interest	Final Balance
$200	4%	2		

Mixed Applications

Solve.

29. On July 1, 2005, the sales tax rate in Chicago, already one of the highest among major U.S. cities, increased a quarter point to 9%. If the sales tax on a computer in Chicago amounted to $162, what was the selling price (before taxes) of the computer? (*Source:* The Tax Foundation)

30. Jonas Salk developed the polio vaccine in 1954. The number of reported polio cases in the United States dropped from 29,000 to 15,000 between 1955 and 1956. What was the percent drop, to the nearest whole percent? (*Source:* U.S. Bureau of the Census)

31. For their fees, one real estate agent charges 11% of a year's rent and another charges the first month's rent. Which agent charges more?

32. A particular community bank makes available loans with simple interest and with no prepayment penalty. How much interest is due on a five-year car loan of $24,000 based on a simple interest rate of 6%?

33. According to a city survey, 49% of respondents approve of how the mayor is handling his job and 31% disapprove. What percent neither approved nor disapproved?

34. Plastics make up about 11% and paper makes up $\frac{9}{25}$ of the solid municipal waste in the United States. Which makes up more of the solid municipal waste? (*Source:* Energy Information Administration)

35. According to a study, 25% of employees do not take all of their vacation time due to the demands of their jobs. Express this percent as a fraction. (*Source:* Families and Work Institute)

36. A typical markup rate in the hobby industry, based on selling price, is 40%. If a hobby shop sells a holiday train set that cost $120 for $220, was the markup rate on the train set above or below the typical markup rate? (*Source:* http://keytaps.com)

37. It takes a worker 50 minutes to commute to work. If he has been traveling for 20 minutes, what percent of his trip has been completed?

38. Approximately 1 of every 10 Americans is left-handed. What percent is this?

39. A couple financed a 30-year mortgage at a fixed interest rate of 6.29%. Express this rate as a decimal.

40. A clothing store places the following ad in a local newspaper:

At the store, what is the sale price of a suit that regularly sells for $230?

41. The following table deals with the oil reserves of two nations that are leading oil producers:

Country	Proven Oil Reserves (in billions of barrels)
Saudi Arabia	262
Canada	179

What percent of the size of Saudi Arabia's reserves is the size of Canada's reserves, rounded to the nearest percent? (***Source:*** *Time Almanac 2006*)

42. The length of a person's thigh bone is usually about 27% of his or her height. Estimate someone's height whose thigh bone is 20 inches long. (***Source:*** *American Journal of Physical Anthropology*)

43. The winner of a men's U.S. Open tennis match got $87\frac{1}{2}\%$ of his first serves in. If he had 72 first serves, how many went in?

44. In a scientific study that relates weight to health, people are considered overweight if their actual weight is at least 20% above their ideal weight. If you weigh 160 pounds and have an ideal weight of 130 pounds, are you considered overweight?

45. An airline oversold a flight to Los Angeles by nine seats, or 5% of the total number of seats available on the airplane. How many seats does the airplane have?

46. When an assistant became editor, the magazine's weekly circulation increased from 50,000 to 60,000. By what percent did the circulation increase?

47. The salary of an executive assistant had been $30,000 before she got a raise of $1,000. If the rate of inflation is 5%, has her salary kept pace with inflation?

48. The cat is the most popular pet in America. According to recent estimates, there are 53 million dogs. If there are 109% as many cats, how many cats are there?

49. At an auction, you bought a table for $150. The auction house also charged a "buyer's premium"—an extra fee—of 10%. How much did you pay in all?

50. According to the news report, 80 tons of food met only 20% of the food needs in the refugee camp. How much additional food was needed?

51. A traveler needs 14,000 more frequent-flier miles to earn a free trip to Hawaii, which is 20% of the total number needed. How many frequent-flier miles in all does this award require?

52. A Sylvania compact fluorescent lightbulb (CFL) has a life of 8,000 hours. A Philips CFL has a life that is 25% longer. How long does a Philips CFL last? (***Source:*** http://Sylvania.com and http://bulbs.com)

53. How much commission does a salesperson make on sales totaling $5,000 at a 20% rate of commission?

54. At the end of the year, the receipts of a retail store amounted to $200,000. Of these receipts, 85% went for expenses; the rest was profit. How much profit did the store make?

55. If a bank customer deposits $7,000 in a bank account that pays a 6.5% rate of interest compounded annually, what will be the balance after 2 years?

56. Suppose that a country's economy expands by 2% per year. By what percent will it expand in 10 years, to the nearest whole percent?

57. Complete the following table, which describes a company's income for the four quarters of last year:

Quarter	Income	Percent of Total Income (rounded to the nearest whole percent)
1	$375,129	
2	289,402	
3	318,225	
4	402,077	
Total		100%

58. The following graph shows the sources from which the federal government received income in a recent year:

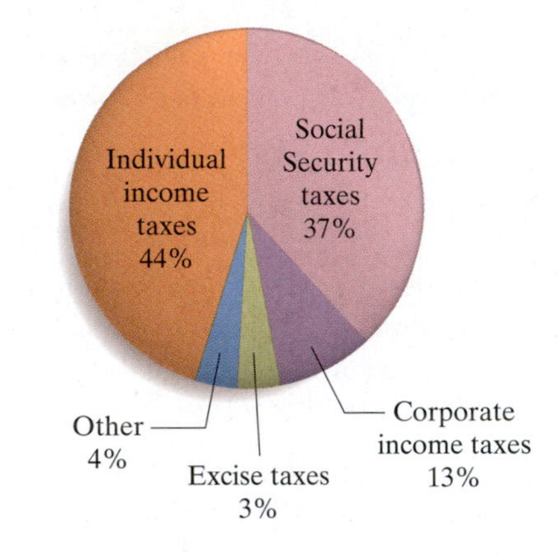

If the total amount of money taken in was $2,154 billion, compute how much money was received from each source, to the nearest billion dollors. (***Source:*** U.S. Office of Management and Budget)

• *Check your answers on page A-11.*

Chapter 6 POSTTEST

FOR EXTRA HELP

Test solutions are found on the enclosed CD.

To see if you have mastered the topics in this chapter, take this test.

Rewrite.

1. 4% as a fraction

2. $27\frac{1}{2}\%$ as a fraction

3. 174% as a decimal

4. 8% as a decimal

5. 0.009 as a percent

6. 10 as a percent

7. $\frac{5}{6}$ as a percent, rounded to the nearest whole percent

8. $2\frac{1}{5}$ as a percent

Solve.

9. What is 25% of 30 miles?

10. Find 120% of 40.

11. Estimate 32% of $20.77.

12. 8% of what number is 16?

13. What percent of 10 is 6?

14. What percent of 4 is 10?

15. To pay for tuition, a college student borrows $2,000 from a relative for 2 years at 5% simple interest. Find the amount of simple interest that is due.

16. In a parking lot that has 150 spaces, 4% are for handicap parking. How many handicap spaces are in the lot?

17. A customer paid $9.95 in sales tax on an iPod nano that cost $199. What was the sales tax rate?

18. Milk is approximately 50% cream. How much milk is needed to produce 2 pints of cream?

19. A department store sells a pair of shoes for $79 at a markup rate of 40% based on the selling price. How much money is the markup?

20. A college ended six straight years of tuition increases by raising its tuition from $3,000 to $3,100. Find the percent increase.

• *Check your answers on page A-12.*

Cumulative Review Exercises

To help you review, solve the following.

1. Divide: $1{,}962 \div 18$

2. Express $\frac{5}{6}$ as a decimal, rounded to the nearest hundredth.

3. Multiply: 0.2×3.5

4. Find the sum of $3\frac{4}{5}$ and $1\frac{9}{10}$.

5. Solve for x: $\frac{x}{3} = 2.5$

6. 20% of what amount is $200?

Solve.

7. The government withdrew $\frac{1}{4}$ million of its 2 million troops. What fraction of the total is this?

8. In a recent survey of college students, 7 out of 10 said they used the Internet every day. At this rate, how many of the 9,160 students at a college would be expected to use the Internet every day?

9. Three FM stations are highlighted on the radio dial shown. These stations have frequencies 99.5 (WBAI), 96.3 (WQXR), and 104.3 (WAXQ). Label the three stations on the dial.

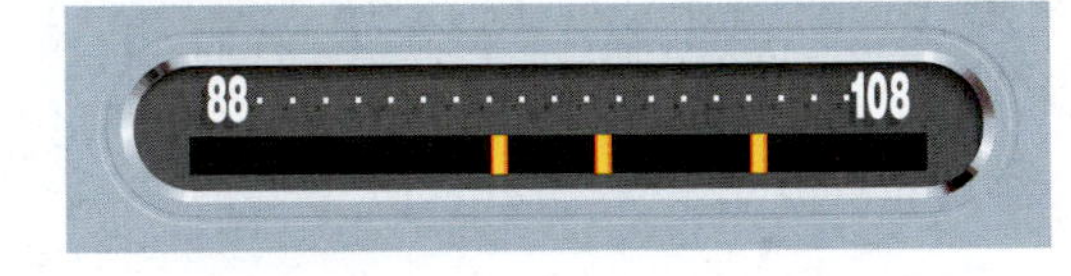

10. In a recent year, about 27% of the 885 thousand American doctors were female. How many female doctors were there, to the nearest thousand? (*Source:* American Medical Association)

• *Check your answers on page A-12.*

CHAPTER 7

Signed Numbers

Signed Numbers and Chemistry

In chemistry, a valence is assigned to each element in a compound. Valences help us study the ways in which the elements combine to form the compound.

The valence is a positive or negative whole number that expresses the combining capacity of the element. For example in the compound H_2O (water), the element hydrogen (H) has a valence of $+1$, whereas the element oxygen (O) has a valence of -2.

The valences in any chemical compound add up to 0. So if you know how to perform signed number computations, you can predict the chemical formula of any compound. (***Source:*** Karen C. Timberlake, *Basic Chemistry*, Prentice Hall, 2005)

Chapter 7 PRETEST

To see if you have already mastered the topics in this chapter, take this test.

1. Which is larger, -23 or $+7$?

2. A negative number has an absolute value of 4. What is the number?

Compute.

3. $-8 + (-9)$

4. $-20 + 20$

5. $34 - 41$

6. $-9 - (-9)$

7. -5×15

8. $-\frac{3}{4} \times \frac{2}{3}$

9. -8^2

10. $\left(-\frac{1}{2}\right)^2$

11. $-18 \div (-9)$

12. $\frac{1}{2} \div (-4)$

13. $-2 + 5 + (-3) + 8$

14. $10 + (-3) - (-1)$

15. $-9 - 3^2 \times (-5)$

16. $8 \div (-2) + 3 \cdot (-1)$

Solve.

17. A dieter on a weight-loss program lost 12 pounds in the first month. Over the next 5 months, he lost an additional 39 pounds. Express as a signed number the total change in weight in 6 months.

18. A fee of \$1.50 is deducted from a student's account each time she uses an ATM on campus. In 1 month, she uses the ATM 7 times. Express as a signed number the impact on her account as a result of the ATM fees.

19. Express as a signed number: The value of a house declined by \$8,000.

20. The closing price of a share of stock was \$34.67. Five trading days later, the closing price was \$31.82. What was the average daily change in the price of the stock?

• *Check your answers on page A-12.*

7.1 Introduction to Signed Numbers

OBJECTIVES

- To find the opposite of a signed number
- To find the absolute value of a signed number
- To compare signed numbers
- To solve word problems involving the comparison of signed numbers

What Signed Numbers Are and Why They Are Important

In this chapter, we discuss negative numbers and show how they relate to positive numbers—the numbers greater than 0. Negative and positive numbers together are referred to as **signed numbers**.

Here are a few applications of signed numbers.

- In football, a positive number represents yards gained; a negative number, yards lost.
- In terms of time, positive applies to a time after an event took place; negative, to a time before that event.
- In the study of electricity, positive represents one kind of electric charge; negative, the opposite kind of electric charge.

These applications can help you develop intuition in working with negative numbers and understand what they represent.

The Number Line

Our previous discussion of the number line on page 0 included only positive numbers. However, the number line can be extended to represent the negative numbers also. If we label the numbers to the right of 0 as positive and extend the line leftward past 0, then we label the numbers to the left of 0 as negative.

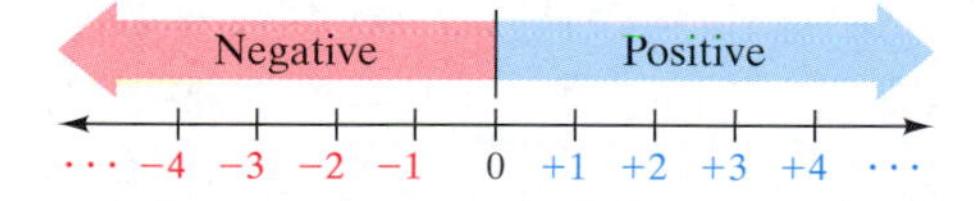

Note that we write "negative two" as -2 and "positive three" as $+3$, or just 3. However, we write no sign before 0 because 0 is neither negative nor positive.

Definition

A **positive number** is a number greater than 0.

A **negative number** is a number less than 0.

A **signed number** is a number with a sign that is either positive or negative.

In drawing the number line, we usually label only the integers, that is, the whole numbers and the corresponding negatives.

Definition

The **integers** are the numbers $\ldots, -4, -3, -2, -1, 0, +1, +2, +3, +4, \ldots$, continuing indefinitely in both directions.

Fractions and decimals and their corresponding negatives can also be represented on the number line. Let's look at how to locate the points on the number line that correspond to the numbers $\frac{3}{4}$, 3.8, and -2.4. The following number line shows these locations.

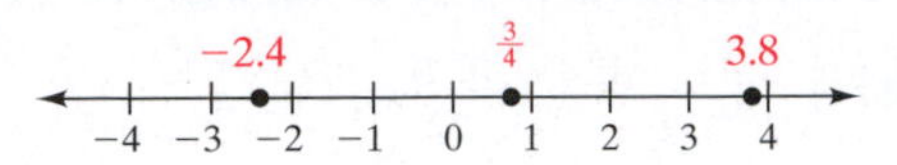

EXAMPLE 1

Locate $\frac{1}{2}$, -2.8, $-\frac{1}{8}$, and 1.2 on the number line.

−4 −3 −2 −1 0 1 2 3 4

Solution

−2.8 $-\frac{1}{8}$ $\frac{1}{2}$ 1.2

−4 −3 −2 −1 0 1 2 3 4

PRACTICE 1

Locate $1\frac{9}{10}$, -1, -3.1, and 0 on the number line.

−4 −3 −2 −1 0 1 2 3 4

On the number line, we say that -1 and $+1$ (or 1) are the opposites of each other. Similarly -50 and $+50$ (or 50) are opposites. What is the opposite of 0?

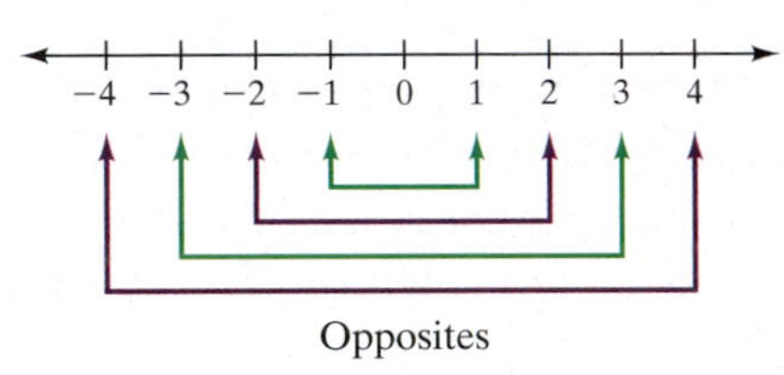

Opposites

Definition

Two numbers that are the same distance from 0 on the number line but on opposite sides of 0 are called **opposites**.

From this definition, we see that opposite numbers have opposite signs.

EXAMPLE 2

Find the opposite of each number in the table.

Solution

Number	Opposite
a. 5	-5
b. $-\frac{1}{2}$	$\frac{1}{2}$
c. 1.5	-1.5
d. -100	100

PRACTICE 2

Find the opposite of each number.

Number	Opposite
a. 9	
b. $-4\frac{9}{10}$	
c. -2.9	
d. 31	

Because the number -2 is negative, it lies 2 units to the left of 0. The number 2, which is positive, lies in the opposite direction: 2 units to the right of 0.

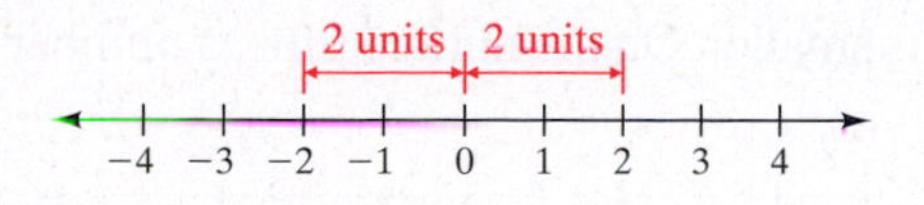

When you locate a number on the number line, the *distance* of that number from 0 is called its *absolute value*. Thus, the absolute value of $+2$ is 2, and the absolute value of -2 is 2.

Definition

The **absolute value** of a number is its distance from 0 on the number line. The absolute value of a number is represented by the symbol $|\ |$.

For example, we write the absolute value of -2 as $|-2|$.

Several properties of absolute value follow from this definition.

- The absolute value of a positive number is the number itself.
- The absolute value of a negative number is its opposite.
- The absolute value of 0 is 0.
- The absolute value of a number is always positive or 0.

These properties help us find the absolute value of any number.

EXAMPLE 3

Compute.

a. $|-8|$ **b.** $|0|$ **c.** $\left|-\frac{1}{2}\right|$ **d.** $|5.3|$

Solution

a. Because -8 is negative, its absolute value is its opposite, or 8.

b. The absolute value of 0 is 0.

c. $\frac{1}{2}$ **d.** 5.3

PRACTICE 3

Compute.

a. $|9|$ **b.** $|1\frac{3}{4}|$ **c.** $|-4.1|$ **d.** $|-5|$

EXAMPLE 4

Determine the sign and the absolute value of the number.

a. 25 **b.** -1.9

Solution

a. Sign: +; absolute value: 25

b. Sign: −; absolute value: 1.9

PRACTICE 4

What are the sign and the absolute value of the number?

	Sign	Absolute value
a. -4		
b. $6\frac{1}{2}$		

Comparing Signed Numbers

The number line helps us compare two signed numbers, that is, to decide which number is larger and which is smaller. On the number line, a number to the right is the larger number.

−4 −3 −2 −1 0 1 2 3 4

So $1 > -2$.

To Compare Signed Numbers

- Locate the points being compared on the number line; a number to the right is larger than a number to the left.

When comparing signed numbers, remember the following:

- Zero is greater than any negative number because all negative numbers lie to the left of 0.
- Zero is less than any positive number because all positive numbers lie to the right of 0.
- Any positive number is greater than any negative number because all positive numbers lie to the right of all negative numbers.

EXAMPLE 5

Which is larger?

a. 2 or 0 **b.** -1 or -3 **c.** 1.4 or -3

Solution

a. Because 2 (or $+2$) is to the right of 0 on the number line, 2 is greater than 0.

−4 −3 −2 −1 0 1 2 3 4

b. Because -1 is to the right of -3, $-1 > -3$, that is, -1 is larger.

−4 −3 −2 −1 0 1 2 3 4

c. Because 1.4 is to the right of -3, $1.4 > -3$, that is, 1.4 is the larger of the two numbers.

−4 −3 −2 −1 0 1 2 3 4

PRACTICE 5

Which is smaller?

a. 0 or $\frac{1}{2}$

b. -5 or -2

c. $2\frac{1}{2}$ or -4

Now, let's try some practical applications of comparing signed numbers. The key is to be able to determine if a number is negative or positive. You should become familiar with the following words that indicate the sign of a number.

Negative	Positive
Loss	Gain
Below	Above
Decrease	Increase
Down	Up
Withdrawal	Deposit
Past	Future
Before	After

EXAMPLE 6

Express as a signed number: Badwater Basin in Death Valley is the lowest elevation in the Western Hemisphere at 282 feet below sea level. (***Source:*** National Park Service)

Solution The number in question represents an elevation below sea level, so we write it as a negative number: -282 feet.

PRACTICE 6

Represent as a signed number: The New York Giants gained 2 yards on a play.

EXAMPLE 7

The following table shows the temperature below which various plants freeze and die.

Plant	Asters	Carnations	Mums
Hardy to	–20°F	–5°F	–30°F

In a very cold climate, which would be planted?
(***Source:*** *The American Horticultural Society A–Z Encyclopedia of Garden Plants*)

Solution First, we compare the temperatures of the asters and the carnations. Because $-20° < -5°$, the asters are hardier than the carnations. Next, we compare the temperatures of the asters and the mums. Because $-20° > -30°$, the mums are hardier. So the mums would be the best of the three to plant.

PRACTICE 7

Apparent visual magnitude is how bright a star appears when viewed from Earth. The lower the apparent magnitude, the brighter the star appears. The table shows the apparent magnitude for various stars.
(***Source:*** *Encyclopedia Britannica Almanac*)

Star	Canopus	Sirius	Alpha Centauri
Apparent Magnitude	–0.72	–1.46	–0.01

Which of these stars is the brightest?

Mathematically Speaking

Fill in each blank with the most appropriate term or phrase from the given list.

larger	smaller	opposites	right
positive number	left	absolute value	integers
signed number	negative number	whole numbers	

1. A number greater than 0 is a(n) _______.

2. A number less than 0 is a(n) _______.

3. A number with a sign that is either positive or negative is called a(n) _______.

4. The _______ are the numbers …, -4, -3, -2, -1, 0, 1, 2, 3, 4, …, continuing indefinitely in both directions.

5. Two numbers that are the same distance from 0 on the number line but on opposite sides of 0 are called _______.

6. The _______ of a number is its distance from 0 on the number line.

7. For two numbers on the number line, the number on the left is _______ than the number on the right.

8. For two numbers on the number line, the number on the _______ is larger.

Mark the corresponding point for each number on the number line.

9. -2

10. 1.1

11. 0

12. $-3\frac{9}{10}$

$-4 \quad -3 \quad -2 \quad -1 \quad 0 \quad 1 \quad 2 \quad 3 \quad 4$

Find the opposite of each number.

13. 8

14. -3

15. 10.2

16. -25

17. -5

18. $-5\frac{1}{2}$

19. $2\frac{1}{3}$

20. $\dfrac{3}{4}$

21. -4.1

22. 0.5

23. -1.2

24. -2.1

Evaluate.

25. $|-6|$

26. $|39|$

27. $\left|-\dfrac{4}{5}\right|$

28. $|-5.8|$

29. $|2|$

30. $|8|$

31. $|-0.6|$

32. $|-1\frac{2}{3}|$

Determine the sign and the absolute value of each number.

	Sign	Absolute Value		Sign	Absolute Value
33. 8			**34.** 11		
35. -4.3			**36.** 9.2		
37. -7			**38.** -30		
39. $\frac{1}{5}$			**40.** $-\frac{3}{4}$		

Solve.

41. How many numbers have an absolute value of 5?

42. How many numbers have an absolute value of 0.5?

43. Is there a number whose absolute value is -1?

44. Are there three different numbers that have the same absolute value?

Circle the larger number in each pair.

45. -4 and -7

46. 4 and -7

47. 12 and 0

48. 0 and -87

49. -3 and 2

50. -3 and -14

51. -4 and $-2\frac{1}{3}$

52. 5.1 and 8

53. -29 and -2

54. -4 and 7

55. 9 and -22

56. -4.9 and -5

57. -8 and -2

58. $+3$ and -14

59. -7 and $-7\frac{1}{4}$

60. 3.888 and 4

61. -8.3 and -8.5

62. -3.9 and -3.4

63. $-3\frac{1}{2}$ and $-3\frac{2}{3}$

64. $-7\frac{1}{2}$ and $-7\frac{1}{4}$

Indicate whether each inequality is true or false.

65. $-5 > -7$

66. $0 < -1$

67. $-1 < 3.4$

68. $-5 > 0$

69. $0 > -2\frac{3}{4}$

70. $-6 < -1$

71. $2 > -2$

72. $-100 < 0$

73. $-3.5 > -3.4$

74. $-1.6 < -1.7$

75. $-4\frac{1}{3} < 0$

76. $-\frac{5}{2} > -\frac{7}{2}$

Arrange the numbers in each group from smallest to largest.

77. 3, -3, 0

78. 3.5, -3.1, -3, 0, 4

79. -9, 9, -4.5

80. $-2\frac{1}{2}$, -2, 3, -2.7

Express each quantity as a signed number.

81. A withdrawal of $150 from an account

82. 6 kilometers below sea level

83. A rise in temperature of 14.5°C

84. A loss of $3\frac{1}{4}$ pounds while on a diet

Mixed Practice

Solve.

85. Locate $2\frac{1}{2}$ and -3.9 on the number line.

-4 -3 -2 -1 0 1 2 3 4

86. Find the opposite of $-2\frac{1}{4}$.

87. Write a withdrawal of \$10.98 from a bank account as a signed number.

88. What are the sign and absolute value of the following numbers?

a. 4 **b.** $-\frac{2}{3}$

	Sign	**Absolute Value**
a.		
b.		

89. Evaluate: **a.** $|0.5|$ **b.** $|-11|$

90. Which number is larger, -4.95 or -4?

91. Complete using the symbol $<$ or $>$.

a. -9 ▢ -6 **b.** 0 ▢ $-8\frac{2}{3}$

92. Rewrite -1.7, -2, and $-\frac{3}{4}$ from largest to smallest.

Applications

Solve.

93. The Mariana Trench, the deepest point in the Pacific Ocean, is 11,033 meters below sea level, and the Puerto Rico Trench, the deepest point in the Atlantic Ocean, is 8,648 meters below sea level. Which trench is deeper? (***Source:*** www.marianatrench.com)

94. A small toy company shows a loss of \$0.3 million for the second quarter of its business and a loss of \$0.9 million for the third quarter. In which quarter did the company show the greater loss?

95. Would a patient be receiving more medication if his dosage is decreased by 50 milligrams or if it is decreased by 25 milligrams?

96. Would a group of passengers be higher if they took the elevator down 2 floors or if they took it down 5 floors?

97. A bone density test is used to determine whether a person has osteoporosis (brittle bone disease). If the result of a bone density test, called the T-score, is below -2.5, then a person has osteoporosis. Does a patient whose T-score is -1.8 have osteoporosis? (***Source:*** http://www.mayoclinic.com)

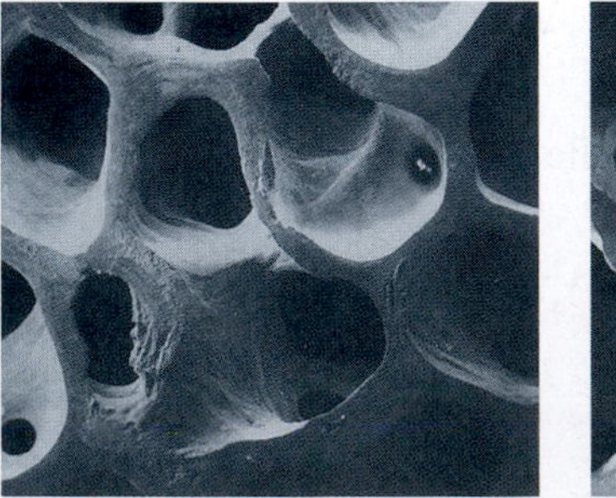

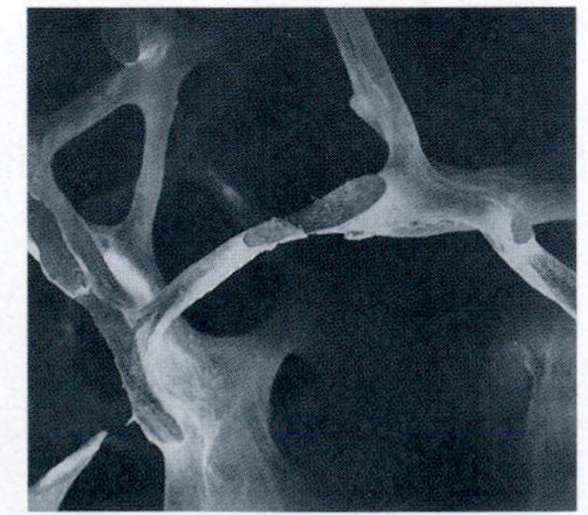

98. A bank customer has a checking account with overdraft privileges. The account is currently overdrawn by \$109.45. If the customer pays off \$100 of the overdraft, will his account still be overdrawn?

99. The average surface temperature on the planet Uranus is $-323°F$; on the planet Mars, $-81°F$, and on the planet Saturn, $-218°F$. Which planet is the warmest? (***Source:*** NASA)

100. Each liquid has its own boiling point—the temperature at which it changes to a gas. Liquid chlorine boils at $-34°C$, liquid fluorine boils at $-188°C$, and liquid bromine boils at 59°C. Which liquid has the lowest boiling point? (***Source:*** *CRC Handbook of Chemistry and Physics*)

101. In golf, scores are given in terms of par; scores above par are positive and scores below par are negative. The table shows Phil Mickelson's scores for each round of a recent PGA Championship tournament. (***Source:*** www.pga.com)

Round	Score
First	−3
Second	−1
Third	−4
Fourth	+2

a. Locate the scores on the number line. What does 0 on the number line represent?

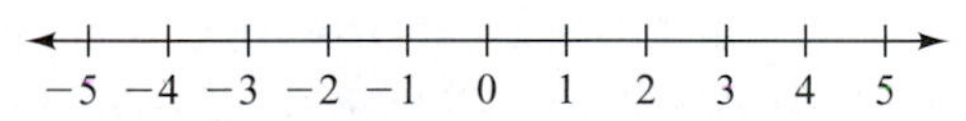

b. In which round did he have the lowest score?

102. The following graph shows the amount of American direct investments in various other countries during a recent year. A positive amount indicated that money is flowing from the United States to the other country, and a negative amount indicates that money is flowing from the other country to the United States. All amounts are rounded to the nearest billion U.S. dollars.

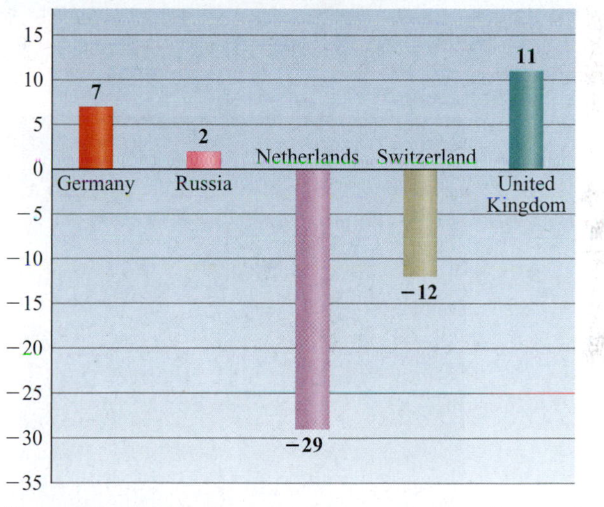

(***Source:*** The U.S. Department of Commerce, Bureau of Economic Analysis, 2005)

a. To which countries is money flowing from the United States?

b. Was more money flowing to the United States from the Netherlands or from Switzerland?

• *Check your answers on page A-12.*

MINDSTRETCHERS

Groupwork

1. a. List several numbers between -2 and -3.

   b. How many numbers are there between -2 and -3?

Mathematical Reasoning

2. On the thermometer at the right, highlight all temperatures within 4 degrees of $-1°$.

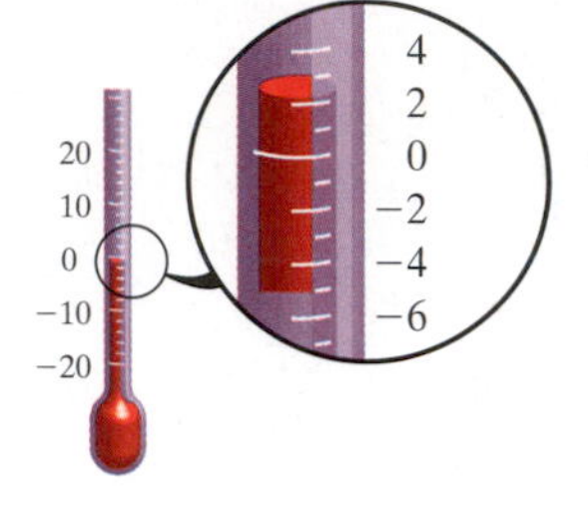

History

3. Negative numbers have not always been accepted as numbers. Either in your college library or on the Web, read about the history of negative numbers. Summarize your findings in a few sentences.

CULTURAL NOTE

Sources:
Lancelot Hogben, *Mathematics in the Making* (London: Galahad Books, 1960).

Henri Michel, *Scientific Instruments in Art and History* (New York: Viking Press, 1966).

Calvin C. Clawson, *The Mathematical Traveler* (New York and London: Plenum Press, 1994).

Up until the work of sixteenth-century Italian physicists, no one was able to measure temperature. Liquid-in-glass thermometers were invented around 1650, when glassblowers in Florence were able to create the intricate shapes that thermometers require. Thermometers from the seventeenth and eighteenth centuries provided a model for working with negative numbers that led to their wider acceptance in the mathematical and scientific communities. Numbers above and below 0 represented temperatures above and below the freezing point of water, just as they do on the Celsius scale today. Before the introduction of these thermometers, a number such as -1 was difficult to interpret for those who believed that the purpose of numbers is to count or to measure.

By contrast, the early Greek mathematicians had rejected negative numbers, calling them absurd. A thousand years later, in the seventh century A.D., the Indian mathematician Brahmagupta argued for accepting negative numbers and put down the first comprehensive rules for computing with them.

7.2 Adding Signed Numbers

OBJECTIVES

- To add signed numbers
- To solve word problems involving the addition of signed numbers

Our previous work in addition, subtraction, multiplication, and division was restricted to positive numbers—whether those positive numbers happened to be whole numbers, fractions, or decimals. Now we consider computations involving *any* signed numbers. Let's first consider the operation of addition.

Suppose that we want to add two negative numbers, say -1 and -2. It is helpful to look at this problem in terms of money. If your hourly wage went down \$1 and down again \$2, altogether it went down \$3. In terms of signed numbers, this example can be expressed as:

$$-1 + (-2) = -3$$

We can also look at this problem on the number line. To add -1 and -2, we start at the point corresponding to the first number, -1. The second number, -2, is *negative*, so we move 2 units to the *left*. We end at -3, which is the answer we expected.

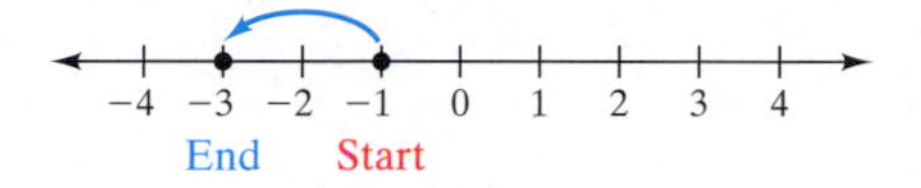

Now let's consider adding the signed numbers -1 and $+3$. Thinking of this problem in terms of money may make it clearer. Suppose that your hourly wage went down \$1 and then up \$3, your total hourly increase is \$2. Using signed numbers, this example can be written as:

$$-1 + 3 = 2$$

We can picture this problem by starting at -1 on the number line. The second number, 3, is *positive*, so we move 3 units to the *right*. We end at 2, which is the answer.

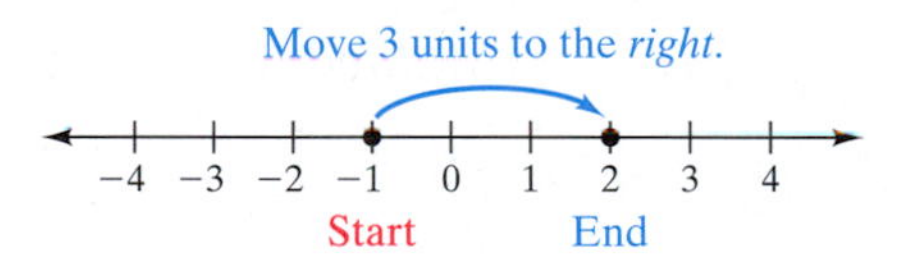

The following rule provides a shortcut for adding signed numbers:

To Add Two Signed Numbers

- If they have the same sign, add the absolute values and keep the sign.
- If they have different signs, subtract the smaller absolute value from the larger and take the sign of the number with the larger absolute value.

EXAMPLE 1

Add: -3 and -2

Solution The sum of the absolute values is 5.

$$|-3| + |-2| = 3 + 2 = 5$$

Both -3 and -2 are negative, so their sum is negative.

$$(-3) + (-2) = -5$$

Check

Move 2 units to the *left*.

−5 −4 −3 −2 −1 0 1 2 3

End Start

PRACTICE 1

Combine: -8 and -17

EXAMPLE 2

Find the sum: $(-3.9) + (-0.5)$

Solution $|-3.9| = 3.9$ and $|-0.5| = 0.5$

$$3.9 + 0.5 = 4.4$$

The sum of two negative numbers is negative, so

$$(-3.9) + (-0.5) = -4.4$$

Check

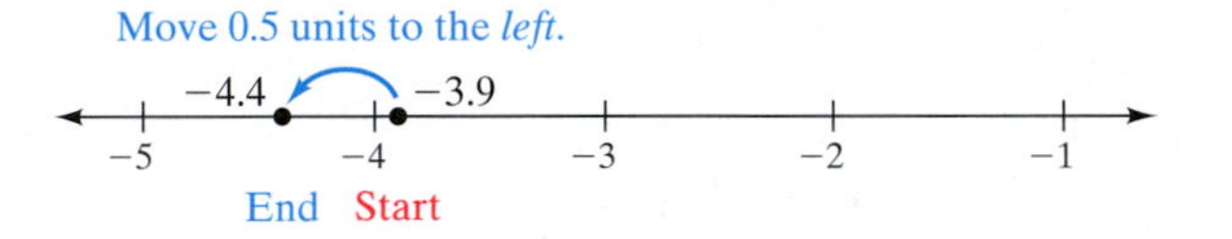

PRACTICE 2

Add: $-3 + (-1\frac{1}{2})$

EXAMPLE 3

Add: $2 + (-1)$

Solution Here, we are adding numbers with different signs. First, we find the absolute values.

$$|2| = 2 \text{ and } |-1| = 1$$

Next, we subtract the smaller absolute value from the larger.

$$2 - 1 = 1$$

Because 2 has the larger absolute value and its sign is positive, the sum is also positive. Our answer is 1, or +1.

$$2 + (-1) = 1$$

Check

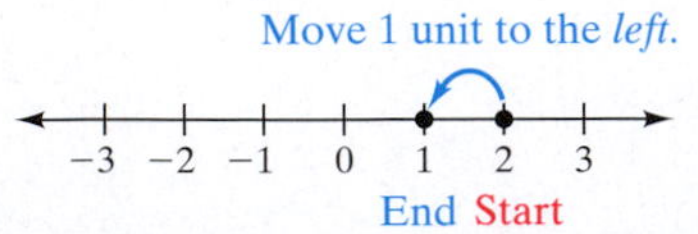

PRACTICE 3

Find the sum of -2 and 9.

Note in Example 3 that, when we added a negative number to 2, we got a smaller result—namely, 1.

EXAMPLE 4

Combine: $(-2) + (+2)$

Solution $|-2| = 2$ and $|+2| = 2$ **Find the absolute values.**

$2 - 2 = 0$ **Subtract the absolute values.**

Zero is neither positive nor negative, for it has no sign.

$$(-2) + (+2) = 0$$

Check Move 2 units to the *right*.

−3 −2 −1 0 1 2 3

Start End

PRACTICE 4

Find the sum: $-35 + 35$

Do you see why the sum of -44 and $+44$ is 0? How about $2.77 + (-2.77)$ or $2\frac{5}{8} + (-2\frac{5}{8})$?

EXAMPLE 5

Add -2.1 to 0.8.

Solution $|-2.1| = 2.1$ and $|0.8| = 0.8$ **Find the absolute values.**

$$\begin{array}{r} 2.1 \\ -0.8 \\ \hline 1.3 \end{array}$$ **Subtract the absolute values.**

Because $|-2.1|$ is greater than $|0.8|$, the answer is negative. So

$$(-2.1) + 0.8 = -1.3$$

PRACTICE 5

Add: $3\frac{4}{5} + (-1\frac{1}{5})$

Some addition problems involve the sum of three or more signed numbers. Rearranging the signed numbers to add the positives and negatives separately can make the addition easier. Note that this rearrangement does not affect the sum because addition is a commutative and associative operation.

EXAMPLE 6

Find the sum: $3 + (-1) + (-8) + 2 + (-11)$

Solution We are adding two positive and three negative numbers. Rearranging the numbers by sign, we get the following.

$$\underbrace{3 + 2}_{\text{Positives}} + \underbrace{(-1) + (-8) + (-11)}_{\text{Negatives}}$$

First, we add the positives. $3 + 2 = 5$

Then, we add the negatives. $(-1) + (-8) + (-11) = -20$

Finally, we combine the positive and the negative subtotals.

$$5 + (-20) = -15$$

So $3 + (-1) + (-8) + 2 + (-11) = -15$.

PRACTICE 6

$-3 + 1 + 8 + (-6) = ?$

EXAMPLE 7

The most famous of all comets is Halley's Comet, which passes by Earth every 76 years. For other comets, however, the length of time between visits is much longer. The Great Comet, for example, comes near Earth only once every 3,000 years. If this comet visited Earth about 1200 B.C., approximately when was its next visit? (***Source:*** Mark R. Kidger, "Some Thoughts on Comet Hale-Bopp")

Solution To help you understand this problem, we draw the number line. Any number line involving time is called a *time line*. On a time line, positive years are A.D. and negative years are B.C.

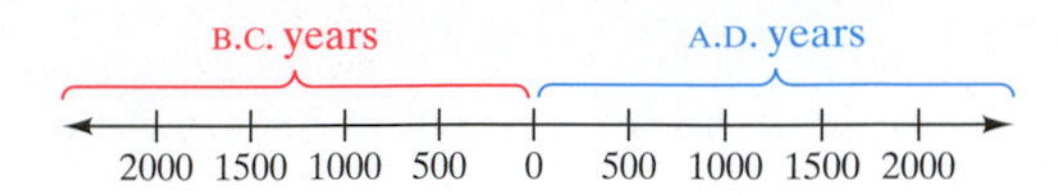

So 1200 B.C. is represented by -1200. We must add -1200 and $+3000$.

$|-1200| = 1200$ and $|3000| = 3000$ **Find the absolute values.**

$3000 - 1200 = 1800$ **Subtract the absolute values.**

The absolute value of 3000 is greater than the absolute value of -1200, so the answer is positive.

$$(-1200) + 3000 = 1800$$

So the Great Comet came near Earth in about A.D. 1800.

Check

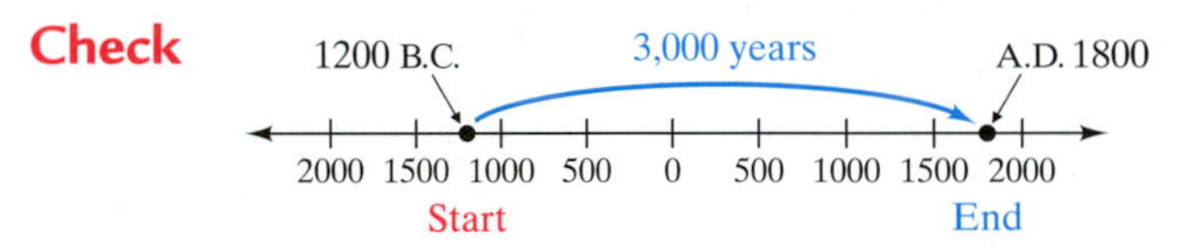

PRACTICE 7

Lake Baikal in Russia is the deepest lake in the world. The deepest point in the lake is 1,187 meters below sea level. If the surface is 1,643 meters above this point, what is the elevation of the surface? (***Source:*** *http://*www.bww.irk.ru)

Signed Numbers on a Calculator

The numbers that we have entered so far on a calculator have been positive numbers. To enter a negative number, we need to hit a special key that indicates that the sign of the number is negative. Some calculators have a negative sign key, [(−)]. Others have a change of sign key, [+/−]. Be careful not to confuse either of these keys with the subtraction key, [−].

EXAMPLE 8

Calculate: $-1.3 + (-5.8)$

Solution

Press	**Display**
[(−)] 1.3 [+] [(] [(−)] 5.8 [)] [ENTER]	−1.3+(−5.8) −7.1

PRACTICE 8

Calculate: $-1.3 + (-5.891) + (4.713)$

7.2 Exercises

FOR EXTRA HELP MyMathLab Math XL PRACTICE WATCH DOWNLOAD READ REVIEW

Mathematically Speaking

Fill in each blank with the most appropriate term or phrase from the given list.

commutative	right	larger
absolute values	left	distributive
smaller	numbers	

1. To add (−6) and (+6) on the number line, start at (−6) and move 6 units to the _______.

2. To find the sum of two signed numbers with the same sign, add the _______ and keep the sign.

3. To find the sum of two signed numbers with different signs, subtract the smaller absolute value from the larger and take the sign of the number with the _______ absolute value.

4. Rearranging signed numbers to add the positives and negatives separately does not affect the sum, because the operation of addition is associative and _______.

Find the sum of each pair of numbers. Use the number line as a visual check.

5. $6 + (-5)$

6. $3 + (-9)$

7. $-2 + 5$

8. $-9 + (-2)$

9. $7 + 0$

10. $3 + (-2)$

11. $7 + (-7)$

12. $-4 + 9$

Find the sum.

13. $67 + (-67)$

14. $23 + (-2)$

15. $-10 + 5$

16. $0 + (-12)$

17. $-100 + 300$

18. $-60 + (-20)$

19. $8 + (-2)$

20. $5{,}000 + (-3{,}000)$

21. $-60 + (-90)$

22. $-5 + (-4)$

23. $-7 + 2$

24. $2 + (-7)$

25. $-27 + 0$

26. $-13 + 13$

27. $-9 + 9$

28. $-2 + (-2)$

29. $5.2 + (-0.3)$

30. $-0.6 + 1$

31. $-0.2 + 0.3$

32. $-5.5 + 0$

33. $60 + (-0.5)$

34. $-0.7 + 0.7$

35. $-9.8 + 3.9$

36. $6.1 + (-5.9)$

37. $(-5.6) + (-8.9)$

38. $(-0.8) + (-0.5)$

39. $\left(-\frac{1}{2}\right) + \left(-5\frac{1}{2}\right)$

40. $-1\frac{1}{3} + \left(-2\frac{2}{3}\right)$

41. $-1\frac{1}{5} + \frac{3}{5}$

42. $2\frac{1}{6} + \left(-\frac{5}{6}\right)$

43. $-\frac{2}{5} + 2$

44. $-14 + \frac{1}{3}$

45. $1\frac{1}{2} + \left(-1\frac{3}{5}\right)$

46. $1\frac{3}{8} + \left(-2\frac{1}{4}\right)$

47. $(-24) + 20 + (-98)$

48. $35 + (-17) + (-18)$

49. $12 + (-7) + (-12\frac{1}{2})$

50. $-8 + (-4) + (-8\frac{1}{4})$

51. $(-7) + 12 + 0 + (-7) + 9$

52. $(-3) + 8 + (-9) + 3 + (-4)$

53. $-0.3 + (-2.6) + (-4)$

54. $-5.25 + (-0.4) + 3$

55. $-12 + 7.58 + 12$

56. $-3.7 + (-1.88) + 5$

57. $8.756 + (-9.08) + (-4.59)$

58. $-5.405 + 6 + (-6.89)$

59. $-3.001 + (-0.59) + 8$

60. $-10 + 5.17 + (-10.002)$

Mixed Practice

Solve.

61. Find the sum of $-6 + 9$ on the number line.

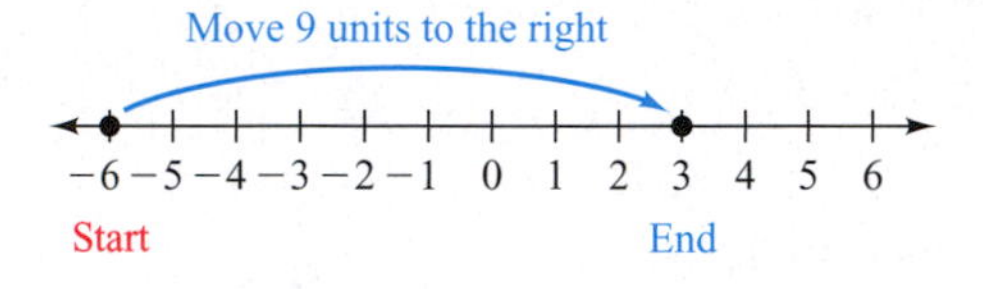

62. Combine 16 and (-24).

63. Add: $9.6 + (-9.6)$

64. Combine: $-4\frac{2}{9} + \left(-2\frac{1}{9}\right)$

65. Add -8, 14, and -10.

66. Find the sum of -1.7, -3.95, and 10.

Applications

Solve. Express each answer as a signed number.

67. The lowest elevation in Africa is Lake Assal at 512 feet below sea level. The highest elevation, Mount Kilimanjaro, is 19,852 feet above Lake Assal. What is the elevation of Mount Kilimanjaro? (***Source:*** *The World Almanac and Book of Facts, 2006*)

68. A student owes \$2,456 on her credit card. After making a payment of \$350, what is the balance on her credit card?

69. During a recession, a manufacturer laid off 182 employees. A year later, another 56 employees were laid off. What was the change in the number of employees working for the manufacturer as a result of the two layoffs?

70. A computer retailer decreases the price of a laptop computer by \$150 during a sale. As a special promotion, the price is decreased another \$75 for customers who trade in their old laptop. What is the total price change for a customer who trades in his old laptop?

71. In a physics class, students study the properties of atomic particles, including protons and electrons. They learn that a proton has an electric charge of $+1$, whereas an electron has an electric charge of -1. What is the total charge of a collection of 3 protons and 4 electrons?

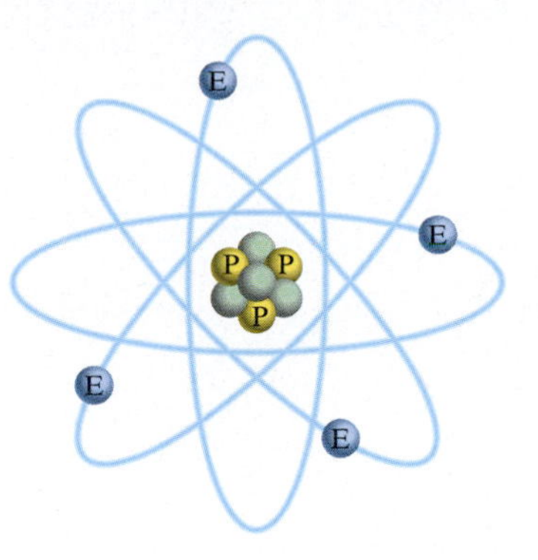

72. In order to conduct an experiment, a chemist cooled a substance to $-10°C$. In the course of this experiment, a chemical reaction took place that raised the temperature of the substance by 15°. What was the final temperature?

73. Ten years ago, a couple got married. Four years later, they got divorced. When was their divorce?

74. Cleopatra became queen of Egypt in 51 B.C. She left the throne 20 years later. In what year was that?

75. A football team gained 5 yards on its first down, lost 7 yards on second down, and lost 4 yards on third down. What was the overall change in position after third down?

76. In the last 4 months, a dieter lost 5 pounds, gained 2 pounds, lost 1 pound, and maintained his weight, respectively. What was his overall change in weight?

77. The table shows the daily change in the price of a share of stock for Texas Instruments, Inc. over a 5-day period in a recent year. (***Source:*** www.marketwatch.com, 2006)

Day	Change in Price
Monday	$+\$0.43$
Tuesday	$-\$0.63$
Wednesday	$+\$0.29$
Thursday	$+\$0.82$
Friday	$-\$0.23$

a. What was the net change in the price of a share of stock over the 5-day period?

b. If the price of the stock was \$29.35 at the start of the 5-day period, what was the closing price on Friday?

78. The following graph shows a company's bottom line (gain or loss) for various quarters, in millions of dollars.

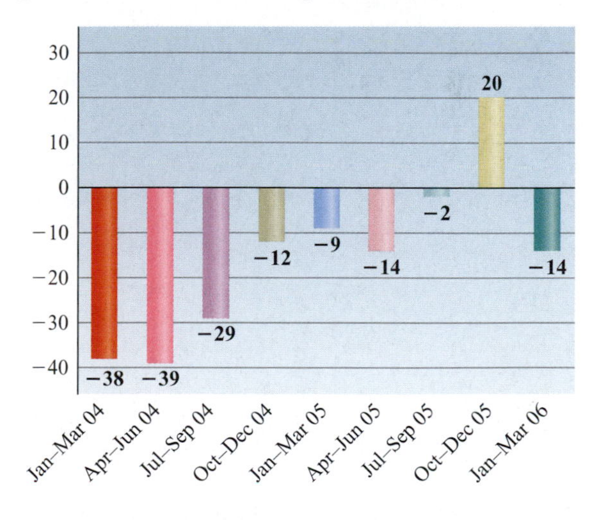

a. What is the company's bottom line for 2005?

b. Was the company's bottom line greater in the last quarter of 2004 or in the first quarter of 2006?

• *Check your answers on page A-13.*

MINDSTRETCHERS

Groupwork

1. Work with a partner on the following.

a. Fill in the following addition table.

+	+3	−2	−1
+4			
−3			
−1			

b. Why do the nine numbers that you entered sum to 0?

Writing

2. For signed numbers, does *adding* always mean *increasing*? Explain.

Patterns

3. Find the missing numbers in the following sequence: −5, −7, −6, −8, −7, −9, −8, _______, _______, _______

7.3 Subtracting Signed Numbers

OBJECTIVES

- To subtract signed numbers
- To solve word problems involving the subtraction of signed numbers

The subtraction of signed numbers is based on two topics previously discussed—adding signed numbers and finding the opposite of a signed number.

Let's first consider a subtraction problem involving money. Suppose that you have \$10 in your bank account and withdraw \$3. Then \$7 will be left in the account. In terms of signed numbers, this can be expressed as

$$10 - (+3) = 7$$

Now suppose that you had started with \$10 in the account and the bank imposed a monthly service charge of \$3. The balance in your account once more would be \$7. Using signed numbers, this can be written as

$$10 + (-3) = 7$$

The answers to the two problems are the same; so

$$10 - (+3) = 7 \text{ and } 10 + (-3) = 7$$

are equivalent problems.

We can change a problem in subtracting signed numbers to an equivalent problem in adding signed numbers by adding the *opposite* of the number that we want to subtract, giving us the following rule.

To Subtract Two Signed Numbers

- Change the operation of subtraction to addition, and change the number being subtracted to its opposite.
- Follow the rule for adding signed numbers.

To see if this rule works when the number we are subtracting is negative, consider $4 - (-1)$. Recall that every subtraction problem has a related addition problem.

$$\underbrace{5 - 3 = 2}_{\text{Subtraction}} \quad \text{because} \quad \underbrace{2 + 3 = 5}_{\text{Related addition}}$$

Therefore, $4 - (-1) = 5$ because $5 + (-1) = 4$. Note that we get the same result using the rule for subtracting signed numbers.

$$4 \underset{\text{Subtract}}{-} \underset{\text{Negative 1}}{(-1)} = 4 \underset{\text{Add}}{+} \underset{\text{Positive 1}}{(+1)} = 5$$

EXAMPLE 1

Find the difference: $-2 - (-4)$

Solution We change the operation of subtraction to addition and also change the number being subtracted from -4 to $+4$.

$$-2 - (-4)$$
$$= -2 + (+4)$$

We already know how to add a negative and a positive number. So we get

$$-2 + 4 = +2, \text{ or } 2$$

PRACTICE 1

Find the difference: $-4 - (-2)$

Note that in Example 1 and Practice 1, the numbers in the differences are the same except for their order. But the answers are quite different: $-2 - (-4) = 2$, whereas $-4 - (-2) = -2$. Why do you think this is so?

EXAMPLE 2

Compute: $3 - (-9)$

Solution Change negative 9 to positive 9.

$$3 - (-9) = 3 + (+9) = +12, \text{ or } 12$$

Change subtraction to addition.

Note that when we subtracted -9 from 3, we got an answer larger than 3.

PRACTICE 2

Subtract: $9 - (-9)$

EXAMPLE 3

Subtract: $-2 - 8\frac{1}{3}$

Solution $-2 - 8\frac{1}{3} = -2 + \left(-8\frac{1}{3}\right) = -10\frac{1}{3}$

PRACTICE 3

Find the difference: $-9 - 12.1$

EXAMPLE 4

Calculate: $5 + (-6) - (-11)$

Solution This problem involves addition and subtraction. According to the order of operations rule, we work from left to right.

$$5 + (-6) - (-11) = -1 - (-11) \quad \text{Add } 5 + (-6).$$
$$= -1 + 11 \quad \text{Subtract } -11.$$
$$= 10 \quad \text{Add } 11.$$

PRACTICE 4

$-2 - 3 + (-5) = ?$

EXAMPLE 5

Normally we think of oxygen as a gas. However, when cooled to $-183°C$ (its boiling point), oxygen becomes a liquid. If it is cooled further to $-218°C$ (its melting point), oxygen becomes a solid. How much higher is the boiling point of oxygen than its melting point? (***Source:*** *Handbook of Chemistry & Physics*)

Solution We need to compute how much greater is -183 than -218.

$$\begin{aligned}(-183) - (-218) &= (-183) + (+218)\\ &= +35\end{aligned}$$

The boiling point of oxygen is 35°C higher than its melting point.

PRACTICE 5

The highest point on the continent of South America is Mt. Aconcagua at an elevation of 22,834 feet above sea level. The lowest point is the Valdes Peninsula at an elevation of 131 feet below sea level. How much higher is Mt. Aconcagua than the Valdes Peninsula?
(***Source:*** National Geographic Society)

Mathematically Speaking

Fill in each blank with the most appropriate term or phrase from the given list.

absolute value	order of operations	addition
sum	multiplication	signed numbers
difference	opposite	

1. To subtract two signed numbers, change the operation of subtraction to addition, and change the number being subtracted to its _______. Then follow the rule for adding signed numbers.

2. Every subtraction problem has a related _______ problem.

3. When a signed number problem involves addition and subtraction, work from left to right according to the _______ rule.

4. When subtracting a negative number, the _______ is greater than the original number.

Find the difference.

5. $5 - (-2)$ **6.** $0 - 1$ **7.** $1 - 0$ **8.** $5 - 9$

9. $-9 - 5$ **10.** $-44 - 2$ **11.** $42 - (-2)$ **12.** $22 - 35$

13. $50 - 75$ **14.** $-44 - (-2)$ **15.** $-20 - (-1)$ **16.** $85 - (-85)$

17. $3 - (-3)$ **18.** $-3 - 3$ **19.** $0 - 38$ **20.** $38 - 0$

21. $-13 - 13$ **22.** $13 - 13$ **23.** $13 - (-13)$ **24.** $23 - 8$

25. $8 - 23$ **26.** $-34 - 7$ **27.** $800 - (-200)$ **28.** $30 - (-10)$

29. $7 - 8.52$ **30.** $9.1 - 10.84$ **31.** $9.2 - (-0.5)$ **32.** $(-3) - (-0.2)$

33. $-5.2 - (-5.2)$ **34.** $0.5 - (-0.5)$ **35.** $8.6 - (-1.9)$ **36.** $-1.9 - 8.6$

37. $-10 - (-9.5)$ **38.** $-6 - 8.7$ **39.** $4\frac{1}{2} - 9\frac{1}{2}$ **40.** $9\frac{1}{2} - 4\frac{1}{2}$

41. $10 - 2\frac{1}{4}$ **42.** $-10 - \left(-2\frac{1}{4}\right)$ **43.** $-7 - \frac{1}{4}$ **44.** $-9 - \frac{1}{8}$

45. $5\frac{3}{4} - \left(-1\frac{1}{2}\right)$ **46.** $-6\frac{1}{2} - \left(-1\frac{1}{3}\right)$

Combine.

47. $4 + (-6) - (-9)$ **48.** $-10 - (-6) + 8$

49. $7 - 7 + (-5)$ **50.** $10 + (-10) - (-5)$

51. $-8 + (-4) - 9 + 7 + (-1)$ **52.** $-5 - (-1) + 6 + (-3) - 4$

53. $7.043 - 9.002 - 1.883$ **54.** $-6.192 - 0.337 - (-23.94)$

55. $-8.722 + (-3.913) - 3.86$ **56.** $2.884 - 0.883 + (-6.125)$

Mixed Practice

Solve.

57. Subtract: $-16 - 9$

58. Find the difference: $8.1 - 10.46$

59. Subtract $-19\frac{3}{4}$ from $-19\frac{3}{4}$.

60. From 6 subtract -5.

61. Combine: $-4 + (-5) + 6 - (-4)$

62. Evaluate: $3 - (-1) - 2 + (-9)$

Applications

Solve. Express each answer as a signed number.

63. Two airplanes take off from the same airport. One flies west and the other east, as shown. How far apart are they?

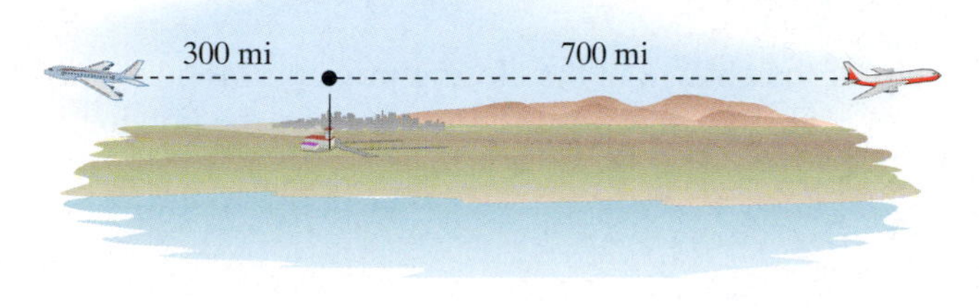

64. Two friends get on different elevators at the same floor. One goes up 2 floors, the other goes down 4 floors. How many floors are they apart?

65. Paper was invented in China in about 100 B.C. About how many years ago was that? (***Source:*** *World of Invention*)

66. Ethiopia was founded around 1,000 B.C., and the United States in A.D. 1789. How much older is Ethiopia than the United States? (***Source:*** *The Concise Columbia Encyclopedia*)

67. In business, net income is calculated by subtracting the costs from the revenue. What is a company's net income if its revenues were \$2.3 million and its costs were \$3.7 million? Express as a signed number.

68. Rapid City, South Dakota, holds the U.S. record for a 2-hour temperature change. On January 12, 1911, the temperature at 6 A.M. was 49°F. If it was 62° colder by 8 A.M., what was the temperature at 8 A.M.? (***Source:*** National Weather Service)

69. A country's trade balance is the difference between the value of its exports and the value of its imports. In 2010, the United States is projected to have \$2,393.7 billion in exports and \$3,282.7 billion in imports. What would be the projected trade balance in 2010? (***Source:*** Bureau of Labor Statistics)

70. The federal deficit in 2006 was \$423 billion and in 2005 was \$318 billion.

a. Write the deficit for each year as a signed number.

b. How much lower was the deficit in 2005? (***Source:*** Office of Management and Budget)

71. The bar graph shows the annual precipitation for Phoenix, Arizona, for the years from 2001 through 2005 (***Source:*** National Weather Service)

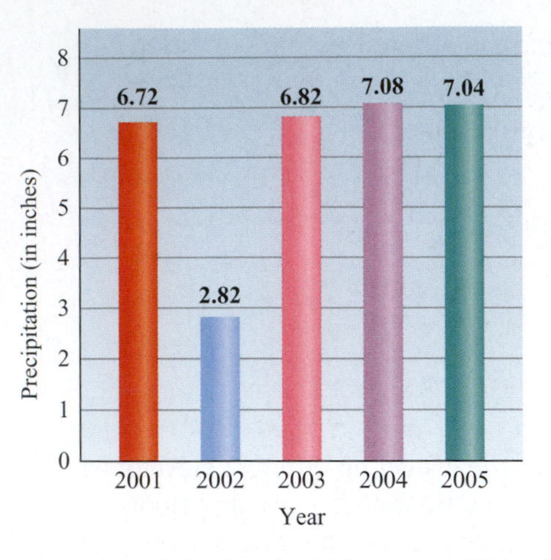

a. The annual precipitation in 2005 was 1.25 inches below the normal annual precipitation. What is the normal annual precipitation?

b. Use the answer from part (a) to calculate the difference between the annual precipitation in 2002 and the normal annual precipitation.

72. The bar graph shows the closing price of a share of Whole Foods Market, Inc. stock for a 5-day period in July of 2006. (***Source:*** http://www.marketwatch.com)

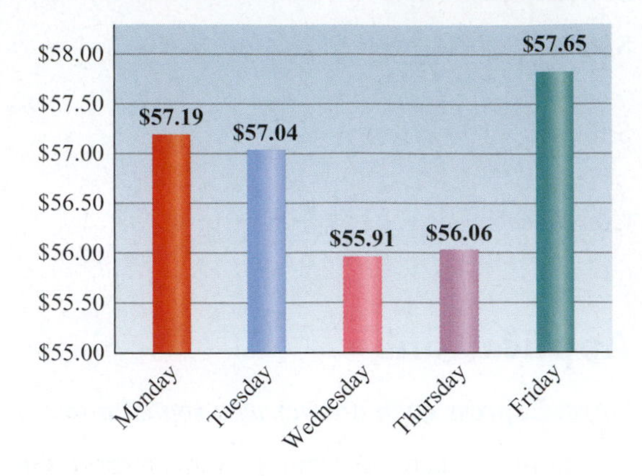

a. Calculate the change in the closing price per share from the previous day for Tuesday through Friday. Express each change as a signed number.

b. If the change in the closing price on Monday was +$0.10, what was the closing price the previous day?

• *Check your answers on page A-13.*

MINDSTRETCHERS

Groupwork

1. Working with a partner, rearrange the numbers in the square on the left so that the sum of every row, column, and diagonal is -6.

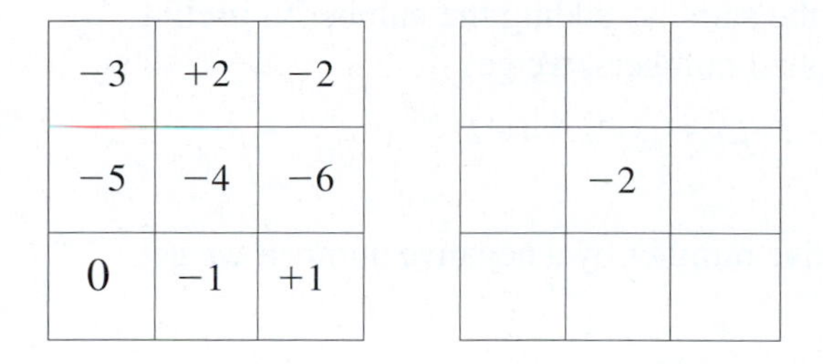

−3	+2	−2
−5	−4	−6
0	−1	+1

	−2	

Mathematical Reasoning

2. The following two columns of numbers add up to the same sum.

3	7
7	?
1	7
9	8
5	9

What number does the question mark represent?

Writing

3. Consider the following two problems.

$$8 - (-2) = 8 + 2 = 10$$

$$8 \div \frac{4}{7} = 8 \times \frac{7}{4} = 14$$

Explain in what way the two problems are similar.

7.4 Multiplying Signed Numbers

OBJECTIVES

- To multiply signed numbers
- To solve word problems involving the multiplication of signed numbers

We now turn to the multiplication of signed numbers. Consider, for example, the problem of finding the product of $+4$ and -2, or $4(-2)$. We know that multiplying a number by 4 means the same as adding the number to itself 4 times. Using the rule for adding signed numbers, we get

$$\begin{aligned} 4(-2) &= -2 + (-2) + (-2) + (-2) \\ &= -8 \end{aligned}$$

Note that when multiplying a positive number by a negative number, we get a negative answer.

Let's take another look at this same problem in practical terms. Suppose that you are on a diet and you lose 2 pounds per month. Compared to your current weight, how much will you weigh 4 months from now? The answer is that you will weigh 8 pounds less. So we write

$$4(-2) = -8$$

Now we examine a different question. Again assume that you lose 2 pounds per month by dieting. Four months ago, you were heavier than you are now. How much heavier were you? To answer this question, note that each month you lost 2 pounds but that you are going back in time 4 months. So you weighed 8 pounds more than you do now, which can be expressed as

$$-4(-2) = +8, \text{ or } 8$$

Note that when we multiply two negative numbers, we get a positive number.

We can use the following rule to multiply signed numbers.

To Multiply Two Signed Numbers

- Multiply their absolute values.
- If the numbers have the same sign, their product is positive; if the numbers have different signs, their product is negative.

Another way to think of multiplying signed numbers is as follows:

Positive · Positive = Positive
Negative · Negative = Positive
Positive · Negative = Negative
Negative · Positive = Negative

EXAMPLE 1

Find the product of -2 and -1.

Solution First, we find the absolute values.

$$|-2| = 2 \text{ and } |-1| = 1$$

Next, we multiply the absolute values.

$$2 \cdot 1 = 2$$

Since the numbers have the same sign, the product is positive. The answer is $+2$, or 2. So we can write

$$(-2)(-1) = 2$$

PRACTICE 1

Compute: $-8(-4)$

EXAMPLE 2

Calculate: $(5)(-10)$

Solution $|5| = 5$ and $|-10| = 10$ — Find the absolute value of each factor.

$5 \cdot 10 = 50$ — Multiply the absolute values.

The factors have different signs, so the product is negative.

$$(5)(-10) = -50$$

PRACTICE 2

Multiply: $(-5)(2)$

EXAMPLE 3

Evaluate: **a.** $(-7)^2$ **b.** -7^2

Solution

a. Recall that $(-7)^2$ means $(-7)(-7)$. Because the two factors have the same sign, their product is positive. So $(-7)^2 = 49$.

b. $-7^2 = -(7 \cdot 7) = -(49)$

$= -49$

PRACTICE 3

Simplify:

a. $(-1)^2$

b. -1^2

Note that in Example 3a, we squared a negative number so that the answer was positive. On the other hand, in Example 3b, squaring a positive number when preceded by a negative sign gives us a negative answer.

EXAMPLE 4

Find the product of $2\frac{1}{5}$ and -5.

Solution $2\frac{1}{5} \cdot 5 = \frac{11}{5} \cdot \frac{5}{1} = 11$ — Multiply the absolute values.

Since the factors $2\frac{1}{5}$ and -5 have different signs, their product is negative.

$$2\frac{1}{5} \cdot (-5) = -11$$

PRACTICE 4

Multiply: $\left(-1\frac{1}{3}\right)\left(-\frac{1}{5}\right)$

EXAMPLE 5

Multiply: $(-1.4)(-0.6)$

Solution $(1.4)(0.6) = 0.84$ — Multiply the absolute values.

Since the factors have the same sign, we get

$$(-1.4)(-0.6) = 0.84$$

PRACTICE 5

Find the product of -2.5 and 8.

EXAMPLE 6

Calculate: $8(-2)(-3)$

Solution We multiply from left to right.

$$\underbrace{8(-2)}(-3) \quad \text{Positive} \cdot \text{Negative} = \text{Negative}$$
$$= \underbrace{-16 \cdot (-3)} \quad \text{Negative} \cdot \text{Negative} = \text{Positive}$$
$$= 48$$

PRACTICE 6

Multiply: $-8(-2)(-3)$

Comparing Example 6 and Practice 6, we note that both problems have the same absolute values but different signs. The product in Example 6 is positive, because there are two negative factors. By contrast, the answer to Practice 6 is negative, because there are three negative factors. Can you explain why a product is positive if it has an even number of negative factors, whereas a product is negative if it has an odd number of negative factors?

EXAMPLE 7

Simplify: $8 - 10(-5)^2$

Solution Use the order of operations rule.

$$8 - 10 \cdot (-5)^2 = 8 - 10 \cdot 25 \quad \text{Square first.}$$
$$= 8 - 250 \quad \text{Multiply.}$$
$$= 8 + (-250) \quad \text{Subtract 250.}$$
$$= -242 \quad \text{Add } -250.$$

PRACTICE 7

Calculate: $-4 + (-2)^2 \cdot 3$

EXAMPLE 8

An oil company is drilling for oil. Each day, the workers drill down 20 feet farther until they hit a pool of oil, as shown. Will they reach oil by the end of the fifth day?

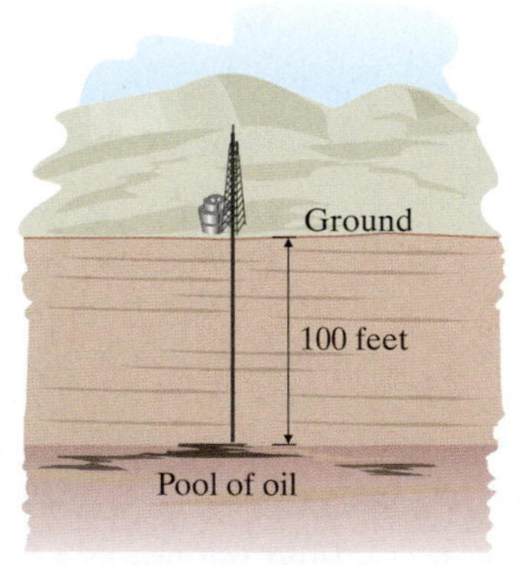

Solution Let's represent movement downward by a negative number. Since each of the 5 days they drill 20 feet farther down, we compute $5 \cdot (-20)$. Using the rule for multiplying signed numbers, we get -100. Therefore, the drill will reach 100 feet below ground level by the fifth day—the depth of the pool of oil.

PRACTICE 8

Alvin is a deep submergence vehicle operated by the Woods Hole Oceanographic Institution for marine research. On a research mission, it descends from the surface to the ocean floor at a rate of 30 meters per minute. Will *Alvin* reach the ocean floor in 60 minutes? (*Source:* American Geophysical Union)

Mathematically Speaking

Fill in each blank with the most appropriate term or phrase from the given list.

odd	positive	negative	even
product	sum	prime	

1. The product of two numbers with the same sign is _______.
2. The _______ of two numbers with different signs is negative.
3. The product of a(n) _______ number of negative factors is positive.
4. The product of a(n) _______ number of negative factors is negative.

Find the product.

5. $(2)(-5)$
6. $-4 \cdot 9$
7. $-2 \cdot 5$
8. $-1(-5)$
9. $-5 \cdot (-5)$
10. $4 \cdot (-3)$
11. $-34(-9)$
12. $8(-100)$
13. $2 \cdot (-8)$
14. $-1 \cdot 5$
15. $907 \cdot (-9)$
16. $-5 \cdot (-812)$
17. $5(-8)$
18. $8 \cdot (-53)$
19. $-88 \cdot 2$
20. $20 \cdot (-30)$
21. $(-200)(-4)$
22. $-4 \cdot (-200)$
23. $-80 \cdot 90$
24. $(-7)(-100)$
25. $(2.5)(-2)$
26. $(0.3)(-0.2)$
27. $(0.2)(-50)$
28. $3 \cdot (-0.3)$
29. $(-1.2)(-4.6)$
30. $(-0.7)(-1.8)$
31. $(5)(-1.6)$
32. $(-40)(2.7)$
33. $-\frac{1}{3} \cdot \frac{5}{9}$
34. $\left(-\frac{5}{6}\right) \cdot \left(-\frac{2}{3}\right)$
35. $1\frac{1}{4}\left(-\frac{2}{3}\right)$
36. $-\frac{1}{5} \cdot 2\frac{1}{2}$

Evaluate.

37. -5^2
38. $(-5)^2$
39. $(-100)^2$
40. $(-300)^2$
41. $(-0.5)^2$
42. $(-0.4)^2$
43. $(-0.1)^3$
44. $(-0.2)^3$
45. $\left(-\frac{3}{4}\right)^2$
46. $\left(-\frac{1}{5}\right)^3$
47. $(-1)^3$
48. $(-4)^4$
49. $(-0.308)^2$
50. $(-7.96)^2$

Multiply.

51. $(9)(12)(-2)$
52. $(2)(-3)(-200)$
53. $(5)(-2)(-1)(3)(-2)$
54. $(-5)(-2)(-1)(3)(-2)$
55. $(-5)(-3)(0)$
56. $(-7)(0)(-10)$
57. $10 \cdot \left(-\frac{1}{2}\right) \cdot (-1)$
58. $\left(-\frac{1}{2}\right)(-4)\left(-\frac{1}{2}\right)$
59. $\frac{4}{5} \cdot \left(-\frac{8}{9}\right) \cdot \frac{1}{3}$
60. $-\frac{3}{4} \cdot \frac{1}{2} \cdot \left(-\frac{5}{7}\right)$
61. $(-2.64)(0.03)(-1.85)$
62. $(5.24)(-0.18)(-2.4)$

Simplify.

63. $(-3)^2 + (-4)$

64. $10^2 - (-5)$

65. $-7 + 3(-3) - 10$

66. $5 - 2(-8) - (-2)$

67. $-3(4) + (-6)(-2)$

68. $8 \cdot (-2) + 3 \cdot (-1)$

69. $2(-8) + 3(-4)$

70. $-2 - 5(-7)$

71. $(-0.5)^2 + 1^2$

72. $(-0.3)^2 + 0.3^2$

73. $\frac{3}{5}(-10) - 6$

74. $\frac{1}{5}(-15) + 32$

75. $-5 \cdot (-3 + 1.2)$

76. $(5 - 0.3) \cdot (-11)$

77. $-2(8-12) + 24 + (-3)^2$

78. $5^2 + 4(-6 - 4) + (-9)(-8)$

Complete each table. Express each answer as a signed number.

79.

Input	Output
a. -2	$(-3)(\mathbf{-2}) - 1 =$
b. -1	$(-3)(\mathbf{-1}) - 1 =$
c. 0	$(-3)(\mathbf{0}) - 1 =$
d. $+1$	$(-3)(\mathbf{+1}) - 1 =$
e. $+2$	$(-3)(\mathbf{+2}) - 1 =$

80.

Input	Output
a. -2	$(-5)(\mathbf{-2}) + 1 =$
b. -1	$(-5)(\mathbf{-1}) + 1 =$
c. 0	$(-5)(\mathbf{0}) + 1 =$
d. $+1$	$(-5)(\mathbf{+1}) + 1 =$
e. $+2$	$(-5)(\mathbf{+2}) + 1 =$

Mixed Practice

Solve.

81. Multiply: $805(-6)$

82. Find the product of $-1\frac{1}{2}$ and $-1\frac{1}{3}$.

83. Calculate: $-(0.01)^2$

84. Compute: $\left(-\frac{2}{3}\right)\left(-\frac{4}{5}\right)\left(-\frac{9}{10}\right)$

85. Simplify: $(-4 + 5) - (-3)^2$

86. Evaluate: $\frac{2}{25}(-10)^2 - 6(4 - 7)$

Applications

Solve. Express each answer as a signed number.

87. Tidal gauge measurements show that the sea level at Kodiak Island in Alaska is dropping at a rate of 12 millimeters per year. At this rate, how much will the sea level change in 6 years? (***Source:*** National Oceanic and Atmospheric Administration)

88. A piece of real estate property dropped \$1,475 in value each month. What was the change in value for 3 months?

89. A patient's dosage of medication is decreased 25 milligrams per day for 1 week. What is the change in her medication at the end of the week?

90. During a drought, the water level in a reservoir fell 2 inches per week for 6 straight weeks. What was the change in the water level in the reservoir at the end of this period?

91. The melting point of a substance is the temperature at which it changes from solid to liquid at standard atmospheric pressure. The melting point of mercury is $-40°C$. Find the melting point of krypton if it is 4 times as great as that of mercury. (*Source:* http://EnvironmentalChemistry.com)

92. In the 10 games the Panthers played this season, they won 3 games by 2 points, won 2 games by 1 point, lost 4 games by 1 point, and tied in the final game. In these games, what was the difference between the number of points they scored and the number scored by the opposing teams?

93. Two seconds after release, the elevation of an object is $\frac{1}{2}(-32)(2)^2$ feet with respect to the point of release. What is this elevation?

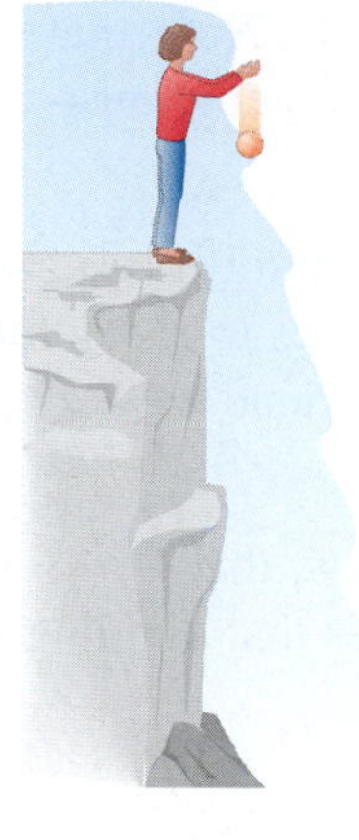

94. Temperatures can be measured in both the Fahrenheit and Celsius scales. To find the Celsius equivalent of the temperature $-4°F$, we need to compute $\frac{5}{9} \cdot (-4 - 32)$ Simplify this expression.

95. The balance in a student's bank account is $1,000. Each month, $150 is withdrawn.

a. What is the change in the account balance after 6 months?

b. What is the balance in the account after 6 months?

96. A man lost 1.8 pounds per week through a diet and exercise program.

a. What was his net change in weight after 15 weeks?

b. If the man weighed 183 pounds at the start of the program, how much did he weigh after 15 weeks?

• *Check your answers on page A-13.*

MINDSTRETCHERS

Groupwork

1. Ask a partner to think of two negative numbers. Then you decide which is larger—the product of these numbers or their sum. Switch roles with your partner and repeat the exercise.

Critical Thinking

2. Fill in the following times table.

×	−1	3	−2
2			
−3			
−2			

Verify that the nine entries sum to 0. Why is this so?

Writing

3. A salesman says that he loses a little money on each item sold but makes it up in volume. Explain if this is possible.

7.5 Dividing Signed Numbers

OBJECTIVES

- To divide signed numbers
- To solve word problems involving the division of signed numbers

Now let's consider an example of division, the last of the four basic operations. Suppose that you and a friend together owe \$8 and you both agree to split the debt evenly. Then each of you will owe \$4.

A debt is considered negative, so this problem requires us to calculate $-8 \div 2$. Recall that every division problem has a related multiplication problem. We see that $-8 \div 2 = -4$ because $-4 \cdot 2 = -8$. Note that when we divide a negative number by a positive number, we get a negative quotient.

Let's look at an example in which we divide one negative number by another negative number. Suppose that your friend owes you \$8 and agrees to repay the debt in installments of \$2 each. How many installments must your friend pay? The answer, of course, is 4.

This problem asks us to calculate $(-8) \div (-2)$. We know that $4 \cdot (-2) = -8$, so it follows that $(-8) \div (-2) = 4$. This example illustrates that dividing two negative numbers gives a positive quotient.

We can use the following rule for dividing signed numbers.

To Divide Two Signed Numbers

- Divide their absolute values.
- If the numbers have the same sign, their quotient is positive; if the numbers have different signs, their quotient is negative.

Another way to think of dividing signed numbers is as follows:

Positive ÷ Positive = Positive　　Positive ÷ Negative = Negative

Negative ÷ Negative = Positive　　Negative ÷ Positive = Negative

EXAMPLE 1

Find the quotient: $-16 \div (-8)$

Solution First, find the absolute values.

$$|-16| = 16 \text{ and } |-8| = 8$$

Next, divide the absolute values.

$$16 \div 8 = 2$$

The numbers have the same sign, so the quotient is positive.

$$-16 \div (-8) = 2$$

PRACTICE 1

Divide: $-24 \div (-2)$

EXAMPLE 2

Simplify: $\frac{-8}{16}$

Solution $|-8| = 8$ and $|16| = 16$ **Find the absolute values.**

$\frac{8}{16} = \frac{1}{2}$ **Express the quotient of the absolute values as a fraction.**

Because the numbers have different signs, the answer is negative.

$$\frac{-8}{16} = -\frac{1}{2}$$

PRACTICE 2

Simplify: $\frac{9}{-15}$

Tip When a fraction has a negative sign in its numerator or denominator, we often rewrite the fraction as a negative number. For instance, we write $\frac{-1}{2}$ as $-\frac{1}{2}$ and $\frac{1}{-2}$ as $-\frac{1}{2}$.

EXAMPLE 3

$7.4 \div (-2) = ?$

Solution $|7.4| = 7.4$ and $|-2| = 2$ **Find the absolute values.**

$7.4 \div 2 = 3.7$ **Divide the absolute values.**

The numbers have different signs, so their quotient is negative.

$$7.4 \div (-2) = -3.7$$

PRACTICE 3

Divide: $-1.5 \div 5$

EXAMPLE 4

Divide: $-8 \div \left(-1\frac{3}{5}\right)$

Solution We divide the absolute values of -8 and $-1\frac{3}{5}$.

$$8 \div 1\frac{3}{5} = 8 \div \frac{8}{5} = \cancel{8} \times \frac{5}{\cancel{8}}$$
$$= 5$$

The quotient of two negative numbers is positive.

$$-8 \div \left(-1\frac{3}{5}\right) = 5$$

PRACTICE 4

Find the quotient: $-\frac{1}{2} \div 3$

We use the order of operations rule to simplify the following expressions.

EXAMPLE 5

Simplify.

a. $-10 + (-8) \div (-2)$

b. $\frac{-7 + (-3)^2}{2}$

PRACTICE 5

Simplify.

a. $6 - (-12) \div (-2)$

b. $\frac{5 - (-1)^2}{-4}$

Solution

a. $-10 + (-8) \div (-2) = -10 + 4$ — Perform division before addition. Divide -8 by -2.

$= -6$ — Add.

b. $\dfrac{-7 + (-3)^2}{2} = \dfrac{-7 + 9}{2}$ — Parentheses are understood to be around the numerator. Square -3.

$= \dfrac{2}{2}$ — Add -7 and 9.

$= 1$

EXAMPLE 6

The federal deficit in 1910 was about \$20 million. Five years later, it was \$60 million. How many times greater was the deficit of 1915 than that of 1910?

Solution The problem asks us to compute $-60 \div (-20)$. The quotient of numbers with the same sign is positive, so the answer is 3. That is, the 1915 deficit was 3 times as great as the deficit of 1910.

PRACTICE 6

A homeowner has her mortgage payment automatically deducted from her checking account. In one year, \$10,500 is deducted from her account. By how much each month did her checking account change, expressed as signed number?

EXAMPLE 7

The table shows the change in the price of a share of a software company's stock each day over a 5-day period.

Day	Change in Price (in cents)
Monday	+32
Tuesday	−18
Wednesday	−21
Thursday	+16
Friday	−54

What was the average daily change in the price of a share of the stocks?

Solution To compute the average change, we add the five changes and divide the sum by 5.

$$\frac{+32 + (-18) + (-21) + (+16) + (-54)}{5}$$

Recall from the order of operations rule that we must find the sum in the numerator before dividing by the denominator.

$$\frac{+48 + (-93)}{5} = \frac{-45}{5} = -9$$

So the average daily change over the 5-day period was down 9 cents per share.

PRACTICE 7

A young girl has a fever. The following chart shows how her temperature changed each day this week.

Monday	Up 2°
Tuesday	Up 1°
Wednesday	Down 1°
Thursday	Up 1°
Friday	Down 3°

What was the average daily change in her temperature?

Mathematically Speaking

Fill in each blank with the most appropriate term or phrase from the given list.

addition	positive	negative
unequal	equal	multiplication

1. The quotient of two numbers with the same signs is _______.
2. The quotient of two numbers with different signs is _______.
3. The fractions $\frac{-2}{3}$, $\frac{2}{-3}$ and $-\frac{2}{3}$, are _______ in value.
4. Every division problem has a related _______ problem.

Find the quotient. Simplify.

5. $-20 \div (-4)$
6. $-7 \div (-1)$
7. $0 \div (-5)$
8. $0 \div 3$
9. $10 \div (-2)$
10. $-9 \div 3$
11. $16 \div (-8)$
12. $-12 \div 4$
13. $-250 \div (-10)$
14. $-300 \div (-3)$
15. $-200 \div 8$
16. $-20 \div 10$
17. $-35 \div (-5)$
18. $8 \div (-4)$
19. $6 \div (-3)$
20. $-8 \div 2$
21. $-17 \div (-1)$
22. $20 \div (-2)$
23. $-72 \div (-12)$
24. $-440 \div (-10)$
25. $-2.4 \div 8$
26. $-0.26 \div 2$
27. $-4 \div 0.2$
28. $9 \div (-0.6)$
29. $-4.8 \div (-0.3)$
30. $-2.6 \div (-0.2)$
31. $\left(-\frac{2}{3}\right) \div \frac{4}{5}$
32. $\left(-\frac{5}{6}\right) \div \left(-\frac{5}{6}\right)$
33. $7 \div \left(-\frac{1}{3}\right)$
34. $-7 \div \left(-\frac{1}{3}\right)$
35. $-40 \div 2\frac{1}{2}$
36. $2\frac{1}{2} \div (-40)$
37. $(-15.1214) \div (-2.45)$
38. $-0.749 \div 0.214$
39. $-12.25 \div 3.5$
40. $50.8369 \div (-7.13)$

Simplify.

41. $\frac{-1}{5}$
42. $\frac{-1}{-5}$
43. $\frac{-11}{-11}$
44. $\frac{-3}{-11}$
45. $\frac{4}{-10}$
46. $\frac{5}{-10}$
47. $\frac{-11}{-2}$
48. $\frac{-2}{-11}$
49. $\frac{-17}{-4}$
50. $\frac{-26}{-5}$
51. $\frac{-9}{-12}$
52. $\frac{-14}{-16}$
53. $-8 \div (-2)(-2)$
54. $-3(-4) \div (-2)$
55. $(3 - 7)^2 \div (-4)$
56. $(4 - 6)^2 \div (1 - 5)^2$

57. $\dfrac{2^2 - (-6)}{2}$

58. $\dfrac{3^2 \cdot (-4)}{-1}$

59. $\dfrac{2^2 + (-6)}{-2}$

60. $\dfrac{3^2 \cdot (-4)^2}{-1}$

61. $\left(\dfrac{-8}{-2}\right)\left(\dfrac{8}{-2}\right)$

62. $\dfrac{-10}{2} \cdot \dfrac{-6}{5}$

63. $\dfrac{-9 - (-3)}{2}$

64. $\dfrac{-5 + (-7)}{2}$

65. $\dfrac{3(-0.2)^2}{-2}$

66. $\dfrac{(-16)(1.5)^2}{-1}$

67. $(-15) + (-3)^2 - 2 \cdot (-1)$

68. $24 \div (-8) + (-5) \cdot 6$

69. $(-13 - 3) \div (-2 - 6)$

70. $-12 \cdot 2 + (-2)^2 - (-5) \cdot 3$

71. $-49 \div (-7)^2 - 4 \cdot (-3)$

72. $10 + (-8) \div (-4)(-5)$

Insert parentheses, if needed, to make the expression on the left equal to the number on the right.

73. $9 \div 1 - 4 = -3$

74. $-2 + 8(-12) = -72$

75. $6 \div 3 - 1 - 4 = -1$

76. $-10 + 8 \div 2 - 5 \cdot 3 = -16$

77. $8 - 10 \cdot 2 - (-5) + 13 \div 4 = -6$

78. $12 \div (-5) + 1 + (-6)(-1) + 2 = -9$

Mixed Practice

Solve.

79. Divide: $-\dfrac{4}{5} \div \dfrac{2}{3}$

80. Divide -0.75 by -0.5.

81. Simplify: $\dfrac{19}{-6}$

82. Find the quotient: $-0.06 \div (-0.3)$

83. Evaluate: $(5 - 3)^2 \div (1 - 4)^3$

84. Simplify: $-4 - 9 \div 3(-5) + 2$

Applications

Solve. Express each answer as a signed number.

85. The population of a certain city decreased by 60,989 in 10 years. Find the average annual change in population.

86. A new computer purchased for $1,800 will have a salvage value of $400. If its value decreases $280 per year, in how many years will it reach its salvage value?

87. In the decade between 1990 and 2000, the population of Washington, D.C. dropped from 607 thousand to 572 thousand, rounded to the nearest thousand. During this decade, what was the change in population per year? (***Source:*** U.S Bureau of the Census.)

88. A football running back lost 4 yards on each of several plays. His total yardage lost was 24 yards. How many plays were involved?

89. The altitude of a plane decreased from 25,000 feet to 19,000 feet in 6 minutes. At what rate did the altitude of the plane change?

90. Over a 5-year period, the height of a cliff eroded by 3.5 feet. By how many feet did it change per year?

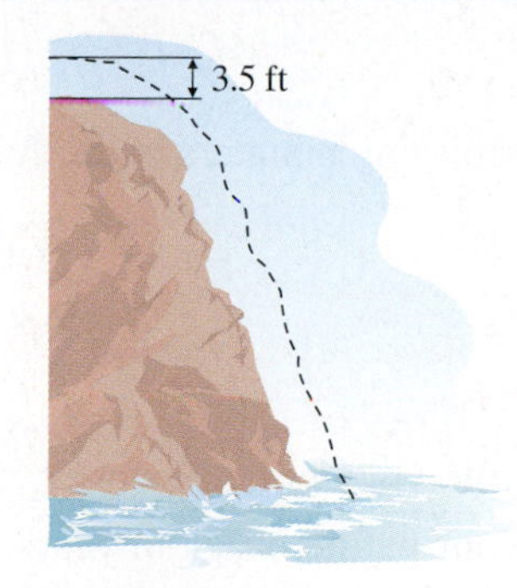

91. A meteorologist is expected to accurately predict the average high temperature for the next 5 days. This week the high temperatures were 3°, 0°, −8°, −11°, and 1°. If her prediction for these days was −3°, was it correct?

92. In a statistics course, a student needs to carry out the following computation.

$$\frac{(-0.5)^2 + (0.3)^2 + (0.2)^2}{3}$$

Find this number, rounded to the nearest hundredth.

93. The bar graph shows the daily high temperature in degrees Fahrenheit in Fairbanks, Alaska, for the first week of January in a recent year. (*Source:* Alaska Climate Research Center, 2006)

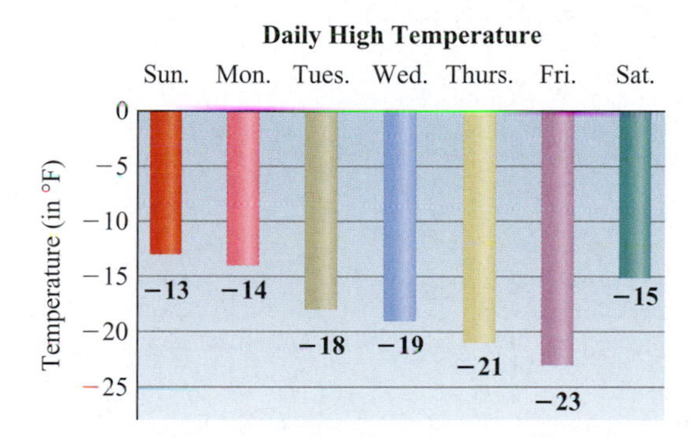

a. Which day of the first week was the coldest? The warmest?

b. To the nearest degree, what was the average daily high temperature that week?

94. The table shows the net income for Zale Corporation for fiscal year 2006.

Quarter	Net Income (in millions)
1	−$23.7
2	+$87.8
3	+$16.8
4	−$26.4

(*Source:* Zale Corporation)

a. In which quarter(s) did the company show a loss?

b. What was the average quarterly net income?

• *Check your answers on page A-13.*

MINDSTRETCHERS

Patterns

1. Find the missing numbers in the following sequence.

$$+1296, +648, -216, -108, +36, +18, -6, \_\_\_\_, \_\_\_\_, \_\_\_\_$$

Groupwork

2. Do the following with a partner.

- Take your partner's age in years.
- Square it.
- Subtract 9.
- Divide the result by 3 less than your partner's age.
- Subtract 53.
- Add your partner's age.
- Divide by 2.
- Add 5^2.

Verify that you wind up where you started—with your partner's age.

Writing

3. Explain the difference between the *opposite* of a number and the *reciprocal* of a number.

KEY CONCEPTS AND SKILLS

CONCEPT SKILL

Concept/Skill	Description	Example
[7.1] **Positive number**	A number greater than 0.	$5, \frac{1}{3}, 2.7$
[7.1] **Negative number**	A number less than 0.	$-5, -\frac{1}{3}, -2.7$
[7.1] **Signed number**	A number with a sign that is either positive or negative.	$5, -5, \frac{1}{3}, -\frac{1}{3}, 2.7, -2.7$
[7.1] **Integers**	The numbers … , −4, −3, −2, −1, 0, 1, 2, 3, 4, … continuing indefinitely in both directions.	$+5, -5$
[7.1] **Opposites**	Two numbers that are the same distance from 0 on the number line but on opposite sides of 0.	$+2$ and -2
[7.1] **Absolute value**	The distance of a number from 0 on the number line, represented by the symbol | |.	2 units, 2 units; number line −4 −3 −2 −1 0 1 2 3 4 $\lvert -2 \rvert = 2, \quad \lvert +2 \rvert = 2$
[7.1] **To compare signed numbers**	• Locate the points being compared on the number line. A number to the right is larger than a number to the left.	number line −4 −3 −2 −1 0 1 2 3 4 $2 > -1$
[7.2] **To add two signed numbers**	• If the numbers have the same sign, add the absolute values and keep the sign. • If the numbers have different signs, subtract the smaller absolute value from the larger and take the sign of the number with the larger absolute value.	$-0.5 + (-1.7) = -2.2$ because $\lvert -0.5 \rvert + \lvert -1.7 \rvert = 0.5 + 1.7 = 2.2$ $3\frac{1}{2} + (-9) = -5\frac{1}{2}$ because $\lvert -9 \rvert > \left\lvert +3\frac{1}{2} \right\rvert$ and $9 - 3\frac{1}{2} = 5\frac{1}{2}$
[7.3] **To subtract two signed numbers**	• Change the operation of subtraction to addition, and change the number being subtracted to its opposite. • Follow the rule for adding signed numbers.	$-2 - (-5) = -2 + 5 = +3$, or 3
[7.4] **To multiply two signed numbers**	• Multiply their absolute values. • If the numbers have the same sign, their product is positive; if the numbers have different signs, their product is negative.	$(-8)\left(-\frac{1}{2}\right) = +4$, or 4 $-0.2 \times 4 = -0.8$
[7.5] **To divide two signed numbers**	• Divide their absolute values. • If the numbers have the same sign, their quotient is positive; if the numbers have different signs, their quotient is negative.	$\frac{-8}{-4} = +2$, or 2 $18 \div (-2) = -9$

Chapter 7 Review Exercises

[7.1] ***Mark the corresponding point for each number on the number line.***

1. -3

2. 1.5

$-4 \quad -3 \quad -2 \quad -1 \quad 0 \quad 1 \quad 2 \quad 3 \quad 4$

Find the opposite signed number.

3. $+6$

4. -4

5. $-7\frac{1}{2}$

6. 10.1

Find the absolute value.

7. $|10|$

8. $|+2.5|$

9. $\left|-1\frac{1}{5}\right|$

10. $|-7|$

Circle the larger number.

11. -11 and -15

12. -15 and 10

13. 9 and $-5\frac{1}{3}$

14. -6.75 and -2

Arrange the numbers in each group from smallest to largest.

15. $-8, 8, -3.5$

16. $9, -6, -9.7$

17. $-2\frac{1}{2}, 0, -2.9$

18. $-4, -1\frac{1}{4}, 0$

Express each quantity as a signed number.

19. Ten feet above sea level

20. A loss of $350 on an investment

[7.2] ***Find the sum.***

21. $-10 + (-10)$

22. $8 + (-10)$

23. $-5\frac{1}{2} + 12$

24. $-\frac{1}{4} + \left(-\frac{3}{4}\right)$

25. $0.9 + (-5)$

26. $-1.2 + (-0.8)$

27. $-8 + 5 + (-4)$

28. $12 + (-12) + \left(-\frac{1}{4}\right)$

[7.3] ***Find the difference.***

29. $-10 - (-10)$

30. $14 - (-14)$

31. $5 - 15$

32. $-2 - 9$

33. $2.5 - (-0.5)$

34. $-\frac{1}{8} - 4$

[7.4] ***Find the product.***

35. $-10(-10)$

36. $-15 \cdot 3$

37. $\frac{-2}{-3}\left(\frac{+10}{-11}\right)$

38. $3.5 \times (-2.1)$

39. $4(-3)(-6)$

40. $-2(-3)(-5)$

Evaluate.

41. $\left(\frac{1}{4}\right)^2$

42. $(-0.7)^2$

43. $(-6)^2$

44. -9^2

[7.5] *Find the quotient.*

45. $-35 \div (-7)$

46. $-80 \div 8$

47. $20 \div (-4)$

48. $-\frac{1}{8} \div (-4)$

49. $15 \div (-0.3)$

50. $\frac{-10}{-5}$

[7.2–7.5] *Simplify.*

51. $-8 - (-3) + 20$

52. $12 \cdot (-3)^2 - (-6)$

53. $(-7 + 3) \cdot (-5)^2$

54. $(20 - 30) \div (-10)$

55. $\frac{(-9.1)(-0.6)}{2}$

56. $\frac{-8 - 5.1}{5}$

57. $10^2 + \frac{-8 - 2}{2} + (-3)^2$

58. $\frac{10}{2} - (5 - 9)^2(-1)$

Mixed Applications

Solve.

59. The Chou dynasty ruled China between 1027 B.C. and 256 B.C. The philosopher Confucius was born in about 551 B.C. and died in about 479 B.C. Was the Chou dynasty in power throughout Confucius's lifetime? (***Source:*** *Asian History on File*)

60. Buddha was born in 563 B.C. and died in 483 B.C. Was he alive in 500 B.C.? (***Source:*** *Compton's Encyclopedia*)

61. A customer has his monthly car payment automatically deducted from his checking account. If his monthly car payment is $235, express the annual change in balance in his checking account as a signed number.

62. An administrative assistant had a balance of $1,498.56 on her credit card. What is her new balance after charges totaling $378.12, a payment of $250, and a finance charge of $23.15 are included?

63. A meteorologist reports that today's low temperature was −5°F and that the normal low for the day is 23°F. How far below the normal low temperature is the low temperature today?

64. An instructor deducts 4 points for each incorrect answer on an exam. If a student received 92 out of a possible 120 points on the exam, how many questions did he answer incorrectly?

65. An investor bought 100 shares of a media company's stock for $3,500. The value of the stock was $3,380 after one month. Express the investor's change in value per share as a signed number.

66. Two of the most influential math books in history were *The Elements*, which Euclid wrote in 323 B.C., and *The Principia*, which Isaac Newton wrote in A.D 1687. To the nearest ten years, how many years apart were these books written? (***Source:*** *Notable Mathematicians from Ancient Times to the Present*)

67. After answering a question incorrectly, a contestant on *Jeopardy* had \$1,000 deducted from his score of \$600. What was his new score?

68. On a diver's first day of scuba diving, he dove to a depth of 30 feet below the surface of the sea. If on the next day he dove to a depth 3 times as great, how deep did he dive on that day?

69. Golf scores are commonly expressed as over or under *par*—the number of expected strokes on each hole. To *birdie* a hole is to take 1 less stroke than par, to *eagle* is to take 2 fewer strokes than par, and to *bogie* is to take 1 more stroke than par. With 3 birdies, 2 eagles, 1 par, and 2 bogies, how far over or under par is the golfer altogether?

70. Physicists have shown that, if an object is thrown upward at a speed of 100 feet per second, its elevation after 5 seconds will be

$$-16 \times 5^2 + 100 \times 5$$

feet relative to the point at which the object was thrown. How far above or below that point will the object be at that time?

71. The following bar graph shows the record low temperatures for selected states.

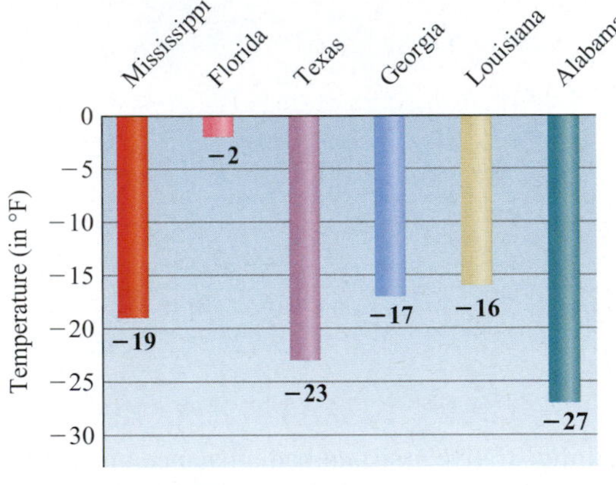

(*Source:* National Climatic Data Center)

a. Which state had the coldest record low temperature?

b. How much higher than the record low temperature in Louisiana was the record low temperature in Florida?

72. The net income for each quarter of fiscal year 2006 for TiVo, Inc. is shown in the chart.

Quarter	Net Income (in millions)
1	\$0.9
2	\$0.2
3	−\$14.2
4	−\$19.5

(*Source:* TiVo, Inc.)

a. What was TiVo's net income for the year?

b. What was the average quarterly net income?

• *Check your answers on page A-13.*

Chapter 7 **POSTTEST**

FOR EXTRA HELP

Test solutions are found on the enclosed CD.

To see if you have mastered the topics in this chapter, take this test.

1. Which is smaller, -10 or -4?

2. A number has an absolute value of $\frac{1}{2}$ and is negative. What is the number?

Evaluate.

3. $-8 + 8$

4. $4.5 + (-5)$

5. $42 - 91$

6. $-12 - (-12)$

7. -23×9

8. -0.5×0.2

9. -12^2

10. $\left(-\frac{1}{4}\right)^2$

11. $-64 \div 16$

12. $-1.8 \div (-0.9)$

13. $-4 + 6 + (-7) + 9$

14. $15 - (-7) + (-1)$

15. $-8 - 4^2 \cdot (-3)$

16. $(2 - 8)^2 \div (-2)$

Solve.

17. The temperature at noon was 74°F. A cold front moved into the region causing the temperature to drop an average of 4°F per hour over the next 4 hours. What was the temperature at 4:00 P.M.?

18. A copier that was purchased new for \$6,900 is worth \$4,700 4 years later. If the copier changes in value by the same amount each year, what is the rate at which its value changes, expressed as a signed number?

19. The Baltimore Orioles were 7 games behind the Boston Red Sox and the Red Sox were $8\frac{1}{2}$ games behind the New York Yankees. What is the standing of the Orioles with respect to the Yankees? Express the answer as a signed number.

20. Chlorine boils at -34.6°C and melts at -100.98°C. How much higher is the boiling point than the melting point? (***Source:*** *Handbook of Chemistry & Physics*)

• *Check your answers on page A-13.*

Cumulative Review Exercises

To help you review, solve the following:

1. Round 2,891 to the nearest thousand.

2. Multiply: $(4)\left(2\frac{1}{2}\right)$

3. Add: $8 + 2.1 + 3.9$

4. Solve for x: $x + 7.5 = 9$

5. What percent of 2.5 is 0.5?

6. Solve for n: $\frac{1.4}{7} = \frac{13}{n}$

7. $(5 - 9)^2 \div (-6 + 2)$

Solve.

8. A patient is to be given a total of 480 milligrams of medication per day. If the medication is to be administered every 4 hours, how much medication should be administered with each dose?

9. Three of the coldest temperature readings ever recorded on Earth were −89°C, −62°C, and −63°C. Of these three temperatures, which was the coldest? (**Source:** *Time Almanac 2006*)

10. When mortgage rates dropped, the number of housing starts rose from 4,000 to 5,000. What percent increase is this?

• *Check your answers on page A-13.*

CHAPTER 8

Basic Statistics

8.1 Introduction to Basic Statistics

8.2 Tables and Graphs

Statistics and the Law

Lawyers make frequent use of statistical evidence to win their cases.

Statistics on the distribution of blood and hair types in the general population are commonly used as evidence in physical assault and robbery trials. Where plaintiffs claim that they are suffering from exposure to a toxic agent, their lawyers often present statistical evidence about the general incidence of their illness. And cases of race and sex discrimination typically focus on such statistics as the proportion of people who are admitted or hired or the average length of time that employees have spent in positions before being promoted.

This use of statistics in the U.S. legal system goes back to a landmark nineteenth-century trial, wherein the claim was made that the signature of Sylvia Howland on her will was forged. The turning point of the case was the testimony of an expert witness—a Harvard mathematician—who developed a system of statistically analyzing the degree of similarity among 42 signatures of the deceased. On the basis of these statistics, he testified that the signature on the will was unreasonably similar to another of Ms. Howland's from which it had probably been traced. (*Source:* Jack B. Weinstein, "Litigation and Statistics," *Statistical Science*, 3, (3), 1988, pp. 286–297)

Chapter 8 PRETEST

To see if you have already mastered the topics in this chapter, take this test.

1. Find the range: 10, 2, 2, 5, 11

2. The following table shows the number of days each month of precipitation for a recent year in Seattle. (*Source:* http://www.beautifulseattle.com)

Jan	Feb	Mar	Apr	May	June
21	6	13	18	16	18
July	**Aug**	**Sept**	**Oct**	**Nov**	**Dec**
11	5	8	24	19	20

What was the median number of days each month that had precipitation?

3. A local fire department tracks its emergency response times. Last month, the response times were: 12 minutes, 7 minutes, 20 minutes, 10 minutes, 6 minutes, 15 minutes, 8 minutes, 12 minutes, and 9 minutes. What was the mean response time?

4. What is the mode of the number of days in a month? (*Reminder:* February has 28 or 29 days; April, June, September, and November have 30 days; and January, March, May, July, August, October, and December have 31 days.)

5. Late in the spring term, your grades were: Spanish I (3 credits)—A; Music (2 credits)—A; Social Science (4 credits)—C; and Physical Education (1 credit)—B. The grades are assigned the following points: A = 4, B = 3, C = 2, D = 1, and F = 0. Calculate your GPA.

In each case use the given table or graph to answer the question.

6. The following mortality table gives estimates for the life spans of individuals (in years), taking into account such factors as year of birth and gender. (*Source:* U.S. National Center for Health Statistics)

Year of Birth	1920	1930	1940	1950	1960	1970	1980	1990	2000	2010 (projected)
Male	53.6	58.1	60.8	65.6	66.6	67.1	70.0	71.8	74.3	75.6
Female	54.6	61.6	65.2	71.1	73.1	74.7	77.5	78.8	79.7	81.4

A female born in 1950 is expected to live how much longer than a male born in 1950?

7. The following graph shows the percent of American cancer patients surviving 5 or more years, during various periods of time.

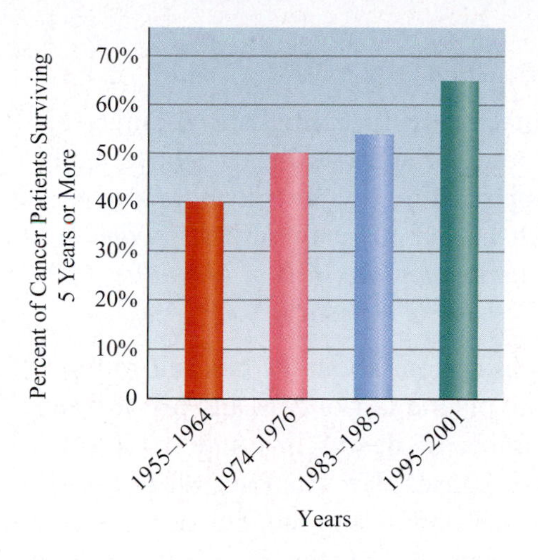

(*Source:* American Cancer Society)

In which period(s) of time shown did more than half the cancer patients survive 5 or more years?

8. The following pictograph shows the daily circulation of some major morning newspapers across the United States for a recent year.

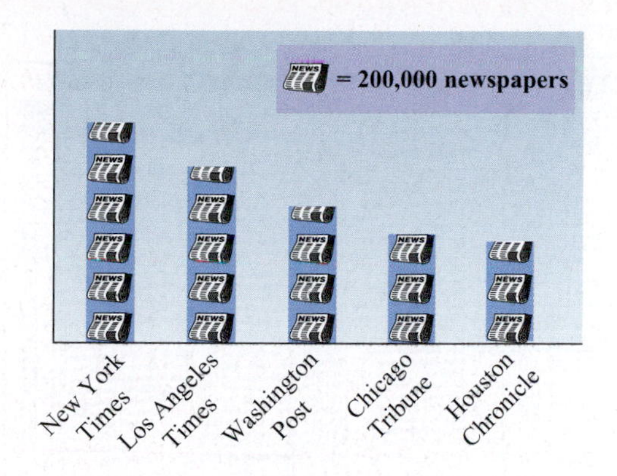

What was the approximate daily circulation of the Los Angeles Times? (*Source: 2005 Editor and Publisher Yearbook*)

9. The first automated teller machine (ATM) in the United States was installed in 1971 at the Citizens & Southern National Bank in Atlanta. Overall, the number of ATMs has grown rapidly. The following graph shows the number in recent years.

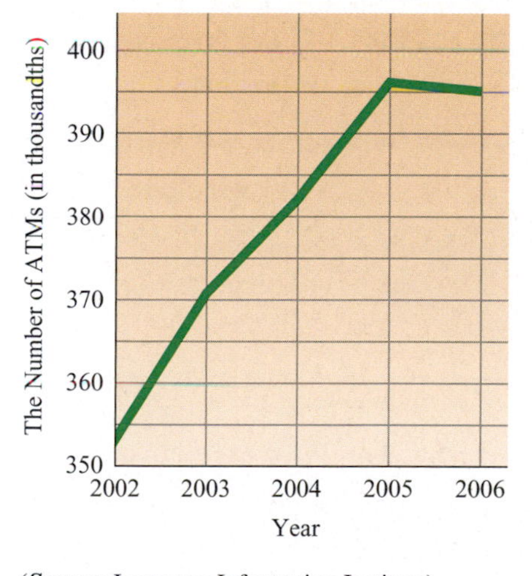

(*Source:* Insurance Information Institute)

In the year 2006, approximately how many ATMs were there in the United States?

10. The graph shows the breakdown of days of school missed in the past 12 months due to illness or injury for U.S. children 5–17 years of age.

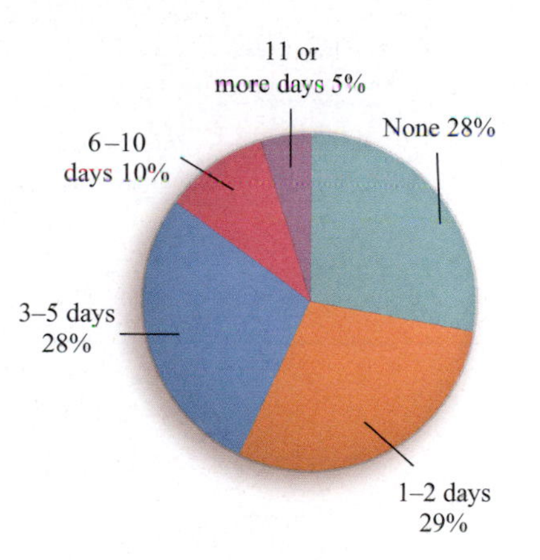

(*Source:* Centers for Disease Control, National Center for Health Statistics, *Summary Health Statistics for U.S. Children: National Health Survey, 2005*)

What percent of children missed 3 or more days of school?

• *Check your answers on page A-14.*

CULTURAL NOTE

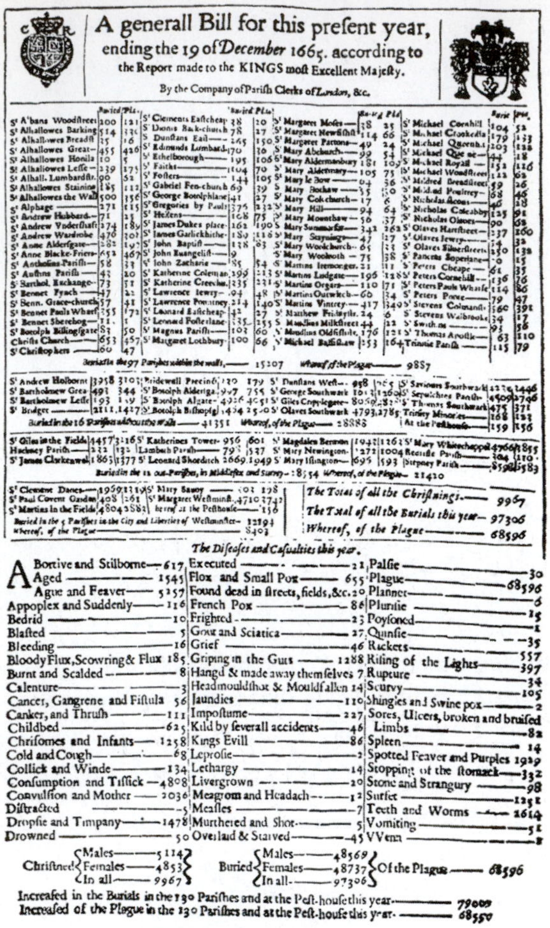

A generall Bill for this preſent year, ending the 19 of *December* 1665. according to the Report made to the KINGS moſt Excellent Majeſty.

By the Company of Pariſh Clerks of *London*, &c.

	Buried	Pla.		Buried	Pla.		Buried	Pla.		Buried	Pla.
St Albans Woodſtreet	200	121	St Clements Eaſtcheap	38	20	St Margaret Moſes	38	25	St Michael Cornhill	104	52
St Alhallowes Barking	514	330	St Dionis Back-church	78	27	St Margaret Newfiſhſt	114	66	St Michael Crookedla	179	133
St Alhallowes Breadſt	35	16	St Dunſtans Eaſt	265	150	St Margaret Pattons	49	24	St Michael Queenhi.	203	122
St Alhallowes Great	455	426	St Edmunds Lumbard.	70	36	St Mary Abchurch	99	54	St Michael Quer[illegible]	44	18
St Alhallowes Honila	10	5	St Ethelborough	195	106	St Mary Aldermanbury	181	109	St Michael Royall	152	116
St Alhallowes Leſſe	239	175	St Faiths	104	70	St Mary Aldermary	105	75	St Michael Woodſtreet	122	62
St Alhall. Lumbardſtr.	90	52	St Foſters	144	105	St Mary le Bow	64	36	St Mildred Breadſtreet	59	26
St Alhallowes Staining	185	112	St Gabriel Fen-church	69	39	St Mary Bothaw	55	30	St Mildred Poultrey	68	46
St Alhallowes the Wall	500	356	St George Botolphlane	41	27	St Mary Colechurch	17	6	St Nicholas Acons	46	28
St Alphage	271	115	St Gregories by Pauls	376	232	St Mary Hill	94	64	St Nicholas Colcabby	125	91
St Andrew Hubbard	71	25	St Hellens	108	75	St Mary Mounthaw	56	37	St Nicholas Olaves	90	62
St Andrew Vnderſhaft	274	189	St James Dukes place	262	190	St Mary Summerſet	342	262	St Olaves Hartſtreet	237	160
St Andrew Wardrobe	476	308	St James Garlickhithe	189	118	St Mary Stayning	47	27	St Olaves Jewry	54	32
St Anne Alderſgate	282	197	St John Baptiſt	138	83	St Mary Woolchurch	65	33	St Olaves Silverſtreet	250	132
St Anne Blacke-Friers	652	467	St John Evangeliſt	9		St Mary Woolnoth	75	38	St Pancras Soperlane	30	15
St Antholins Pariſh	58	33	St John Zacharie	85	54	St Martins Iremonger.	21	11	St Peters Cheape	61	35
St Auſtins Pariſh	43	20	St Katherine Coleman	299	213	St Martins Ludgate	196	128	St Peters Cornehill	136	76
St Barthol. Exchange	73	51	St Katherine Creechu.	335	231	St Martins Orgars	110	71	St Peters Pauls Wharfe	114	86
St Bennet Fynch	47	22	St Lawrence Jewry	94	48	St Martins Outwich	60	34	St Peters Poore	79	47
St Benn. Grace-church	57	41	St Lawrence Pountney	214	140	St Martins Vintrey	417	349	St Stevens Colmanſt.	560	391
St Bennet Pauls Wharf	355	172	St Leonard Eaſtcheap	42	27	St Matthew Fridayſtr.	24	6	St Stevens Walbrooke	34	17
St Bennet Sherehog	11	1	St Leonard Foſterlane	335	255	St Maudlins Milkſtreet	44	22	St Swithins	93	56
St Botolph Billingſgate	83	50	St Magnus Pariſh	103	60	St Maudlins Oldfiſhſtr.	176	121	St Thomas Apoſtle	163	110
Chriſts Church	653	457	St Margaret Lothbury	100	66	St Michael Baſſiſhaw	253	164	Trinitie Pariſh	115	79
St Chriſtophers	60	47									

Buried in the 97 Pariſhes within the walls, 15207 *Whereof, of the Plague* 9887

St Andrew Holborne	3958	3103	Bridewell Precinct	230	179	St Dunſtans Weſt	958	665	St Saviours Southwark	4235	3446
St Bartholmew Great	493	344	St Botolph Alderſga.	997	755	St George Southwark	1613	1260	St Sepulchres Pariſh	4509	2746
St Bartholmew Leſſe	193	139	St Botolph Algate	4926	4051	St Giles Cripplegate	8069	4838	St Thomas Southwark	475	371
St Bridget	2111	1427	St Botolph Biſhopſg.	3464	2500	St Olaves Southwark	4793	2785	Trinity Minories	168	123
									At the Peſthouſe	159	156

Buried in the 16 Pariſhes without the walls 41351 *Whereof, of the Plague* 28888

St Giles in the Fields	4457	3216	St Katherines Tower	956	601	St Magdalen Bermon.	1943	1363	St Mary Whitechappel	4766	3855
Hackney Pariſh	232	132	Lambeth Pariſh	798	537	St Mary Newington	1272	1004	Rederiff Pariſh	304	210
St James Clarkenwel	1863	1377	St Leonard Shorditch	2669	1949	St Mary Iſlington	696	593	Stepney Pariſh	8598	6583

Buried in the 12 out-Pariſhes, in Middleſex and Surrey 28554 *Whereof, of the Plague* 21420

St Clement Danes	1969	1319	St Mary Savoy	303	198
St Paul Covent Garden	408	261	St Margaret Weſtminſt.	4710	3742
St Martins in the Fields	4804	2883	whereof at the Peſthouſe		156

Buried in the 5 Pariſhes in the City and Liberties of Weſtminſter 12194
Whereof, of the Plague 8403

The Total of all the Chriſtnings 9967
The Total of all the Burials this year 97306
Whereof, of the Plague 68596

The Diſeaſes and Caſualties this year.

Abortive and Stilborne	617	Executed	21	Palſie	30
Aged	1545	Flox and Small Pox	655	Plague	68596
Ague and Feaver	5257	Found dead in ſtreets, fields, &c.	20	Planner	6
Appoplex and Suddenly	116	French Pox	86	Pluriſie	15
Bedrid	10	Frighted	23	Poyſoned	1
Blaſted	5	Gowt and Sciatica	27	Quinſie	35
Bleeding	16	Grief	46	Rickets	557
Bloody Flux, Scowring & Flux	185	Griping in the Guts	1288	Riſing of the Lights	397
Burnt and Scalded	8	Hangd & made away themſelves	7	Rupture	34
Calenture	3	Headmouldſhot & Mouldfallen	14	Scurvy	105
Cancer, Gangrene and Fiſtula	56	Jaundies	110	Shingles and Swine pox	2
Canker, and Thruſh	111	Impoſtume	227	Sores, Ulcers, broken and bruiſed Limbs	82
Childbed	625	Kild by ſeverall accidents	46	Spleen	14
Chriſomes and Infants	1258	Kings Evill	86	Spotted Feaver and Purples	1929
Cold and Cough	68	Leproſie	2	Stopping of the ſtomack	332
Collick and Winde	134	Lethargy	14	Stone and Strangury	98
Conſumption and Tiſſick	4808	Livergrown	20	Surfet	1251
Convulſion and Mother	2036	Meagrom and Headach	12	Teeth and Worms	2614
Diſtracted	5	Meaſles	7	Vomiting	51
Dropſie and Timpany	1478	Murthered and Shot	5	VVenn	8
Drowned	50	Overlaid & Starved	45		

Chriſtned: Males 5114, Females 4853, In all 9967
Buried: Males 48569, Females 48737, In all 97306. Of the Plague 68596

Increaſed in the Burials in the 130 Pariſhes and at the Peſt-houſe this year 79009
Increaſed of the Plague in the 130 Pariſhes and at the Peſt-houſe this year 68590

A seventeenth-century English clothing salesman named John Graunt had the insight to apply a numerical approach to major social problems. In 1662, he published a book entitled *Natural and Political Observations upon the Bills of Mortality*, and so founded the science of statistics.

Graunt was curious about the periodic outbreaks of the bubonic plague in London, and his book analyzed the number of deaths in London each week due to various causes. He was the first to discover that, at least in London, the number of male births exceeded the number of female births. He also found that there was a higher death rate in urban areas than in rural areas and that more men than women died violent deaths. Graunt summarized large amounts of information to make it understandable and made conjectures about large populations based on small samples. Graunt was also a pioneer in examining expected life span—a statistic that became vital to the insurance companies formed at the end of the seventeenth century.

Sources:

Morris Kline, *Mathematics, a Cultural Approach* (Reading, Mass.: Addison-Wesley Publishing Company, 1962), p. 614.

F. N. David, *Games, Gods and Gambling* (New York: Hafner Publishing Company, 1962).

8.1 Introduction to Basic Statistics

OBJECTIVES

- To find the mean, median, and mode(s) of a set of numbers
- To find the range of a set of numbers
- To interpret these statistics and use them to solve word problems

What Basic Statistics Is and Why It Is Important

Statistics is the branch of mathematics that deals with ways of handling large quantities of information. The goal is to make this information easier to interpret.

With unorganized data, spotting trends and making comparisons is difficult. The study of statistics teaches you how to organize data in various ways in order to make the data more understandable.

One approach is to calculate special numbers, also called statistics, which describe the data. In this section, we consider four statistics: the mean, the median, the mode, and the range.

You have already seen that another way to organize data is to display the information in the form of a table or graph. We will discuss tables and graphs in greater detail in the next section of this chapter.

Many situations lend themselves to the application of statistical techniques. Wherever there are large quantities of information—from sports to business—statistics can help us to find meaning where, at first glance, there seems to be none, and to become more quantitatively literate.

Averages

We begin our introduction to statistics by revisiting the meaning of "average." Previously, we defined the average of a set of numbers to be the sum of the numbers divided by however many numbers are in the set. This statistic, which is more precisely called the *arithmetic mean*, or just the **mean**, is what most people think of as the average. However, it is not the only kind of average used to represent the numbers in a set.

A second average, the **median**, may describe the numbers better than the mean when there is an unusually large or unusually small number in the set to be averaged. The third average, the **mode**, has a special property—unlike the mean and the median, it is always in the set of numbers being averaged.

Mean

Let's look at an example of the mean.

EXAMPLE 1

The area of the United States is about 3,800,000 square miles.

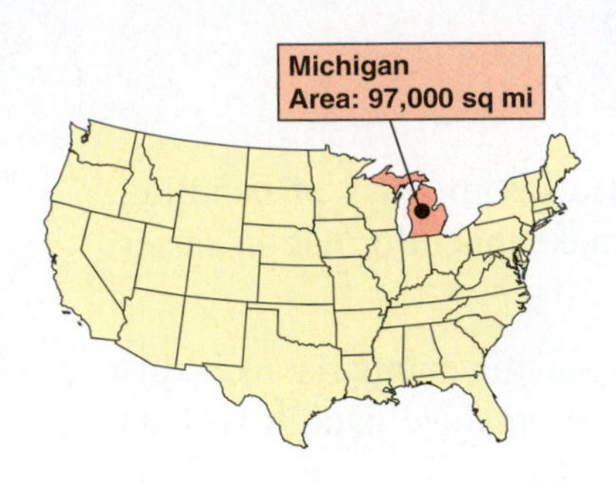

a. Approximately what is the average area of each of the 50 states?

b. Is Michigan above or below average in area?

Solution

a. The mean area of a state is $\frac{3{,}800{,}000}{50}$, or 76,000 square miles.

b. Since 97,000 is greater than 76,000, Michigan is above average in area.

PRACTICE 1

Reggie Jackson hit five home runs in the 1977 World Series, which lasted six games. By contrast, Lou Gehrig hit four home runs in the 1928 World Series, a four-game series. On the average, which baseball player hit fewer home runs per game?

Note that a property of the mean is that it is changed substantially if even a single number in a set of numbers is replaced by one much larger or much smaller. For instance, if five people each make $10 per hour, then their mean hourly wage is $10. However, if the hourly wage of one of these individuals jumps to $500, then the mean wage skyrockets to $108, more than 10 times the previous mean.

Another kind of mean, called the **weighted average**, is used when some numbers in a set count more heavily than others. Weighted average comes into play if you want to compute the average of your test scores in a class and the final exam counts twice as much as any of the other tests. Or if you are computing your grade point average (GPA), and some courses carry more credits than others.

EXAMPLE 2

Last term, a student's grades were as follows.

Course	Credits	Grade	Grade Equivalent
Psychology	4	A	4
English	4	C	2
Art	3	B	3
Physical Education	1	B	3

Compute the student's GPA for the term.

PRACTICE 2

The following table shows the test scores that a classmate earned.

Exam	Score
1	95
2	80
3	80
Final	90

If the final exam is equivalent to two other exams, did the classmate earn an exam average above or below 85?

Solution To calculate the GPA, we first multiply the number of credits each course carries by the numerical grade equivalent received. We then add these products to find the total number of grade points. Finally, we divide this sum by the total number of credits.

Number of credits for the first course

Grade equivalent of the first course

$$\text{GPA} = \frac{4 \cdot 4 + 4 \cdot 2 + 3 \cdot 3 + 1 \cdot 3}{12}$$

Total number of credits

$$= \frac{16 + 8 + 9 + 3}{12} = \frac{36}{12} = 3$$

The GPA of 3 is exactly equivalent to a B.

Median

As we have seen, a very large number can affect the mean of a set of numbers to such an extent that it is not representative of the set. Another kind of average, the **median**, is used when we wish to reduce the impact of an extreme number in the set, for instance, in computing average salary.

Definition

In a set of numbers arranged in numerical order, the **median** of the numbers is the number in the middle. If there are two numbers in the middle, the median is the mean of the two middle numbers.

EXAMPLE 3

Find the median.

a. 6, 8, 2, 1, and 5 **b.** 6, 8, 2, and 5

Solution

a. We arrange the numbers from smallest to largest.

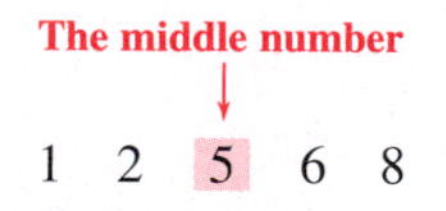

So the median is 5. If we arrange the numbers from largest to smallest, we get the same answer.

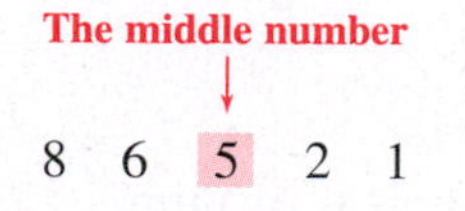

The median is still 5.

PRACTICE 3

Compute the median.

a. 7, 2, 8, 5, 10, 7, 9, 10, 2, 5, 8, and 6

b. 0, 4, 1, 5, 7, 2, 5, 9, and 3

b. We order the numbers from smallest to largest.

2 5 6 8

Since four numbers are on the list and four is even, no single number is in the middle. In this case, the median is the mean of the two middle numbers.

Two middle numbers

2 **5 6** 8

$$\frac{5+6}{2} = 5.5$$

So 5.5 is the median.

EXAMPLE 4

The first 11 justices on the U.S. Supreme Court served the following numbers of years: 5, 8, 1, 20, 5, 9, 0, 13, 4, 15, and 30. (*Source: Time Almanac 2006*)

a. What was the median number of years that they served on the Court?

b. The next appointed justice served on the Court for 3 years. With the addition of this new justice, by how much did the median number of years that the justices served change?

Solution

a. To compute the median of the years of service, let's first arrange these numbers in increasing order:

The middle number

0 1 4 5 5 **8** 9 13 15 20 30

Because 8 is the middle number, the median number of years of service was 8.

b. Let's arrange the 12 numbers in increasing order:

Two middle numbers

0 1 3 4 5 **5 8** 9 13 15 20 30

The numbers 5 and 8 are in the middle. So the median is $\frac{5+8}{2}$, or 6.5. Since the median decreased from 8 to 6.5, it dropped by 1.5, that is, by 1.5 years.

PRACTICE 4

The weekend box office (in millions of dollars) for five leading movies was:

25 16 9 12 26

a. What was the median box office for these movies?

b. A sixth movie also took in $9 million at the box office. How does the median box office of the six movies compare with that of the five movies? (*Source: Variety*, Nov. 14–16, 2005)

Mode

The last type of average that we will consider is the **mode**. Note that a set of numbers can have one mode, more than one mode, or even no mode.

Definition

The **mode** of a set of numbers is the number (or numbers) occurring most frequently in the set.

EXAMPLE 5

Compute the mode(s).

a. 8, 6, 10, 8, 10, 8, 9, and 6

b. 2, 9, 3, 5, 7, 12, 3, 2, 18, 12, 2, and 3

c. 4, 10, 1, 5, 12, and 7

Solution

a. When we count how often each number occurs on this list, we see that 6 occurs twice, 8 occurs three times, 9 once, and 10 twice. Because there are more 8's than any other number, 8 is the mode.

b. Here, 2 occurs three times, 3 occurs three times, 5 once, 7 once, 9 once, 12 twice, and 18 once. So both 2 and 3, occurring most frequently, are modes.

c. No number occurs more than once. There is no mode.

PRACTICE 5

Find the mode(s).

a. 7, 2, 5, 1, 2, 5, and 2

b. 9, 1, 0, 4, 9, 4, 1, 5, 9, and 4

c. 8, 13, 9, and 2

EXAMPLE 6

Of the 46 states in the United States with maximum allowable speed limits for driving, 15 states have a limit of 55 miles per hour, 30 have a limit of 65 miles per hour, and 1 has a limit of 70 miles per hour. What is the mode of these speed limits? (*Source: Time Almanac 2006*)

Solution The speed limit 65 miles per hour occurs more frequently than any other limit (30 times). So 65 miles per hour is the mode of the state speed limits.

PRACTICE 6

Students in a class discussed the number of hours in their college schedules. One student had a weekly schedule consisting of 3 hours of classes, 2 students had 6 hours, 15 had 12 hours, 1 had 13 hours, and 1 had 14 hours. Find the mode of the number of hours in these schedules.

Range

The last statistic that we consider is called the **range**. The range is not an average because it does not represent a typical number in the set. Instead, the range is a measure of the spread of the numbers in the set.

Definition

The **range** of a set of numbers is the difference between the largest and the smallest number in the set.

EXAMPLE 7

Find the range of the numbers 3, 13, 2, 5, 9, and 2.

Solution The largest number in the set is 13, and the smallest is 2. So the range is $13 - 2$, or 11.

PRACTICE 7

What is the range of 8, 10, 3, and 8?

EXAMPLE 8

The monthly profits on 5 months of investments were as follows:

$680 $760 $1,135 $725 −$2,500

For the following 5 months, the monthly profits were:

$330 $600 $840 $620 −$1,070

In which of the two 5-month periods was the spread on monthly profits greater? (*Source:* CashFlow Avenue, January–May and June–October, 2006)

Solution During the first 5-month period, the largest monthly profit was $1,135 and the smallest was −$2,500. The range of monthly profits was $1,135 – (−$2,500), or $3,635. For the second 5-month period, the largest monthly profit was $840 and the smallest was −$1,070. So the range was $840 – (−$1,070), or $1,910. Since the first 5-month period had a larger range, its spread was greater.

PRACTICE 8

Each of the 43 states in the United States with a state minimum hourly wage rate has one of the following rates: $2.65, $4.25, $5.15, $6.15, $6.25, $6.75, $7.05, $7.10, $7.15, and $7.16. What is the range of these rates? (*Source: Time Almanac 2006*)

8.1 Exercises

Mathematically Speaking

Fill in each blank with the most appropriate term or phrase from the given list.

range	weighted	mean
median	statistics	arithmetic
mode	algebra	

1. The branch of mathematics that deals with handling large quantities of information is called _______.

2. The sum of the numbers in a set divided by however many numbers are in the set is called the _______ mean.

3. When some numbers in a set count more heavily than other numbers in the set, the average is said to be _______.

4. The middle number in a set of numbers arranged in numerical order is called the _______.

5. The _______ is the number (or numbers) occurring most frequently in a set of numbers.

6. The _______ is the difference between the largest number and the smallest number in a set of numbers.

Compute the indicated statistics. Round to the nearest tenth, where necessary.

7.

Numbers	Mean	Median	Mode(s)	Range
a. 8, 2, 9, 4, 8				
b. 3, 0, 0, 3, 10				
c. 6.5, 9, 8.5, 6.5, 8.1				
d. $3\frac{1}{2}, 3\frac{3}{4}, 4, 3\frac{1}{2}, 3\frac{1}{4}$				
e. 4, −2, −1, 0, −1				

8.

Numbers	Mean	Median	Mode(s)	Range
a. 5, 3, 5, 5, 3				
b. 8, 0, 7, 5, 0				
c. 2.1, 2.6, 2.4, 2.5, 2.4				
d. $4\frac{1}{2}, 3\frac{3}{4}, 4, 4\frac{1}{2}, 4\frac{1}{4}$				
e. −1, −3, −3, −2, −2				

9. Calculate the mean, rounded to the nearest cent.
\$9,125.88 \$11,724.87 \$12,705 \$11,839.75
\$13,500.79 \$14,703.71

10. Find the mean, rounded to the nearest foot, of the following measurements.
3,725 ft 3,719 ft 3,740 ft 3,726 ft 3,729 ft
3,734 ft 3,725 ft

Applications

Solve and check.

11. Here are a student's grades last term: A in College Skills (2 credits), B in World History (4 credits), C in Music (2 credits), A in Spanish (3 credits), and B in Physical Education (1 credit). Did the student make the Dean's List, which requires a GPA of 3.5? Explain. (*Reminder:* A = 4, B = 3, C = 2, and D = 1.)

12. On a test, 9 students earned 80, 10 students earned 70, and 1 student earned 75. Was the grade of 75 below the class average (mean), exactly average, or above the class average? Explain.

13. A woman leaves $1,000,000 to her 10 heirs. What is the mean amount left to each heir? Can you compute the median amount with the given information? Explain.

14. A woman and four men are riding in an elevator. Two men are taller than the woman, and two are shorter. Who has the median height of the people in the elevator?

15. In the U.S. House of Representatives, 435 members of Congress represent the 50 states. The table below shows the number of representatives of 8 states.

State	Number of Representatives
Maine	2
Indiana	9
Missouri	9
Hawaii	2
Colorado	7
North Carolina	13
Tennessee	9
Nebraska	3

(*Source:* U.S. Bureau of the Census, 2007)

Which of these 8 states has representation that is above the average for all 50 states?

16. The table shows the quarterly revenues (in billions) for Hewlett-Packard in a recent year.

Quarter	Revenue (in billions)
1	$21.5
2	$21.6
3	$20.8
4	$22.9

(*Source:* Hewlett-Packard, 2005)

What was the median quarterly revenue?

17. The table shows the salary of six teachers based on the number of years of service in a local town.

Years of Service	Salary
6	$44,424
10	$57,418
1	$37,925
4	$42,656
13	$58,358
18	$70,852

a. Find the median salary.

b. What is the range?

18. The table shows the federal minimum wage rates for various years.

Year(s)	Minimum Wage
1981–1989	$3.35
1990	$3.80
1991–1995	$4.25
1996	$4.75
1997–2006	$5.15

(*Source:* U.S. Employment Standards Administration)

a. What is the mode of the federal minimum wage for the given years?

b. Find the range

19. The diameters for the eight planets of the solar system, rounded to the nearest 1,000 miles, are as follows:

Planet	Miles (in thousands)
Mercury	3
Venus	8
Earth	8
Mars	4
Jupiter	89
Saturn	75
Uranus	32
Neptune	31

(*Source: Encyclopedia Americana*)

Find each of the following distances, rounded to the nearest 1,000 miles:

a. mean diameter

b. median diameter

c. mode(s) of the diameters

d. range of the diameters

20. Consider the following utility bills for the past 10 months:

Jan	Feb	Mar	Apr	May	Jun	Jul	Aug	Sep	Oct
$90	$80	$90	$70	$100	$110	$140	$140	$100	$90

Find each of the following:

a. the mean bill

b. the median bill

c. the mode(s) of the bills

d. the range of the bills

Using a calculator, solve each problem, giving (a) the operation(s) carried out in the solution, (b) the exact answer, and (c) an estimate of the answer.

21. In the year 1990, when the number of U.S. residents was about 249 million, the U.S. Postal Service delivered some 166 billion pieces of mail. By 2003, when the population had grown to 292 million, the Service delivered approximately 202 billion pieces of mail. On the average, how many more pieces of mail did a resident receive in 2003 than in 1990? (*Source:* U.S. Bureau of the Census; *The Washington Post*, September 25, 2004)

22. In the 20 years from 1985 through 2004, the number of American workers (in thousands) who were involved in strikes was as follows:

324 533 174 118 452 185 392 364 182 322
192 273 339 387 73 394 99 46 129 320

Find the mean number of strikers in a year, to the nearest thousand workers. (*Source:* U.S. Department of Labor, Bureau of Labor Statistics)

Check your answers on page A-14.

MINDSTRETCHERS

Groupwork

1. Working with a partner, construct an example of a set of 10 numbers
   - **a.** whose mean, median, and mode are equal.
   - **b.** whose mean is less than its median.
   - **c.** that has two modes.

Mathematical Reasoning

2. Can the range of a set of numbers be equal to a negative number? Explain.

Investigation

3. In your college library or on the Web, research the legal drinking age in each of 10 countries. Then determine the mean, median, mode, and range of these ages.

8.2 Tables and Graphs

OBJECTIVES

To read and interpret tables

To read and interpret pictographs, bar graphs, histograms, line graphs, and circle graphs

What Tables and Graphs Are and Why They Are Important

We frequently present data in the form of tables or graphs. A **table** is a rectangular display of data. A **graph** is a picture or diagram of the data.

Organizing data in a table or graph makes it easier for a reader to make comparisons, to understand relationships, to spot trends, and to gain a sense of the data.

Graphs generally provide less accurate information than tables because we often have to read a graph by estimating. However, graphs are pictorial, so they make a more lasting impression than a table.

Tables and graphs are used in many different situations. Train schedules, insurance premium charts, and accountants' spreadsheets are common examples of tables. Graphs, such as a bar graph of a changing population, a line graph of fluctuating stock prices, and a circle graph of budget allocations, appear regularly in newspapers, magazines, and reports.

Tables

A table consists of rows and columns. Rows run horizontally, and columns run vertically. The nature of the entries in a row or column is described with labels called *headings*.

To read a table, first identify a particular row and column and then locate the entry at their intersection. Consider the following table that shows the typical heartbeat rates for people of various ages:

Column

Heading →

Person's Age	Beats per Minute
Newborn	135
2	110
6	95
10	87
20	71
40	72
60	74

Row → (Newborn row)

Entry → (60)

As the headings indicate, the entries in the first column are a person's age. The second column gives the number of times per minute that the heart of a person of that age typically beats.

As given in the table, the heart of a 20-year-old beats 71 times a minute. Does this table suggest that a child's heart beats more quickly than the heart of an adult? How many times per minute would you estimate that the heart of an 8-year-old beats?

Example 1 illustrates both reading and drawing conclusions from tables.

EXAMPLE 1

The following is the schedule of math classes this term:

Course	Section	Day/Time	Room	Professor
010	090	Tu W Th 9–10:50	516	Einstein
011	091	M W Th 9–9:50	518	von Neumann
011	611	Tu Th 6–7:15	516	Kovalevski
051	111	M Tu W Th 11–11:50	523	Noether
051	711	M W 7–8:40	518	Hilbert
056	131	M W Th 1–2:50	516	von Neumann
100	121	M Tu W Th 12–12:50	511	Hilbert
104	081	M W Th 8–8:50	511	Einstein
104	111	M W Th 11–11:50	511	Newton
150	091	M Tu W Th 9–9:50	511	Newton
150	511	M W 5:25–7:05	511	Kovalevski
206	131	M Tu W Th 1–1:50	523	Noether
301	141	M Tu W Th 2–2:50	523	Hilbert
302	511	Tu Th 5:25–7:05	520	Gauss

a. In what room does Math 301, Section 141 meet?

b. Is Professor Kovalevski teaching a section of Math 056 this term?

c. Today is Monday, and a student needs to speak to Professor Einstein. When and where is he teaching?

Solution

a. Math 301, Section 141 meets in room 523.

b. No, Professor Kovalevski is not teaching Math 056.

c. Professor Einstein is in room 511 from 8 to 8:50.

PRACTICE 1

A mail-order catalog contains the following chart for determining shipping and handling (S&H) charges:

Amount of Merchandise	Up to $5	$5.01–$15	$15.01–$25*
Charges	$2.95	$3.95	$4.95

*Add $0.10 for each additional $1 of merchandise over $25.

a. What are the S&H charges on $23.45 worth of merchandise?

b. How much must a customer pay in all for merchandise that, excluding S&H charges, sells for $3?

c. How much are the S&H charges on merchandise selling for $30?

Graphs

Now let's discuss displaying data in the form of graphs. We deal with five kinds of graphs: pictographs, bar graphs, histograms, line graphs, and circle graphs.

Pictographs

A **pictograph** is a kind of graph in which images of people, books, coins, and so on are used to represent and to compare quantities. A *key* is given to explain what each image represents.

Pictographs are visually appealing. However, they make it difficult to distinguish between small differences—say, between a half and a third of an image.

EXAMPLE 2

The following graph shows the number of degrees awarded in the United States in a recent year:

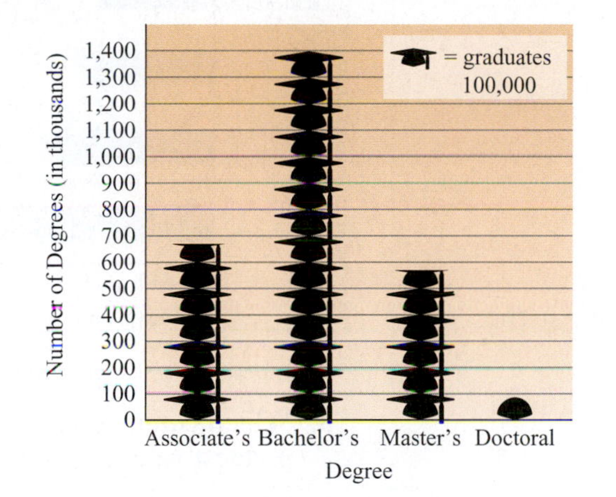

(*Source:* The U.S. Bureau of the Census, 2004)

a. What does the symbol mean in the key at the top of the graph?

b. About how many master's degrees were awarded?

c. About how many more bachelor's degrees than associate's degrees were awarded?

Solution

a. According to the key, the symbol represents 100,000 graduates.

b. The number of master's degrees awarded was about $5\frac{1}{2}(100{,}000)$, or 550,000.

c. About 14(100,000), or 1,400,000, bachelor's degrees were awarded, in contrast to about $6\frac{1}{2}(100{,}000)$, or 650,000, associate's degrees. So there were approximately 750,000 more bachelor's degrees awarded.

PRACTICE 2

The following pictograph shows the number of passengers in a recent year who took off or landed at four busy U.S. airports.

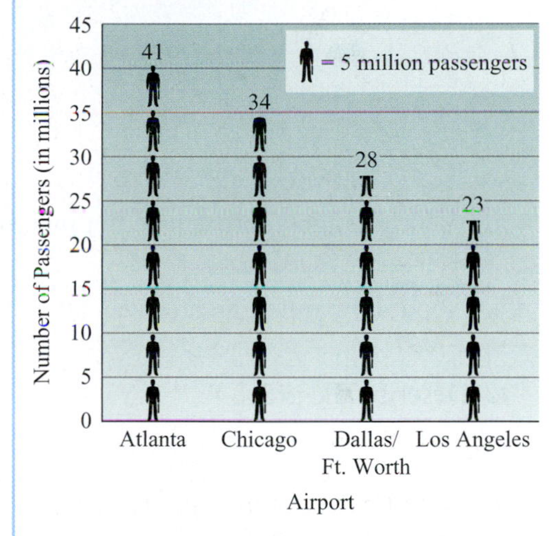

(*Source:* U.S. Bureau of Transportation Statistics, 2004)

a. What does the symbol represent?

b. Which of the four airports was the busiest in terms of passengers?

c. Approximately how many passengers did the Los Angeles airport serve?

Bar Graphs

On a **bar graph**, quantities are represented by thin, parallel rectangles called bars. The length of each bar is proportional to the quantity that it represents.

On some graphs, the bars extend to the right. On others, they extend upward or downward. Sometimes, bar lengths are labeled. Other times, bar lengths are read against an *axis*—a straight line parallel to the bars and similar to a number line.

Bar graphs are especially useful for making comparisons or contrasts among a few quantities, as the following example illustrates.

EXAMPLE 3

The following graph shows the net income of U.S. Airways in recent years.

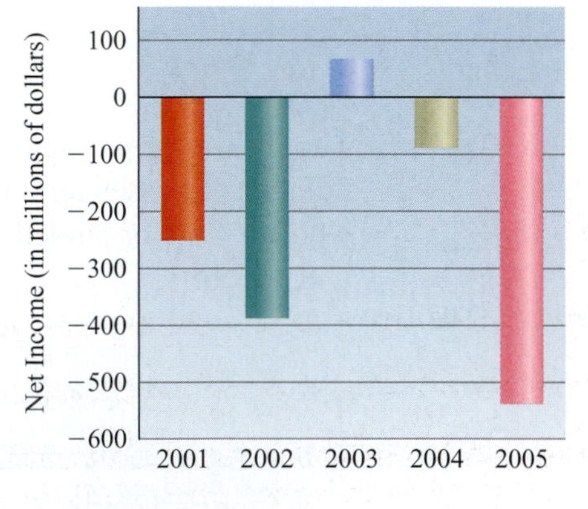

(*Source:* U.S. Airways Group)

a. What was the approximate net income of the company in the year 2005?

b. About how much greater was the net income in 2003 than in 2004?

c. Describe the graph.

Solution

a. In 2005, the net income was about $-\$540$ million, that is, a loss of about \$540 million.

b. In 2003, the net income was approximately \$70 million. The next year, it was about $-\$90$ million. So the net income for 2003 was about \$160 million greater than in 2004.

c. The company operated at a profit in the year 2003. In other years, however, it operated at a loss, especially in 2005.

PRACTICE 3

The following graph shows the value of the top five agricultural commodities in the United States in a recent year.

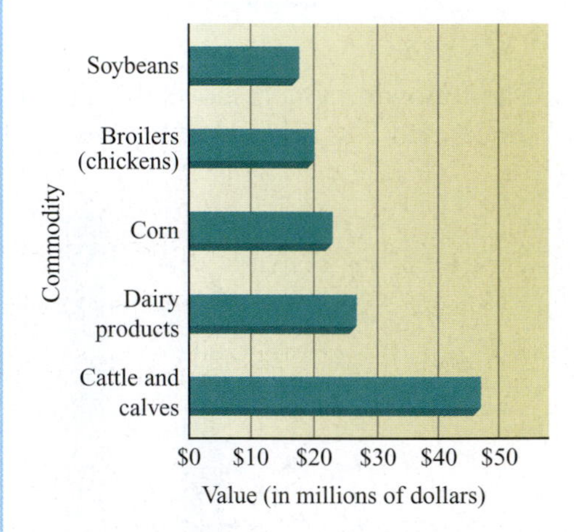

(*Source:* Department of Agriculture)

a. Which commodity had the greatest value?

b. What was the approximate value of dairy products?

c. About how much greater was the value of corn than of broilers?

The next graph is an example of a *double-bar graph*. This kind of graph is used to compare two sets of data in various ways, as the following example illustrates.

EXAMPLE 4

The following graph shows the number of male and female physicians in the United States for various years.

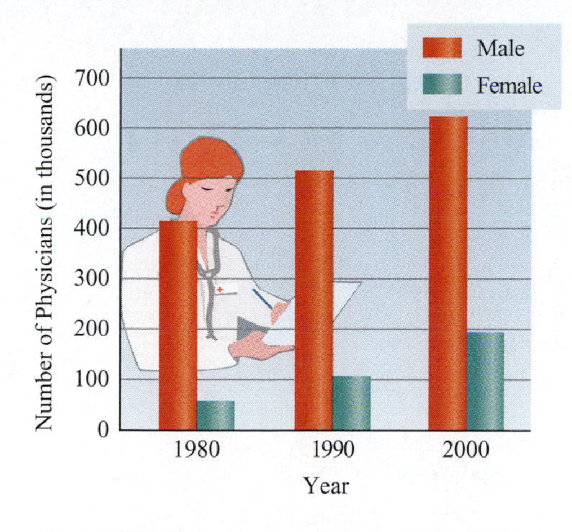

(*Source: Physician Characteristics and Distribution in the U.S.,* as reported in the *Statistical Abstract of the United States: 2006*)

a. In the year 2000, how many more male physicians were there than female physicians, to the nearest hundred thousand?

b. To the nearest hundred thousand, how many physicians were there in 1990?

Solution

a. In 2000, there were about 600,000 male physicians and 200,000 female physicians. So there were approximately 400,000 more male physicians.

b. In 1990, there were about 500,000 + 100,000, or 600,000, physicians.

PRACTICE 4

The following graph shows, among households with Internet connections, the percent of households in a recent year with a particular type of Internet connection as related to family income.

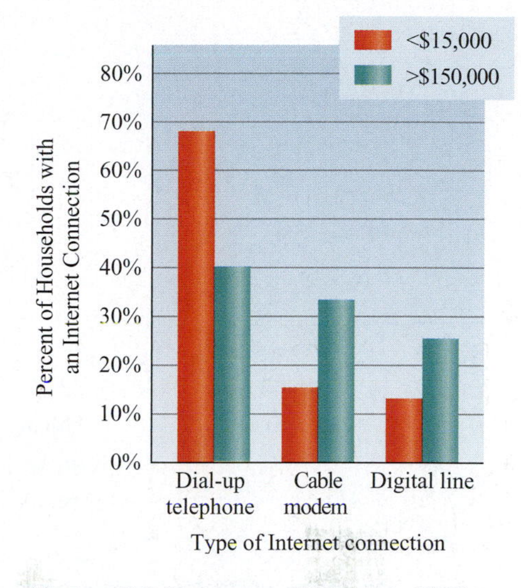

(*Source:* U.S. Department of Commerce, 2003)

a. Among households with Internet connections and with income greater than $150,000, about what fraction had dial-up telephone access?

b. Among households with Internet connections, approximately how much greater was the percent of wealthier households with cable modem than the percent of less wealthy households?

Histograms

Now let's consider another kind of bar graph called a **histogram**. To understand what a histogram is, we consider an example.

Suppose that in a math class, 24 students take a final exam. The results are organized into a *frequency table*, as shown to the right. Note that in the left column, the scores are grouped into *class intervals* all of the same width and that these class intervals are written in increasing order. The right column shows the *class frequencies*, that is, the number of students who earned scores that fall into the class interval on the left. How would you have predicted the sum of the class frequencies in the right column?

Score (Class Interval)	Frequency (Class Frequency)
40–49	2
50–59	2
60–69	6
70–79	4
80–89	6
90–99	4

A histogram is a graph of a frequency table. In a histogram, adjacent bars touch. For each bar, the width represents a class interval, and the height stands for the corresponding class frequency. Consider the following histogram, which corresponds to the frequency table shown on the previous page.

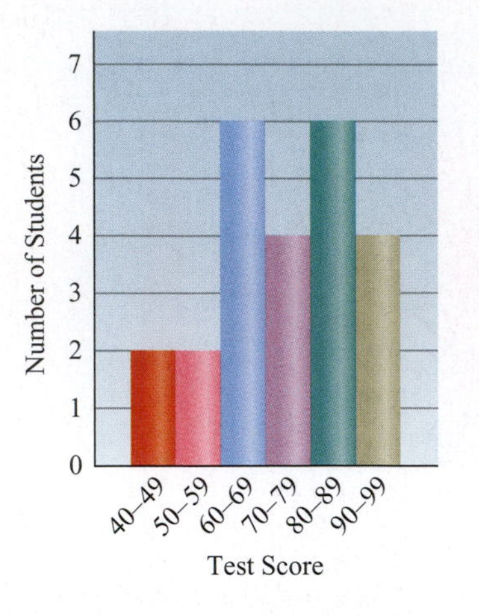

Note that according to the histogram, 4 students scored between 70 and 79, as given in the frequency table. Can you explain how the histogram shows that 10 students scored 80 or above?

EXAMPLE 5

The following graph shows the number of earthquakes worldwide in a recent year with magnitude 1.0 or greater:

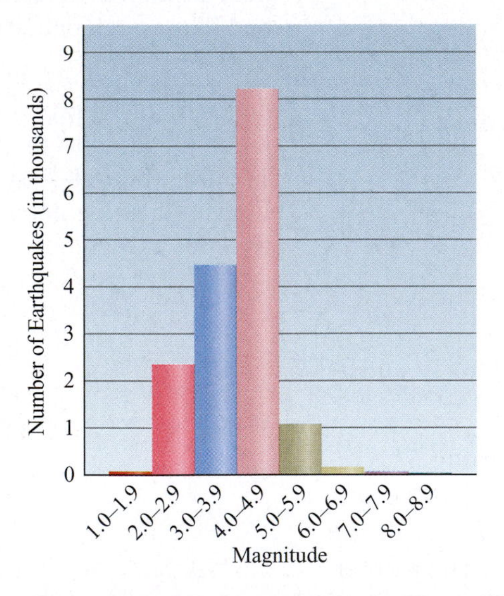

(*Source:* National Earthquake Information Center, 2003)

a. Approximately how many earthquakes were there with magnitude between 5.0 and 5.9?

b. To the nearest thousand, how many earthquakes were there with magnitude 4.9 or below?

c. What is the approximate ratio in simplified form of the number of earthquakes with magnitude 4.0–4.9 to those with magnitude 3.0–3.9?

PRACTICE 5

The following histogram summarizes the ages of the first 43 U.S. presidents at the time of their initial inauguration.

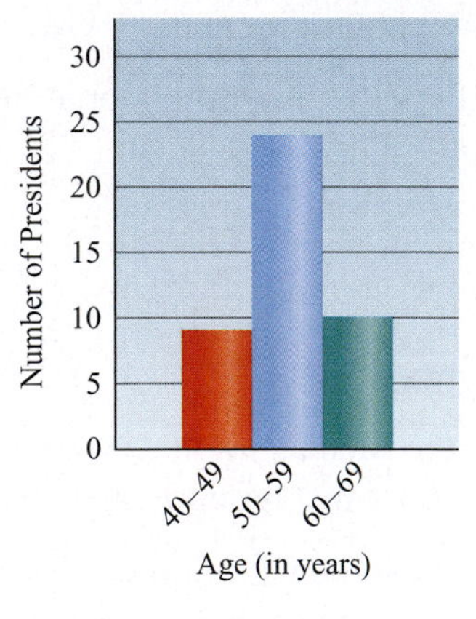

(*Source: Time Almanac 2006*)

a. Approximately how many presidents were in their fifties when they were first inaugurated?

b. Were any presidents younger than 40 at their initial inauguration?

c. About how many presidents were 59 or younger when they were first inaugurated?

Solution

a. The height of the bar for the class interval 5.0–5.9 is approximately 1,000. So there were about 1,000 earthquakes with magnitude between 5.0 and 5.9.

b. The number of earthquakes with magnitude 4.9 or below is the sum of the height of the bar for the class interval 4.0–4.9, as well as the height of all bars to the left. To the nearest thousand, this sum is 15,000.

c. The ratio of the number of earthquakes with magnitude 4.0–4.9 to those with magnitude 3.0–3.9 is about 8 to 4, which simplifies to 2 to 1.

Line Graphs

On a **line graph**, quantities are represented as points connected by straight-line segments. The height of any point on a line is read against the vertical axis.

A line graph, also called a **broken-line graph**, is commonly used to highlight changes and trends over a period of time. Especially when we have data for many points in time, we are more likely to use a line graph than a bar graph.

EXAMPLE 6

The following graph shows the number of Americans 65 years of age and older during the twentieth century.

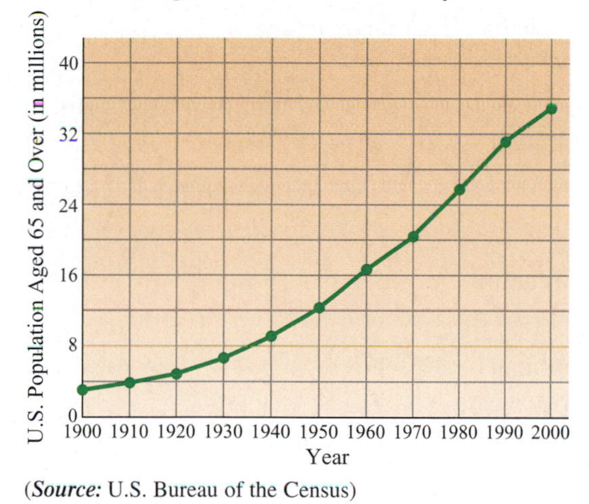

(*Source:* U.S. Bureau of the Census)

a. Approximately how big was this population in the year 2000?

b. In what year did this population number about 21 million?

c. In the year 2000, the U.S. population overall was approximately 4 times as large as it had been in the year 1900. Did the population shown in the graph grow more quickly?

Solution

a. In 2000, there were about 35 million Americans aged 65 and above.

b. There were approximately 21 million Americans aged 65 and above in the year 1970.

c. In 2000, the overall U.S. population was 4 times what it had been in 1900. But the population shown in the graph grew by a factor of about 10 and so grew more quickly.

PRACTICE 6

The following graph shows the mean temperatures in Chicago over a 30-year period for each month of the year.

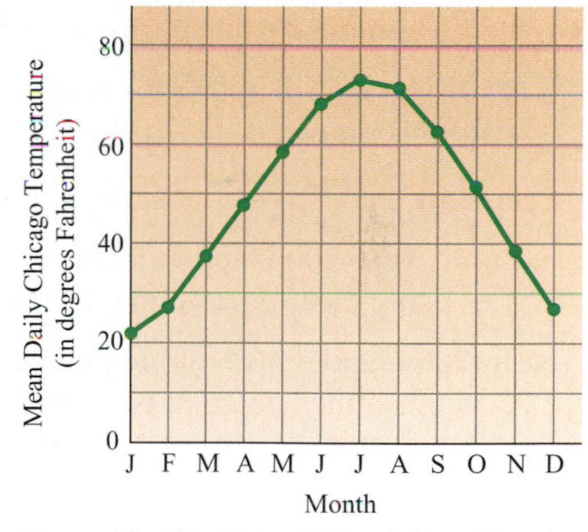

(*Source:* The U.S. National Climatic Data Center)

a. Which month in Chicago has the highest mean temperature?

b. Approximately what is the mean temperature in February?

c. What trend does the graph show?

Comparison line graphs show two or more changing quantities, as Example 7 illustrates.

EXAMPLE 7

The following graph shows the percent of American voters, by year, in counties using various types of voting machines.

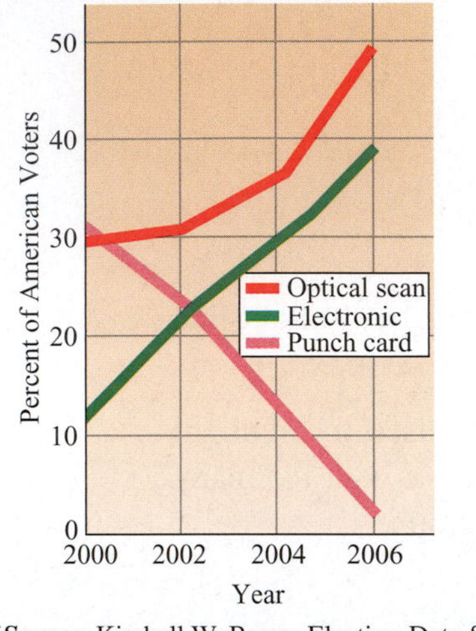

(*Source:* Kimball W. Brace, Election Data Services)

a. Approximately what percent of voters in the year 2006 were in counties using electronic voting machines?

b. In the year 2004, was the percent of voters in counties using optical scan voting machines higher than the percent of voters in counties using electronic voting machines?

c. Describe the trend that the graph shows in the use of punch card machines.

Solution

a. In the year 2006, approximately 39% of voters were in counties using electronic voting machines.

b. In 2004, about 36% of voters were in counties with optical scan machines, in contrast to approximately 29% of voters in counties with electronic machines. So the figure for optical scan machines was higher.

c. The use of punch card ballots dropped every year between 2000 and 2006. In 2006, the percent of voters in counties using punch card ballots was close to 0.

PRACTICE 7

The following graph shows the number of children in New York State who are living in foster care and the number of foster children who have been adopted, for the years between 1995 and 2006.

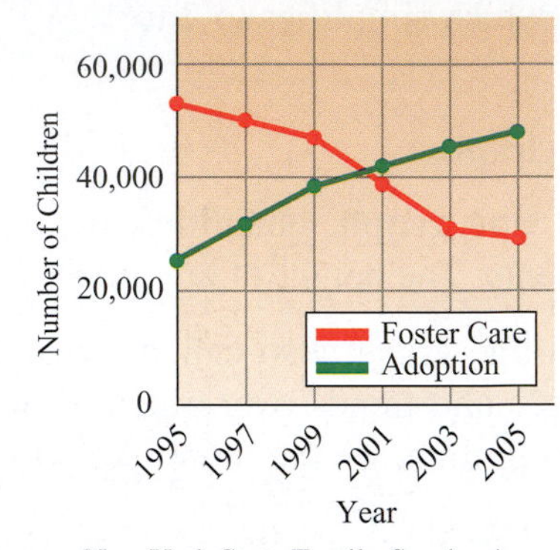

(*Source:* New York State Family Services)

a. Approximately how many children were adopted in the year 2005?

b. In what year did the number of adopted children exceed the number of children in foster care by about 14,000?

c. What trend does this graph show?

Circle Graphs

Circle graphs are commonly used to show how a whole amount—say, an entire budget or population—is broken into its parts. The graph resembles a pie (the whole amount) that has been cut into slices (the parts).

Each slice (or *sector*) is proportional in size to the part of the whole that it represents. Each slice is appropriately labeled with either its actual count or the percent of the whole that it represents.

Example 8 on the next page illustrates how to read and interpret the information given by a circle graph.

EXAMPLE 8

The following graph shows the percents of American households that own a single kind of pet:

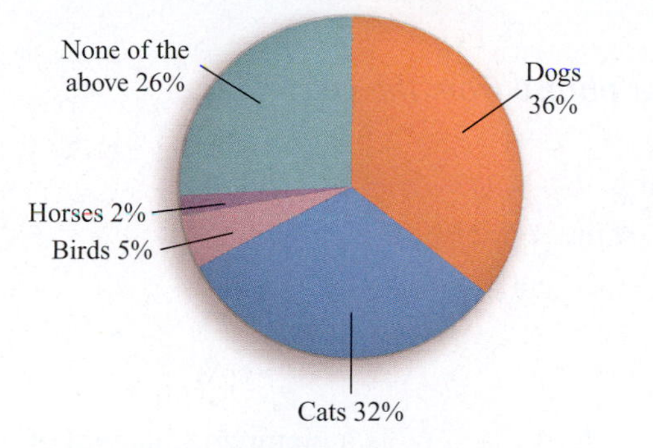

(*Source: Statistical Abstract of the United States, 2006*)

a. What is the difference between the percent of households owning dogs and the percent owning cats?

b. What fraction of the households owns birds?

c. How many times as great is the percent of households owning cats as the percent owning horses?

Solution

a. Dog owners comprise 36% of the households, in contrast to 32% for cats. So the difference is 4%.

b. 5%, or 5 out of every 100, of the households own birds, which is equivalent to $\frac{5}{100}$, or $\frac{1}{20}$.

c. 32% of the households own cats, and 2% own horses. So the percent of households that owns cats is 16 times the percent that owns horses.

PRACTICE 8

The following graph shows the distribution of ages, in a recent year, of the 25 million living veterans.

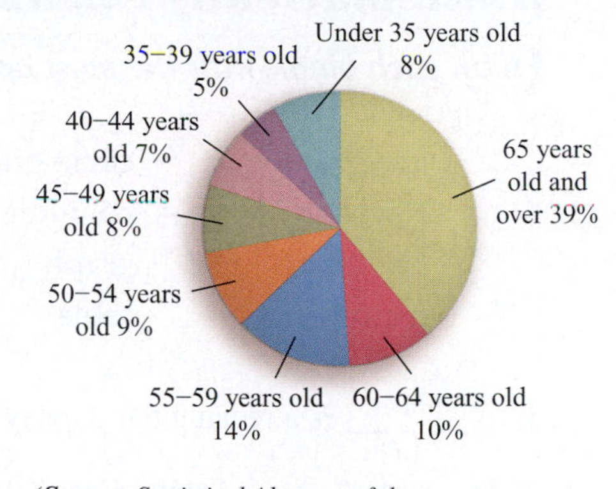

(*Source: Statistical Abstract of the United States, 2006*)

a. What fraction of the living veterans were under 35 years of age?

b. What percent of the living veterans were 60 years old or over?

c. How many living veterans were 35–39 years of age?

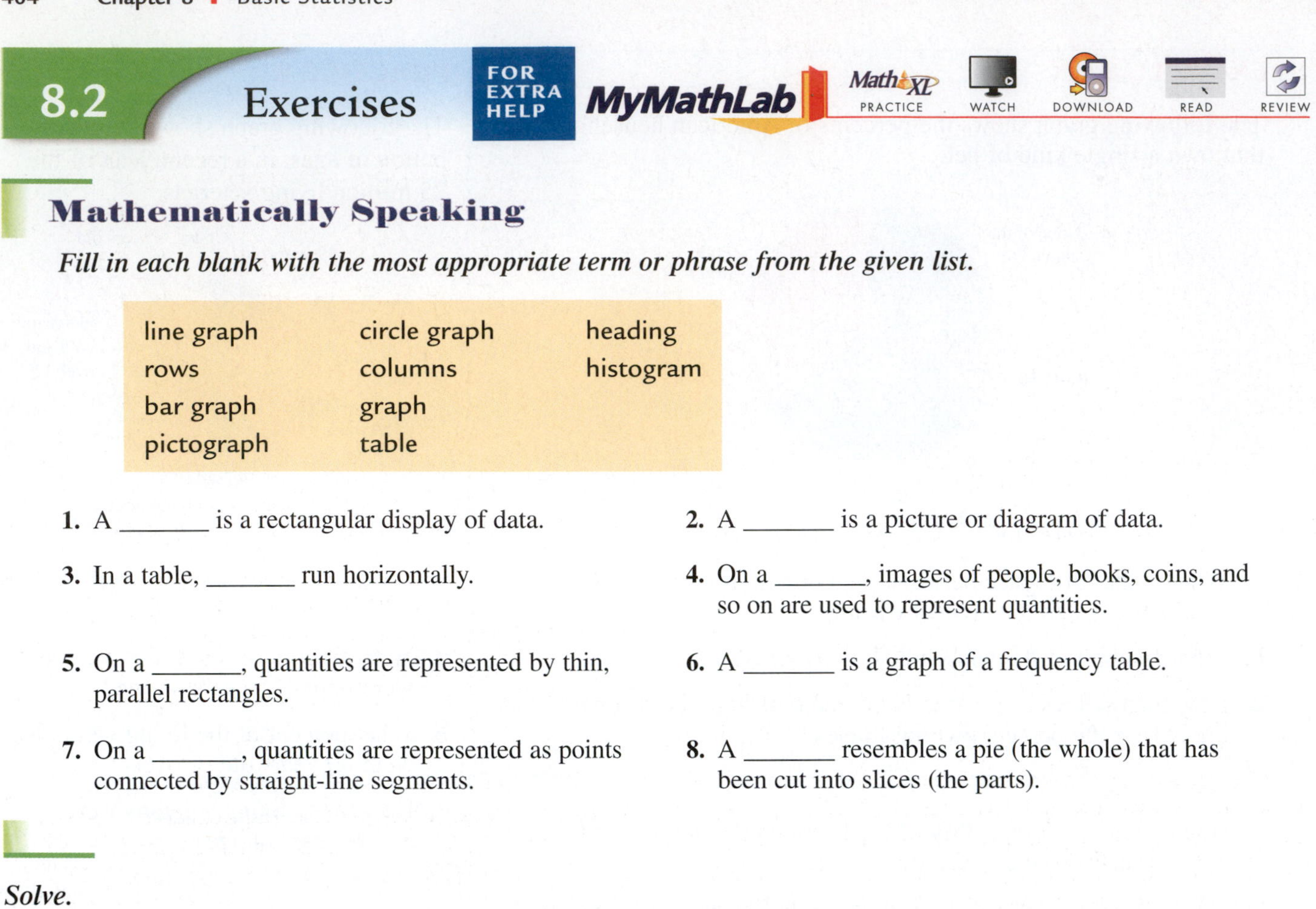

8.2 Exercises

FOR EXTRA HELP MyMathLab Math XL PRACTICE WATCH DOWNLOAD READ REVIEW

Mathematically Speaking

Fill in each blank with the most appropriate term or phrase from the given list.

line graph	circle graph	heading
rows	columns	histogram
bar graph	graph	
pictograph	table	

1. A _______ is a rectangular display of data.

2. A _______ is a picture or diagram of data.

3. In a table, _______ run horizontally.

4. On a _______, images of people, books, coins, and so on are used to represent quantities.

5. On a _______, quantities are represented by thin, parallel rectangles.

6. A _______ is a graph of a frequency table.

7. On a _______, quantities are represented as points connected by straight-line segments.

8. A _______ resembles a pie (the whole) that has been cut into slices (the parts).

Solve.

9. The following table shows how to determine a stockbroker's commission in a stock transaction. The commission depends on both the number of shares sold and the price per share.

	Number of Shares				
Price per Share	**100**	**200**	**300**	**400**	**500**
\$1–\$20	\$40	\$50	\$60	\$70	\$80
>\$20	\$40	\$60	\$80	\$90	\$100

a. What is the broker's commission on a sale of 300 shares of stock at \$15.75 a share?

b. What is the commission on a sale of 500 shares of stock at \$30 a share?

c. Will an investor pay her broker a lower commission if she sells 400 shares of stock in a single deal or 200 shares of stock in each of two deals?

10. The table shows the 2006 federal income tax schedule for single filers.

If taxable income is over—	But not over—	The tax is
\$0	\$7,550	10% of the amount over \$0
\$7,550	\$30,650	\$755 plus 15% of the amount over \$7,550
\$30,650	\$74,200	\$4,220 plus 25% of the amount over \$30,650
\$74,200	\$154,800	\$15,107.50 plus 28% of the amount over \$74,200
\$154,800	\$336,550	\$37,675.50 plus 33% of the amount over \$154,800
\$336,550	no limit	\$97,653.00 plus 35% of the amount over \$336,550

(*Source:* http://www.irs.gov)

a. What is the tax for a person whose taxable income was \$30,650?

b. Compute the tax for a person whose taxable income is \$25,000.

11. The pictograph shows the projected populations of five countries in the year 2015. (*Source:* U.S. Bureau of the Census, International Database)

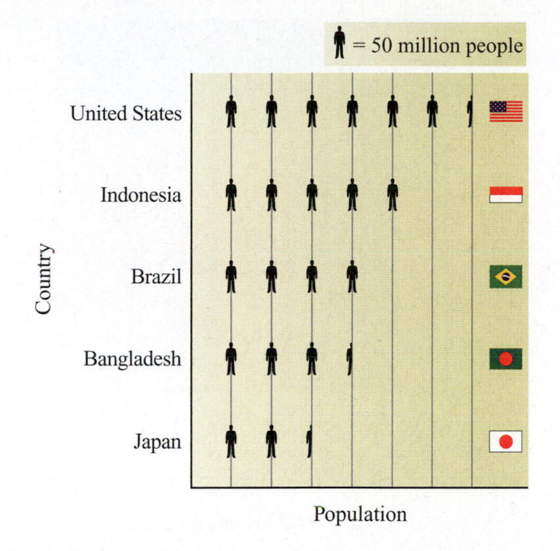

a. Which of the countries will have the largest population in 2015?

b. Approximate the population of Japan in 2015.

12. The following pictograph shows the number of various types of books sold in the United States in a recent year. (*Source:* Book Industry Study Group)

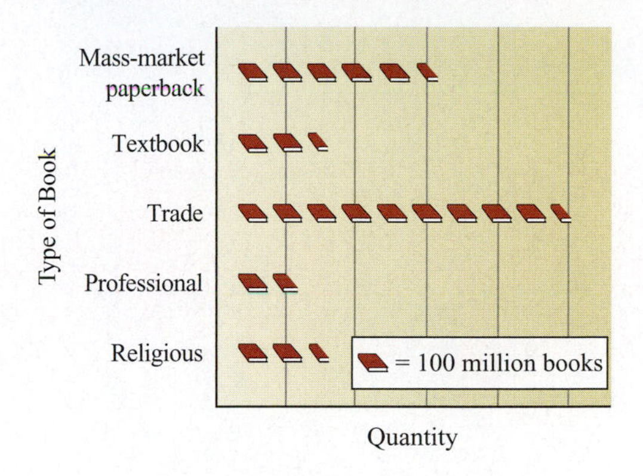

a. About how many professional books were sold?

b. Estimate the ratio of the number of textbooks sold to the number of trade books sold.

13. In chemistry, the pH scale measures how acidic or basic a solution is. A solution with a pH of 7 is considered neutral. Solutions with a pH less than 7 are acids, and solutions with a pH greater than 7 are bases. The graph below shows the pH of various solutions.

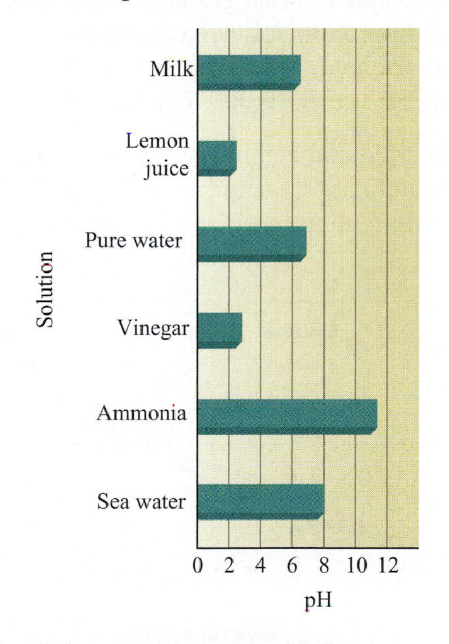

a. Which of the solutions are acids?

b. Approximate the pH of sea water.

c. Which solution is neutral?

14. The following graph shows the percent change in population between the years 2000 and 2020 projected for young Americans by age group and by race/ethnicity:

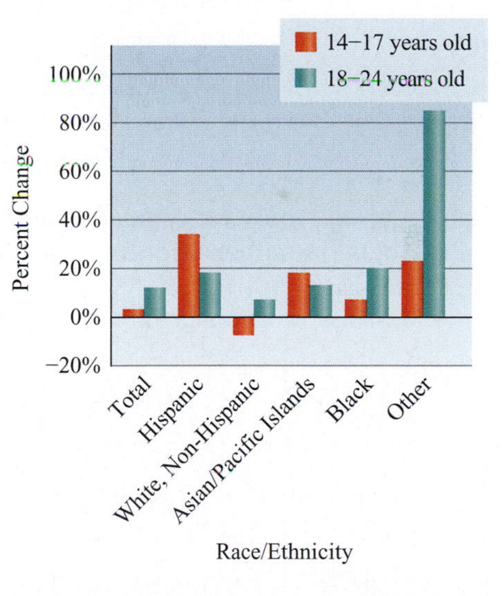

(*Source:* The National Center of Educational Statistics)

a. Estimate the predicted percent increase between the years 2000 and 2020 in the total population of 14–17-year-olds.

b. Among Hispanics, is a higher percent change predicted for 14–17-year-olds or for 18–24-year-olds?

c. According to the projection, which of the populations will have the greatest percent increase?

15. The following histogram shows the waiting time for students at a college while registering:

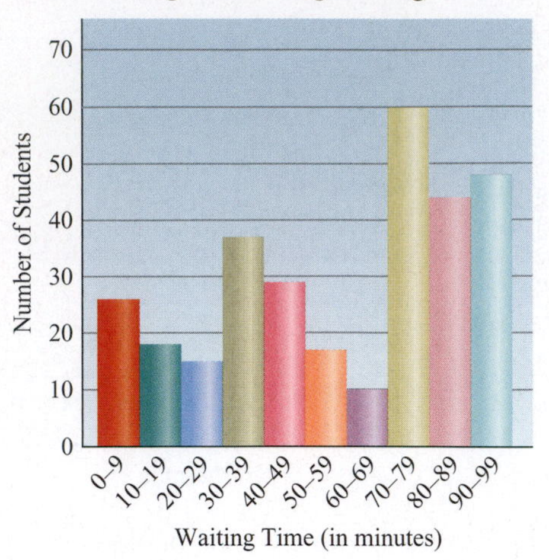

a. Did more students wait less than 10 minutes or more than 89 minutes?

b. Approximately how many students waited 80 minutes or more?

c. About how many students waited between 60 and 79 minutes?

16. The following histogram shows the total annual income of U.S. households for households with incomes less than $100,000 in a recent year:

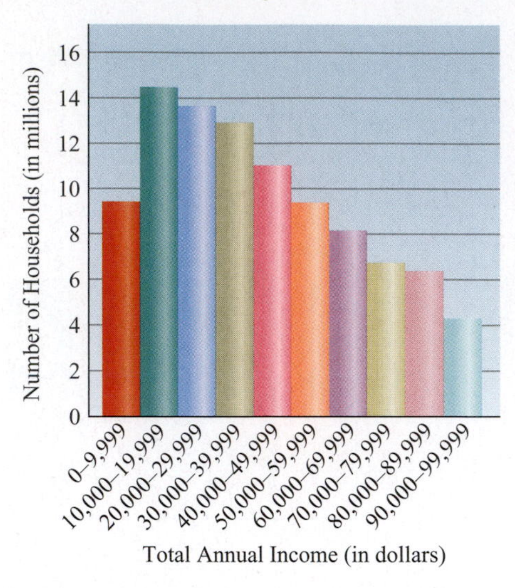

(*Source:* U.S. Bureau of the Census, 2005)

a. About how many households had incomes less than $30,000?

b. Describe the trend in this histogram.

c. The number of households with incomes $100,000 or more, not shown in the graph, was approximately 20 million. Approximately what is the ratio of this number of households to the number of households with incomes between $90,000 and $99,999?

17. The graph shows the mid-year estimated number of cell phone subscribers (in millions) for the years 2000 through 2006. (*Source:* CTIA)

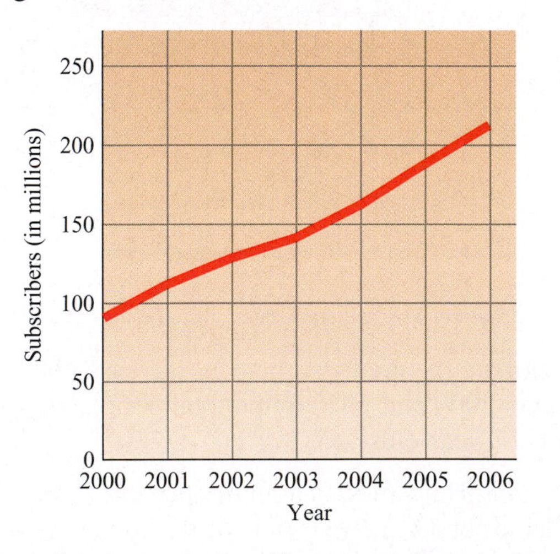

a. About how many subscribers were there in the year 2000?

b. In what year did the number of subscribers reach 150 million?

c. Describe the trend shown in the graph.

18. A human child and a chimp were raised together. Scientists graphed the number of words that the child and the chimp understood at different ages.

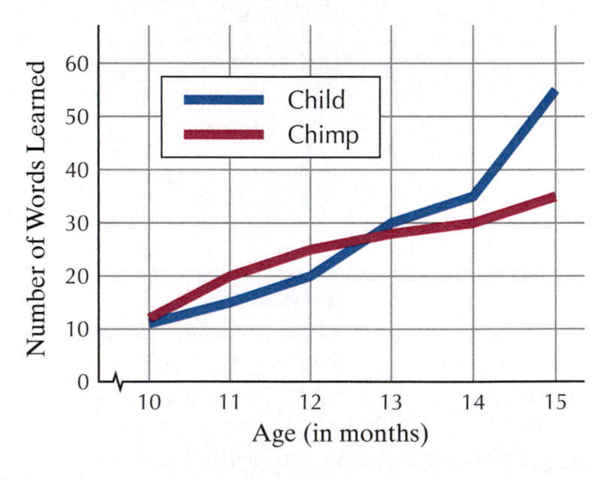

(*Source:* A. H. Kritz, *Problem solving in the Sciences*: W. N. Kellogg and L. A. Kellogg. *The Ape and the Child*)

a. At about what age was the child's vocabulary first better than that of the chimp?

b. At age 15 months, about how many more words did the child know than the chimp?

19. The graph shows the distribution of prescriptions filled by sales outlet in a recent year. (*Source:* National Association of Chain Drug Stores)

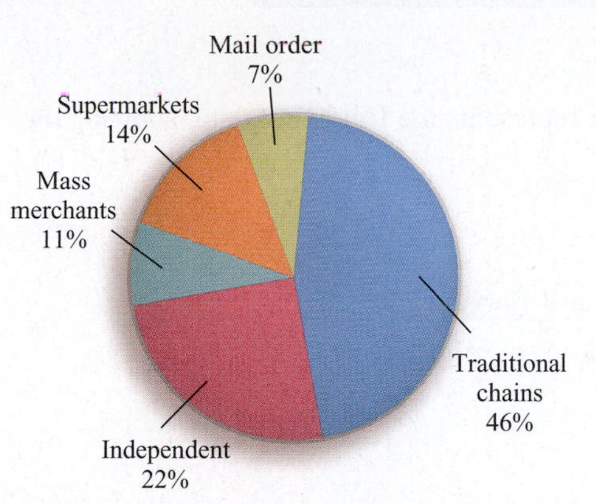

a. What percent of the prescriptions were filled by mail-order outlets?

b. Which outlet filled half the number of prescription drugs that independent outlets filled?

c. If traditional chains filled about 1.5 billion prescriptions, how many prescriptions, to the nearest 10 million, were filled by supermarkets?

20. Readers of a newspaper were surveyed as to how they get the news. The results are reported in the following circle graph:

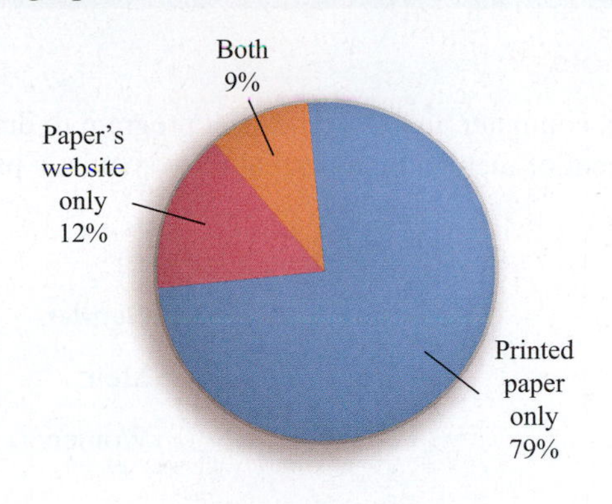

(*Source:* Pew Research Center for the People and the Press)

a. What percent of the readers read both the printed paper and the paper's website?

b. What percent of the readers read the printed paper?

c. Is it true that the number of readers who read the printed paper only was more than 6 times as great as the number of readers who read the paper's website only?

Check your answers on page A-14.

MINDSTRETCHERS

Technology

1. On a computer, use a spreadsheet program to draw a circle graph that represents the following data, showing the percent of men and the percent of women at a party:

Gender	Number of Guests at a Party
Men	12
Women	8

Writing

2. A *stacked bar graph* not only allows comparisons between quantities but also shows how each quantity is divided into parts. For example, the following stacked bar graph deals with how American mothers and fathers, on the average, spend their time each week. In a few sentences, describe some of the main trends that this graph implies. (***Source:*** Suzanne Bianchi, *Changing Rythms of American Family Life*, published by the Russell Sage Foundation and the American Sociological Association, as reported in the *New York Times* on October 17, 2006)

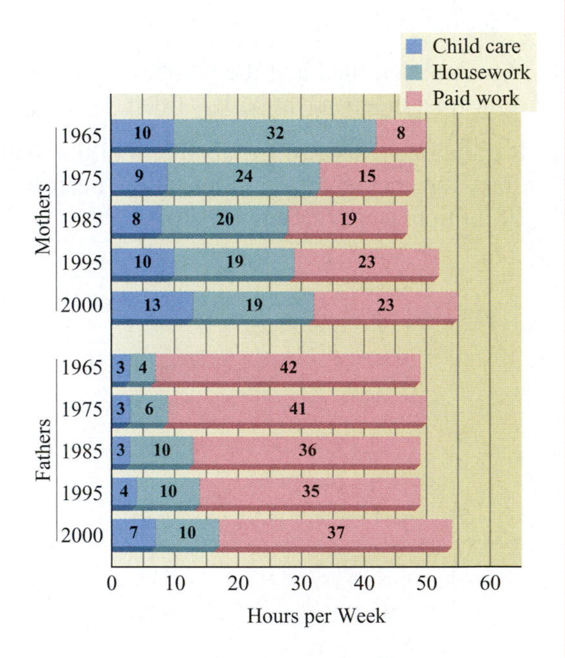

Mathematical Reasoning

3. Consider the following two bar graphs. They both represent the same data, namely, the percent of voters in the presidential election of 2000 who voted either Democratic or Republican. (***Source:*** Center for Political Studies, University of Michigan, as presented in *Statistical Abstract of the United States: 2006*)

a. Explain the difference between the impression that the top graph gives and the impression the bottom graph gives.

b. How is this difference in impression achieved?

c. Why would someone prefer to give one impression more than the other?

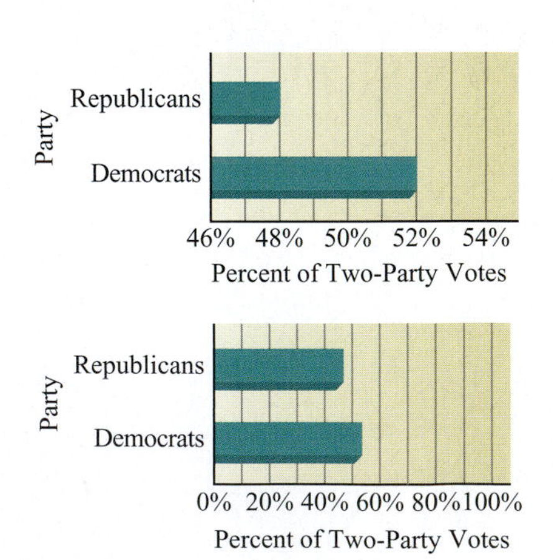

KEY CONCEPTS AND SKILLS

CONCEPT SKILL

Concept/Skill	Description	Example
[8.1] **Mean**	Given a set of numbers, the sum of the numbers divided by however many numbers are in the set.	For 0, 0, 1, 3, and 5 the mean is $\frac{0+0+1+3+5}{5} = \frac{9}{5} = 1.8$
[8.1] **Median**	Given a set of numbers arranged in numerical order, the number in the middle. If there are two numbers in the middle, the mean of the two middle numbers.	For 0, 0, 1, 3, and 5, the median is 1.
[8.1] **Mode**	Given a set of numbers, the number (or numbers) occurring most frequently in the set.	For 0, 0, 1, 3, and 5, the mode is 0.
[8.1] **Range**	Given a set of numbers, the difference between the largest and the smallest number in the set of numbers.	For 0, 0, 1, 3, and 5, the range is $5 - 0$, or 5.
[8.2] **Table**	A rectangular display of data.	Column ↓ |	Dems.	Reps. | Row →	Men	37	45
Women	14	15	
[8.2] **Pictograph**	A graph in which images of people, books, coins, and so on are used to represent the quantities.	= 10 robots A B C D Factory
[8.2] **Bar graph**	A graph in which quantities are represented by thin, parallel rectangles called bars. The length of each bar is proportional to the quantity that it represents.	Decade Ending in: 1990, 1970, 1950, 1930, 1910 0 2,000 4,000 6,000 8,000 Fatalities

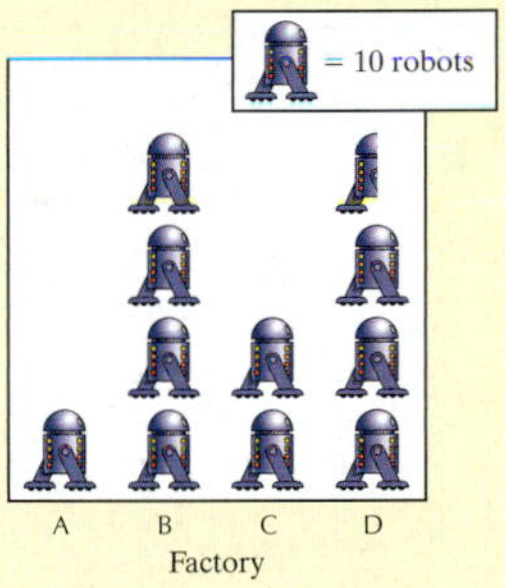

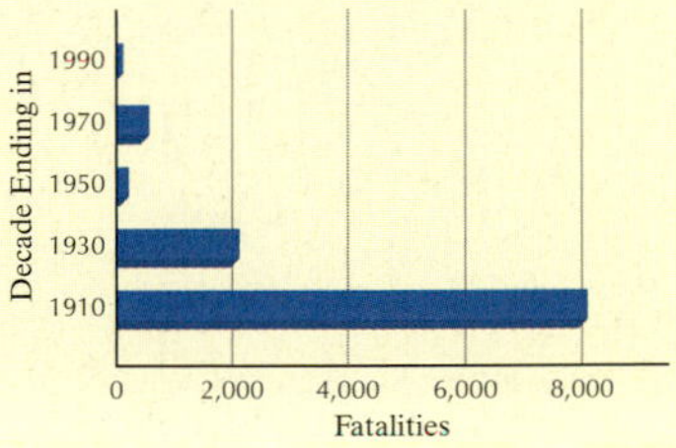

continued

CONCEPT SKILL

Concept/Skill	Description	Example
[8.2] **Histogram**	A graph of a frequency table.	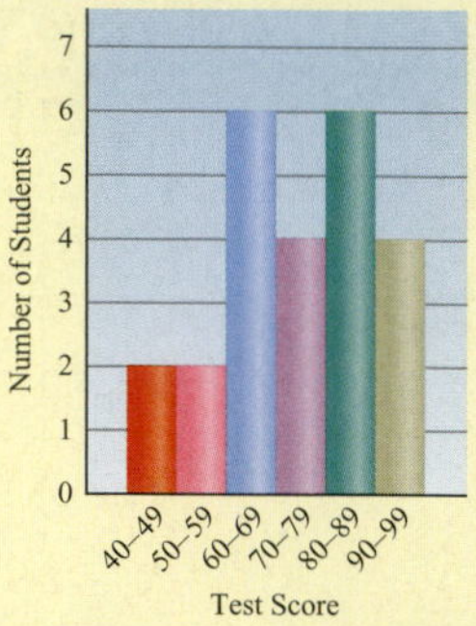
[8.2] **Line graph**	A graph in which quantities are represented as points connected by straight-line segments. The height of any point on a line is read against the vertical axis.	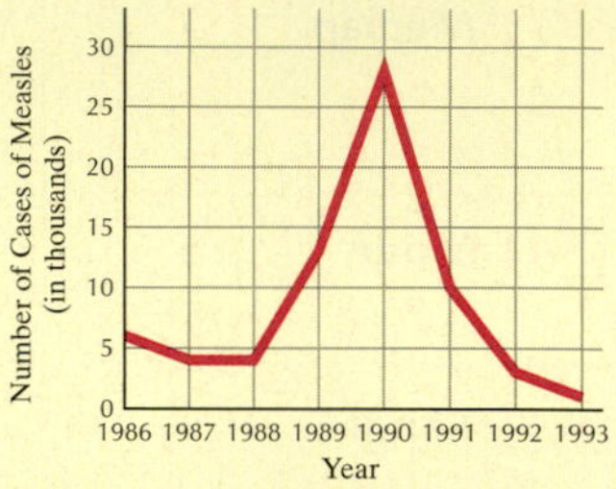
[8.2] **Circle graph**	A graph that resembles a pie (a whole amount) that has been cut into slices (the parts).	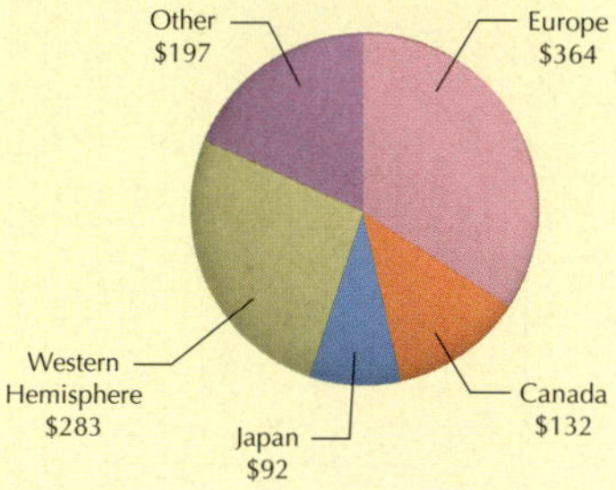

Chapter 8 Review Exercises

To help you review this chapter, solve these problems.

[8.1] *Compute the desired statistic for each list of numbers.*

1.

	List of Numbers	Mean	Median	Mode	Range
a.	6, 7, 4, 10, 4, 5, 6, 8, 7, 4, 5				
b.	1, 3, 4, 4, 2, 3, 1, 4, 5, 1				

Mixed Applications

2. The following table shows how long the first five American presidents and their wives lived:

President	Age	President's Wife	Age
George Washington	67	Martha Washington	70
John Adams	90	Abigail Adams	74
Thomas Jefferson	83	Martha Jefferson	34
James Madison	85	Dolley Madison	81
James Monroe	73	Eliza Monroe	62

(*Source: Presidents, First Ladies, and Vice Presidents*)

a. Using the median as the average, did the husbands or the wives live longer?

b. By how many years?

c. What was the range of the ages of the presidents?

3. According to the census of 2000, the median age in the United States was 35.3. (*Source:* U.S. Bureau of the Census)

a. Explain what this statement means.

b. In 2000, by how many years was your age above or below average?

4. A soda machine is considered reliable if the range of the amounts of soda that it dispenses is less than 2 fluid ounces (fl oz). In 10 tries, a particular machine dispensed the following amounts (in fl oz).

8.1 7.8 8.6 8.1 8.4 7.8 8 7.7 6.9 8.4

Is the machine reliable? Explain.

5. According to its producer, a Broadway show would make a profit if it averaged at least 1,000 paying customers a night. For the past 10 nights, the number of paying customers, rounded to the nearest hundred, was as follows:

900 700 1,500 800 1,100 800 700 1,600 800 1,100

Using the mean as the average, determine whether the show was making a profit.

6. In a recent year, the amount of cargo (in tons) that the five busiest U.S. ports handled was as follows:

	Cargo (in tons)		
Port	**Total**	**Domestic**	**Foreign**
Port of South Louisiana	224,187,322	119,416,519	104,770,803
Houston, Texas	202,047,327	64,510,816	137,536,511
New York, New York	152,377,503	70,177,949	82,199,554
Beaumont, Texas	91,697,948	20,823,732	70,874,216
Long Beach, California	79,708,424	17,193,016	62,515,408

(*Source:* U.S. Army Corps of Engineers)

a. To the nearest 10,000 tons, how much domestic cargo did Houston handle?

b. Which of these five ports handled the least foreign cargo?

c. To the nearest 10,000 tons, how much more domestic cargo did New York handle than Beaumont?

d. Which of these ports handled more foreign than domestic cargo?

7. The following pictograph shows the number of successful launches of rockets carrying satellites for the first 6 months of 2006:

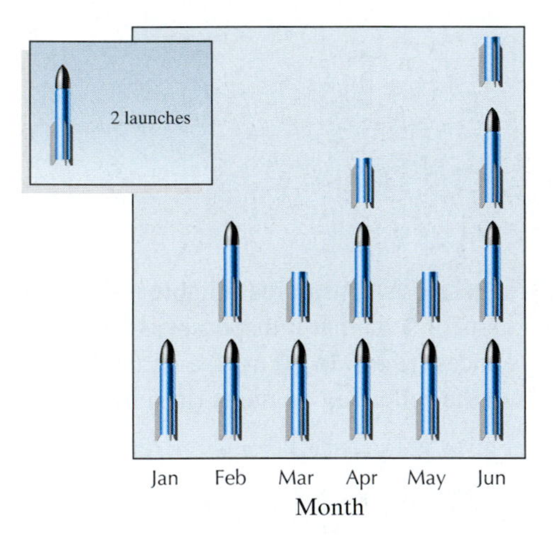

(*Source:* spaceflightnow.com)

a. In January, February, and March, about how many total launches were there?

b. In which month were there the most launches?

c. About what fraction of the launches in the first 6 months occurred in May?

8. The following graph shows the number of movie screens—indoor and drive-in—in the United States between 2000 and 2005. (*Source:* National Association of Theater Owners)

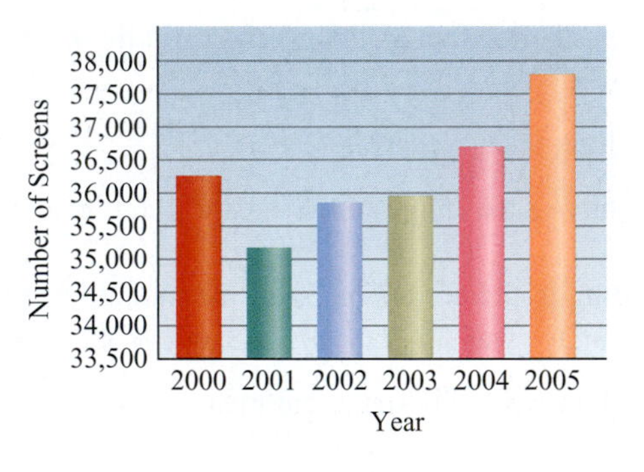

a. Approximately how many movie screens were there in 2003?

b. About how many more movie screens were there in 2005 than in 2000?

c. Describe the trend the graph illustrates.

9. The following graph displays the number of U.S. licensed drivers in a recent year, aged 20 through 84, according to their age.

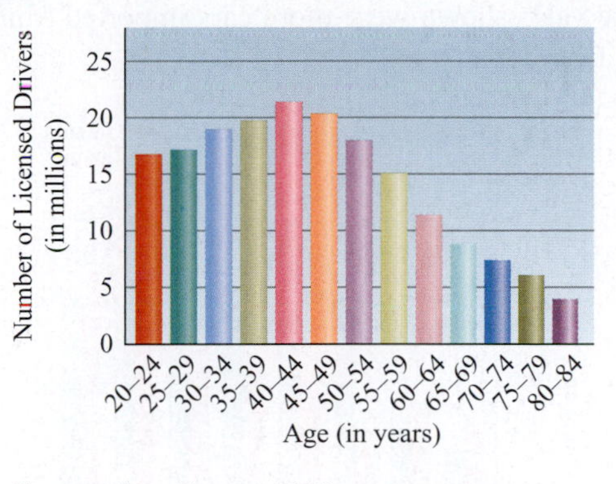

(*Source: Time Almanac 2006*)

a. About how many licensed drivers were between 35 and 49 years of age?

b. If the total number of licensed drivers was about 200 million, approximately what percent of these were between the ages of 35 and 39?

c. Describe the trend in this graph.

10. Consider the following learning curve that shows how long a rat running through a maze takes on each run:

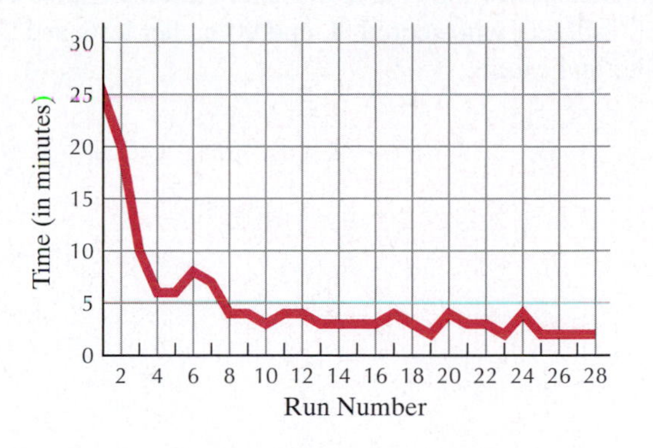

a. On which run does the rat run through the maze in 10 minutes?

b. How long does the rat take on the tenth run?

c. What general conclusion can you draw from this learning curve?

11. The graph shows the health care expenditures from out-of-pocket and insurance sources.

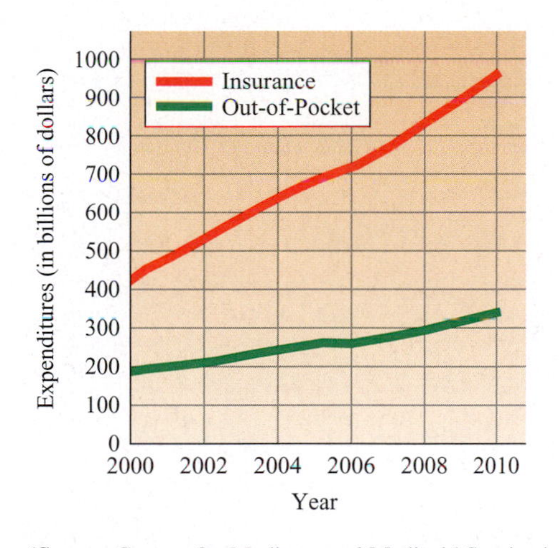

(*Source:* Centers for Medicare and Medicaid Services)

a. In what year were the insurance expenditures approximately \$700 billion?

b. In the year 2000, what was the approximate ratio of out-of-pocket expenditures to insurance expenditures?

c. Express in words the trend the graph is illustrating.

12. The following graph shows the number of organ transplants performed in the United States for a recent year.

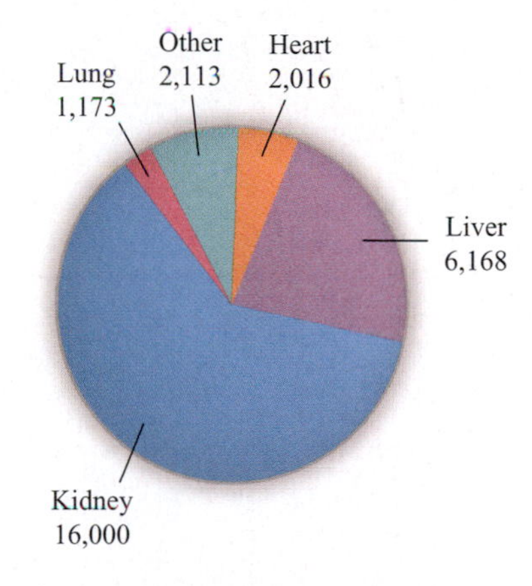

(*Source:* U.S. Department of Health and Human Services)

a. How many transplants in all were performed?

b. How many more liver transplants than lung transplants were performed?

c. What percent of all transplants performed were kidney transplants, to the nearest whole percent?

13. In a college algebra course, each of two tests counts 20% of a student's course average, whereas the final exam counts 60% of the course average. Find the course average of a student who scored 80 and 90 on her tests and 70 on the final exam.

14. The following double-bar graph shows the number of new passenger cars imported in various years into the United States from Japan or from Canada. In how many of the years shown were more cars imported from Canada than from Japan?

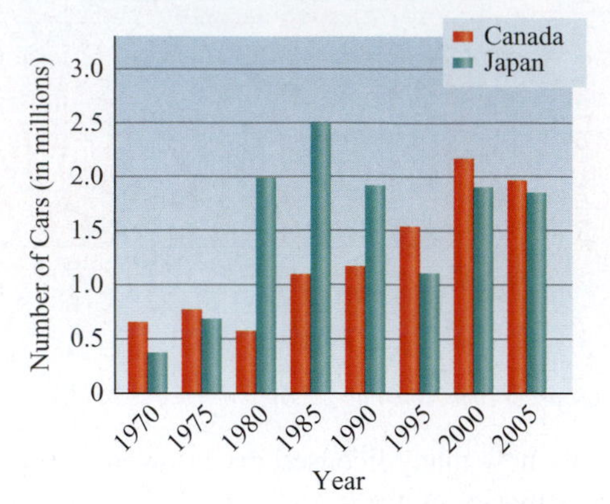

(***Source:*** Bureau of the Census, Foreign Trade Division)

Check your answers on page A-14.

Chapter 8 POSTTEST

FOR EXTRA HELP  Test solutions are found on the enclosed CD.

To see whether you have mastered the topics in this chapter, take this test.

1. Late in a term, your grades were as follows: English Composition I (4 credits)—A; Freshman Orientation (1 credit)—A; College Skills (2 credits)—C; and Physical Education (1 credit)—B. If A = 4, B = 3, C = 2, D = 1, and F = 0, calculate your GPA.

2. A local hospital kept track of the number of babies born each month last year.

Jan	Feb	Mar	Apr	May	June
106	115	138	165	189	202
July	Aug	Sept	Oct	Nov	Dec
208	216	190	172	138	105

What was the mean number of babies born at the hospital in a month last year?

3. A math instructor records the following exam scores for students in her Math 110 course:

86	78	96	82	74	56	72	76
88	60	48	76	100	98	64	80

What was the median exam score?

4. The heights (in inches) of the 19 players on the roster for the Indiana Pacers during a recent season were as follows:

73	82	78	84	83	81	76	81	84	76
80	76	78	79	83	81	75	79	81	

(*Source:* http://www.nba.com/pacers, 2006)

What are the mode and range of the heights?

5. Shown in the table is the fuel economy of the seven most fuel-efficient vehicles for the model year 2007.

Vehicle	City (mpg)	Highway (mpg)
Toyota Prius	60	51
Honda Civic Hybrid	49	51
Toyota Camry Hybrid	40	38
Ford Escape Hybrid FWD	36	31
Toyota Yaris (manual)	34	40
Toyota Yaris (automatic)	34	39
Honda Fit (manual)	33	38

(*Source:* U.S. Environmental Protection Agency, U.S. Department of Energy, *Fuel Economy Guide*)

Which vehicle(s) have better fuel efficiency for city driving than for highway driving?

6. The following bar graph displays the projected carbon dioxide emissions (in billions of metric tons) from energy consumption in North America.

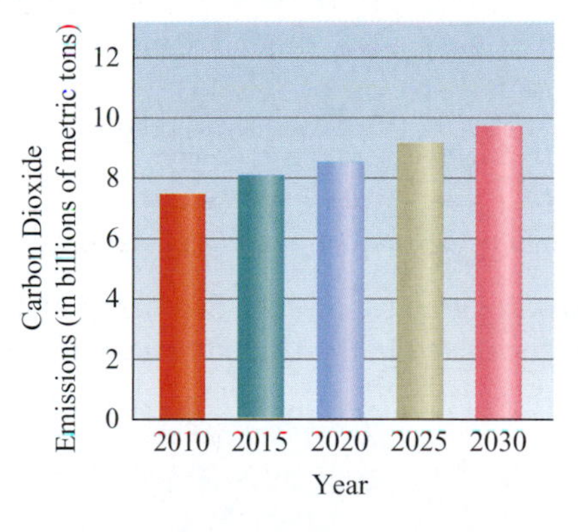

(*Source:* Energy Information Administration)

Approximately how much greater are the projected carbon dioxide emissions in 2030 than in 2015?

7. The following line graph shows the gross seasonal ticket sales for Broadway shows.

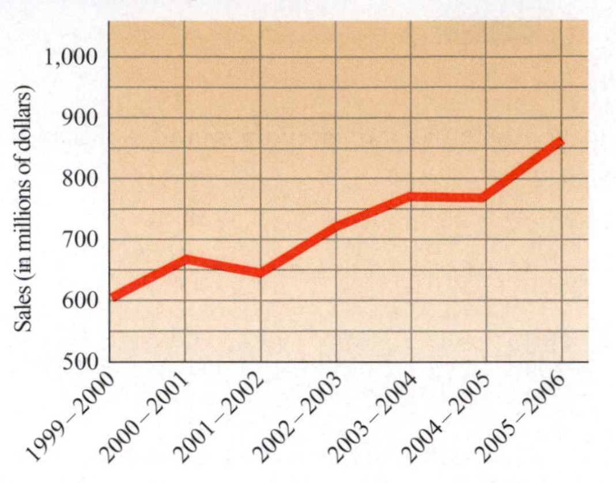

(***Source:*** The League of American Theaters and Producers, Inc.)

About how much higher than in the 1999–2000 season were the gross sales in the 2004–2005 season?

8. The following graph shows the enrollments in public and private colleges in the United States.

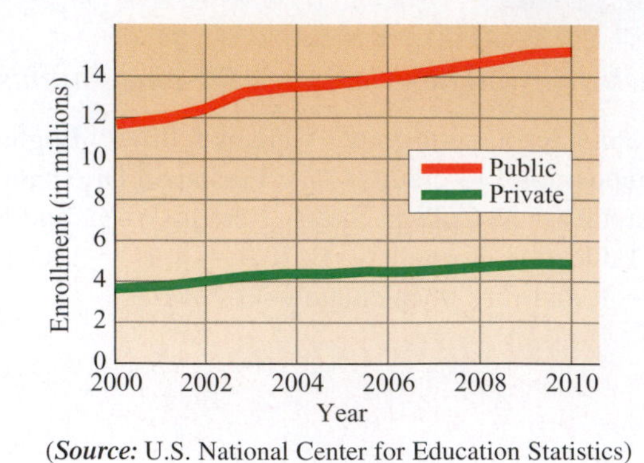

(***Source:*** U.S. National Center for Education Statistics)

About how many students were enrolled in public colleges in 2004?

9. The following circle graph shows the breakdown of the U.S. coastline (in miles) by region:

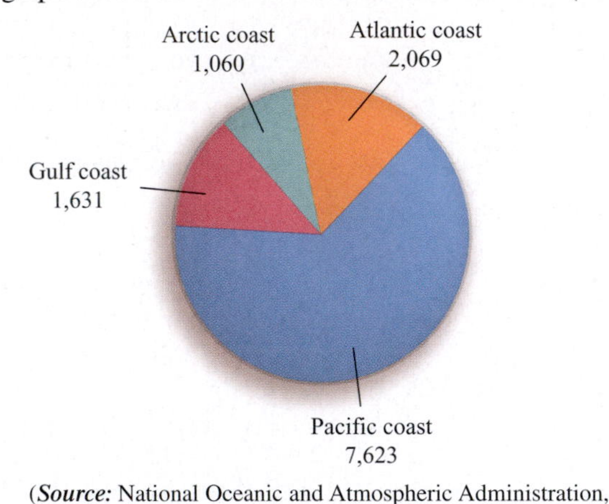

(***Source:*** National Oceanic and Atmospheric Administration, U.S. Department of Commerce)

To the nearest 10%, what percent of the U.S. coastline is on the Pacific coast?

10. The following graph shows the number of Nobel prizes awarded in the sciences for research conducted in the United States and in the United Kingdom.

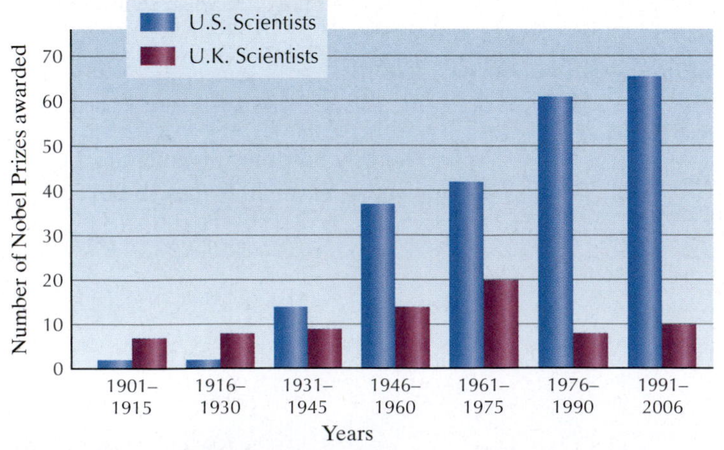

(***Source:*** *Who's Who of Nobel Prize Winners*)

In which period of time did scientists in the United States receive about twice as many prizes as scientists in the United Kingdom?

Check your answers on page A-14.

Cumulative Review Exercises

To help you review, solve the following.

1. Simplify: $\frac{20}{25}$

2. Estimate the following product: $8\frac{1}{10} \quad 4\frac{9}{10}$

3. $3.01 \quad 1{,}000 \quad\quad ?$

4. Solve for x: $x \quad 7\frac{1}{4} \quad 10$

5. Solve for n: $\frac{7}{10} \quad \frac{n}{30}$

6. What is $12\frac{1}{2}\%$ of 16?

7. Simplify: $3 \quad 4(2 \quad 7)$

Solve.

8. The Great Pyramid of Khufu—the last surviving wonder of the ancient world—was built around 2680 B.C. To the nearest thousand years, how long ago was this pyramid built? (***Source:*** *The Concise Columbia Encyclopedia*)

9. According to the Recording Industry Association of America, about 290 million digital singles were sold in the first 6 months of 2006, whereas 170 million were sold for the same period in 2005. What was the percent increase from 2005 to 2006 in the number of digital singles sold, to the nearest percent? (***Source:*** Recording Industry Association of America)

10. The graph below shows the Super Bowl game scores for the American Football Conference (AFC) team and the National Football Conference (NFC) team for the years from 2000 to 2006.

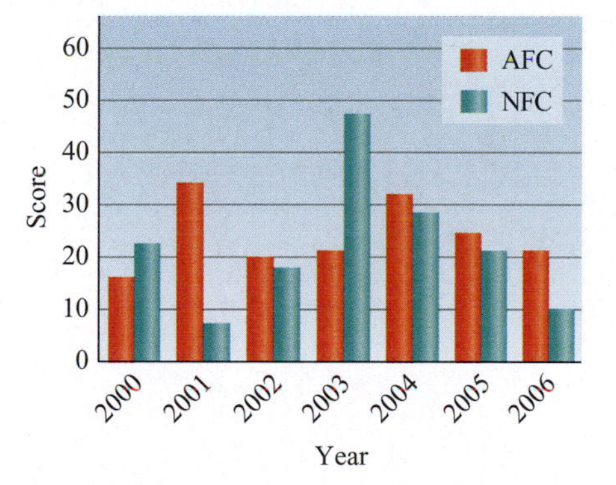

(***Source:*** *The World Almanac and Book of Facts 2007*)

In which year(s) did the NFC team win the Super Bowl?

Check your answers on page A-14.

CHAPTER 9

More on Algebra

Equations and Accounting

Accountants commonly compute financial quantities—say tax or profit—using a computer program called a *spreadsheet*.

Like a piece of accounting paper, a spreadsheet has the appearance of a grid, with rows and columns and numbers in the cells where the rows and columns intersect. For example, B3 is the cell at the intersection of the B column and the third row.

	A	B
1	Software Budget	20,000
2	Office budget	B2
3	TOTAL BUDGET	B3

We might have the spreadsheet compute the number in B3 by adding the quantity in B2 to that in B1 (20,000), getting the following equation.

$$B3 = 20{,}000 + B2$$

This equation shows the relationship between the variables B2 and B3. (***Source:*** Philip E. Fess, et al., *Accounting*, Thomson South-Western, 2004)

Chapter 9 PRETEST

To see if you have already mastered the topics in this chapter, take this test.

Solve and check.

1. $y + 8 = 2$

2. $x - 6 = -8$

3. $-6x = 24$

4. $-1 = \frac{a}{5}$

5. $3x + 1 = 10$

6. $4 = 5y - 1$

7. $3 - 2c = 7$

8. $\frac{a}{2} - 7 = -12$

9. $2x + 3x = 10$

10. $4y + 3 = y + 9$

11. $2x - 4 = x - 5$

12. $4x = 2(x - 4)$

Write each relationship as an equation or a formula.

13. Three times the sum of a number and 7 is 30.

14. The length, l, of a woman's radius bone—the longest bone in the forearm—is about $\frac{1}{7}$ of her height, h.

(***Source:*** Colin Evans, *Casebook of Forensic Detection*)

Write an equation for each word problem. Solve and check.

15. A dog groomer worked 4 hours last Saturday. Her boss paid her \$45 for the day, which included a \$5 tip. How much money did she earn per hour?

16. A disc jockey charges \$250 to provide 3 hours of music for a party. For each additional hour, he charges \$75. If he received \$475 for working at a graduation party, how many hours did he work?

17. Each used book at a library sale sold for the same amount of money. A student bought 6 short-story and 8 mystery books. Then she bought a college poster for \$5. If she spent \$19 in all, how much did each book cost?

18. A publishing company ordered 15 cases of copy paper. The total cost was \$795, which included a shipping fee of \$75. How much did each case of paper cost?

19. Lumber is frequently sold by the board foot, where the number of board feet, B, for a piece of lumber is $\frac{1}{12}$ the product of the lumber's thickness t, width w, and length l. (The thickness and width are given in inches, and the length is given in feet.)

a. Express this relationship as a formula.

b. How many board feet would a piece of lumber that measures 2 inches $\times$ 4 inches $\times$ 20 feet contain?

20. In marketing, the sale price, s, equals the regular price, r, minus the discount price, d.

a. Write a formula for this relationship.

b. Find the sale price of an item when $r = \$400$ and $d = \$99$.

• *Check your answers on page A-15.*

9.1 Solving Equations

OBJECTIVES

- To solve one-step equations involving signed numbers
- To solve two-step equations
- To solve word problems involving equations with signed numbers or two steps

In Chapter 4, we solved one-step equations such as the following.

$$x + 3 = 5, \qquad x - 3 = 7, \qquad 3x = 6, \qquad \text{and} \qquad \frac{x}{4} = 3$$

In this chapter, we extend the discussion to include equations that involve

- both positive and negative numbers or
- more than a single operation.

Solving One-Step Equations Involving Signed Numbers

Recall that in solving one-step equations, the key is to isolate the variable, that is, to get the variable alone on one side of the equation. We do this by performing the appropriate opposite operation.

Definition

A **solution** to an equation is a value of the variable that makes the equation a true statement. To **solve** an equation means to find all solutions of the equation.

We have already discussed rules for solving addition, subtraction, multiplication, and division equations involving positive numbers. Let's now apply these rules to simple equations involving both positive and negative numbers.

EXAMPLE 1

Solve and check: $y + 5 = 3$

Solution

$y + 5 = 3$

$y + 5 - 5 = 3 - 5$ **Subtract 5 from each side of the equation.**

$y + 0 = -2$ **$5 - 5 = 0$ and $3 - 5 = 3 + (-5) = -2$**

$y = -2$ **$y + 0 = y$**

Check

$y + 5 = 3$

$-2 + 5 \stackrel{?}{=} 3$ **Substitute -2 for y in the original equation.**

$3 \stackrel{\checkmark}{=} 3$

The solution is -2.

PRACTICE 1

Solve and check: $x + 7 = 4$

EXAMPLE 2

Solve and check: $n - 5 = -11$

Solution

$$n - 5 = -11$$
$$n - 5 + 5 = -11 + 5 \quad \text{Add 5 to each side of the equation.}$$
$$n + 0 = -6$$
$$n = -6$$

Check

$$n - 5 = -11$$
$$-6 - 5 \stackrel{?}{=} -11 \quad \text{Substitute } -6 \text{ for } n \text{ in the original equation.}$$
$$-11 \stackrel{\checkmark}{=} -11$$

The solution is -6.

PRACTICE 2

Solve and check: $m - 7 = -19$

EXAMPLE 3

Solve and check: $-7x = 21$

Solution

$$-7x = 21$$
$$\frac{-7x}{-7} = \frac{21}{-7} \quad \text{Divide each side of the equation by } -7.$$
$$x = -3$$

Check

$$-7x = 21$$
$$-7(-3) \stackrel{?}{=} 21 \quad \text{Substitute } -3 \text{ for } x \text{ in the original equation.}$$
$$21 \stackrel{\checkmark}{=} 21$$

The solution is -3.

PRACTICE 3

Solve and check: $9y = -18$

EXAMPLE 4

Solve and check: $-1 = \frac{x}{6}$

Solution

$$-1 = \frac{x}{6}$$
$$-1 \cdot 6 = \frac{x}{\cancel{6}} \cdot \cancel{6} \quad \text{Multiply each side of the equation by 6.}$$
$$-6 = x$$
$$x = -6$$

Check

$$-1 = \frac{x}{6}$$
$$-1 \stackrel{?}{=} \frac{-6}{6} \quad \text{Substitute } -6 \text{ for } x \text{ in the original equation.}$$
$$-1 \stackrel{\checkmark}{=} -1$$

The solution is -6.

PRACTICE 4

Solve and check: $\frac{y}{-3} = -2$

EXAMPLE 5

A student lost 8 points for each incorrect answer on an exam. How many incorrect answers did he get if he lost a total of 32 points?

Solution We can represent a loss of 8 points as -8 and a loss of 32 points as -32. To find the number of incorrect answers, we write the following equation, letting x represent the number of incorrect answers:

$$-8x = -32$$

We solve this equation for x.

$$-8x = -32$$
$$\frac{-8x}{-8} = \frac{-32}{-8}$$
$$x = 4$$

Check

$$-8x = -32$$
$$-8(4) \stackrel{?}{=} -32$$
$$-32 \stackrel{\checkmark}{=} -32$$

PRACTICE 5

A patient's daily dosage of a medication was decreased by 15 milligrams per week. In how many weeks did the dosage decrease by 105 milligrams?

Solving Two-Step Equations

We now turn our attention to solving equations that involve two operations, such as

$$3y + 5 = 26 \quad \text{and} \quad \frac{c}{3} - 4 = 7$$

which are commonly referred to as two-step equations.

To solve a two-step equation

- First use the rule for solving addition or subtraction equations.
- Then use the rule for solving multiplication or division equations.

EXAMPLE 6

Solve and check: $3y + 11 = 26$

Solution

$$3y + 11 = 26$$
$$3y + 11 - 11 = 26 - 11 \quad \text{Subtract 11 from each side of the equation.}$$
$$3y = 15$$
$$\frac{3y}{3} = \frac{15}{3} \quad \text{Divide each side of the equation by 3.}$$
$$y = 5$$

PRACTICE 6

Solve and check: $2x + 8 = -6$

Check $3y + 11 = 26$

$$3(5) + 11 \stackrel{?}{=} 26$$ Substitute 5 for y in the original equation.

$$15 + 11 \stackrel{?}{=} 26$$

$$26 \stackrel{\checkmark}{=} 26$$

The solution is 5.

EXAMPLE 7

Solve and check: $\dfrac{c}{3} - 4 = 7$

Solution

$$\frac{c}{3} - 4 = 7$$

$$\frac{c}{3} - 4 + 4 = 7 + 4$$ Add 4 to each side of the equation.

$$\frac{c}{3} = 11$$

$$3 \cdot \frac{c}{3} = 3 \cdot 11$$ Multiply each side of the equation by 3.

$$c = 33$$

Check

$$\frac{c}{3} - 4 = 7$$

$$\frac{33}{3} - 4 \stackrel{?}{=} 7$$ Substitute 33 for c in the original equation.

$$11 - 4 \stackrel{?}{=} 7$$

$$7 \stackrel{\checkmark}{=} 7$$

The solution is 33.

PRACTICE 7

Solve and check: $\dfrac{k}{5} - 6 = -3$

EXAMPLE 8

Solve and check: $1 - 2x = 5$

Solution

$$1 - 2x = 5$$

$$1 - 1 - 2x = 5 - 1$$ Subtract 1 from each side of the equation.

$$-2x = 4$$

$$\frac{-2x}{-2} = \frac{4}{-2}$$ Divide each side of the equation by -2.

$$x = -2$$

PRACTICE 8

Solve and check: $8 - 3d = -4$

Check $1 - 2x = 5$

$$1 - 2(-2) \stackrel{?}{=} 5 \quad \text{Substitute } -2 \text{ for } x \text{ in the original equation.}$$

$$1 + 4 \stackrel{?}{=} 5$$

$$5 \stackrel{\checkmark}{=} 5$$

The solution is -2.

Note that in Example 8 we solved by subtracting before dividing. Can you think of any other way to solve this equation?

EXAMPLE 9

Solve and check: $12 - \frac{a}{2} = 10$

Solution $12 - \frac{a}{2} = 10$

$$12 - 12 - \frac{a}{2} = 10 - 12 \quad \text{Subtract 12 from each side of the equation.}$$

$$-\frac{a}{2} = -2$$

$$-\frac{a}{2}(-2) = -2(-2) \quad \text{Multiply each side of the equation by } -2.$$

$$a = 4$$

Check $12 - \frac{a}{2} = 10$

$$12 - \frac{4}{2} \stackrel{?}{=} 10 \quad \text{Substitute 4 for } a \text{ in the original equation.}$$

$$12 - 2 \stackrel{?}{=} 10$$

$$10 \stackrel{\checkmark}{=} 10$$

The solution is 4.

PRACTICE 9

Solve and check: $1 - \frac{x}{8} = -9$

EXAMPLE 10

A customer paid $80 for monthly cable service and $3.50 for each pay-per-view movie he ordered.

a. Write an equation to determine how many pay-per-view movies the customer ordered if his monthly bill was $101.

b. Solve this equation.

Solution

a. If we let x represent the number of pay-per-view movies ordered, then $3.5x$ represents the total amount paid for the pay-per-view movies.

PRACTICE 10

Suppose that an empty crate for oranges weighs 2 kilograms. A typical orange weighs about 0.2 kilogram and the total weight of the crate with oranges is 10 kilograms.

a. Write an equation to determine how many oranges the crate contains.

b. Solve this equation.

Next, we write a sentence for the problem and then translate it to an algebraic equation.

Total amount paid for pay-per-view movies	plus	Amount paid for cable service	equals	Total monthly bill
↓	↓	↓	↓	↓
$3.5x$	$+$	80	$=$	101

b. Now, we solve the equation: $3.5x + 80 = 101$.

$$3.5x + 80 - 80 = 101 - 80$$
$$3.5x = 21$$
$$x = 6$$

Check

$$3.5x + 80 = 101$$
$$3.5(6) + 80 \stackrel{?}{=} 101$$
$$21 + 80 \stackrel{?}{=} 101$$
$$101 \stackrel{\checkmark}{=} 101$$

So the customer ordered 6 pay-per-view movies.

9.1 Exercises

FOR EXTRA HELP MyMathLab MathXL PRACTICE WATCH DOWNLOAD READ REVIEW

Solve and check.

1. $a - 7 = -21$

2. $x - 6 = -9$

3. $b + 4 = -7$

4. $y + 12 = -12$

5. $-11 = z - 4$

6. $-15 = m - 20$

7. $x + 21 = 19$

8. $a + 12 = 10$

9. $c + 33 = 14$

10. $d + 27 = 15$

11. $z + 2.4 = -5.3$

12. $t + 2.3 = -6.7$

13. $2.3 = x - 5.9$

14. $4.1 = d - 6.9$

15. $y - 2\frac{1}{3} = -3$

16. $s - 4\frac{1}{2} = -8$

17. $n + \frac{1}{3} = \frac{1}{2}$

18. $\frac{1}{4} + t = -\frac{1}{6}$

19. $-5 = t + 1\frac{1}{4}$

20. $-3 = 1\frac{2}{3} + c$

21. $39 = z + 51$

22. $33 = c + 49$

23. $-5x = 30$

24. $-8y = 8$

25. $-36 = -9n$

26. $-125 = -25x$

27. $\frac{m}{-1.5} = 1$

28. $\frac{x}{-7} = 1.3$

29. $\frac{w}{10} = -24$

30. $\frac{a}{5} = -40$

31. $-6 = \frac{x}{-2}$

32. $-8 = \frac{y}{-4}$

33. $1.7t = -51$

34. $-1.5x = 45$

35. $\frac{y}{9} = -\frac{5}{3}$

36. $\frac{z}{3} = \frac{-4}{3}$

37. $-10y = 4$

38. $-15a = 3$

39. $4n - 20 = 36$

40. $3a - 13 = 11$

41. $3x + 1 = 7$

42. $7m + 1 = 22$

43. $6k + 23 = 5$

44. $2x + 21 = 7$

45. $3x + 20 = 20$

46. $4x + 28 = 28$

47. $31 = 3 - 4h$

48. $-68 = 10 - 3x$

49. $34 = 13 - 4p$

50. $36 = 25 - 3c$

51. $-7b + 8 = -6$

52. $-2x + 15 = -9$

53. $21 + \frac{a}{3} = 10$

54. $25 + \frac{w}{5} = 15$

55. $\frac{1}{2}y + 5 = -13$

56. $\frac{x}{5} + 15 = 0$

57. $5 - \frac{x}{12} = 1$

58. $16 - \frac{a}{2} = 15$

59. $\frac{c}{3} + 3 = -4$

60. $\frac{m}{4} + 1 = -5$

61. $\frac{4}{9}x - 13 = -5$

62. $\frac{5}{4}y - 19 = 26$

63. $-8 - x = 11$

64. $-24 = 2 - x$

65. $-7 - t = 0$

66. $10 = -3 - x$

Solve. Round each solution to the nearest tenth. Check.

67. $8{,}950 = -6.24n$

68. $-1{,}458 = 20.9p$

69. $-2.57 = \frac{x}{5.91}$

70. $-4.6 = \frac{z}{-2.78}$

71. $58.3r + 23.58 = 2.79$

72. $-51.5 = 29m - 4.06$

73. $\frac{x}{2.4} - 0.03 = -0.14$

74. $\frac{a}{2.7} + 11.9 = 0.02$

Mixed Practice

Solve and check.

75. $\frac{t}{-8} = -1.2$

76. $-10 = m + 6$

77. $6n + 21 = 15$

78. $1.9 + \frac{x}{4} = 2.1$

79. $32 = 27 - 2c$

80. $-1.4a = 42$

81. $-3 = y - 1\frac{1}{5}$

82. $\frac{2}{3}n - 5 = 4$

Applications

Write an equation. Solve and check.

83. On a particular day, the price of one share of a video gaming company's stock dropped by \$1.50. If a stockholder lost \$750 that day, how many shares of stock did he own?

84. The value of a networked color printer purchased new decreased by the same amount each year. After 4 years, the value of the printer dropped by \$1,300. By how much did the value decrease each year?

85. A small company lost a total of \$15,671 in the first two quarters of last year. If the company lost \$9,046 in the first quarter, how much did it lose in the second quarter?

86. If you were to write a check for \$350, your account would be overdrawn by \$200. What is the present balance of your account?

87. Fifteen members of the drama club at a college were given a special group discount of \$25 off the total price of their tickets. If they paid \$155 with the discount, what was the original price of each ticket?

88. A car rents for \$35 per day plus 30¢ per mile. How many miles were driven if the rental fee for a day was \$65?

89. At a graduation dinner, an equal number of guests were seated at each of 8 large tables, and 3 late-arriving guests were seated at a small table. If there were 43 guests in all, how many guests were seated at each of the large tables?

90. A toy maker has daily manufacturing costs of \$890, plus \$3 for each action figure it produces. How many action figures does the toy maker produce each day if the total daily manufacturing cost is \$5,390?

Use a calculator to solve the following problems, giving (a) the equation, (b) the exact answer, and (c) an estimate of the answer.

91. A homeowner's electric bill was \$67.35 last month. How many kilowatt hours were used if the electric company charges \$0.0625 per kilowatt hour plus a service fee of \$12.25 per month?

92. A long-distance service provider charges a flat monthly rate of \$2.99 plus \$0.05 per minute for long-distance phone calls. If a customer's monthly bill (excluding taxes and other fees) for long-distance service was \$21.04, how many minutes of long-distance calls was she charged for this month?

• Check your answers on page A-15.

MINDSTRETCHERS

Groupwork

1. In the following magic square, the sum of every row, column, and diagonal is 15. Working with a partner, solve for a, b, and c.

6	1	$\frac{c}{4} + 2$
7	$2b + 1$	3
$3a - 1$	9	4

Mathematical Reasoning

2. Suppose that there are three objects to weigh: **a**, **b**, and **c**. It is given that two of the three objects are equal in weight but that the third object weighs more than either of the other two. On the balance scale shown, indicate how to identify which object is the heaviest one by *only one weighing*.

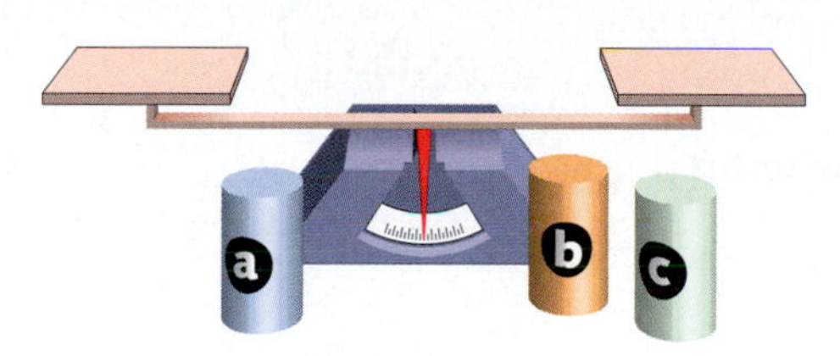

Writing

3. For each of the following equations, write two different situations that the equation models.

a. $12x + 500 = 24{,}500$

•

b. $\frac{x}{2} - 40 = 60$

•

9.2 More on Solving Equations

In this section, we extend the discussion to include equations that involve *like terms* or contain parentheses.

OBJECTIVES

- To combine like terms
- To solve equations involving like terms
- To solve equations involving parentheses
- To solve word problems with like terms or parentheses

Solving Equations Involving Like Terms

In Chapter 4, we considered algebraic expressions such as $2x$ or $x - 1$ that consist of one or two terms. Some algebraic expressions, such as $2x + x + 1$, have three terms.

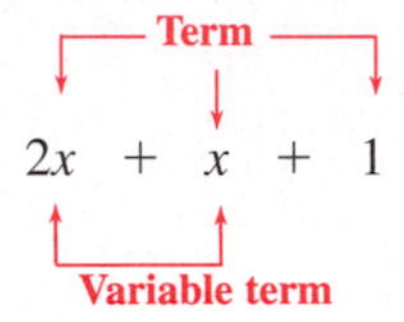

The terms $2x$ and x are like terms.

Definition

Like terms are terms that have the same variables with the same exponents. Terms that are not like are called **unlike terms.**

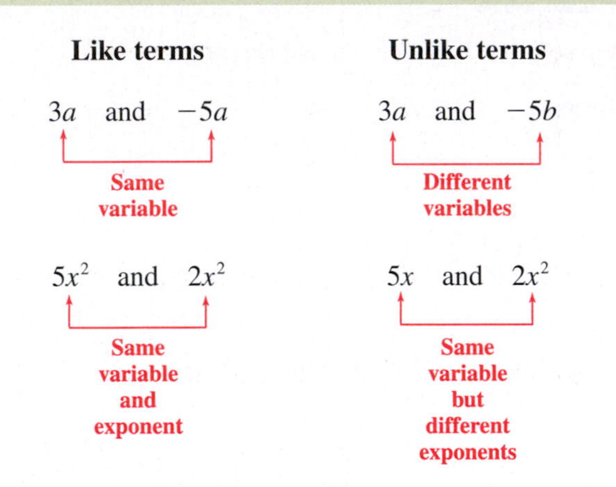

We cannot combine unlike terms, such as $3a$ and $-5b$. However, we can combine like terms using the following rule.

To combine like terms

- Use the Distributive Property in reverse.
- Add or subtract.

EXAMPLE 1

Combine like terms.

a. $7x + 3x$ **b.** $6y - 8y$ **c.** $4a + a + 9$

Solution

a. $7x + 3x = (7 + 3)x$ Use the Distributive Property in reverse.

$= 10x$ Add 7 and 3.

PRACTICE 1

Simplify.

a. $5x + 7x$

b. $7y - y$

c. $4z - 5z + 6$

b. $6y - 8y = 6y + (-8y)$

$= (6 + (-8))y$ Use the Distributive Property in reverse.

$= (6 - 8)y$

$= -2y$

c. $4a + a + 9 = (4 + 1)a + 9$ Use the Distributive Property in reverse. Recall that $a = 1a$.

$= 5a + 9$ Add 4 and 1.

Some equations have like terms on one side. To solve this type of equation, we begin by combining all like terms.

EXAMPLE 2

Solve and check: $3x - 5x = 10$

Solution $3x - 5x = 10$

$-2x = 10$ Combine like terms: $3x - 5x = (3 - 5)x = -2x$

$\frac{-2x}{-2} = \frac{10}{-2}$ Divide each side of the equation by -2.

$x = -5$

Check $3x - 5x = 10$

$3(-5) - 5(-5) \stackrel{?}{=} 10$ Substitute -5 for x in the original equation.

$-15 - (-25) \stackrel{?}{=} 10$

$10 \stackrel{\checkmark}{=} 10$

The solution is -5.

PRACTICE 2

Solve and check: $2y - 3y = 8$

Some equations have like terms on both sides. To solve this type of equation, we use the addition or subtraction rule to get like terms on the same side so that they can be combined.

EXAMPLE 3

Solve and check: $8y - 11 = 5y - 2$

Solution $8y - 11 = 5y - 2$

$8y - 5y - 11 = 5y - 5y - 2$ Subtract $5y$ from each side of the equation.

$3y - 11 = -2$ Combine like terms.

$3y - 11 + 11 = -2 + 11$ Add 11 to each side of the equation.

$3y = 9$

$y = 3$

Check $8y - 11 = 5y - 2$

$8(3) - 11 \stackrel{?}{=} 5(3) - 2$ Substitute 3 for y in the original equation.

$24 - 11 \stackrel{?}{=} 15 - 2$

$13 \stackrel{\checkmark}{=} 13$

The solution is 3.

PRACTICE 3

Solve and check: $10x - 1 = 2x + 7$

Recall that in Chapter 4 we translated to algebraic expressions phrases, such as *four more than n* and *twice n*. Now we look at phrases that involve more than one operation, such as *four more than twice n*, which translates to $2n + 4$. Some translations lead to multistep equations, as shown in Example 4.

EXAMPLE 4

The product of 5 and a number is 24 less than twice that number.

a. Write an equation to find the number.

b. Solve this equation.

Solution

a. If we let x represent the number, $5x$ represents the product of 5 and that number. Then we represent 24 less than twice that number by $2x - 24$.

We can now translate the given problem to an equation: $5x = 2x - 24$.

b. Next, we solve the equation $5x = 2x - 24$.

$$5x = 2x - 24$$
$$5x - 2x = 2x - 2x - 24$$
$$3x = -24$$
$$x = -8$$

Check $5x = 2x - 24$

$$5(-8) \stackrel{?}{=} 2(-8) - 24$$
$$-40 \stackrel{?}{=} -16 - 24$$
$$-40 \stackrel{\checkmark}{=} -40$$

The number is -8.

PRACTICE 4

The sum of a number and 3 times that number is 9 more than the number.

a. Write an equation to find the number.

b. Solve this equation.

Solving Equations Containing Parentheses

Some equations contain parentheses. To solve this type of equation, we first remove the parentheses, using the Distributive Property.

EXAMPLE 5

Solve and check: $2(6 - x) = -8$

Solution $2(6 - x) = -8$

$$12 - 2x = -8 \quad \text{Use the Distributive Property.}$$
$$12 - 12 - 2x = -8 - 12 \quad \text{Add } -12 \text{ to each side of the equation.}$$
$$-2x = -20$$
$$x = 10$$

Check $2(6 - x) = -8$

$$2(6 - 10) \stackrel{?}{=} -8 \quad \text{Substitute 10 for } x \text{ in the original equation.}$$
$$2(-4) \stackrel{?}{=} -8$$
$$-8 \stackrel{\checkmark}{=} -8$$

PRACTICE 5

Solve and check: $3(1-2x) = 9$

EXAMPLE 6

Solve and check: $6x = -4(x + 5)$

Solution

$$6x = -4(x + 5)$$
$$6x = (-4)(x) + (-4)(5) \quad \text{Use the Distributive Property.}$$
$$6x = -4x - 20$$
$$6x + 4x = -4x + 4x - 20 \quad \text{Add } 4x \text{ to each side of the equation.}$$
$$10x = -20 \quad \text{Combine like terms.}$$
$$x = -2$$

Check

$$6x = -4(x + 5)$$
$$6(-2) \stackrel{?}{=} -4(-2 + 5) \quad \text{Substitute } -2 \text{ for } x \text{ in the original equation.}$$
$$-12 \stackrel{?}{=} -4(3)$$
$$-12 \stackrel{\checkmark}{=} -12$$

The solution is -2.

PRACTICE 6

Solve and check: $-3(x - 4) = x$

EXAMPLE 7

A parking garage charges $6 for the first hour and $1.50 for each additional hour of parking. If a driver paid $13.50 to park in the garage, how many hours did she park?

Solution Let h represent the number of hours she parked. Then $h-1$ represents the number of hours for which she paid the hourly rate of $1.50.

Amount for first hour	plus	Amount for each additional hour	equals	Total cost of parking
↓	↓	↓	↓	↓
6	$+$	$1.5(h - 1)$	$=$	13.5

$$6 + 1.5(h - 1) = 13.5$$
$$6 + 1.5h - 1.5 = 13.5$$
$$1.5h + 4.5 = 13.5$$
$$1.5h = 9$$
$$h = 6$$

So the driver parked in the garage for 6 hours.

PRACTICE 7

A taxi fare is $3.50 for the first mile and $2.00 for each additional mile. If a passenger's total fare was $9.50, how far did he travel in the taxi?

9.2 Exercises

FOR EXTRA HELP MyMathLab MathXL PRACTICE WATCH DOWNLOAD READ REVIEW

Simplify.

1. $4x + 3x$
2. $y + 5y$
3. $4a - a$
4. $3d - 2d$
5. $6y - 9y$
6. $2x - 7x$
7. $2n - 3n$
8. $4z - 7z$
9. $2c + c + 12$
10. $5a + a - 1$
11. $8 + x - 7x$
12. $5 - a + 4a$
13. $-y + 5 + 3y$
14. $2x - 4 - x$

Solve and check.

15. $5m + 4m = 36$
16. $3y + 8y = 22$
17. $18 = 4y - 2y$
18. $24 = 3x - x$
19. $2a - 3a = 0$
20. $x - 7x = 12$
21. $7 = -5b + b$
22. $5 = -y + 3y$
23. $n + n - 13 = 13$
24. $y + 2y - 3 = 18$
25. $6 = 7x - 3x - 6$
26. $11 = s - 4s + 5$
27. $n + 3n - 7 = 29$
28. $6m + m - 9 = 30$
29. $6a - a + 4a = -6$
30. $2y + 3y - y = -8$
31. $5x = 2x + 12$
32. $3t + 8 = t$
33. $4p + 1 = 3p - 1$
34. $10w + 1 = 3w + 1$
35. $8x + 1 = x - 6$
36. $5x - 7 = 2x + 2$
37. $3p - 2 = -p + 4$
38. $2y - 3 = y + 1$
39. $4n - 6 = 3n + 6$
40. $2x + 1 = -x - 4$
41. $3 - 6t = 5t - 19$
42. $5 - 8r = 2r - 25$
43. $-7y + 2 = 3y - 8$
44. $-6s - 2 = -8s - 4$
45. $3x - 2\frac{1}{2} = 3\frac{1}{3} - 3x - x$
46. $4y - 7 = 4\frac{2}{3} - y - 2y$
47. $2(n + 3) = 12$
48. $3(b + 4) = 24$
49. $8(x - 1) = -24$
50. $3(r - 5) = -18$
51. $\frac{1}{2}(x + 12) = 7$
52. $0.4(10 - x) = 8$
53. $6n = 5(n + 7)$
54. $3x = 2(x + 4)$
55. $3(y - 5) = 2y$
56. $5(m - 3) = 2m$
57. $5y = -3(y + 1)$
58. $6x = -4(x - 1)$
59. $-4n = 7(9 - n)$
60. $2(3 - x) = -x$
61. $6r + 2(r - 1) = 14$
62. $3t + 2(t + 4) = 13$

Write an equation to find the number. Solve and check.

63. Five times a number subtracted from 3 times the number is equal to -20.

64. Twenty-one less than 5 times a number is equal to 28 more than -2 times the number.

65. Three times the sum of a number and 5 equals twice the number.

66. Eleven more than a number equals -2 times 1 less than the number.

Solve. Round each solution to the nearest tenth. Check.

67. $60r - 17r + 23.58 = 2.79$

68. $1.02m + 3.007m = 50.1$

69. $3.61n = 4 + 2.135n$

70. $0.138a = -4.667a + 2.931a - 4.625$

71. $1.72y = 3.16(y - 8.72)$

72. $6.19t + 3.81(1 - t) = 2.72$

Mixed Practice

Simplify.

73. $-x + 6 + 5x$

74. $-1 + a - 9a$

Solve and check.

75. $6 = -5y + y$

76. $7n - 5 = 2n - 5$

77. $2x + 3(x + 1) = -x$

78. $-8 = 8 - t + 5t$

79. $-5m - 1 = 1 + 5m$

80. $-2(y - 3) = 5.4$

Applications

Write an equation. Solve and check.

81. A homeowner sold her house through a broker who charged 6% of the selling price. After the commission was paid, the homeowner received \$211,500. Find the selling price.

82. A student purchased 8 tickets for a monster truck rally through an online ticket agency. A service charge of \$5 was added to the price of each ticket. If the total cost of the tickets was \$208, what was the price of one ticket?

83. The perimeter of a rectangle 70 feet long is 200 feet. Find the width, w.

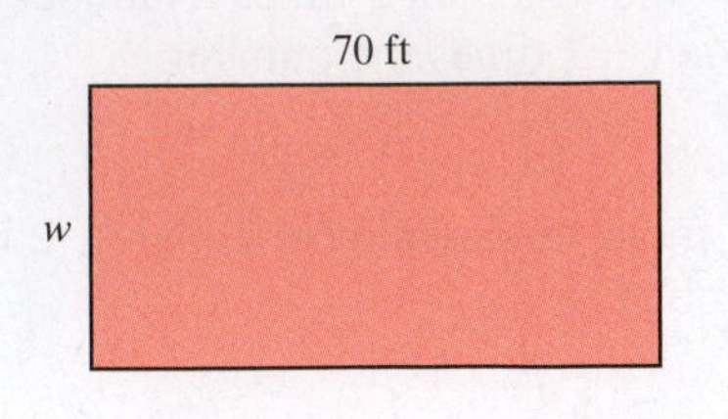

84. The sum of the measures of the angles of a triangle is 180°. In the triangle shown, what is the measure of the two equal angles, x?

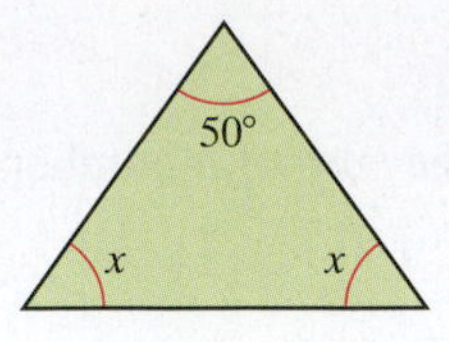

85. One-third of the proceeds from a fund-raiser was donated to a local charity, whereas one-half was donated to a national charity. If a total of $24,000 was donated, how much was donated to each charity?

86. A family's budget allows $\frac{1}{3}$ of their monthly income for housing and $\frac{1}{4}$ for food. If a total of $1,050 a month is budgeted for housing and food, what is their monthly income?

87. Two car dealerships offer lease options on comparable midsize sedans. One dealership requires $3,000 down and $120 per month. The other dealership requires $2,400 down and $145 per month. In how many months will a customer spend the same amount for each lease?

88. In any polygon with n sides, the sum of the measures of the interior angles is $180(n - 2)$ degrees. If the sum of the measures of the angles of a polygon is 540 degrees, how many sides does the polygon have?

89. In the spreadsheet shown, $C1 = A1 \cdot B1$, $B3 = B1 + B2$, $C2 = A2 \cdot B2$, and $C3 = C1 + C2$. Find B1.

	A	B	C
1	2		
2	3		
3		20	40

90. A company bought a copier for $10,000. After n years of depreciation, the copier is valued at $10{,}000\left(1 - \frac{n}{20}\right)$ dollars. After how many years will the copier be valued at $5,500?

MINDSTRETCHERS

Mathematical Reasoning

1. The *algebra tiles* pictured represent $3x + 1$ and $2x - 3$, respectively.

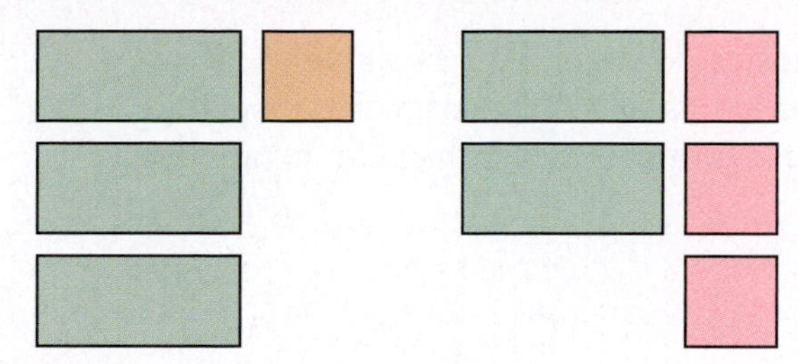

Represent each of the following expressions by algebra tiles.

a. $2x - 1$

b. $x + 4$

Use algebra tiles to represent the sum of the expressions in parts (a) and (b).

Writing

2. Write a list of steps for solving equations with variables on both sides.

Critical Thinking

3. Suppose that you work through a two-step equation and obtain $6x = 5x$. Can you solve this equation by dividing both sides by x? Explain.

9.3 Using Formulas

Translating Rules to Formulas

OBJECTIVES

- To translate a rule to a formula
- To evaluate formulas
- To solve word problems involving formulas

A **formula** is an equation that indicates how a number of variables are related to one another. In a formula, letters and mathematical symbols are used to represent words. We often use formulas as shorthand for expressing a stated rule or relationship.

EXAMPLE 1

To convert a temperature expressed in Celsius degrees, C, to Fahrenheit degrees, F, we multiply the Celsius temperature by $\frac{9}{5}$ and then add 32. Write a formula for this relationship.

Solution Stating the rule briefly in words, we get

Fahrenheit equals nine-fifths times Celsius plus 32.

Now we can easily translate the rule to mathematical symbols.

$$F = \frac{9}{5}C + 32$$

PRACTICE 1

To predict the temperature, t, at a particular altitude, a, scientists who study weather conditions subtract $\frac{1}{200}$ of the altitude from the temperature on the ground, g. Here t and g are in degrees Fahrenheit and a is in feet. Write this relationship as a formula.

Evaluating Formulas

In business and the health and physical sciences, as well as in many other areas of life, we often must evaluate a formula.

EXAMPLE 2

The formula $r = \frac{72}{y}$, known as the Rule of 72, gives the approximate rate of compound interest r (expressed as a percent) on an investment that doubles in y years. Find the approximate rate on an investment that doubles in 5 years.

Solution

$$r = \frac{72}{y}$$

$$= \frac{72}{5} \quad \text{Replace } y \text{ with 5.}$$

$$= 14.4$$

So r is 14.4%

PRACTICE 2

The formula for finding simple interest is $I = Prt$, where I is the interest, P is the principal, r is the rate of interest, and t is the time in years that the principal has been on deposit. Evaluate $I = Prt$ when $P = \$3{,}000$, $r = 0.06$, and $t = 2$ years.

EXAMPLE 3

A dietitian used the formula $I = 1.7h - 53$, where h is height in inches, to calculate the ideal body weight, I, in kilograms for women. According to this formula, what is the ideal body weight for a woman who is 65 inches tall?

Solution We use the formula $I = 1.7h - 53$ and substitute for h.

$$\begin{aligned} I &= 1.7h - 53 \\ &= 1.7(65) - 53 \\ &= 110.5 - 53 \\ &= 57.5 \end{aligned}$$

So her ideal body weight is 57.5 kilograms.

PRACTICE 3

When a cricket chirps n times per minute, the temperature outside in degrees Fahrenheit, F, can be found by using the following formula.

$$F = \frac{n}{4} + 37$$

What is the temperature outside when a cricket chirps 10 times per minute?

EXAMPLE 4

The top of a can of soda is a circle with radius 1 inch. Find the area of the top of the can. Use 3.14 for π.

Solution We use the formula $A = \pi r^2$ and substitute for r.

$$\begin{aligned} A &= \pi r^2 \\ &\approx 3.14 \times (1)^2 \\ &\approx 3.14 \times 1 \\ &\approx 3.14 \end{aligned}$$

So the top of the can has area 3.14 square inches. Note that since the radius of the circle is measured in inches, the area of the circle is measured in square inches.

PRACTICE 4

The body mass index (BMI), a ratio used by doctors who study obesity, can be found using the following formula.

$$BMI = \frac{w}{h^2}$$

where w is the weight in kilograms and h is the height in meters. Find the BMI, measured in kilograms per square meter, for a person 2 meters tall who weighs 83 kilograms. Round to the nearest whole number.

Translate each stated rule or relationship to a formula.

1. To figure out how far away in kilometers a bolt of lightning hits the ground, d, we divide the number of seconds, t, between the flash of lightning and the associated sound of thunder by 3.

2. The length, l, of a certain spring in centimeters is 25 more than 0.4 times the weight, w, in grams of the object hanging from it.

3. In baseball, a batting average, a, is computed by dividing the number of hits made, h, by the number of times up at bat, n.

4. In a particular state, the sales tax, t, can be computed by multiplying 0.0625 by the price of the item sold, p.

5. The total surface area, A, of a cube equals six times the square of the length of one of its edges, e.

6. The equivalent energy, E, of a mass equals the product of the mass, m, and the square of the speed of light, c.

Evaluate each formula for the given quantity in Exercises 7–14.

	Formula	Given	Find
7.	$F = \frac{9}{5}C + 32$	$C = -5°$	F
8.	$C = K - 273$	$K = 270°$	C
9.	$y = mx + b$	$m = -\frac{1}{2}$, $x = -3$, and $b = -4$	y
10.	$S = \frac{n(n+1)}{2}$	$n = 13$	S
11.	$P = a + b + c$	$a = 7$ inches, $b = 5.8$ inches, and $c = 6.1$ inches	P
12.	$A = \frac{a+b+c}{3}$	$a = -8$, $b = 6$, and $c = -4$	A
13.	$S = \frac{ad + bc}{bd}$	$a = 1$, $b = 5$, $c = 3$, and $d = 8$	S
14.	$R = \frac{s^2}{A}$	$s = 10$ feet and $A = 25$ square feet	R

Applications

Solve.

15. The formula for finding the present value of an item that depreciates yearly is $v = c - crt$. In this formula, v is the present value, c is the original cost, r is the rate of depreciation per year, and t is the number of years that have passed. After 5 years, what is the value of a car originally costing \$27,000 that depreciated at a rate of 0.1 per year?

16. The formula $A = P(1 + rt)$ is used for calculating the current amount in an account earning simple interest. Here, A is the current amount, P is the original amount deposited into the account, r is the annual interest rate, and t is the time in years. What is the current amount in the account after 3 years if \$500 was deposited into an account earning 4% annually?

17. One of the formulas for calculating the correct dosage for a child is as follows.

$$C = \frac{a}{a + 12} \cdot A$$

Here, C is the child's dosage in milligrams, a is the age of the child, and A is the adult dosage in milligrams. What is the prescribed dosage of a certain medicine for a child 6 years old, if the adult dosage of the medicine is 180 milligrams?

18. A *regular* polygon has equal sides and equal interior angles. The measure of an interior angle of a regular polygon is given by the formula $a = \frac{180(n - 2)}{n}$, where a is in degrees and n is the number of sides of the polygon. What is the measure of an interior angle of the regular polygon shown?

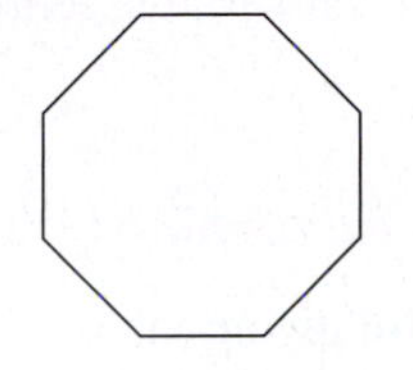

Use a calculator to solve the following problems, giving (a) the equation, (b) the exact answer, and (c) an estimate of the answer.

19. Scientists studying the American buffalo use formulas to predict the number of male, m, and female, f, calves that will be born in a year.

$$m = 0.48A \qquad \text{and} \qquad f = 0.42A$$

This prediction is based on the number of adult females, A, in the herd. If there are 328 adult females in a particular herd, how many calves will be born during the coming year?

20. A formula for the braking distance, d, in feet of a car traveling on dry pavement is $d = \frac{s^2}{25}$, where s is the speed in miles per hour. The corresponding formula for wet pavements is $d = \frac{s^2}{15}$. At a speed of 74 mph, how much greater is the braking distance on wet pavement than on dry pavement, to the nearest foot?

• *Check your answers on page A-15.*

MINDSTRETCHERS

Writing

1. A formula used in mathematics is

$$S = \frac{a}{1 - r}$$

For which value of r is there no value of S? Explain.

Groupwork

2. Working with a partner, give an example of a situation that the following formula might describe.

$$d = bc - a$$

Explain what each variable represents in your example.

Mathematical Reasoning

3. Consider the following table.

Counting Number, C	1	2	3	4	…	1,000
Odd Number, O	1	3	5	7	…	1,999

Write a formula expressing O in terms of C.

CULTURAL NOTE

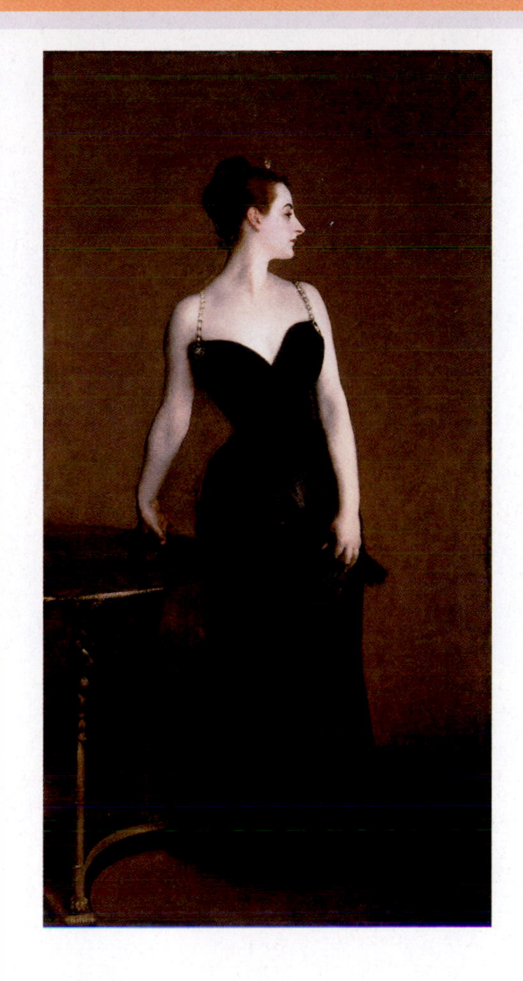

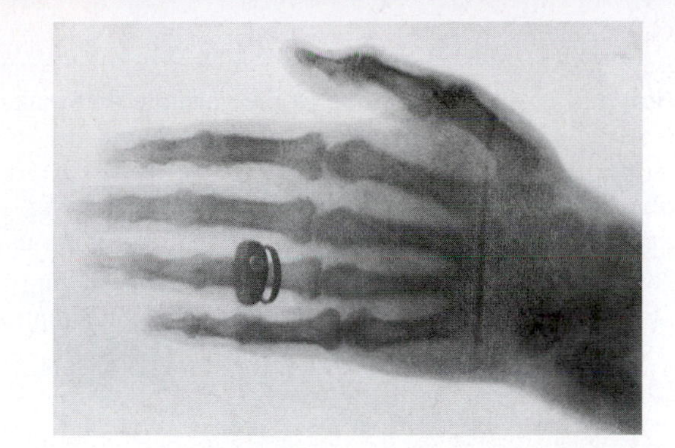

These two pictures both reflect the mathematical tradition of representing an unknown by the letter x. The painting of an unnamed woman, entitled *Madame X* (1884), is by the American artist John Singer Sargent. About a dozen films made throughout the twentieth century dealing with enigmatic women bear the same name.

The small photograph is one of the first taken by the German physicist Wilhelm Roentgen in 1895 after he discovered strange rays that he called *X-rays* since they were a mysterious phenomenon. Roentgen's discovery allowed doctors to see inside the body for the first time without surgery and earned him the first Nobel prize awarded for physics.

The practice of using an x and other letters from the end of the alphabet to represent mathematical unknowns goes back to the seventeenth-century French mathematician René Descartes, who made major contributions to the development of algebra.

Source: Florian Cajori, *A History of Mathematical Notations* (Chicago: The Open Court Publishing Company, 1929).

KEY CONCEPTS AND SKILLS

CONCEPT SKILL

Concept/Skill	Description	Example
[9.1] **Solution**	A value of the variable that makes an equation a true statement	-2 is a solution of $x + 5 = 3$ because $-2 + 5 = 3$.
[9.1] **To solve a two-step equation**	• First use the rule for solving addition or subtraction equations. • Then use the rule for solving multiplication or division equations.	$2y - 7 = 13$ $2y - 7 + 7 = 13 + 7$ $2y = 20$ $\frac{2y}{2} = \frac{20}{2}$ $y = 10$
[9.2] **Like terms**	Terms that have the same variables with the same exponents	$2x$ and $3x$
[9.2] **Unlike terms**	Terms that are not like	$4a^2$ and $3a$
[9.2] **To combine like terms**	• Use the Distributive Property in reverse. • Add or subtract.	$2x + 3x = (2 + 3)x$ $= 5x$
[9.3] **Formula**	An equation that indicates how a number of variables are related to one another	$I = \frac{V}{R}$

Chapter 9 Review Exercises

To help you review this chapter, solve these problems.

[9.1] *Solve and check.*

1. $x + 2 = -4$

2. $y + 3 = -6$

3. $d + 9 = 0$

4. $w + 11 = 0$

5. $8 = x + 17$

6. $4 = y + 20$

7. $p + 11 = 11$

8. $x - 9 = -9$

9. $a + 7.4 = -2.3$

10. $b + 3.2 = -5.8$

11. $-10 = y - 4\frac{1}{2}$

12. $-12 = d - 5\frac{2}{3}$

13. $-4x = 28$

14. $-9c = 9$

15. $4 = -16p$

16. $6 = -9a$

17. $-1.6y = -32$

18. $-2.5w = -75$

19. $\frac{b}{6} = -30$

20. $\frac{x}{10} = -45$

21. $\frac{x}{-5} = 1.5$

22. $\frac{a}{-2} = 2.7$

23. $3x + 1 = 13$

24. $2a + 3 = 7$

25. $4y - 3 = 17$

26. $5w - 1 = 9$

27. $2y - 1 = 6$

28. $3y - 5 = 3$

29. $\frac{a}{3} + 1 = 9$

30. $\frac{w}{4} + 3 = 7$

31. $\frac{x}{5} + 4 = -1$

32. $\frac{b}{2} + 2 = -3$

33. $\frac{c}{7} - 1 = -1$

34. $\frac{d}{8} - 2 = -2$

35. $-c - 6 = 0$

36. $-a - 9 = 9$

[9.2] *Solve and check.*

37. $4y - 2y = 18$

38. $-2b + 7b = 30$

39. $2c + c = -6$

40. $-8x + 3x = -11$

41. $y + y + 2 = 18$

42. $5 - t - t = -1$

43. $3x - 4x + 6 = -2$

44. $7 = 4m - 2m + 1$

45. $0 = -7n + 4 - 5n$

46. $a - 5a - 6 = 30$

47. $4r + 2 = 3r - 6$

48. $-5y + 8 = -3y + 10$

49. $2x - 8 = x + 1$

50. $4y - 1 = 2y - 3$

51. $7s + 4 = 5s + 8$

52. $2 + 8x = 3x - 8$

53. $5m - 9 = 9 - 4m$

54. $x = -x + 12$

55. $s + s - 6\frac{2}{3} = 4\frac{1}{3} + s$

56. $5z - 6\frac{1}{2} = z - 4z + \frac{1}{2}$

57. $2(8 + w) = 22$

58. $-3(z + 5) = -15$

59. $6x + 12(10 - x) = 84$

60. $3(y - 1) + y = 37$

61. $6(y + 4) = 2y - 8$

62. $5(m - 1) = 11 - m$

63. $0 = \frac{1}{3}(b + 9) + 2$

64. $-\frac{1}{5}(d - 5) = 9$

Write an equation to find the number. Solve and check.

65. The sum of 6 times a number and 3 is equal to 17 less than the number.

66. The sum of 3 times a number and twice 1 less than the number is 33.

[9.3] ***Write each relationship as a formula.***

67. Engineers use the Rankin temperature scale, where a Rankin temperature, R, is 460° more than the corresponding Fahrenheit temperature, F.

68. The speed of a sound wave, s, is given by the frequency, f, multiplied by the wavelength, w.

69. In baseball, a pitcher's earned run average, A, is calculated by multiplying 9 by the ratio of the number of earned runs, E, to the number of innings pitched, I.

70. The midrange, m, of a collection of numbers is one-half the sum of the smallest number, s, and the largest number, l.

Evaluate each formula for the given quantity.

	Formula	Given	Find
71.	$F = ma$	$m = 3.6$ and $a = 14$	F
72.	$d = 16t^2$	$t = 3$	d
73.	$P = n(p - c)$	$n = 8$, $p = 350$, and $c = 240$	P
74.	$C = \frac{5}{9}(F - 32)$	$F = -4$	C

Mixed Applications

Solve.

75. In marketing, the cost, c, of a certain number of units of an item can be found by using the formula $c = p \cdot n$, where p is the price per unit and n is the number of units. If the total cost of 50 units of an item is \$269, what is the price per unit?

76. A video artist plans to make a down payment of \$45 on a digital camcorder and then make weekly payments of \$50. How many weeks will it take him to pay for the camcorder if it sells for \$895?

77. The perimeter of a rectangular rug is 42 feet. If the length of the rug is 12 feet, what is the width?

78. An auto repair shop charges a customer \$96 for parts and \$40 per hour for labor. How many hours were charged for labor if the bill for the repair was \$256?

79. The density, D, of an object can be found by using the formula $D = \frac{m}{V}$, where m is its mass and V is its volume. Find the density of a piece of ice if $m = 10$ grams and $V = 10.9$ cubic centimeters. Round the answer to the nearest tenth.

80. When police discover a human thigh bone, police doctors need to be able to predict the height of the entire skeleton. A formula that they use is $h = 2.38t + 61.41$, where h is the height of the skeleton and t is the length of the thigh bone, both in centimeters. What length thigh bone, rounded to the nearest centimeter, would lead to a predicted height of 150 centimeters? (***Source:*** Colin Evans, *Casebook of Forensic Detection*)

• *Check your answers on page A-15.*

Chapter R POSTTEST

FOR EXTRA HELP

Test solutions are found on the enclosed CD.

To see whether you have mastered the topics in this chapter, take this test.

Solve and check.

1. $x + 5 = 0$

2. $y - 6 = -6$

3. $27 = -9a$

4. $-2 = \frac{b}{5}$

5. $2x + 3 = 4$

6. $-9 = 4y + 1$

7. $2 - y = 11$

8. $\frac{a}{5} + 6 = -6$

9. $3x - 5x = 12$

10. $-4c + 3 = 7 - 2c$

11. $-3y - 4 = 8 - y$

12. $-2x = 4(x - 3)$

Write each relationship as an equation or formula.

13. According to Newton's second law of motion, force, F, is equal to the mass, m, of an object times its acceleration, a.

14. Four times the sum of a number and 3 is 5 times that number.

Write an equation for each word problem. Then solve and check.

15. The cost of a long-distance telephone call to your friend is \$0.35 for the first minute and \$0.16 for each additional minute. What was the total length of a call that cost \$2.75?

16. An automobile dealer sells cars at a price that is $\frac{8}{9}$ the suggested retail price, plus a \$100 handling fee. What is the suggested retail price of a car that the dealer sells for \$12,906?

17. A fitness club charges an initial fee of \$200 plus \$30 per month for membership. In how many months will a member pay a total of \$500 for membership?

18. An investor put the same amount of money into a high-risk fund and a low-risk fund. After 1 year, the high-risk fund had a return of 6% and the low-risk fund had a return of 4%. If the total return on her investments was \$80, how much did she invest in each fund?

Solve

19. The *reaction distance* is the distance that a car travels during the time that the driver is getting his or her foot on the brake. Studies have shown that the reaction distance, d, is approximately 2.2 times the rate, r, of the car. Here, the reaction distance is in feet, and the rate is in miles per hour. (***Source:*** National Highway Traffic Safety Administration)

a. Write a formula for the reaction distance.

b. Find the distance when $r = 55$ miles per hour.

20. To calculate the speed, f, of an object in feet per second, we multiply its speed, m, expressed in miles per hour by $\frac{22}{15}$.

a. Write this relationship as a formula.

b. If an object is moving 45 miles per hour, what is its speed in feet per second?

• *Check your answers on page A-15.*

Cumulative Review Exercises

To help you review, solve the following.

1. Subtract: $5\frac{1}{3} - 2\frac{1}{2}$

2. Find the median: 5, 9, 3, and 8

3. Compute: $(0.1)^3$

4. What percent of 30 is 25?

5. Calculate: $(-1)^2 - 3$

6. You serve on a large military ship on which there are 75 officers and 1,275 enlisted personnel. What is the ratio of officers to enlisted personnel among the ship's crew?

7. Solve: $3x - 5 = 13$

8. A doctor changed a patient's dosage of thyroxine from 0.075 milligrams to 0.1 milligrams. Explain whether this change represented an increase or a decrease.

9. The following graph shows the growth of BlackBerry subscribers in recent years.

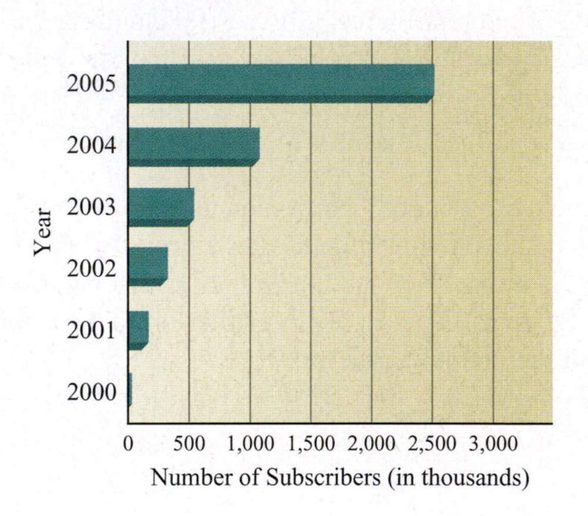

Approximately how many more BlackBerry subscribers were there in 2005 than in 2003?

10. According to some medical research, women with a body mass index between 22.0 and 23.4 are likely to have the longest life span. The formula for body mass index (represented by the letter i) is as follows.

$$i = \frac{697.5w}{h^2}$$

In this formula, the weight, w, is in pounds and the height, h, is in inches. Estimate i if $w = 100$ pounds and $h = 60$ inches. (***Source:*** *The New England Journal of Medicine*)

• *Check your answers on page A-15.*

CHAPTER 10

Measurement and Units

10.1 U.S. Customary Units

10.2 Metric Units and Metric/U.S. Customary Unit Conversions

Units and International Trade

Although American consumers have been reluctant to change, the United States is increasingly "going metric." The federal government is replacing inches with centimeters and gallons with liters in its business dealings. The largest U.S. car manufacturer began building car parts in metric units (everything except the speedometer) in the early 1970s. Most packaged goods are now labeled in grams and liters as well as ounces and quarts.

Industry leaders may have to switch over to the metric system if they want to survive in the global economy. One major U.S. corporation had a shipment of appliances rejected by Saudi Arabia because their electrical cords were 6 feet long instead of the required standard of 2 meters (approximately 6.6 feet).

However, a complete switch to the metric system would be costly, entailing endless changes in machines, tools, dials, containers, signs, contracts, and laws. (*Source:* United States Metric Association)

Chapter 10 PRETEST

To see if you have already mastered the topics in this chapter, take this test.

Solve.

1. 8 qt = ________ gal

2. 5 tons = ________ lb

3. Add: 7 ft 11 in. + 4 ft 7 in.

4. Which of the following units is a measure of length?

a. a gram **b.** a meter **c.** a liter **d.** a second

5. The width of a dime is about ______.

a. a meter **b.** a kilometer

c. a millimeter **d.** a centimeter

6. 3.5 kg = ________ g

7. 2,100 mm = ________ m

8. Which is larger: 2.3 m or 700 cm?

9. A teaching assistant leased a car for 36 months. For how many years did she lease the car?

10. The IMAX/IWERKS movie *Everest* was shot on 65-mm-wide film. Convert this width to centimeters. (*Source:* http://www.pbs.org)

• *Check your answers on page A-16.*

10.1 U.S. Customary Units

OBJECTIVES

- To identify units in the U.S. customary system
- To change a measurement from one U.S. customary unit to another
- To add and subtract measurements expressed in U.S. customary units

What Measurement and Units Are and Why They Are Important

When taking measurements, we express quantities in standardized units such as pounds and yards, which enable us to compare characteristics of physical objects. With standardized units, we can decide which of two packages is heavier or how many times longer one room is than another.

Throughout this chapter, we focus on four kinds of measure: length, weight, capacity, and time. The units that we use to express these measurements come from two systems of measurement: the U.S. customary system and the metric system.

U.S. Customary Units of Length, Weight, Capacity, and Time

We begin with U.S. customary units (also called U.S. units). These are the units that we use most often in everyday situations.

Sometimes, we need to change the unit in which a measurement is expressed. To convert units, we must know how different units are related.

In measuring *length*, the main U.S. units are inches (in.), feet (ft), yards (yd), and miles (mi). The key conversion relationships among these units are as follows.

$$12 \text{ in.} = 1 \text{ ft}$$
$$3 \text{ ft} = 1 \text{ yd}$$
$$5{,}280 \text{ ft} = 1 \text{ mi}$$

The main U.S. units of *weight* are ounces (oz), pounds (lb), and tons. Note the following relationships.

$$16 \text{ oz} = 1 \text{ lb}$$
$$2{,}000 \text{ lb} = 1 \text{ ton}$$

Next, let's consider the main U.S. units of *capacity* (*liquid volume*): fluid ounces (fl oz), cups (c), pints (pt), quarts (qt), and gallons (gal). Here is how these units are related.

$$8 \text{ fl oz} = 1 \text{ c}$$
$$2 \text{ c} = 1 \text{ pt}$$
$$2 \text{ pt} = 1 \text{ qt}$$
$$4 \text{ qt} = 1 \text{ gal}$$

Finally, there are the main U.S. units of *time*: seconds (sec), minutes (min), hours (hr), days, weeks (wk), months (mo), and years (yr). Some of the key relationships among these units of time are as follows.

60 sec = 1 min
60 min = 1 hr
24 hr = 1 day
7 days = 1 wk
52 wk = 1 yr
12 mo = 1 yr
365 days = 1 yr

Study these relationships so that you can use them to solve problems involving units.

Note that the abbreviations of the units are the same regardless of whether they are singular or plural. For example, the abbreviation for foot (ft) is the same as the abbreviation for feet (ft).

Changing Units

Suppose that you are ordering the rug pictured from a catalog. If you know that the length of the space to be covered is 2 ft, how long is it in inches? To change a length given in feet to inches, we need to know that there are 12 in. in 1 ft.

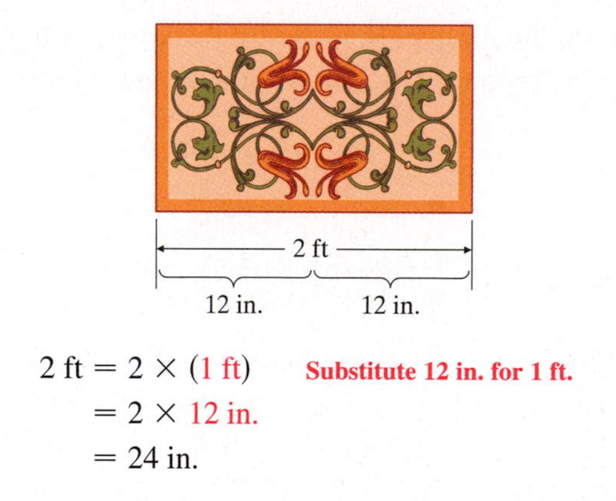

$$\begin{aligned} 2 \text{ ft} &= 2 \times (1 \text{ ft}) \\ &= 2 \times 12 \text{ in.} \\ &= 24 \text{ in.} \end{aligned}$$

Substitute 12 in. for 1 ft.

The rug is 24 in. long.

> **Tip** When we change *a large unit to a small unit,* the numerical part of the answer is larger than the original number. But when we change *a small unit to a large unit*, the numerical part of the answer is smaller than the original number.

Another way of solving this problem is to multiply the original measurement by the *unit factor* $\frac{12 \text{ in.}}{1 \text{ ft}}$. The unit factor method, commonly used in the physical and health sciences, is particularly helpful in solving complex conversion problems.

Because the numerator 12 in. and the denominator 1 ft represent the same length, the unit factor $\frac{12 \text{ in.}}{1 \text{ ft}}$ is equivalent to 1.

$$2 \text{ ft} = 2 \cancel{\text{ft}} \frac{12 \text{ in}}{1 \cancel{\text{ft}}}$$

$$= \frac{2 \times 12 \text{ in.}}{1} = 24 \text{ in.}$$

Note how we simplified the answer by canceling common units, as if the units were numbers.

Or suppose that we wanted to change 24 in. to feet. We can solve this problem by multiplying the original measurement by the unit factor $\frac{1 \text{ ft}}{12 \text{ in.}}$.

$$24 \text{ in.} = \overset{2}{\cancel{24}} \cancel{\text{in.}} \times \frac{1 \text{ ft}}{\underset{1}{\cancel{12}} \cancel{\text{in.}}}$$

$$= 2 \text{ ft}$$

Let's look back at the two problems that we just solved. Both involved inches and feet. In the first problem, we multiplied by the unit factor $\frac{12 \text{ in.}}{1 \text{ ft}}$; in the second problem, we multiplied by its reciprocal, $\frac{1 \text{ ft}}{12 \text{ in.}}$. (Note that both unit factors are equivalent to 1.)

In general, when converting from one unit to another unit, we multiply the original measurement by the unit factor that has the desired unit in its numerator and the original unit in its denominator.

EXAMPLE 1

5 qt = _____ pt

Solution From the capacity chart on page 451 we know that 2 pt = 1 qt.

We want to change from *quarts* to *pints*, so we use the unit factor $\frac{2 \text{ pt}}{1 \text{ qt}}$.

$$5 \text{ qt} = 5 \cancel{\text{qt}} \times \frac{2 \text{ pt}}{1 \cancel{\text{qt}}}$$

$$= 10 \text{ pt}$$

PRACTICE 1

32 oz = ____ lb

EXAMPLE 2

Express 30 min in hours.

Solution $30 \text{ min} = \overset{1}{\cancel{30}} \cancel{\text{min}} \times \frac{1 \text{ hr}}{\underset{2}{\cancel{60}} \cancel{\text{min}}}$

$$= \frac{1}{2} \text{ hr}$$

PRACTICE 2

Change 5 ft to yards.

EXAMPLE 3

How many cups are equivalent to 3 gal?

Solution To solve this problem, we must use several steps.

Step 1. Change gallons to quarts.

Step 2. Change quarts to pints.

Step 3. Change pints to cups.

To combine these three steps, we multiply the original measurement by a chain of appropriate unit factors to get *cups* in the final answer.

$$3 \text{ gal} = 3 \cancel{\text{gal}} \times \frac{4 \cancel{\text{qt}}}{1 \cancel{\text{gal}}} \times \frac{2 \cancel{\text{pt}}}{1 \cancel{\text{qt}}} \times \frac{2 \text{ c}}{1 \cancel{\text{pt}}}$$
$$= \frac{3 \times 4 \times 2 \times 2}{1} \text{ c}$$
$$= 48 \text{ c}$$

PRACTICE 3

How many seconds are there in a day?

EXAMPLE 4

The following sign is displayed on a bridge:

If a driver's car weighs about 3,000 lb, can she cross the bridge safely?

Solution To compare the weight limit of the bridge and the weight of the car, let's express the two measurements in the same unit.

$$2 \text{ tons} = 2 \cancel{\text{tons}} \times \frac{2{,}000 \text{ lb}}{1 \cancel{\text{ton}}}$$
$$= 4{,}000 \text{ lb}$$

Because the bridge can support 4,000 lb—a lot more than the car weighs—the driver can safely cross the bridge.

PRACTICE 4

A grocery store had 3 pt of onion dip in stock. If a shopper needed 2 qt of dip for a party, was the stock in the grocery store sufficient?

In Example 4, we used the unit factor $\frac{2{,}000 \text{ lb}}{1 \text{ ton}}$ instead of $\frac{1 \text{ ton}}{2{,}000 \text{ lb}}$. Explain why.

Adding and Subtracting Mixed Units

Some measurements involve more than a single unit. For instance, we may express in *mixed units* the height of a child as 3 ft 6 in. or the weight of a package as 10 lb 7 oz.

When a measurement is expressed in mixed units, we write the larger unit with the largest possible whole number. For instance, the length of a movie is written as 3 hr 5 min and not as 2 hr 65 min, even though the two lengths of time are equal.

EXAMPLE 5

Convert 2 lb 15 oz to ounces.

Solution
$$2 \text{ lb } 15 \text{ oz} = 2 \cancel{\text{lb}} \times \frac{16 \text{ oz}}{1 \cancel{\text{lb}}} + 15 \text{ oz}$$
$$= (32 + 15) \text{ oz}$$
$$= 47 \text{ oz}$$

PRACTICE 5

Write 1 ft 7 in. in inches.

EXAMPLE 6

Change 271 min to hours and minutes.

Solution
$$271 \text{ min} = 4 \times 60 \text{ min} + 31 \text{ min}$$
$$= \left(4 \times 60 \cancel{\text{min}} \times \frac{1 \text{ hr}}{60 \cancel{\text{min}}}\right) + 31 \text{ min}$$
$$= 4 \text{ hr } 31 \text{ min}$$

PRACTICE 6

What is 20 mo expressed in years and months?

There are many practical situations in which we need to add or subtract measurements written in mixed units—for instance, when we want to find out how much longer one car is than another or how much paint there will be if we combine the contents of several cans.

In addition or subtraction problems, only quantities having the same unit can be added or subtracted.

$$\begin{array}{r} 1 \text{ ft } 4 \text{ in.} \\ +1 \text{ ft } 2 \text{ in.} \\ \hline 2 \text{ ft } 6 \text{ in.} \end{array} \qquad \begin{array}{r} 2 \text{ hr } 10 \text{ min} \\ -1 \text{ hr } 3 \text{ min} \\ \hline 1 \text{ hr } 7 \text{ min} \end{array}$$

Often addition or subtraction problems involve changing units.

EXAMPLE 7

Find the sum:
$$\begin{array}{r} 7 \text{ ft } 9 \text{ in.} \\ +2 \text{ ft } 5 \text{ in.} \\ \hline \end{array}$$

Solution

$$\begin{array}{r} 7 \text{ ft } 9 \text{ in.} \\ +2 \text{ ft } 5 \text{ in.} \\ \hline 14 \text{ in.} \end{array}$$

Start with the inches column. (The smaller unit is always in the column on the right.) Add the numbers in the inches column.

$$\begin{array}{r} 1 \text{ ft} \phantom{\text{ 9 in.}} \\ 7 \text{ ft } 9 \text{ in.} \\ +2 \text{ ft } 5 \text{ in.} \\ \hline \cancel{14 \text{ in.}} \\ 2 \text{ in.} \end{array}$$

Because 14 in. = 1 ft 2 in., replace the 14 in. by 2 in. and carry the 1 ft to the feet column.

$$\begin{array}{r} 1 \text{ ft} \phantom{\text{ 9 in.}} \\ 7 \text{ ft } 9 \text{ in.} \\ +2 \text{ ft } 5 \text{ in.} \\ \hline 10 \text{ ft } 2 \text{ in.} \end{array}$$

Add the numbers in the feet column.

The sum is 10 ft 2 in.

PRACTICE 7

Add 3 lb 10 oz and 1 lb 14 oz.

EXAMPLE 8

A local theater was showing a double feature of two of your favorite films: *Harry Potter and the Sorcerer's Stone* and *Harry Potter and the Chamber of Secrets*. The first film runs 2 hr 32 min, and the second, 2 hr 41 min. How long was the double feature?

Solution

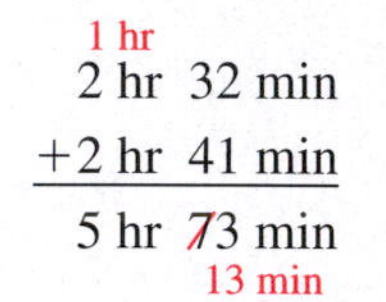

Start with the minutes column. Because 73 min = 1 hr 13 min, replace the 73 min with 13 min and carry the 1 hr to the hours column. Add the number in the hours column.

So the double feature ran 5 hr 13 min.

PRACTICE 8

The first leg of a cross-country flight from Boston to Los Angeles took 2 hr 45 min. The second leg took 4 hr 25 min. How long was the flight time from Boston to Los Angeles?

Now let's look at some examples of subtraction.

EXAMPLE 9

Subtract 1 yd 2 ft from 3 yd 1 ft.

Solution

$$\begin{array}{r} \overset{2}{\cancel{3}} \text{ yd } \overset{4}{\cancel{1}} \text{ ft} \\ -1 \text{ yd } 2 \text{ ft} \\ \hline \end{array}$$

Start with the feet column. Because 2 ft is larger than 1 ft, replace the 3 yd with 2 yd, and the borrowed yard with 3 ft, which when added to 1 ft gives 4 ft.

$$\begin{array}{r} \overset{2}{\cancel{3}} \text{ yd } \overset{4}{\cancel{1}} \text{ ft} \\ -1 \text{ yd } 2 \text{ ft} \\ \hline 1 \text{ yd } 2 \text{ ft} \end{array}$$

Subtract the numbers within each column.

Check Remember that we can check a subtraction by using addition.

$$\begin{array}{r} \overset{1 \text{ yd}}{1 \text{ yd }} 2 \text{ ft} \\ +1 \text{ yd } 2 \text{ ft} \\ \hline 3 \text{ yd } \underset{1 \text{ ft}}{\cancel{4}} \text{ ft} \end{array}$$

Adding, we get 3 yd 1 ft, so our answer checks.

PRACTICE 9

How much greater is 6 gal than 5 gal 1 qt?

EXAMPLE 10

The *Mona Lisa* is one of the best known paintings in the world.

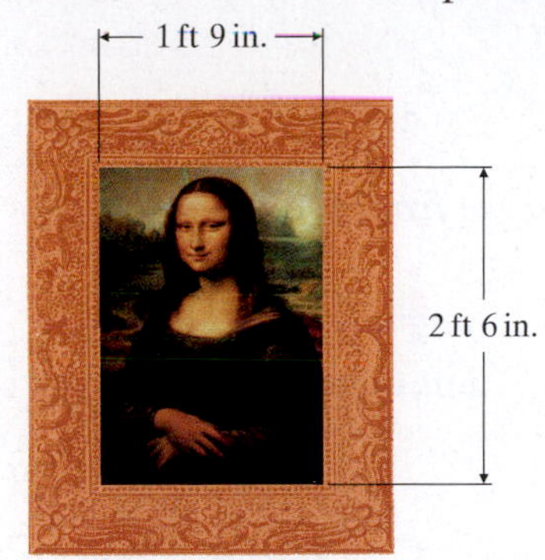

What is the difference between the height and width of the painting?

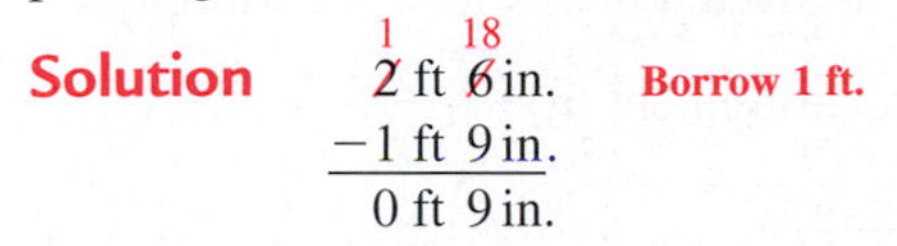

Solution

$$\begin{array}{r} \overset{1}{\cancel{2}}\text{ ft } \overset{18}{\cancel{6}}\text{ in.} \\ -1\text{ ft } 9\text{ in.} \\ \hline 0\text{ ft } 9\text{ in.} \end{array}$$

Borrow 1 ft.

The difference between the height and the width is 9 in.

PRACTICE 10

Two of the most famous horses in history were Citation and Secretariat. In 1948, Citation won the Kentucky Derby with a time of 2 min 5 sec. Twenty-five years later, Secretariat won in 1 min 59 sec. What was the difference between the two times?

(***Source:*** *Facts and Dates of American Sports*)

10.1 Exercises

Mathematically Speaking

Fill in each blank with the most appropriate term or phrase from the given list.

smaller	pound	gallon	weight
larger	numerator	denominator	unit factor
sign	length	unit	

1. In the U.S. customary system, the yard is a unit of _______.

2. In the U.S. customary system, the _______ is a unit of capacity.

3. When we change a large unit to a small unit, the numerical part of the answer is _______ than the original number.

4. When changing from miles to feet, we multiply the original measurement by $\frac{5{,}280 \text{ ft}}{1 \text{ mi}}$, which is called a _______.

5. In addition or subtraction problems, only quantities having the same _______ can be added or subtracted.

6. When converting from one unit to another unit, multiply the original measurement by the unit factor that has the desired unit in its _______.

Change each quantity to the indicated unit.

7. 48 in. = _____ ft

8. 6 pt = _____ qt

9. 9 ft = _____ yd

10. 48 mo = _____ yr

11. 60 ft = _____ in.

12. 8 qt = _____ pt

13. 7 min = _____ sec

14. 4 lb = _____ oz

15. 10 yd = _____ ft

16. 2 yr = _____ mo

17. 32 oz = _____ lb

18. 30 sec = _____ min

19. 2 mi = _____ yd

20. 5 hr = _____ min

21. 32 pt = _____ fl oz

22. $\frac{1}{4}$ day = _____ hr

23. $\frac{1}{2}$ gal = _____ qt

24. 32 fl oz = _____ pt

25. $2\frac{1}{2}$ qt = _____ pt

26. $1\frac{1}{2}$ qt = _____ gal

27. 7 pt = _____ qt

28. 36 hr = _____ day

29. $1\frac{1}{2}$ tons = _____ lb

30. 7,000 lb = _____ tons

31. 45 min = _____ hr

32. 7,920 yd = _____ mi

33. $\frac{1}{2}$ day = _____ hr

34. $\frac{1}{2}$ hr = _____ day

35. 5 min 10 sec = _____ sec

36. 1 lb 2 oz = _____ oz

37. 90 in. = ___ ft ___ in.

38. 50 hr = ___ da ___ hr

Complete each table.

	Length	Inches	Feet	Yards
39.	Giraffe (height)		16	
40.	Baseball (diameter)	3		
41.	Dog pen (width)			5
42.	Pond (depth)		12	

	Weight	Ounces	Pounds	Tons
43.	Ostrich egg		3	
44.	Baseball bat	29		
45.	Tongue of a blue whale			4
46.	Pony		200	

	Capacity	Fluid Ounces	Pints	Quarts
47.	Can of paint			1
48.	Case of cream		24	
49.	Bottle of mouthwash	16		
50.	Container of milk			2

	Time	Seconds	Minutes	Hours
51.	Rocket blast	50		
52.	Baseball game			3
53.	News report		22	
54.	Standing ovation		8	

Compute.

55. 4 lb 7 oz
$-$2 lb 9 oz

56. 2 hr
$-$1 hr 2 min

57. 20 lb 5 oz
$+$ 9 lb 10 oz

58. 5 lb 10 oz
$+$1 lb 8 oz

59. 5 yr 7 mo $+$ 3 yr 11 mo

60. 4 yr $-$ 2 yr 3 mo

61. 5 gal 1 qt $-$ 2 gal 2 qt

62. 1 pt 10 fl oz $+$ 3 pt 8 fl oz

63. 2 qt 1 pt $+$ 1 qt 1 pt

64. 2 ft $-$ 5 in.

65. 6 min 2 sec
70 sec
$+$1 min 3 sec

66. 20 ft 5 in.
5 ft 7 in.
$+$ 9 ft 10 in.

Mixed Practice

Solve.

67. Find the sum of 10 lb 12 oz and 3 lb 5 oz.

68. 2 gal = ________ pt

69. Express 4 min 7 sec in seconds.

70. Compute: 4 yr − 1 yr 3 mo

71. Convert 15 ft to yards.

72. How many pints are equivalent to 20 fl oz?

Applications

Solve.

73. The longest field goal made by Mike Vanderjagt while playing for the Indianapolis Colts was 54 yd. The longest field goal made by Jay Feely while playing for the Atlanta Falcons was 165 ft. Who made the longer field goal? (*Source:* http://www.nfl.com)

74. A person sheds about 40 lb of skin in a lifetime. How many ounces is this? (*Source: Webster's New World Book of Facts*)

75. As part of a kitchen remodel, a homeowner bought a new refrigerator.

a. The refrigerator is 36 in. wide and 30 in. deep. Express the width and depth in feet.

b. Using the answer from part (a), calculate the area of floor space the refrigerator will occupy.

76. A recipe calls for 8 oz of chicken broth.

a. How many pints of chicken broth are needed for the recipe?

b. If 1 qt of chicken broth is available, will there be enough to triple the recipe?

77. The record for a person holding his or her breath under water without special equipment is 823 sec. Express this time in minutes and seconds. (*Source: Atlantic Monthly*)

78. One of the tallest women who ever lived was an American named Sandy Allen, who, at age 22, was 91 in. tall. What was her height in feet and inches? (*Source: Guinness Book of World Records*)

79. In 1940, U.S. athlete Cornelius Warmerdam used a bamboo pole to vault 15 ft 8 in., setting a record. In 1962, U.S. athlete Dave Tork, using a fiberglass pole, vaulted 16 ft 2 in. How much higher was Tork's vault than Warmerdam's? (*Source: Facts and Dates of American Sports*)

80. In an Olympic marathon, an athlete runs 26 mi plus an additional 385 yd. How many total yards does an athlete run in a marathon?

81. Born in 1934, the Dionne sisters were Canadians who became world famous as the first quintuplets to survive beyond infancy. At birth, the tiniest of these babies weighed 1 lb 15 oz, and the largest weighed 3 lb 4 oz. What was the difference between their weights? (*Source: Encyclopaedia Brittanica*)

82. Abraham Lincoln spoke of "four score and seven years ago." If a score is 20, how many months are there in four score and seven years?

83. The lease on a tenant's apartment runs for 3 yr. If he has lived in the apartment already for 1 yr 2 mo, does he have more or less than $1\frac{1}{2}$ yr left on the lease?

84. A driver parks her car next to a meter that will expire in 1 hr 55 min. If she pays for an additional $1\frac{1}{2}$ hr of parking, is there enough time on the meter for her to park for $3\frac{1}{4}$ hr?

Use a calculator to solve the following problems, giving (a) the operation(s) carried out in your solution, (b) the exact answer, and (c) an estimate of the answer.

85. The highest mountain in the world is Mt. Everest. If its peak is 29,035 ft above sea level, find the height of the mountain to the nearest tenth of a mile.

86. The Earth is made up of three main layers: the crust, the mantle, and the core. The core, which is composed of a liquid outer core and a solid inner core, is about 2,156 mi thick. How many feet thick is the core, rounded to the nearest million? (*Source:* U.S. Geological Survey)

• *Check your answers on page A-16.*

MINDSTRETCHERS

Writing

1. Not all units are standardized. For example, the term *city block* varies in meaning from town to town. Explain the consequences of this lack of standardization.

Mathematical Reasoning

2. In measuring, we often introduce errors. Suppose that each of two measurements could be as much as an inch off. If we then add the two measurements, how far from the truth could our sum be? Explain.

History

3. The foot is not the only body part used as a measure. For instance, the ancient Egyptians used the *mouthful* as a unit of measure of volume. In your college library or on the Web, investigate other examples.

CULTURAL NOTE

Joseph Louis Lagrange (1736–1813) was a mathematician who, as chairman of the French commission on weights and measures, was influential during the years following the French Revolution of 1789 in developing the metric system of measures based on decimals and powers of 10. Since then, the United States has been resistant to adopting the metric system, although in 1790, Thomas Jefferson—then secretary of state and later president—argued that the country should adopt a decimal system of weights and measures.

Source: Gullberg, *Mathematics: From the Birth of Numbers* (New York: W. W. Norton, 1997), p. 52.

10.2 Metric Units and Metric/U.S. Customary Unit Conversions

OBJECTIVES

- To identify units in the metric system
- To change a measurement from one metric unit to another
- To change a measurement from a metric unit to a U.S. customary unit, and vice versa

Metric Units of Length, Weight, and Capacity

Now we turn to metric units. Developed by French scientists over 200 years ago, the metric system (formally known as the International System of Units, or SI) has become standard in most countries of the world. Even in the United States, metric units predominate in many important fields, including scientific research, medicine, the film industry, food and drink packaging, sports, and the import–export industry.

As in Section 10.1, which dealt with the U.S. customary system, we consider measurements of length, weight, and capacity. Time units are identical in both systems, so we do not discuss them in this section. Again, abbreviations of units in the singular and plural are the same. For example, the abbreviation for meter (m) is the same as the abbreviation for meters (m).

We begin this discussion of the metric system by considering the basic metric units:

- the **meter**, a unit of length (which gives the metric system its name);
- the **gram**, a unit of weight (technically, a unit of mass); and
- the **liter**, a unit of capacity (liquid volume).

There are quite a few other metric units as well. The names for many of the other units are formed by combining a basic unit with one of the metric prefixes. These prefixes are listed in the following table.

METRIC PREFIXES

Prefix	Symbol	Meaning
Milli-	m	One thousandth $\left(\frac{1}{1,000}\right)$
Centi-	c	One hundredth $\left(\frac{1}{100}\right)$
Deci-	d	One tenth $\left(\frac{1}{10}\right)$
Deka-	da	Ten (10)
Hecto-	h	Hundred (100)
Kilo-	k	One thousand (1,000)

Next, let's see how the three basic metric units combine with the metric prefixes to form new units. We begin with units of length.

Length

The table below shows the four most commonly used metric units of length: millimeters (mm), centimeters (cm), meters (m), and kilometers (km). Memorize the following table, noting what each unit means as well as its symbol.

METRIC UNITS OF LENGTH

Unit	Symbol	Meaning
Millimeter	mm	$\frac{1}{1{,}000}$ meter
Centimeter	cm	$\frac{1}{100}$ meter
Meter	m	1 meter
Kilometer	km	1,000 meters

In this table, the first unit of length, the *millimeter*, is the smallest—about the thickness of a dime. We use the millimeter to measure short lengths—say, the dimensions of an insect.

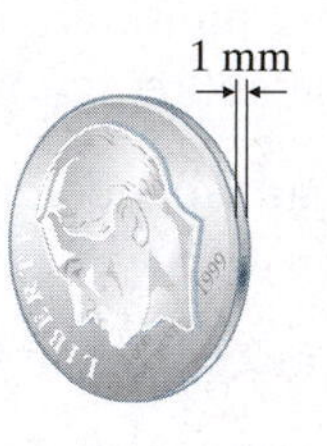

The next unit of length, the *centimeter*, is approximately the width of your little finger, or somewhat less than half an inch. In the metric system, the width of an envelope is expressed in centimeters.

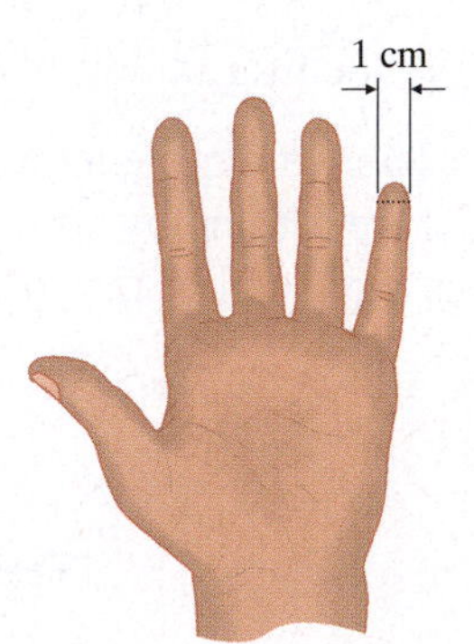

The *meter*, the basic metric unit of length, is a little longer than a yard, or about the width of a twin bed. In the metric system, we use meters to measure medium-size lengths—say, the length of a room.

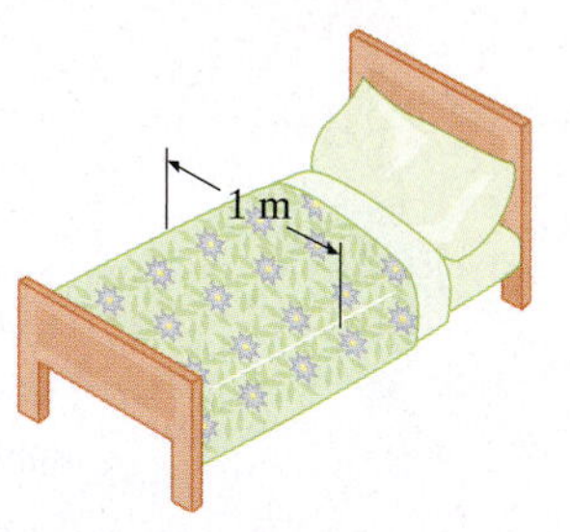

The largest unit of length in the table is the *kilometer*. A kilometer is a little more than half a mile, or about 3 times the height of the Empire State Building. Great lengths, such as the distance between two cities, are expressed in kilometers.

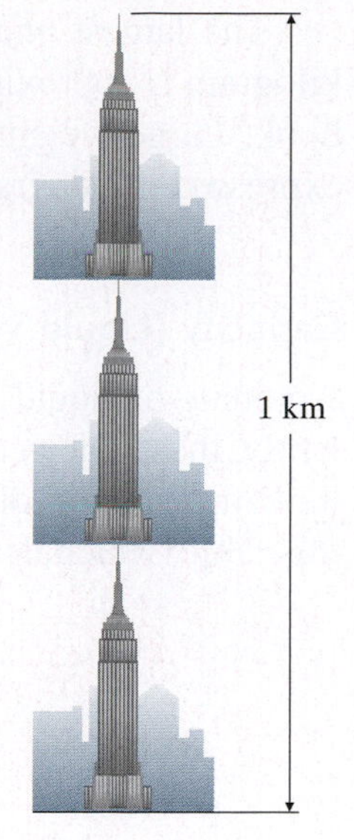

Weight

Now, we turn to the metric units of weight, shown in the following table: milligrams (mg), grams (g), and kilograms (kg). Memorize this table, which deals with the three most commonly used metric units of weight.

METRIC UNITS OF WEIGHT

Unit	Symbol	Meaning
Milligram	mg	$\frac{1}{1,000}$ gram
Gram	g	1 gram
Kilogram	kg	1,000 grams

The smallest unit, the *milligram*, is tiny—about the weight of a hair. It is therefore used to measure light weights—say, that of a small pill.

The next unit, the *gram*, is larger but is still only about $\frac{1}{30}$ oz, or about the weight of a raisin.

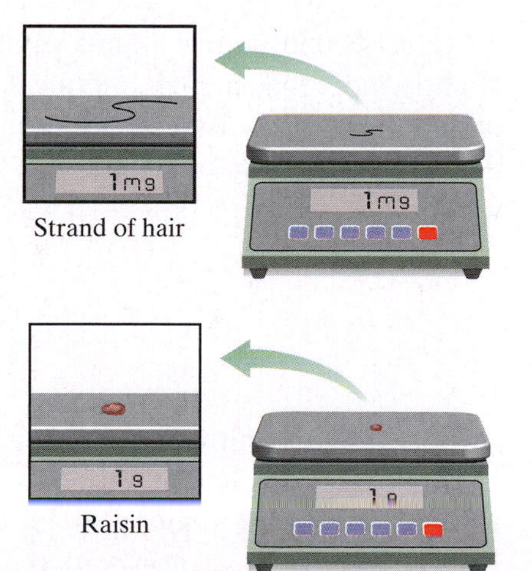

Strand of hair

Raisin

The largest unit of weight in the table is the *kilogram*. A kilogram is approximately 2 lb, or about the weight of this textbook. Large weights—say, that of a car or of a person—are expressed in kilograms.

This textbook

Capacity (Liquid Volume)

Amounts of liquid are commonly measured in terms of liquid volume, or equivalently, the capacity of containers that hold the liquid. The following table deals with three metric units of capacity: milliliters (ml), liters (L), and kiloliters (kl). Memorize this table, which describes the three primary metric units of capacity.

METRIC UNITS OF CAPACITY

Unit	Symbol	Meaning
Milliliter	ml	$\frac{1}{1{,}000}$ liter
Liter	L	1 liter
Kiloliter	kl	1,000 liters

In this table, the first unit, the *milliliter*, represents a very small amount of liquid—about as much as an eyedropper contains. Milliliters are used in measuring small volumes of liquid—say, the amount of perfume in a tiny bottle.

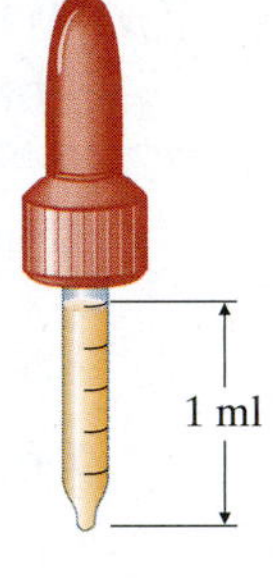

The second unit of liquid volume, the *liter*, is slightly more than a quart, a typical size for a bottle of soda. Liters are used in measuring larger quantities of liquid, such as the amount of water that a sink will hold.

The largest unit of capacity in the table is the *kiloliter*. The amount of water that a typical collapsible swimming pool holds is about 1 kiloliter. Kiloliters are used to measure large volumes of liquid, for instance, the capacity of an oil barge or the amount of soda that a factory produces annually.

Changing Units

As we have already seen, sometimes we need to change the unit in which a measurement is expressed. One reason the metric system is widely used is that unit conversions in this system are much easier than those in the U.S. customary system.

Such conversions simply involve multiplying or dividing by a power of 10, such as 100 or 1,000. The following table shows several metric conversion relationships.

Length	Weight	Capacity
1,000 mm = 1 m	1,000 mg = 1 g	1,000 ml = 1 L
100 cm = 1 m	1,000 g = 1 kg	1,000 L = 1 kl
1,000 m = 1 km		

From these relationships, we can set up unit factors to carry out unit conversions.

EXAMPLE 1

1.5 g = ________ mg

Solution Because 1,000 mg = 1 g and we want to convert to *milligrams*, we use the unit factor $\frac{1{,}000 \text{ mg}}{1 \text{ g}}$ to solve this problem.

$$1.5 \text{ g} = 1.5 \not{\text{g}} \times \frac{1{,}000 \text{ mg}}{1 \not{\text{g}}}$$
$$= 1.5 \times 1{,}000 \text{ mg}$$
$$= 1{,}500 \text{ mg}$$

To multiply 1.5 by 1,000, move the decimal point in 1.5 three places to the right.

PRACTICE 1

3,100 mg = _______ g

As illustrated in Example 1, a part of a metric unit is usually expressed as a decimal, not as a fraction. That is, we write 1.5 g, not $1\frac{1}{2}$ g.

EXAMPLE 2

500 m = ________ km

Solution Because we want *kilometers*, we solve this problem by multiplying the original measurement by the unit factor $\frac{1 \text{ km}}{1{,}000 \text{ m}}$.

$$500 \text{ m} = 500 \not{\text{m}} \times \frac{1 \text{ km}}{1000 \not{\text{m}}}$$
$$= \frac{500}{1{,}000} \text{ km}$$
$$= 0.5 \text{ km}$$

To divide 500 by 1,000, move the decimal point in 500 three places to the left.

As a quick check, note that the number of the larger unit (0.5 km) is less than the number of the smaller unit (500 m) for an equivalent measurement.

PRACTICE 2

2,500 cm = _______ m

EXAMPLE 3

Express 3 km in millimeters.

Solution Because the table of lengths on page 464 does not indicate how many millimeters are equivalent to a kilometer, we need to solve the problem in steps.

Step 1. Change kilometers to meters.

Step 2. Change meters to millimeters.

To combine these two steps, we multiply the original measurement by a chain of appropriate unit factors to get *millimeters* for the final answer.

$$3 \text{ km} = 3 \cancel{\text{km}} \times \frac{1{,}000 \cancel{\text{m}}}{1 \cancel{\text{km}}} \times \frac{1{,}000 \text{ mm}}{1 \cancel{\text{m}}}$$
$$= 3 \times 1{,}000 \times 1{,}000 \text{ mm}$$
$$= 3{,}000{,}000 \text{ mm}$$

Note that because a kilometer is larger than a millimeter, the number of kilometers (3) is less than the number of millimeters (3,000,000).

PRACTICE 3

Change 5,000,000 mm to kilometers.

EXAMPLE 4

For a 20-year-old female, the U.S. recommended dietary allowances (RDAs) for calcium and iron are 1 g and 18 mg, respectively. Which RDA is higher? (*Source: The National Institutes of Health*)

Solution When comparing quantities expressed in different units, we convert them to the same unit, usually the smaller unit. Here, we change 1 g to milligrams.

$$1 \text{ g} = 1 \cancel{\text{g}} \times \frac{1{,}000 \text{ mg}}{1 \cancel{\text{g}}}$$
$$= 1{,}000 \text{ mg}$$

The RDA for calcium is 1,000 mg, which is higher than the 18-mg RDA for iron.

PRACTICE 4

The small intestine is 6 m long, and the large intestine is 150 cm long. Which is shorter? (*Source: Webster's New World Book of Facts*)

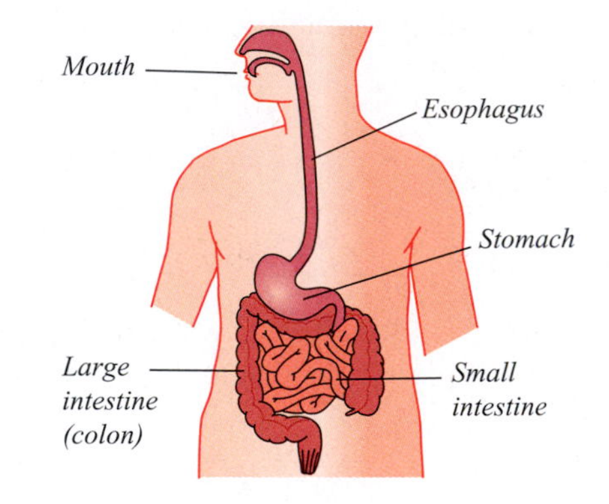

Metric/U.S. Customary Unit Conversions

In some situations, we need to change a measurement expressed in a U.S. unit to a metric unit, or vice versa. For example, if we were driving in Canada and saw a road sign giving the distance to the next town in kilometers, we might want to express that distance in miles.

Or suppose that we had gone shopping to buy mouthwash and wondered how many pint bottles are equal in capacity to a 750-ml bottle. Here, we might want to change pints to milliliters.

To convert, we must either have memorized or have access to metric/U.S. unit conversion relationships. The following table shows some of these key relationships.

METRIC/U.S. UNIT CONVERSION RELATIONSHIPS

Length	Weight	Capacity
2.5 cm ≈ 1 in.	28 g ≈ 1 oz	470 ml ≈ 1 pt
30 cm ≈ 1 ft	450 g ≈ 1 lb	2.1 pt ≈ 1 L
39 in. ≈ 1 m	2.2 lb ≈ 1 kg	1.1 qt ≈ 1 L
3.3 ft ≈ 1 m	910 kg ≈ 1 ton	3.8 L ≈ 1 gal
3,300 ft ≈ 1 km		260 gal ≈ 1 kl
1,600 m ≈ 1 mi		
1.6 km ≈ 1 mi		

EXAMPLE 5

Express 2 oz in grams.

Solution According to the conversion table, 1 oz ≈ 28 g. Because we want *grams*, we multiply 2 oz by the unit factor $\frac{28\text{ g}}{1\text{ oz}}$.

$$2\text{ oz} \approx 2\text{ oz} \times \frac{28\text{ g}}{1\text{ oz}}$$
$$\approx 56\text{ g}$$

So 2 ounces is about 56 grams.

PRACTICE 5

Express 10 gal in terms of liters.

Note that our answer in Example 5 is only an approximation because the unit factor is not exact. Also note that the number of grams is more than the number of ounces because an ounce is larger than a gram.

EXAMPLE 6

Different kinds of barrels have different capacities. Some contain 31 gal, others contain 31.5 gal. Since 1866, the capacity of an oil barrel in the United States has been standardized at 42 gal. How many liters of oil does an oil barrel hold, rounded to the nearest liter? (*Source:* Daniel Yergin, *The Prize*)

Solution The conversion table indicates that 3.8 L ≈ 1 gal. To convert to liters, we use the conversion factor $\frac{3.8\text{ L}}{1\text{ gal}}$.

$$42\text{ gal} \approx 42\text{ gal} \times \frac{3.8\text{ L}}{1\text{ gal}}$$
$$\approx 159.6\text{ L}$$

So an oil barrel holds approximately 160 L of oil.

PRACTICE 6

The distance from Las Vegas to Seattle is 608 km. Express this distance in miles, rounded to the nearest mile.

Mathematically Speaking

Fill in each blank with the most appropriate term or phrase from the given list.

kilo-	centi-	milli-
weight	deci-	quart
liter	length	

1. In the metric system, the gram is a unit of _______.

2. In the metric system, the _______ is a unit of capacity.

3. The prefix _______ means one thousand.

4. The prefix _______ means one-thousandth.

5. The prefix _______ means one-hundredth.

6. The prefix _______ means one-tenth.

Choose the unit that would most likely be used to measure each quantity.

7. The volume of liquid in a test tube

a. millimeter **b.** milligram **c.** milliliter

8. The weight of a television set

a. milligram **b.** gram **c.** kilogram

9. The width of a street

a. millimeter **b.** meter **c.** kilometer

10. The length of a river

a. kilometer **b.** kilogram **c.** millimeter

Choose the best estimate in each case.

11. The capacity of a large bottle of soda

a. 1 ml **b.** 1 L **c.** 1 kg

12. The width of film for slides

a. 35 mm **b.** 35 cm **c.** 35 m

13. The height of the Washington Monument

a. 170 cm **b.** 170 m **c.** 170 km

14. The length of a pencil

a. 20 mm **b.** 20 cm **c.** 20 m

15. The capacity of a bottle of hydrogen peroxide

a. 400 ml **b.** 400 L **c.** 400 g

16. The weight of an adult

a. 70 mg **b.** 70 g **c.** 70 kg

Change each quantity to the indicated unit.

17. 1,000 mg = ____ g

18. 253 mm = ____ m

19. 750 g = ____ kg

20. 2 L = ____ ml

21. 0.08 kl = ____ L

22. 4.3 kg = ____ g

23. 3.5 m = ____ mm

24. 900 m = ____ km

25. 5 ml = ____ L

26. 250 mg = ____ g

27. 4,000 mm = ____ m

28. 5 L = ____ ml

29. 7,000 L = ____ kl

30. 2,500 mg = ____ g

31. 413 cm = ____ m

32. 2.8 m = ____ cm

33. 0.002 kg = ____ mg

34. 3,000 mm = ____ cm

35. 7,500 ml = ____ kl

36. 2.1 km = ______ cm

Complete each table.

	Length	Millimeters	Centimeters	Meters
37.	Swan wingspan		238	
38.	Strip of cloth width		5	
39.	Layer of soil depth	10		
40.	Tubeworm height			3

	Weight	Milligrams	Grams	Kilograms
41.	Capsule	300		
42.	Human liver		1,560	
43.	Pastry		450	
44.	Kangaroo cub			1

	Capacity	Milliliters	Liters	Kiloliters
45.	Container of tile cleaner	709		
46.	Bottle of spring water		3	
47.	Gas station fuel tank			17
48.	Aquarium		110	

Compute.

49. 3 km + 250 m

50. 5 L − 600 ml

51. 98 kg + 25.6 g

52. 30 cm + 2 m

Change each quantity to the indicated unit. If needed, round the answer to the nearest tenth of the unit.

53. 30 oz ≈ ___ g

54. 4 mi ≈ ____ km

55. 10 cm ≈ ___ in.

56. 900 g ≈ ___ lb

57. 48 in. ≈ ____ m

58. 6 qt ≈ ____ L

59. 5 pt ≈ ____ L

60. 6 ft ≈ _____ cm

Mixed Practice

Solve.

61. Combine 3 m, and 50 cm.

62. Change 500 g to kilograms.

63. 2.5 m = __________ mm

64. Express 2,000 ml in liters.

65. The distance between your home and your college would most likely be measured in

a. kilometers **b.** kilograms **c.** millimeters

66. Express 60 cm in feet.

67. Change 3 L to quarts.

68. The best estimate for the capacity of a bottle of olive oil is

a. 500 L **b.** 500 kg **c.** 500 ml

Applications

Solve. If needed, round the answer to the nearest tenth of the unit.

69. According to a medical journal, the average daily U.S. diet contains 6,000 mg of sodium. How many grams is this? (***Source:*** *Journal of the American Medical Association*)

70. The speed limit on many European highways is 100 kilometers per hour. Express this speed in miles per hour.

71. A nurse must administer a 4-ml dose of a drug daily to a patient. If there is 1 L of this drug on hand, will it last the patient 120 days?

72. Vitamin C commonly comes in pills with a strength of 500 mg. How many of these pills will an adult need to take if she wants a dosage of half a gram?

73. A student in a physics lab measured the length of a pendulum string as 7.5 cm. Express this length in inches.

74. In the Summer Olympics, a major track-and-field event is the 100-m dash. How long is this race in kilometers?

75. The diameter of the primary mirror of the Hubble Space Telescope is 2.4 m. Express this diameter in millimeters. (***Source:*** http://hubblesite.org)

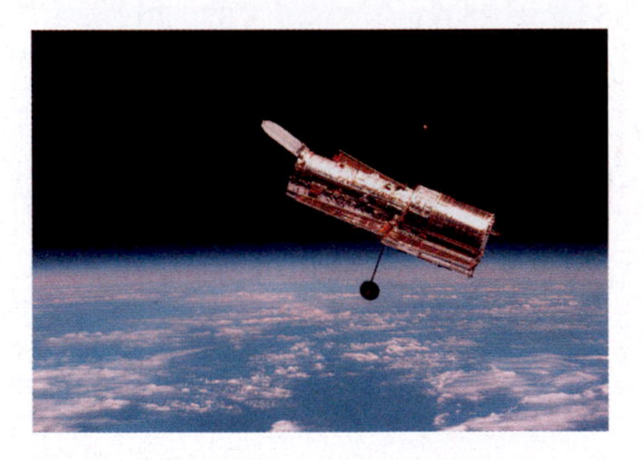

76. The Sears Tower in Chicago is 442 m tall. Express this height in kilometers. (***Source:*** Emporis Buildings)

77. The side of a square tile is 75 mm long. If 100 of these tiles are placed on the floor side by side, what is their total length in centimeters?

78. Two hours before surgery, a patient was told to drink 300 ml of a clear fluid. Express this amount of fluid in quarts.

79. One of the heaviest babies ever born was an Italian boy who at birth weighed 360 oz. What was the baby's weight in kilograms? (***Source:*** *Guinness Book of World Records*)

80. A prehistoric bird had a wingspan of 8 m. Express this wingspan in feet. (***Source:*** *Guinness Book of World Records*)

81. The average weight of an adult human's brain is about 3 lb. Express this weight in grams. (***Source:*** *The Top 10 of Everything, 2006*)

82. A passenger car is generally considered small if the distance between its front and back wheels is less than 95 inches. What is this distance expressed in meters?

83. A chemistry professor mixes the contents of two beakers containing 2.5 L and 700 ml of a liquid in liters. What is the combined amount?

84. A can contains 355 ml of soda. Express in liters the amount of soda in a six-pack of these cans.

• *Check your answers on page A-16.*

MINDSTRETCHERS

History

1. Either in your college library or on the Web, investigate the history of decimal coinage in the United States. In what respect are decimal coinage and the metric system similar? Explain.

Writing

2. Consider the following measurement expressed in the metric system.

$$\underbrace{37}_{\text{km}},\ \underbrace{568}_{\text{m}}.\ \underbrace{251}_{\text{mm}}\ \text{meters}$$

Note how we can split this measurement in meters into other metric units. Would this work with U.S. units? Explain.

Groupwork

3. A liter is about 5% more than a quart. Work with a partner to answer the following questions.

a. Do you think that, as the United States goes metric, containers will increase in size? Explain.

b. Suppose containers were to increase in size. What do you think the economic consequence of this increase would be?

KEY CONCEPTS AND SKILLS

CONCEPT SKILL

Quantity	U.S. Customary Units	Relationships
[10.1] Length	Inch (in.), foot (ft), yard (yd), and mile (mi)	12 in. = 1 ft 3 ft = 1 yd 5,280 ft = 1 mi
[10.1] Weight	Ounce (oz), pound (lb), and ton	16 oz = 1 lb 2,000 lb = 1 ton
[10.1] Capacity (Liquid Volume)	Fluid ounce (fl oz), cup (c), pint (pt), quart (qt), and gallon (gal)	8 fl oz = 1 c 2 c = 1 pt 2 pt = 1 qt 4 qt = 1 gal
[10.1] Time	Second (sec), minute (min), hour (hr), day, week (wk), month (mo), and year (yr)	60 sec = 1 min 60 min = 1 hr 24 hr = 1 day 7 days = 1 wk 52 wk = 1 yr 12 mo = 1 yr 365 days = 1 yr

METRIC PREFIXES

Prefix	Symbol	Meaning
Milli-	m	One thousandth $\left(\frac{1}{1{,}000}\right)$
Centi-	c	One hundredth $\left(\frac{1}{100}\right)$
Deci-	d	One tenth $\left(\frac{1}{10}\right)$
Deka-	da	Ten (10)
Hecto-	h	Hundred (100)
Kilo-	k	One thousand (1,000)

Quantity	Metric Units	Relationships
[10.2] Length	Millimeter (mm), centimeter (cm), meter (m), and kilometer (km)	1,000 mm = 1 m 100 cm = 1 m 1,000 m = 1 km
[10.2] Weight	Milligram (mg), gram (g), and kilogram (kg)	1,000 mg = 1 g 1,000 g = 1 kg
[10.2] Capacity (Liquid Volume)	Milliliter (ml), liter (L), and kiloliter (kl)	1,000 ml = 1 L 1,000 L = 1 kl
[10.2] Time	Same as U.S. customary units.	

continued

CONCEPT SKILL

Key Metric/U.S. Unit Conversion Relationships

[10.2] **Length**	2.5 cm ≈ 1 in. 30 cm ≈ 1 ft 39 in. ≈ 1 m 3.3 ft ≈ 1 m 3,300 ft ≈ 1 km 1,600 m ≈ 1 mi 1.6 km ≈ 1 mi
[10.2] **Weight**	28 g ≈ 1 oz 450 g ≈ 1 lb 2.2 lb ≈ 1 kg 910 kg ≈ 1 ton
[10.2] **Capacity (Liquid Volume)**	470 ml ≈ 1 pt 2.1 pt ≈ 1 L 1.1 qt ≈ 1 L 3.8 L ≈ 1 gal 260 gal ≈ 1 kl

Chapter 10 Review Exercises

To help you review this chapter, solve these problems.

[10.1] *Change each quantity to the indicated unit.*

1. 5 yd = ____ ft

2. 20 mo = ____ yr

3. 32 oz = ____ lb

4. 10 ft = ____ yd

5. $1\frac{1}{2}$ tons = ____ lb

6. $8\frac{1}{2}$ lb = ____ oz

7. 3 pt = ____ fl oz

8. 150 sec = ____ min

9. 7 hr 15 min = ____ min

10. 50 in. = ____ ft ____ in.

11. 10,560 ft = ____ mi

12. 2,000 oz = ____ lb

Compute the given sum or difference.

13.
```
  4 hr 20 min
+3 hr 50 min
```

14.
```
  20 ft
−  1 ft 3 in.
```

15. 3 gal 2 qt − 1 gal 3 qt

16. 3 lb 6 oz + 2 lb 9 oz + 1 lb 3 oz

[10.2] *Choose the unit that you would most likely use to measure each quantity.*

17. The weight of a car

a. milligrams **b.** grams **c.** kilograms

18. The width of a pencil's point

a. millimeters **b.** centimeters **c.** meters

19. The capacity of an oil barrel

a. milliliters **b.** liters **c.** meters

20. The distance a commuter drives

a. millimeters **b.** centimeters **c.** kilometers

Choose the best estimate in each case.

21. The width of a piece of typing paper

a. 16 mm **b.** 16 cm **c.** 16 km

22. The capacity of a bottle of mouthwash

a. 100 ml **b.** 100 L **c.** 100 g

23. The weight of an aspirin pill

a. 200 mg **b.** 200 g **c.** 200 kg

24. The length of an athlete's long jump

a. 6.72 mm **b.** 6.72 cm **c.** 6.72 m

Change each quantity to the indicated unit.

25. 37 mg = ____ g

26. 4 kl = ____ L

27. 8 m = ____ cm

28. 2.1 km = ____ m

29. 600 mm = ____ m

30. 5,100 g = ____ kg

Change each quantity to the indicated unit, rounding to the nearest unit.

31. 4 oz ≈ ____ g

32. 5 cm ≈ ____ in.

33. 32 km ≈ ____ mi

34. 4 gal ≈ ____ L

Mixed Applications

Solve.

35. A DVD plays for 72 min. Express this playing time in hours.

36. In a recent year, a typical U.S. resident used about 1,600 gal of water a day for residential, agricultural, and industrial purposes. How many pints is this? (*Source:* U.S. Geological Survey)

37. *Frankenstein* (130 minutes) and *Dracula* (1 hr 15 min) are two classic horror films made in 1931. Which film is longer?

38. Some doctors recommend that athletes drink about 600 ml of fluid each hour. Express this amount in liters.

39. A teaspoon of common table salt contains about 2,000 mg of sodium. How many grams of sodium is this?

40. In a factory, a chemical process produced 3 mg of a special compound each hour. How many grams were produced in 24 hr?

41. The average level of cholesterol in U.S. children is 160 mg per 100 ml of blood. Express this ratio in terms of grams per liter. (*Source: The Peoplepedia*)

42. According to a journal article, 750 kg of pesticide is sprayed on a typical U.S. golf course each year. How many grams is this?

43. A computer virus checker took 349 min to scan each file on a 50-gigabyte hard drive for any viruses. Express this length of time to the nearest hour.

44. A daily reference value (DRV) is a reference point that serves as a general guideline for a healthy diet. For a 2,000-calorie diet, the DRV for fiber is 25 g. Express this DRV in milligrams. (*Source:* U.S. Food and Drug Administration)

45. In Olympic gymnastics, the floor exercise is performed on a square mat measuring 12 m on a side. What is the length of one side to the nearest foot?

46. The weight of a precious stone is given in carats, where 1 carat is equal to 200 mg. The Hope Diamond weights 45.52 carats. Express this weight in grams. (*Source:* Smithsonian Institution)

47. In pairs figure skating, the free skate is 1 min 40 sec longer than the short program. If the short program is 2 min 50 sec, how long is the free skate?

48. The following diagram shows the heights of an average U.S. woman and an average 10-year-old U.S. girl.

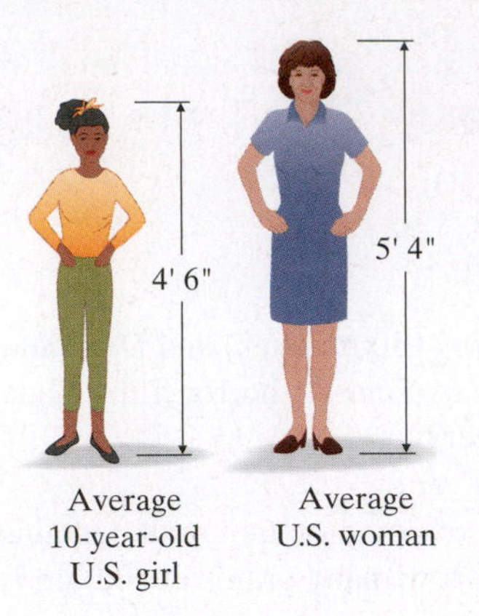

What is the difference in their heights? (*Source: Archives of Pediatrics and Adolescent Medicine*)

49. One of the shortest dinosaurs that ever lived was only 60 cm long when fully grown. What was the length of this dinosaur in inches? (*Source: Encarta Learning Zone Encyclopedia*)

50. There are about 6 qt of blood in an average-sized man. Express this amount to the nearest tenth of a liter.

51. The table shows the average gestation, in days, for various mammals. (*Source: The World Almanac and Book of Facts, 2006*)

Mammal	Gestation
Polar bear	240
Cow	284
Hippopotamus	238
Gorilla	258
Sea lion	350

How many more weeks is the gestation for a sea lion than a hippopotamus?

52. The table shows the Saffir-Simpson Hurricane Scale, which is used to rate a hurricane based on its present intensity. (*Source:* National Weather Service, NOAA)

Category	Wind Speed (mph)
1	74–95
2	96–110
3	111–130
4	131–155
5	156 and above

Express the range of wind speeds for a category 3 hurricane to the nearest kilometer per hour.

● *Check your answers on page A-16.*

Chapter 10 POSTTEST

FOR EXTRA HELP

Test solutions are found on the enclosed CD.

To see if you have mastered the topics in this chapter, take this test.

1. 120 sec = _____ min

2. 7 yd = _____ ft

3. Subtract 1 hr 25 min from 3 hr 10 min.

4. Which of the following units is a measure of capacity?

a. a gram **b.** a meter **c.** a liter **d.** an hour

5. The weight of a baby is measured in

a. milligrams **b.** grams **c.** kilograms

6. 400 cm = _____ m

7. 500 ml = _____ L

8. Which is larger: 4 mm or 2 km?

9. In order for a satellite to be in a geosynchronous orbit, it must be approximately 35,786 km above the surface of the Earth. Express this distance to the nearest thousand miles. (*Source:* Marshall Space Flight Center, NASA)

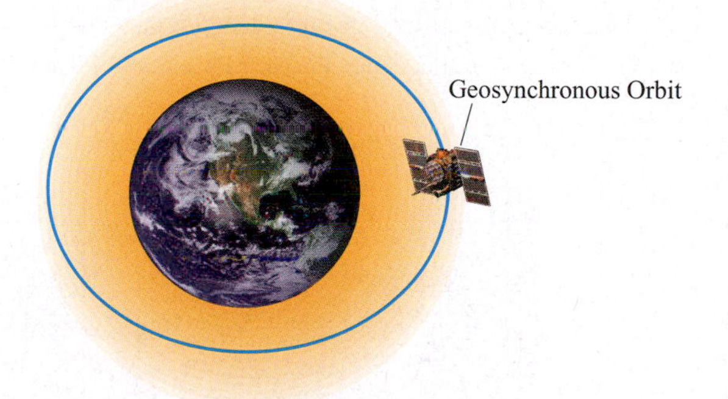

10. An adult humpback whale weighs about 60,000 lb and an adult finback whale weights about 40 tons. Which whale weighs more? (*Source:* Whale Center of New England)

• *Check your answers on page A-16.*

Cumulative Review Exercises

To help you review, solve the following:

1. Compute: 9^3

2. Express as a mixed number: $\frac{11}{2}$

3. Divide: $\frac{2.8}{0.2}$

4. Solve: $3(x - 2) = -1$

5. Find the price of a book that normally sells for $24 but is on sale at a 25% discount.

6. Find the range: 8, 6, 2, 9, 1, and 6

7. Fill in the blank: 7 ft = _____ in.

8. The closing price of a share of technology stock on Monday was $23.86. On Tuesday, the closing price was $22.39. What was the change in the price of the stock?

9. The bar graph shows the number of American Kennel Club registrations for various breeds in a recent year.

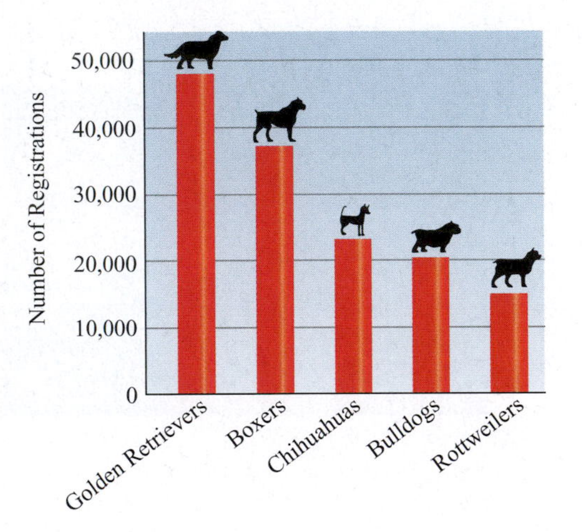

About how many more golden retrievers were registered than bulldogs? (***Source:*** American Kennel Club)

10. In Finland, police levied traffic fines proportional to an offender's income. For speeding, a wealthy driver was fined about $70,000, and a student, with a monthly income of approximately $700, was fined about $100. Estimate, to the nearest hundred thousand dollars, the monthly income of the wealthy driver. (***Source:*** Steve Stecklow, "Helsinki on Wheels: Fast Finns Find Fines Fit Their Finances," *Wall Street Journal*)

• *Check your answers on page A-16.*

CHAPTER 11

Basic Geometry

Geometry and Architecture

Students of geometry study abstract figures in space, whereas architects design real structures in space. The two fields, geometry and architecture, are, therefore, closely related.

The simplest architectural structures have basic geometric shapes. An igloo in the far north and a dome that graces a state capitol are shaped like hemispheres. A tepee is in the shape of a cone, and the peak of a roof is triangular.

The rectangle plays an especially important role in architectural design. Bricks, windows, doors, rooms, buildings, lots, city blocks, and street grids are all based on the rectangle—one of the most adaptable shapes for human needs.

Of all the rectangles with a given area, the square has the smallest perimeter. As a result, warehouses are often built in the form of squares. On the other hand, houses, hotels, and hospitals—for which daylight and a long perimeter are more important—are seldom square shaped. (*Source:* William Blackwell, *Geometry in Architecture,* John Wiley and Sons, 1984)

Chapter 11 PRETEST

To see if you have already mastered the topics in this chapter, take this test.

1. Sketch and label an example of each figure.
   a. Obtuse $\angle PQR$ **b.** Right triangle ABC

2. What is the square root of each perfect square?
   a. $\sqrt{36}$ **b.** $\sqrt{121}$

3. Find the supplement of 100°.

4. Find the complement of 36°.

Find each perimeter or circumference. Use $\pi \approx$ 3.14 when needed.

5.

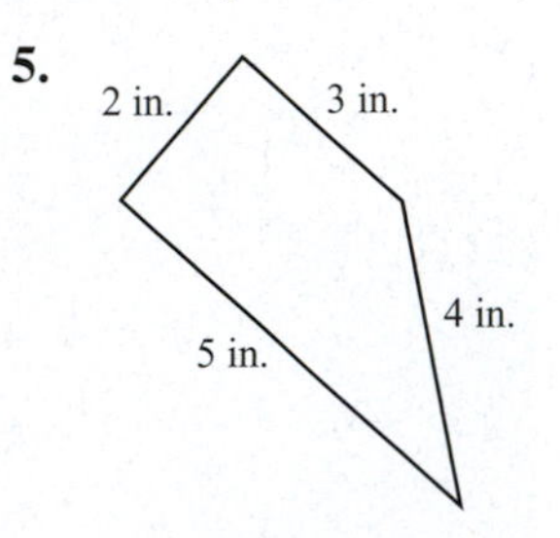

6.

8 ft
2 ft

7. A circle with a diameter of 4 inches.

8. A square with side 2.6 meters.

Find each area. Use $\pi \approx$ 3.14 when needed.

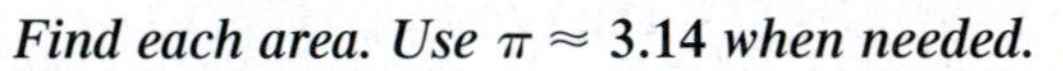

9. **10.**

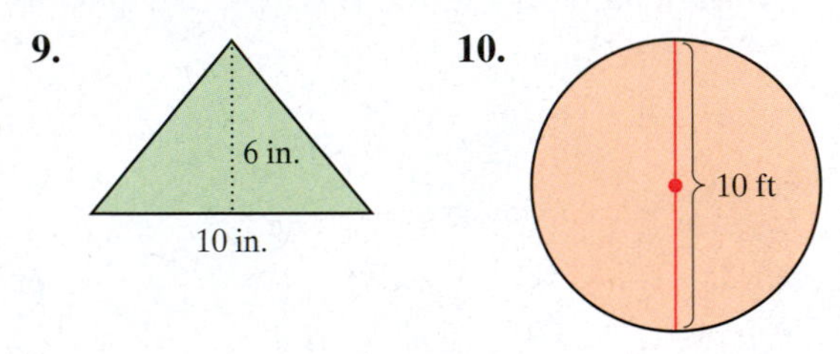

Find each volume. Use $\pi \approx$ 3.14 when needed.

11. **12.**

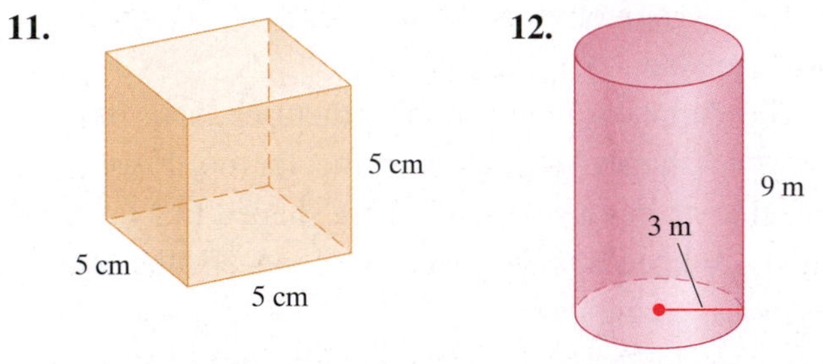

Find the unknown measure(s) in each figure.

13. **14.**

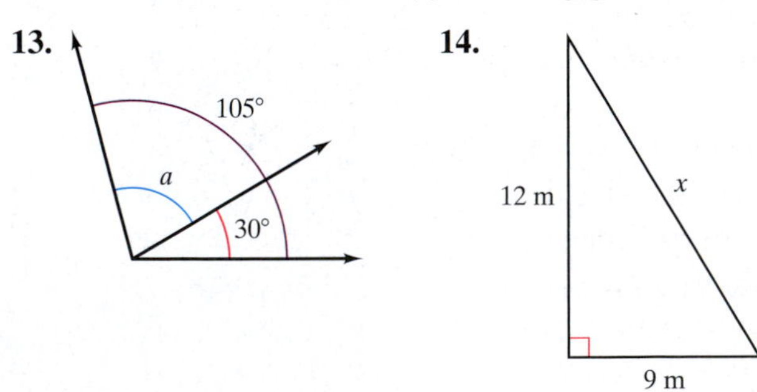

15.

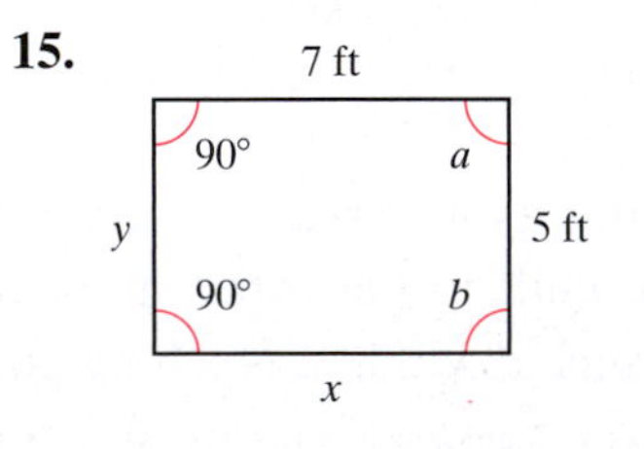

16.

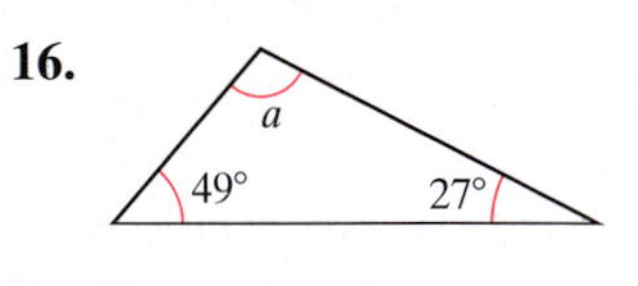

17. For the diagrams shown, $\triangle ABC$ is similar to $\triangle DEF$. Find y.

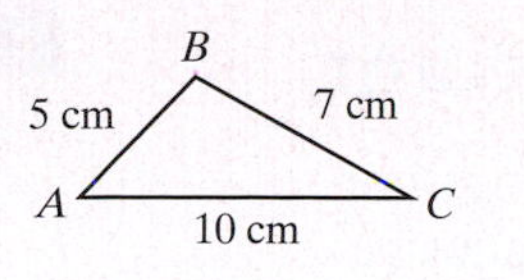

E
15 cm
y
D
F
30 cm

Solve.

18. Rescuers are searching for survivors of a shipwreck that took place within a mile of a rock. What is the area of the region that would be most appropriate to search for survivors? Round to the nearest square mile.

19. What is the length of the loading ramp shown in the diagram? Round the answer to the nearest tenth.

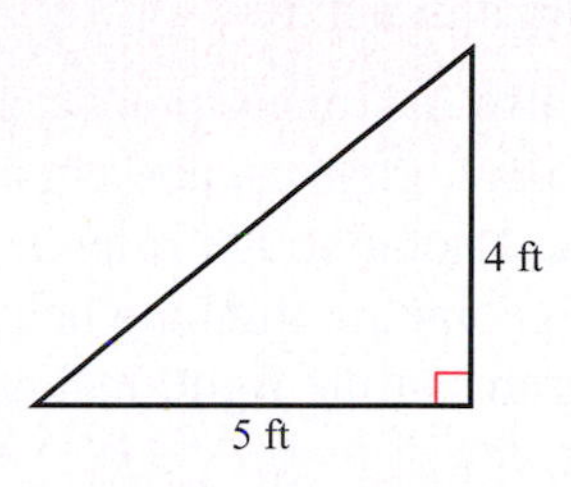

20. In constructing the foundation for a house, a contractor digs a hole 6 feet deep, 54 feet long, and 25 feet wide. How many cubic feet of earth are removed?

• *Check your answers on page A-16.*

CULTURAL NOTE

A *tessellation* is any repeating pattern of interlocking shapes. Some shapes, depending on their geometric properties, will tessellate, that is, go on indefinitely, covering the plane without overlapping and without gaps. These shapes include squares and equilateral triangles. Many other shapes will not tessellate. Tessellations are commonly found in the home—for instance, in the design of wall, ceiling, and floor coverings. More elaborate tessellations are found in mosaics that survive from ancient times. The tile mosaic shown to the left is found at the Alhambra, the summer residence of Moorish kings built in the fourteenth century in Granada, Spain.

11.1 Introduction to Basic Geometry

OBJECTIVES

- To identify basic geometric concepts
- To identify basic geometric figures
- To solve word problems involving basic geometric concepts and figures

What Geometry Is and Why It Is Important

The word *geometry*, which dates back thousands of years, means "measurement of the Earth." Today, we use the term to mean the branch of mathematics that deals with concepts such as point, line, angle, perimeter, area, and volume.

Ancient peoples, including the Egyptians, used the principles of geometry in their construction projects. They understood these principles because of observations they made in their daily lives and their studies of the physical forms in nature.

Geometry also has many practical applications in such diverse fields as art and design, architecture, physics, and engineering. In city planning, geometric concepts, relationships, and notation are often used when designing the layout of a city. Note how the use of geometric thinking helps to transform the street plan on the left to the geometric diagram on the right, making it easier to focus on the key features of the street plan.

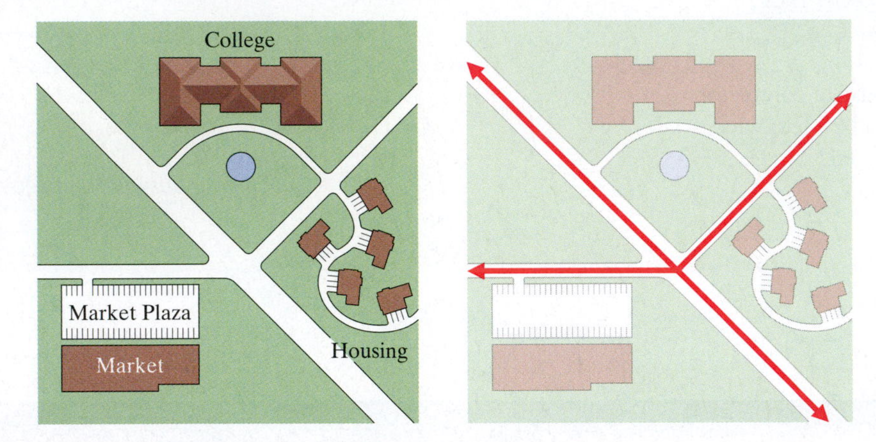

Basic Geometric Concepts

Let's first consider some of the basic concepts that underlie the study and application of geometry. The following table gives the definitions of some basic geometric terms illustrated in the preceding street plan. We use these terms throughout this chapter.

Definition	Example
A **point** is an exact location in space. A point has no dimension.	A • (read "point A")
A **line** is a collection of points along a straight path that extends endlessly in both directions. A line has only one dimension—its length.	$\overleftrightarrow{CB}$ (read "line CB")
A **line segment** is a part of a line having two endpoints. Every line segment has a length.	$\overline{AB}$ (read "line segment AB") The length of $\overline{AB}$ is denoted AB.

A **ray** is a part of a line having only one endpoint.	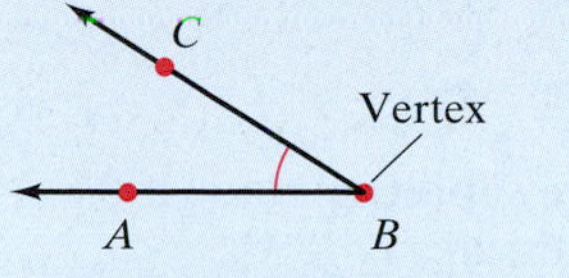 $\overrightarrow{CD}$ (read "ray *CD*") (The endpoint is always the first letter.)
An **angle** consists of two rays that have a common endpoint called the **vertex** of the angle.	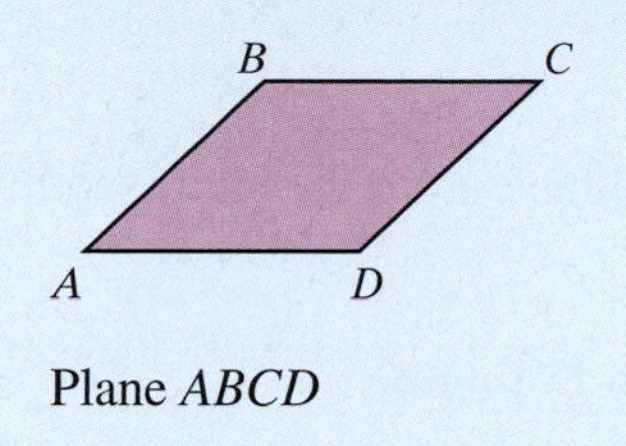 $\angle ABC$ (read "angle *ABC*") (The vertex is always the middle letter.) $\angle ABC$ can also be written as $\angle CBA$ or just $\angle B$.
A **plane** is a flat surface that extends endlessly in all directions.	Plane *ABCD*

The unit in which angles are commonly measured is the degree (°). Angles are classified according to their measures. To indicate the measure of $\angle ABC$, we write $m\angle ABC$.

Definition	Example
A **straight angle** is an angle whose measure is 180°.	$\angle ABC$ is a straight angle.
A **right angle** is an angle whose measure is 90°.	Symbol for right angle. $\angle DEF$ is a right angle.
An **acute angle** is an angle whose measure is less than 90°.	$\angle XYZ$ is an acute angle.
An **obtuse angle** is an angle whose measure is more than 90° and less than 180°.	$\angle CDE$ is an obtuse angle.

continued

Two angles are complementary if the sum of their measures is 90°.	$m\angle A + m\angle B = 25° + 65° = 90°$ $\angle A$ and $\angle B$ are complementary angles.
Two angles are supplementary if the sum of their measures is 180°.	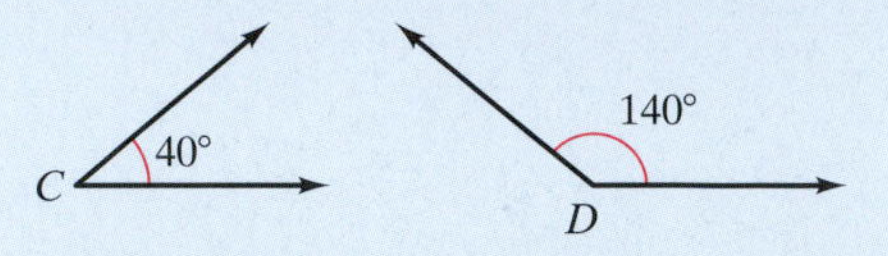 $m\angle C + m\angle D = 40° + 140° = 180°$ $\angle C$ and $\angle D$ are supplementary angles.

Lines in a plane are either intersecting or parallel.

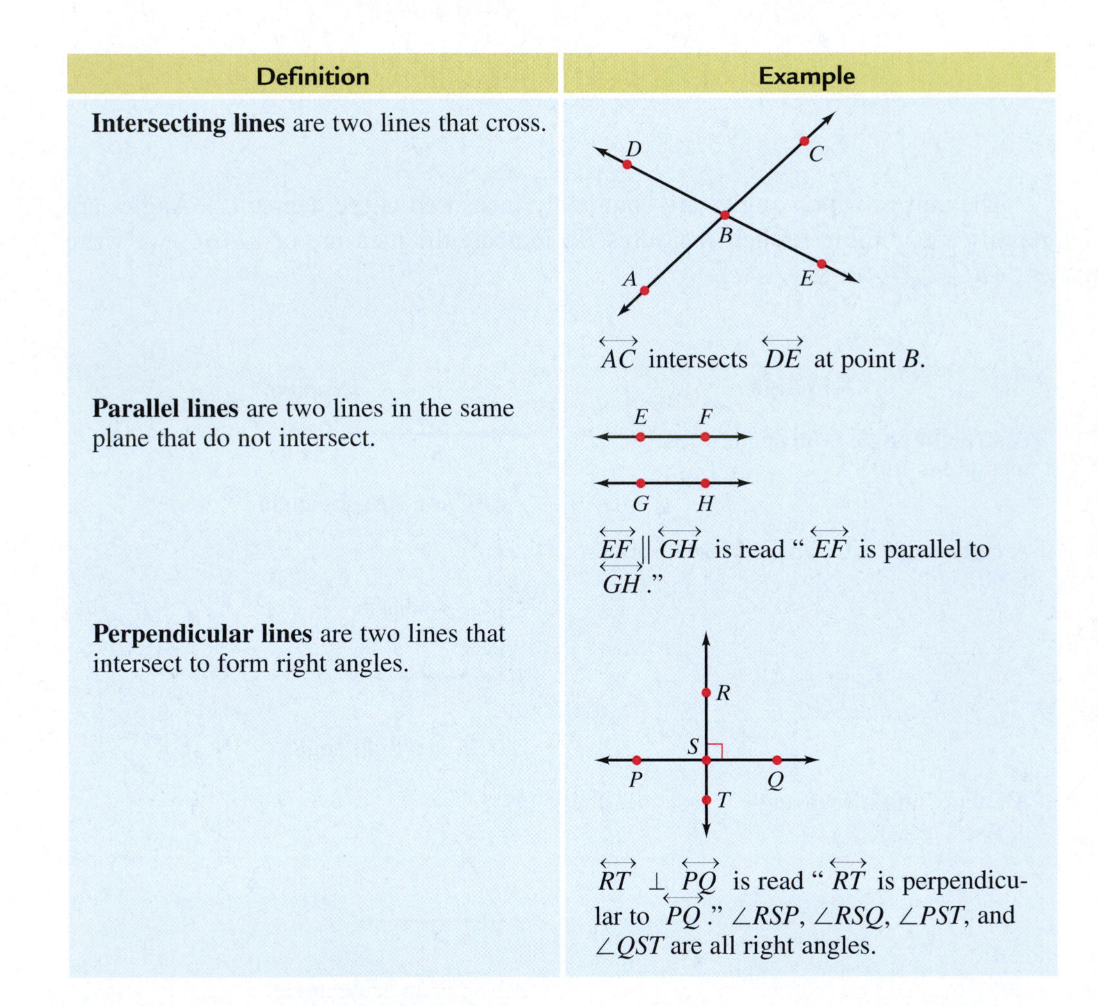

Definition	Example
Intersecting lines are two lines that cross.	$\overleftrightarrow{AC}$ intersects $\overleftrightarrow{DE}$ at point B.
Parallel lines are two lines in the same plane that do not intersect.	$\overleftrightarrow{EF} \parallel \overleftrightarrow{GH}$ is read "$\overleftrightarrow{EF}$ is parallel to $\overleftrightarrow{GH}$."
Perpendicular lines are two lines that intersect to form right angles.	$\overleftrightarrow{RT} \perp \overleftrightarrow{PQ}$ is read "$\overleftrightarrow{RT}$ is perpendicular to $\overleftrightarrow{PQ}$." $\angle RSP$, $\angle RSQ$, $\angle PST$, and $\angle QST$ are all right angles.

When two lines intersect, two special pairs of angles are formed.

Definition	Example
Vertical angles are two angles with equal measure formed by two intersecting lines.	$\angle BAE$ and $\angle DAC$ are vertical angles. $\angle BAD$ and $\angle EAC$ are vertical angles.

Try drawing another pair of vertical angles. Do you think that they are equal? Can you describe the pair of angles formed by intersecting lines that are not vertical angles?

Now let's consider more examples involving these basic geometric terms.

EXAMPLE 1

Sketch and label $\overline{EF}$.

Solution First sketch a line segment.

Then label the line segment.

This line segment is written $\overline{EF}$ and is read "line segment EF."

PRACTICE 1

Draw $\angle ABC$.

EXAMPLE 2

$\angle A$ and $\angle B$ are complementary angles. Find the measure of $\angle A$ if $m\angle B = 69°$.

Solution Because $\angle A$ and $\angle B$ are complementary angles, $m\angle A + m\angle B = 90°$.

$$m\angle A + m\angle B = 90°$$
$$m\angle A + 69° = 90°$$
$$m\angle A + 69° - 69° = 90° - 69° \quad \text{Subtract } 69° \text{ from each side.}$$
$$m\angle A = 21°$$

PRACTICE 2

What is the measure of the angle complementary to 37°?

EXAMPLE 3

Find the measure of the angle that is supplementary to 89°.

Solution To find the measure of the angle that is supplementary to 89°, we write the following equation:

$$89° + x = 180°$$

where x represents the measure of the supplementary angle.

$$89° + x = 180°$$
$$89° - 89° + x = 180° - 89°$$
$$x = 91°$$

So an angle with measure 91° is supplementary to one with measure 89°.

PRACTICE 3

What is the measure of the angle supplementary to 15°?

EXAMPLE 4

In the following diagram, $\angle ABC$ is a straight angle. Find y.

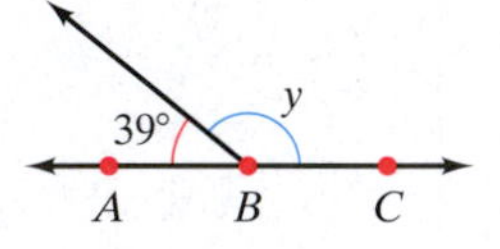

Solution Because $\angle ABC$ is a straight angle, $y + 39° = 180°$. We solve this equation for y.

$$y + 39° = 180°$$
$$y + 39° - 39° = 180° - 39°$$
$$y = 141°$$

PRACTICE 4

In the diagram shown, find x.

EXAMPLE 5

Find the values of x and y in the diagram shown at the right.

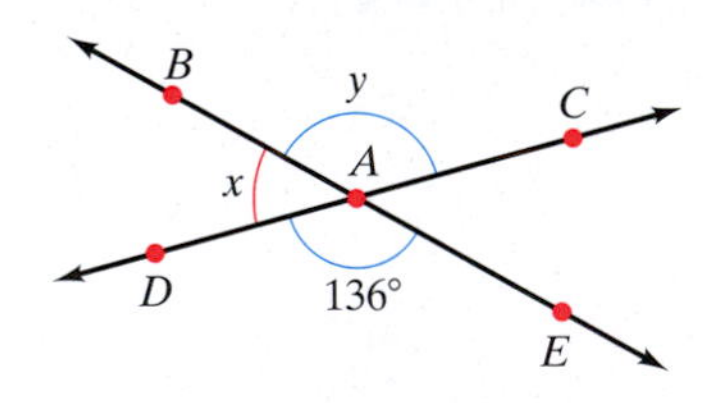

Solution Because $\angle BAC$ and $\angle DAE$ are vertical angles and $\angle DAE = 136°$, $\angle BAC = 136°$, or $y = 136°$. Because $\angle DAC$ is a straight angle, the sum of x and y is 180°.

$$x + y = 180°$$
$$x + 136° = 180°$$
$$x + 136° - 136° = 180° - 136°$$
$$x = 44°$$

So $x = 44°$ and $y = 136°$.

PRACTICE 5

In the following diagram, what are the values of a and b?

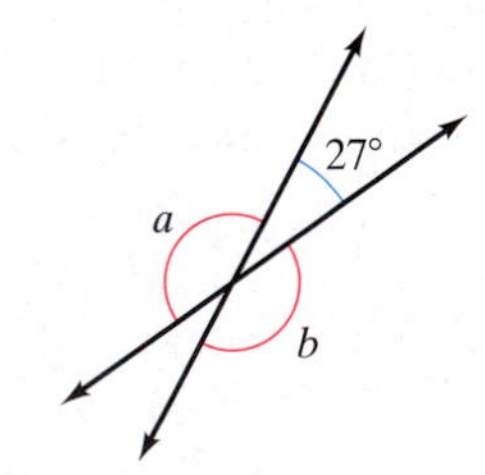

Basic Geometric Figures

Here we use the concepts just discussed to define some basic geometric figures: triangles, trapezoids, parallelograms, rectangles, squares, and circles. Except for circles, these figure are *polygons*.

> **Definition**
>
> A **polygon** is a closed plane figure made up of line segments.

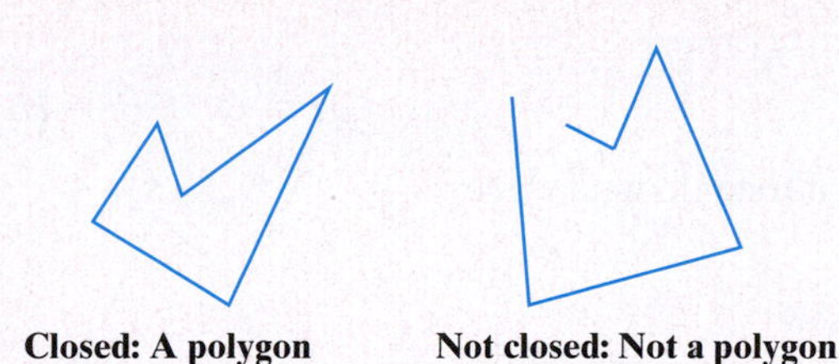

Polygons are classified according to the number of their sides. Here we examine two types of polygons—triangles and quadrilaterals.

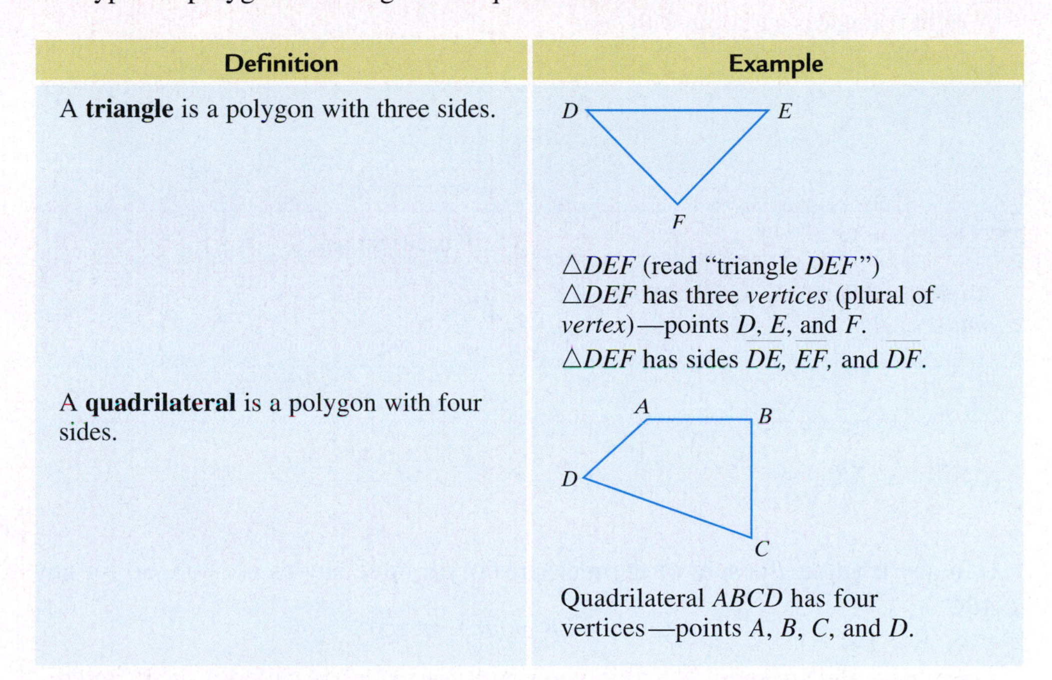

Definition	Example
A **triangle** is a polygon with three sides.	$\triangle DEF$ (read "triangle DEF") $\triangle DEF$ has three *vertices* (plural of *vertex*)—points D, E, and F. $\triangle DEF$ has sides $\overline{DE}$, $\overline{EF}$, and $\overline{DF}$.
A **quadrilateral** is a polygon with four sides.	Quadrilateral $ABCD$ has four vertices—points A, B, C, and D.

Triangles are classified according to the measures of either their sides or their angles.

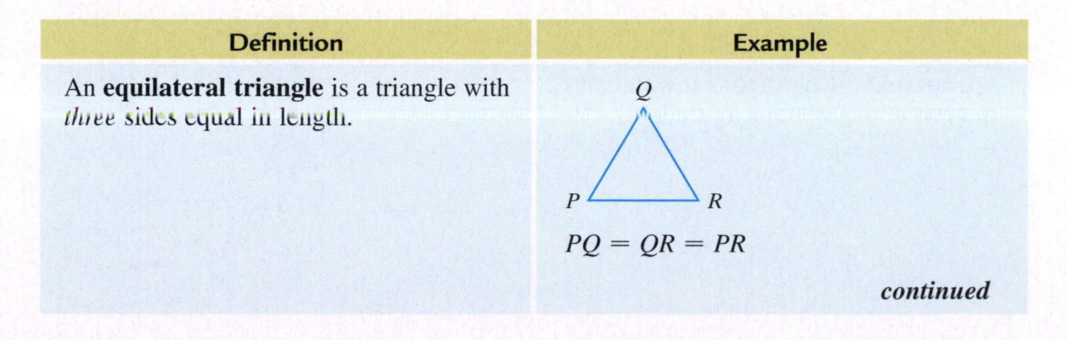

Definition	Example
An **equilateral triangle** is a triangle with *three* sides equal in length.	$PQ = QR = PR$

continued

An **isosceles triangle** is a triangle with *two* sides equal in length.	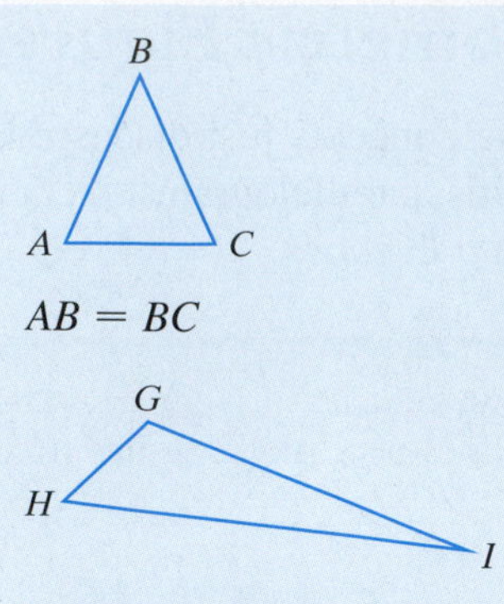 $AB = BC$
A **scalene triangle** is a triangle with *no* sides equal in length.	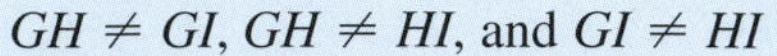 $GH \neq GI$, $GH \neq HI$, and $GI \neq HI$
An **acute triangle** is a triangle with *three acute* angles.	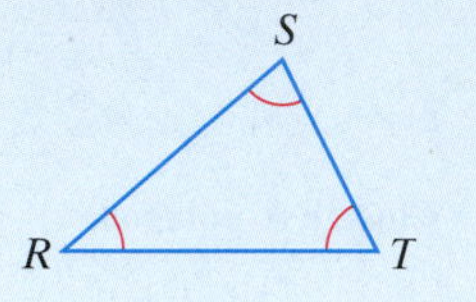 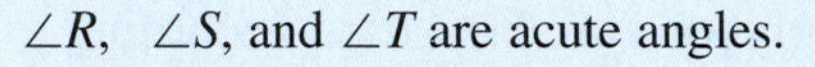 $\angle R$, $\angle S$, and $\angle T$ are acute angles.
A **right triangle** is a triangle with *one right* angle.	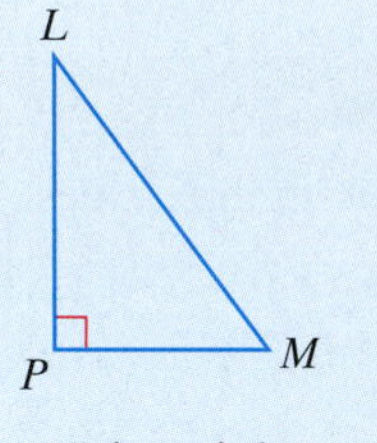 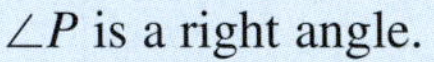 $\angle P$ is a right angle.
An **obtuse triangle** is a triangle with *one obtuse* angle.	$\angle Y$ is an obtuse angle.

In any triangle, the sum of the measures of all three angles is 180°. So for any $\triangle ABC$,

$$m\angle A + m\angle B + m\angle C = 180°$$

We have already seen that a polygon with four sides is called a *quadrilateral*. Let's consider special types of quadrilaterals.

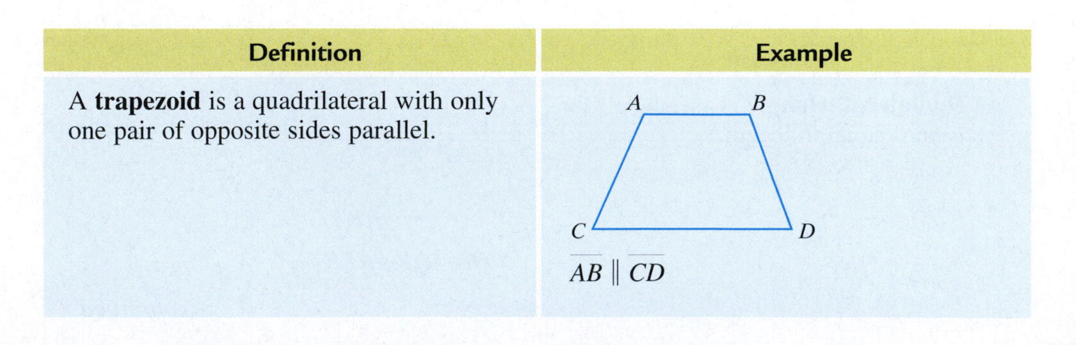

Definition	Example
A **trapezoid** is a quadrilateral with only one pair of opposite sides parallel.	$\overline{AB} \parallel \overline{CD}$

Definition	Example
A **parallelogram** is a quadrilateral with both pairs of opposite sides parallel. Opposite sides are equal in length, and opposite angles have equal measures.	$\overline{LM} \parallel \overline{PO}$ and $\overline{LP} \parallel \overline{MO}$ $LM = PO$ and $LP = MO$ $m\angle L = m\angle O$ and $m\angle P = m\angle M$
A **rectangle** is a parallelogram with four right angles.	$m\angle R = m\angle T = m\angle U = m\angle S = 90°$
A **square** is a rectangle with four sides equal in length.	$DE = EG = FG = DF$

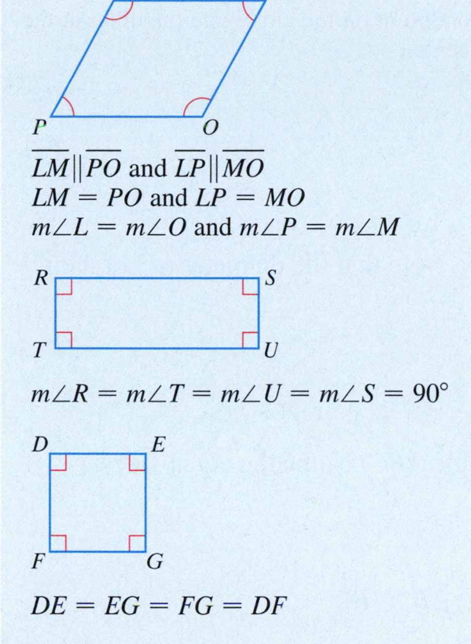

In any quadrilateral, the sum of the measures of the angles is 360°. So for any quadrilateral *ABCD*,

$$m\angle A + m\angle B + m\angle C + m\angle D = 360°$$

We can see why this is true by cutting a quadrilateral into two triangles. In each triangle, the sum of the measures of the three angles is 180°.

The last basic geometric figure we consider here is the circle.

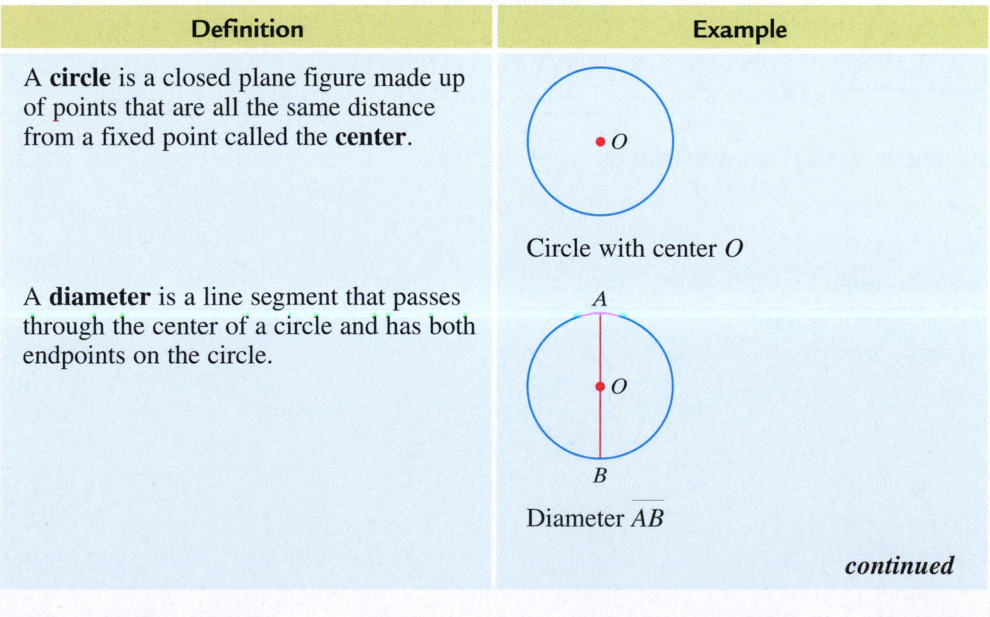

Definition	Example
A **circle** is a closed plane figure made up of points that are all the same distance from a fixed point called the **center**.	Circle with center *O*
A **diameter** is a line segment that passes through the center of a circle and has both endpoints on the circle.	Diameter $\overline{AB}$

continued

A **radius** is a line segment with one endpoint on the circle and the other at the center.

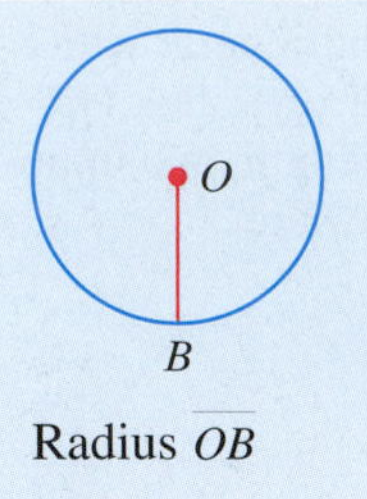

Radius $\overline{OB}$

Note that the diameter (d) of a circle is twice the radius (r), or $d = 2r$.

EXAMPLE 6

Sketch and label isosceles triangle ABC. Name the equal sides.

Solution

B

A C

$AB = BC$

PRACTICE 6

Draw and label quadrilateral $ABCD$ that has at least one right angle with opposite sides equal and parallel. Name both pairs of parallel sides.

EXAMPLE 7

In $\triangle DEF$, $m\angle D = 45°$ and $m\angle E = 65°$. Find the measure of $\angle F$.

Solution First we draw a diagram.

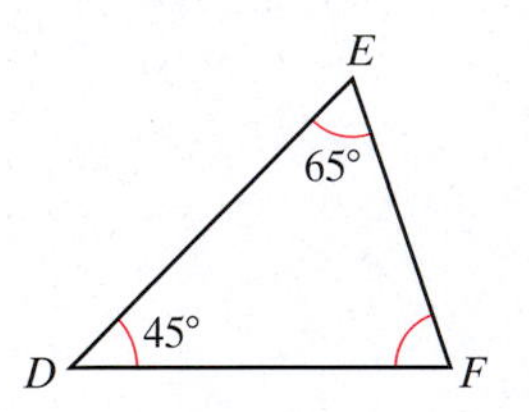

The sum of the measures of the angles is 180°, so we write the following.

$$m\angle D + m\angle E + m\angle F = 180°$$
$$45° + 65° + m\angle F = 180°$$
$$110° + m\angle F = 180°$$
$$110° - 110° + m\angle F = 180° - 110°$$
$$m\angle F = 70°$$

PRACTICE 7

In triangle RST, where $\angle S$ is a right angle and $m\angle T = 30°$, what is $m\angle R$?

EXAMPLE 8

In the quadrilateral shown, what is $m\angle D$?

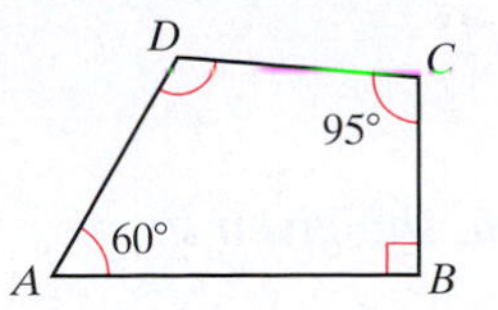

Solution The sum of the measures of the four angles is 360°. Note that $\angle B$ is a right angle, so $m\angle B = 90°$. We write the following.

$$\begin{aligned} m\angle A + m\angle B + m\angle C + m\angle D &= 360° \\ 60° + 90° + 95° + m\angle D &= 360° \\ 245° + m\angle D &= 360° \\ 245° - 245° + m\angle D &= 360° - 245° \\ m\angle D &= 115° \end{aligned}$$

PRACTICE 8

In the trapezoid shown, find the measure of $\angle U$?

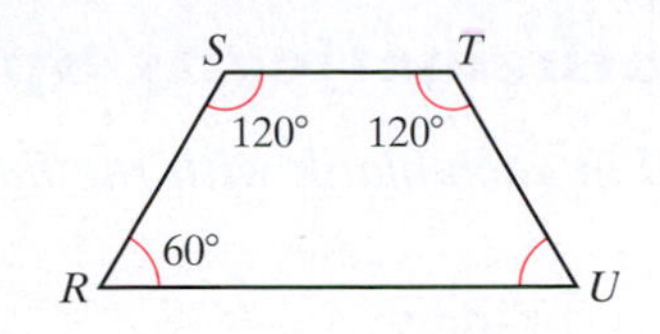

EXAMPLE 9

A dartboard has a diameter of 18 inches. What is the radius of the dartboard?

Solution The diameter of the dartboard is 18 inches. The distance from the center to a point on the edge of the board is the radius. To find the radius, we divide the diameter by 2.

$$18 \div 2 = 9$$

The radius of the dartboard is 9 inches.

PRACTICE 9

A manhole cover has a radius of 12 inches. What is the diameter of the manhole cover?

Mathematically Speaking

Fill in each blank with the most appropriate term or phrase from the given list.

radius	perpendicular	line segment
scalene	parallelogram	obtuse
acute	isosceles	trapezoid
supplementary	ray	parallel
vertical	complementary	diameter

1. A(n) _______ is a part of a line having two endpoints.

2. A(n) _______ is an angle whose measure is more than 90° and less than 180°.

3. Two angles are _______ if the sum of their measures is 90°.

4. Two angles are _______ if the sum of their measures is 180°.

5. Lines that intersect to form right angles are called _______.

6. Angles with equal measure formed by two intersecting lines are called _______.

7. Lines in the same plane that do not intersect are called _______.

8. A(n) _______ triangle has two sides equal in length.

9. A(n) _______ triangle has no sides equal in length.

10. A(n) _______ is a quadrilateral with only one pair of opposite sides parallel.

11. A(n) _______ is a quadrilateral with both pairs of opposite sides parallel.

12. A(n) _______ is a line segment that passes through the center of a circle and has both endpoints on the circle.

Sketch and label each geometric object. Where appropriate, use symbols to express your answer.

13. Point P

14. Line $\overleftrightarrow{AB}$

15. Line segment $\overline{BC}$

16. Ray $\overrightarrow{AB}$

17. Parallel lines $\overleftrightarrow{MN}$ and $\overleftrightarrow{ST}$

18. Perpendicular lines $\overleftrightarrow{UV}$ and $\overleftrightarrow{WX}$

19. Equilateral $\triangle ABC$

20. Isosceles $\triangle PQR$

21. Circle

22. Trapezoid

23. Scalene $\triangle ABC$

24. Right $\triangle WXY$

25. Acute $\angle FGH$

26. Vertical angles

Solve.

27. Find x.

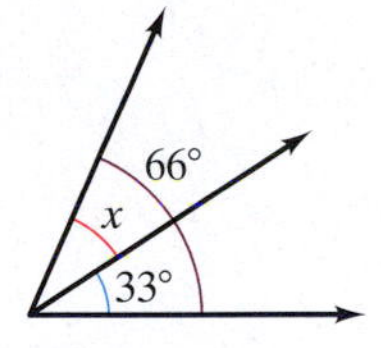

28. $\angle PQR$ is a straight angle.
Find the measure of $\angle PQS$.

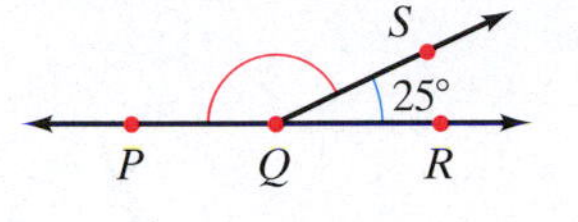

29. $\angle DEF$ is a right angle.
Find the measure of $\angle DEG$.

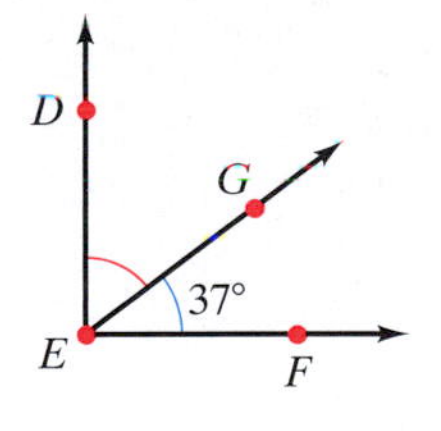

30. Solve for x and y.

In the diagram shown, $\overleftrightarrow{AB} \perp \overleftrightarrow{CD}$ and $m\angle CPE = 35°$. Find the measure of each angle.

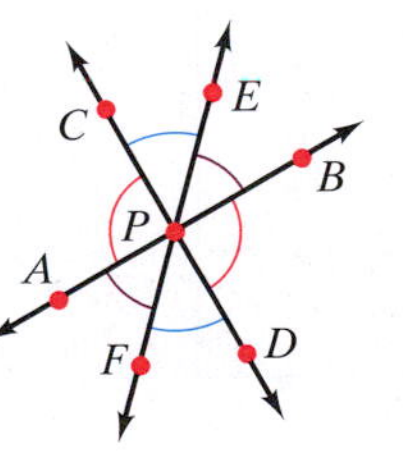

31. $\angle CPD$

32. $\angle APD$

33. $\angle BPD$

34. $\angle CPB$

35. $\angle APB$

36. $\angle BPE$

37. $\angle FPD$

38. $\angle APF$

In each figure, find the measure of the unknown angle(s).

39. **40.**

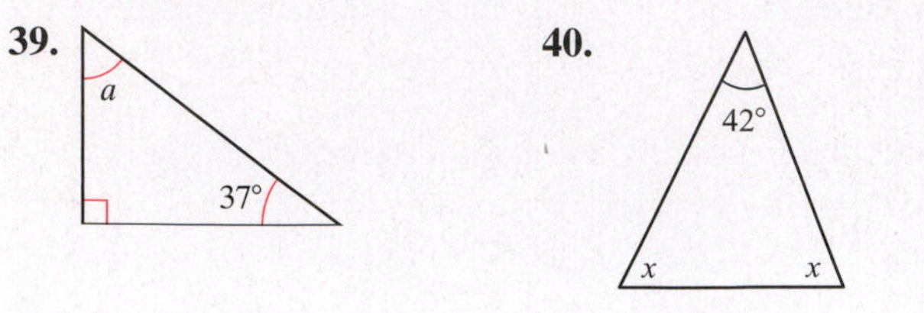

41.

y
40°

42.

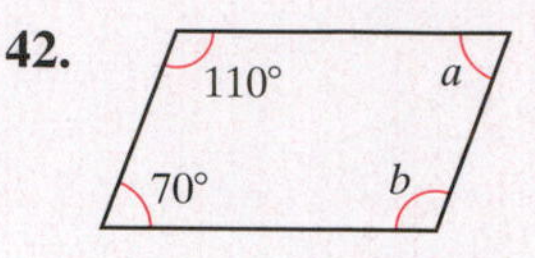

Solve.

43. Find the complement of 35°.

44. What is the measure of an angle that is complementary to itself?

45. Find the supplement of 105°.

46. What is the measure of an angle that is supplementary to 88°?

47. In $\triangle ABC$, $m\angle A = 35°$ and $m\angle B = 75°$. Find the measure of $\angle C$.

48. In $\triangle DEF$, $m\angle E = 90°$ and $m\angle F = 19°$. Find the measure of $\angle D$.

49. In a parallelogram, the sum of three of the angles is 275°. What is the measure of the fourth angle?

50. In a triangle in which all angles are equal, what is the measure of each angle?

Mixed Practice

Solve.

51. Sketch and label the diameter of a circle

52. In the following diagram, find the measure of $\angle DBE$.

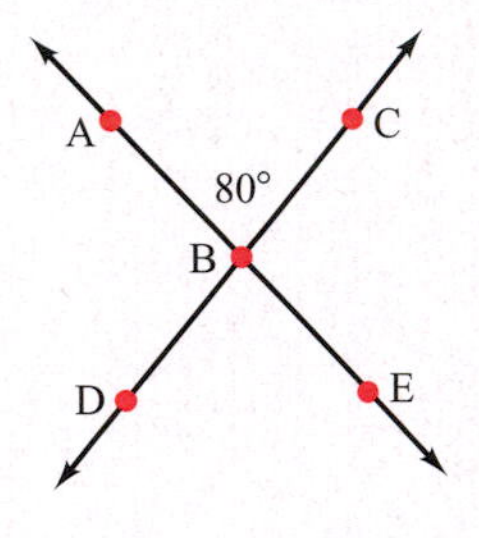

53. Find x.

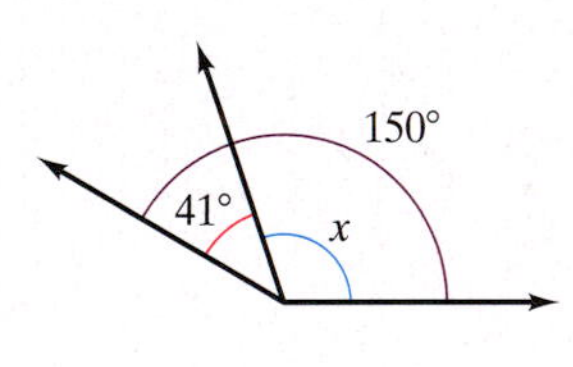

54. Find x.

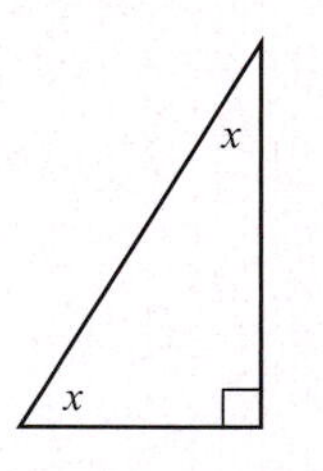

55. Sketch and label obtuse $\triangle RST$.

56. What is the measure of an angle that is complementary to 20°?

Applications

Solve.

57. An ancient circular medicine wheel made with rocks, as shown in the diagram, was built by Native Americans in Wyoming. What is the wheel's radius? (*Source:* Works Projects Administration, *Wyoming: A Guide to Its History, Highways, and People*)

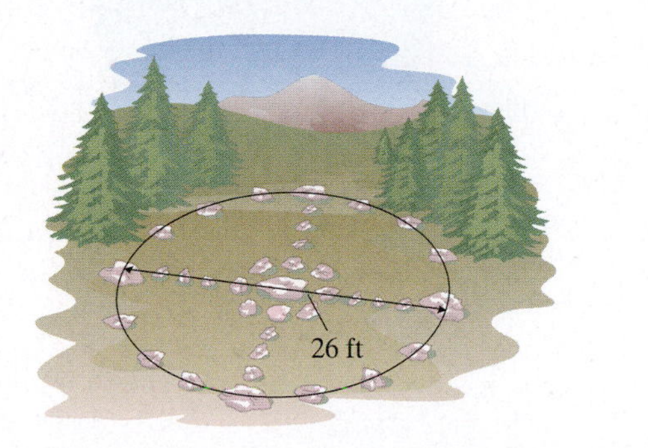

58. A circular access road is to be constructed around an office building. The distance between the center of the office building and the road is 350 feet. Find the diameter of the circle formed by the road.

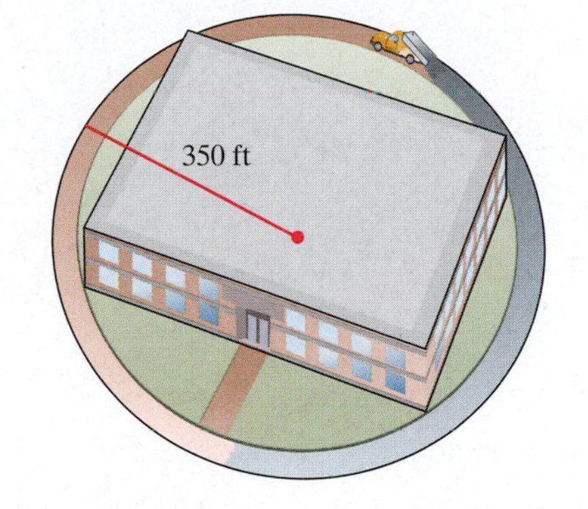

59. In a physics class, the instructor connects two blocks with a cord that passes over a small, frictionless pulley, as shown. What is the measure of the missing angle?

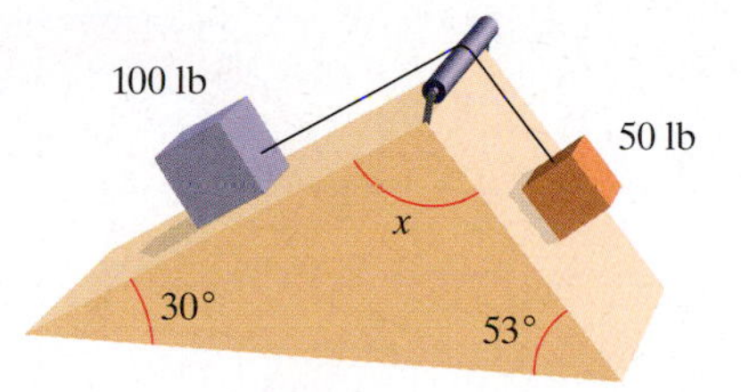

60. A solar eclipse occurs when the Moon passes between the Sun and Earth. In the following diagram of a solar eclipse, $m\angle RST = 140°$ and $m\angle SRT = 5°$. Find the measure of $\angle STR$.

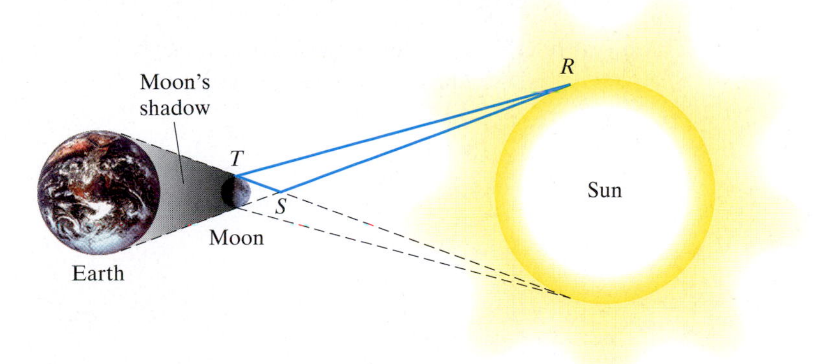

61. The Bermuda triangle, mapped below, is a region of the Atlantic Ocean famous because many people, aircraft, and ships have disappeared within its bounds.

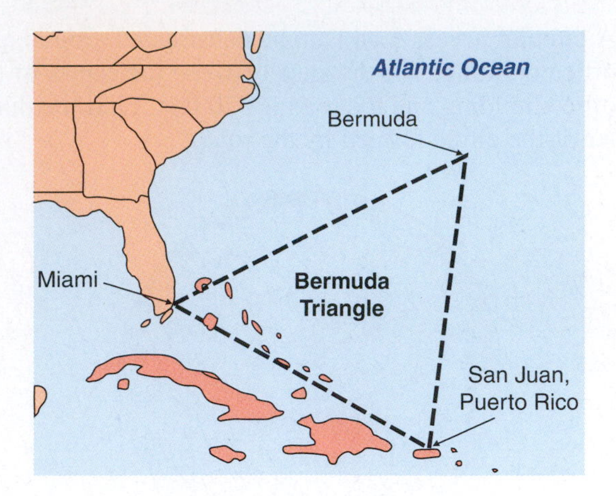

Is this triangle acute, right, or obtuse?

62. The following circle graph shows the relative annual egg production of three Midwest states. Which state's production is not represented by an acute angle?

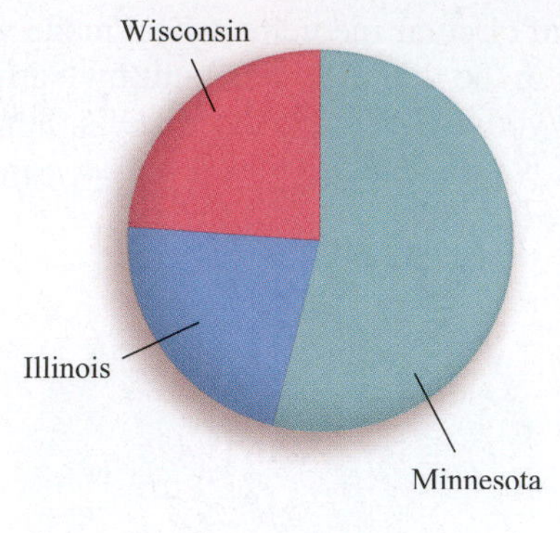

Source: http://www.usda.gov

63. A ceramic tile in the shape of a parallelogram used for a bathroom remodeling project is shown in the diagram.

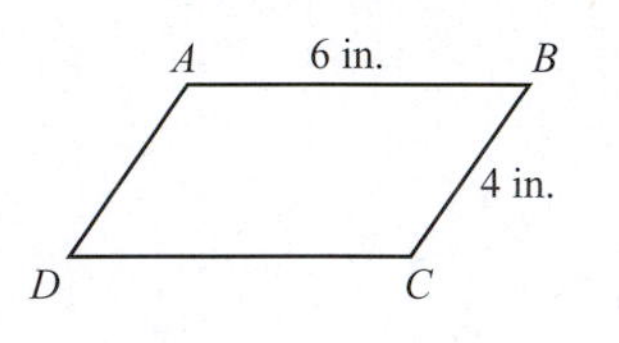

a. What is the length of the side parallel to $\overline{AB}$?

b. If $m\angle C = 120°$, then what is the measure of $\angle D$?

64. A circular mirror shown in the diagram has a radius of 10 inches.

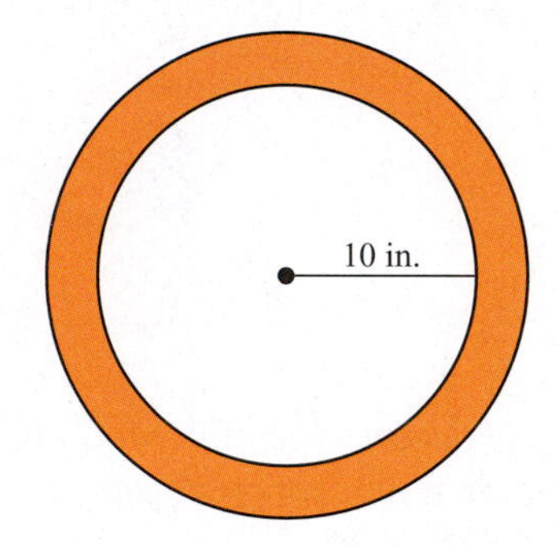

a. What is the diameter of the mirror, *excluding* the circular frame?

b. The diameter of the mirror, *including* the frame, is 28 inches. How wide is the frame?

MINDSTRETCHERS

Patterns

1. How many rectangles can you find in the diagram below?

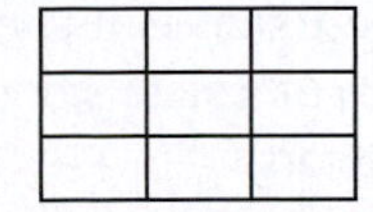

Writing

2. Can a triangle contain one right angle and one obtuse angle? Explain.

Groupwork

3. A *hexagon* is a polygon with 6 sides and 6 angles. Working with a partner, show how you can find the sum of the measures of the angles in the hexagon.

11.2 Perimeter and Circumference

The Perimeter of a Polygon

OBJECTIVES

- To find the perimeter of a polygon
- To find the circumference of a circle
- To find the perimeter or circumference of a composite geometric figure
- To solve word problems involving perimeter or circumference

One of the most basic features of a plane geometric figure is its *perimeter*. The length of a fence around a plot of land, the length of a state's border, and the length of a picture frame are examples of perimeters.

Definition

The **perimeter** of a polygon is the distance around it.

To find the perimeter of any polygon, we add the lengths of its sides. Note that perimeters are measured in linear units such as feet or meters.

Suppose that we want to build a fence around the garden shown.

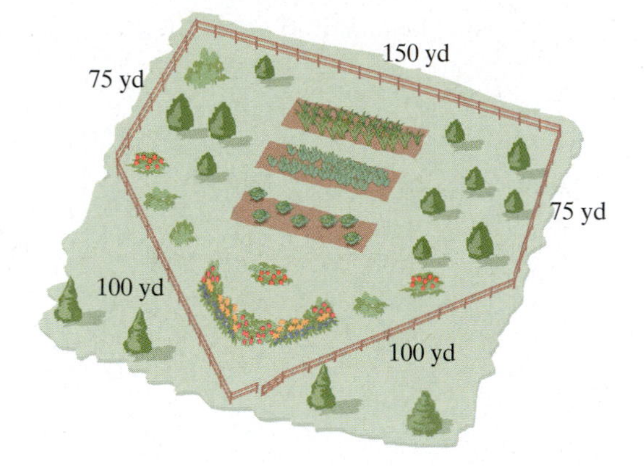

How much fencing do we need? Using the definition of perimeter, we obtain the distance around this garden.

$$75 + 150 + 75 + 100 + 100 = 500$$

So we need 500 yards of fencing.

For some polygons, we can also use a *formula* to find the perimeter. Let's consider the formulas for the perimeter of a triangle, a rectangle, and a square.

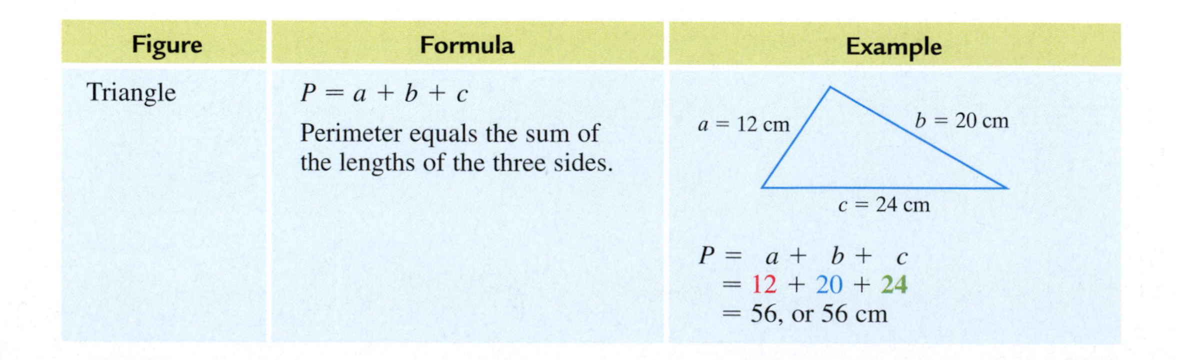

Figure	Formula	Example
Triangle	$P = a + b + c$ Perimeter equals the sum of the lengths of the three sides.	$a = 12$ cm, $b = 20$ cm, $c = 24$ cm $P = a + b + c$ $= 12 + 20 + 24$ $= 56$, or 56 cm

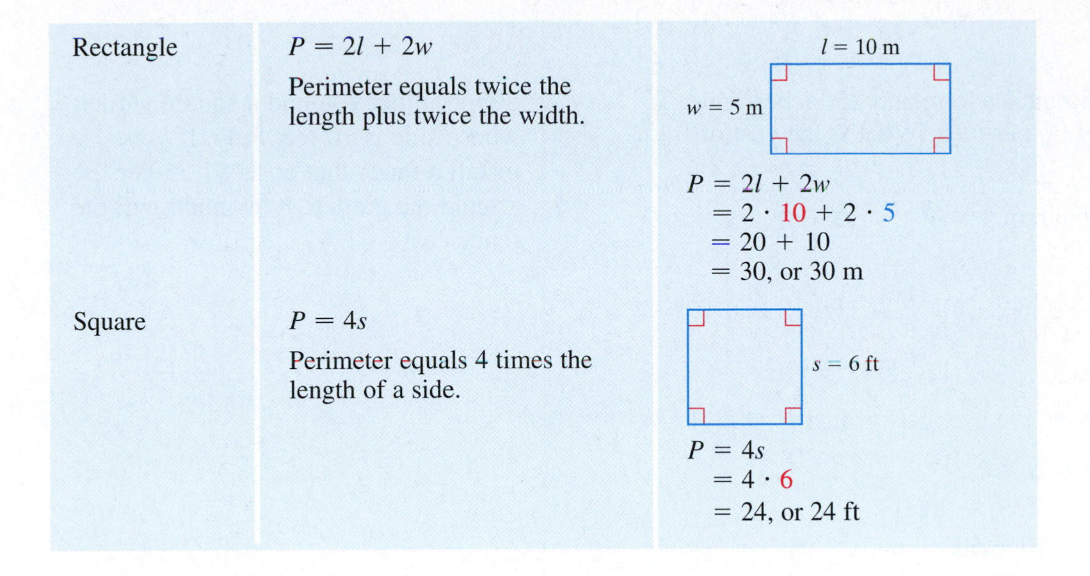

Rectangle	$P = 2l + 2w$ Perimeter equals twice the length plus twice the width.	$l = 10$ m, $w = 5$ m $P = 2l + 2w$ $= 2 \cdot 10 + 2 \cdot 5$ $= 20 + 10$ $= 30$, or 30 m
Square	$P = 4s$ Perimeter equals 4 times the length of a side.	$s = 6$ ft $P = 4s$ $= 4 \cdot 6$ $= 24$, or 24 ft

EXAMPLE 1

Find the perimeter of the polygon shown.

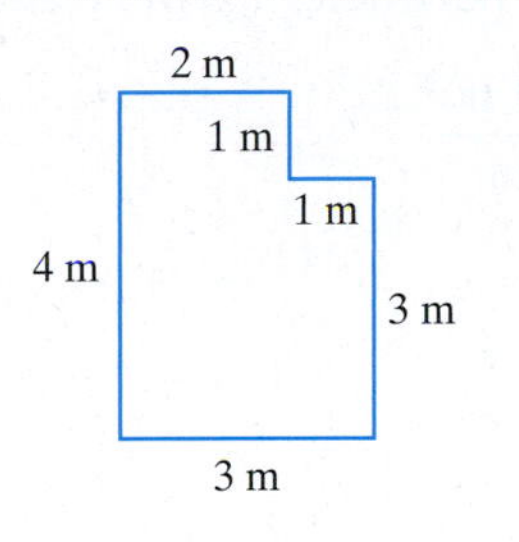

Solution To find the perimeter, we add the lengths of the sides.

$$2 + 1 + 1 + 3 + 3 + 4 = 14$$

So the perimeter is 14 meters.

PRACTICE 1

What is the perimeter of this polygon?

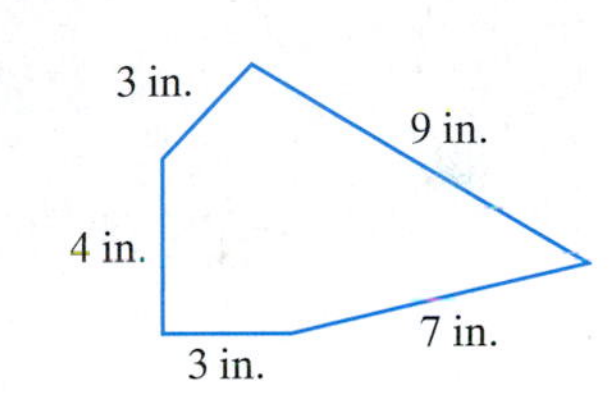

EXAMPLE 2

Find the perimeter of an equilateral triangle with side 1.4 meters long.

Solution Recall that all three sides of an equilateral triangle are equal.

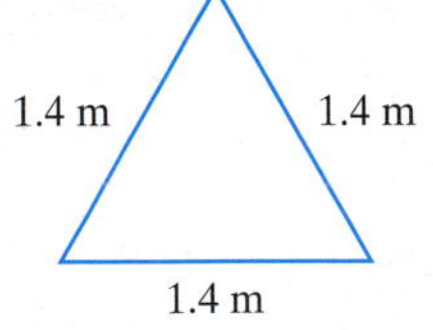

We use the formula for the perimeter of a triangle.

$$P = a + b + c$$
$$= \mathbf{1.4} + \mathbf{1.4} + \mathbf{1.4}$$
$$= 4.2$$

Therefore the perimeter of the triangle is 4.2 meters. Because all three sides are equal in length, we could have used the formula

$$P = 3s = 3(1.4) = 4.2, \text{ or } 4.2 \text{ meters}$$

PRACTICE 2

Find the perimeter of a square with side $\frac{3}{4}$ miles long.

EXAMPLE 3

A rectangular picture is 35 inches long and 25 inches wide. To frame the picture costs $1.00 per inch. What is the cost of framing the picture?

Solution Let's draw a diagram.

35 in.

25 in.

The picture is rectangular, so we use the formula $P = 2l + 2w$ to find its perimeter.

$$\begin{aligned} P &= 2l + 2w \\ &= 2(35) + 2(25) \\ &= 70 + 50 \\ &= 120 \end{aligned}$$

The distance around the picture is 120 inches. To find the cost of framing the picture, we multiply this perimeter by the cost per inch.

$$\begin{aligned} \text{Cost} &= 120\ \cancel{\text{in.}} \times \frac{\$1.00}{\cancel{\text{in.}}} \\ &= \$120 \end{aligned}$$

So the cost of framing the picture is $120.

PRACTICE 3

Suppose that you had a square garden whose side is 10 feet long. If you install a fence that costs $1.75 per foot around the garden, how much will the fence cost?

The Circumference of a Circle

Just as we speak of the perimeter of a polygon, we refer to the circumference of a circle.

Definition

The distance around a circle is called its **circumference.**

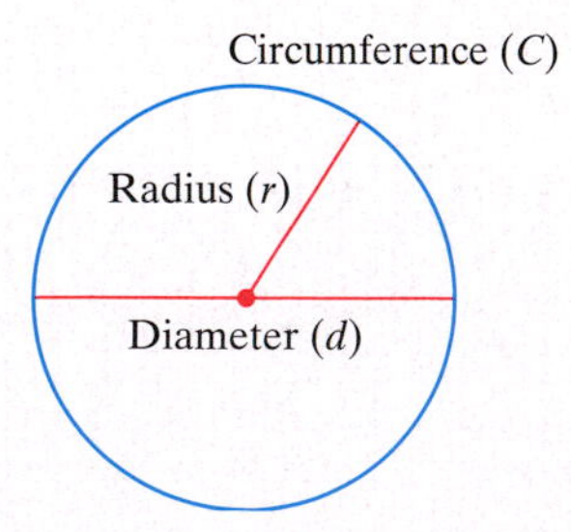

For every circle, the ratio of the circumference, C, to the diameter, d, is the same number, which is written as π (read "pi"). This relationship, $\frac{C}{d} = \pi$, can also be written as $C = \pi d$ or $C = 2\pi r$. Do you see why πd and $2\pi r$ are equal? Explain.

The number π equals 3.1415926. . . . It is an *irrational number*, so that when π is expressed as a decimal, the digits go on indefinitely without any pattern being repeated. For convenience, we often use approximate values of π, such as 3.14 and $\frac{22}{7}$, when calculating circumferences by hand.

Figure	Formula	Example
Circumference	$C = \pi d$, or $C = 2\pi r$ Circumference equals π times the diameter, or 2 times π times the radius r.	10 cm $C = \pi d$ $\approx 3.14(\mathbf{10})$ ≈ 31.4, or 31.4 cm

EXAMPLE 4

Find the circumference of the circle shown. Use $\pi \approx 3.14$.

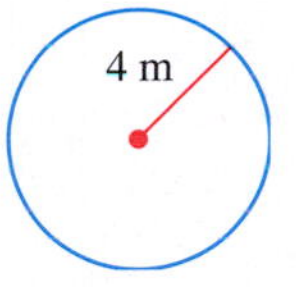

Solution The radius of the circle is 4 meters. We use the formula for the circumference of a circle in terms of the radius.

$$\begin{aligned} C &= 2\pi r \\ &\approx 2(3.14)(4) \quad \text{Substitute 4 for } r. \\ &\approx 25.12 \end{aligned}$$

Therefore, the circumference is approximately 25.12 meters.

PRACTICE 4

What is the circumference of the circle shown? Use $\pi \approx \frac{22}{7}$.

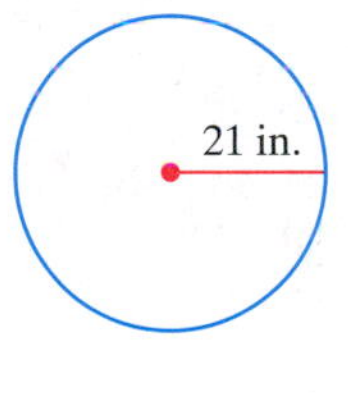

EXAMPLE 5

Suppose that the diameter of a rolling wheel is 20 inches. How far does it travel in one complete turn?

Solution First let's draw a diagram.

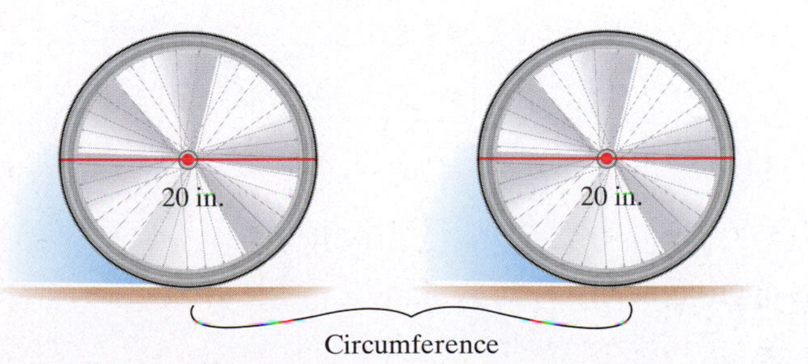

PRACTICE 5

A circular swimming pool has a radius of 18 feet. If a metal rail is to be placed around the edge of the pool, how many feet of railing are needed?

The wheel makes one complete turn, so we know that it travels a distance equal to its circumference. We use the formula for the circumference in terms of the diameter.

$$C = \pi d$$
$$\approx 3.14(20) \quad \text{Substitute 20 for } d.$$
$$\approx 62.8$$

Thus the wheel travels approximately 62.8 feet in one turn. Do you see how to solve this problem using the formula $C = 2\pi r$?

Composite Figures

Two or more basic geometric figures may be combined to form a **composite figure.**

EXAMPLE 6

Find the perimeter of the following figure, which consists of a semicircle and a rectangle.

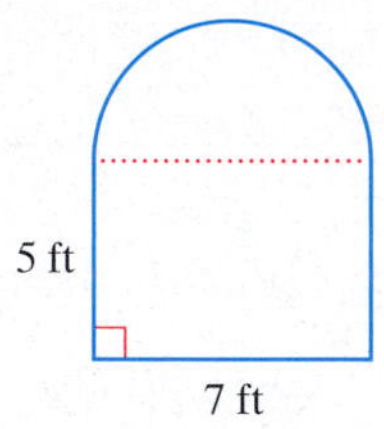

Solution The upper part of the figure is a semicircle with a diameter of 7 feet. The distance around the semicircle is $\frac{1}{2}$ the circumference of the entire circle. So let's compute this circumference and then divide by 2.

$$C = \pi d$$
$$\approx \frac{22}{7}(7)$$
$$\approx 22$$

The circumference of the semicircle is approximately $\frac{22}{2}$, or 11 feet.

Now we find the perimeter of three sides of the rectangle at the lower part of the figure.

$$P = 5 + 7 + 5$$
$$= 17$$

The perimeter of the lower part is 17 feet.

$$\text{Total perimeter} = \text{Circumference of top} + \text{Perimeter of bottom}$$
$$\approx 11 + 17$$
$$\approx 28$$

Therefore, the perimeter of the composite figure is approximately 28 feet.

PRACTICE 6

What is the perimeter of the figure shown, which is composed of a square and a semicircle?

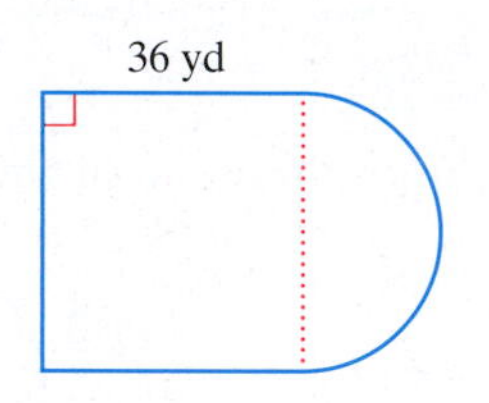

EXAMPLE 7

To prevent a cellar from being flooded, a plumber puts drainage pipes around the outside of a building complex. The outline of the building, consisting of two rectangles and a square, is shown.

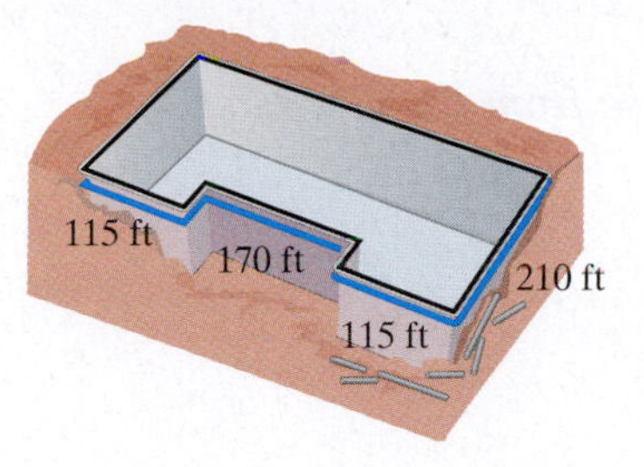

Ignoring the distance between the pipes and the walls, what is the length of the pipes required to go around the complex?

Solution We want to find the perimeter of the building. So we break the composite figure into three figures—two rectangles and one square.

We find the length of the indented part of each rectangular shape by subtracting 170 feet from 210 feet to get 40 feet.

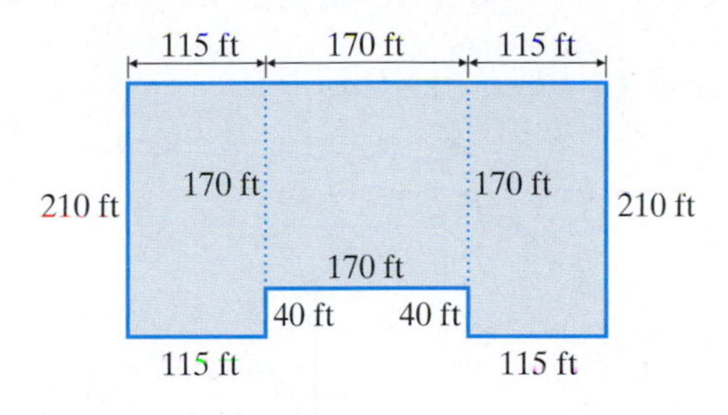

Now we know all the lengths that make up the building complex, so we can find its perimeter.

$$P = \underbrace{(115 + 210 + 115 + 40)}_{\text{Left rectangle}} + \underbrace{(170 + 170)}_{\text{Center square}}$$
$$+ \underbrace{(115 + 210 + 115 + 40)}_{\text{Right rectangle}}$$
$$= 1{,}300$$

Note that the dotted lines are not part of the perimeter because they are not part of the outside of the figure.

So 1,300 feet of drainage pipe is needed to go around the building complex.

PRACTICE 7

An interior decorator plans to place a wallpaper border at the top of the walls of the room shown in the diagram. The room consists of two rectangles. How many feet of the wallpaper border are needed?

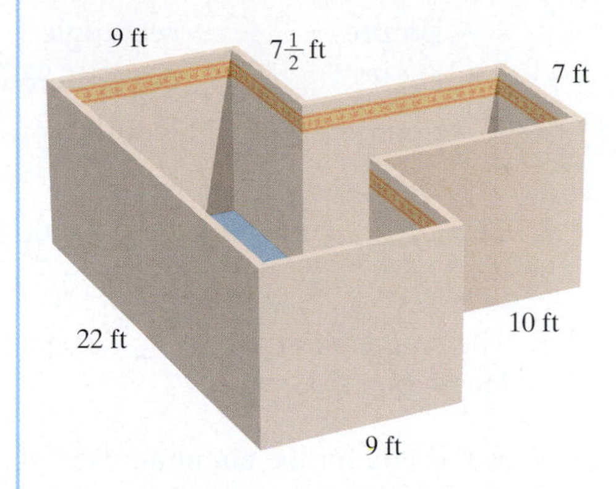

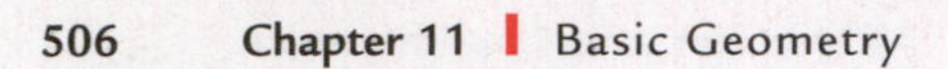

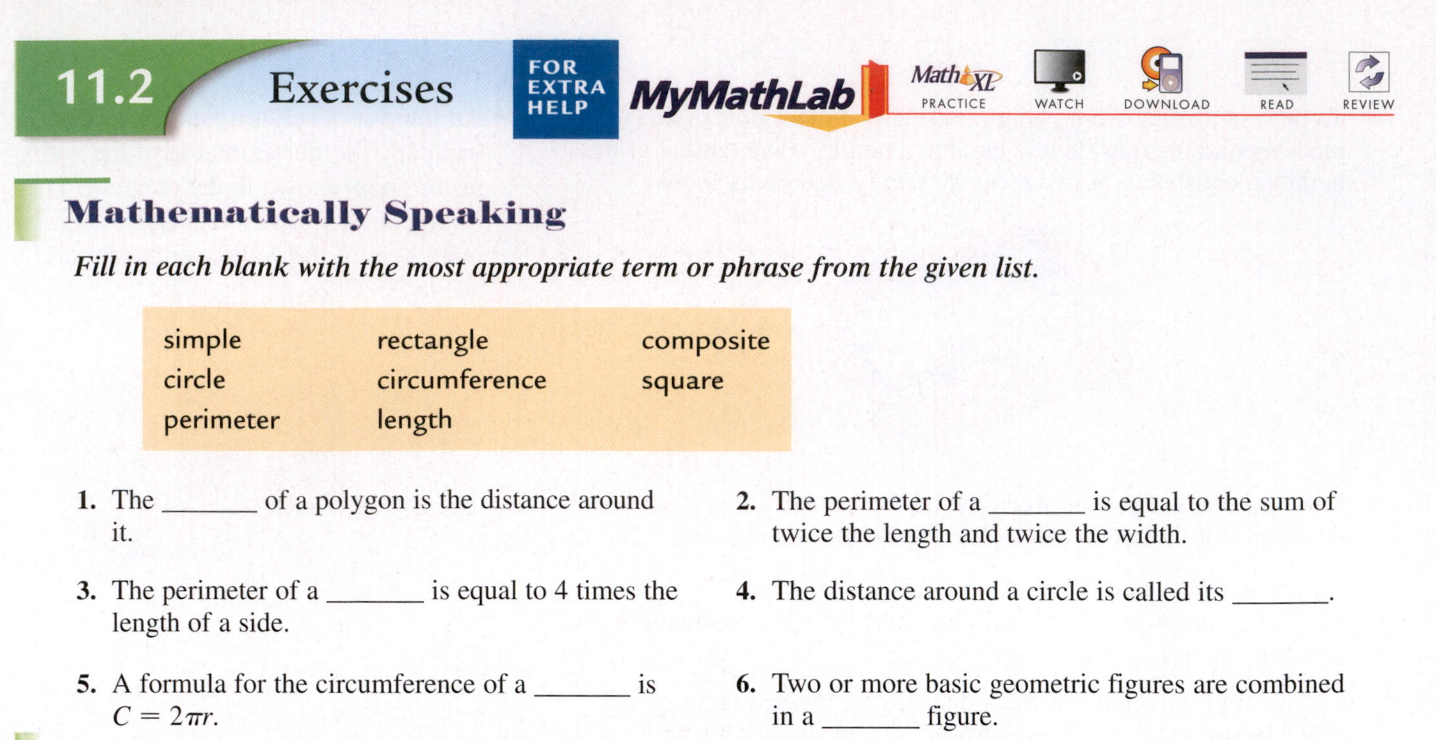

11.2 Exercises

FOR EXTRA HELP MyMathLab MathXL PRACTICE WATCH DOWNLOAD READ REVIEW

Mathematically Speaking

Fill in each blank with the most appropriate term or phrase from the given list.

simple	rectangle	composite
circle	circumference	square
perimeter	length	

1. The _______ of a polygon is the distance around it.
2. The perimeter of a _______ is equal to the sum of twice the length and twice the width.
3. The perimeter of a _______ is equal to 4 times the length of a side.
4. The distance around a circle is called its _______.
5. A formula for the circumference of a _______ is $C = 2\pi r$.
6. Two or more basic geometric figures are combined in a _______ figure.

Find the perimeter or circumference of each figure. Use $\pi \approx 3.14$ *when needed.*

7. 8. 9. 10. 11. 12. 13. 14.

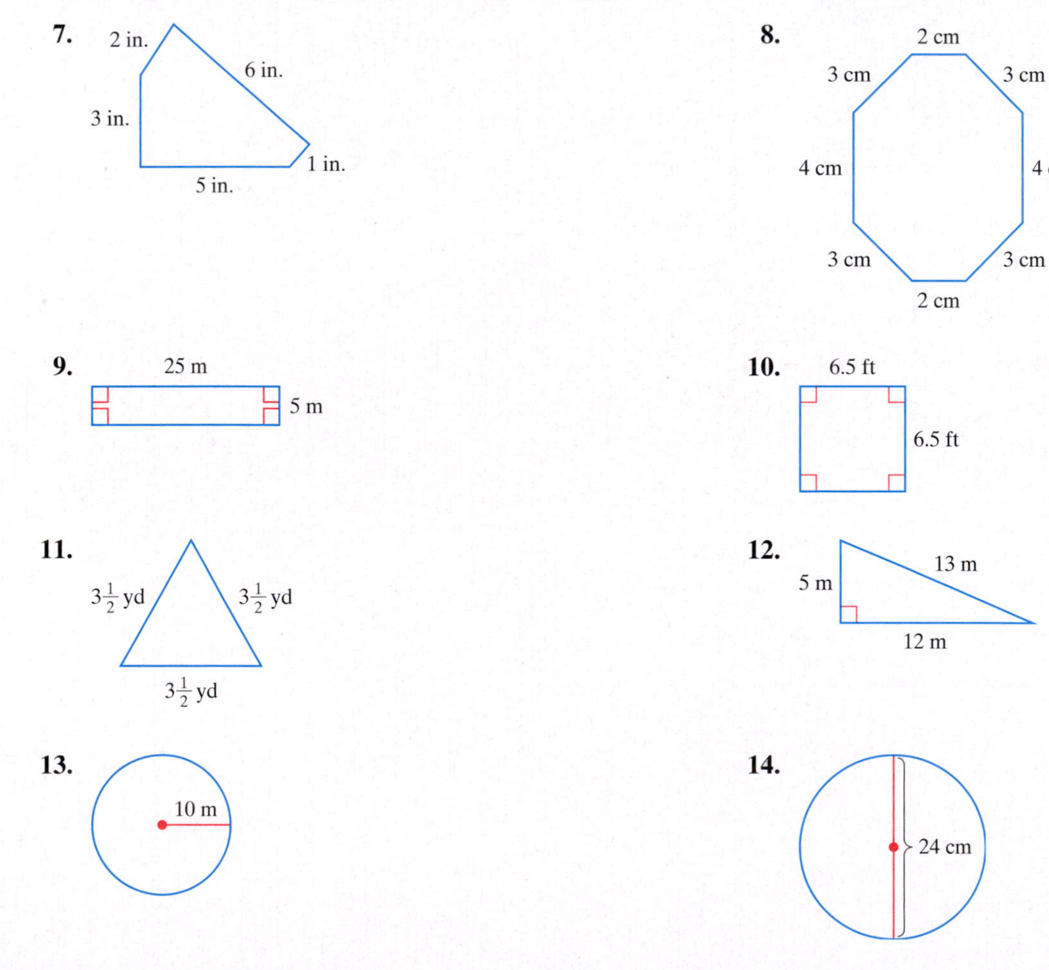

15.

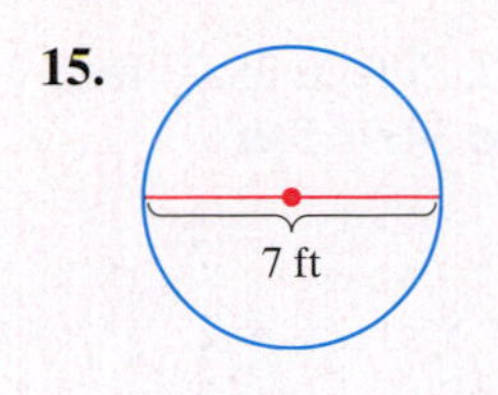

16.

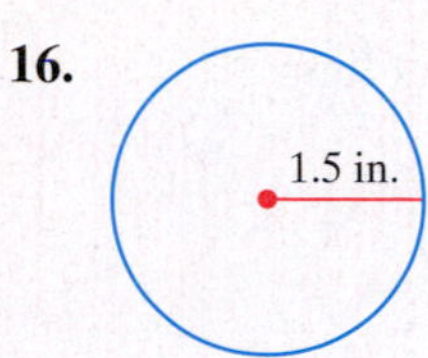

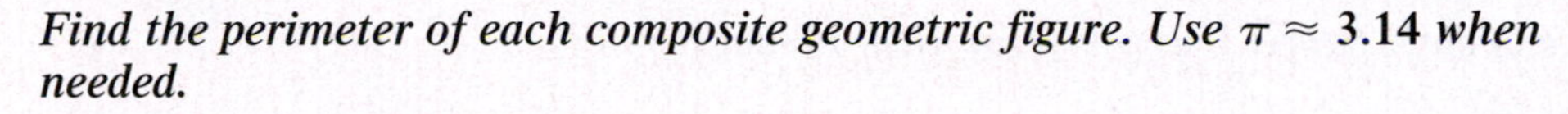

Find the perimeter of each composite geometric figure. Use $\pi \approx 3.14$ when needed.

17. **18.** **19.** **20.** **21.** **22.**

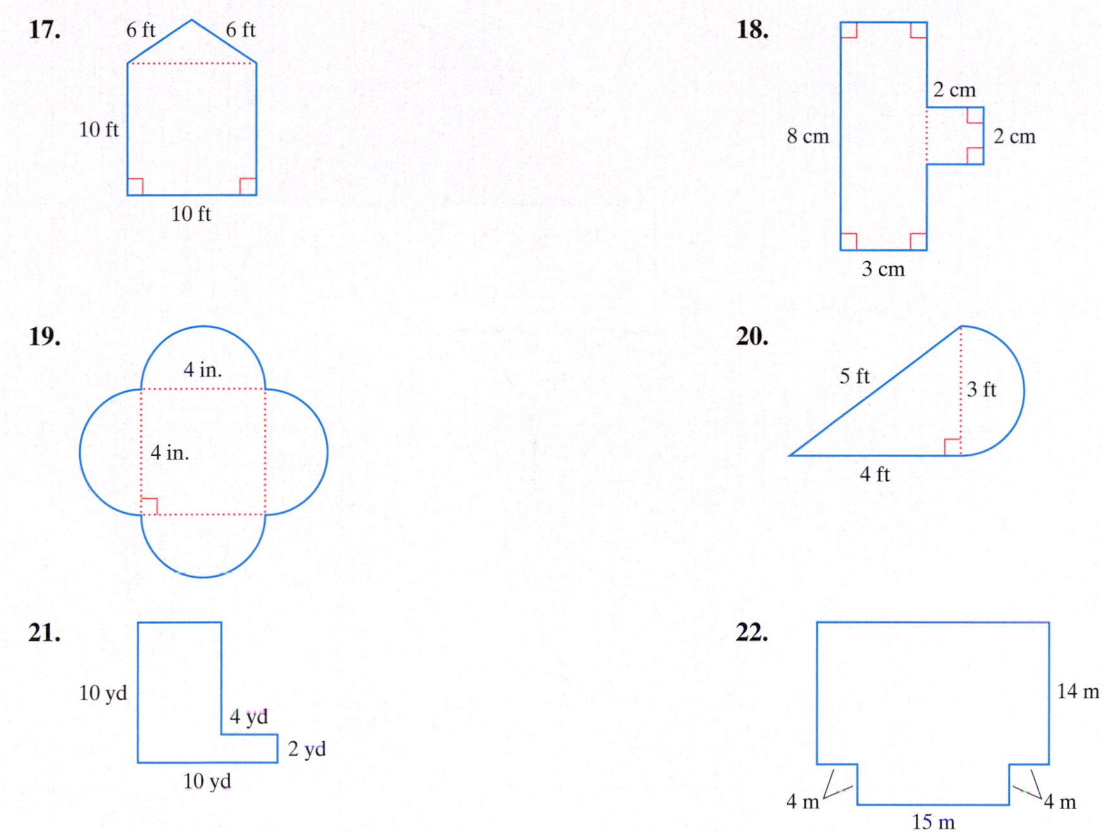

Find the perimeter or circumference. Use $\pi \approx 3.14$ when needed.

23. A square with side $5\frac{1}{4}$ yards long

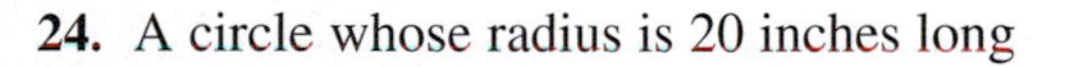

24. A circle whose radius is 20 inches long

25. A rectangle of length $5\frac{3}{4}$ feet and width $3\frac{1}{4}$ feet

26. A triangle whose side lengths are 2 inches, $1\frac{1}{2}$ inches, and $\frac{7}{8}$ inches.

27. An isosceles triangle whose equal sides are $7\frac{1}{2}$ centimeters long and whose third side is 4 centimeters long

28. A rectangle with length 8 meters and width $4\frac{1}{2}$ meters

29.

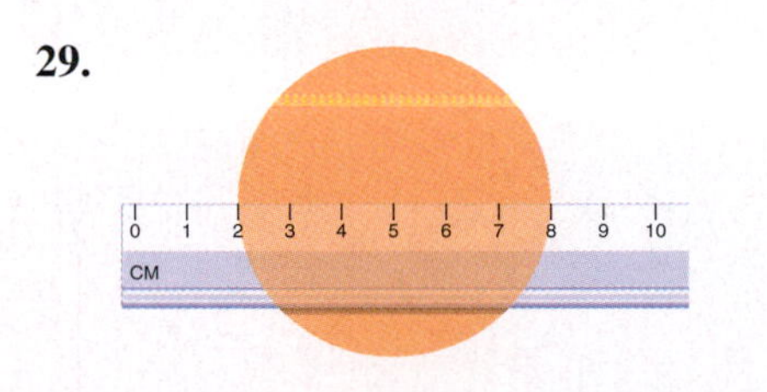

30.

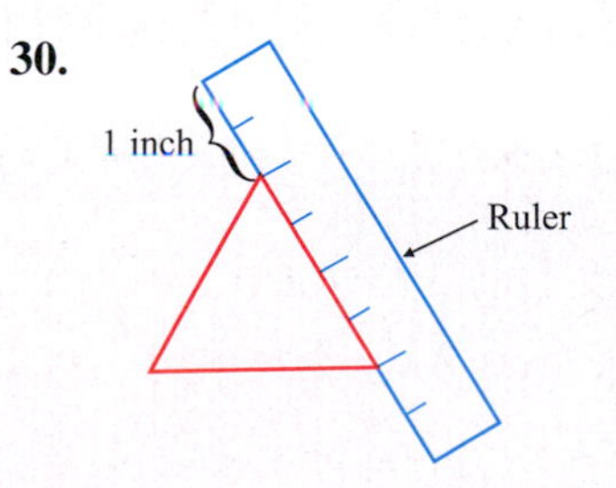

31. A circle whose diameter is 3.54 meters long

32. A polygon whose side lengths are 22.75 feet, 25.73 feet, 15.94 feet, 18.23 feet, 21.65 feet, and 34.98 feet

Mixed Practice

Find the perimeter or circumference of each figure. Use $\pi \approx$ **3.14** *when needed.*

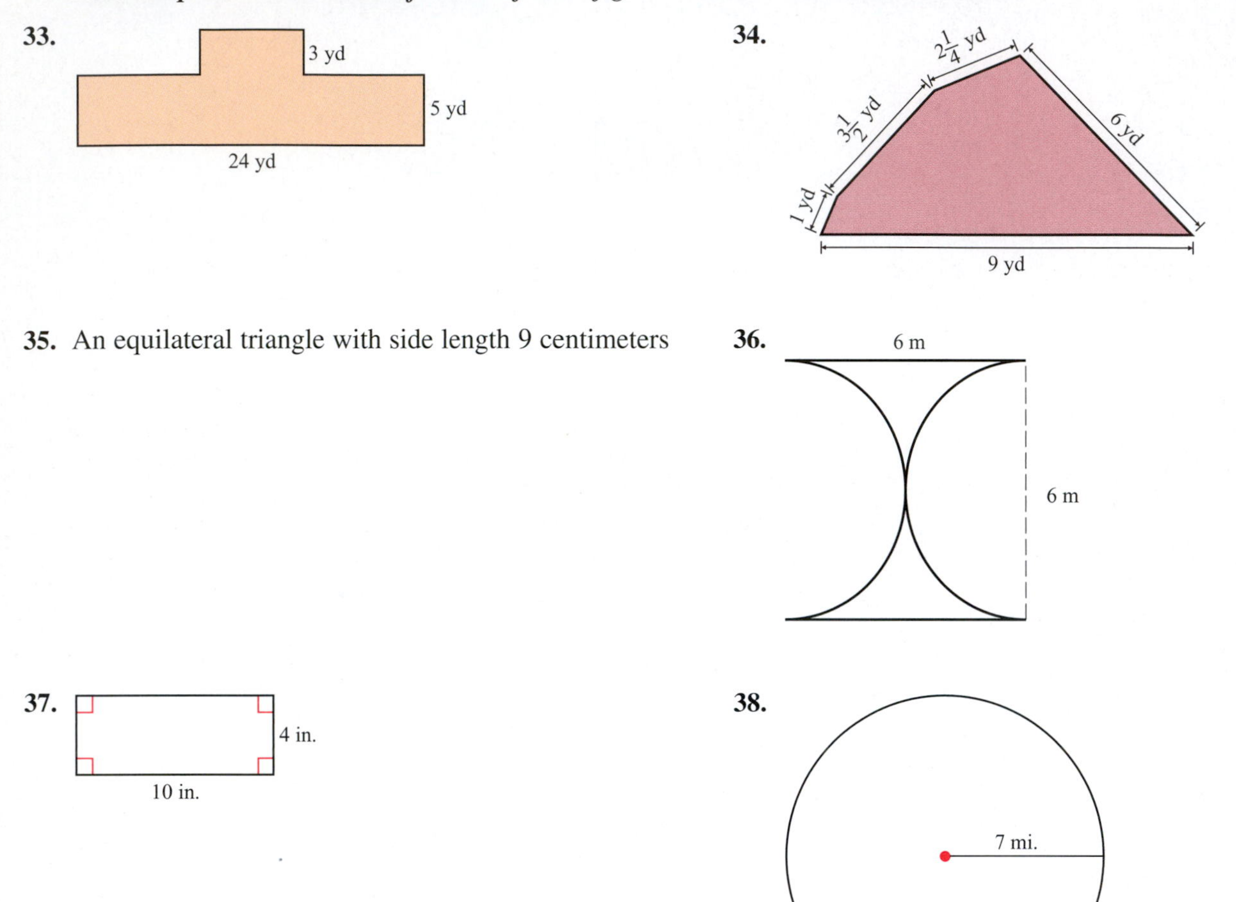

Applications

Solve.

39. Find the perimeter of the doubles tennis court shown below.

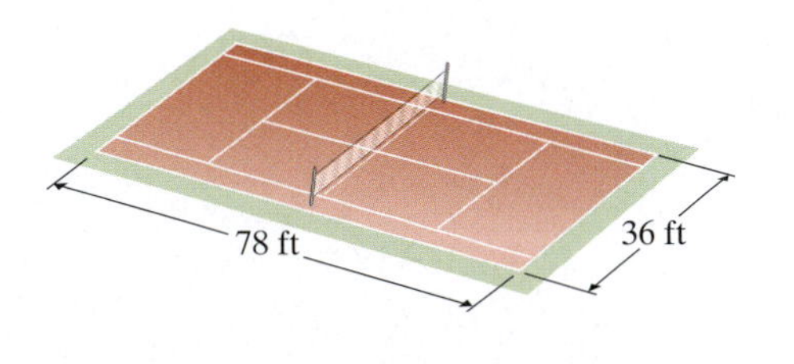

40. If a student drives from Atlanta to New York City to Chicago and back to Atlanta, what is the total mileage?

41. As the following diagram shows, bicycle wheels come in different diameters.

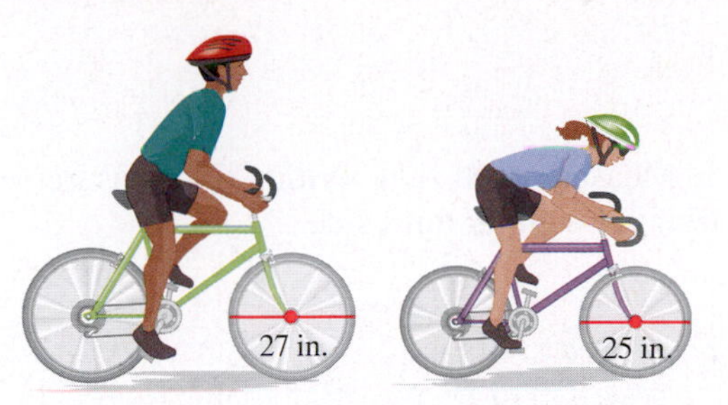

In one wheel rotation, how much farther to the nearest inch does the 27-inch bicycle wheel go than the 25-inch bicycle wheel?

42. The Texas Star at Fair Park in Dallas is the largest ferris wheel in North America. The diameter of the wheel is 212 feet. How many feet does a rider travel in one revolution of the wheel? (*Source:* http://www.bigtex.com)

43. A field 50 meters wide and 100 meters long is to be enclosed with a fence. If fence posts are placed every 10 meters, how many posts are needed?

44. If rug binding costs \$1.95 per foot, what is the cost of binding a rectangular rug that is 21 feet long and 12 feet wide?

45. Find the length of line needed for the clothesline pulley.

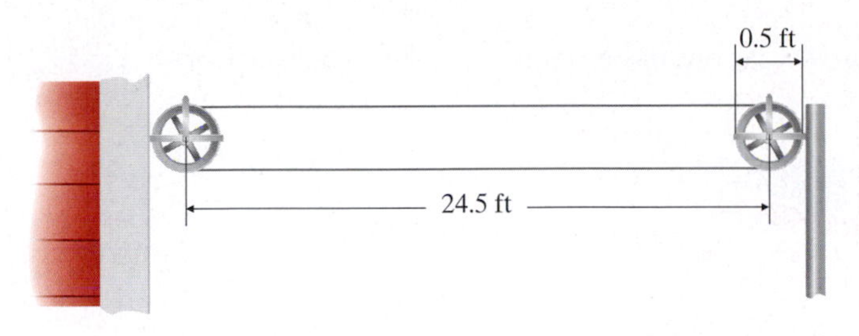

46. A carpenter plans to lay a wood molding in the room shown. If the room has three doors, each 3 feet wide, what is the total length of floor molding required?

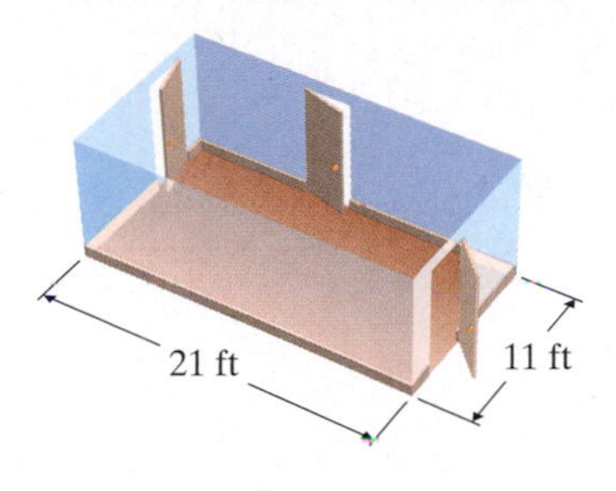

Use a calculator to solve the following problems, giving (a) the operation(s) carried out in your solution, (b) the rounded answer, and (c) an estimate of the answer.

47. The radius of Earth is about 6,400 kilometers. If a satellite is orbiting 400 kilometers above Earth, find the distance to the nearest hundred kilometers that the satellite travels in one orbit.

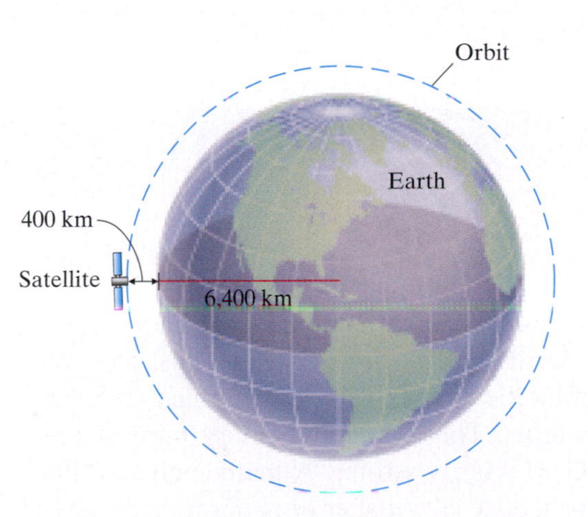

48. A circular crater on the Moon has a circumference of about 214.66 miles. What is the radius of the crater to the nearest mile?

● *Check your answers on page A-17.*

MINDSTRETCHERS

Investigation

1. Draw three triangles. Label the sides of each triangle a, b, and c. Measure each side, writing the measurements in the following table. Compare the sum of any two sides of a triangle with its third side.

	a	b	c	$a + b$	$a + c$	$b + c$
Triangle 1						
Triangle 2						
Triangle 3						

How does the length of the side of a triangle compare to the sum of the lengths of the other two sides?

Mathematical Reasoning

2. Consider the cart pictured. Which wheel do you think will wear out more quickly? Justify your answer.

Groupwork

3. Explain how you can approximate the circumference of a circular room with a ruler. Compare your method with those of other members of the group.

CULTURAL NOTE

Shown on the left are mural decorations from Southern Africa called *litema.* They illustrate how simple geometric shapes can be combined to create diverse patterns. The twentieth-century pure mathematician G. H. Hardy wrote, "A mathematician, like a painter or a poet, is a maker of patterns."

Source: Paulus Gerdes, *Women, Art and Geometry in Southern Africa* (Trenton, New Jersey: Africa World Press, Inc.), pp. 117 and 147.

11.3 Area

The Area of a Polygon and a Circle

OBJECTIVES

- To find the area of a polygon or a circle
- To find the area of a composite figure
- To solve word problems involving area

Area is a measure of the size of a plane geometric figure. The size of a piece of paper, the size of a volleyball court, and the size of a lawn are all examples of areas.

To find the area of the rectangle shown, we split it into little squares, each representing 1 square inch.

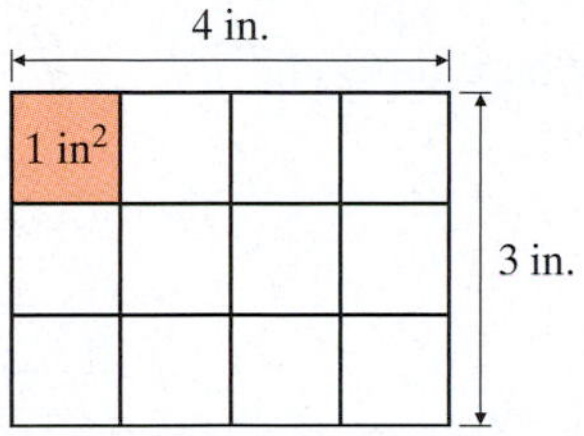

Then we count the number of square inches within the rectangle, which is 12 square inches.

Each row of the rectangle contains 4 square inches, and the rectangle has 3 rows. So a shortcut to counting the total number of square inches is to multiply 3×4, getting 12 square inches in all. Note that areas are measured in square units, such as square inches (sq in. or in^2), square miles (sq mi or mi^2), or square meters (m^2).

Definition

Area is the number of square units that a figure contains.

In this section, we focus on finding the area of certain polygons and also circles by using the following formulas. First, we consider the areas of polygons.

<table>
<tr><th>Figure</th><th>Formula</th><th>Example</th></tr>
<tr><td>Rectangle</td><td>$A = lw$
Area equals the length times the width.
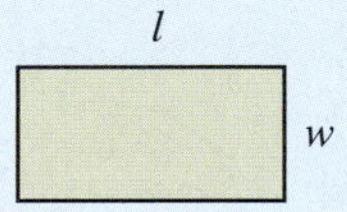
</td><td>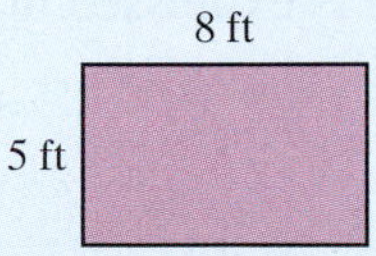

$A = lw$
$= 8 \cdot 5$
$= 40$, or 40 ft^2</td></tr>
<tr><td>Square</td><td>$A = s \cdot s = s^2$
Area equals the square of a side.
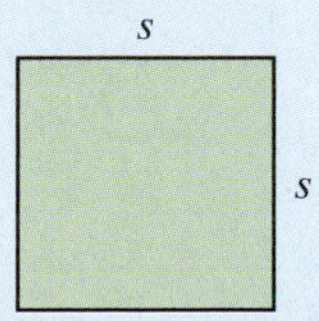
</td><td>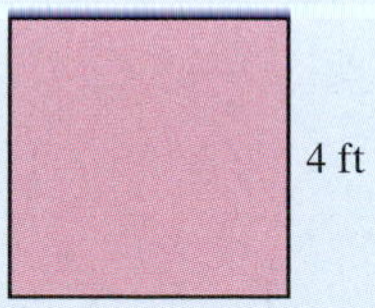

$A = s^2$
$= (4)^2$
$= 4 \cdot 4$
$= 16$, or 16 ft^2</td></tr>
</table>

continued

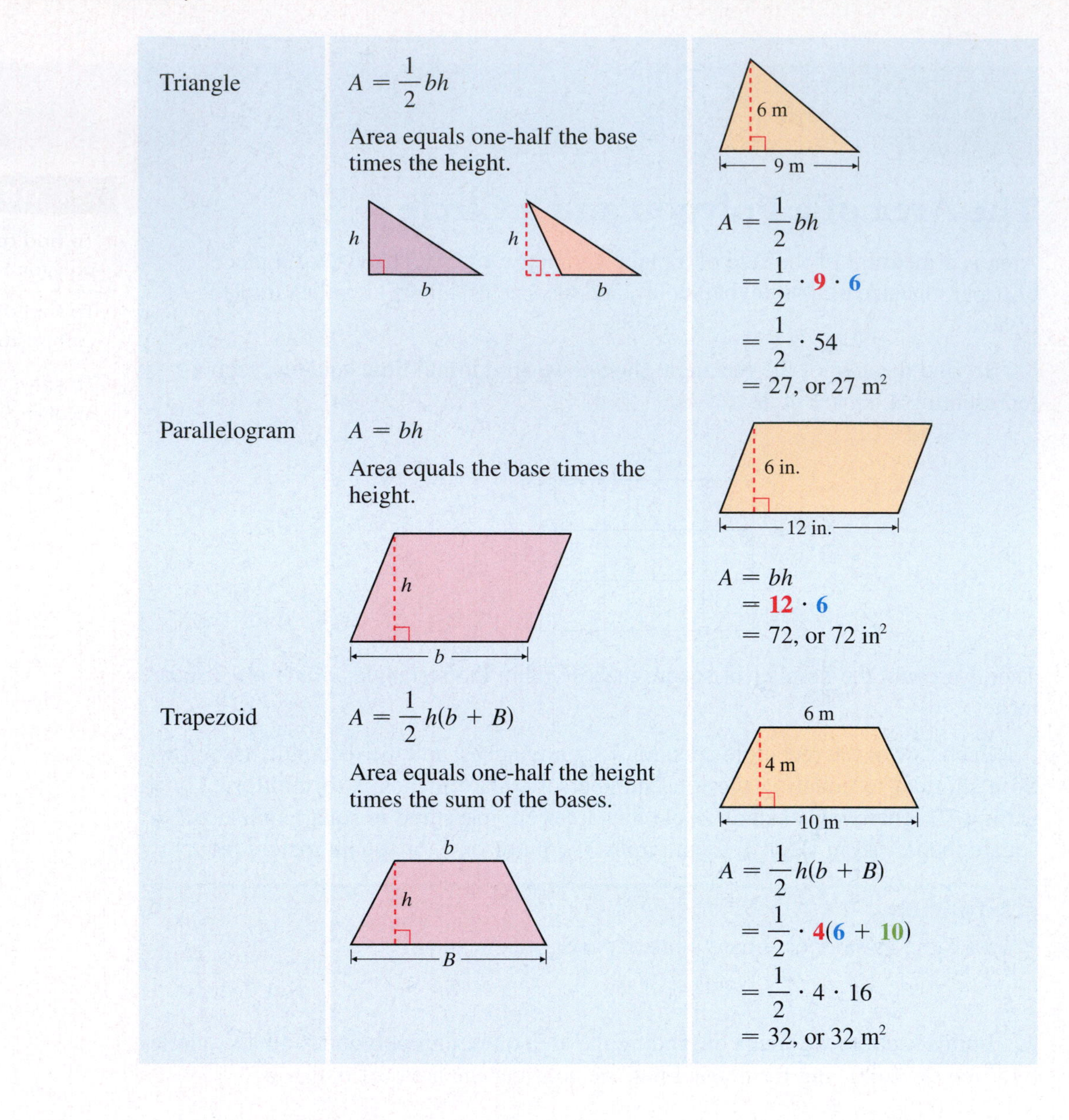

Figure	Formula	Example
Triangle	$A = \frac{1}{2}bh$ Area equals one-half the base times the height.	$A = \frac{1}{2}bh$ $= \frac{1}{2} \cdot 9 \cdot 6$ $= \frac{1}{2} \cdot 54$ $= 27$, or $27\ m^2$
Parallelogram	$A = bh$ Area equals the base times the height.	$A = bh$ $= 12 \cdot 6$ $= 72$, or $72\ in^2$
Trapezoid	$A = \frac{1}{2}h(b + B)$ Area equals one-half the height times the sum of the bases.	$A = \frac{1}{2}h(b + B)$ $= \frac{1}{2} \cdot 4(6 + 10)$ $= \frac{1}{2} \cdot 4 \cdot 16$ $= 32$, or $32\ m^2$

Now, let's consider the area of a circle. As in the case of the circumference, the area of a circle is expressed in terms of π. Recall that π is approximately 3.14 or $\frac{22}{7}$.

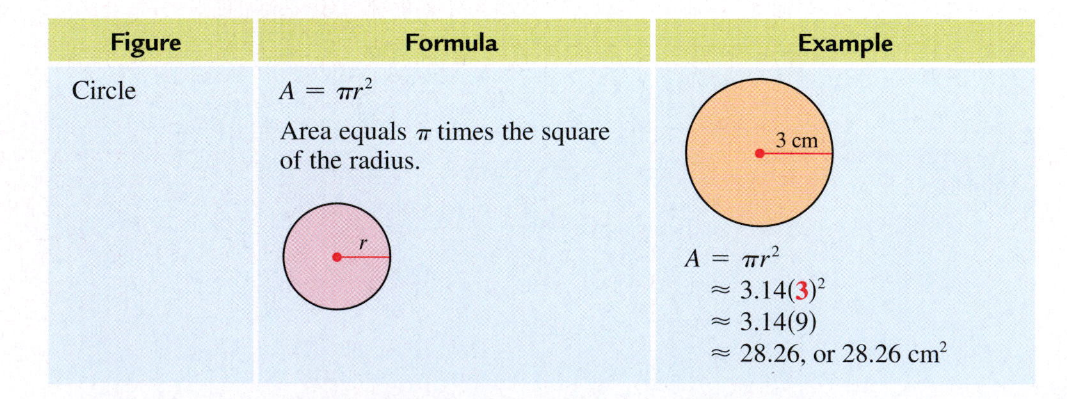

Figure	Formula	Example
Circle	$A = \pi r^2$ Area equals π times the square of the radius.	$A = \pi r^2$ $\approx 3.14(3)^2$ $\approx 3.14(9)$ ≈ 28.26, or $28.26\ cm^2$

EXAMPLE 1

Find the area of a rectangle whose length is 5 feet and whose width is 3 feet.

Solution First, we draw a diagram to visualize the problem.

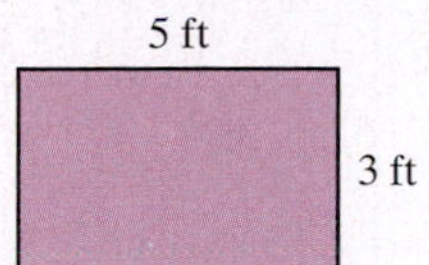

Then, we use the formula for the area of a rectangle.

$$\begin{aligned} A &= lw \\ &= (5)(3) \\ &= 15 \end{aligned}$$

So the area of the rectangle is 15 square feet.

PRACTICE 1

A rectangle has length 6 centimeters and width 2 centimeters. Find its area.

EXAMPLE 2

Find the area of the square.

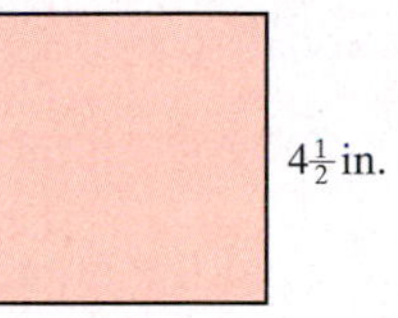

Solution We use the formula for the area of a square.

$$\begin{aligned} A &= s^2 \\ &= (4\tfrac{1}{2})^2 \\ &= (4\tfrac{1}{2})(4\tfrac{1}{2}) \\ &= \frac{9}{2} \cdot \frac{9}{2} \\ &= \frac{81}{4}, \text{ or } 20\tfrac{1}{4} \end{aligned}$$

The area of the square is $20\frac{1}{4}$ square inches.

PRACTICE 2

What is the area of a square with side 3.6 cm?

EXAMPLE 3

Find the area of a triangle with base 8 centimeters and height 5.9 centimeters.

Solution First, we draw a diagram.

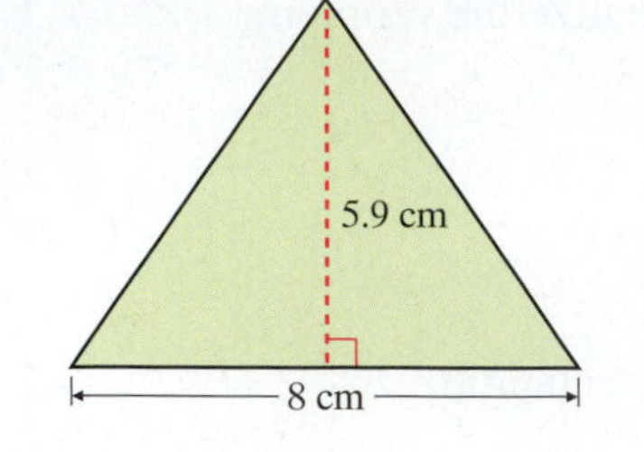

Next, we use the formula for finding the area of a triangle.

$$A = \frac{1}{2}bh$$
$$= \frac{1}{\underset{1}{2}}(\overset{4}{8})(5.9)$$
$$= 23.6$$

The area of the triangle is 23.6 square centimeters.

PRACTICE 3

A triangle has a height of 3 inches and a base of 5 inches. What is its area?

EXAMPLE 4

What is the area of a parallelogram with base $6\frac{1}{2}$ meters and height 3 meters?

Solution We draw a diagram and then use the formula for the area of a parallelogram.

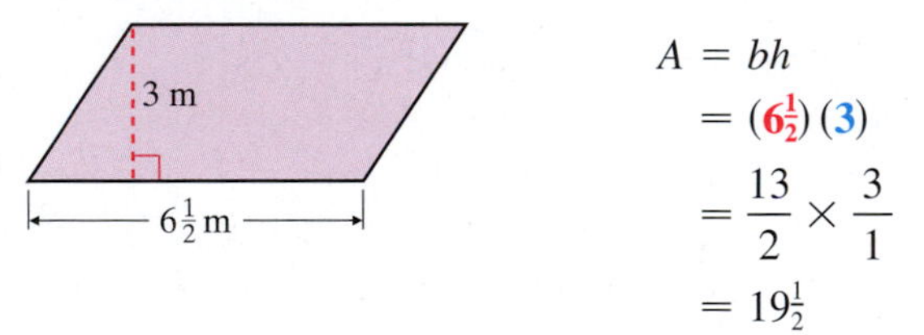

$$A = bh$$
$$= (6\tfrac{1}{2})(3)$$
$$= \frac{13}{2} \times \frac{3}{1}$$
$$= 19\tfrac{1}{2}$$

The area of the parallelogram is $19\frac{1}{2}$ square meters.

PRACTICE 4

Find the area of a parallelogram whose base is 5 feet and height is $2\frac{1}{2}$ feet.

EXAMPLE 5

What is the area of the trapezoid shown?

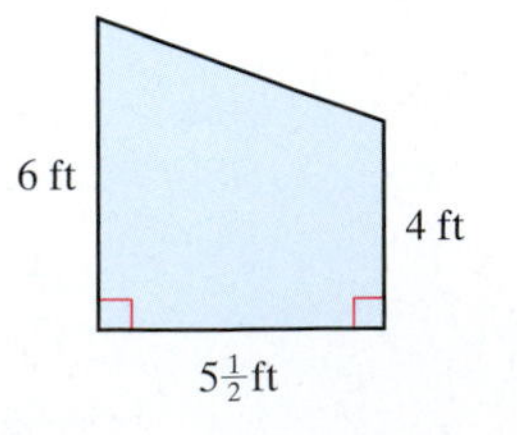

PRACTICE 5

Find the area of the following trapezoid.

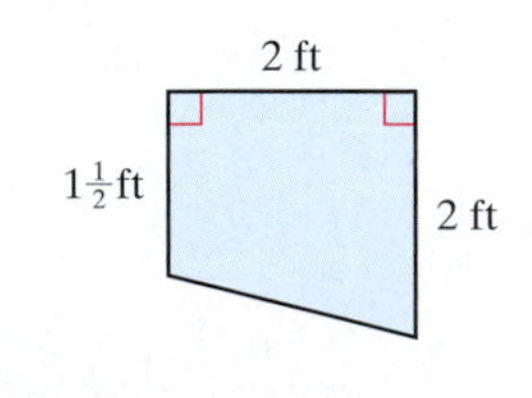

Solution This polygon is a trapezoid, so we use the following formula to find its area.

$$\begin{aligned} A &= \frac{1}{2}h(b + B) \\ &= \frac{1}{2} \cdot 5\tfrac{1}{2}(6 + 4) \\ &= \frac{1}{2} \cdot \frac{11}{\cancel{2}_1} \cdot \cancel{10}^{5} \\ &= \frac{55}{2}, \text{ or } 27\tfrac{1}{2} \end{aligned}$$

The area of the trapezoid is $27\frac{1}{2}$ square feet.

EXAMPLE 6

What is the area of a circle whose diameter is 8 meters?

Solution First, we draw a diagram.

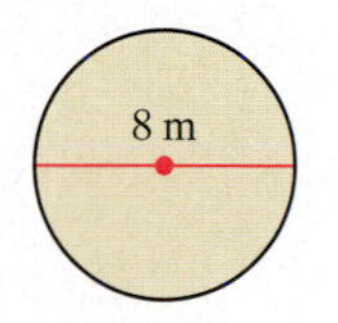

We know that the radius is one-half of 8 meters, or 4 meters, which we substitute in the formula for the area of a circle.

$$\begin{aligned} A &= \pi r^2 \\ &\approx 3.14(4)^2 \\ &\approx 3.14(16) \\ &\approx 50.24 \end{aligned}$$

The area of the circle is approximately 50.24 square meters.

PRACTICE 6

Find the area of a circle whose radius is 5 yards.

EXAMPLE 7

An artist wants to buy an ad in a magazine that charges $1,000 per square inch for advertising space. How much will this ad cost?

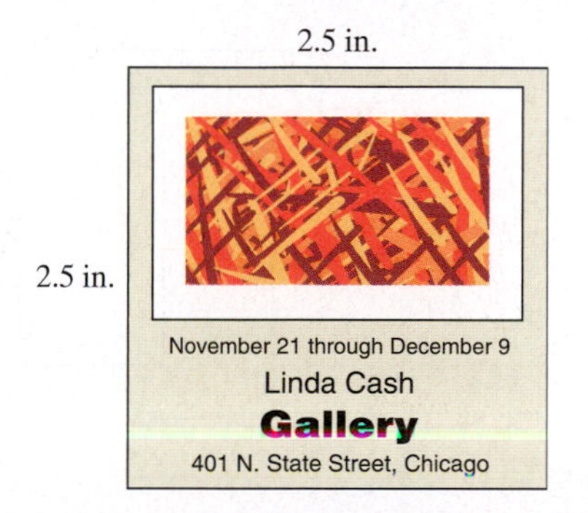

2.5 in.
2.5 in.
November 21 through December 9
Linda Cash
Gallery
401 N. State Street, Chicago

Solution First, we need to find the area of the ad, which is square.

$$\begin{aligned} A &= s^2 \\ &= (2.5)^2 \\ &= (2.5)(2.5) \\ &= 6.25, \text{ or } 6.25 \text{ square inches} \end{aligned}$$

To find the cost of the ad, we multiply 6.25 by 1,000, getting 6,250. So the cost of the ad is $6,250.

PRACTICE 7

In a flooring store, a customer wants to tile a 9 foot × 12 foot room with 1 ft^2 tiles that sell for $4.99 apiece. Will $500 be enough to pay for the tiles?

EXAMPLE 8

In a certain town, only students living outside a 2-mile radius of their school must pay a fee for bus transportation. To the nearest square mile, what is the area of the region in which students do not pay a fee for bus transportation?

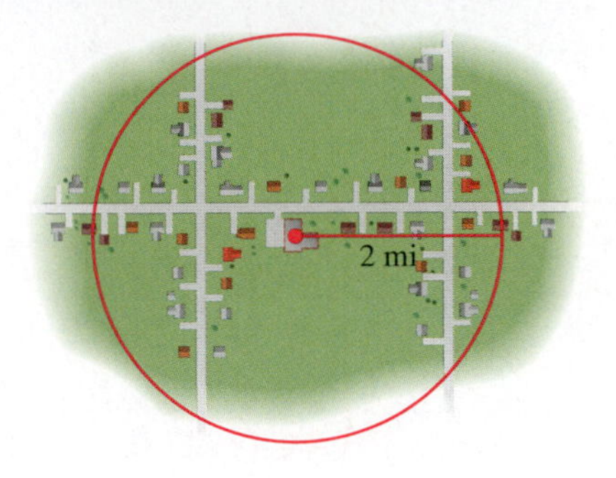

Solution We need to find the area of the region, which is a circle.

$$\begin{aligned} A &= \pi r^2 \\ &\approx \pi(2)^2 \\ &\approx 3.14(4) \\ &\approx 12.56 \end{aligned}$$

The area of the region is approximately 13 square miles.

PRACTICE 8

The following diagram shows the region in which the beam of light from a lighthouse can be seen in any direction in the fog. To the nearest square mile, what is the area of the region in which the light is visible?

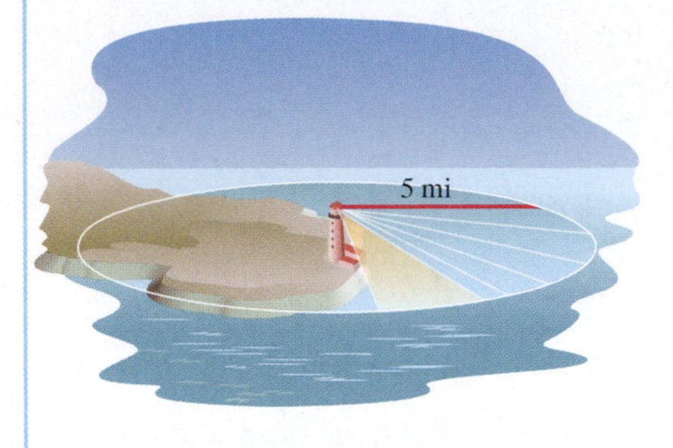

Composite Figures

Recall that a composite figure comprises two or more simple figures. Let's consider finding areas of such figures.

EXAMPLE 9

Find the area of the shaded portion of the figure.

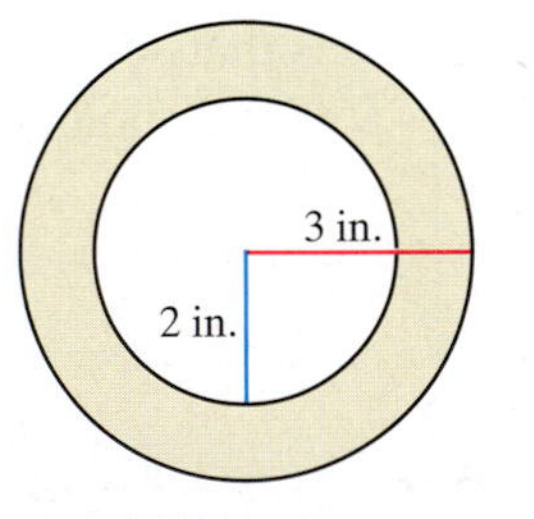

Solution To find the area of the shaded portion, we subtract the area of the small (inner) circle from the area of the large (outer) circle.

$$\begin{aligned} \text{Shaded Area} &= \textbf{Area of large circle} - \textbf{Area of small circle} \\ &\approx 3.14(3)^2 - 3.14(2)^2 \\ &\approx 3.14(9) - 3.14(4) \\ &\approx 28.26 - 12.56 \\ &\approx 15.70 \end{aligned}$$

The area of the shaded figure is approximately 15.7 square inches. How could the distributive property be used to solve this problem?

PRACTICE 9

Find the shaded area.

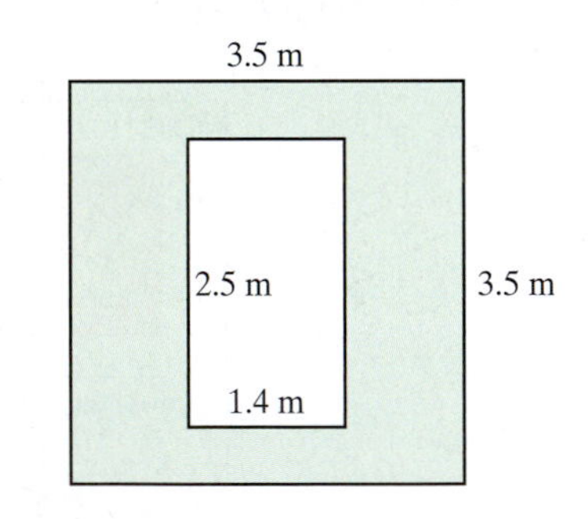

EXAMPLE 10

At $19 per square foot, how much will it cost to carpet the bedroom pictured?

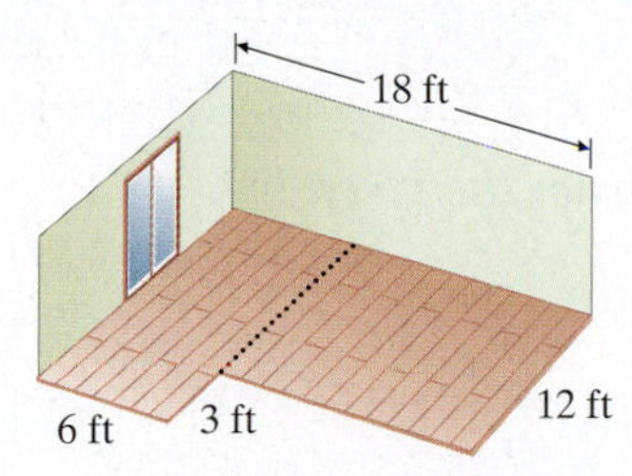

Solution First, we must find the area of the room. Note that the room consists of a 15 foot $\times$ 6 foot rectangle, and a square 12 feet on a side.

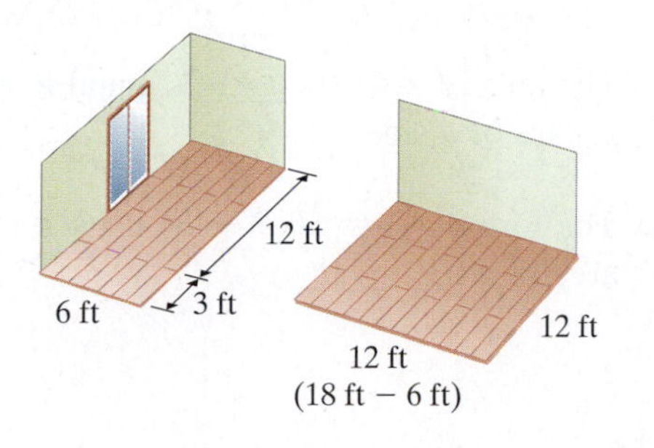

$$\begin{aligned}\text{Total area} &= \textbf{Area of rectangle} + \textbf{Area of square}\\ &= l \cdot w + s^2\\ &= 15 \cdot 6 + (12)^2\\ &= 90 + 144\\ &= 234, \text{ or } 234 \text{ square feet}\end{aligned}$$

The total area of the room is 234 square feet. The carpet costs $19 per square foot, so we calculate the total cost as follows:

$$234 \text{ ft}^2 \times \frac{\$19}{\text{ft}^2} = \$4{,}446$$

So carpeting the bedroom costs $4,446.

PRACTICE 10

A coating of polyurethane is applied to the central circle on the gymnasium floor shown below. What is the area of the part of the floor that still needs coating?

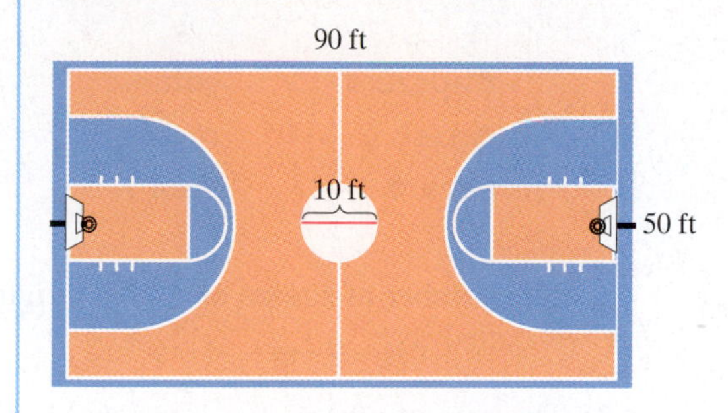

Mathematically Speaking

Fill in each blank with the most appropriate term or phrase from the given list.

trapezoid	volume	square meters
circle	meters	triangle
square	area	

1. The number of square units that a figure contains is called its _______.

2. Areas are measured in square units, such as _______.

3. The area of a(n) _______ is equal to one-half the product of the base and the height.

4. The area of a(n) _______ is equal to equal to the square of a side.

5. A formula for the area of a(n) _______ is $A = \pi r^2$.

6. The formula $A = \frac{1}{2}h(b + B)$ is used to find the area of a(n) _______.

Find the area of each figure. Use $\pi \approx 3.14$ when needed.

7.

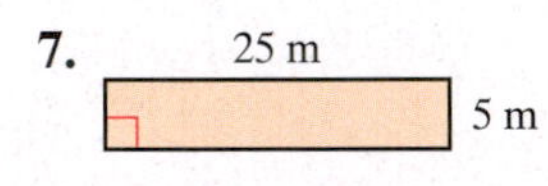

8.

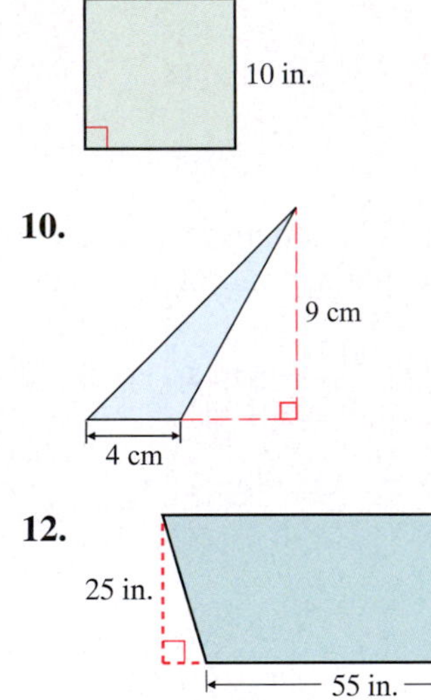

9.

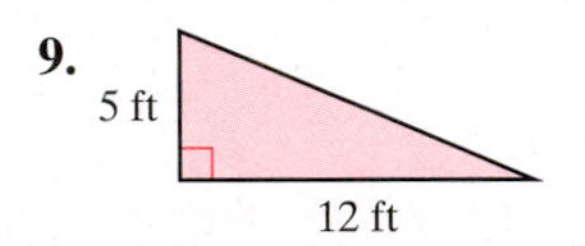

10.

11.

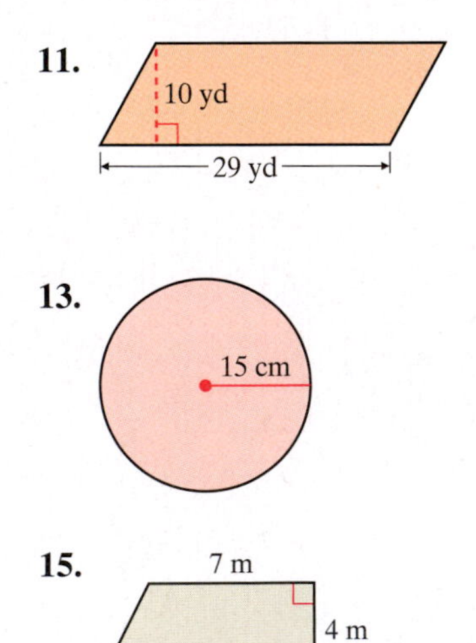

12.

13.

14.

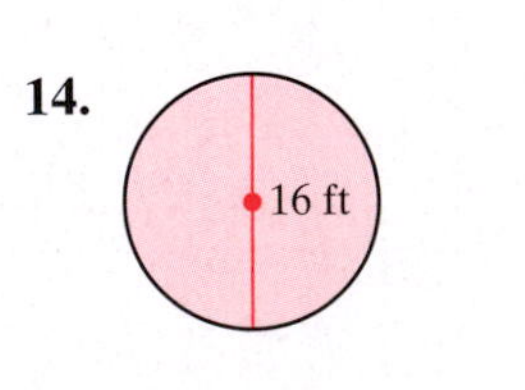

15.

16.

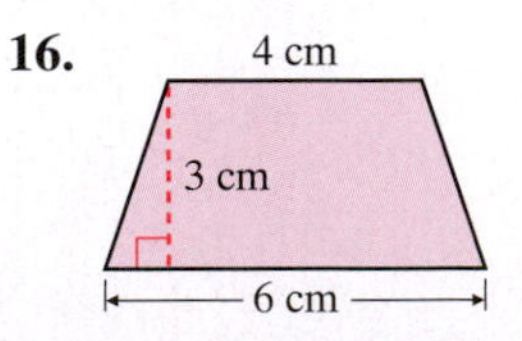

17. A parallelogram with base 4 meters and height 3.9 meters

18. A parallelogram with base 6.5 inches and height 4 inches

19. A circle with diameter 20 inches

20. A circle with radius 100 feet

21. A triangle with height 2.5 feet and base 5 feet

22. A triangle with base 8 inches and height $6\frac{1}{2}$ inches

23. A trapezoid with height 4.2 yards and bases 7 yards and 14 yards

24. A trapezoid with height 3.5 meters and bases 4 meters and 6.5 meters

25. A rectangle with length 2.6 meters and width 1.4 meters

26. A rectangle with length $\frac{1}{2}$ foot and width $\frac{2}{3}$ foot

27. A square with side $\frac{1}{4}$ yard long

28. A square with side 15.5 centimeters long

29.

4 yd
10 yd
3 yd

30.

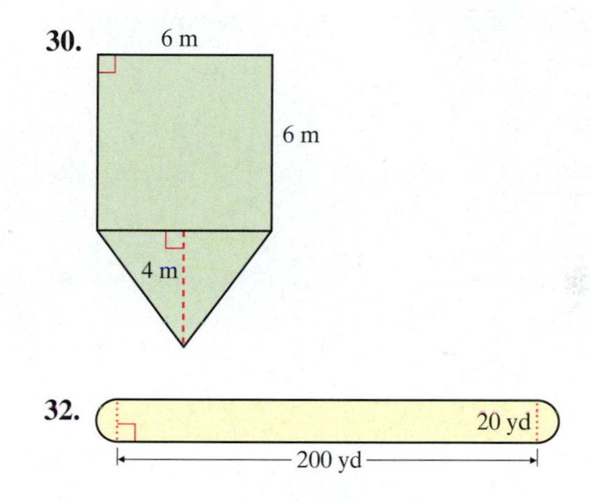

31.

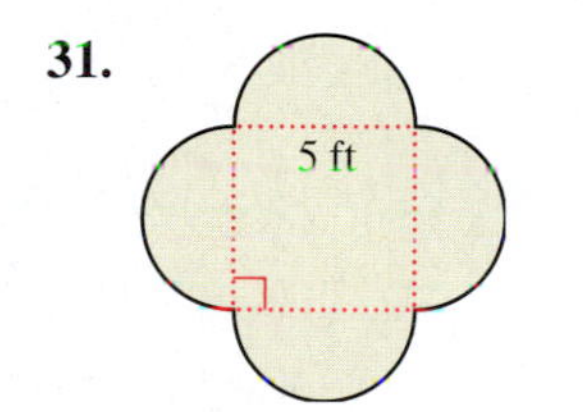

32.

20 yd
200 yd

Find the shaded area.

33.

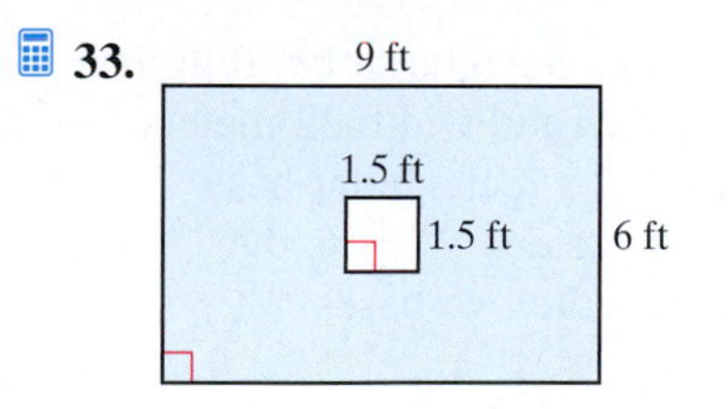

34.

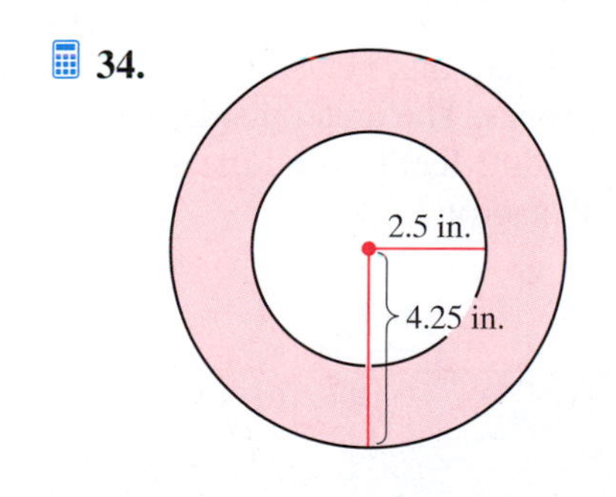

Mixed Practice

Find the area of each figure. Use $\pi \approx 3.14$ when needed.

35.

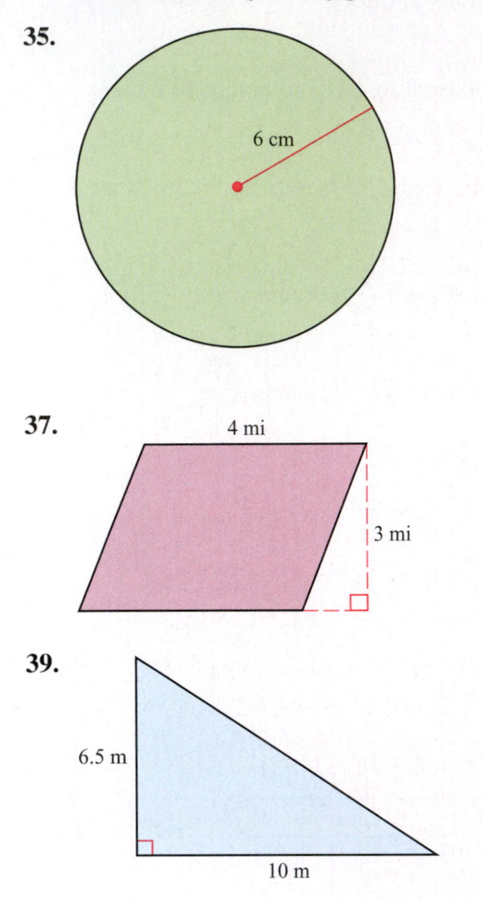

36. The shaded region.

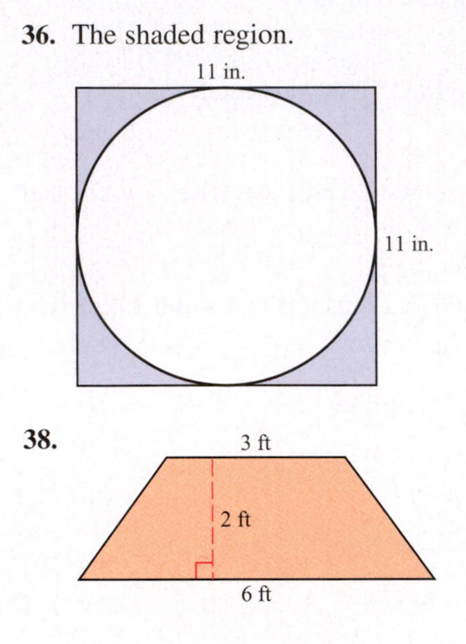

37.

38.

39.

40. A parallelogram with base 8.5 yards and height 7 yards

Applications

Solve. Use $\pi \approx 3.14$ when needed.

41. The boxing ring below is a square that measures 18 feet on a side inside the ropes. The area outside the ropes, called the apron, extends 2 feet beyond the ropes. What is the area of the apron?

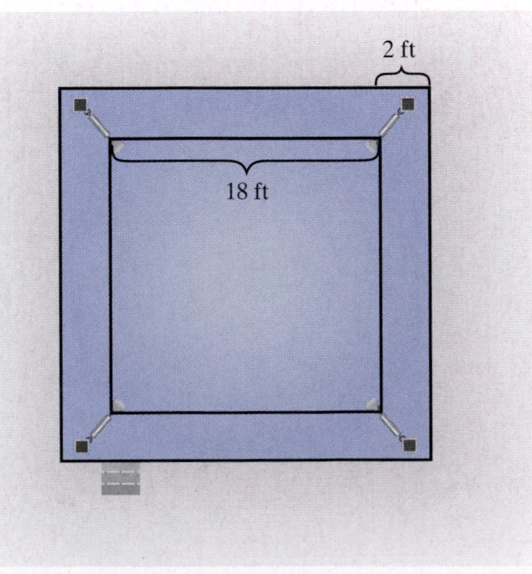

42. The base of the United Nations Secretariat building is a rectangle with length 88 meters and width 22 meters. The Empire State Building has a rectangular base measuring 129 meters by 57 meters. What is the difference in the area between the two bases?

43. A microscope allows a scientist to see a circular region that is 0.25 millimeters in diameter. What is the area of this region?

44. An air-traffic control tower can identify an airplane within 10 miles of the tower in any direction. What area does the tower cover?

45. Suppose that an L-shaped house is located on the rectangular lot shown. How much yard space is there?

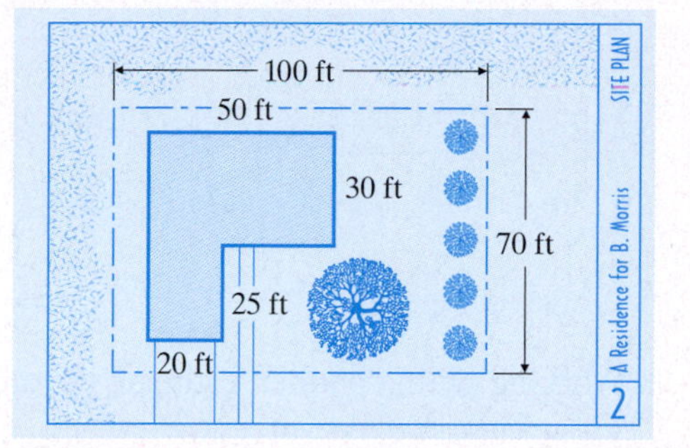

46. A walkway 2 yards wide, shown below, is built around the entire building below. Find the area of the walkway.

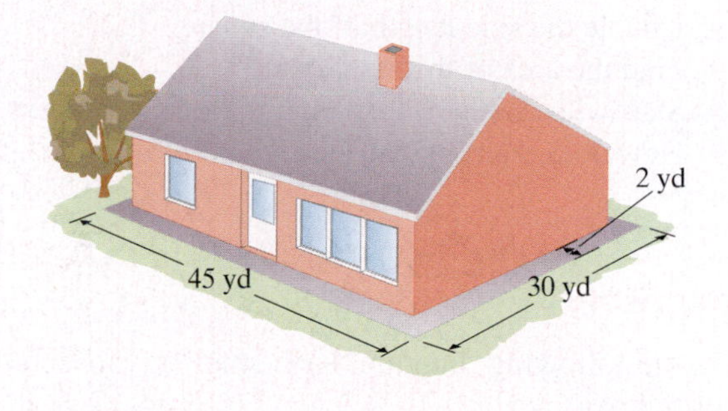

Use a calculator to solve the following problems, giving (a) the operation(s) carried out in the solution, (b) the answer, and (c) an estimate of the answer.

47. Even though an LP record is larger than a CD, the CD holds twice as much music.

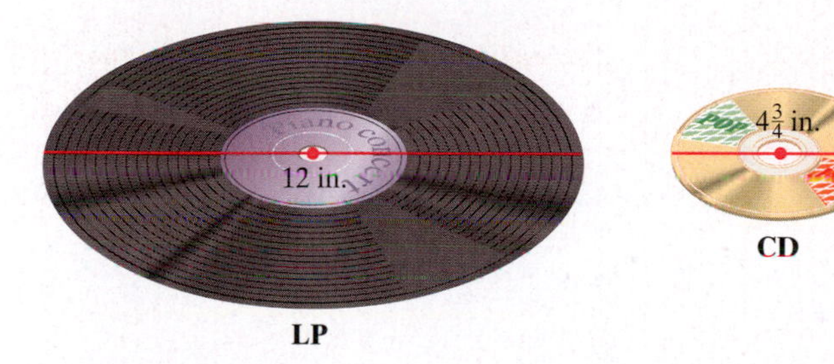

How much larger in area is an LP record than a CD? Round the answer to the nearest 10 square inches.

48. Consider the two tables pictured. How much larger in area to the nearest square meter is the semicircular table than the rectangular table?

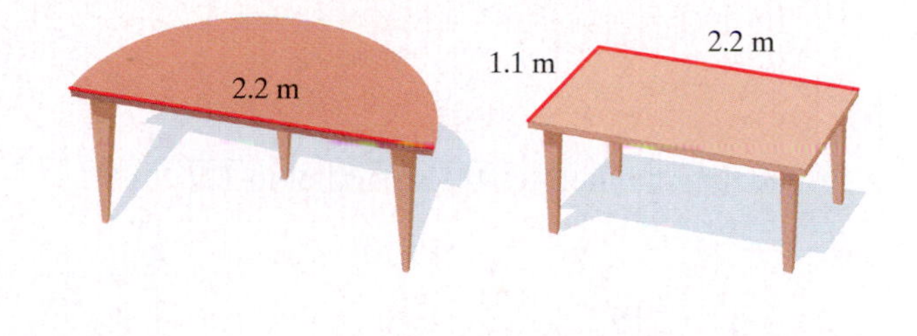

• *Check your answers on page A-17.*

MINDSTRETCHERS

Investigation

1. • Draw a square.
   • Measure its side lengths.
   • Find its area.
   • Double the side length of the square.
   • Find the area of the new square.
   • Start with another square and repeat this process several times.
   • How does doubling the side length of a square affect its area?

Groupwork

2. In the following diagram, each small square represents 1 square inch. Working with a partner, estimate the area of the oval.

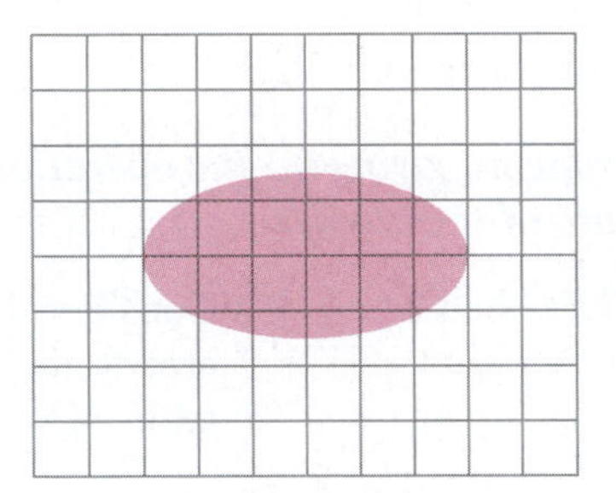

Mathematical Reasoning

3. In the diagram below, $\overline{AC}$ is parallel to $\overline{ED}$.

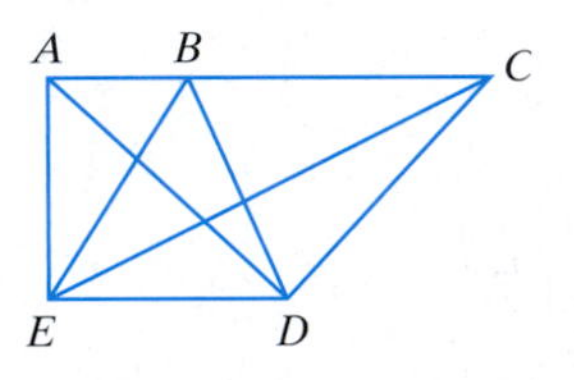

What relationship do you see between the areas of $\triangle EAD$, $\triangle EBD$, and $\triangle ECD$? Justify your answer.

11.4 Volume

The Volume of a Geometric Solid

OBJECTIVES

- To find the volume of a geometric solid
- To find the volume of a composite geometric solid
- To solve word problems involving volume

Volume is a measure of the amount of space inside a three-dimensional figure. The amount of water in an aquarium, the amount of juice in a can, or the amount of grain in a bin are all examples of volumes.

To find the volume of the box shown, we can split it into little cubes, each representing 1 cubic inch. Then we count the number of cubic inches within the box, which is 24 cubic inches.

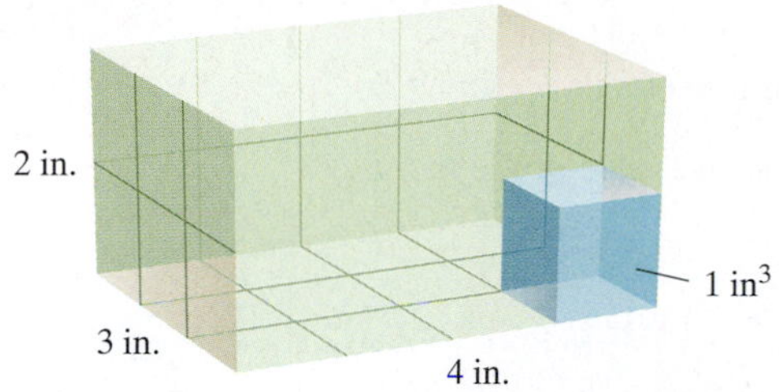

A shortcut to counting the total number of cubic inches is to multiply the length, the width, and the height: $2 \times 3 \times 4$, getting 24, or 24 cubic inches. Note that volumes are measured in cubic units, such as cubic inches (cu in. or in^3), cubic miles (cu mi or mi^3), or cubic meters (cu m or m^3).

Definition

Volume is the number of cubic units required to fill a three-dimensional figure.

In this section, we consider basic three-dimensional objects and find their volume by using the following formulas.

Definition	Formula	Example
A **rectangular solid** is a solid in which all six faces are rectangles.	$V = lwh$ Volume equals length times width times height. 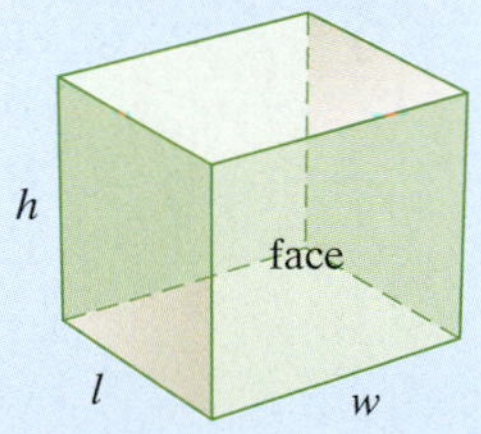 	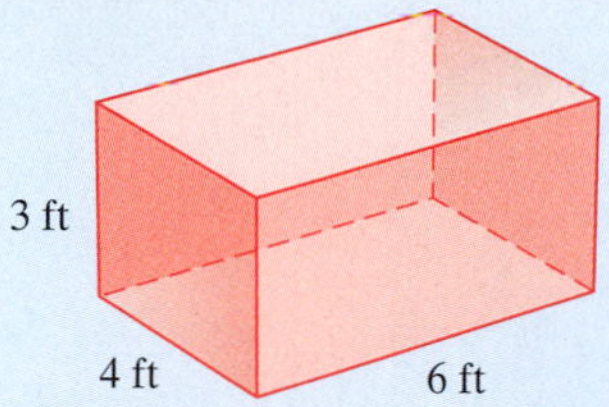 $V = lwb$ $= 6 \cdot 4 \cdot 3$ $= 72$, or $72\ ft^3$
A **cube** is a solid in which all six faces are squares.	$V = e^3$ Volume equals the cube of the edge. 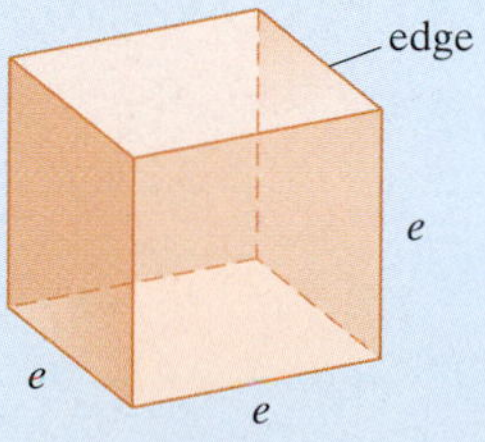 	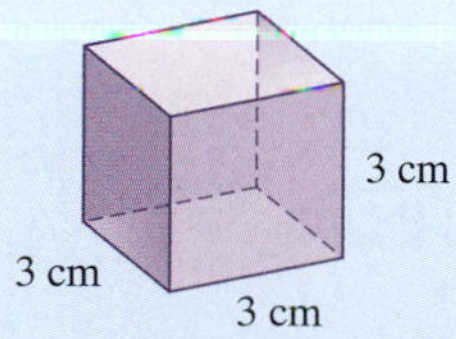 $V = e^3$ $= (3)^3$ $= 27$, or $27\ cm^3$

continued

A **cylinder** is a solid in which the bases are circles and are perpendicular to the height.

$V = \pi r^2 h$

Volume equals π times the square of the radius times the height.

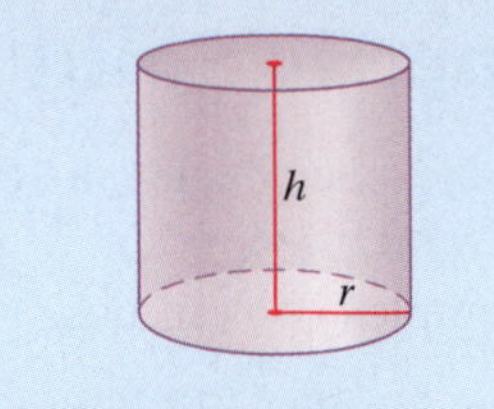

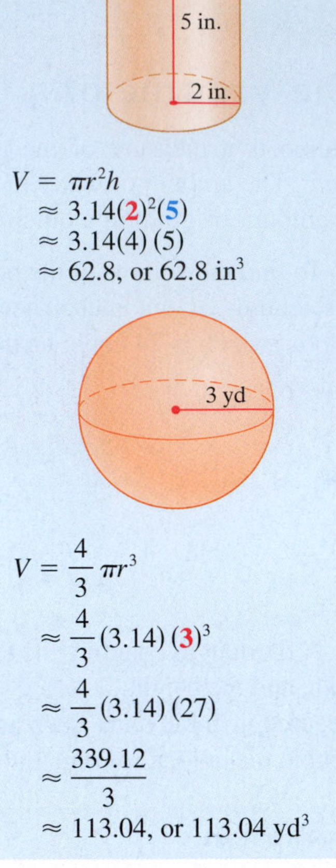

$$\begin{aligned} V &= \pi r^2 h \\ &\approx 3.14(2)^2(5) \\ &\approx 3.14(4)(5) \\ &\approx 62.8, \text{ or } 62.8 \text{ in}^3 \end{aligned}$$

A **sphere** is a three-dimensional figure made up of all points a given distance from the center.

$V = \frac{4}{3}\pi r^3$

Volume equals $\frac{4}{3}$ times π times the cube of the radius.

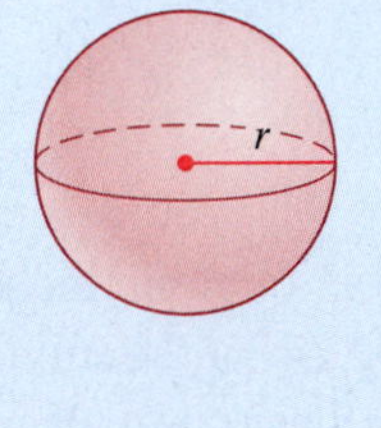

$$\begin{aligned} V &= \frac{4}{3}\pi r^3 \\ &\approx \frac{4}{3}(3.14)(3)^3 \\ &\approx \frac{4}{3}(3.14)(27) \\ &\approx \frac{339.12}{3} \\ &\approx 113.04, \text{ or } 113.04 \text{ yd}^3 \end{aligned}$$

EXAMPLE 1

Find the volume of the rectangular solid shown.

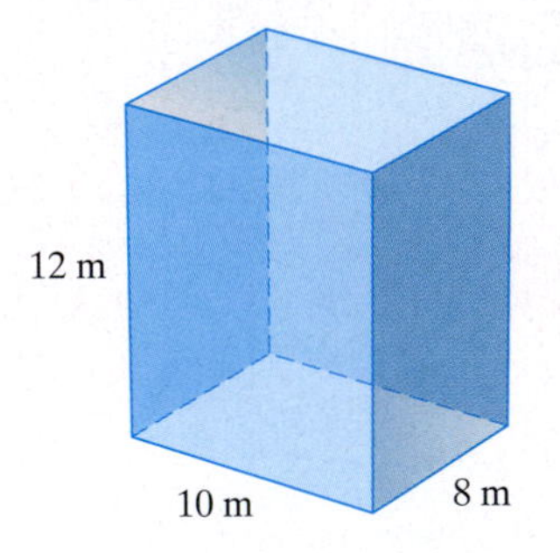

Solution To find the volume of a rectangular solid, use the formula $V = lwh$.

$$\begin{aligned} V &= lwh \\ &= 8 \cdot 10 \cdot 12 \\ &= 960 \end{aligned}$$

The volume of the rectangular solid is 960 cubic meters.

PRACTICE 1

What is the volume of this box?

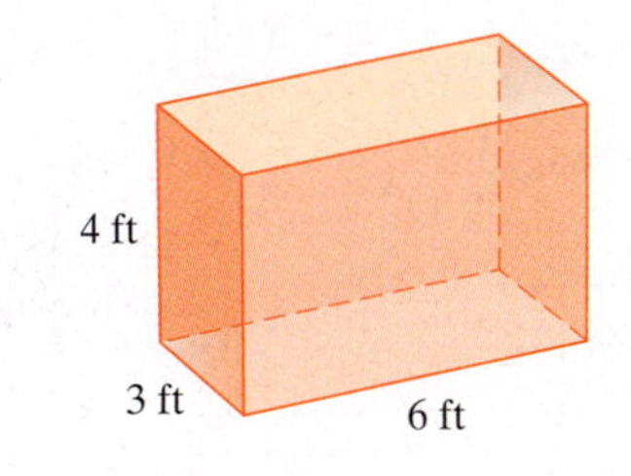

EXAMPLE 2

Find the volume of a cube with length 11 inches on a side.

Solution Use the formula for the volume of a cube, $V = e^3$.

$$\begin{aligned} V &= e^3 \\ &= (11)^3 \\ &= 11 \cdot 11 \cdot 11 \\ &= 1{,}331 \end{aligned}$$

So the volume is therefore 1,331 cubic inches.

PRACTICE 2

What is the volume of a cube with an edge 15 centimeters long?

EXAMPLE 3

What is the volume of the cylinder shown?

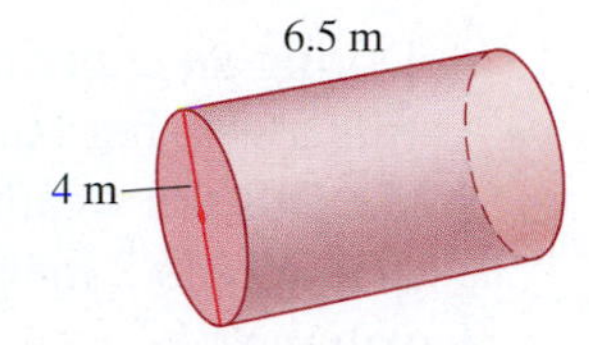

Solution In this cylinder the radius of the base is 2 meters and the height is 6.5 meters. We substitute these quantities into the formula for the volume of a cylinder.

$$\begin{aligned} V &= \pi r^2 h \\ &\approx 3.14(2)^2(6.5) \\ &\approx 3.14(4)(6.5) \\ &\approx 81.64 \end{aligned}$$

The volume is approximately 82 cubic meters.

PRACTICE 3

Find the volume of this pipe.

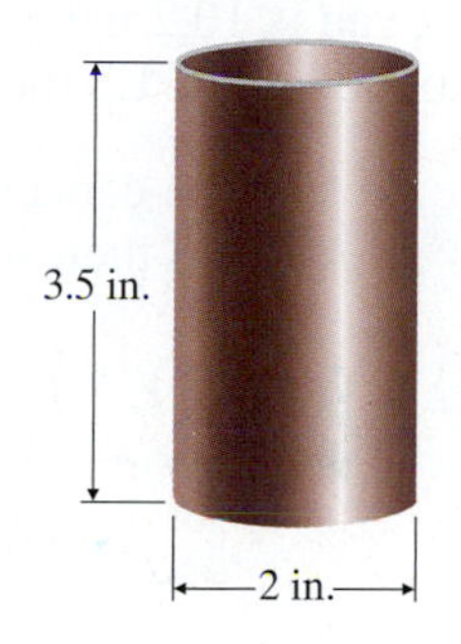

EXAMPLE 4

What is the volume of the sphere shown? Use $\pi \approx \frac{22}{7}$.

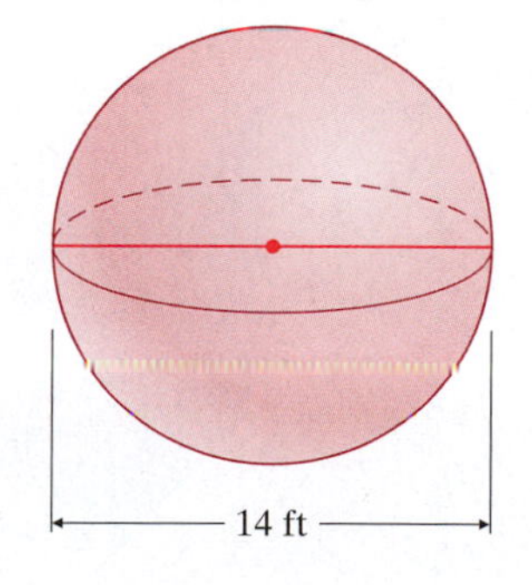

PRACTICE 4

What is the volume of a ball whose diameter is 6 inches?

Solution We use the formula for the volume of a sphere.
The diameter of the sphere is 14 feet, so the radius is 7 feet.

$$V = \frac{4}{3}\pi r^3$$
$$V \approx \frac{4}{3} \cdot \frac{22}{7}(7)^3$$
$$\approx \frac{4}{3} \cdot \frac{22}{7} \cdot \frac{7 \cdot 7 \cdot 7}{1}$$
$$\approx \frac{4{,}312}{3}, \text{ or } 1{,}437\tfrac{1}{3}$$

So the volume is about 1,437 cubic feet.

EXAMPLE 5

A shipping van is 10.2 meters long and 2.5 meters wide. If it can be filled to a depth of 1.8 meters, what is the capacity of the van?

Solution To find the capacity or volume of the van, we use the formula $V = lwh$ and then substitute the given values.

$$V = lwh$$
$$= (10.2)(2.5)(1.8)$$
$$= 45.9$$

The capacity of the van is 45.9 cubic meters.

PRACTICE 5

During an experiment, a meteorologist fills a weather balloon with helium. If she fills the weather balloon until its diameter is 2 meters, what is its volume?

Composite Geometric Solids

Now, let's find the volume of composite geometric solids made up of two or more basic solid figures.

EXAMPLE 6

Find the volume of the solid pictured, which is a cube with a cylinder deleted from its center.

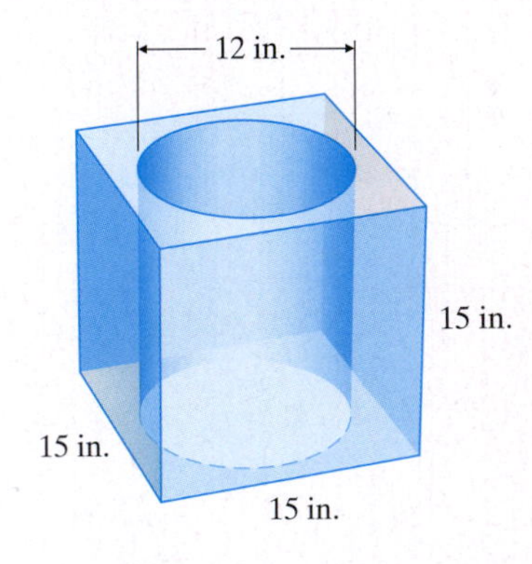

PRACTICE 6

A ball is packaged in a cube-shaped box touching all its sides, as pictured.

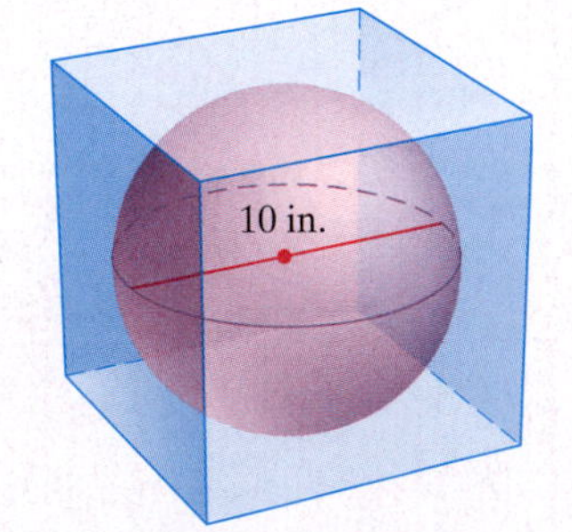

Does the ball occupy more or less than one-half the volume of the box? Explain.

Solution To find the volume of this solid, we subtract the volume of the cylinder from the volume of the cube.

$$\begin{aligned}\text{Volume of the solid} &= \textbf{Volume of cube} - \textbf{Volume of cylinder}\\ &= e^3 - \pi r^2 h\\ &\approx (15)^3 - (3.14)(6)^2(15)\\ &\approx 3{,}375 - (3.14)(36)(15)\\ &\approx 3{,}375 - 1{,}695.6\\ &\approx 1{,}679.4\end{aligned}$$

The volume of the solid is approximately 1,679 cubic inches.

EXAMPLE 7

A pharmaceutical company produces a medicine capsule that is shaped like a cylinder with half a sphere at each end. What is the volume of the capsule?

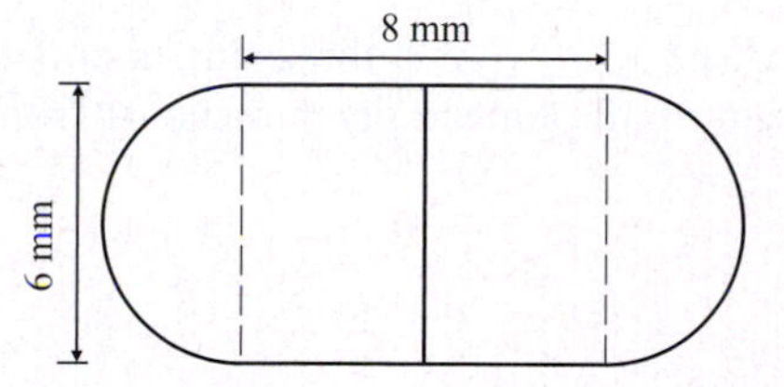

Solution To find the volume of the capsule, we add the volume of the cylinder to the volume of the sphere formed by the two ends.

$$\text{Volume of capsule} = \text{Volume of cylinder} + \text{Volume of sphere}$$

$$\begin{aligned}V &= \pi r^2 h + \frac{4}{3}\pi r^3\\ &\approx 3.14(3)^2(8) + \frac{4}{3}(3.14)(3)^3\\ &\approx 3.14(9)(8) + \frac{4}{3}(3.14)(27)\\ &\approx 226.08 + 113.04\\ &\approx 339.12\end{aligned}$$

The volume of the capsule is approximately 339 cubic millimeters.

PRACTICE 7

Each tier of a wedding cake is a 4-inch-high rectangular solid. What is the total volume of the cake?

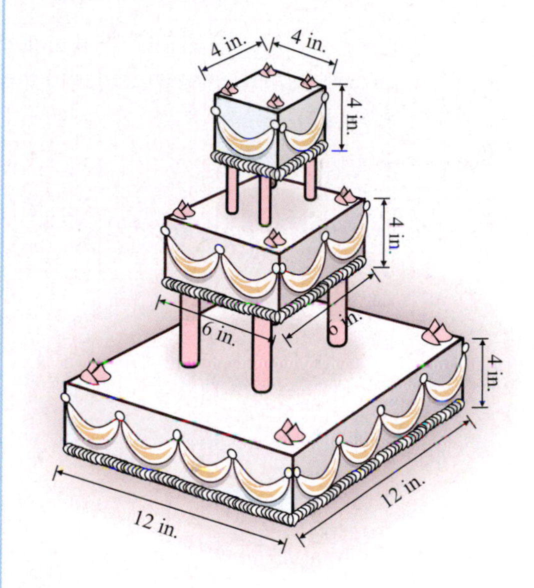

Mathematically Speaking

Fill in each blank with the most appropriate term or phrase from the given list.

circle	composite	volume
cube	rectangular solid	simple
cylinder	area	sphere

1. The number of cubic units required to fill a three-dimensional figure is called its _______.

2. Combining two or more basic solid figures results in a(n) _______ geometric solid.

3. The formula $V = lwh$ is used the find the volume of a(n) _______.

4. A formula for the volume of a(n) _______ is $V = e^3$.

5. A(n) _______ is a solid in which the bases are circles and are perpendicular to the height.

6. A(n) _______ is a three-dimensional figure made up of all points a given distance from the center.

Find the volume of each solid. Use $\pi \approx 3.14$ when needed.

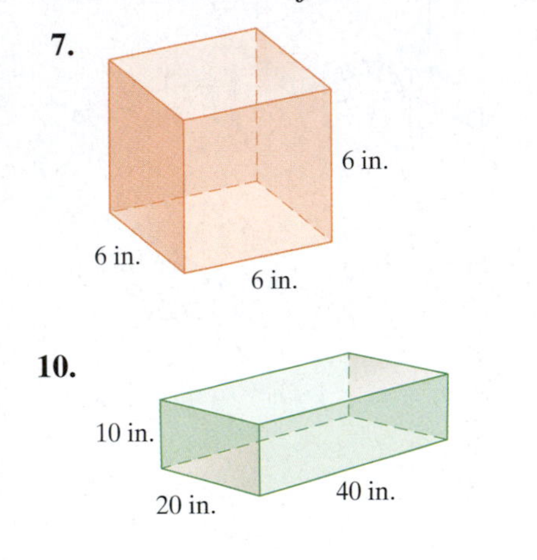

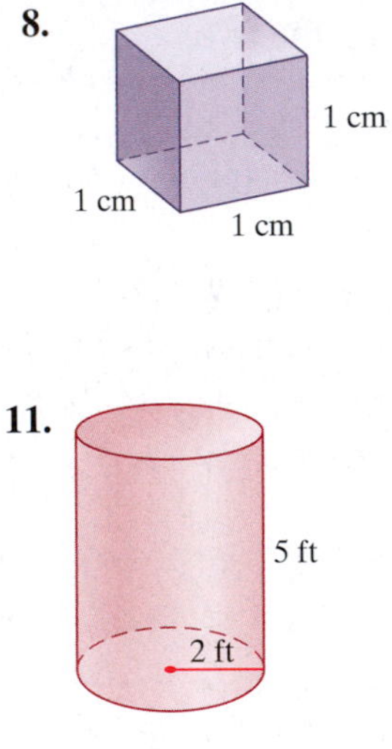

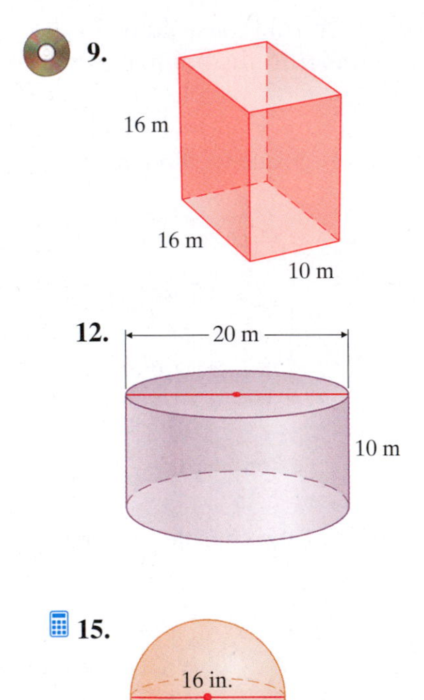

13. A rectangular solid with length 3.5 feet, width 5.5 feet, and height 6.5 feet

14. A cube with length 45 meters on a side

15. (sphere, 16 in.)

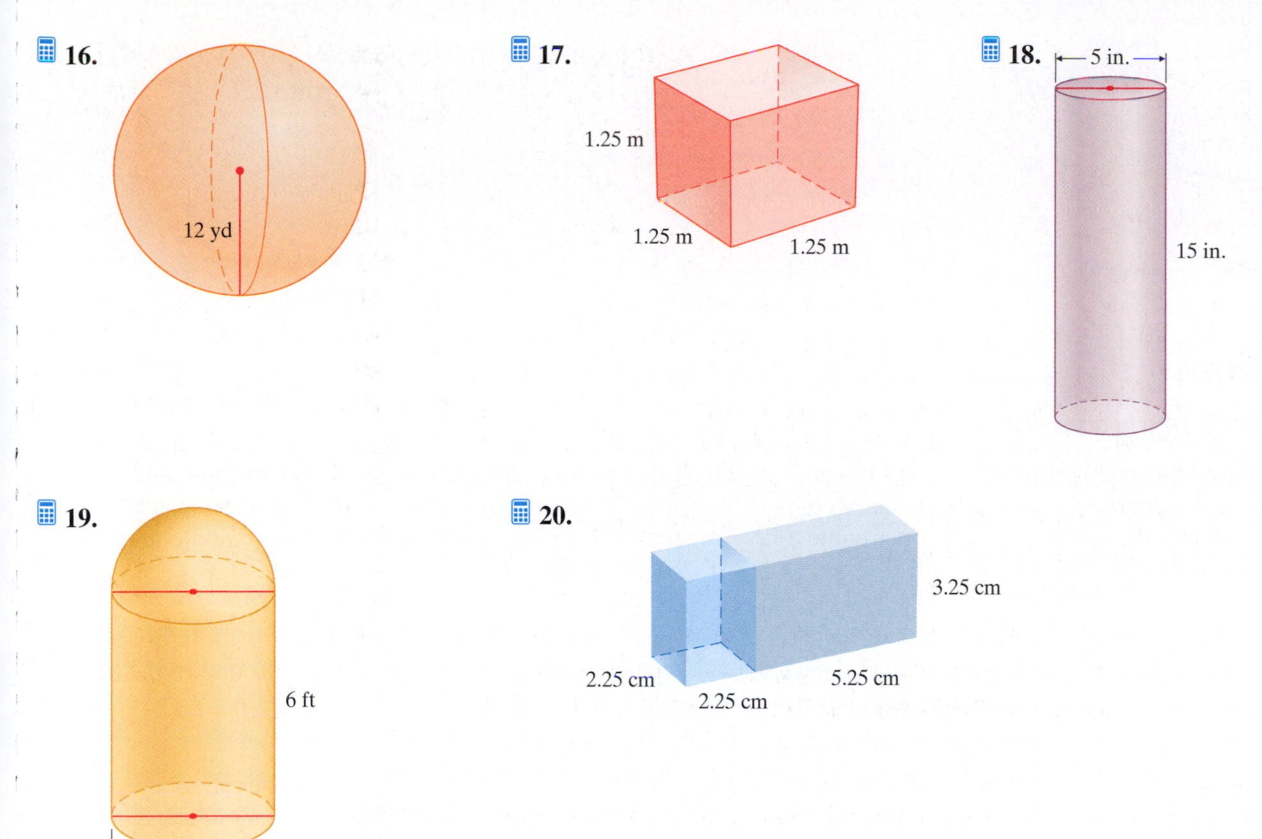

Mixed Practice

Find the volume of each solid. Use $\pi \approx 3.14$ *when needed.*

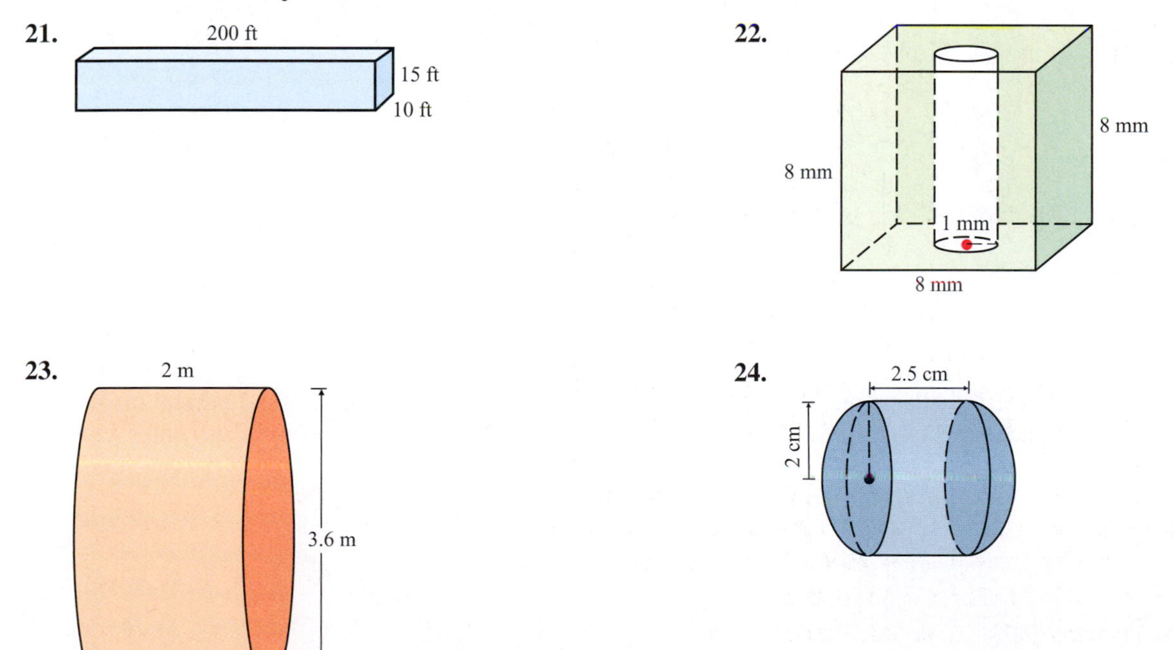

25.

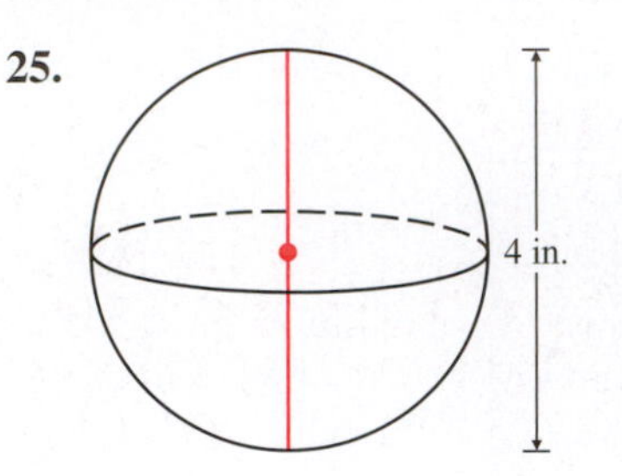

26. A cube with side length 9 inches.

Applications

Solve. Use $\pi \approx 3.14$ *when needed.*

27. A rectangular gold bar is 20 centimeters long, 10 centimeters wide, and 6 centimeters high. If the bar weighs 13 kilograms, what is the weight of the bar per cubic centimeter? Round to the nearest hundredth.

28. A fish tank is 12 inches wide, 18 inches long, and 16 inches high. How much water, to the nearest tenth of a gallon, will fill the tank if 1 gallon equals 231 cubic inches?

29. The crew of a spaceship began a search for a missing shuttlecraft 1,000 miles in every direction. How many cubic miles did the crew search? Round to the nearest million.

30. What is the volume of a hot air balloon, in the shape of a sphere, with a diameter of 30 feet? Round to the nearest hundred cubic feet.

31. The displacement of a car engine's cylinder is the volume of gas and air forced up by the piston. This volume is the product of the distance that the top of the piston travels and the area of the base of the cylinder. Find the displacement of the cylinder shown, rounded to the nearest cubic inch.

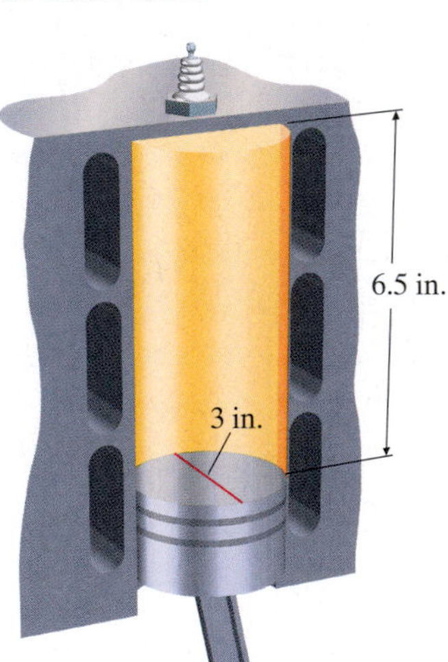

32. A board foot is a special measure of volume used in the lumber industry. If a board foot contains 144 cubic inches of wood, how many board feet rounded to the nearest tenth are there in the board shown?

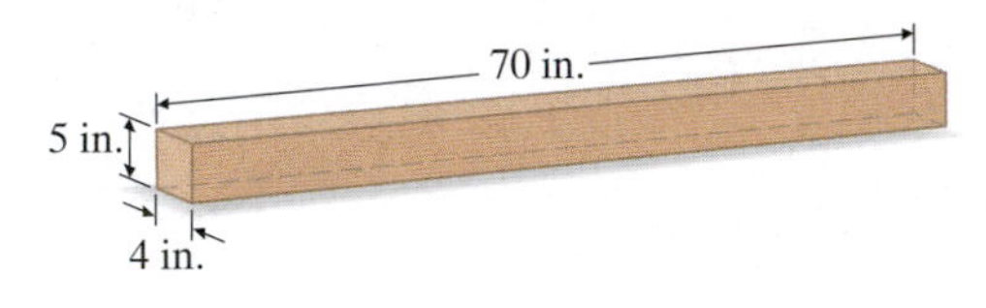

33. A large box is placed in a delivery truck, as pictured. What is the volume of the space remaining in the truck?

34. The metal machine part shown is cylindrical in shape, with a smaller cylinder drilled out of its center. How many cubic millimeters of metal does the machine part contain?

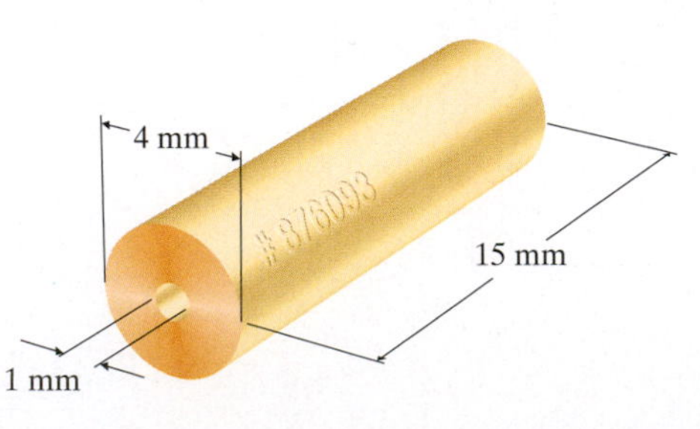

Use a calculator to solve the following problems, giving (a) the operation(s) carried out in the solution, (b) the exact answer, and (c) an estimate of the answer. Use $\pi \approx 3.14$ when needed.

35. In a chemistry course, students learn that a substance's density is found by dividing its weight by its volume. If a 300-gram block of wood is 7 centimeters by 5 centimeters by 3 centimeters, find the density of the wood. Round to the nearest tenth.

36. A cylindrical wheat silo has a height of 35 yards and a diameter of 5 yards. What is the maximum amount of wheat that can be stored in this silo? Round to the nearest tenth.

• *Check your answers on page A-17.*

MINDSTRETCHERS

Groupwork

1. Working with a partner, decide which three-dimensional figure can be formed by folding each pattern.

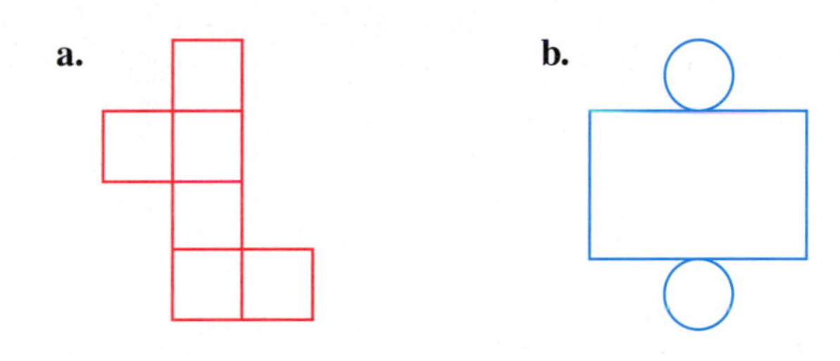

Mathematical Reasoning

2. Geometric solids have not only volume but also *surface area*. Find the total surface area of a cube with volume x^3.

Patterns

3. Consider the following table:

Figure	Measure	Number of Dimensions	Formula
Circle	Circumference	1	$2\pi r$
Circle	Area	2	πr^2
Sphere	Volume	3	$\frac{4}{3}\pi r^3$

Describe the pattern that you observe.

11.5 Similar Triangles

Identifying Corresponding Sides of Similar Triangles

OBJECTIVES

- To identify corresponding sides of similar triangles
- To find the missing sides of similar triangles
- To solve word problems involving similar triangles

When discussing ratio and proportion earlier, we looked at figures that have the same shape but different size. For example, in Section 5.2, we noted that everything in the enlargement of a rectangular photo is the same shape as in the original—only larger. In this section, we focus on triangles that have this relationship, which are called *similar triangles.*

Definition

Similar triangles are triangles that have the same shape but not necessarily the same size.

When two triangles are similar, for each angle of the first triangle there corresponds an angle of the second triangle with the same measure. The sides opposite these *corresponding angles* are called *corresponding sides*.

In similar triangles, the measures of corresponding angles are equal and corresponding sides are in proportion. For example, the following triangles ABC and DEF are similar:

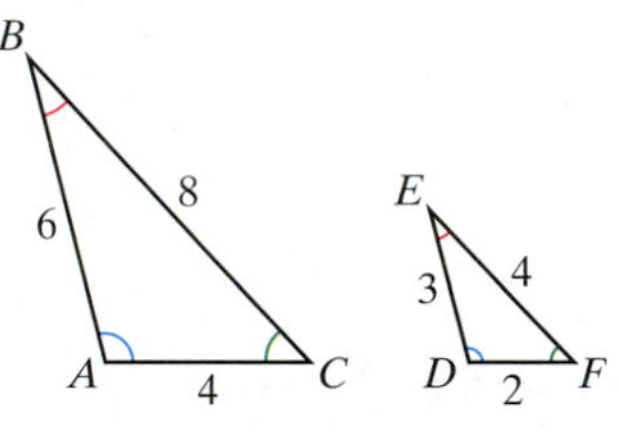

Since these triangles are similar, the measures of their corresponding angles are equal. So we write:

$$m\angle A = m\angle D$$
$$m\angle B = m\angle E$$
$$m\angle C = m\angle F$$

Also the lengths of the corresponding sides are in proportion, that is:

$$\frac{AB}{DE} = \frac{BC}{EF} = \frac{AC}{DF}$$
$$\frac{6}{3} = \frac{8}{4} = \frac{4}{2} = \frac{2}{1}$$

The ratio of the corresponding sides is $\frac{2}{1}$.

When we write that two triangles are similar, we name them so that the order of corresponding angles in both triangles is the same. In this case,

$$\triangle ABC \sim \triangle DEF$$

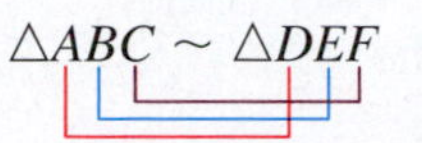

EXAMPLE 1

$\triangle RST \sim \triangle XYZ$. Name the corresponding sides of these triangles.

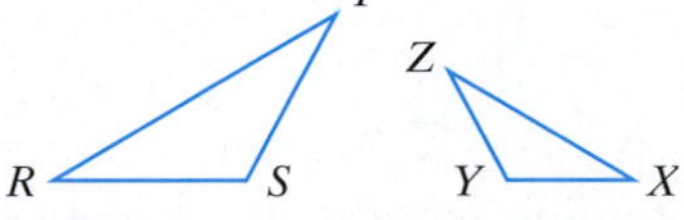

Solution Because $\triangle RST \sim \triangle XYZ$, $m\angle R = m\angle X$, $m\angle S = m\angle Y$, and $m\angle T = m\angle Z$. We know that the corresponding sides are opposite angles with equal measure. So we write the following:

Because $m\angle R = m\angle X$, $\overline{ST}$ corresponds to $\overline{YZ}$. — **$\overline{ST}$ is opposite $\angle R$, and $\overline{YZ}$ is opposite $\angle X$.**

Because $m\angle S = m\angle Y$, $\overline{RT}$ corresponds to $\overline{XZ}$. — **$\overline{RT}$ is opposite $\angle S$, and $\overline{XZ}$ is opposite $\angle Y$.**

Because $m\angle T = m\angle Z$, $\overline{RS}$ corresponds to $\overline{XY}$. — **$\overline{RS}$ is opposite $\angle T$, and $\overline{XY}$ is opposite $\angle Z$.**

PRACTICE 1

$\triangle ABC \sim \triangle GHI$. List the corresponding sides of these triangles.

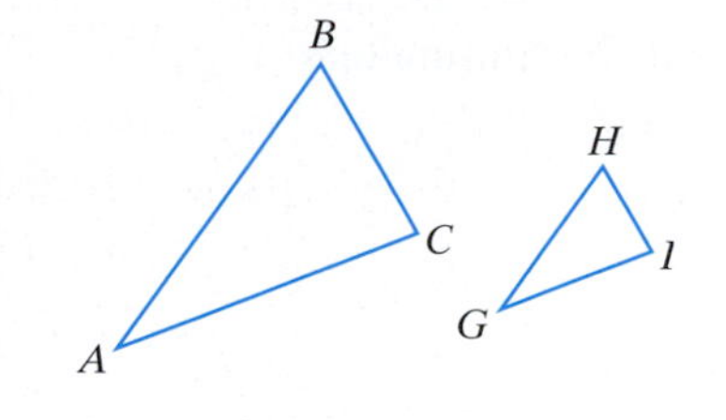

Finding the Missing Sides of Similar Triangles

Since corresponding sides in similar triangles are in proportion, we can use these proportions to find the length of a missing side.

To Find a Missing Side of Similar Triangles

- Write the ratios of the lengths of the corresponding sides.
- Write a proportion using a ratio with known terms and a ratio with an unknown term.
- Solve the proportion for the unknown term.

EXAMPLE 2

In the following diagram, $\triangle TAP \sim \triangle RUN$. Find x.

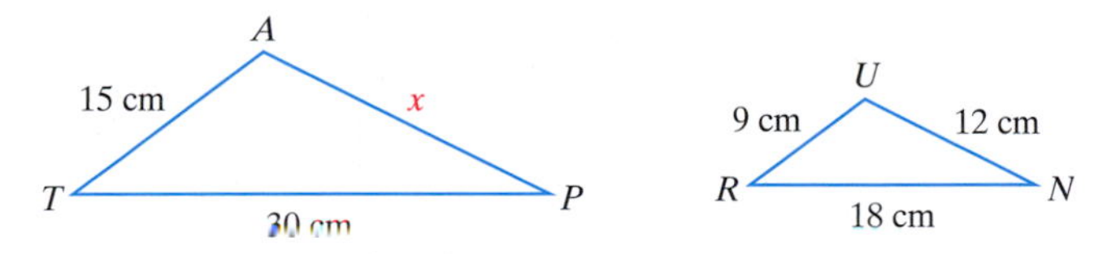

Solution Because $\triangle TAP \sim \triangle RUN$, we write the ratios of the lengths of the corresponding sides.

$$\frac{TA}{RU} = \frac{AP}{UN} = \frac{TP}{RN}$$

$$\frac{15}{9} = \frac{x}{12} = \frac{30}{18}$$

PRACTICE 2

In the following diagram, $\triangle DOT \sim \triangle PAN$. Find y.

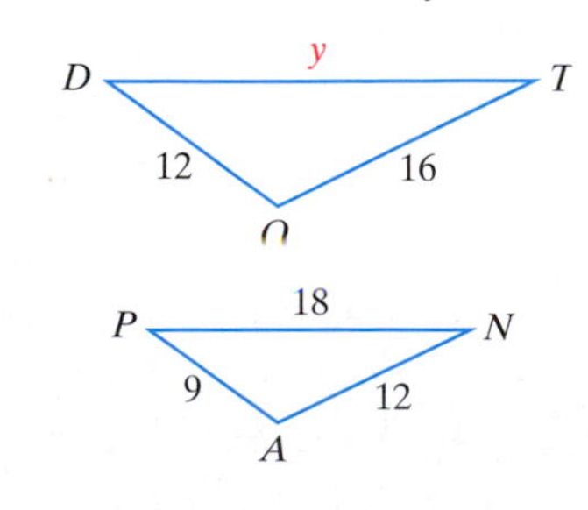

To solve for x, we can consider either the proportion $\frac{15}{9} = \frac{x}{12}$ or the proportion $\frac{x}{12} = \frac{30}{18}$. Note that each proportion contains the unknown term x. If we choose the first proportion and solve for x, we get the following.

$$\frac{15}{9} = \frac{x}{12}$$
$$9x = 180$$
$$x = 20$$

So x is 20 centimeters.

Similar triangles are useful in finding lengths that cannot be measured directly, as Example 3 shows.

EXAMPLE 3

A surveyor took the measurements shown. If $\triangle ABC \sim \triangle EFC$, find d, the distance across the river.

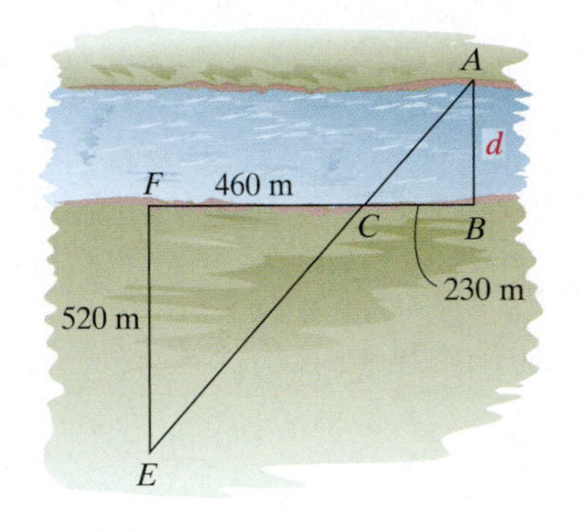

Solution Here, $\triangle ABC \sim \triangle EFC$, so we write the proportion $\frac{AB}{EF} = \frac{BC}{FC}$.

Then we substitute the values given in the diagram.

$$\frac{d}{520} = \frac{230}{460}$$

$$460d = (230)(520) \quad \text{Cross multiply.}$$

$$\frac{\overset{1}{\cancel{460}}d}{\underset{1}{\cancel{460}}} = \frac{(\overset{1}{\cancel{230}})(520)}{\underset{2}{\cancel{460}}} \quad \text{Divide both sides by 460.}$$

$$d = 260$$

The distance across the river is 260 meters.

PRACTICE 3

The height of a man and his shadow form a triangle similar to that formed by a nearby tree and its shadow. What is the height of the tree?

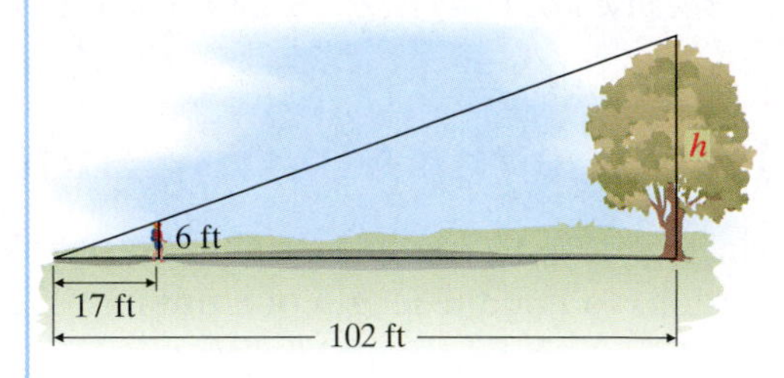

Mathematically Speaking

Fill in each blank with the most appropriate term or phrase from the given list.

equal	shape	similar
area	in proportion	corresponding

1. The symbol ~ is used to indicate that triangles are ________.

2. Similar triangles have the same ________ but not necessarily the same size.

3. In similar triangles, ________ sides are opposite angles with equal measure.

4. Corresponding sides of similar triangles are ________.

Find each missing side.

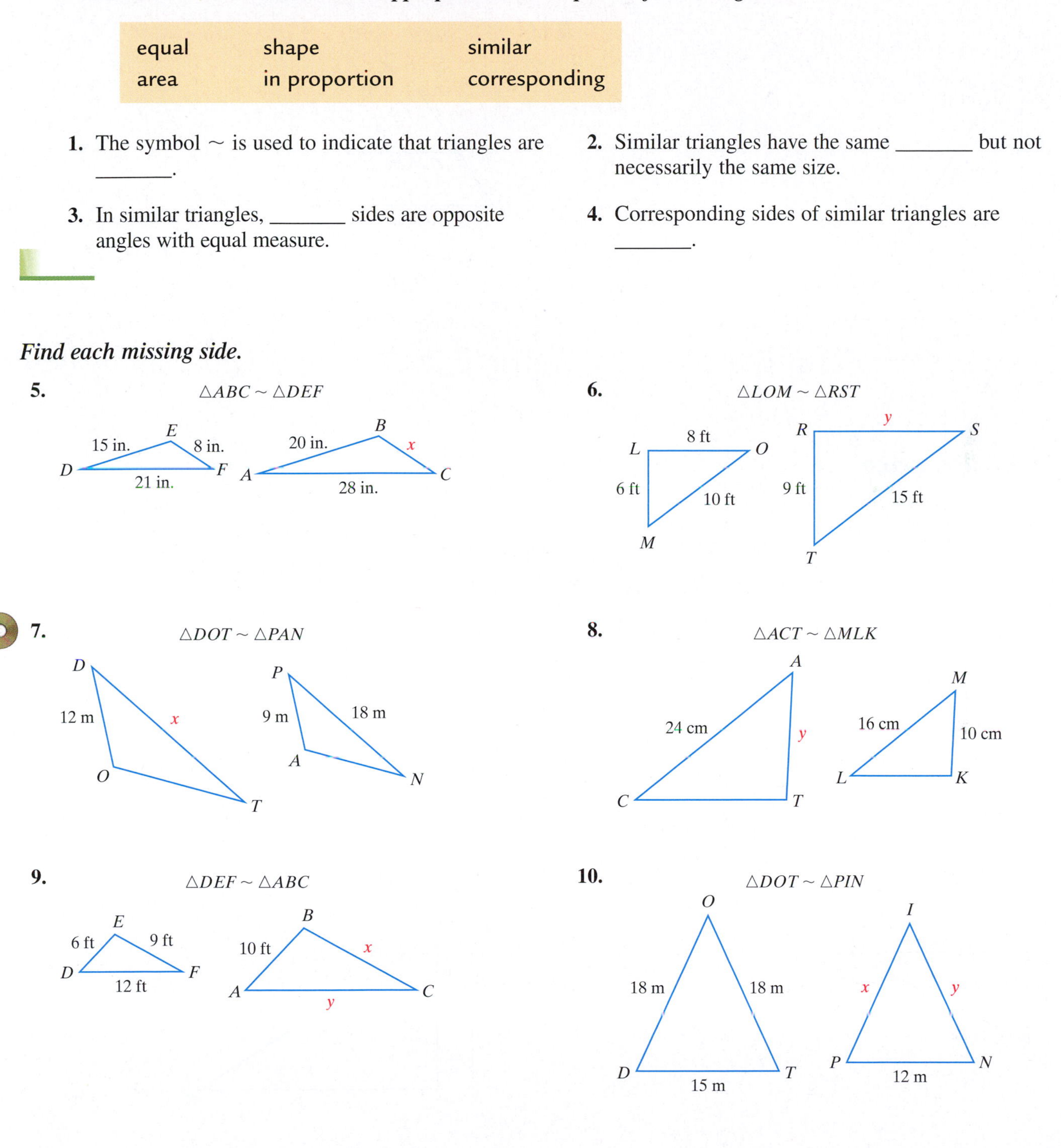

11.

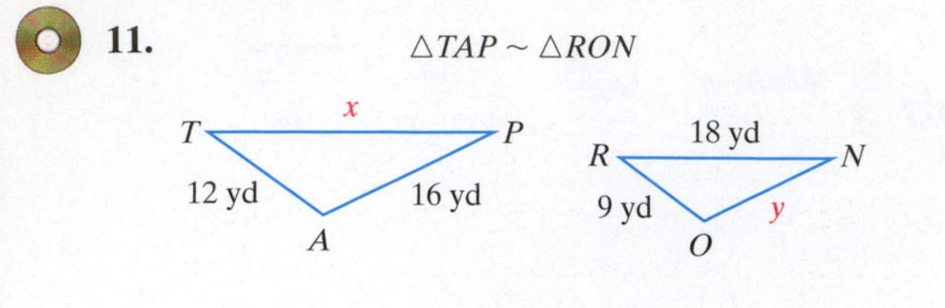

12.

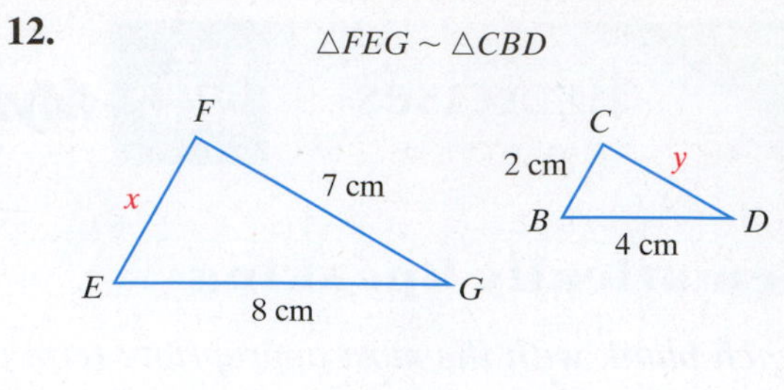

13.

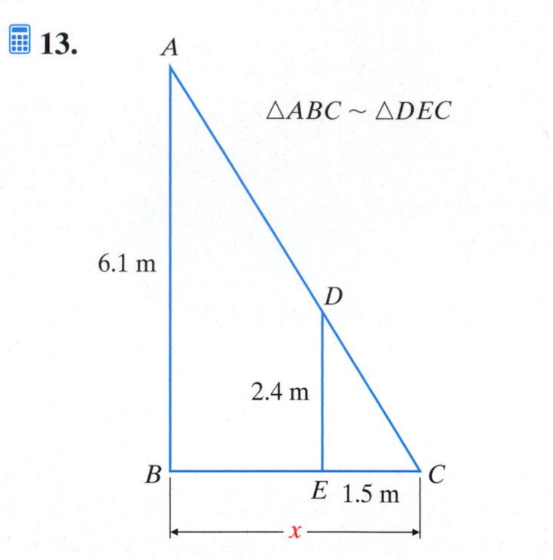

14.

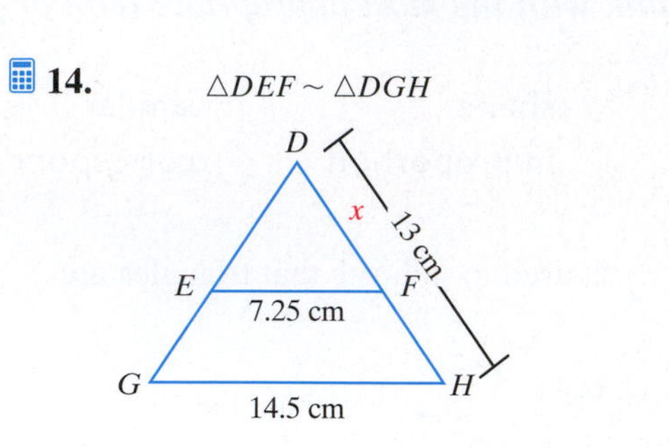

Mixed Practice

Find each missing side.

15. $\triangle ABC \sim \triangle DEF$

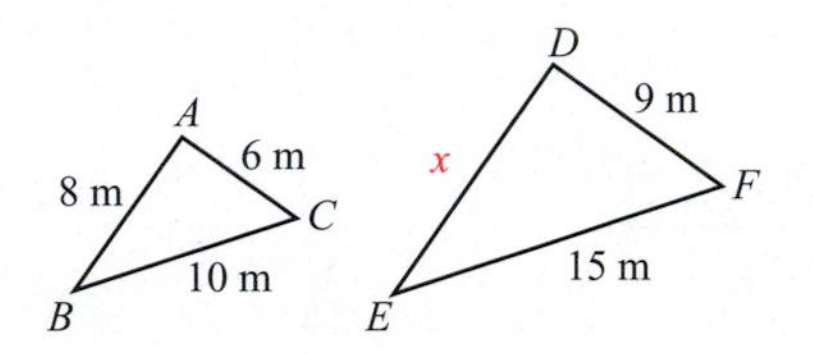

16. $\triangle PQR \sim \triangle STU$

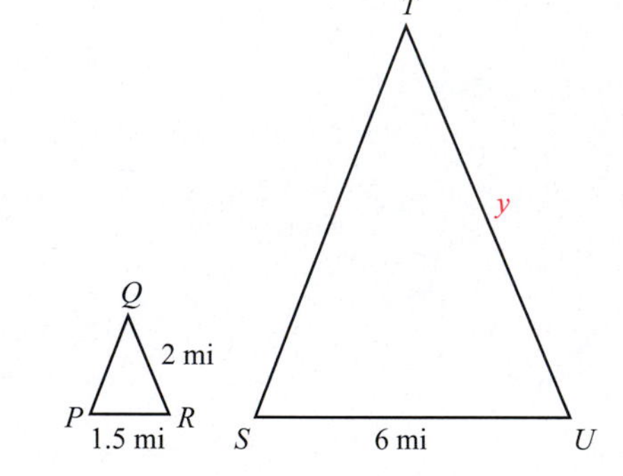

17. $\triangle PQR \sim \triangle TSR$

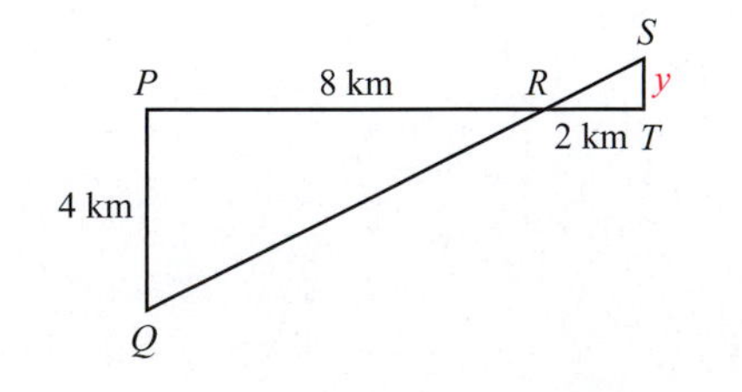

18. $\triangle ABC \sim \triangle ADE$

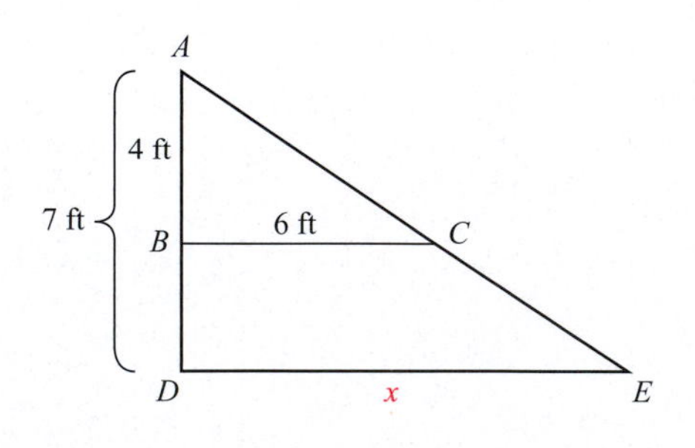

Applications

Solve. Assume the triangles are similar.

19. The tallest land animal is the giraffe. How tall is the giraffe pictured here?

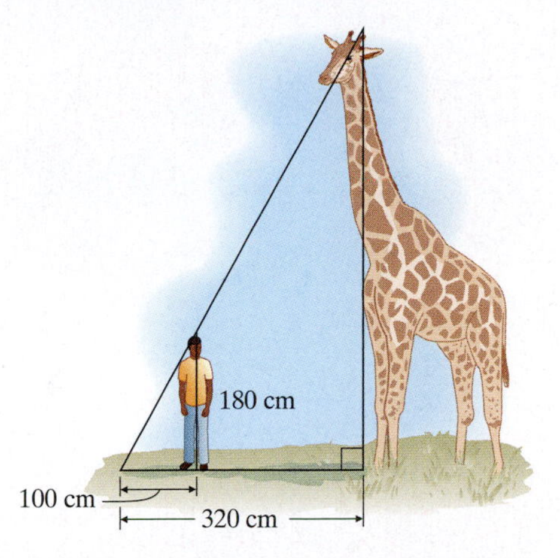

20. Light from a flashlight shines through a transparent dragon puppet onto a screen behind it, as shown below. Find the height of the puppet's image.

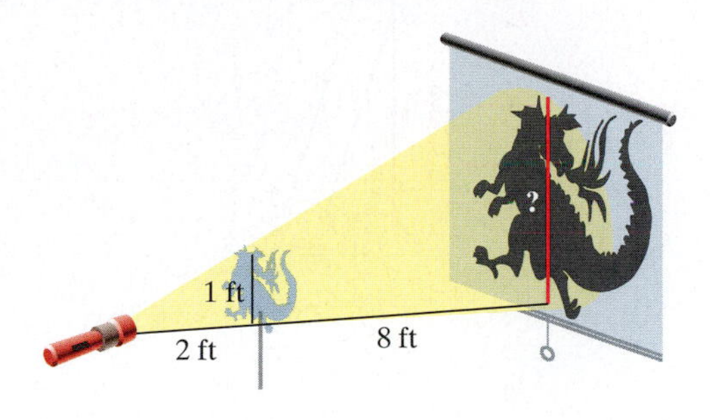

21. A Coast Guard observer sees a ship out on the ocean and wants to know how far it is from the shore. Use the diagram to find that distance.

22. One way to measure the height of a building is to position a mirror on the ground so that the top of the building's reflection can be seen. Find the height of the building.

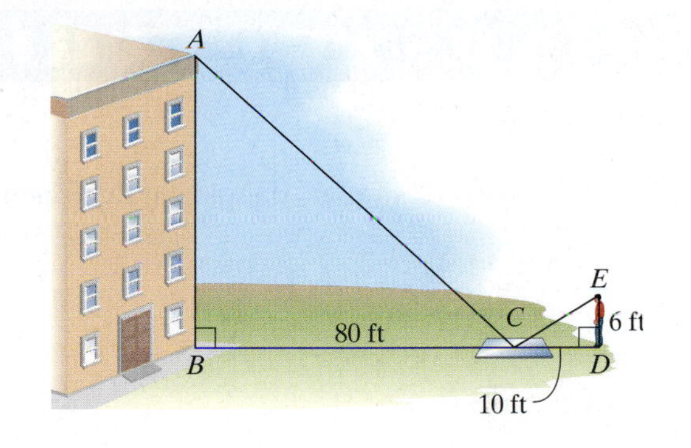

23. Two support wires are attached to a utility pole as shown in the diagram. If $\triangle ABC \sim \triangle ADE$, find AD.

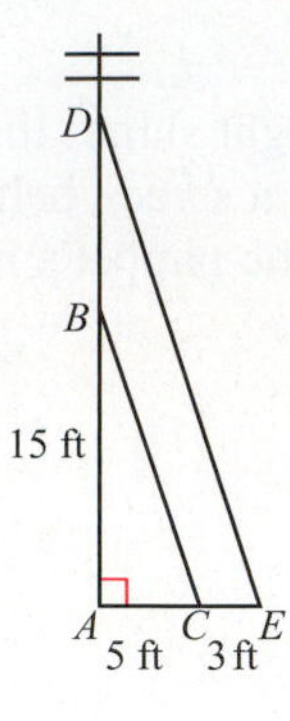

24. To measure AB, the distance across a certain lakes, distances BC, DC, and ED were "staked out," as shown in the diagram below. If $\triangle ABC \sim \triangle EDC$, how wide is the lake?

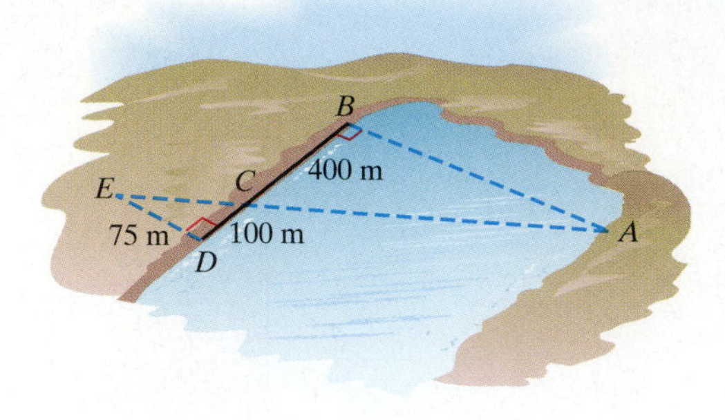

• *Check your answers on page A-17.*

MINDSTRETCHERS

Patterns

1. List 10 pairs of similar triangles in the following square.

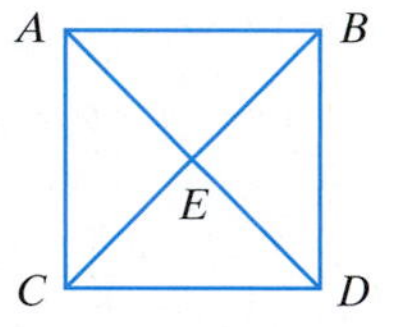

Writing

2. When you enlarge a photograph, the new image is similar to the old one. Give some other everyday examples of similarity.

Groupwork

3. Working with a partner, decide whether this statement is true: If two quadrilaterals have corresponding angles equal, the two quadrilaterals must have the same shape. Draw a diagram to support your answer.

11.6 Square Roots and the Pythagorean Theorem

In Section 11.3, we found the area of a square by squaring the length of one of its sides. In this section, we look at the opposite of squaring a number, that is, finding its *square root.*

OBJECTIVES

- To find the square root of a number
- To find the unknown side of a right triangle, using the Pythagorean theorem
- To solve word problems involving a square root or the Pythagorean theorem

Finding the Square Root of a Number

Consider the following problem. Suppose that a square has an area of 25 ft^2. What is the length, s, of each side?

s s $A = 25\text{ ft}^2$ s s

Recall that, for the area of a square, $A = s^2$. So for $25 = s^2$, we need to determine what whole number when multiplied by itself equals 25. Because $25 = 5 \cdot 5$, the whole number must be 5. So a square with an area of 25 ft^2 has sides of length 5 feet.

Squaring the whole number 5 gives 25. We say that 25 is the *square* of 5, or that 5 is the (principal) *square root* of 25 (written $\sqrt{25} = 5$).

Since 25 is the square of a whole number, it is called a *perfect square*. Perfect squares play a special role in the discussion of square roots.

Definitions

A **perfect square** is a number that is the square of a whole number.

The (principal) **square root** of a number n, written $\sqrt{n}$, is the positive number whose square is n.

The square root of 36 is 6, because the square of 6 is 36. And $\sqrt{4} = 2$, because $2^2 = 4$. Squaring and taking a square root are opposite operations, since one operation undoes the other.

EXAMPLE 1

Find the square root.

a. $\sqrt{25}$ **b.** $\sqrt{100}$

Solution In each case, we need to find the whole number that when squared gives us the number under the square root sign.

a. $\sqrt{25} = 5$ because $5 \cdot 5 = 25$

b. $\sqrt{100} = 10$ because $10 \cdot 10 = 100$

PRACTICE 1

What is the square root of each perfect square?

a. $\sqrt{49}$

b. $\sqrt{144}$

Many numbers are not perfect squares. For instance, 28 is not a perfect square, because there is no whole number that when multiplied by itself equals 28.

If a number is not a perfect square, we can either estimate or use a calculator to find its square root.

EXAMPLE 2

$\sqrt{28}$ lies between which two consecutive whole numbers?

Solution To begin, let's find the two consecutive perfect squares that 28 lies between. The number 28 is more than 25 and less than 36, which are consecutive perfect squares.

Because 28 lies between 5^2 and 6^2, $\sqrt{28}$ lies between 5 and 6.

PRACTICE 2

Between which two consecutive whole numbers does $\sqrt{47}$ lie?

EXAMPLE 3

Using a calculator, find each square root to the nearest tenth.

a. $\sqrt{75}$ **b.** $\sqrt{21}$

Solution

a.

Press	Display
2nd √ 75 ENTER	√(75 8.660254038

So $\sqrt{75} = 8.7$, rounded to the nearest tenth.

b.

Press	Display
2nd √ 21 ENTER	√(21 4.582575695

Therefore, $\sqrt{21} = 4.6$, rounded to the nearest tenth.

PRACTICE 3

Using a calculator, find each square root to the nearest hundredth.

a. $\sqrt{56}$

b. $\sqrt{12}$

The Pythagorean Theorem

Recall that a right triangle is a triangle that has one 90° angle. In a right triangle, the side opposite the right angle is called the **hypotenuse**. The other two sides are called **legs**.

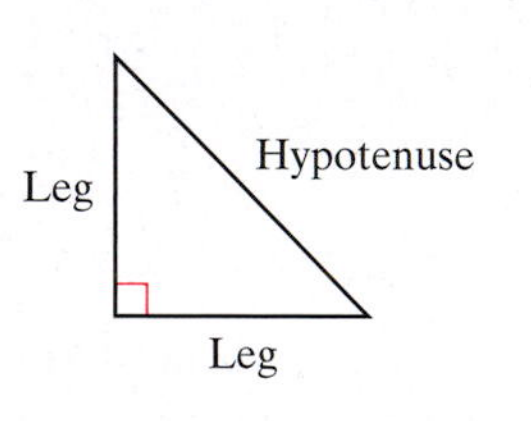

The lengths of the three sides of a right triangle are related in a special way. To understand this relationship, consider the areas of the squares on the legs and on the hypotenuse, as in the following example.

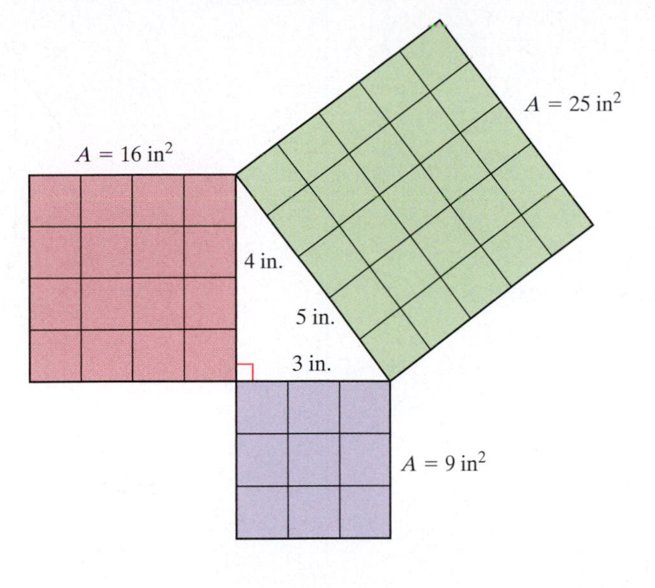

$$9 \text{ in}^2 + 16 \text{ in}^2 = 25 \text{ in}^2$$

Area of the square on one leg + Area of the square on the other leg = Area of the square on the hypotenuse

In general, if we let a and b represent the lengths of the legs and c represent the length of the hypotenuse, then $a^2 + b^2 = c^2$.

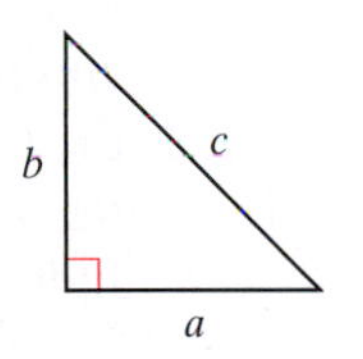

This relationship is called the *Pythagorean theorem.*

The Pythagorean theorem

For every right triangle, the sum of the squares of the two legs equals the square of the hypotenuse.

We can use the Pythagorean theorem to find the third side of a right triangle if we know the other two sides.

EXAMPLE 4

Find the length of the hypotenuse.

Solution To find the length of the hypotenuse, we use the Pythagorean theorem.

$$a^2 + b^2 = c^2$$
$$5^2 + 12^2 = c^2$$
$$25 + 144 = c^2$$
$$169 = c^2$$
$$\sqrt{169} = c$$
$$13 = c, \text{ or } c = 13$$

The hypotenuse is 13 centimeters long.

PRACTICE 4

Find the length of the unknown side in $\triangle ABC$.

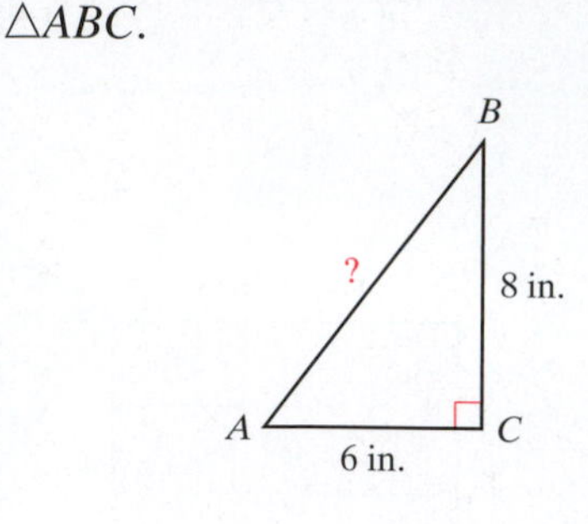

EXAMPLE 5

If b equals 1 centimeter and c equals 2 centimeters, what is the length of a in a right triangle, where a and b are legs and c is the hypotenuse? Round the answer to the nearest tenth of a centimeter.

Solution First we draw a diagram.

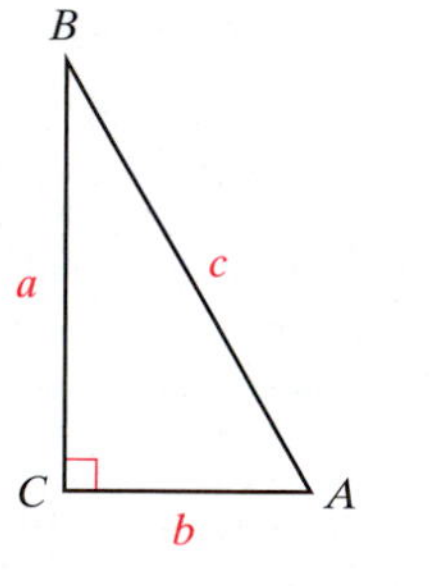

Then we use the Pythagorean theorem and substitute the given values to obtain the following.

$$a^2 + b^2 = c^2$$
$$a^2 + 1^2 = 2^2$$
$$a^2 + 1 = 4$$
$$a^2 + 1 - 1 = 4 - 1$$
$$a^2 = 3$$
$$a = \sqrt{3}$$

To express the answer as a decimal, we use a calculator. Rounding to the nearest tenth, we find that $\sqrt{3}$ is approximately 1.7. So $a \approx 1.7$ centimeters.

PRACTICE 5

In a right triangle, one leg equals 2 feet and the hypotenuse equals 4 feet. Find the length of the missing leg. Round the answer to the nearest tenth of a foot.

EXAMPLE 6

A baseball diamond is a square with sides 90 feet long. How far, to the nearest foot, must the third baseman throw the ball to reach the first baseman?

Solution The sides of the diamond together with the diagonal from third base to first base form a right triangle. To find the distance from third base to first base, we use the Pythagorean theorem.

$$a^2 + b^2 = c^2$$
$$90^2 + 90^2 = c^2$$
$$8{,}100 + 8{,}100 = c^2$$
$$\sqrt{16{,}200} = c^2$$
$$\sqrt{16{,}200} = c$$
$$c = \sqrt{16{,}200}$$

We use a calculator and round to find that $\sqrt{16{,}200}$ is approximately 127 feet. So the third baseman must throw the ball approximately 127 feet to reach the first baseman.

PRACTICE 6

What is the distance across the intersection shown in the diagram, to the nearest foot?

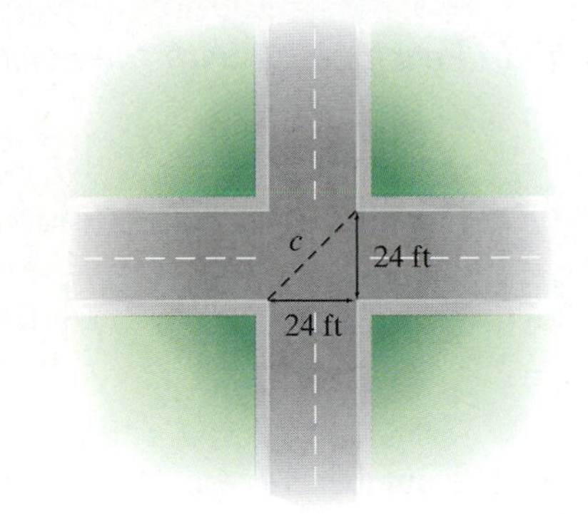

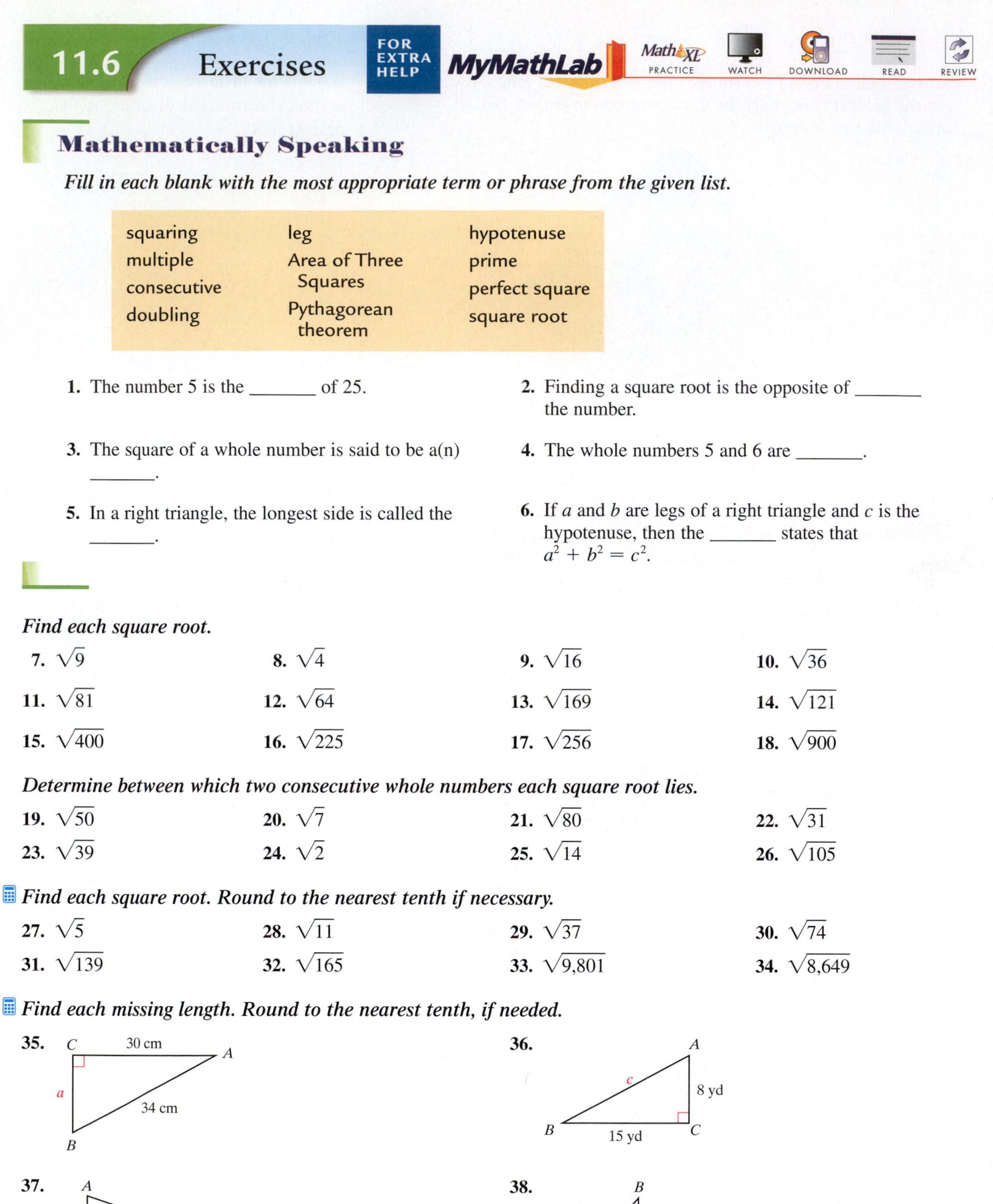

11.6 Exercises

FOR EXTRA HELP MyMathLab MathXL PRACTICE WATCH DOWNLOAD READ REVIEW

Mathematically Speaking

Fill in each blank with the most appropriate term or phrase from the given list.

squaring	leg	hypotenuse
multiple	Area of Three Squares	prime
consecutive		perfect square
doubling	Pythagorean theorem	square root

1. The number 5 is the _______ of 25.

2. Finding a square root is the opposite of _______ the number.

3. The square of a whole number is said to be a(n) _______.

4. The whole numbers 5 and 6 are _______.

5. In a right triangle, the longest side is called the _______.

6. If a and b are legs of a right triangle and c is the hypotenuse, then the _______ states that $a^2 + b^2 = c^2$.

Find each square root.

7. $\sqrt{9}$
8. $\sqrt{4}$
9. $\sqrt{16}$
10. $\sqrt{36}$
11. $\sqrt{81}$
12. $\sqrt{64}$
13. $\sqrt{169}$
14. $\sqrt{121}$
15. $\sqrt{400}$
16. $\sqrt{225}$
17. $\sqrt{256}$
18. $\sqrt{900}$

Determine between which two consecutive whole numbers each square root lies.

19. $\sqrt{50}$
20. $\sqrt{7}$
21. $\sqrt{80}$
22. $\sqrt{31}$
23. $\sqrt{39}$
24. $\sqrt{2}$
25. $\sqrt{14}$
26. $\sqrt{105}$

Find each square root. Round to the nearest tenth if necessary.

27. $\sqrt{5}$
28. $\sqrt{11}$
29. $\sqrt{37}$
30. $\sqrt{74}$
31. $\sqrt{139}$
32. $\sqrt{165}$
33. $\sqrt{9{,}801}$
34. $\sqrt{8{,}649}$

Find each missing length. Round to the nearest tenth, if needed.

35. C, 30 cm, A, a, 34 cm, B

36. A, c, 8 yd, B, 15 yd, C

37. A, 1 m, c, C, 3 m, B

38.

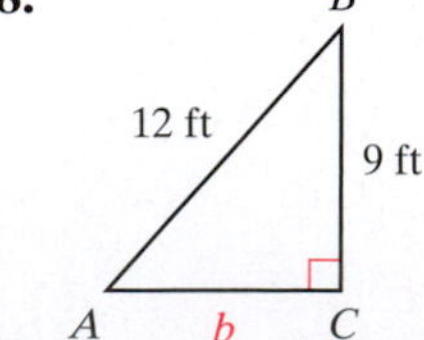

Given a right triangle with legs a and b, and hypotenuse c, find the missing side. Round to the nearest tenth, if needed.

	a	*b*	*c*
39.	24 m		25 m
40.	5 in.	12 in.	
41.	6 ft		10 ft
42.		4 cm	5 cm
43.	12 m	16 m	
44.		9 in.	15 in.
45.	7 cm	9 cm	
46.	2 yd	5 yd	
47.		18 ft	20 ft
48.	2 in.	2 in.	

Mixed Practice

Solve.

49. Find the missing length.

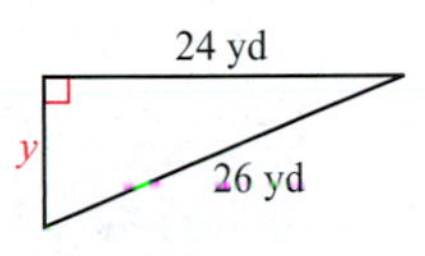

50. Determine between which two consecutive whole numbers $\sqrt{95}$ lies.

51. Find the missing length.

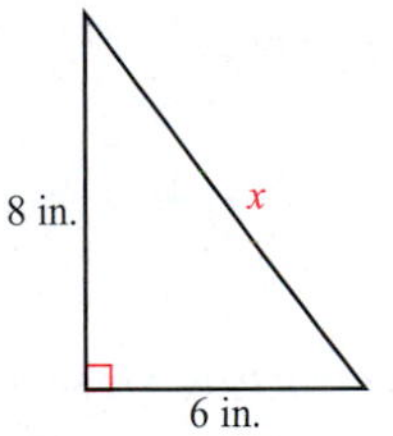

52. Find $\sqrt{196}$.

53. Find the missing length. Round to the nearest tenth.

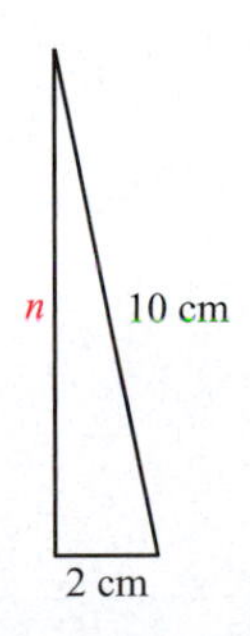

54. Find $\sqrt{41}$ to the nearest tenth.

55. A contractor leans a ladder against the side of a building. How high up the building does the ladder reach?

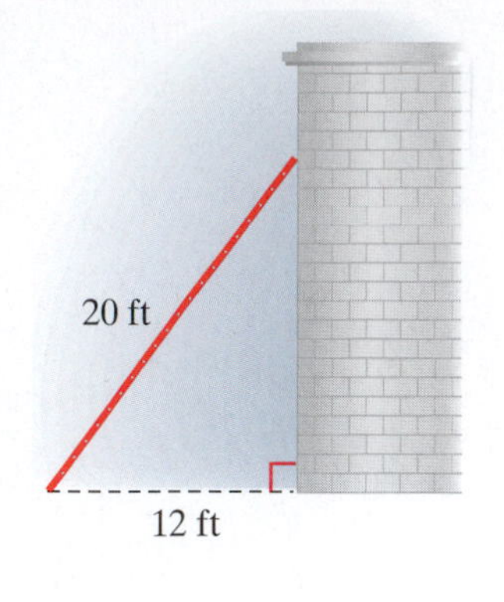

56. A scuba diver swims away from the boat and then dives, as shown. How far from the boat, to the nearest foot, will he be?

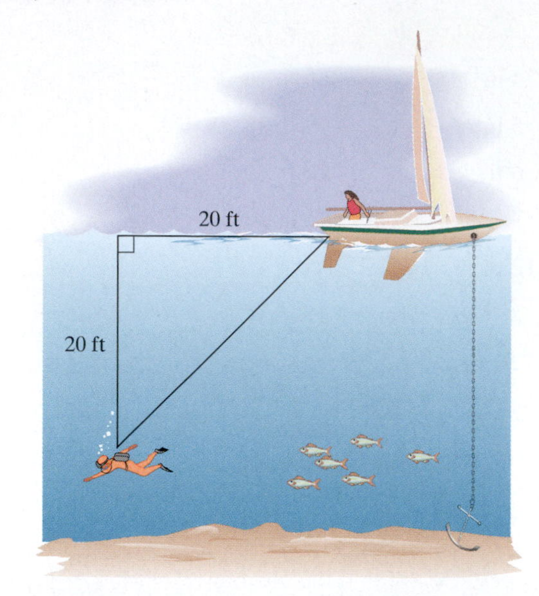

57. What is the length of the rectangular plot of land shown?

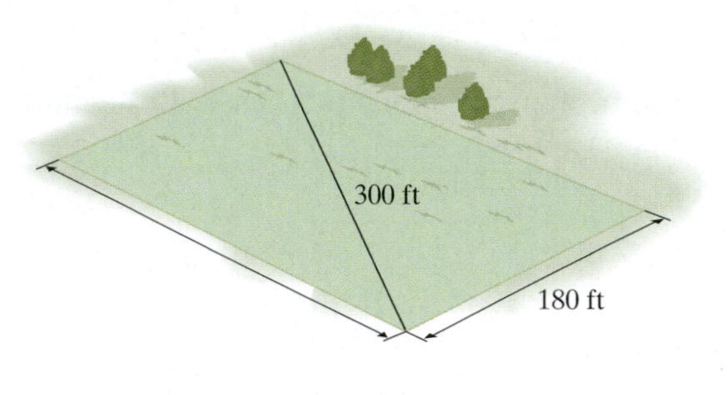

58. When Ronald Reagan was president, his helicopter and a private plane were involved in a near-miss incident, as pictured. How close were the president's helicopter and the plane? (***Source:*** *New York Times*, August 15, 1987)

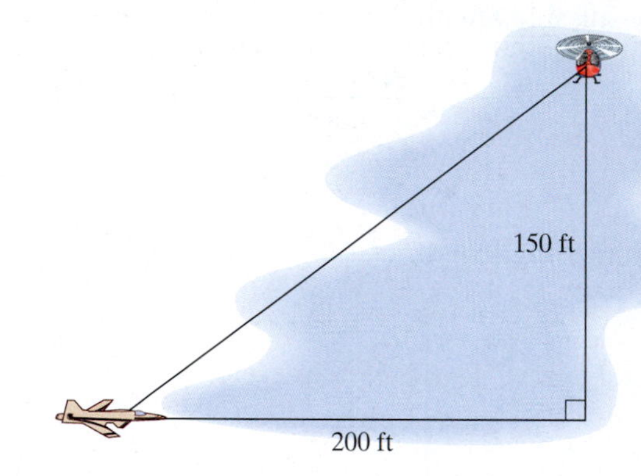

59. A builder constructed a roof of wooden beams. According to the diagram, what is the length of the sloping beam?

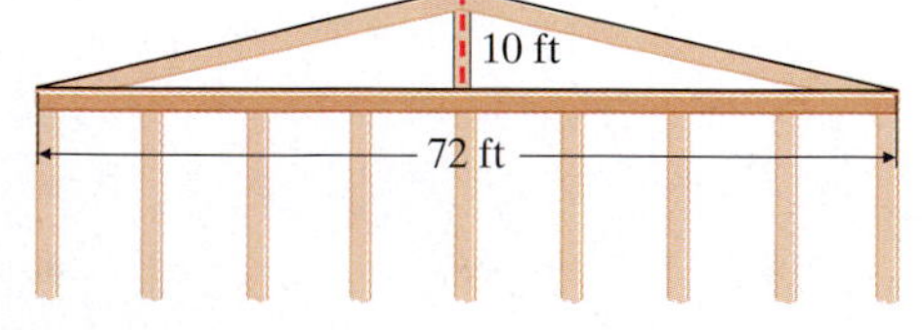

60. A college is constructing an access ramp for the handicapped to a door in one of its buildings, as shown. Find the length of the ramp.

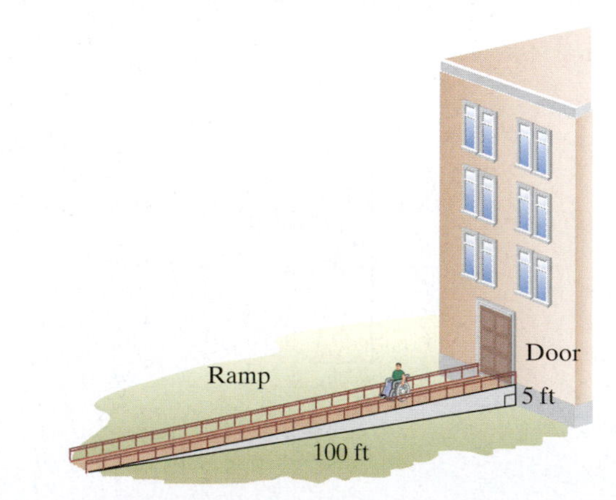

● *Check your answers on page A-17.*

MINDSTRETCHERS

Mathematical Reasoning

1. Give an example of a number that is smaller than its square root.

Writing

2. Thousands of years ago, the ancient Egyptians had a clever way of creating a right angle for their construction projects. They would use a rope, tying it in a circle with 12 equally spaced knots, as shown.

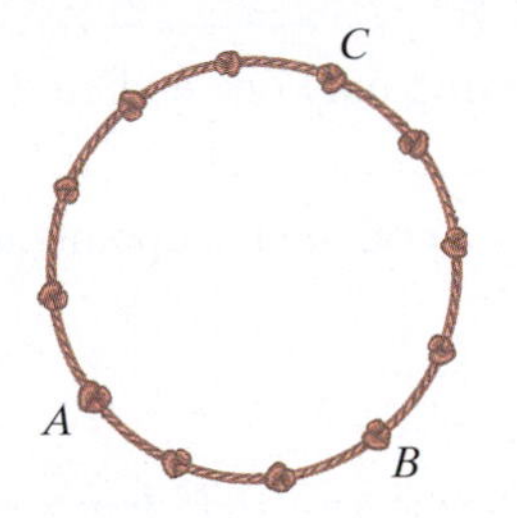

They would then pull on the knots labeled A, B, and C. Explain why doing so would create a right triangle.
(*Source:* Peter Tompkins, *Secrets of the Great Pyramids*)

Investigation

3. Choose a whole number. Use a calculator to determine whether it is a perfect square.

KEY CONCEPTS AND SKILLS

CONCEPT SKILL

Concept/Skill	Description	Example
[11.1] **Point**	An exact location in space, with no dimension.	•A
[11.1] **Line**	A collection of points along a straight path, that extends endlessly in both directions.	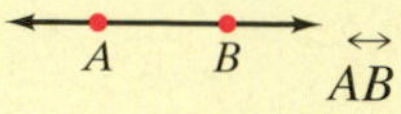 A B $\overleftrightarrow{AB}$
[11.1] **Line segment**	A part of a line having two endpoints.	 B C $\overline{BC}$
[11.1] **Ray**	A part of a line having only one endpoint.	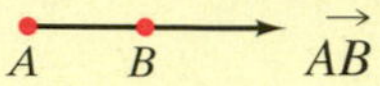 A B $\overrightarrow{AB}$
[11.1] **Angle**	Two rays that have a common endpoint called the *vertex* of the angle.	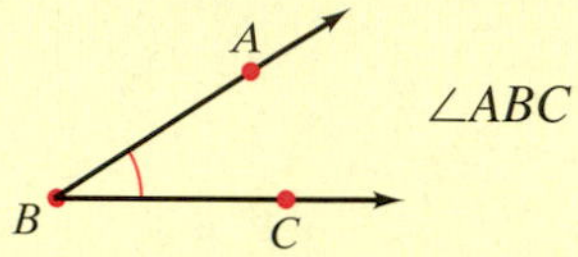 A B C $\angle ABC$
[11.1] **Plane**	A flat surface that extends endlessly in all directions.	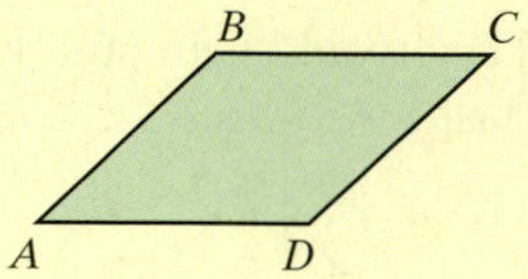 B C A D
[11.1] **Straight angle**	An angle whose measure is 180°.	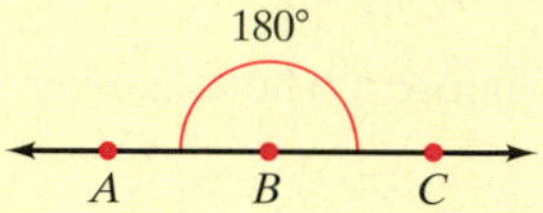 180° A B C
[11.1] **Right angle**	An angle whose measure is 90°.	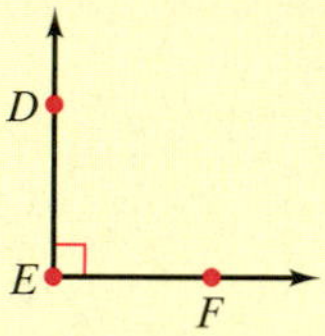 D E F
[11.1] **Acute angle**	An angle whose measure is less than 90°.	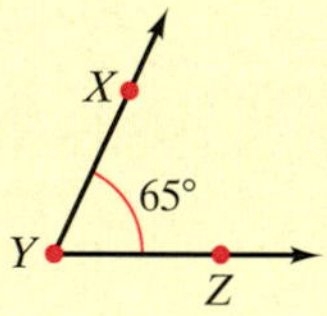 X 65° Y Z
[11.1] **Obtuse angle**	An angle whose measure is more than 90° and less than 180°.	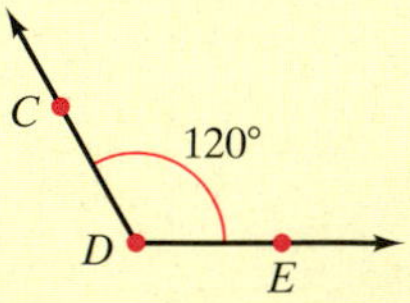 C 120° D E
[11.1] **Complementary angles**	Two angles the sum of whose measures is 90°.	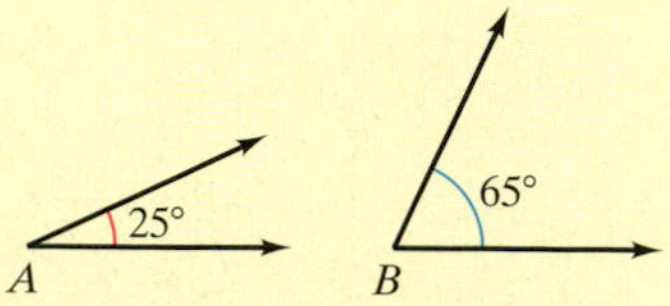 25° A 65° B
[11.1] **Supplementary angles**	Two angles the sum of whose measures is 180°.	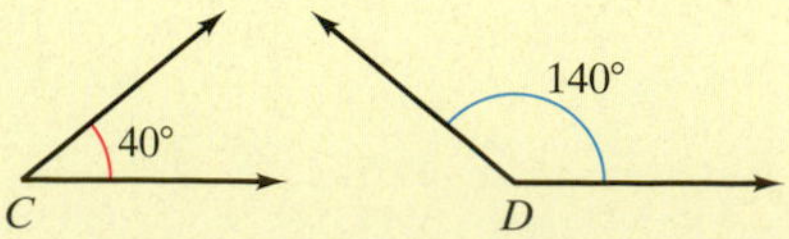 40° C 140° D

Concept/Skill	Description	Example
[11.1] **Intersecting lines**	Two lines that cross.	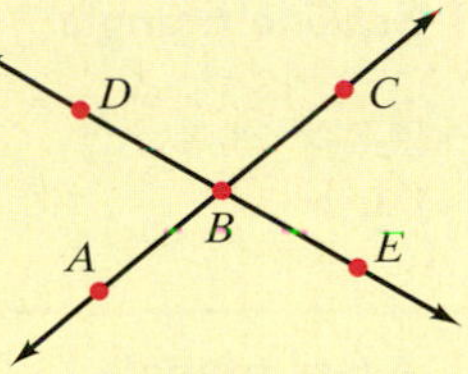 D, C, B, A, E
[11.1] **Parallel lines**	Two lines on the same plane that do not intersect.	 E, F, G, H; $\overleftrightarrow{EF} \parallel \overleftrightarrow{GH}$
[11.1] **Perpendicular lines**	Two lines that intersect to form right angles.	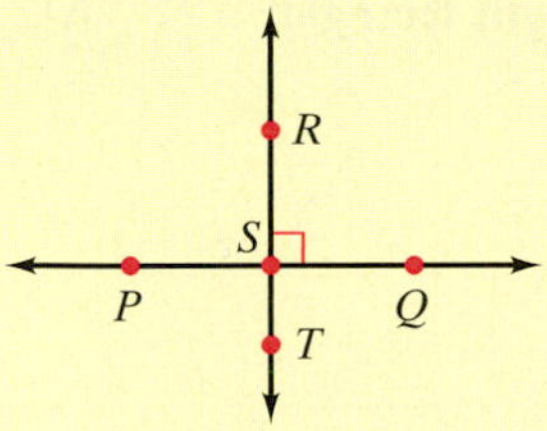 R, S, P, Q, T; $\overleftrightarrow{RT} \perp \overleftrightarrow{PQ}$
[11.1] **Vertical angles**	Two angles with equal measure formed by two intersecting lines.	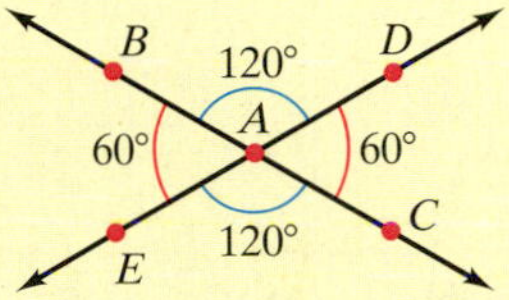 B, D, A, C, E; 120°, 60°, 60°, 120°
[11.1] **Polygon**	A closed plane figure made up of line segments.	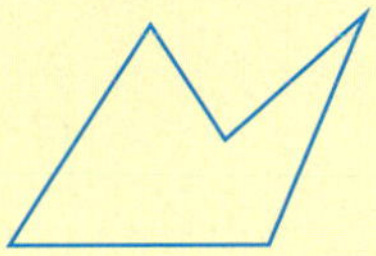
[11.1] **Triangle**	A polygon with three sides.	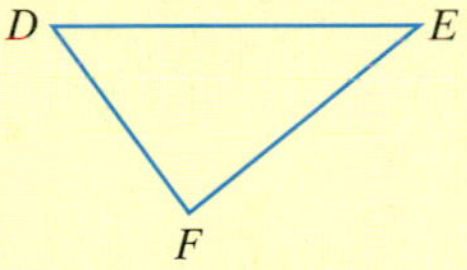 D, E, F
[11.1] **Quadrilateral**	A polygon with four sides.	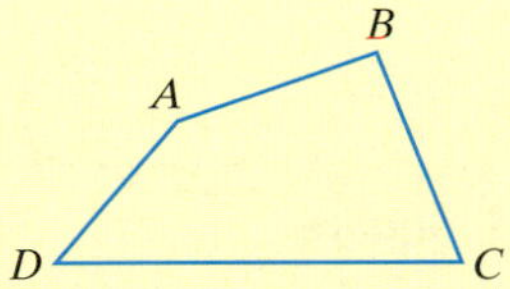 A, B, C, D
[11.1] **Equilateral triangle**	A triangle with three sides equal in length.	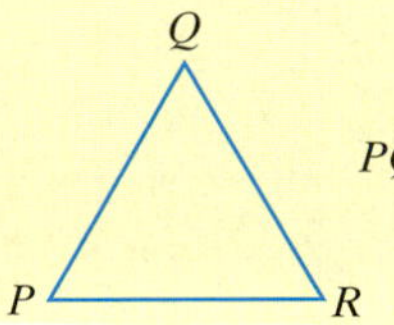 Q, P, R; $PQ = QR = PR$
[11.1] **Isosceles triangle**	A triangle with two sides equal in length.	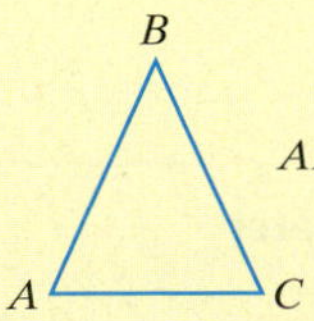 B, A, C; $AB = BC$

continued

CONCEPT SKILL

Concept/Skill	Description	Example
[11.1] **Scalene triangle**	A triangle with no sides equal in length.	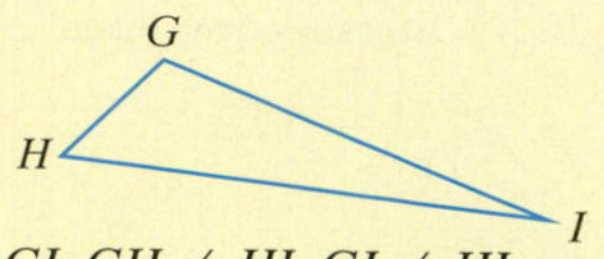 $GH \neq GI$, $GH \neq HI$, $GI \neq HI$
[11.1] **Acute triangle**	A triangle with three acute angles.	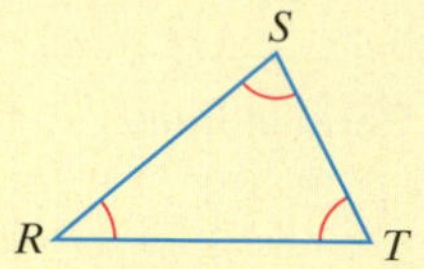
[11.1] **Right triangle**	A triangle with one right angle.	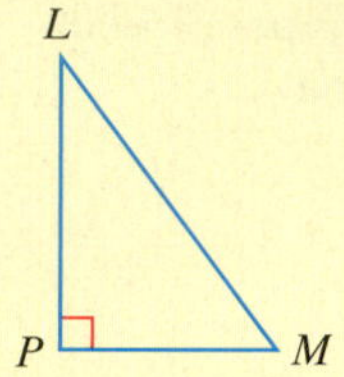
[11.1] **Obtuse triangle**	A triangle with one obtuse angle.	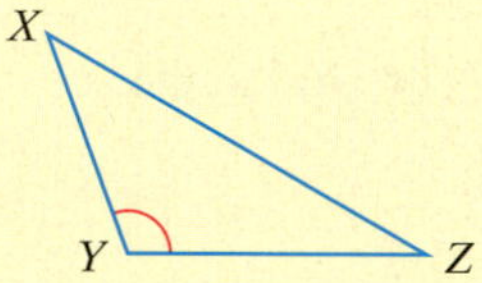
[11.1] **Trapezoid**	A quadrilateral with only one pair of opposite sides parallel.	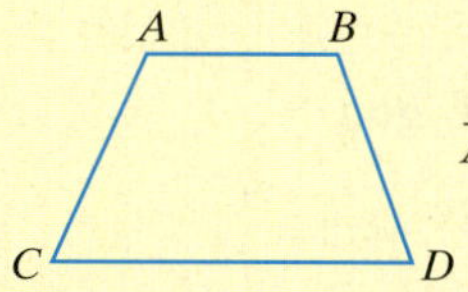 $\overline{AB} \parallel \overline{CD}$
[11.1] **Parallelogram**	A quadrilateral with both pairs of opposite sides parallel. Opposite sides are equal in length, and opposite angles have equal measures.	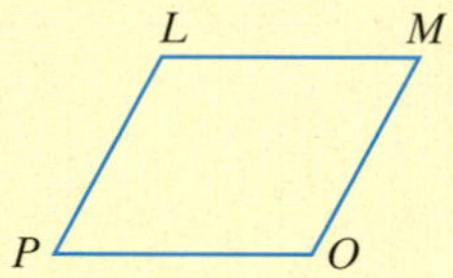 $\overline{LM} \parallel \overline{PO}$ $\overline{LP} \parallel \overline{MO}$
[11.1] **Rectangle**	A parallelogram with four right angles.	
[11.1] **Square**	A rectangle with four sides equal in length.	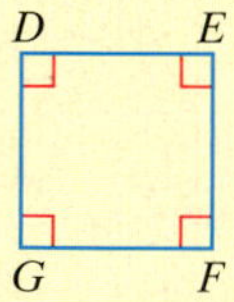 $DE = EF = FG = GD$
[11.1] **Circle**	A closed plane figure made up of points that are all the same distance from a fixed point called the center.	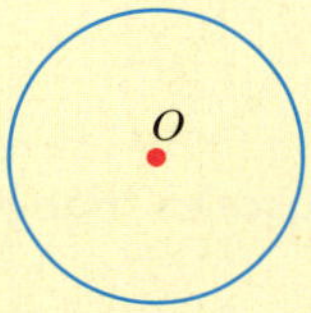
[11.1] **Diameter**	A line segment that passes through the center of a circle and has both endpoints on the circle.	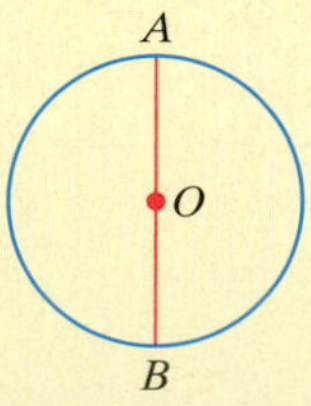

Concept/Skill	Description	Example
[11.1] **Radius**	A line segment with one endpoint on the circle and the other at the center.	O, B
[11.2] **Perimeter**	The distance around a polygon.	
[11.2] **Circumference**	The distance around a circle.	
[11.3] **Area**	The number of square units that a figure contains.	
[11.4] **Volume**	The number of cubic units required to fill a three-dimensional figure.	
[11.5] **Similar triangles**	Triangles that have the same shape but not necessarily the same size.	$\triangle ABC \sim \triangle DEF$
[11.5] **Corresponding sides**	In similar triangles, the sides opposite the equal angles.	In the similar triangles pictured, $\overline{AB}$ corresponds to $\overline{DE}$, $\overline{BC}$ corresponds to $\overline{EF}$, and $\overline{AC}$ corresponds to $\overline{DF}$.
[11.5] **To find the missing sides of similar triangles**	• Write the ratios of the lengths of the corresponding sides. • Write a proportion using a ratio with known terms and a ratio with an unknown term. • Solve the proportion for the unknown term.	$\triangle TRS \sim \triangle XYW$ Find a. S, 6 in., R, 4 in., 8 in., T; W, 9 in., Y, 6 in., a, X $\frac{ST}{WX} = \frac{TR}{XY}$ $\frac{4}{6} = \frac{8}{a}$ $4a = 48$ $a = 12$, or 12 in.

continued

CONCEPT SKILL

Concept/Skill	Description	Example
[11.6] Perfect square	A number that is the square of a whole number.	49 and 144
[11.6] (Principal) square root of *n*	The positive number, written $\sqrt{n}$, whose square is n.	$\sqrt{36}$ and $\sqrt{8}$
[11.6] Pythagorean theorem	For every right triangle, the sum of the squares of the two legs equals the square of the hypotenuse, that is, $$a^2 + b^2 = c^2$$ where a and b are legs, and c is the hypotenuse.	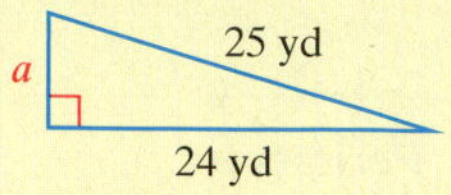 Find a. $a^2 + b^2 = c^2$ $a^2 + (24)^2 = (25)^2$ $a^2 + 576 = 625$ $a^2 + 576 - 576 = 625 - 576$ $a^2 = 49$ $a = \sqrt{49}$ $= 7$, or 7 yd

KEY FORMULAS

CONCEPT SKILL

Figure	Formula	Example
[11.2]–[11.3] Triangle	*Perimeter* $$P = a + b + c$$ Perimeter equals the sum of the lengths of the three sides. *Area* $$A = \frac{1}{2}bh$$ Area equals one-half the base times the height.	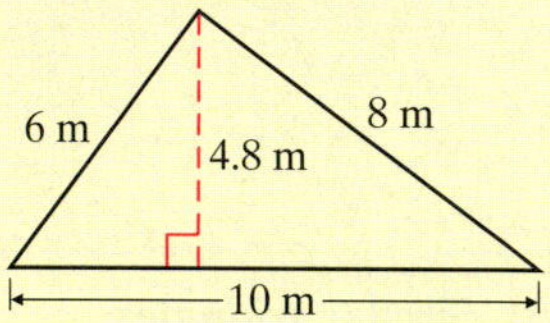 $P = a + b + c$ $= 6 + 10 + 8$ $= 24$, or 24 m $A = \frac{1}{2}bh$ $= \frac{1}{\cancel{2}_{1}} \cdot \cancel{10}^{5} \cdot 4.8$ $= 24$, or 24 m^2
[11.2]–[11.3] Rectangle	*Perimeter* $$P = 2l + 2w$$ Perimeter equals twice the length plus twice the width. *Area* $$A = lw$$ Area equals the length times the width.	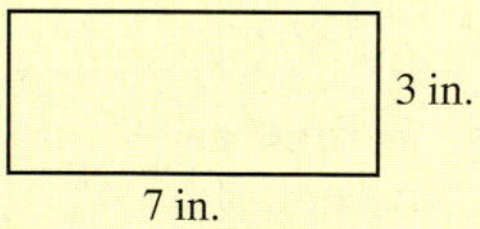 $P = 2l + 2w$ $= 2(7) + 2(3)$ $= 14 + 6$ $= 20$ or 20 in. $A = lw$ $= 7 \cdot 3$ $= 21$, or 21 in^2

Figure	Formula	Example
[11.2]–[11.3] **Square**	*Perimeter* $P = 4s$ Perimeter equals four times the length of a side. *Area* $A = s^2$ Area equals the square of a side.	$\frac{1}{2}$ in. $P = 4s$ $= 4 \cdot \frac{1}{2}$ $= 2$, or 2 in. $A = s^2$ $= \left(\frac{1}{2}\right)^2$ $= \frac{1}{4}$, or $\frac{1}{4}$ in^2
[11.3] **Parallelogram**	*Area* $A = bh$ Area equals the base times the height.	3 ft 6 ft $A = bh$ $= 6 \cdot 3$ $= 18$, or 18 ft^2
[11.3] **Trapezoid**	*Area* $A = \frac{1}{2}h(b + B)$ Area equals one-half the height times the sum of the bases.	3 in. 4 in. 5 in. $A = \frac{1}{2}h(b + B)$ $= \frac{1}{2} 4 (3 + 5)$ $= \frac{1}{\cancel{2}_1} \cdot \cancel{4}^2 \cdot 8$ $= 16$, or 16 in^2
[11.2]–[11.3] **Circle**	*Circumference* $C = \pi d$, or $C = 2\pi r$ Circumference equals π times the diameter, or 2 times π times the radius. *Area* $A = \pi r^2$ Area equals π times the square of the radius.	8 cm $C = \pi d$ $\approx 3.14(8)$ ≈ 25.12, or 25.12 cm $A = \pi r^2$ $\approx 3.14(4)^2$ $\approx 3.14(16)$ ≈ 50.24, or 50.24 cm^2 Note: $d = 8$ cm, so $r = 4$ cm.

continued

CONCEPT SKILL

Concept/Skill	Description	Example
[11.4] Rectangular solid	*Volume* $V = lwh$ Volume equals length times width times height.	5 cm, 7 cm, 15 cm. $V = lwh$ $= 15 \cdot 7 \cdot 5$ $= 525$, or 525 cm^3
[11.4] Cube	*Volume* $V = e^3$ Volume equals the cube of the edge.	2 in. $V = e^3$ $= (2)^3$ $= 2 \cdot 2 \cdot 2$ $= 8$, or 8 in^3
[11.4] Cylinder	*Volume* $V = \pi r^2 h$ Volume equals π times the square of the radius times the height.	12 m, 4 m. $V = \pi r^2 h$ $\approx 3.14(4)^2(12)$ $\approx 3.14(16)(12)$ ≈ 602.88, or 602.88 m^3
[11.4] Sphere	*Volume* $V = \frac{4}{3}\pi r^3$ Volume equals $\frac{4}{3}$ times π times the cube of the radius.	2 ft. $V = \frac{4}{3}\pi r^3$ $\approx \frac{4}{3}(3.14)(2)^3$ $\approx \frac{4}{3}(3.14)(8)$ $\approx \frac{100.48}{3}$ ≈ 33.5, or 33.5 ft^3 to the nearest tenth of a cubic foot

Chapter 11 Review Exercises

To help you review this chapter, solve these problems.

[11.1] *Sketch and label an example of each of the following.*

1. $\overline{AB}$

2. $\angle PQR$

3. Parallel lines $\overleftrightarrow{ST}$ and $\overleftrightarrow{UV}$

4. Obtuse $\triangle ABC$

Find each missing angle.

5. **6.** **7.** **8.**

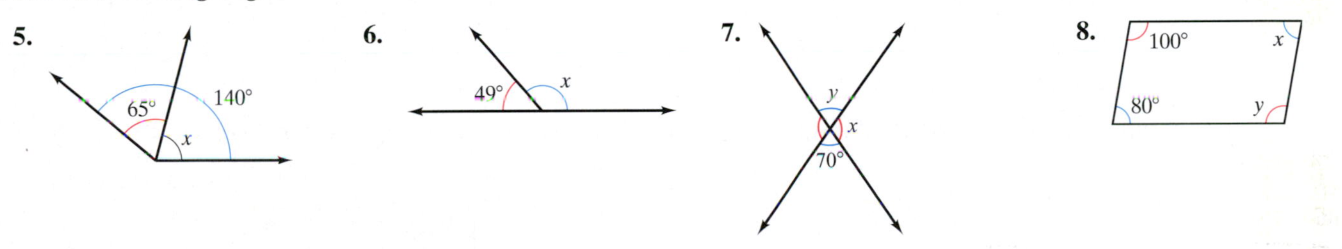

[11.2] *Find each perimeter or circumference. Use $\pi \approx 3.14$ when needed.*

9. An equilateral triangle with side 1.8 meters

10. A polygon whose side lengths are 4.5 feet, 9 feet, 7.5 feet, 3 feet, and 6 feet.

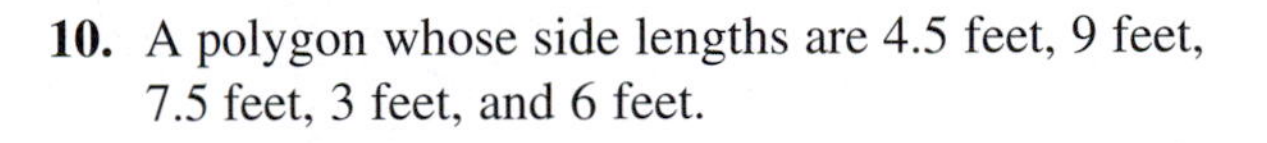

11.

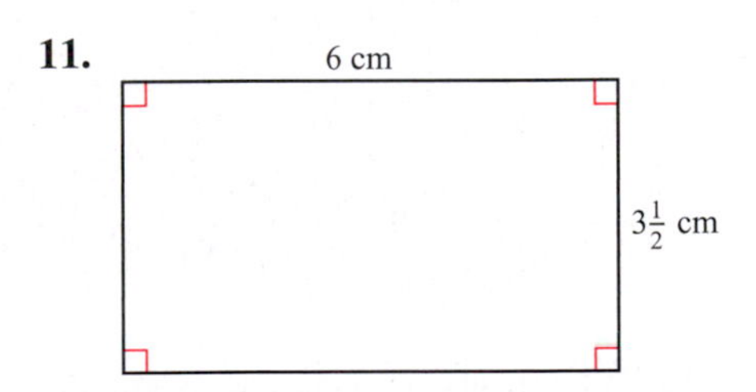

12.

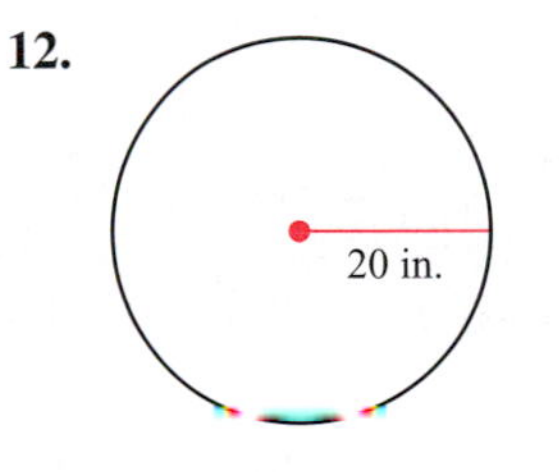

[11.3] *Find the area of each figure. Use $\pi \approx 3.14$ when needed.*

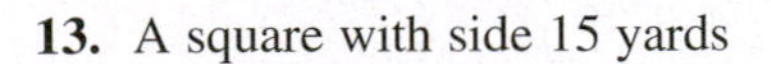

13. A square with side 15 yards

14. A circle with radius 14 feet

15.

8 in.

16 in.

16.

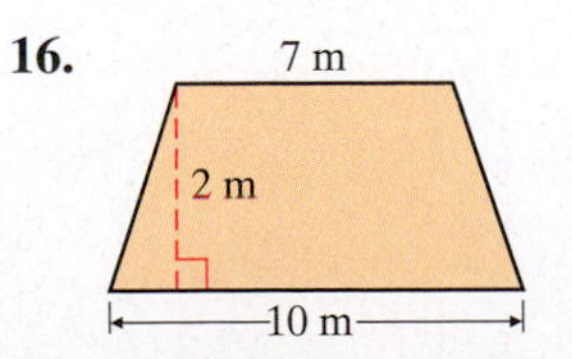

[11.4] *Find the volume of each figure. Use $\pi \approx 3.14$ when needed.*

17. A cylinder with radius 10 inches and height 4.2 inches

18. A rectangular solid with length 16 feet, width $4\frac{1}{2}$ feet, and height 3 feet

19. A cube with edge 1.25 meters

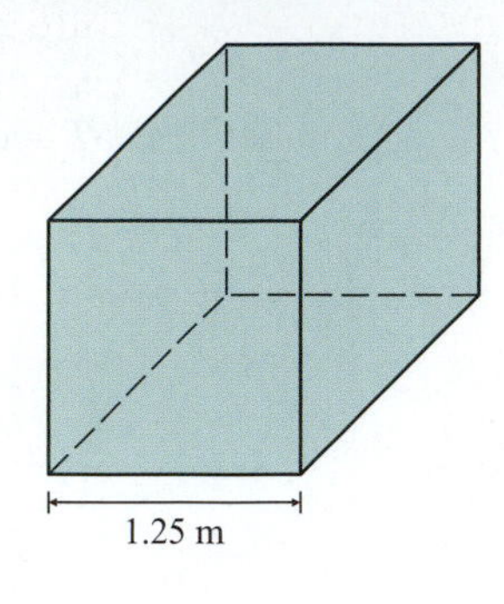

20. A sphere with diameter 2.5 centimeters

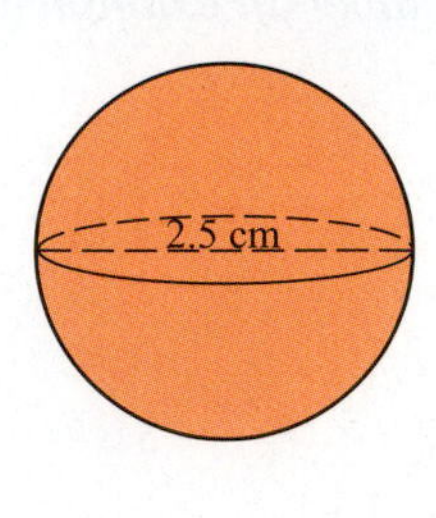

[11.2]–[11.4]

Solve.

21. Find the perimeter of the figure shown, which is made up of a semicircle and a trapezoid. Use $\pi \approx 3.14$.

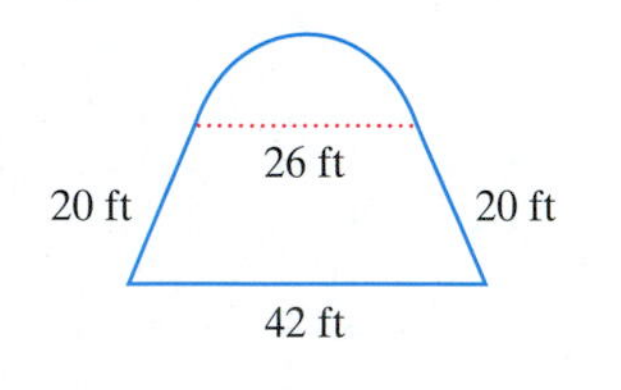

22. Find the area of the shaded portion of the figure. Use $\pi \approx 3.14$.

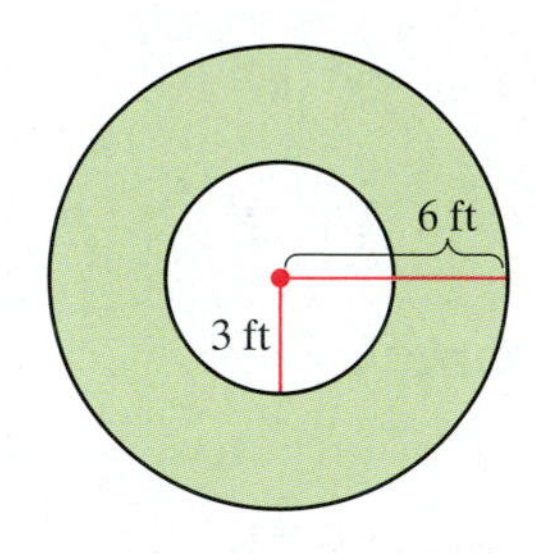

23. What is the area of the figure that consists of a square and two semicircles?

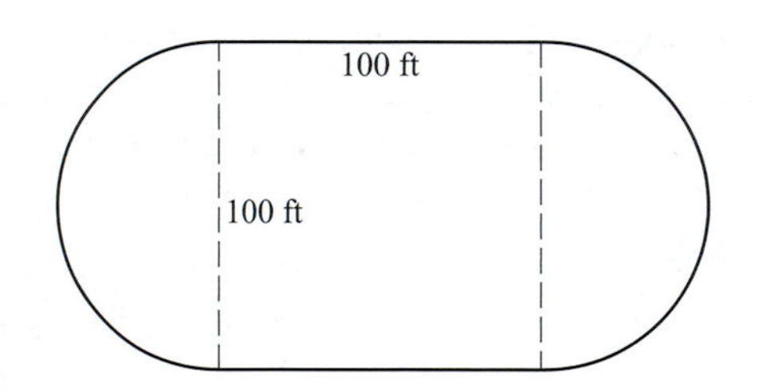

24. Find the volume of the shaded region between the sphere and the cube.

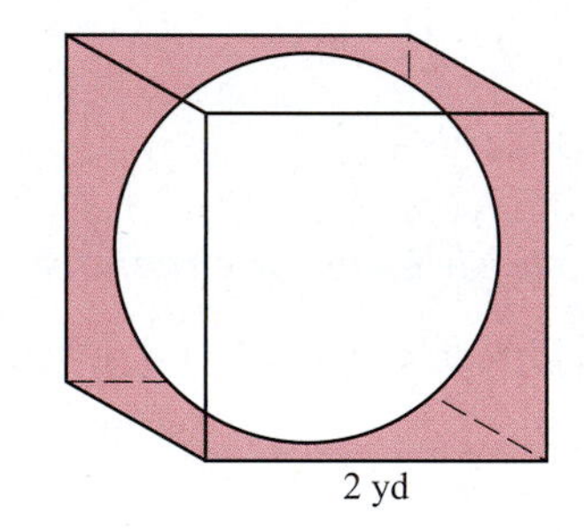

[11.5] *Find each missing length.*

25.

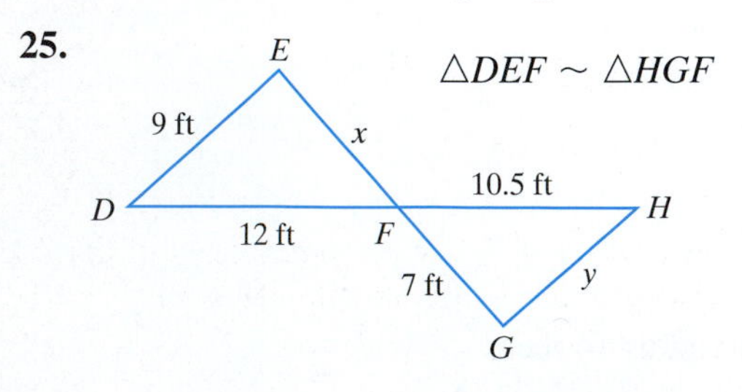

26.

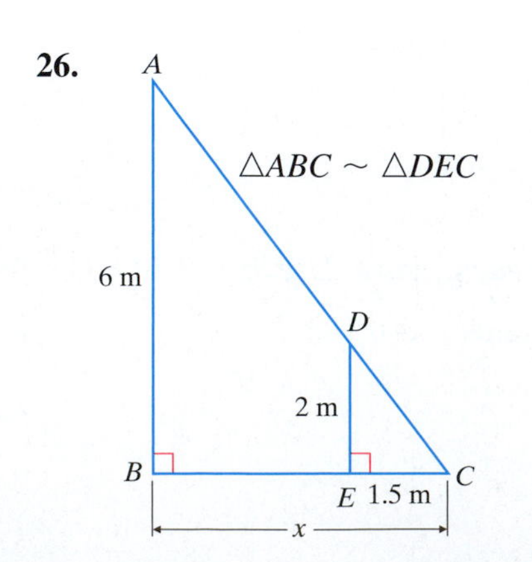

[11.6] *Find the square root.*

27. $\sqrt{9}$ **28.** $\sqrt{64}$ **29.** $\sqrt{121}$ **30.** $\sqrt{900}$

Determine between which two consecutive whole numbers each square root lies.

31. $\sqrt{3}$ **32.** $\sqrt{84}$ **33.** $\sqrt{40}$ **34.** $\sqrt{10}$

Find the square root. Round to the nearest hundredth.

35. $\sqrt{8}$ **36.** $\sqrt{1{,}235}$ **37.** $\sqrt{195}$ **38.** $\sqrt{29}$

For a given triangle with legs a and b and hypotenuse c, find the missing side. Round to the nearest tenth, if needed.

	a	*b*	*c*
39.	9 ft		15 ft
40.		24 in.	26 in.
41.	8 yd	5 yd	
42.	2 ft	2 ft	

Applications

Solve.

43. A roll of aluminum foil is 12 inches wide and 2,400 inches long. Find the area of the roll of aluminum foil.

44. Following an explosion, poisonous gases spread in all directions for 2 miles. How big was the affected area, to the nearest square mile?

45. In a couple's apartment, an air conditioner can cool a room up to 3,000 cubic feet in volume. Based on the floor plan of their living room and a ceiling height of 10 feet, can the air conditioner cool the room?

5 ft
8 ft
10 ft
15 ft

46. A high-definition LCD television has a screen with dimensions approximately 20 inches by 35 inches. Another has dimensions approximately 23 inches by 40 inches. How much greater is the area of the larger screen?

47. A pilot flies 12 miles west from city A to city B. Then he flies 5 miles south from city B to city C. What is the straight-line distance from city A to city C?

48. How high up on a wall will a 12-foot ladder reach if the bottom of the ladder is placed 6 feet from the wall? Round to the nearest foot.

49. On the pool table shown, a player hits the ball at point E. It ricochets off point C and winds up in the pocket at point A. If $\triangle ABC \sim \triangle EDC$, find CD.

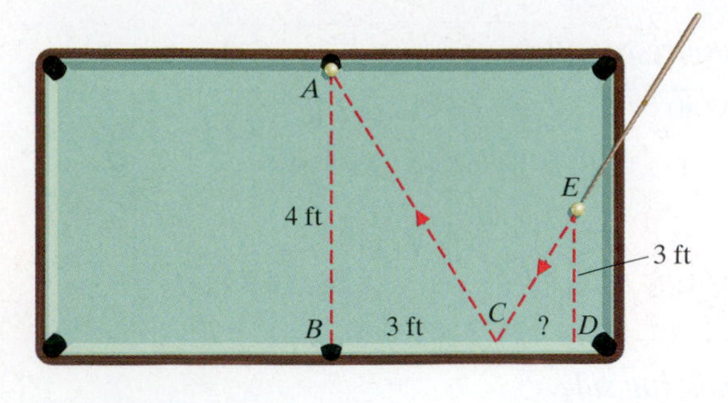

50. Use the campus map below to find the distance between the Athletic Center and the Student Center.

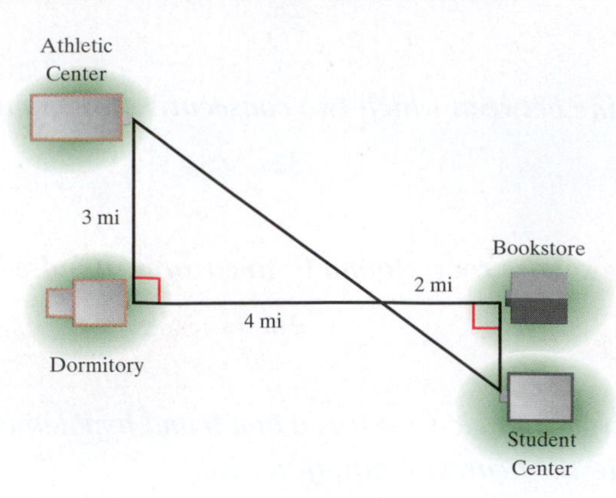

51. From the following drawing, find the total length of the building's walls.

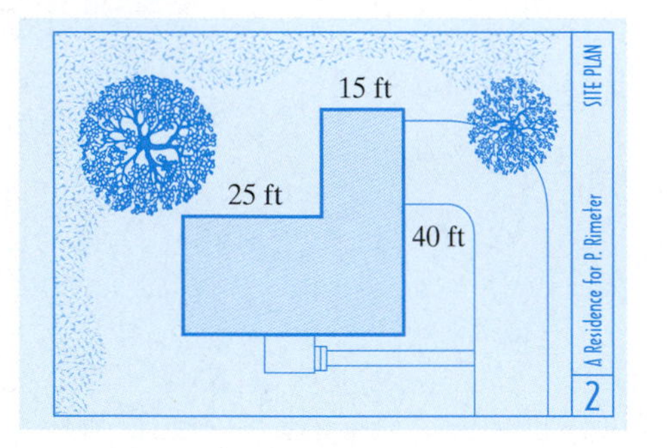

52. According to interior designers, the distances between the refrigerator, stove, and sink usually form a work triangle. To be efficient, the perimeter of a work triangle must be no more than 22 feet. Determine whether the model kitchen shown is efficient.

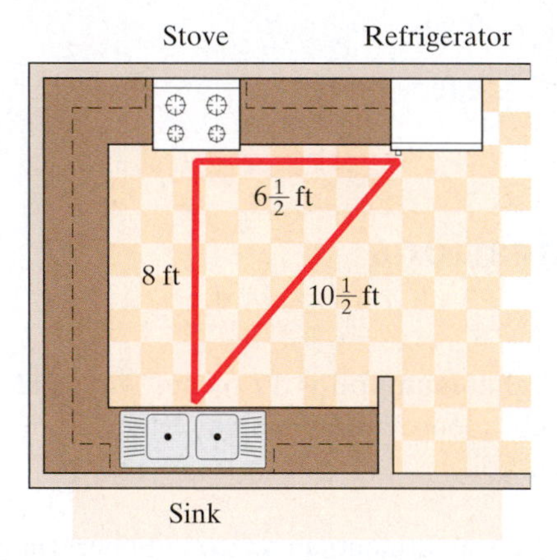

53. The coffee in this cylindrical can weighs 13 ounces.

What is the weight of a cubic inch of coffee, to the nearest tenth of an ounce?

54. How much soil is needed to fill the flower box shown to 1 centimeter from the top? Round to the nearest cubic centimeter.

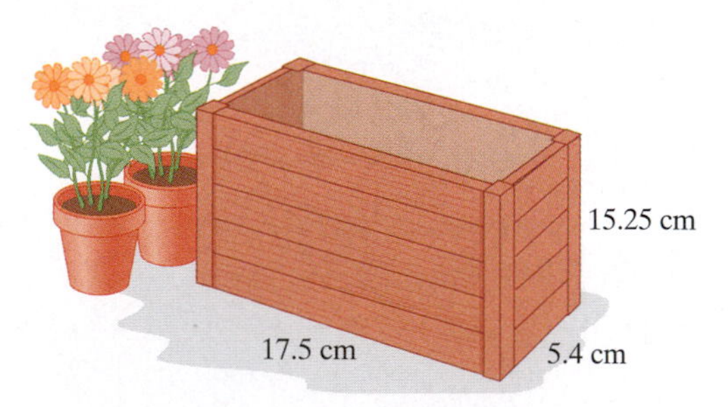

55. Find the area of the picture matting shown.

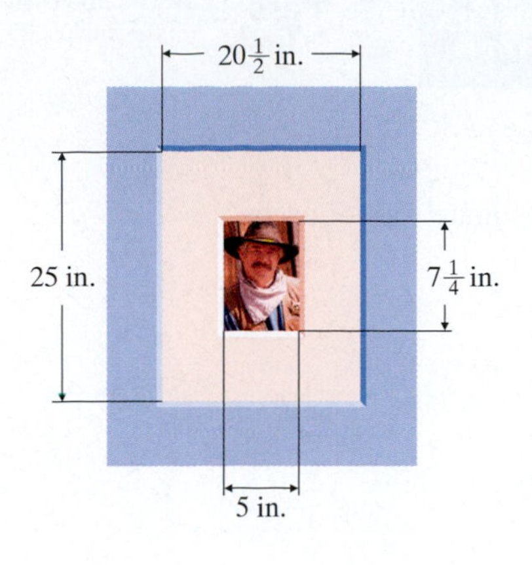

56. The game of racquetball is played with a small, hollow rubber ball, as depicted.

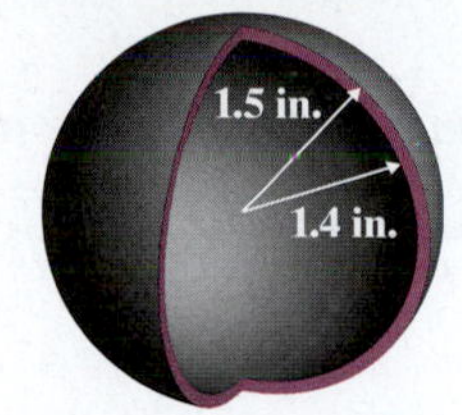

How much rubber, to the nearest tenth of a cubic inch, does a racquetball contain?

• *Check your answers on page A-18.*

Chapter 11 POSTTEST

Test solutions are found on the enclosed CD.

To see if you have mastered the topics in this chapter, take this test.

1. Sketch and label an example of each.

a. Acute $\angle PQR$

b. $\overrightarrow{AB}$

2. Find the square root of each.

a. $\sqrt{49}$

b. $\sqrt{225}$

3. Find the complement of 25°.

4. What is the measure of an angle that is supplementary to 91°?

Find each perimeter or circumference. Use $\pi \approx$ 3.14 when needed.

5. A square with side $3\frac{1}{2}$ feet

6. An equilateral triangle with side 1.5 meters

7.

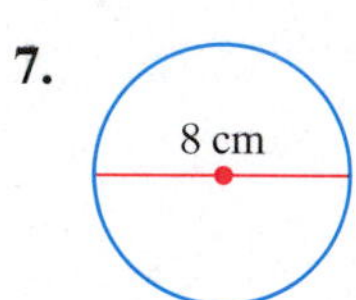

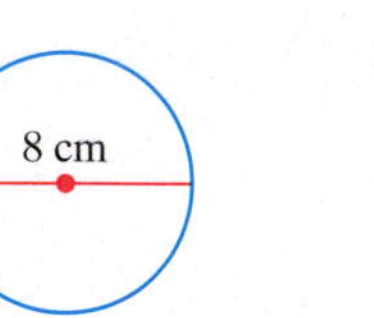

8.

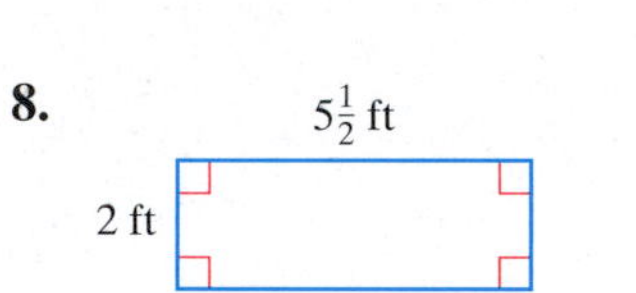

Find each area. Use $\pi \approx$ 3.14 when needed.

9.

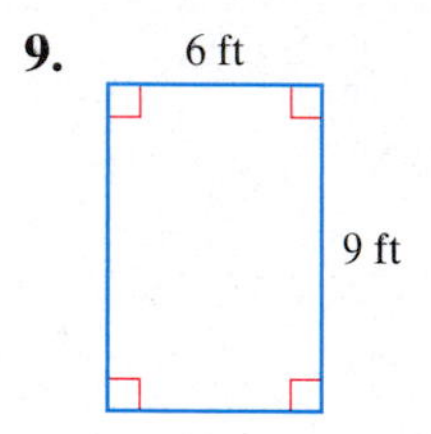

10.

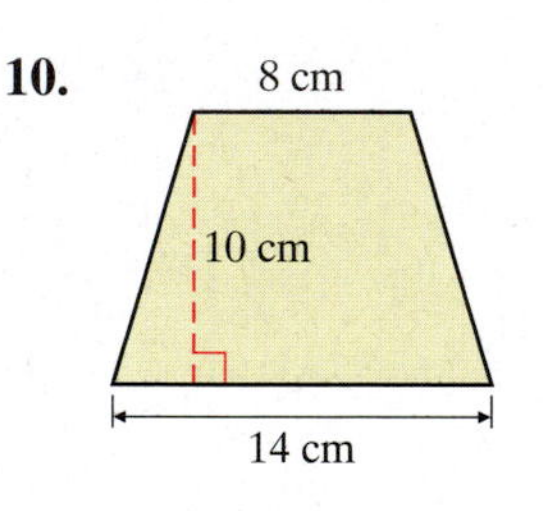

Find each volume. Use $\pi \approx$ 3.14 when needed.

11.

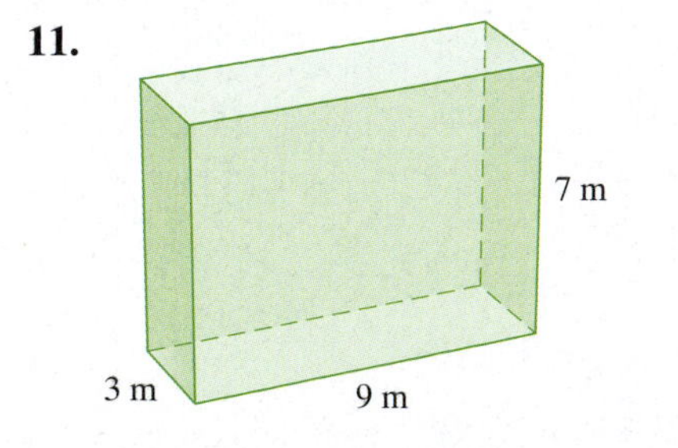

12.

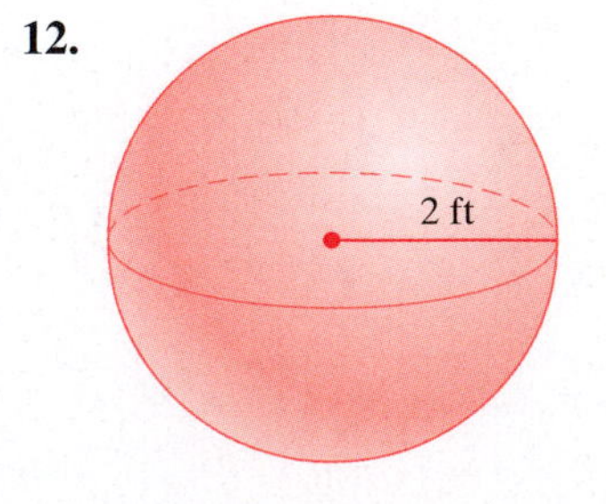

Find each unknown.

13.

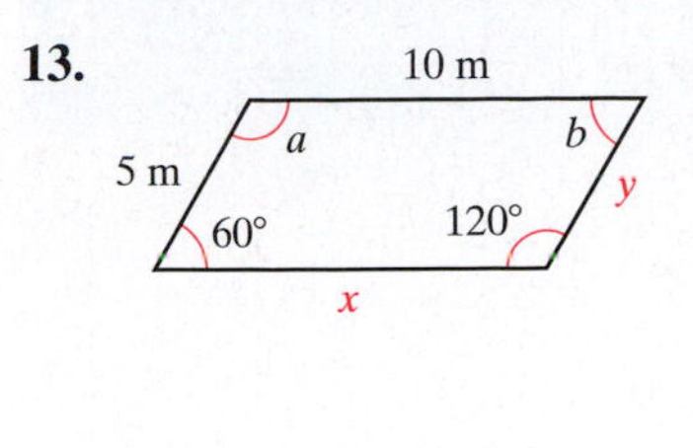

14.

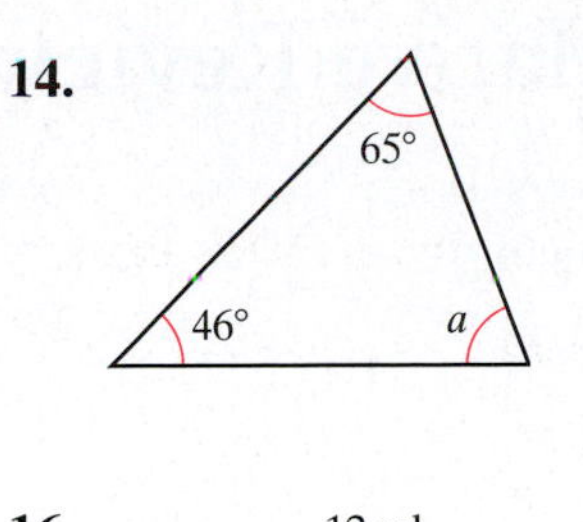

15.

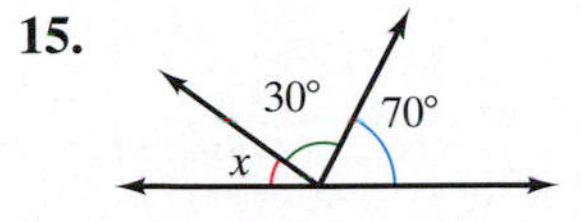

16.

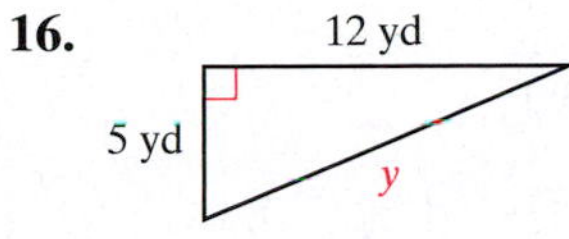

17. In the following diagram, $\triangle ABC \sim \triangle DEF$. Find x and y.

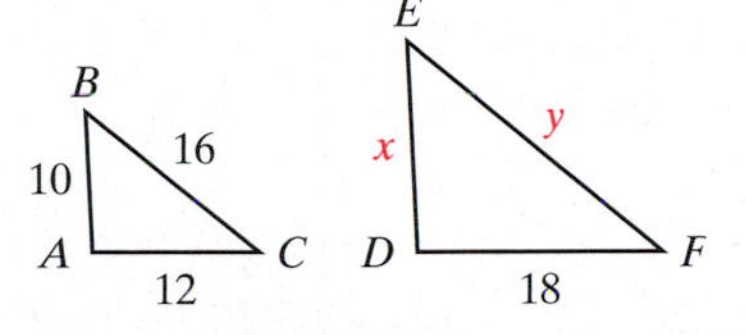

Solve.

18. In constructing the foundation for a house, a contractor digs a hole 6 feet deep, 54 feet long, and 25 feet wide. How many cubic feet of earth are removed?

19. Suppose that the chain on a dog is 5 meters long. To the nearest square meter, what is the area of the dog's "run"? Use $\pi \approx 3.14$.

20. An airplane flying due north is 200 miles from the airport. At the same time, another airplane flying due east is 150 miles from the airport. How far apart are the two airplanes at that time?

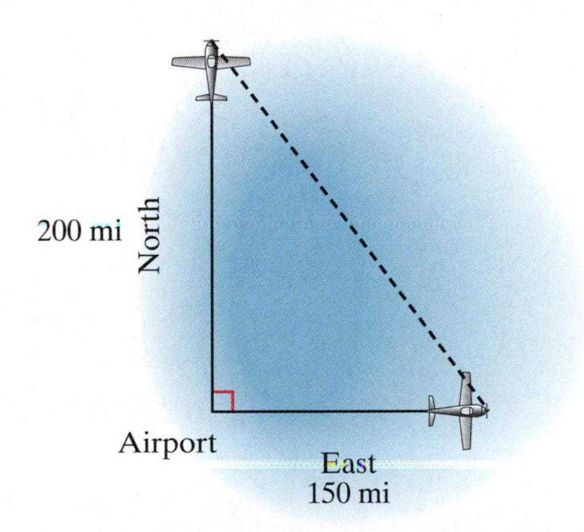

• *Check your answers on page A-18.*

Cumulative Review Exercises

To help you review, solve the following.

1. Estimate: $23{,}802 \div 396$

2. Simplify: $\dfrac{24}{30}$

3. Round to the nearest tenth: 3.061

4. 40% of what number is 20?

5. Evaluate: $(-4)(-2) + 7$

6. Solve for y: $\dfrac{y}{12} = \dfrac{2}{3}$

7. A typical hurricane has a calm, circular center, called the eye, which is approximately 30 miles across. What is the area that the eye covers?

8. Find the width of the river pictured. The triangles are similar.

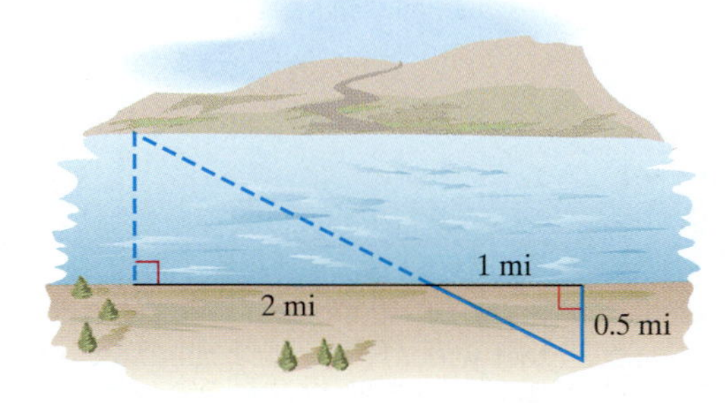

9. A common liquid insecticide for houseplants requires mixing 50 parts of water with every part of insecticide. What fraction of the mixture is insecticide?

10. The graph shows the number of identity-theft complaints reported in the United States for the years from 2001 to 2005. About how many more complaints were there in 2005 than in 2001? (***Source:*** Federal Trade Commission)

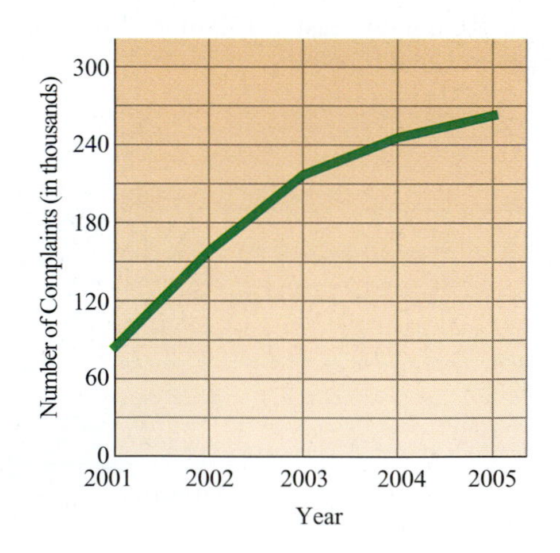

• *Check your answers on page A-18.*

APPENDIX

Scientific Notation

Frequently, scientists deal with numbers that are either very large or very small. For instance, in astronomy, they study the distance to the nearest star; in biology, the length of a virus.

40,000,000,000,000 kilometers ← **The distance between the Sun and Proxima Centauri**

0.0000000000001 meters ← **The length of a virus**

Scientists commonly write such numbers not in standard notation, but rather in *scientific notation*. Also, scientific and graphing calculators generally show answers in scientific notation that are too long to fit in their display.

Scientific notation is based on powers of 10. ***A number is said to be in scientific notation if it is written as the product of a decimal between 1 and 10 (not including 10) and a power of 10.***

For instance, the number 7.35×10^5 is written in scientific notation, because the decimal factor 7.35 is between 1 and 10, and 10^5 is a power of 10. However, the number 81.45×2^6 is not in scientific notation, because the base of the exponent is not 10, and the decimal factor 81.45 is not less than 10.

Tip A number written in scientific notation has one nonzero digit in its decimal factor.

Let's consider how to change a number from scientific notation to standard notation and vice versa. First, we will look at *large numbers*.

EXAMPLE 1

Change the number 7.35×10^5 from scientific notation to standard notation.

Solution To express 7.35×10^5 in standard notation, we need to multiply 7.35 by 10^5 or 100,000.

$$7.35 \times 10^5 = 7.35 \times 100{,}000 = 735{,}000.$$

So 7.35×10^5 written in standard notation is 735,000.

PRACTICE 1

Express 2.539×10^7 in standard notation.

In Example 1, note that the power of 10 is *positive* and that the decimal point is moved five places *to the right*. So a shortcut for writing 7.35×10^5 in standard notation is to move the decimal point in 7.35 five places to the right.

$$7.35 \times 10^5 = 7\underbrace{35000} = 735{,}000$$

To change a number from standard notation to scientific notation, the process is reversed.

EXAMPLE 2

Rewrite 37,000,000,000 in scientific notation.

Solution We know that for a number to be written in scientific notation, it must be the product of a decimal number between 1 and 10 and a power of 10. Recall that 37,000,000,000 and 37,000,000,000. are the same. We move the decimal point *to the left* so that there is one nonzero digit to its left. The number of places moved is the power of 10 by which we need to multiply.

$$37{,}000{,}000{,}000. = 3.\underbrace{7000000000} \times 10^{10} = 3.7 \times 10^{10}$$

Note that we dropped the extra zeros in 3.7000000000. So 37,000,000,000 expressed in scientific notation is 3.7×10^{10}.

PRACTICE 2

Write 8,000,000,000,000 in scientific notation.

Now, let's turn our attention to writing *small numbers* in scientific notation. The key is an understanding of *negative exponents*. Until now, we have only considered exponents that were either positive integers or 0. What meaning should we attach to a negative exponent? The following pattern, in which each number is $\frac{1}{10}$ of the previous number, suggests an answer.

$$10^3 = 1{,}000$$
$$10^2 = 100$$
$$10^1 = 10$$
$$10^0 = 1$$
$$10^{-1} = \frac{1}{10}$$
$$10^{-2} = \frac{1}{100}$$
$$10^{-3} = \frac{1}{1{,}000}$$

Notice that 10^{-1}, or $\frac{1}{10}$, is the reciprocal of 10^1 or 10. So in general, ***a number raised to a negative exponent is defined to be the reciprocal of that number raised to the corresponding positive exponent***.

When written in scientific notation, large numbers have positive powers of 10, whereas small numbers have negative powers of 10. For instance, 3×10^5 is large, but 3×10^{-5} is small.

Next, let's look at how we change *small* numbers from scientific notation to standard notation and vice versa.

EXAMPLE 3

Convert 3×10^{-5} to standard notation.

Solution Using the meaning of negative exponents, we get:

$$3 \times 10^{-5} = 3 \times \frac{1}{10^5}, \quad \text{or} \quad \frac{3}{10^5}$$

Since $10^5 = 100{,}000$, dividing 3 by 10^5 gives us:

$$\frac{3}{10^5} = \frac{3}{100{,}000} = 0.00003$$

So 3×10^{-5} written in standard notation is 0.00003.

PRACTICE 3

Change 4.3×10^{-9} to standard notation.

In Example 3, note that the power of 10 is *negative* and the decimal point, which is understood to be at the right end of a whole number, was moved five places *to the left*. So just as with 7.35×10^5, there is a shortcut for expressing 3×10^{-5} in standard notation. To do this, we move the decimal point five places *to the left*:

$$3 \times 10^{-5} = 3. \times 10^{-5} = .00003 = \mathbf{.00003}, \text{ or } 0.00003.$$

Tip When converting a number from scientific notation to standard notation, move the decimal point to the *left* if the power of 10 is *negative* and to the *right* if the power of 10 is *positive*.

EXAMPLE 4

Write 0.00000000000000002 in scientific notation.

Solution To write 0.00000000000000002 in scientific notation, we move the decimal point *to the right* until there is one nonzero digit to the left of the decimal point. The number of places moved, preceded by a *negative* sign, is the power of 10 that we need.

$$0.00000000000000002 = \mathbf{00000000000000002.} \times 10^{-17}$$
$$= 2 \times 10^{-17}$$

PRACTICE 4

Express 0.000000000071 in scientific notation.

Computation Involving Scientific Notation

Now we consider how to perform calculations on numbers written in scientific notation. We focus on the operations of multiplication and division.

Multiplying and dividing numbers written in scientific notation can best be understood in terms of two *laws of exponents*—the *product rule* and the *quotient rule*.

- The *product rule of exponents* states that when we multiply a base raised to a power by the same base raised to another power, we add the exponents and leave the base the same. For example,

$$10^3 \cdot 10^2 = 10^{3+2} = 10^5$$

Add the exponents.

Keep the base.

This result is reasonable, since $10^3 \times 10^2 = 1{,}000 \times 100 = 100{,}000 = 10^5$.

- The *quotient rule of exponents* states that when we divide a base raised to a power by the same base to another power, we subtract the second power from the first power, and leave the base the same. For instance,

Subtract the exponents.

$$10^5 \div 10^2 = 10^{5-2} = 10^3$$

Keep the base.

We would have expected this result even if we did not know the quotient rule, since $\dfrac{10^5}{10^2} = \dfrac{100{,}000}{100} = \dfrac{1{,}000}{1} = 1{,}000 = 10^3$.

EXAMPLE 5

Calculate, writing the result in scientific notation:

a. $(4 \times 10^{-1})(2.1 \times 10^6)$

b. $(1.2 \times 10^5) \div (2 \times 10^{-4})$

Solution

a. $(4 \times 10^{-1})(2.1 \times 10^6) = (4 \times 2.1)(10^{-1} \times 10^6)$ Regroup the factors.

$= (8.4)(10^{-1} \times 10^6)$ Multiply the decimal factors.

$= 8.4 \times 10^{-1+6}$ Use the product rule of exponents.

$= 8.4 \times 10^5$ Simplify.

b. $(1.2 \times 10^5) \div (2 \times 10^{-4}) = \dfrac{1.2 \times 10^5}{2 \times 10^{-4}}$

$= \dfrac{1.2}{2} \times \dfrac{10^5}{10^{-4}}$ Write as the product of fractions.

$= 0.6 \times \dfrac{10^5}{10^{-4}}$ Divide the decimal factors.

$= 0.6 \times 10^{5-(-4)}$ Use the quotient rule of exponents.

$= 0.6 \times 10^9$ Simplify.

Note that 0.6×10^9 is not written in scientific notation, because 0.6 is not between 1 and 10, that is, it does not have one nonzero digit to the left of the decimal point. To write 0.6×10^9 in scientific notation, we convert 0.6 to scientific notation and simplify the product.

$0.6 \times 10^9 = 6 \times 10^{-1} \times 10^9$

$= 6 \times 10^{-1+9}$ Use the product rule of exponents.

$= 6 \times 10^8$

So the quotient, written in scientific notation, is 6×10^8.

PRACTICE 5

Calculate, expressing the answer in scientific notation:

a. $(7 \times 10^{-2})(3.52 \times 10^3)$

b. $(5.01 \times 10^3) \div (6 \times 10^{-9})$

Exercises

FOR EXTRA HELP MyMathLab Math XL PRACTICE WATCH DOWNLOAD READ REVIEW

Express in scientific notation.

1. 400,000,000
2. 10,000,000
3. 0.0000035
4. 0.00017
5. 0.00000000031
6. 218,000,000,000

Express in standard notation.

7. 3.17×10^8
8. 9.1×10^5
9. 1×10^{-6}
10. 8.013×10^{-4}
11. 4.013×10^{-5}
12. 2.1×10^{-3}

Multiply, and write the result in scientific notation.

13. $(3 \times 10^2)(3 \times 10^5)$
14. $(5 \times 10^6)(1 \times 10^3)$
15. $(2.5 \times 10^{-2})(8.3 \times 10^{-3})$
16. $(2.1 \times 10^4)(8 \times 10^{-4})$

Divide, and write the result in scientific notation.

17. $(2.5 \times 10^8) \div (2 \times 10^{-2})$
18. $(3.0 \times 10^4) \div (1 \times 10^3)$
19. $(1.2 \times 10^5) \div (3 \times 10^3)$
20. $(4.88 \times 10^{-3}) \div (8 \times 10^2)$

Answers

Chapter 1

Pretest: Chapter 1, *p. 2*

1. Two hundred five thousand, seven **2.** 1,235,000 **3.** Hundred thousands **4.** 8,100 **5.** 8,226 **6.** 4,714 **7.** 185 **8.** 29,124 **9.** 260 **10.** 308 R6 **11.** 2^3 **12.** 36 **13.** 5 **14.** 43 **15.** 75 years old **16.** $675 **17.** 69 **18.** 156 sec **19.** $27 **20.** Room C, which measures 126 sq ft

Practices: Section 1.1, *pp. 4–8*

1, ***p. 4:*** **a.** Thousands **b.** Hundred thousands **c.** Ten millions **2,** ***p. 4:*** Eight billion, three hundred seventy-six thousand, fifty-two **3,** ***p. 4:*** $7,372,050 Seven million, three hundred seventy-two thousand, fifty dollars **4,** ***p. 5:*** $95,000,003 **5,** ***p. 5:*** $375,000 **6,** ***p. 6:*** **a.** 2 ten thousands + 7 thousands + 0 hundreds + 1 ten + 3 ones = 20,000 + 7,000 + 0 + 10 + 3 **b.** 1 million + 2 hundred thousands + 7 ten thousands + 9 tens + 3 ones = 1,000,000 + 200,000 + 70,000 + 90 + 3 **7,** ***p. 7:*** **a.** 52,000 **b.** 50,000 **8,** ***p. 8:*** 420,000,000 **9,** ***p. 8:*** **a.** Two hundred forty-eight thousand, seven hundred eighty-eight **b.** 400,000

Exercises 1.1, *pp. 9–13*

1. whole numbers **3.** odd **5.** standard form **7.** placeholder **9.** expanded form **11.** 4,867 **13.** 316 **15.** 28,461,013 **17.** Hundred thousands **19.** Hundreds **21.** Billions **23.** Four hundred eighty-seven thousand, five hundred **25.** Two million, three hundred fifty thousand **27.** Nine hundred seventy-five million, one hundred thirty-five thousand **29.** Two billion, three hundred fifty-two **31.** One billion **33.** 10,120 **35.** 150,856 **37.** 6,000,055 **39.** 50,600,195 **41.** 400,072 **43.** 3 ones = 3 **45.** 8 hundreds + 5 tens + 8 ones = 800 + 50 + 8 **47.** 2 millions + 5 hundred thousands + 4 ones = 2,000,000 + 500,000 + 4 **49.** 670 **51.** 7,100 **53.** 30,000 **55.** 700,000 **57.** 30,000

59.

To the nearest	135,842	2,816,533
Hundred	135,800	2,816,500
Thousand	136,000	2,817,000
Ten thousand	140,000	2,820,000
Hundred thousand	100,000	2,800,000

61. 1 ten thousand + 2 thousands + 5 tens + 1 one = 10,000 + 2,000 + 50 + 1 **63.** 40,059 **65.** 1,056,100; one million, fifty-six thousand, one hundred **67.** Nine hundred thousand **69.** Eight thousand, nine hundred fifty-nine **71.** Thirty-seven thousand, eight hundred forty-two **73.** 100,000,000,000 **75.** 3,288 **77.** 2,908,000 **79.** 150 ft **81.** 20,000 mi **83.** 500 g **85.** **a.** One million, three hundred ninety-nine thousand, five hundred forty-two **b.** 700,000

Practices: Section 1.2, *pp. 15–22*

1, ***p. 15:*** 385 **2,** ***p. 16:*** 10,436 **3,** ***p. 17:*** 16 mi **4,** ***p. 18:*** 651 **5,** ***p. 19:*** 4,747 **6,** ***p. 19:*** 765 plant species **7,** ***p. 21:*** **a.** 128,000 **b.** 233,000 **c.** Less **8,** ***p. 22:*** 9,477 **9,** ***p. 22:*** 2,791 **10,** ***p. 22:*** 87,000 mi

Calculator Practices, *p. 23*

11, ***p. 23:*** 49,532 **12,** ***p. 23:*** 31,899 **13,** ***p. 23:*** 2,499 ft

Exercises 1.2, *pp. 24–30*

1. right **3.** sum **5.** Associative Property of Addition **7.** subtrahend **9.** 177,778 **11.** 14,710 **13.** 14,002 **15.** 56,188 **17.** 6,978 **19.** 4,820 **21.** 413 **23.** 14,865 **25.** 15,509 m **27.** 82 hr **29.** $104,831 **31.** $12,724 **33.** 31,200 tons **35.** 13,296,657 **37.** 22,912,891

39.

+	400	200	1,200	300	Total
300	700	500	1,500	600	3,300
800	1,200	1,000	2,000	1,100	5,300
Total	1,900	1,500	3,500	1,700	8,600

41.

+	389	172	1,155	324	Total
255	644	427	1,410	579	3,060
799	1,188	971	1,954	1,123	5,236
Total	1,832	1,398	3,364	1,702	8,296

43. a; possible estimate: 12,800 **45.** a; possible estimate: $900,000 **47.** 217 **49.** 90 **51.** 362 **53.** 68,241 **55.** 2,285 **57.** 52,999 **59.** 2,943 **61.** 203,465 **63.** 368 **65.** 4,996 **67.** 982 **69.** 1,995 mi **71.** $669 **73.** $3,609 **75.** 273 books **77.** 209 m **79.** 2,001,000 **81.** 813,429 **83.** c; possible estimate: 40,000,000 **85.** a; possible estimate: $200,000 **87.** 7,065 **89.** 1,676 **91.** 5,186 **93.** 281,000,000 **95.** 2,600,000 sq mi **97. a.** Austria: 23; Canada: 24; Germany: 29; Russia: 22; United States: 25 **b.** Germany **99.** About 43 years old **101.** No, the elevator is not overloaded. The total weight of passengers is 963 lb. **103.** 180°F **105.** 12 mi **107.** 36 hr **109.** $2,951 **111. a.** No **b.** Yes **113. a.** 347,000,000 lb **b.** Possible estimate: 2,300,000,000 lb **c.** 2,260,000,000 lb **115.** (a) Addition; (b) yes, $1,563; (c) possible estimate: $1,600

Practices: Section 1.3, *pp. 34–37*

1, ***p. 34:*** 608 **2,** ***p. 34:*** 4,230 **3,** ***p. 35:*** 480,000 **4,** ***p. 35:*** 205,296 **5,** ***p. 36:*** 107 sq ft **6,** ***p. 36:*** 112,840 **7,** ***p. 37:*** No; possible estimate = 20,000

Calculator Practices, *p. 38*

8, ***p. 38:*** 1,026,015; **9,** ***p. 38:*** 345,546;

Exercises 1.3, *pp. 39–42*

1. product **3.** Identity Property of Multiplication **5.** addition **7.** 400 **9.** 142,000 **11.** 170,000 **13.** 7,000,000 **15.** 12,700 **17.** 418 **19.** 3,248,000 **21.** 65,268 **23.** 817 **25.** 34,032 **27.** 3,003 **29.** 3,612 **31.** 57,019 **33.** 243,456 **35.** 220,120 **37.** 149,916 **39.** 144,500 **41.** 123,830 **43.** 3,312 **45.** 2,106 **47.** 40,000 **49.** 23,085 **51.** 3,274,780 **53.** 54,998,850 **55.** c; possible estimate: 480,000 **57.** b; possible estimate: 80,000 **59.** 2,880 **61.** 230,520 **63.** 1,071,000 **65.** 300,000 **67.** 3,300 yr **69. a.** 3,000,000 **b.** 1,000,000 **71.** Yes **73.** 5,775 sq in. **75.** 1,750 mi **77.** $442 **79. a.** 294 mi **b.** 1,470 mi **81.** (a) Multiplication (b) Colorado; area ≈ 106,700 sq mi (c) possible estimate: 120,000 sq mi

Practices: Section 1.4, *pp. 47–51*

1, ***p. 47:*** 807 **2,** ***p. 47:*** 7,002 **3,** ***p. 48:*** 5,291 R1 **4,** ***p. 49:*** 79 R1 **5,** ***p. 49:*** 94 R10 **6,** ***p. 50:*** 607 R3 **7,** ***p. 50:*** 200 **8,** ***p. 51:*** 967 **9,** ***p. 51:*** 5 times

Calculator Practice, *p. 52:* 603

Exercises 1.4, *p. 53*

1. divisor **3.** multiplication **5.** 400 **7.** 560 **9.** 301 **11.** 3,003 **13.** 202 **15.** 500 **17.** 30 **19.** 14 **21.** 42 **23.** 400 **25.** 159 **27.** 5,353 **29.** 1,002 **31.** 6,944 **33.** 1,001 **35.** 3,050 **37.** 651 R2 **39.** 11 R7 **41.** 116 R83 **43.** 700 R2 **45.** 723 R19 **47.** 428 R8 **49.** 721 **51.** 155 **53.** a; possible estimate: 7,000 **55.** a; possible estimate: 400 **57.** 907 R1 **59.** 2,000 **61.** 2,400 **63.** 370 **65.** $135 **67.** 2 times **69.** 300 people per square mile **71.** 6 calories **73. a.** 304 tiles **b.** 26 boxes **c.** $468 **75.** (a) Division, (b) more than 4 times, since the quotient of 1,306,313,800 and 295,734,100 is 4 with a remainder, (c) possible estimate of the quotient: 4 with a remainder, so that China's population is more than 4 times that of the United States.

Practices: Section 1.5, *pp. 57–62*

1, ***p. 57:*** $5^5 \cdot 2^2$ **2,** ***p. 58:*** **a.** 1 **b.** 1,331 **3,** ***p. 58:*** 784 **4,** ***p. 58:*** 10^9 **5,** ***p. 59:*** 28 **6,** ***p. 60:*** 146 **7,** ***p. 60:*** 4 **8,** ***p. 60:*** 23 **9,** ***p. 61:*** 60 ft **10,** ***p. 61:*** $40 **11,** ***p. 62:*** **a.** 46 fatalities **b.** 2004 and 2005

Calculator Practices, *p. 63*

12, ***p. 63:*** 140,625; **13,** ***p. 63:*** 131

Exercises 1.5, *pp. 64–68*

1. base **3.** adding

5.

n	0	2	4	6	8	10	12
n^2	0	4	16	36	64	100	144

7.

n	0	2	4	6	8
n^3	0	8	64	216	512

9. 10^2 **11.** 10^4 **13.** 10^6 **15.** $2^2 \cdot 3^2$ **17.** $4^3 \cdot 5^1$ **19.** 900 **21.** 1,568 **23.** 18 **25.** 2 **27.** 35 **29.** 343 **31.** 250 **33.** 36 **35.** 8 **37.** 92 **39.** 60 **41.** 28 **43.** 6 **45.** 99 **47.** 99 **49.** 4 **51.** 39 **53.** 16 **55.** 93 **57.** 18 **59.** 419 **61.** 137,088

63. $\boxed{4} \cdot 3 + \boxed{6} \cdot 5 + \boxed{8} \cdot 7 = 98$

65. $(\boxed{8})(3 + \boxed{4}) - 2 \cdot \boxed{6} = 44$

67. $\boxed{8} + 10 \times \boxed{4} - \boxed{6} \div 2 = 45$

69. $(5 + 2) \cdot 4^2 = 112$ **71.** $(5 + 2 \cdot 4)^2 = 169$ **73.** $(8 - 4) \div 2^2 = 1$ **75.** 242 sq cm **77.** 3,120 sq in.

79.

Input	Output
0	$21 + 3 \times \mathbf{0} = 21$
1	$21 + 3 \times \mathbf{1} = 24$
2	$21 + 3 \times \mathbf{2} = 27$

81. 25 **83.** 40 **85.** 4 **87.** 2,412 mi **89.** 8 **91.** 10^8 **93.** 289 **95.** 48 **97.** 8 **99.** 625 sq ft **101.** $5^2 + 12^2 = 13^2$; $25 + 144 = 169$ **103.** 10^6 **105. a.** $21,500 **b.** $1,050 **107. a.** 69 **b.** At home; the average score for home games was higher than the average score for away games. **109. a.** Broadcast television:

791 hr; cable and satellite television: 976 hr **b.** Cable and satellite television; by 185 hr. **111.** (a) Addition, division, and subtraction. (b) The daily average circulation of newspaper B was greater by 11,553. (c) possible estimate: 10,000

Practices: Section 1.6, *pp. 71–73*

1, *p. 71:* 10,670 employees **2, *p. 72:*** 2 yr **3, *p. 72:*** 1,551 students **4, *p. 73:*** 180 lb

Exercises 1.6, *pp. 74–75*

1. $2,150 **3.** 27 mi **5.** 75 times **7.** 5,882 mi **9.** 528,179 immigrants **11.** 300¢, or $3 **13.** $17,000 **15.** $6,036 **17.** $1,458 **19.** 8 extra pens **21.** 1952 was closer by 31 votes. **23.** (a) Subtraction and division, (b) $983, (c) possible estimate: $1,000

Review Exercises: Chapter 1, *pp. 79–83*

1. Ones **2.** Ten thousands **3.** Hundred millions **4.** Ten billions **5.** Four hundred ninety-seven **6.** Two thousand, fifty **7.** Three million, seven **8.** Eighty-five billion **9.** 251 **10.** 9,002 **11.** 14,000,025 **12.** 3,000,003,000 **13.** 2 millions + 5 hundred thousands = 2,000,000 + 500,000 **14.** 4 ten thousands + 2 thousands + 7 hundreds + 7 ones = 40,000 + 2,000 + 700 + 7 **15.** 600 **16.** 1,000 **17.** 380,000 **18.** 70,000 **19.** 9,486 **20.** 65,692 **21.** 173,543 **22.** 150,895 **23.** 1,957,825 **24.** $223,067 **25.** 445 **26.** 10,016 **27.** 11,109 **28.** 5,510 **29.** 11,042,223 **30.** $2,062,852 **31.** 11,006 **32.** 2,989 **33.** 432 **34.** 1,200 **35.** 149,073 **36.** 12,000,000 **37.** 477,472 **38.** 1,019,000 **39.** 1,397,508 **40.** 188,221,590 **41.** 39 **42.** 307 R3 **43.** 37 R10 **44.** 680 R8 **45.** 25,625 **46.** 957 **47.** 343 **48.** 1 **49.** 72 **50.** 300,000 **51.** 5 **52.** 169 **53.** 5 **54.** 19 **55.** 12 **56.** 18 **57.** 10,833,312 **58.** 2,694 **59.** $7^2 \cdot 5^2$ **60.** $2^2 \cdot 5^3$ **61.** 39 **62.** 7 **63.** 6 **64.** 5 **65.** Two million, four hundred thousand **66.** 150,000,000 **67.** $307 per week **68.** 1758 **69.** 300,000 sq mi **70.** 22,000,000 iPods **71.** 9 **72.** 272 legs **73.** 509 m **74.** 23 flats **75.** $604,015,000,000

76.

Net sales	$430,000
− Cost of merchandise sold	− 175,000
Gross margin	$255,000
− Operating expenses	− 135,000
Net profit	$120,000

77. 6,675 sq m **78.** Possible answer: 20 **79.** 1968 to 1972 (15,379,754 votes) **80.** 4,341 points **81. a.** 1,832 km **b.** 1,800 km **82. a.** 12,677,500 **b.** The average would increase by 239,475 **83.** 29 sq mi **84.** 162 cm

Posttest: Chapter 1, *p. 84*

1. 225,067 **2.** 1,768,405 **3.** One million, two hundred five thousand, seven **4.** 200,000 **5.** 1,894 **6.** 607 **7.** 147 **8.** 297,496 **9.** 509 **10.** 622 R19 **11.** 625 **12.** $4^3 \cdot 5^2$ **13.** 84 **14.** 2 **15.** 5,700,000 sq mi **16.** 46,848,000 acres **17.** $469 **18.** Below; $1,000,000 is smaller than the average, which was $1,380,468. **19.** $1,380 **20.** 12 mg

Chapter 2

Pretest: Chapter 2, *p. 86*

1. 1, 2, 4, 5, 10, 20 **2.** $2 \times 2 \times 2 \times 3 \times 3$, or $2^3 \times 3^2$ **3.** $\frac{2}{5}$ **4.** $\frac{61}{3}$ **5.** $1\frac{1}{30}$ **6.** $\frac{3}{4}$ **7.** 20 **8.** $\frac{1}{8}$ **9.** $1\frac{1}{5}$ **10.** $12\frac{5}{6}$ **11.** $2\frac{1}{4}$ **12.** $4\frac{5}{8}$ **13.** $3\frac{1}{2}$ **14.** 60 **15.** $\frac{2}{3}$ **16.** $3\frac{2}{3}$ **17.** $\frac{1}{8}$ **18.** 6 students **19.** $20\frac{7}{8}$ ft **20.** 66 g

Practices: Section 2.1, *pp. 87–93*

1, *p. 87:* 1, 7 **2, *p. 88:*** 1, 3, 5, 15, 25, 75 **3, *p. 89:*** 1, 2, 3, 5, 6, 9, 10, 15, 18, 30, 45, 90 **4, *p. 89:*** Yes; 24 is a multiple of 3. **5, *p. 90:*** **a.** Prime **b.** Composite **c.** Prime **d.** Composite **e.** Prime **6, *p. 91:*** $2^3 \times 7$ **7, *p. 91:*** 3×5^2 **8, *p. 92:*** 18 **9, *p. 93:*** 66 **10, *p. 93:*** 12 **11, *p. 93:*** 6 yr

Exercises 2.1, *pp. 94–95*

1. factors **3.** prime **5.** prime factorization **7.** 1, 3, 7, 21 **9.** 1, 17 **11.** 1, 2, 3, 4, 6, 12 **13.** 1, 31 **15.** 1, 2, 3, 4, 6, 9, 12, 18, 36 **17.** 1, 29 **19.** 1, 2, 4, 5, 10, 20, 25, 50, 100 **21.** 1, 2, 4, 7, 14, 28 **23.** Prime **25.** Composite (2, 4, 8) **27.** Composite (7) **29.** Prime **31.** Composite (3, 9, 27) **33.** 2^3 **35.** 7^2 **37.** $2^3 \times 3$ **39.** 2×5^2 **41.** 7×11 **43.** 3×17 **45.** 5^2 **47.** 2^5 **49.** 3×7 **51.** $2^3 \times 13$ **53.** 11^2 **55.** 2×71 **57.** $2^2 \times 5^2$ **59.** 5^3 **61.** $3^3 \times 5$ **63.** 15 **65.** 40 **67.** 90 **69.** 110 **71.** 72 **73.** 360 **75.** 300 **77.** 84 **79.** 105 **81.** 60 **83.** 3×5^2 **85.** 1, 2, 3, 4, 6, 8, 9, 12, 18, 24, 36, and 72 **87. a.** No, because 1995 is not a multiple of 10 **b.** Yes, because 1990 is a multiple of 10 **89.** No **91.** 99 students **93.** 30 days

Practices: Section 2.2, *pp. 98–106*

1, *p. 98:* $\frac{5}{8}$ **2, *p. 98:*** $\frac{7}{30}$ **3, *p. 98:*** $\frac{41}{101}$

4, *p. 99:*

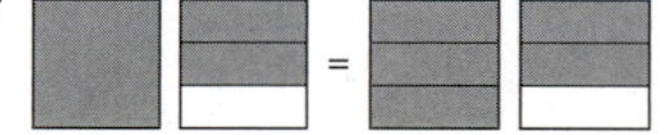

5, *p. 99:* **a.** $\frac{16}{3}$ **b.** $\frac{102}{5}$ **6, *p. 100:*** **a.** 2 **b.** $5\frac{5}{9}$ **c.** $2\frac{2}{3}$ **7, *p. 102:*** Possible answer: $\frac{4}{10}, \frac{6}{15}, \frac{8}{20}$ **8, *p. 102:*** $\frac{45}{72}$ **9, *p. 103:*** $\frac{2}{3}$ **10, *p. 103:*** $\frac{7}{3}$ **11, *p. 104:*** $\frac{5}{16}$ **12, *p. 105:*** $\frac{11}{16}$ **13, *p. 106:*** $\frac{8}{15}, \frac{23}{30}, \frac{9}{10}$ **14, *p. 106:*** Country stations

Exercises 2.2, *pp. 107–112*

1. proper fraction **3.** equivalent **5.** like fractions **7.** $\frac{1}{3}$ **9.** $\frac{3}{6}$ **11.** $1\frac{1}{4}$ **13.** $3\frac{2}{4}$ **15.**

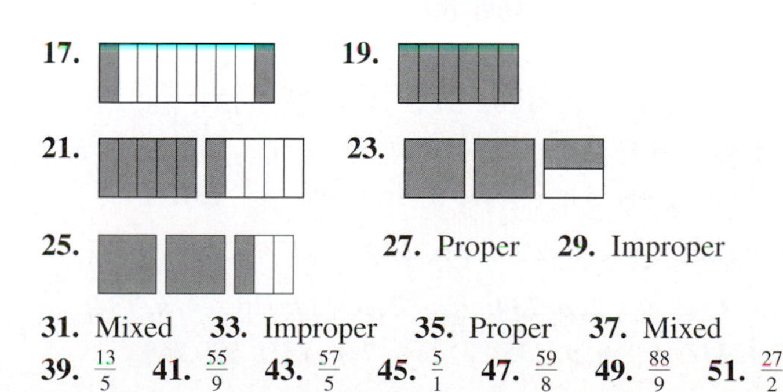

17. **19.** **21.** **23.** **25.** **27.** Proper **29.** Improper **31.** Mixed **33.** Improper **35.** Proper **37.** Mixed **39.** $\frac{13}{5}$ **41.** $\frac{55}{9}$ **43.** $\frac{57}{5}$ **45.** $\frac{5}{1}$ **47.** $\frac{59}{8}$ **49.** $\frac{88}{9}$ **51.** $\frac{27}{2}$

53. $\frac{98}{5}$ **55.** $\frac{14}{1}$ **57.** $\frac{54}{11}$ **59.** $\frac{115}{14}$ **61.** $\frac{202}{25}$ **63.** $1\frac{1}{3}$
65. $1\frac{1}{9}$ **67.** 3 **69.** 1 **71.** $19\frac{4}{5}$ **73.** $9\frac{1}{9}$ **75.** 1
77. $8\frac{2}{9}$ **79.** $13\frac{1}{2}$ **81.** $11\frac{1}{9}$ **83.** 27 **85.** 8
87. Possible answers: $\frac{2}{16}, \frac{3}{24}$ **89.** Possible answers: $\frac{4}{22}, \frac{6}{33}$
91. Possible answers: $\frac{6}{8}, \frac{9}{12}$ **93.** Possible answers: $\frac{2}{18}, \frac{3}{27}$
95. 9 **97.** 15 **99.** 40 **101.** 36 **103.** 40 **105.** 54
107. 36 **109.** 42 **111.** 6 **113.** 49 **115.** 32
117. 30 **119.** $\frac{2}{3}$ **121.** 1 **123.** $\frac{1}{3}$ **125.** $\frac{9}{20}$ **127.** $\frac{1}{4}$
129. $\frac{1}{8}$ **131.** $\frac{5}{4}$, or $1\frac{1}{4}$ **133.** $\frac{33}{16}$, or $2\frac{1}{16}$ **135.** $\frac{9}{16}$
137. $\frac{7}{24}$ **139.** 3 **141.** $\frac{1}{7}$ **143.** $3\frac{2}{3}$ **145.** 3 **147.** $<$
149. $>$ **151.** $=$ **153.** $<$ **155.** $\frac{1}{4}, \frac{1}{3}, \frac{1}{2}$ **157.** $\frac{7}{12}, \frac{2}{3}, \frac{5}{6}$
159. $\frac{3}{5}, \frac{2}{3}, \frac{8}{9}$ **161.** $\frac{5}{6}$ **163.** Possible answers: $\frac{4}{18}, \frac{6}{27}$
165. $\frac{12}{15}$ **167.** $2\frac{1}{5}$ hr per day **169. a.** $\frac{33}{758}$ **b.** $\frac{725}{758}$
171. $\frac{50}{103}$ **173.** No; $\frac{1}{4} = \frac{25}{100}$, which is greater than $\frac{23}{100}$
175. a. Petroleum products **b.** Natural gas **177.** $190\frac{1}{3}$ lb

Practices: Section 2.3, *pp. 113–127*

1, *p. 113:* $\frac{2}{3}$ **2, *p. 114:*** $1\frac{7}{40}$ **3, *p. 114:*** $\frac{2}{5}$
4, *p. 114:* **a.** $\frac{3}{5}$ g **b.** $\frac{2}{5}$ g **5, *p. 116:*** $1\frac{2}{3}$ **6, *p. 116:*** $\frac{3}{10}$
7, *p. 116:* $\frac{71}{72}$ **8, *p. 117:*** 2 mi **9, *p. 118:*** $34\frac{2}{5}$
10, *p. 118:* $7\frac{1}{2}$ **11, *p. 118:*** 4 lengths **12, *p. 119:*** $7\frac{5}{8}$
13, *p. 119:* $11\frac{5}{24}$ **14, *p. 120:*** $4\frac{2}{5}$ **15, *p. 121:*** $2\frac{1}{8}$ in.
16, *p. 121:* $4\frac{7}{12}$ **17, *p. 122:*** $6\frac{1}{4}$ mi **18, *p. 123:*** $1\frac{2}{7}$
19, *p. 124:* $5\frac{1}{6}$ **20, *p. 124:*** $13\frac{1}{2}$ **21, *p. 125:*** No, there will be only $3\frac{5}{8}$ yd left. **22, *p. 126:*** $10\frac{11}{20}$ **23, *p. 126:*** $1\frac{3}{8}$
24, *p. 127:* $6\frac{3}{4}$

Exercises 2.3, *pp. 128–132*

1. numerators **3.** borrow **5.** $1\frac{1}{4}$ **7.** $1\frac{1}{2}$ **9.** $\frac{4}{5}$ **11.** $\frac{3}{5}$
13. $1\frac{1}{6}$ **15.** $\frac{7}{8}$ **17.** $\frac{77}{100}$ **19.** $\frac{37}{40}$ **21.** $1\frac{5}{18}$ **23.** $1\frac{17}{100}$
25. $\frac{3}{4}$ **27.** $\frac{53}{80}$ **29.** $1\frac{7}{72}$ **31.** $1\frac{13}{40}$ **33.** $3\frac{1}{3}$ **35.** $15\frac{2}{5}$
37. $14\frac{1}{5}$ **39.** 15 **41.** $10\frac{5}{12}$ **43.** $3\frac{11}{15}$ **45.** $13\frac{13}{15}$
47. $6\frac{19}{24}$ **49.** $20\frac{1}{4}$ **51.** $10\frac{3}{100}$ **53.** $11\frac{3}{8}$ **55.** $36\frac{3}{50}$
57. $91\frac{7}{12}$ **59.** $6\frac{1}{2}$ **61.** $10\frac{33}{40}$ **63.** $11\frac{3}{8}$ **65.** $\frac{1}{5}$ **67.** $\frac{2}{5}$
69. $\frac{4}{25}$ **71.** $\frac{1}{2}$ **73.** 2 **75.** $\frac{1}{12}$ **77.** $\frac{5}{18}$ **79.** $\frac{1}{20}$ **81.** $\frac{1}{14}$
83. $\frac{5}{72}$ **85.** $\frac{1}{4}$ **87.** $4\frac{2}{7}$ **89.** $1\frac{3}{4}$ **91.** 20 **93.** $4\frac{1}{10}$
95. $3\frac{1}{3}$ **97.** $3\frac{3}{10}$ **99.** $6\frac{1}{3}$ **101.** $5\frac{1}{2}$ **103.** $4\frac{1}{2}$
105. $3\frac{1}{4}$ **107.** $11\frac{4}{5}$ **109.** $6\frac{2}{3}$ **111.** $7\frac{5}{6}$ **113.** $3\frac{13}{24}$
115. $15\frac{7}{18}$ **117.** $2\frac{29}{30}$ **119.** $\frac{1}{4}$ **121.** $5\frac{5}{12}$ **123.** $13\frac{39}{40}$
125. $\frac{3}{8}$ **127.** $1\frac{11}{40}$ **129.** $16\frac{23}{30}$ **131.** $5\frac{1}{5}$ **133.** $18\frac{11}{20}$
135. $4\frac{1}{8}$ **137.** $8\frac{1}{3}$ **139.** $2\frac{14}{15}$ **141.** $\frac{1}{8}$ in. **143. a.** $1\frac{1}{2}$ mi **b.** $\frac{1}{4}$ mi **145.** 5 hr **147.** $5\frac{5}{6}$ ft **149.** $\frac{1}{10}$ **151.** 1 lb

Practices: Section 2.4, *pp. 134–143*

1, *p. 134:* $\frac{15}{28}$ **2, *p. 134:*** $\frac{81}{100}$ **3, *p. 134:*** 20 **4, *p. 134:*** $\frac{7}{22}$
5, *p. 135:* $\frac{2}{9}$ **6, *p. 135:*** $5\frac{1}{4}$ hr **7, *p. 135:*** \$20,769
8, *p. 136:* $7\frac{7}{8}$ **9, *p. 136:*** 28 **10, *p. 137:*** $25\frac{1}{2}$ sq in.
11, *p. 138:* $18\frac{1}{4}$ **12, *p. 139:*** 6 **13, *p. 140:*** 8
14, *p. 140:* $2\frac{2}{3}$ yr **15, *p. 141:*** $1\frac{3}{5}$ **16, *p. 141:*** $\frac{7}{16}$
17, *p. 141:* 6 lb **18, *p. 142:*** 6 **19, *p. 143:*** $4\frac{1}{2}$

Exercises 2.4, *pp. 144–148*

1. multiply **3.** reciprocal **5.** $\frac{2}{15}$ **7.** $\frac{5}{12}$ **9.** $\frac{9}{16}$ **11.** $\frac{8}{25}$
13. $\frac{35}{32} = 1\frac{3}{32}$ **15.** $\frac{45}{16} = 2\frac{13}{16}$ **17.** $\frac{2}{9}$ **19.** $\frac{7}{12}$ **21.** $\frac{3}{40}$
23. $\frac{31}{30} = 1\frac{1}{30}$ **25.** $\frac{40}{3} = 13\frac{1}{3}$ **27.** $\frac{40}{3} = 13\frac{1}{3}$ **29.** 16
31. 4 **33.** 4 **35.** $\frac{35}{4} = 8\frac{3}{4}$ **37.** $1\frac{5}{16}$ **39.** $2\frac{1}{8}$ **41.** $\frac{25}{27}$
43. $2\frac{2}{3}$ **45.** 1 **47.** $\frac{7}{8}$ **49.** $1\frac{13}{35}$ **51.** $4\frac{41}{100}$ **53.** $7\frac{4}{5}$
55. 375 **57.** 8 **59.** 3 **61.** $41\frac{2}{3}$ **63.** $113\frac{1}{3}$ **65.** $1\frac{1}{6}$
67. $\frac{7}{12}$ **69.** $\frac{77}{100}$ **71.** $3\frac{3}{8}$ **73.** $\frac{9}{10}$ **75.** $\frac{32}{35}$ **77.** $3\frac{1}{2}$
79. $4\frac{4}{9}$ **81.** $1\frac{1}{2}$ **83.** $2\frac{1}{3}$ **85.** $1\frac{1}{5}$ **87.** $\frac{1}{4}$ **89.** $\frac{2}{21}$
91. $\frac{1}{9}$ **93.** 40 **95.** $16\frac{1}{3}$ **97.** $13\frac{1}{3}$ **99.** 7 **101.** $6\frac{11}{18}$
103. $1\frac{2}{3}$ **105.** $9\frac{22}{27}$ **107.** $100\frac{1}{2}$ **109.** $\frac{7}{90}$ **111.** $\frac{5}{26}$
113. $3\frac{1}{5}$ **115.** $\frac{21}{200}$ **117.** $\frac{35}{44}$ **119.** $1\frac{47}{115}$ **121.** $\frac{14}{27}$
123. $2\frac{1}{4}$ **125.** $1\frac{7}{18}$ **127.** $4\frac{13}{15}$ **129.** $\frac{87}{160}$ **131.** $3\frac{19}{27}$
133. $4\frac{1}{5}$ **135.** $3\frac{1}{8}$ **137.** $11\frac{1}{6}$ **139.** $20\frac{13}{16}$ **141.** $2\frac{5}{22}$
143. $\frac{21}{40}$ **145.** 8 **147.** $\frac{7}{12}$ **149.** \$12,000 **151.** $6\frac{1}{4}$
153. \$$191\frac{1}{4}$ **155.** $\frac{27}{64}$ **157.** 7 times **159. a.** The scented candle **b.** The unscented candle

Review Exercises: Chapter 2, *pp. 152–157*

1. 1, 2, 3, 5, 6, 10, 15, 25, 30, 50, 75, 150 **2.** 1, 2, 3, 4, 5, 6, 9, 10, 12, 15, 18, 20, 30, 36, 45, 60, 90, 180
3. 1, 3, 19, 57 **4.** 1, 2, 5, 7, 10, 14, 35, 70 **5.** Prime
6. Composite **7.** Composite **8.** Prime **9.** $2^2 \times 3^2$
10. 3×5^2 **11.** $3^2 \times 11$ **12.** 2×3^3 **13.** 42 **14.** 10
15. 72 **16.** 60 **17.** $\frac{2}{4}$ **18.** $\frac{6}{12}$ **19.** $1\frac{1}{6}$ **20.** $2\frac{3}{5}$
21. Mixed **22.** Proper **23.** Improper **24.** Improper
25. $\frac{23}{3}$ **26.** $\frac{9}{5}$ **27.** $\frac{91}{10}$ **28.** $\frac{59}{7}$ **29.** $6\frac{1}{2}$ **30.** $4\frac{2}{3}$
31. $2\frac{3}{4}$ **32.** 1 **33.** 84 **34.** 4 **35.** 5 **36.** 27
37. $\frac{1}{2}$ **38.** $\frac{5}{7}$ **39.** $\frac{2}{3}$ **40.** $\frac{3}{4}$ **41.** $5\frac{1}{2}$ **42.** $8\frac{2}{3}$ **43.** $6\frac{2}{7}$
44. $8\frac{5}{7}$ **45.** $>$ **46.** $>$ **47.** $<$ **48.** $>$ **49.** $>$
50. $>$ **51.** $>$ **52.** $>$ **53.** $\frac{2}{7}, \frac{3}{8}, \frac{1}{2}$ **54.** $\frac{2}{15}, \frac{1}{5}, \frac{1}{3}$
55. $\frac{3}{4}, \frac{4}{5}, \frac{9}{10}$ **56.** $\frac{13}{18}, \frac{7}{9}, \frac{7}{8}$ **57.** $\frac{6}{5} = 1\frac{1}{5}$ **58.** $\frac{3}{4}$ **59.** $\frac{15}{8} = 1\frac{7}{8}$
60. $\frac{3}{5}$ **61.** $\frac{11}{15}$ **62.** $1\frac{17}{24}$ **63.** $1\frac{4}{5}$ **64.** $1\frac{37}{40}$ **65.** $5\frac{7}{8}$
66. $9\frac{1}{2}$ **67.** $10\frac{3}{5}$ **68.** 8 **69.** $12\frac{1}{3}$ **70.** $4\frac{3}{10}$ **71.** $5\frac{7}{10}$
72. $17\frac{13}{24}$ **73.** $23\frac{5}{12}$ **74.** $46\frac{3}{8}$ **75.** $20\frac{3}{4}$ **76.** $56\frac{1}{24}$ **77.** $\frac{1}{4}$
78. $\frac{2}{3}$ **79.** 1 **80.** 0 **81.** $\frac{1}{4}$ **82.** $\frac{3}{8}$ **83.** $\frac{7}{20}$ **84.** $\frac{7}{30}$
85. $7\frac{1}{2}$ **86.** $2\frac{3}{10}$ **87.** $3\frac{3}{4}$ **88.** $18\frac{1}{2}$ **89.** $6\frac{1}{2}$ **90.** $1\frac{7}{10}$
91. $2\frac{2}{3}$ **92.** $\frac{1}{5}$ **93.** $1\frac{4}{5}$ **94.** $\frac{3}{4}$ **95.** $2\frac{1}{2}$ **96.** $3\frac{1}{3}$ **97.** $\frac{3}{10}$
98. $2\frac{7}{8}$ **99.** $\frac{7}{12}$ **100.** $3\frac{8}{9}$ **101.** $\frac{2}{3}$ **102.** $9\frac{9}{20}$ **103.** $\frac{3}{16}$
104. $\frac{7}{16}$ **105.** $\frac{5}{8}$ **106.** $\frac{1}{6}$ **107.** $5\frac{1}{3}$ **108.** $\frac{7}{10}$ **109.** $\frac{1}{125}$
110. $\frac{8}{27}$ **111.** $\frac{1}{4}$ **112.** $\frac{7}{120}$ **113.** $\frac{24}{25}$ **114.** $1\frac{5}{9}$ **115.** $2\frac{2}{3}$
116. $\frac{2}{3}$ **117.** 6 **118.** $18\frac{5}{12}$ **119.** $8\frac{7}{16}$ **120.** $21\frac{1}{4}$

121. $\frac{9}{20}$ **122.** $1\frac{9}{16}$ **123.** $37\frac{1}{27}$ **124.** $3\frac{3}{8}$ **125.** $3\frac{1}{8}$
126. $1\frac{41}{90}$ **127.** $2\frac{1}{10}$ **128.** $7\frac{1}{5}$ **129.** $\frac{3}{2}$ **130.** $\frac{2}{3}$ **131.** $\frac{1}{8}$
132. 4 **133.** $\frac{7}{40}$ **134.** $\frac{5}{81}$ **135.** $\frac{2}{15}$ **136.** $\frac{1}{200}$
137. $\frac{3}{4}$ **138.** $1\frac{1}{3}$ **139.** 30 **140.** $8\frac{3}{4}$ **141.** $1\frac{1}{6}$ **142.** $1\frac{4}{5}$
143. 2 **144.** 4 **145.** $1\frac{3}{4}$ **146.** $\frac{4}{7}$ **147.** $1\frac{7}{12}$
148. $\frac{12}{19}$ **149.** $5\frac{1}{2}$ **150.** $2\frac{11}{20}$ **151.** 2 **152.** 3 **153.** $9\frac{3}{4}$
154. $1\frac{3}{10}$ **155.** $5\frac{1}{3}$ **156.** $7\frac{5}{9}$ **157.** No **158.** 50¢
159. $\frac{1}{4}$ **160.** $\frac{2}{9}$ **161.** The Filmworks camera **162.** $\frac{39}{40}$
163. The patient got back more than $\frac{1}{3}$, because $\frac{275}{700} = \frac{11}{28} = \frac{33}{84}$ which is greater than $\frac{1}{3} = \frac{28}{84}$. **164.** Yes it should, because $\frac{23}{32}$ is greater than $\frac{2}{3}$. $\frac{23}{32} = \frac{69}{96}$, whereas $\frac{2}{3} = \frac{64}{96}$ **165. a.** $\frac{12}{23}$ **b.** $\frac{3}{4}$
166. a. Lisa Gregory **b.** Monica Yates **167.** $\frac{3}{4}$ **168.** $\frac{11}{12}$ oz
169. $\frac{1}{4}$ carat **170.** $\frac{3}{5}$ **171.** 12 women **172.** 2,685 undergraduate students **173.** $1,050 **174.** 7 lb **175.** $18
176. 2 times **177.** 19 fish **178.** $11\frac{3}{4}$ mi **179.** $62\frac{5}{6}$ ft
180. $\frac{5}{6}$ hr **181.** 1,500 fps **182.** 500 lb/sq in.
183. $281\frac{1}{4}$ lb **184.** $46,000 **185.** 8 orbits
186. $29\frac{5}{9}$ sq yd
187.

Employee	Saturday	Sunday	Total
L. Chavis	$7\frac{1}{2}$	$4\frac{1}{4}$	$11\frac{3}{4}$
R. Young	$5\frac{3}{4}$	$6\frac{1}{2}$	$12\frac{1}{4}$
Total	$13\frac{1}{4}$	$10\frac{3}{4}$	24

188.

Worker	Hours per Day	Days Worked	Total Hours	Wage per Hour	Gross Pay
Maya	5	3	15	$7	$105
Noel	$7\frac{1}{4}$	4	29	$10	$290
Alisa	$4\frac{1}{2}$	$5\frac{1}{2}$	$24\frac{3}{4}$	$9	$\$222\frac{3}{4}$

189. $10\frac{10}{11}$ lb **190.** $22\frac{1}{2}$ cups

Posttest: Chapter 2, *p. 158*

1. 1, 3, 7, 9, 21, 63 **2.** 2×3^3 **3.** $\frac{4}{9}$ **4.** $\frac{12}{1}$
5. $10\frac{1}{4}$ **6.** $\frac{7}{8}$ **7.** $\frac{5}{10}$ **8.** 24 **9.** $1\frac{13}{24}$ **10.** $8\frac{7}{40}$
11. $4\frac{2}{7}$ **12.** $5\frac{23}{30}$ **13.** $\frac{1}{81}$ **14.** 12 **15.** $\frac{7}{9}$ **16.** $7\frac{5}{6}$
17. $\frac{5}{6}$ **18.** $19\frac{1}{5}$ mi **19.** $90\frac{2}{3}$ sq ft **20.** $16\frac{1}{4}$ in.

Cumulative Review: Chapter 2, *p. 159*

1. Five million, three hundred fifteen **2.** 581,400
3. 908 **4.** $\frac{3}{4}$ **5.** $6\frac{2}{5}$ **6.** 7 **7.** $\frac{3}{8}$ **8.** 1 million times
9. 549 copies **10.** $\frac{1}{3}$

Chapter 3

Pretest: Chapter 3, *p. 162*

1. Hundredths **2.** Four and twelve thousandths **3.** 3.1
4. 0.0029 **5.** 21.52 **6.** 7.3738 **7.** 11.69 **8.** 9.81
9. 8,300 **10.** 18.423 **11.** 0.0144 **12.** 7.1 **13.** 0.00605
14. 32.7 **15.** 0.875 **16.** 2.83 **17.** One with a pH value of 2.95 **18.** $39.788 billion **19.** 3 times
20. $3.74

Practices: Section 3.1, *pp. 164–170*

1, *p. 164:* **a.** The tenths place **b.** The ten-thousandths place **c.** The thousandths place **2, *p. 165:*** $\frac{7}{8}$ **3, *p. 165:*** $2\frac{3}{100}$
4, *p. 165:* **a.** $5\frac{3}{5}$ **b.** $5\frac{3}{5}$ **5, *p. 166:*** **a.** $7\frac{3}{1,000}$ **b.** $4\frac{1}{10}$
6, *p. 166:* **a.** Sixty-one hundredths **b.** Four and nine hundred twenty-three thousandths **c.** Seven and five hundredths
7, *p. 166:* **a.** 0.043 **b.** 10.26 **8, *p. 167:*** 3.14
9, *p. 167:* 0.8297 **10, *p. 167:*** 3.51, 3.5, 3.496
11, *p. 168:* The one with the rating of 8.1, because $9 > 8.2 > 8.1$ **12, *p. 169:*** **a.** 748.1 **b.** 748.08
c. 748.077 **d.** 748 **e.** 700 **13, *p. 170:*** 7.30
14, *p. 170:* 11.4 m

Exercises 3.1, *pp. 171–175*

1. right **3.** hundredths **5.** greater **7.** $2.\underline{7}8$
9. $2.001\underline{7}5$ **11.** $3\underline{5}8.02$ **13.** $0.7\underline{7}2$ **15.** Tenths
17. Hundredths **19.** Thousandths **21.** Ones **23.** $\frac{3}{5}$
25. $\frac{39}{100}$ **27.** $1\frac{1}{2}$ **29.** 8 **31.** $5\frac{3}{250}$ **33.** Fifty-three hundredths **35.** Three hundred five thousandths
37. Six tenths **39.** Five and seventy-two hundredths
41. Twenty-four and two thousandths **43.** 0.8
45. 1.041 **47.** 60.01 **49.** 4.107 **51.** 3.2 m **53.** $>$
55. $<$ **57.** $>$ **59.** $=$ **61.** $<$ **63.** 7, 7.07, 7.1
65. 4.9, 5.001, 5.2 **67.** 9.1 mi, 9.38 mi, 9.6 mi **69.** 17.4
71. 3.591 **73.** 37.1 **75.** 0.40 **77.** 7.06 **79.** 9 mi
81.

To the Nearest	8.0714	0.9916
Tenth	8.1	1.0
Hundredth	8.07	0.99
Ten	10	0

83. $0.\underline{0}24$ **85.** 870.06 **87.** 2.04 m, 2.14 m, 2.4 m
89. Twenty-three and nine hundred thirty-four thousandths
91. Eighteen and seven tenths; eighteen and eight tenths
93. Three hundred one and three tenths, Fifty-five and nine tenths, Two hundred sixty-eight and two tenths, Forty-six and six tenths, Forty-three and six tenths **95.** One hundred-thousandth; eight hundred-thousandths **97.** 1.2 acres
99. 74.59 mph **101.** 14.7 lb **103.** $0.005
105. 352.1 kWh **107.** Less **109.** Last winter
111. Yes **113.** No **115.** $57.03 **117.** 0.001 **119.** 1.8

Practices: Section 3.2, *pp. 177–182*

1, *p. 177:* 10.387 **2, *p. 177:*** 39.3 **3, *p. 178:*** 102.1°F
4, *p. 178:* 46.2125 **5, *p. 179:*** $485.43
6, *p. 179:* 13.5 mi **7, *p. 179:*** 22.13 mi **8, *p. 180:*** 0.863
9, *p. 180:* 0.079 **10, *p. 180:*** 0.5744
11, *p. 181:* Possible estimate: $480
12, *p. 181:* Possible estimate: $2 million

Calculator Practices, *p. 182*

13, ***p. 182*****:** 79.23; **14,** ***p. 182*****:** 0.00002

Exercises 3.2, *pp. 183–186*

1. decimal points **3.** sum **5.** 9.33 **7.** 0.9 **9.** 8.13 **11.** 21.45 **13.** 7.67 **15.** $77.21 **17.** 1.08993 **19.** 24.16 **21.** 44.422 **23.** 20.32 mm **25.** 16.682 kg **27.** 23.30595 **29.** 0.7 **31.** 16.8 **33.** 18.41 **35.** 75.63 **37.** 22.324 **39.** 0.17 **41.** 0.1142 **43.** 6.2 **45.** 15.37 **47.** 5.9 **49.** 6.21 **51.** 1.85 lb **53.** 4.9°F **55.** 39.752 **57.** 27.9 mg **59.** 3.205 **61.** 21.19896 **63.** c; possible estimate: 0.083 **65.** b; possible estimate: 0.06 **67.** 7.771 **69.** 7.75 lb **71.** 11.6013 **73.** $1.03 **75.** 56.8 centuries **77.** $1.7 million **79.** 6.84 in. **81.** Yes; 2.8 + 2.9 + 2.6 + 1.6 = 9.9

83. a.

Gymnast	VT	UB	BB	FX	AAS
Madeline Whiteman	9.2	9.275	8.6	8.05	35.125
Jordyn Stengel	9	9	8.65	8.45	35.1

b. Madeline Whiteman **85.** (a) Addition, subtraction; (b) total 13.2 mg iron; no, she needs 4.8 mg more; (c) possible estimates: 1 + 2 + 0 + 2 + 1 + 1 + 1 + 1 + 1 + 2 + 1 + 0 = 13; 18 − 13 = 5

Practices: Section 3.3, *pp. 188–192*

1, ***p. 188:*** 9.835 **2,** ***p. 189:*** 1.4 **3,** ***p. 189:*** 0.01 **4,** ***p. 189:*** 0.024 **5,** ***p. 189:*** 9.91 **6,** ***p. 190:*** 325 **7,** ***p. 190:*** 327,000 **8,** ***p. 190:*** **a.** 18.015 **b.** 18 **9,** ***p. 191:*** 0.0003404; possible estimate: 0.004 × 0.09 = 0.00036 **10,** ***p. 191:*** 3.6463 **11,** ***p. 192:*** Possible answer: 1,200 mi

Calculator Practices, *p. 192*

12, ***p. 192:*** 815.6 **13,** ***p. 192:*** 9.261

Exercises 3.3, *pp. 193–195*

1. multiplication **3.** two **5.** square **7.** 2.99212 **9.** 204.360 **11.** 2,492.0 **13.** 0.0000969 **15.** 2,870.00 **17.** $0.73525 **19.** 0.54 **21.** 0.4 **23.** 0.02 **25.** 0.0028 **27.** 0.765 **29.** 2.016 **31.** 7.602 **33.** 0.5 **35.** 5.852 **37.** 151.14 **39.** 3.7377 **41.** 1.7955 **43.** 8,312.7 **45.** 23 **47.** 0.09 **49.** 1.05

51. 0.000000001 **53.** 42.5 ft **55.** 1.4 mi **57.** 3.29025 **59.** 272,593.75 **61.** 70 **63.** 25.75 **65.** 1.09 **67.** 2.86

69.

Input	Output
1	3.8 × **1** − 0.2 = 3.6
2	3.8 × **2** − 0.2 = 7.4
3	3.8 × **3** − 0.2 = 11.2
4	3.8 × **4** − 0.2 = 15

71. a; possible estimate: 50 **73.** b; possible estimate: 0.014 **75.** 8.75 **77.** 0.068 **79.** 4.48 **81.** 2,900 fps **83.** 57,900,000 km **85.** 19.6 sq ft **87.** 1.25 mg **89.** 1,308 calories

91. a.

Purchase	Quantity	Unit Price	Price
Belt	1	$11.99	$11.99
Shirt	3	$16.95	$50.85
Total Price			$62.84

b. $17.16 **93.** (a) Multiplication and addition; (b) 88.81 in.; (c) possible estimate: 90 in.

Practices: Section 3.4, *pp. 198–205*

1, ***p. 198:*** 0.375 **2,** ***p. 198:*** 7.625 **3,** ***p. 199:*** 83.3 **4,** ***p. 199:*** 0.8 **5,** ***p. 201:*** 18.04 **6,** ***p. 201:*** 2,050 **7,** ***p. 202:*** 73.4 **8,** ***p. 202:*** 0.0341 **9,** ***p. 203:*** 0.00086 **10,** ***p. 203:*** 1.5 **11,** ***p. 204:*** 21.1; possible estimate: 20 **12,** ***p. 204:*** 295.31 **13,** ***p. 205:*** 8 times as great

Calculator Practices, *pp. 205–206*

14, ***p. 205:*** 0.2 **15,** ***p. 206:*** 4.29

Exercises 3.4, *pp. 207–210*

1. decimal **3.** right **5.** quotient **7.** 0.5 **9.** 0.25 **11.** 3.7 **13.** 1.625 **15.** 2.875 **17.** 21.03 **19.** 4.25 **21.** 4.2 **23.** 1.375 **25.** 8.5 **27.** 0.67 **29.** 0.78 **31.** 3.11 **33.** 5.06 **35.** 3.286 **37.** 0.273 **39.** 6.571 **41.** 70.077 **43.** 58.82 **45.** 0.0663 **47.** 2.8875 **49.** 0.286 **51.** 4.3 **53.** 0.0015 **55.** 1.73 **57.** 2.875 **59.** 4 **61.** 70.4 **63.** 94 **65.** 12.5 **67.** 0.3 **69.** 0.2 **71.** 0.952 **73.** 0.00082 **75.** 383.88 **77.** 0.01 **79.** 9.23 **81.** 9,666.67 **83.** 1,952.38 **85.** 325.18 **87.** 67.41 **89.** 41.61 **91.** 0.17136 **93.** 0.13 **95.** 52.2 **97.** 4.05

99.

Input	Output
1	**1** ÷ 5 − 0.2 = 0
2	**2** ÷ 5 − 0.2 = 0.2
3	**3** ÷ 5 − 0.2 = 0.4
4	**4** ÷ 5 − 0.2 = 0.6

101. c; possible estimate: 50 **103.** b; possible estimate: 0.2 **105.** 0.8 **107.** 1.17 **109.** 0.45 **111.** 0.0037 in. per yr **113. a.** 0.6 **b.** 0.55 **c.** The women's team has a better record. The team won $\frac{3}{5}$, or 0.6, of the games played, and the men's team won $\frac{11}{20}$, or 0.55, of the games played.

115. a.

SUVs	Distance Driven (in miles)	Gasoline Used (in gallons)	Miles per Gallon
A	17.4	1.2	14.5
B	8.4	0.6	14
C	23.4	1.2	19.5

b. SUV C **117.** 2,000 shares **119.** 13 times **121.** 0.4 lb **123.** (a) Division; (b) .366; (c) .4

Review Exercises: Chapter 3, *pp. 214–216*

1. Hundredths **2.** Tenths **3.** Tenths **4.** Ten-thousandths **5.** $\frac{7}{20}$ **6.** $8\frac{1}{5}$ **7.** $4\frac{7}{1{,}000}$ **8.** 10 **9.** Seventy-two hundredths **10.** Five and six tenths **11.** Three and nine ten-thousandths **12.** Five hundred ten and thirty-six thousandths **13.** 0.007 **14.** 2.1 **15.** 0.09 **16.** 7.041 **17.** $<$ **18.** $>$ **19.** $>$ **20.** $>$ **21.** 1.002, 0.8, 0.72 **22.** 0.004, 0.003, 0.00057 **23.** 7.3 **24.** 0.039 **25.** 4.39 **26.** $899 **27.** 12.11 **28.** 52.75 **29.** $24.13 **30.** 12 m **31.** 28.78 **32.** 87.752 **33.** 1.834 **34.** 48.901 **35.** 98.2033 **36.** $90,948.80 **37.** 2.912 **38.** 1,008 **39.** 0.00001 **40.** 13.69 **41.** 2,710 **42.** 0.034 **43.** 5.75 **44.** 13.5 **45.** 1,569.36846 **46.** 441.760662 **47.** 0.625 **48.** 90.2 **49.** 4.0625 **50.** 0.045 **51.** 0.17 **52.** 0.29 **53.** 8.33 **54.** 11.22 **55.** 0.65 **56.** 1.6 **57.** 0.175 **58.** 0.277 **59.** 5.2 **60.** 3.2 **61.** 23.7 **62.** 16,358.3 **63.** 1.9 **64.** 360.7 **65.** 3.0 **66.** 0.3 **67.** 1.18 **68.** 117 **69.** 34.375 **70.** 1.4 **71.** Four ten-millionths **72.** $57.86 **73.** 54.49 sec **74.** 1.647 in. **75.** No, it would have traveled 0.585 mi, which is less than 0.75 mi. **76.** 4.35 times **77.** $0.06 **78.** Possible estimate: 15 in. **79.** 7.19 g **80.** 3.5°C **81.** 36,162.45

82.

Quarter	Google	Yahoo!
1st	1.257	1.174
2nd	1.384	1.253
3rd	1.578	1.330
4th	1.919	1.501
Total:	6.138	5.258

$0.88 billion, or $880,000,000

Posttest: Chapter 3, *p. 217*

1. 6 **2.** Five and one hundred two thousandths **3.** 320.15 **4.** 0.00028 **5.** $3\frac{1}{25}$ **6.** 0.004 **7.** 4.354 **8.** $5.66 **9.** 20.9 **10.** 5.72 **11.** 0.001 **12.** 3.36 **13.** 0.0029 **14.** 32.7 **15.** 0.375 **16.** 4.17 **17.** 0.01 lb **18.** 2.6 ft **19.** Belmont Stakes **20.** $2,807.21

Cumulative Review: Chapter 3, *p. 218*

1. 1,000,000 **2.** 32 **3.** $1\frac{1}{2}$ **4.** 27,403 **5.** $2\frac{2}{3}$ **6.** $\frac{17}{30}$ **7.** 610 **8.** 325 **9.** 26,000 mi **10.** $193.86

Chapter 4

Pretest: Chapter 4, *p. 220*

1. Possible answer: four less than t **2.** Possible answer: quotient of y and three **3.** $m + 8$ **4.** $2n$ **5.** 4 **6.** $1\frac{1}{2}$ **7.** $x + 3 = 5$ **8.** $4y = 12$ **9.** $x = 6$ **10.** $t = 10$ **11.** $n = 13$ **12.** $a = 12$ **13.** $m = 6.1$ **14.** $n = 30$ **15.** $m = \frac{1}{2}$, or 0.5 **16.** $n = 15$ **17.** $63 = x + 36$; $x = 27$ moons **18.** $6.75 = x - 2.75$; $x = \$9.50$ **19.** $\frac{2}{5}x = 39{,}900$; $x = 99{,}750$ sq mi **20.** $40 = 10x$; $x = 4$ mg

Practices: Section 4.1, *pp. 222–225*

1, ***p. 222:*** Answers may vary **a.** One-half of p **b.** x less than 5 **c.** y divided by 4 **d.** 3 more than n **e.** $\frac{3}{5}$ of b **2,** ***p. 223:*** **a.** $x + 9$ **b.** $10y$ **c.** $n - 7$ **d.** $p \div 5$ **e.** $\frac{2}{5}v$ **3,** ***p. 223:*** **a.** $q + 12$ **b.** $\frac{9}{a}$ **c.** $\frac{2}{7}c$ **4,** ***p. 223:*** $\frac{h}{4}$ hr **5,** ***p. 223:*** $s - 3$ **6,** ***p. 224:*** **a.** 25 **b.** 0.38 **c.** 4.8 **d.** 26.6 **7,** ***p. 224:*** $\frac{1}{5}p$; $69,800 **8,** ***p. 225:*** The total amount is $(15.45 + t)$ dollars; $18.45 for $t = \$3$

Exercises 4.1, *pp. 226–229*

1. variable **3.** algebraic **5.** 9 more than t; t plus 9 **7.** c minus 12; 12 subtracted from c **9.** c divided by 3; the quotient of c and 3; **11.** 10 times s; the product of 10 and s **13.** y minus 10; 10 less than y **15.** 7 times a; the product of 7 and a **17.** x divided by 6; the quotient of x and 6 **19.** x minus $\frac{1}{2}$; $\frac{1}{2}$ less than x **21.** $\frac{1}{4}$ times w; $\frac{1}{4}$ of w **23.** 2 minus x; the difference between 2 and x **25.** 1 increased by x; x added to 1 **27.** 3 times p; the product of 3 and p **29.** n decreased by 1.1; n minus 1.1 **31.** y divided by 0.9; the quotient of y and 0.9 **33.** $x + 10$ **35.** $n - 1$ **37.** $y + 5$ **39.** $t \div 6$ **41.** $10y$ **43.** $w - 5$ **45.** $n + \frac{4}{5}$ **47.** $z \div 3$ **49.** $\frac{2}{7}x$ **51.** $k - 6$ **53.** $n + 12$ **55.** $n - 5.1$ **57.** 26 **59.** 2.5 **61.** 15 **63.** $1\frac{1}{6}$ **65.** 1.1 **67.** $\frac{1}{5}$

69.

x	$x + 8$
1	9
2	10
3	11
4	12

71.

n	$n - 0.2$
1	0.8
2	1.8
3	2.8
4	3.8

73.

x	$\frac{3}{4}x$
4	3
8	6
12	9
16	12

75.

z	$\frac{z}{2}$
2	1
4	2
6	3
8	4

77. $x - 7$ **79.** Possible answers: n over 2; n divided by 2 **81.** $3.5t$ **83.** Possible answers: 6 more than x; the sum of x and 6 **85.** $(m - 25)$ mg **87.** $30° + 90° + d°$, or $120° + d°$ **89.** 220 mi **91.** **a.** $1.5w$ dollars **b.** $13.50

Practices: Section 4.2, *pp. 231–235*

1, ***p. 231:*** **a.** $n - 5.1 = 9$ **b.** $y + 2 = 12$ **c.** $n - 4 = 11$ **d.** $n + 5 = 7\frac{3}{4}$ **2,** ***p. 231:*** $p - 6 = 49.95$, where p is the regular price. **3,** ***p. 232:*** $x = 9$; **4,** ***p. 233:*** $t = 2.7$; **5,** ***p. 233:*** $m = 5\frac{1}{4}$; **6,** ***p. 234:*** **a.** $11 = m - 4$; $m = 15$

b. $12 + n = 21; n = 9$ **7, *p. 234:*** $x + 3.99 = 27.18;$ $x = \$23.19$ **8, *p. 235:*** $x - 262{,}000 = 308{,}000;$ $x = 570{,}000$ sq mi

Exercises 4.2, *pp. 236–239*

1. equation **3.** subtract **5.** $z - 9 = 25$ **7.** $7 + x = 25$ **9.** $t - 3.1 = 4$ **11.** $\frac{3}{2} + y = \frac{9}{2}$ **13.** $n - 3\frac{1}{2} = 7$ **15. a.** Yes **b.** No **c.** Yes **d.** No **17.** Subtract 4. **19.** Add 11. **21.** Add 7. **23.** Subtract 2. **25.** $a = 31$ **27.** $y = 2$ **29.** $x = 12$ **31.** $n = 4$ **33.** $m = 2$ **35.** $y = 90$ **37.** $z = 2.9$ **39.** $n = 8.9$ **41.** $y = 0.9$ **43.** $x = 8\frac{2}{3}$ **45.** $m = 5\frac{1}{3}$ **47.** $x = 3\frac{3}{4}$ **49.** $c = 47\frac{1}{5}$ **51.** $x = 13$ **53.** $y = 6\frac{1}{4}$ **55.** $a = 3\frac{5}{12}$ **57.** $x = 8.2$ **59.** $y = 19.91$ **61.** $x = 4.557$ **63.** $y = 10.251$ **65.** $n + 3 = 11; n = 8$ **67.** $y - 6 = 7; y = 13$ **69.** $n + 10 = 19; n = 9$ **71.** $x + 3.6 = 9; x = 5.4$ **73.** $n - 4\frac{1}{3} = 2\frac{2}{3}; n = 7$ **75.** Equation c **77.** Equation a **79.** $a = 14.5$ **81.** Equation b **83.** Yes **85.** Subtract 2. **87.** Equation a **89.** $621{,}000 = x - 13{,}000; x = \$634{,}000$ **91.** $40° + x = 90°; 50°$ **93.** $x + 12 = 96; \$84$ **95.** $45 = x - 20;$ 65 mph **97. a.** $x + 794{,}000{,}000 = 1{,}324{,}089{,}000$ **b.** \$530,089,000 **c.** \$500,000,000

Practices: Section 4.3, *pp. 241–246*

1, *p. 241:* **a.** $2x = 14$ **b.** $\frac{a}{6} = 1.5$ **c.** $\frac{n}{0.3} = 1$ **d.** $10 = \frac{1}{2}n$ **2, *p. 241:*** $15 = 3w$ **3, *p. 243:*** $x = 5$ **4, *p. 243:*** $a = 6$ **5, *p. 243:*** $x = 4$ **6, *p. 244:*** $a = 2.88$ **7, *p. 244:*** $x = 16$ **8, *p. 245:*** **a.** $12 = \frac{z}{6}, z = 72; 12 \stackrel{?}{=} \frac{72}{6}, 12 \stackrel{\checkmark}{=} 12$ **b.** $16 = 2x,$ $8 = x$, or $x = 8; 16 \stackrel{?}{=} 2(8), 16 \stackrel{\checkmark}{=} 16$ **9, *p. 245:*** $1.6 = 5x;$ $x = 0.32$ km **10, *p. 246:*** $\frac{3}{8}x = 525;$ \$1,400

Exercises 4.3, *pp. 247–251*

1. divide **3.** substituting **5.** equation **7.** $\frac{3}{4}y = 12$ **9.** $\frac{x}{7} = \frac{7}{2}$ **11.** $\frac{1}{3}x = 2$ **13.** $\frac{n}{3} = \frac{1}{3}$ **15.** $9a = 27$ **17. a.** Yes **b.** No **c.** No **d.** No **19.** Divide by 3. **21.** Multiply by 2. **23.** Divide by $\frac{3}{4}$ or multiply by $\frac{4}{3}$ **25.** Divide by 1.5. **27.** $x = 6$ **29.** $x = 18$ **31.** $n = 4$ **33.** $x = 91$ **35.** $y = 4$ **37.** $b = 20$ **39.** $m = 157.5$ **41.** $t = 0.4$ **43.** $x = \frac{3}{2}$, or $1\frac{1}{2}$ **45.** $x = 36$ **47.** $t = 3$ **49.** $y = \frac{2}{5}$ **51.** $n = 700$ **53.** $x = 12.5$ **55.** $x = \frac{1}{2}$ **57.** $m = 6$ **59.** $x = 6.8$ **61.** $x = 4.9$ **63.** $8n = 56;$ $n = 7$ **65.** $\frac{3}{4}y = 18; y = 24$ **67.** $\frac{x}{5} = 11; x = 55$ **69.** $2x = 36; x = 18$ **71.** $\frac{1}{2}a = 4; a = 8$ **73.** $\frac{n}{5} = 1\frac{3}{5};$ $n = 8$ **75.** $\frac{n}{2.5} = 10; n = 25$ **77.** Equation d **79.** Equation a **81.** $x = 5.5$ **83.** Equation d **85.** $\frac{y}{3} = 6$ **87.** $2x = 5$ **89.** Yes **91.** $4s = 60;$ $s = 15$ units **93.** $56 = \frac{1}{2}x;$ 112 mi **95.** $12x = 119.88; \$9.99$ **97. a.** $\frac{2}{5}x = 60;$ 150 ml **b.** 90 ml **99. a.** $79.6 = \frac{x}{3{,}537{,}441}$ **b.** 281,580,304 people **c.** 280,000,000 people

Review Exercises: Chapter 4, *pp. 253–255*

1. x plus 1 **2.** Four more than y **3.** w minus 1 **4.** Three less than s **5.** c divided by 7 **6.** The quotient of a and 10 **7.** Two times x **8.** The product of 6 and y **9.** y divided by 0.1 **10.** The quotient of n and 1.6 **11.** One-third of x **12.** One-tenth of w **13.** $m + 9$ **14.** $b + \frac{1}{2}$ **15.** $y - 1.4$ **16.** $z - 3$ **17.** $\frac{3}{x}$ **18.** $n \div 2.5$ **19.** $3n$ **20.** $12n$ **21.** 12 **22.** 19 **23.** 0 **24.** 6 **25.** 0.3 **26.** 6.5 **27.** $1\frac{1}{2}$ **28.** $\frac{5}{12}$ **29.** 0.4 **30.** $4\frac{1}{2}$ **31.** 1.6 **32.** 9 **33.** $x = 9$ **34.** $y = 9$ **35.** $n = 26$ **36.** $b = 20$ **37.** $a = 3.5$ **38.** $c = 7.5$ **39.** $x = 11$ **40.** $y = 2$ **41.** $w = 1\frac{1}{2}$ **42.** $s = \frac{1}{3}$ **43.** $c = 6\frac{3}{4}$ **44.** $p = 11\frac{2}{3}$ **45.** $m = 5$ **46.** $n = 0$ **47.** $c = 78$ **48.** $y = 90$ **49.** $n = 11$ **50.** $x = 25$ **51.** $x = 31.0485$ **52.** $m = 26.6225$ **53.** $n - 19 = 35$ **54.** $a - 37 = 234$ **55.** $9 + n = 5\frac{1}{2}$ **56.** $s + 26 = 30\frac{1}{3}$ **57.** $2y = 16$ **58.** $25t = 175$ **59.** $34 = \frac{n}{19}$ **60.** $17 = \frac{z}{13}$ **61.** $\frac{1}{3}n = 27$ **62.** $\frac{2}{5}n = 4$ **63. a.** No **b.** Yes **c.** Yes **d.** No **64. a.** Yes **b.** No **c.** No **d.** Yes **65.** $x = 5$ **66.** $t = 2$ **67.** $a = 105$ **68.** $n = 54$ **69.** $y = 9$ **70.** $r = 10$ **71.** $w = 90$ **72.** $x = 100$ **73.** $y = 20$ **74.** $a = 120$ **75.** $n = 32$ **76.** $b = 32$ **77.** $m = 3.15$ **78.** $z = 0.57$ **79.** $x = \frac{2}{5}$, or 0.4 **80.** $t = \frac{1}{2}$, or 0.5 **81.** $m = 1.2$ **82.** $b = 9.8$ **83.** $x = 12.5$ **84.** $x = 1.4847$ **85.** $2h$ degrees; 6 degrees **86.** $\frac{d}{20}$ dollars per hr; \$9.55 per hr **87.** $89p$ cents; \$2.67 **88.** $(3{,}000 + d)$ dollars; \$3,225 **89.** $x + 238 = 517;$ \$279 **90.** $\frac{1}{4}x = 500{,}000;$ $x = 2{,}000{,}000$ **91.** $177 = 2.5x;$ 71 L **92.** $225 = x + 50; x = 175$ **93.** $\frac{x}{6} = 30;$ 180 lb **94.** $1.8x = 6{,}696; x = 3{,}720$ km **95.** $98.6 + x = 101;$ $x = 2.4$°F **96.** $x - 256 = 8{,}957;$ 9,213 applications

Posttest: Chapter 4, *p. 256*

1. Possible answer: x plus $\frac{1}{2}$ **2.** Possible answer: the quotient of a and 3 **3.** $n - 10$ **4.** $\frac{8}{p}$ **5.** 0 **6.** $\frac{1}{4}$ **7.** $x - 6 = 4\frac{1}{2}$ **8.** $\frac{y}{8} = 3.2$ **9.** $x = 0$ **10.** $y = 12$ **11.** $n = 27$ **12.** $a = 738$ **13.** $m = 7.8$ **14.** $n = 50$ **15.** $x = \frac{11}{20}$ **16.** $n = 760$ **17.** $1\frac{3}{4} + x = 2\frac{1}{4}; x = \frac{1}{2}$ lb **18.** $\frac{1}{3}x = 30{,}000; x = 90{,}000$ elephants **19.** $6 = \frac{2}{3}x;$ 9 billion people **20.** $x - 19.8 = 7.6;$ 27.4 degrees Celsius

Cumulative Review: Chapter 4, *p. 257*

1. $5\frac{3}{8}$ **2.** 0.0075 **3.** Yes **4.** 23,316 **5.** 3.14 **6.** $n = 7.8$ **7.** $x = 32$ **8.** 7,200 cartoons **9.** 55,000 beehives **10.** He got back $\frac{2}{7}$ of his money, which is less than $\frac{1}{3}$.

Chapter 5

Pretest: Chapter 5, *p. 260*

1. $\frac{3}{4}$ **2.** $\frac{2}{5}$ **3.** $\frac{5}{3}$ **4.** $\frac{19}{51}$ **5.** $\frac{16 \text{ gal}}{5 \text{ min}}$ **6.** $\frac{5 \text{ mg}}{3 \text{ hr}}$ **7.** $\frac{2 \text{ dental assistants}}{1 \text{ dentist}}$ **8.** $\frac{1 \text{ calculator}}{1 \text{ student}}$ **9.** $\frac{\$230}{\text{box}}$ **10.** $\frac{\$0.50}{\text{bottle}}$ **11.** True **12.** False **13.** $x = 9$ **14.** $x = 31\frac{1}{2}$ **15.** $x = 16$ **16.** $x = 160$ **17.** $\frac{4}{5}$ **18.** 200 lb/min **19.** \$264,000 **20.** 87 mi

Practices: Section 5.1, *pp. 261–265*

1, ***p. 261:*** $\frac{2}{3}$ **2,** ***p. 262:*** $\frac{9}{5}$ **3,** ***p. 262:*** $\frac{1}{3}$
4, ***p. 263:*** **a.** $\frac{5\text{ ml}}{2\text{ min}}$ **b.** $\frac{3\text{ lb}}{2\text{ wk}}$ **5,** ***p. 263:*** **a.** 48 ft/sec
b. 0.375 hit per time at bat **6,** ***p. 264:*** $\frac{500\text{ beats}}{1\text{ min}}$
7, ***p. 264:*** **a.** \$174/flight **b.** \$2.75/hr **c.** \$0.99/download
8, ***p. 265:*** The 39-oz can

Exercises 5.1, *pp. 266–271*

1. quotient **3.** simplest form **5.** denominator
7. $\frac{2}{3}$ **9.** $\frac{2}{3}$ **11.** $\frac{11}{7}$ **13.** $\frac{3}{2}$ **15.** $\frac{1}{4}$ **17.** $\frac{4}{3}$ **19.** $\frac{1}{1}$ **21.** $\frac{5}{3}$
23. $\frac{7}{24}$ **25.** $\frac{20}{1}$ **27.** $\frac{8}{7}$ **29.** $\frac{4}{5}$ **31.** $\frac{5\text{ calls}}{2\text{ days}}$ **33.** $\frac{36\text{ cal}}{5\text{ min}}$
35. $\frac{1\text{ million hits}}{3\text{ mo}}$ **37.** $\frac{17\text{ baskets}}{30\text{ attempts}}$ **39.** $\frac{37\text{ points}}{2\text{ games}}$ **41.** $\frac{100\text{ sq ft}}{\$329}$
43. $\frac{16\text{ males}}{3\text{ females}}$ **45.** $\frac{8\text{ Democrats}}{7\text{ Republicans}}$ **47.** $\frac{1\text{ lb}}{8\text{ servings}}$ **49.** $\frac{307\text{ flights}}{3\text{ days}}$
51. $\frac{1\text{ lb}}{200\text{ sq ft}}$ **53.** 225 revolutions/min **55.** 8 gal/day
57. 0.3 tank/acre **59.** 1.6 yd/dress **61.** 2 hr/day
63. 0.25 km/min **65.** 70 fat calories/tbsp **67.** \$0.45/bar
69. \$2.95/roll **71.** \$66.67/plant **73.** \$99/night

75.

Number of Units	Total Price	Unit Price
125	\$6.69	\$0.05
500	\$15.49	\$0.03

500 envelopes

77.

Number of Units	Total Price	Unit Price
180	\$12.99	\$0.072
250	\$17.49	\$0.070

250 tablets

79.

Number of Units	Total Price	Unit Price
25	\$14.99	\$0.60
50	\$26.55	\$0.53
100	\$54.99	\$0.55

50 discs

81. \$0.16/oz **83.** 2 tutors/15 students **85.** $\frac{5}{1}$ **87.** $\frac{2}{3}$
89. 170 cal/oz **91.** 25 times/min **93.** $\frac{1}{2}$ **95.** 8.4 people/sq km **97.** Lower **99.** **a.** $\frac{62}{67}$ **b.** $\frac{7}{8}$ **101.** (a) Division; (b) 0.5131495; (c) possible estimate: 0.5

Practices: Section 5.2, *pp. 272–276*

1, ***p. 272:*** Yes **2,** ***p. 272:*** Not a true proportion
3, ***p. 273:*** No **4,** ***p. 274:*** $x = 8$ **5,** ***p. 274:*** $x = 12$
6, ***p. 274:*** 64,000 flowers **7,** ***p. 275:*** 810 mi
8, ***p. 276:*** 160 ft

Exercises 5.2, *pp. 277–280*

1. proportion **3.** as **5.** True **7.** False **9.** True
11. False **13.** True **15.** True **17.** $x = 20$ **19.** $x = 38$
21. $x = 4$ **23.** $x = 13$ **25.** $x = 8$ **27.** $x = 4$
29. $x = 20$ **31.** $x = 15$ **33.** $x = 21$ **35.** $x = 13\frac{1}{3}$
37. $x = 100$ **39.** $x = 1.8$ **41.** $x = 21$ **43.** $x = 280$
45. $x = 300$ **47.** $x = 20$ **49.** $x = 10$
51. $x = \frac{1}{5}$, or 0.2 **53.** $x = 0.005$ **55.** $x = \frac{2}{5}$
57. $x = 1\frac{3}{5}$ **59.** False **61.** Not the same **63.** $1\frac{7}{8}$ gal
65. 54.5 g **67.** 100 hydrogen atoms **69.** $41\frac{2}{3}$ in.
71. 0.25 ft **73.** \$600 **75.** 12,000 fish **77.** 280 times
79. 90 mg and 50 mg **81.** **a.** 92 g **b.** 4 g **83.** (a) Multiplication, division; (b) 835,312.5 gal; (c) possible estimate: 880,000 gal

Review Exercises: Chapter 5, *pp. 283–284*

1. $\frac{2}{3}$ **2.** $\frac{1}{2}$ **3.** $\frac{3}{4}$ **4.** $\frac{25}{8}$ **5.** $\frac{8}{5}$ **6.** $\frac{3}{4}$ **7.** $\frac{44\text{ ft}}{5\text{ sec}}$
8. $\frac{9\text{ applicants}}{2\text{ positions}}$ **9.** 0.0025 lb/sq ft **10.** 500,000,000 calls/day
11. 8 yd/down **12.** 400 sq ft/gal **13.** 10,500,000 vehicles/yr **14.** 76,000 commuters/day **15.** \$118.75/night **16.** \$3.89/rental **17.** \$1,250/station
18. \$93.64/share

19.

Number of Units	Total Price	Unit Price
47	\$11.95	\$0.25
92	\$29.90	\$0.33

47 issues

20.

Number of Units	Total Price	Unit Price
300	\$59.99	\$0.20
525	\$74.99	\$0.14

525 checks

21.

Number of Units	Total Price	Unit Price
90	\$7.19	\$0.08
180	\$7.43	\$0.04
360	\$17.91	\$0.05

180 capsules

22.

Number of Units (Fluid Ounces)	Total Price	Unit Price
4	\$1.89	\$0.47
14	\$3.59	\$0.26
20	\$4.69	\$0.23

20 fl ounce-bottle

23. True **24.** False **25.** False **26.** True
27. $x = 6$ **28.** $x = 3$ **29.** $x = 32$ **30.** $x = 30$
31. $x = 2$ **32.** $x = 8$ **33.** $x = \frac{7}{10}$ **34.** $x = \frac{2}{5}$
35. $x = 67\frac{1}{2}$ **36.** $x = 45$ **37.** $x = \frac{3}{4}$, or 0.75
38. $x = \frac{3}{20}$, or 0.15 **39.** $x = 28$ **40.** $x = 0.14$
41. $\frac{1}{15}$ **42.** $\frac{23}{45}$ **43.** \$90/day **44.** 0.125 in./mo **45.** $\frac{2}{3}$
46. 50,000 books **47.** No **48.** $2\frac{1}{7}$ hr **49.** 55 cc
50. 1.25 in. **51.** 0.68 g/cc **52.** 1,175.4 people/sq mi

Posttest: Chapter 5, *p. 285*

1. $\frac{2}{3}$ **2.** $\frac{5}{14}$ **3.** $\frac{55}{31}$ **4.** $\frac{12}{1}$ **5.** $\frac{13 \text{ revolutions}}{12 \text{ sec}}$ **6.** $\frac{1 \text{ cm}}{25 \text{ km}}$ **7.** 68 mph **8.** 8 m/sec **9.** \$136/day **10.** \$0.80/greeting card **11.** False **12.** True **13.** $x = 25$ **14.** $x = 6$ **15.** $x = 28$ **16.** $x = 1$ **17.** 5 million e-mail addresses **18.** $\frac{3}{19}$ **19.** 25 ft **20.** 48 beats/min

Cumulative Review: Chapter 5, *p. 286*

1. $\frac{2}{5}$ **2.** 8,200 **3.** $x = 2.5$ **4.** $\frac{1}{4}$ **5.** \$4 per yd **6.** Possible answer: 3 **7.** $x = \frac{3}{4}$ **8.** 2,106 sq ft **9.** 7.5 in. **10.** \$180

Chapter 6

Pretest: Chapter 6, *p. 288*

1. $\frac{1}{20}$ **2.** $\frac{3}{8}$ **3.** 2.5 **4.** 0.03 **5.** 0.7% **6.** 800% **7.** 67% **8.** 110% **9.** $37\frac{1}{2}$ ft **10.** 55 **11.** Possible estimate: \$48 **12.** 250 **13.** 40% **14.** 250% **15.** \$14 **16.** 8% **17.** $\frac{6}{25}$ **18.** 25% **19.** \$61.11 **20.** \$10,000

Practices: Section 6.1, *pp. 290–296*

1, *p. 290:* $\frac{21}{100}$ **2, *p. 291:*** $\frac{9}{4}$, or $2\frac{1}{4}$ **3, *p. 291:*** $\frac{1}{8}$ **4, *p. 291:*** $\frac{43}{50}$ **5, *p. 292:*** 0.31 **6, *p. 292:*** 0.05 **7, *p. 293:*** 0.482 **8, *p. 293:*** 0.6225 **9, *p. 293:*** 1.12 **10, *p. 294:*** 2.5% **11, *p. 294:*** 9% **12, *p. 294:*** 70% **13, *p. 294:*** 300% **14, *p. 294:*** 71% **15, *p. 295:*** Nitrogen; 78% > 0.93%, or 0.78 > 0.0093. **16. *p. 295:*** 16% **17. *p. 296:*** True. $\frac{2}{3} \approx 67\% > 60\%$ **18. *p. 296:*** 27%

Exercises 6.1, *pp. 297–301*

1. percent **3.** left **5.** $\frac{2}{25}$ **7.** $2\frac{1}{2}$ **9.** $\frac{33}{100}$ **11.** $\frac{9}{50}$ **13.** $\frac{7}{50}$ **15.** $\frac{13}{20}$ **17.** $\frac{3}{400}$ **19.** $\frac{3}{1{,}000}$ **21.** $\frac{3}{40}$ **23.** $\frac{1}{7}$ **25.** 0.06 **27.** 0.72 **29.** 0.001 **31.** 1.02 **33.** 0.425 **35.** 5 **37.** 1.069 **39.** 0.035 **41.** 0.009 **43.** 0.0075 **45.** 31% **47.** 17% **49.** 30% **51.** 4% **53.** 12.5% **55.** 129% **57.** 290% **59.** 287% **61.** 101.6% **63.** 900% **65.** 30% **67.** 10% **69.** 16% **71.** 90% **73.** 6% **75.** $55\frac{5}{9}\%$ **77.** $11\frac{1}{9}\%$ **79.** 600% **81.** 150% **83.** $216\frac{2}{3}\%$ **85.** < **87.** < **89.** 44% **91.** 225%

93.

Fraction	Decimal	Percent
$\frac{1}{3}$	0.333 ...	$33\frac{1}{3}\%$
$\frac{2}{3}$	0.666 ...	$66\frac{2}{3}\%$
$\frac{1}{4}$	0.25	25%
$\frac{3}{4}$	0.75	75%
$\frac{1}{5}$	0.2	20%
$\frac{2}{5}$	0.4	40%
$\frac{3}{5}$	0.6	60%

95. $1\frac{1}{25}$ **97.** $316\frac{2}{3}\%$ **99.** 0.275 **101.** 310% **103.** 254% **105.** 0.79 **107.** $\frac{9}{10}$ **109.** $\frac{1}{10}$ **111.** 900% **113.** $1\frac{7}{20}$ **115.** 0.845 **117.** Among men; $\frac{1}{3} = 33\frac{1}{3}\% < 40\%$ **119. a.** 0.4% **b.** 99.6% **121. a.** Division, multiplication; **b.** 55.27...%; **c.** possible estimate: 50%

Practices: Section 6.2, *pp. 303–309*

1, *p. 303:* **a.** $x = 0.7 \cdot 80$ **b.** $0.5 \cdot x = 10$ **c.** $x \cdot 40 = 20$ **2, *p. 303:*** 8 **3, *p. 303:*** 12 **4, *p. 304:*** Possible estimate: 200 **5, *p. 304:*** 51 workers **6, *p. 305:*** 50 **7, *p. 305:*** 7.2 **8, *p. 305:*** 2,500,000 sq ft **9, *p. 306:*** $83\frac{1}{3}\%$ **10, *p. 306:*** $112\frac{1}{2}\%$ **11, *p. 307:*** 30% **12, *p. 307:*** 270 **13, *p. 308:*** 1,080 **14, *p. 308:*** $33\frac{1}{3}\%$ **15, *p. 308:*** \$98 **16, *p. 309:*** \$340,000 **17, *p. 309:*** 105% **18, *p. 309:*** \$5.97

Exercises 6.2, *pp. 310–313*

1. base **3.** percent **5.** 6 **7.** 23 **9.** 2.87 **11.** \$140 **13.** 0.62 **15.** 0.1 **17.** 4 **19.** \$18.32 **21.** 32 **23.** \$120 **25.** 2.5 **27.** \$250 **29.** 45 **31.** 1.75 **33.** 4,600 **35.** \$49,230.77 **37.** 50% **39.** 75% **41.** $83\frac{1}{3}\%$ **43.** 25% **45.** 150% **47.** $112\frac{1}{2}\%$ **49.** 62.5% **51.** 31% **53.** 60 **55.** $66\frac{2}{3}\%$ **57.** 175 mi **59.** 5% **61.** \$500 **63.** 10 **65.** $66\frac{2}{3}\%$ **67.** 10.2 gal **69.** \$600 **71.** 25% **73.** 54 tables **75.** 30% **77.** 40 questions **79.** 6.8 million **81. a.** Approximately 4 million **b.** 75% **83.** \$30,000,000 **85.** 5,100 employees **87.** \$9,000

Practices: Section 6.3, *pp. 314–320*

1, *p. 314:* 300% **2, *p. 315:*** 1929 **3, *p. 316:*** \$555 **4, *p. 316:*** **a.** \$1,125 **b.** \$2,625 **5, *p. 317:*** \$69.60 **6, *p. 317:*** 50% **7, *p. 318:*** \$1,450 **8, *p. 318:*** \$1,792 **9, *p. 320:*** \$2,524.95

Exercises 6.3, *pp. 321–326*

1. original **3.** discount

5.

Original Value	New Value	Percent Increase or Decrease
\$10	\$12	20% increase
\$10	\$8	20% decrease
\$6	\$18	200% increase
\$35	\$70	100% increase
\$14	\$21	50% increase
\$10	\$1	90% decrease
\$8	\$6.50	$18\frac{3}{4}\%$ decrease
\$6	\$5.25	$12\frac{1}{2}\%$ decrease

7.

Selling Price	Rate of Sales Tax	Sales Tax
\$30.00	5%	\$1.50
\$24.88	3%	\$0.75
\$51.00	$7\frac{1}{2}\%$	\$3.83
\$196.23	4.5%	\$8.83

9.

Sales	Rate of Commission	Commission
$700	10%	$70.00
$450	2%	$9.00
$870	$4\frac{1}{2}$%	$39.15
$922	7.5%	$69.15

11.

Original Price	Rate of Discount	Discount	Sale Price
$700.00	25%	$175.00	$525.00
$18.00	10%	$1.80	$16.20
$43.50	20%	$8.70	$34.80
$16.99	5%	$0.85	$16.14

13.

Selling Price	Rate of Markup	Markup	Cost
$10.00	50%	$5.00	$5.00
$23.00	70%	$16.10	$6.90
$18.40	10%	$1.84	$16.56
$13.55	60%	$8.13	$5.42

15.

Principal	Interest Rate	Time (in years)	Interest	Final Balance
$300	4%	2	$24.00	$324.00
$600	7%	2	$84.00	$684.00
$500	8%	2	$80.00	$580.00
$375	10%	4	$150.00	$525.00
$1,000	3.5%	3	$105.00	$1,105.00
$70,000	6.25%	30	$131,250.00	$201,250.00

17.

Principal	Interest Rate	Time (in years)	Final Balance
$500	4%	2	$540.80
$6,200	3%	5	$7,187.50
$300	5%	8	$443.24
$20,000	4%	2	$21,632.00
$145	3.8%	3	$162.17
$810	2.9%	10	$1,078.05

19.

Original Value	New Value	Percent Decrease
$5	$4.50	10%

21.

Original Price	Rate of Discount	Discount	Sale Price
$87.33	40%	$34.93	$52.40

23.

Selling Price	Rate of Sales Tax	Sales Tax
$200	7.25%	$14.50

25.

Principal	Interest Rate	Kind of Interest	Time (in years)	Interest	Final Balance
$3,000	5%	simple	5	$750.00	$3,750.00

27. 28% **29.** 550% **31.** 300% **33.** $84.95 **35.** 6.5% **37.** $2,700 **39.** $9 **41.** $23.98 **43.** $53\frac{1}{3}$% **45.** $259.35 **47.** $150 **49.** $250 **51. a.** $144 **b.** $152.64 **53.** $3,370.80 **55.** 5,856

Review Exercises: Chapter 6, *pp. 329–332*

1.

Fraction	Decimal	Percent
$\frac{1}{4}$	0.25	25%
$\frac{7}{10}$	0.7	70%
$\frac{3}{400}$	0.0075	$\frac{3}{4}$%
$\frac{5}{8}$	0.625	62.5%
$\frac{41}{100}$	0.41	41%
$1\frac{1}{100}$	1.01	101%
$2\frac{3}{5}$	2.6	260%
$3\frac{3}{10}$	3.3	330%
$\frac{3}{25}$	0.12	12%
$\frac{2}{3}$	0.66 ...	$66\frac{2}{3}$%
$\frac{1}{6}$	0.166 ...	$16\frac{2}{3}$%

2.

Fraction	Decimal	Percent
$\frac{3}{8}$	0.375	37.5%
$\frac{49}{100}$	0.49	49%
$\frac{1}{1,000}$	0.001	0.1%
$1\frac{1}{2}$	1.5	150%
$\frac{7}{8}$	0.875	87.5%
$\frac{5}{6}$	0.833 ...	$83\frac{1}{3}$%
$2\frac{3}{4}$	2.75	275%
$1\frac{1}{5}$	1.2	120%
$\frac{3}{4}$	0.75	75%
$\frac{1}{10}$	0.1	10%
$\frac{1}{3}$	0.33 ...	$33\frac{1}{3}$%

3. 12 **4.** 120% **5.** 50% **6.** 20 **7.** 43.75% **8.** 5.5 **9.** \$6 **10.** 20% **11.** 0.3 **12.** 460 **13.** \$70 **14.** 1,000 **15.** 2000% **16.** 25% **17.** \$200 **18.** $44\frac{4}{9}\%$ **19.** \$12 **20.** \$1,600 **21.** 17% **22.** 42.42

23.

Original Value	New Value	Percent Decrease
24	16	$33\frac{1}{3}\%$

24.

Selling Price	Rate of Sales Tax	Sales Tax
\$50	6%	\$3.00

25.

Sales	Rate of Commission	Commission
\$600	4%	\$24

26.

Original Price	Rate of Discount	Discount	Sale Price
\$200	15%	\$30	\$170

27.

Selling Price	Rate of Markup (based on the selling price)	Markup	Cost
\$51	50%	\$25.50	\$25.50

28.

Principal	Interest Rate	Time (in years)	Simple Interest	Final Balance
\$200	4%	2	\$16	\$216

29. \$1,800 **30.** 48% **31.** The agent that charges 11% **32.** \$7,200 **33.** 20% **34.** Paper **35.** $\frac{1}{4}$ **36.** Above the typical markup **37.** 40% **38.** 10% **39.** 0.0629 **40.** \$207 **41.** 68% **42.** Possible estimate: 80 in. **43.** 63 first serves **44.** Yes **45.** 180 **46.** 20% **47.** No **48.** 57,770,000 cats **49.** \$165 **50.** 320 tons **51.** 70,000 mi **52.** 10,000 hr **53.** \$1,000 **54.** \$30,000 **55.** \$7,939.58 **56.** 22%

57.

Quarter	Income	Percent of Total Income (rounded to the nearest whole precent)
1	\$375,129	27%
2	289,402	21%
3	318,225	23%
4	402,077	29%
Total	\$1,384,833	100%

58. Individual income taxes: \$948 billion; Social Security taxes: \$797 billion; corporate income taxes: \$280 billion; excise taxes: \$65 billion; other: \$86 billion

Posttest: Chapter 6, *p. 333*

1. $\frac{1}{25}$ **2.** $\frac{11}{40}$ **3.** 1.74 **4.** 0.08 **5.** 0.9% **6.** 1,000% **7.** 83% **8.** 220% **9.** 7.5 mi **10.** 48 **11.** Possible estimate: \$7 **12.** 200 **13.** 60% **14.** 250% **15.** \$200 **16.** 6 spaces **17.** 5% **18.** 4 pt **19.** \$31.60 **20.** $3\frac{1}{3}\%$

Cumulative Review: Chapter 6, *p. 334*

1. 109 **2.** 0.83 **3.** 0.7 **4.** $5\frac{7}{10}$ **5.** 7.5 **6.** \$1,000 **7.** $\frac{1}{8}$ **8.** 6,412 students

9.

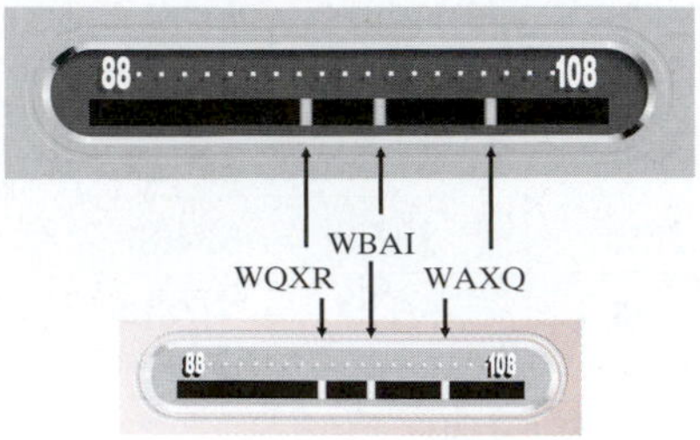

10. 239 thousand

Chapter 7

Pretest: Chapter 7, *p. 336*

1. +7 **2.** −4 **3.** −17 **4.** 0 **5.** −7 **6.** 0 **7.** −75 **8.** $-\frac{1}{2}$ **9.** −64 **10.** $\frac{1}{4}$ **11.** 2 **12.** $-\frac{1}{8}$ **13.** 8 **14.** 8 **15.** 36 **16.** −7 **17.** −51 lb **18.** −\$10.50 **19.** −\$8,000 **20.** −\$0.57

Practices: Section 7.1, *pp. 338–341*

1, *p. 338:* 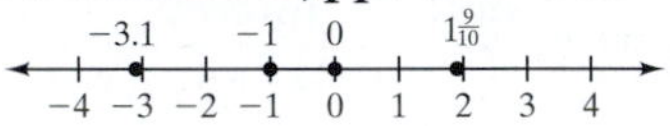

2, *p. 338:* **a.** −9 **b.** $4\frac{9}{10}$ **c.** 2.9 **d.** −31 **3, *p. 339:*** **a.** 9 **b.** $1\frac{3}{4}$ **c.** 4.1 **d.** 5 **4, *p. 339:*** **a.** Sign: −; absolute value: 4 **b.** Sign: +; absolute value: $6\frac{1}{2}$ **5, *p. 340:*** **a.** 0 **b.** −5 **c.** −4 **6, *p. 341:*** +2 yd **7, *p. 341:*** Sirius

Exercises 7.1, *pp. 342–346*

1. positive number **3.** signed number **5.** opposites **7.** smaller **9.** & **11.** 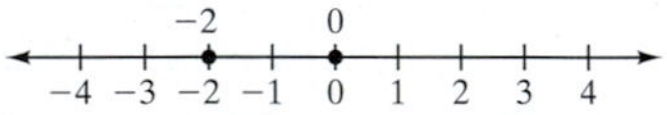

13. −8 **15.** −10.2 **17.** 5 **19.** $-2\frac{1}{3}$ **21.** 4.1 **23.** 1.2 **25.** 6 **27.** $\frac{4}{5}$ **29.** 2 **31.** 0.6 **33.** Sign: +; Absolute value: 8 **35.** Sign: −; Absolute value: 4.3 **37.** Sign: −; Absolute value: 7 **39.** Sign: +; Absolute value: $\frac{1}{5}$ **41.** Two; −5 and 5 **43.** No **45.** −4 **47.** 12 **49.** 2 **51.** $-2\frac{1}{3}$ **53.** −2 **55.** 9 **57.** −2 **59.** −7 **61.** −8.3 **63.** $-3\frac{1}{2}$ **65.** T **67.** T **69.** T **71.** T **73.** F **75.** T **77.** −3, 0, 3 **79.** −9, −4.5, 9 **81.** −\$150 **83.** +14.5°C

85. 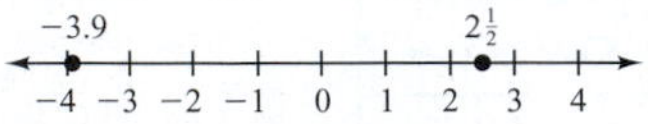

87. −\$10.98

89. a. 0.5 **b.** 11 **91. a.** < **b.** > **93.** Mariana Trench **95.** Decreased by 25 mg **97.** No **99.** Mars

101. a. Par score

Third First Second Fourth
−5 −4 −3 −2 −1 0 1 2 3 4 5

b. The third round

Practices: Section 7.2, *pp. 348–350*

1, ***p. 348:*** −25 **2,** ***p. 348:*** $-4\frac{1}{2}$ **3,** ***p. 348:*** 7
4, ***p. 349:*** 0 **5,** ***p. 349:*** $2\frac{3}{5}$ **6,** ***p. 349:*** 0
7, ***p. 350:*** 456 m **8,** ***p. 350:*** −2.478

Exercises 7.2, *pp. 351–354*

1. right **3.** larger **5.** 1 **7.** 3 **9.** 7 **11.** 0 **13.** 0 **15.** −5 **17.** 200 **19.** 6 **21.** −150 **23.** −5 **25.** −27 **27.** 0 **29.** 4.9 **31.** 0.1 **33.** 59.5 **35.** −5.9 **37.** −14.5 **39.** −6 **41.** $-\frac{3}{5}$ **43.** $1\frac{3}{5}$ **45.** $-\frac{1}{10}$ **47.** −102 **49.** $-7\frac{1}{2}$ **51.** 7 **53.** −6.9 **55.** 7.58 **57.** −4.914 **59.** 4.409
61. 3

Move 9 units to the right
−6 −5 −4 −3 −2 −1 0 1 2 3 4 5 6
Start End

63. 0 **65.** −4 **67.** 19,340 ft **69.** −238 employees **71.** −1 **73.** −6 yr (6 yr ago) **75.** −6 yd **77. a.** +\$0.68 **b.** \$30.03

Practices: Section 7.3, *pp. 356–357*

1, ***p. 356:*** −2 **2,** ***p. 356:*** 18 **3,** ***p. 356:*** −21.1
4, ***p. 356:*** −10 **5,** ***p. 357:*** 22,965 ft

Exercises 7.3, *pp. 358–361*

1. opposite **3.** order of operations **5.** 7 **7.** 1 **9.** −14 **11.** 44 **13.** −25 **15.** −19 **17.** 6 **19.** −38 **21.** −26 **23.** 26 **25.** −15 **27.** 1,000 **29.** −1.52 **31.** 9.7 **33.** 0 **35.** 10.5 **37.** −0.5 **39.** −5 **41.** $7\frac{3}{4}$ **43.** $-7\frac{1}{4}$ **45.** $7\frac{1}{4}$ **47.** 7 **49.** −5 **51.** −15 **53.** −3.842 **55.** −16.495 **57.** −25 **59.** 0 **61.** 1 **63.** 1,000 mi **65.** About 2,100 yr **67.** −\$1.4 million **69.** −\$889 billion **71. a.** 8.29 in. **b.** −5.47 in.

Practices: Section 7.4, *pp. 362–364*

1, ***p. 362:*** 32 **2,** ***p. 363:*** −10 **3,** ***p. 363:*** **a.** 1 **b.** −1
4, ***p. 363:*** $\frac{4}{15}$ **5,** ***p. 363:*** −20 **6,** ***p. 364:*** −48
7, ***p. 364:*** 8 **8,** ***p. 364:*** No

Exercises 7.4, *pp. 365–367*

1. positive **3.** even **5.** −10 **7.** −10 **9.** 25 **11.** 306 **13.** −16 **15.** −8,163 **17.** −40 **19.** −176 **21.** 800 **23.** −7,200 **25.** −5 **27.** −10 **29.** 5.52 **31.** −8 **33.** $-\frac{5}{27}$ **35.** $-\frac{5}{6}$ **37.** −25 **39.** 10,000 **41.** 0.25 **43.** −0.001 **45.** $\frac{9}{16}$ **47.** −1 **49.** 0.094864 **51.** −216 **53.** −60 **55.** 0 **57.** 5 **59.** $-\frac{32}{135}$ **61.** 0.14652 **63.** 5 **65.** −26 **67.** 0 **69.** −28 **71.** 1.25 **73.** −12 **75.** 9 **77.** 41 **79. a.** 5 **b.** 2 **c.** −1 **d.** −4 **e.** −7 **81.** −4,830 **83.** −0.0001 **85.** −8 **87.** −72 mm **89.** −175 mg **91.** −160° C **93.** −64 ft **95. a.** −\$900 **b.** \$100

Practices: Section 7.5, *pp. 368–370*

1, ***p. 368:*** 12 **2,** ***p. 369:*** $-\frac{3}{5}$ **3,** ***p. 369:*** −0.3
4, ***p. 369:*** $-\frac{1}{6}$ **5,** ***p. 369:*** **a.** 0 **b.** −1 **6,** ***p. 370:*** −\$875
7, ***p. 370:*** 0 degrees

Exercises 7.5, *pp. 371–374*

1. positive **3.** equal **5.** 5 **7.** 0 **9.** −5 **11.** −2 **13.** 25 **15.** −25 **17.** 7 **19.** −2 **21.** 17 **23.** 6 **25.** −0.3 **27.** −20 **29.** 16 **31.** $-\frac{5}{6}$ **33.** −21 **35.** −16 **37.** 6.172 **39.** −3.5 **41.** $-\frac{1}{5}$ **43.** 1 **45.** $-\frac{2}{5}$ **47.** $5\frac{1}{2}$ **49.** $4\frac{1}{4}$ **51.** $\frac{3}{4}$ **53.** −8 **55.** −4 **57.** 5 **59.** 1 **61.** −16 **63.** −3 **65.** −0.06 **67.** −4 **69.** 2 **71.** 11 **73.** $9 \div (1 - 4) = -3$
75. $6 \div (3 - 1) - 4 = -1$
77. $(8 - 10) \cdot 2 - (-5 + 13) \div 4 = -6$
79. $-\frac{6}{5}$ **81.** $-3\frac{1}{6}$ **83.** $-\frac{4}{27}$ **85.** −6,098.9 **87.** −3.5 thousand **89.** −1,000 ft/min **91.** Yes. **93. a.** Friday; Sunday **b.** −18°F

Review Exercises: Chapter 7, *pp. 376–378*

1. & **2.**

−3 1.5
−4 −3 −2 −1 0 1 2 3 4

3. −6 **4.** 4 **5.** $7\frac{1}{2}$ **6.** −10.1 **7.** 10 **8.** 2.5 **9.** $1\frac{1}{5}$ **10.** 7 **11.** −11 **12.** 10 **13.** 9 **14.** −2 **15.** −8, −3.5, 8 **16.** −9.7, −6, 9 **17.** −2.9, $-2\frac{1}{2}$, 0 **18.** −4, $-1\frac{1}{4}$, 0 **19.** +10 ft **20.** −\$350 **21.** −20 **22.** −2 **23.** $6\frac{1}{2}$ **24.** −1 **25.** −4.1 **26.** −2 **27.** −7 **28.** $-\frac{1}{4}$ **29.** 0 **30.** 28 **31.** −10 **32.** −11 **33.** 3 **34.** $-4\frac{1}{8}$ **35.** 100 **36.** −45 **37.** $-\frac{20}{33}$ **38.** −7.35 **39.** 72 **40.** −30 **41.** $\frac{1}{16}$ **42.** 0.49 **43.** 36 **44.** −81 **45.** 5 **46.** −10 **47.** −5 **48.** $\frac{1}{32}$ **49.** −50 **50.** 2 **51.** 15 **52.** 114 **53.** −100 **54.** 1 **55.** 2.73 **56.** −2.62 **57.** 104 **58.** 21 **59.** Yes **60.** Yes **61.** −\$2,820 **62.** \$1,649.83 **63.** 28°F **64.** 7 questions **65.** −\$1.20 **66.** 2,010 years apart **67.** −\$400 **68.** 90 feet below the surface **69.** 5 under par **70.** 100 ft above the point of release (+100 ft) **71. a.** Alabama **b.** 14°F **72. a.** −\$32.6 million **b.** −\$8.15 million

Posttest: Chapter 7, *p. 379*

1. −10 **2.** $-\frac{1}{2}$ **3.** 0 **4.** −0.5 **5.** 49 **6.** 0 **7.** −207 **8.** −0.1 **9.** −144 **10.** $\frac{1}{16}$ **11.** −4 **12.** 2 **13.** 4 **14.** 21 **15.** 40 **16.** −18 **17.** 58°F **18.** −\$550/yr **19.** $-15\frac{1}{2}$ games **20.** 66.38°C

Cumulative Review: Chapter 7, *p. 380*

1. 3,000 **2.** 10 **3.** 14 **4.** 1.5 **5.** 20% **6.** $n = 65$ **7.** −4 **8.** 80 mg **9.** −89°C **10.** 25%

Chapter 8

Pretest: Chapter 8, *pp. 382–383*

1. 9 **2.** 17 **3.** 11 min. **4.** 31 **5.** 3.1 **6.** 5.5 yr longer **7.** 1983–1985 and 1995–2001 **8.** 900,000 newspapers **9.** 395 thousand **10.** 43%

Practices: Section 8.1, *pp. 386–390*

1, *p. 386:* Reggie Jackson **2, *p. 386:*** Above **3, *p. 387:*** **a.** 7 **b.** 4 **4, *p. 388:*** **a.** $16 million **b.** The median for the six movies was $2 million less. **5, *p. 389:*** **a.** 2 **b.** 4 and 9 **c.** No mode **6, *p. 389:*** 12 hours **7, *p. 389:*** 7 **8, *p. 390:*** $4.51

Exercises 8.1, *p. 391*

1. statistics **3.** weighted **5.** mode

7.

Numbers	Mean	Median	Mode(s)	Range
a. 8, 2, 9, 4, 8	6.2	8	8	7
b. 3, 0, 0, 3, 10	3.2	3	0 and 3	10
c. 6.5, 9, 8.5, 6.5, 8.1	7.7	8.1	6.5	2.5
d. $3\frac{1}{2}, 3\frac{3}{4}, 4, \frac{1}{2}, 3\frac{1}{4}$	$3\frac{3}{5}$	$3\frac{1}{2}$	$3\frac{1}{2}$	$\frac{3}{4}$
e. 4, −2, −1, 0, −1	0	−1	−1	6

9. $12,266.67 **11.** No; the GPA was 3.25. **13.** The mean amount is $100,000. We can't compute the median amount because we don't know the actual amounts given to each heir. **15.** Indiana, Missouri, North Carolina, and Tennessee **17. a.** $50,921 **b.** $32,927 **19. a.** 31,000 mi **b.** 20,000 mi **c.** 8,000 mi **d.** 86,000 mi **21.** (a) Division and subtraction; (b) 25.11; (c) possible estimate: 13

Practices: Section 8.2, *pp. 396–403*

1, *p. 396:* **a.** The S&H charges are $4.95. **b.** The customer must pay $5.95. **c.** The S&H charges total $5.45. **2, *p. 397:*** **a.** 5 million passengers **b.** Atlanta **c.** 23 million passengers **3, *p. 398:*** **a.** Cattle and calves **b.** $27 million **c.** $2 million **4, *p. 399:*** **a.** $\frac{2}{5}$ **b.** 17% **5, *p. 400:*** **a.** 24 presidents **b.** No **c.** 33 presidents **6, *p. 401:*** **a.** July **b.** 27°F **c.** In Chicago, mean temperatures increase from January through July and then decrease from July through December. **7, *p. 402:*** **a.** 47,000 children.: **b.** 2003 **c.** Possible answer: Between 1995 and 2005, the number of children in foster care declined, and the number of children who were adopted increased. **8, *p. 403:*** **a.** $\frac{2}{25}$ **b.** 49% **c.** 1,250,000 living veterans

Exercises 8.2, *p. 404*

1. table **3.** rows **5.** bar graph **7.** line graph **9. a.** $60 **b.** $100 **c.** She will pay a lower commission if she sells 400 shares in a single deal; 400 shares in a single deal will cost her $70 or $90, whereas two deals of 200 shares will cost her between $100 and $120. **11. a.** United States **b.** 125 million **13. a.** Milk, lemon juice, and vinegar **b.** 8 **c.** Pure water **15. a.** More than 89 minutes **b.** 92 **c.** 70 **17. a.** 90 million **b.** 2003 **c.** Every year the number of cell phone subscribers increased. **19. a.** 7% **b.** Mass merchants **c.** 460 million

Review Exercises: Chapter 8, *p. 411*

1.

List of Numbers	Mean	Median	Mode	Range
6, 7, 4, 10, 4, 5, 6, 8, 7, 4, 5	6	6	4	6
1, 3, 4, 4, 2, 3, 1, 4, 5, 1	2.8	3	1 and 4	4

2. a. The husbands (83 yr) lived longer than the wives (70 yr). **b.** 13 yr **c.** 23 **3. a.** Half of the people were younger than 35.3 and half were older. **b.** Answers will vary. **4.** This machine is reliable because the range is 1.7 fl oz. **5.** The average was 1,000; therefore the show was making a profit. **6. a.** 64,510,000 tons **b.** Long Beach, 62,515,408 **c.** 49,350,000 tons **d.** All except the Port of South Louisiana **7. a.** 9 **b.** June **c.** $\frac{1}{8}$ **8. a.** 36,000 **b.** 1,500 **c.** The number of movie screens in the United States decreased from 2000 to 2001 and then increased each year from 2001 to 2005. **9. a.** 61 million **b.** 10% **c.** It increases for younger age groups, peaks for 40–44, and decreases for older age groups. **10. a.** Run number 3 **b.** Approximately 3 min **c.** With practice, the rat ran through the maze more quickly. **11. a.** 2005 **b.** $\frac{4}{9}$ **c.** Both out-of-pocket expenditures (with the exception of the period 2005–2006) and insurance expenditures increased and are expected to continue to increase each year from 2000 through 2010. **12. a.** 27,470 **b.** 4,995 **c.** 58% **13.** 76 **14.** Five years

Posttest: Chapter 8, *p. 415*

1. 3.375 **2.** 162 **3.** 77 **4.** Mode: 81 in., range: 11 in. **5.** Toyota Prius, Toyota Camry Hybrid, and Ford Escape Hybrid **6.** 2,000,000,000 metric tons **7.** About $175,000,000 **8.** 13,500,000 students **9.** 60% **10.** 1961–1975; U.S. scientists received approximately 40 and U.K. scientists received approximately 20 Nobel prizes.

Cumulative Review: Chapter 8, *p. 417*

1. $\frac{4}{5}$ **2.** 40 **3.** 3,010 **4.** $x = 17\frac{1}{4}$ **5.** $n = 21$ **6.** 2 **7.** −23 **8.** 5,000 yr **9.** 71% **10.** 2000 and 2003

Chapter 9

Pretest: Chapter 9, *p. 420*

1. $y = -6$ **2.** $x = -2$ **3.** $x = -4$ **4.** $a = -5$ **5.** $x = 3$ **6.** $y = 1$ **7.** $c = -2$ **8.** $a = -10$ **9.** $x = 2$ **10.** $y = 2$ **11.** $x = -1$ **12.** $x = -4$ **13.** $3(n + 7) = 30$ **14.** $l = \frac{1}{7}h$ **15.** $45 = 4x + 5$; \$10/hr. **16.** $475 = 250 + 75x$; 3, so he worked a total of 6 hr. **17.** $14x + 5 = 19$; \$1 **18.** $15x + 75 = 795$; \$48 **19. a.** $B = \frac{1}{12}twl$ **b.** $B = 13\frac{1}{3}$ board feet **20. a.** $s = r - d$ **b.** $s = \$301$

Practices: Section 9.1, *pp. 421–426*

1, *p. 421:* $x = -3$ **2, *p. 422:*** $m = -12$ **3, *p. 422:*** $y = -2$ **4, *p. 422:*** $y = 6$ **5, *p. 423:*** $-15w = -105$; 7 weeks **6, *p. 423:*** $x = -7$ **7, *p. 424:*** $k = 15$ **8, *p. 424:*** $d = 4$ **9, *p. 425:*** $x = 80$ **10, *p. 425:*** **a.** $2 + 0.2x = 10$ **b.** $x = 40$ oranges

Exercises 9.1, *p. 427*

1. $a = -14$ **3.** $b = -11$ **5.** $z = -7$ **7.** $x = -2$ **9.** $c = -19$ **11.** $z = -7.7$ **13.** $x = 8.2$ **15.** $y = -\frac{2}{3}$ **17.** $n = \frac{1}{6}$ **19.** $t = -6\frac{1}{4}$ **21.** $z = -12$ **23.** $x = -6$ **25.** $n = 4$ **27.** $m = -1.5$ **29.** $w = -240$ **31.** $x = 12$ **33.** $t = -30$ **35.** $y = -15$ **37.** $y = -\frac{2}{5}$ **39.** $n = 14$ **41.** $x = 2$ **43.** $k = -3$ **45.** $x = 0$ **47.** $h = -7$ **49.** $p = -5\frac{1}{4}$ **51.** $b = 2$ **53.** $a = -33$ **55.** $y = -36$ **57.** $x = 48$ **59.** $c = -21$ **61.** $x = 18$ **63.** $x = -19$ **65.** $t = -7$ **67.** $n = -1{,}434.3$ **69.** $x = -15.2$ **71.** $r = -0.4$ **73.** $x = -0.3$ **75.** $t = 9.6$ **77.** $n = -1$ **79.** $c = -2\frac{1}{2}$ **81.** $y = -1\frac{4}{5}$ **83.** $-1.5x = -750$; 500 shares **85.** $-9{,}046 + x = -15{,}671$; a loss of \$6,625 **87.** $15x - 25 = 155$; \$12 **89.** $8x + 3 = 43$; 5 people **91.** **(a)** $0.0625x + 12.25 = 67.35$; **(b)** exact answer: 881.6 kWh; **(c)** possible estimate: 1,000 kWh

Practices: Section 9.2, *pp. 430–433*

1, *p. 430:* **a.** $12x$ **b.** $6y$ **c.** $-z + 6$ **2, *p. 431:*** $y = -8$ **3, *p. 431:*** $x = 1$ **4, *p. 432:*** **a.** $x + 3x = x + 9$ **b.** $x = 3$ **5, *p. 432:*** $x = -1$ **6, *p. 433:*** $x = 3$ **7, *p. 433:*** 4 mi

Exercises 9.2, *p. 434*

1. $7x$ **3.** $3a$ **5.** $-3y$ **7.** $-n$ **9.** $3c + 12$ **11.** $8 - 6x$ **13.** $2y + 5$ **15.** $m = 4$ **17.** $y = 9$ **19.** $a = 0$ **21.** $b = -1\frac{3}{4}$ **23.** $n = 13$ **25.** $x = 3$ **27.** $n = 9$ **29.** $a = -\frac{2}{3}$ **31.** $x = 4$ **33.** $p = -2$ **35.** $x = -1$ **37.** $p = 1\frac{1}{2}$ **39.** $n = 12$ **41.** $t = 2$ **43.** $y = 1$ **45.** $x = \frac{5}{6}$ **47.** $n = 3$ **49.** $x = -2$ **51.** $x = 2$ **53.** $n = 35$ **55.** $y = 15$ **57.** $y = -\frac{3}{8}$ **59.** $n = 21$ **61.** $r = 2$ **63.** $3x - 5x = -20$; $x = 10$ **65.** $3(x + 5) = 2x$; $x = -15$ **67.** $r = -0.5$ **69.** $n = 2.7$ **71.** $y = 19.1$ **73.** $4x + 6$ **75.** $y = -1\frac{1}{2}$ **77.** $x = -\frac{1}{2}$ **79.** $m = -\frac{1}{5}$ **81.** $s - 0.06s = 211{,}500$; \$225,000 **83.** $140 + 2w = 200$; 30 ft **85.** $\frac{1}{3}x + \frac{1}{2}x = 24{,}000$; \$9,600 was donated to the local charity, and \$14,400 was donated to the national charity. **87.** $3{,}000 + 120x = 2{,}400 + 145x$; 24 mo **89.** Let $x = $ B1; $2x + 3(20 - x) = 40$; 20

Practices: Section 9.3, *pp. 438–439*

1, *p. 438:* $t = g - \frac{1}{200}a$ **2, *p. 438:*** $I = \$360$ **3, *p. 439:*** $F = 39.5°$ **4, *p. 439:*** 21 kg per sq m

Exercises 9.3, *p. 440*

1. $d = \frac{t}{3}$ **3.** $a = \frac{h}{n}$ **5.** $A = 6e^2$ **7.** 23° **9.** $-2\frac{1}{2}$ **11.** 18.9 in **13.** $\frac{23}{40}$ **15.** $v = \$13{,}500$ **17.** $C = 60$ mg **19.** **(a)** $T = 0.48A + 0.42A$, where T represents the total number of calves born; **(b)** exact answer: 295 calves; **(c)** possible estimate: 270 calves

Review Exercises: Chapter 9, *p. 445*

1. $x = -6$ **2.** $y = -9$ **3.** $d = -9$ **4.** $w = -11$ **5.** $x = -9$ **6.** $y = -16$ **7.** $p = 0$ **8.** $x = 0$ **9.** $a = -9.7$ **10.** $b = -9$ **11.** $y = -5\frac{1}{2}$ **12.** $d = -6\frac{1}{3}$ **13.** $x = -7$ **14.** $c = -1$ **15.** $p = -\frac{1}{4}$ **16.** $a = -\frac{2}{3}$ **17.** $y = 20$ **18.** $w = 30$ **19.** $b = -180$ **20.** $x = -450$ **21.** $x = -7.5$ **22.** $a = -5.4$ **23.** $x = 4$ **24.** $a = 2$ **25.** $y = 5$ **26.** $w = 2$ **27.** $y = 3\frac{1}{2}$ **28.** $y = 2\frac{2}{3}$ **29.** $a = 24$ **30.** $w = 16$ **31.** $x = -25$ **32.** $b = -10$ **33.** $c = 0$ **34.** $d = 0$ **35.** $c = -6$ **36.** $a = -18$ **37.** $y = 9$ **38.** $b = 6$ **39.** $c = -2$ **40.** $x = 2\frac{1}{5}$ **41.** $y = 8$ **42.** $t = 3$ **43.** $x = 8$ **44.** $m = 3$ **45.** $n = \frac{1}{3}$ **46.** $a = -9$ **47.** $r = -8$ **48.** $y = -1$ **49.** $x = 9$ **50.** $y = -1$ **51.** $s = 2$ **52.** $x = -2$ **53.** $m = 2$ **54.** $x = 6$ **55.** $s = 11$ **56.** $z = \frac{7}{8}$ **57.** $w = 3$ **58.** $z = 0$ **59.** $x = 6$ **60.** $y = 10$ **61.** $y = -8$ **62.** $m = 2\frac{2}{3}$ **63.** $b = -15$ **64.** $d = -40$ **65.** $6x + 3 = x - 17$; -4 **66.** $3x + 2(x - 1) = 33$; 7 **67.** $R = F + 460$ **68.** $s = fw$ **69.** $A = 9(\frac{E}{I})$ **70.** $m = \frac{1}{2}(s + l)$ **71.** 50.4 **72.** 144 **73.** 880 **74.** -20 **75.** $p = \$5.38$/unit **76.** 17 wk **77.** 9 ft **78.** 4 hr **79.** $D = 0.9$ g/cc **80.** $t = 37.2$ cm

Posttest: Chapter 9, *p. 447*

1. $x = -5$ **2.** $y = 0$ **3.** $a = -3$ **4.** $b = -10$ **5.** $x = \frac{1}{2}$ **6.** $y = -2\frac{1}{2}$ **7.** $y = -9$ **8.** $a = -60$ **9.** $x = -6$ **10.** $c = -2$ **11.** $y = -6$ **12.** $x = 2$ **13.** $F = ma$ **14.** $4(n + 3) = 5n$ **15.** $0.35 + 0.16x = 2.75$; 15 min, so the total length of the call was 16 min. **16.** $100 + \frac{8}{9}P = 12{,}906$; \$14,406.75. **17.** $200 + 30x = 500$; 10 months **18.** $0.06x + 0.04x = 80$; \$800 **19. a.** $d = 2.2r$ **b.** $d = 121$ ft **20. a.** $f = \frac{22}{15}m$ **b.** $f = 66$ fps

Cumulative Review: Chapter 9, *p. 448*

1. $2\frac{5}{6}$ **2.** 6.5 **3.** 0.001 **4.** $83\frac{1}{3}\%$ **5.** -2 **6.** $\frac{1}{17}$ **7.** $x = 6$ **8.** This change represented an increase because 0.1 is larger than 0.075 (by 0.025). **9.** 2,000,000 subscribers **10.** Possible answer: $i = 20$

Chapter 10

Pretest: Chapter 10, *p. 450*

1. 2 **2.** 10,000 **3.** 12 ft 6 in. **4.** b **5.** d **6.** 3,500 **7.** 2.1 **8.** 700 cm **9.** 3 yr **10.** 6.5 cm

Practices: Section 10.1, *pp. 453–457*

1, *p. 453:* 2 **2, *p. 453:*** $1\frac{2}{3}$ yd **3, *p. 454:*** 86,400 sec **4, *p. 454:*** No; **5, *p. 455:*** 19 in **6, *p. 455:*** 1 yr 8 mo **7, *p. 455:*** 5 lb 8 oz **8, *p. 456:*** 7 hr 10 min **9, *p. 456:*** 3 qt **10, *p. 457:*** 6 sec

Exercises 10.1, *p. 458–461*

1. length **3.** larger **5.** unit **7.** 4 **9.** 3 **11.** 720 **13.** 420 **15.** 30 **17.** 2 **19.** 3,520 **21.** 512 **23.** 2 **25.** 5 **27.** $3\frac{1}{2}$ **29.** 3,000 **31.** $\frac{3}{4}$ **33.** 12 **35.** 310 **37.** 7 ft 6 in.

39.	192	$5\frac{1}{3}$
41.	180	15
43.	48	2,000
45.	128,000	8,000
47.	32	2
49.	1	$\frac{1}{2}$
51.	$\frac{5}{6}$	$\frac{1}{72}$
53.	1,320	$\frac{11}{30}$

55. 1 lb 14 oz **57.** 29 lb 15 oz **59.** 9 yr 6 mo **61.** 2 gal 3 qt **63.** 4 qt **65.** 8 min 15 sec **67.** 14 lb 1 oz **69.** 247 sec **71.** 5 yd **73.** Jay Feely **75. a.** Width: 3 ft, depth: $2\frac{1}{2}$ ft **b.** $7\frac{1}{2}$ sq ft **77.** 13 min 43 sec **79.** 6 in. **81.** 1 lb 5 oz **83.** More; he has1 yr 10 mo left on the lease. **85. a.** Division; **b.** 5.5 mi; **c.** possible estimate: 6 mi

Practices: Section 10.2, *pp. 467–469*

1, *p. 467:* 3.1 **2, *p. 467:*** 25 **3, *p. 468:*** 5 km **4, *p. 468:*** Large intestine **5, *p. 469:*** 38 L **6, *p. 469:*** 380 mi

Exercises 10.2, *p. 478*

1. weight **3.** kilo- **5.** centi- **7.** c **9.** b **11.** b **13.** b **15.** a **17.** 1 **19.** 0.75 **21.** 80 **23.** 3,500 **25.** 0.005 **27.** 4 **29.** 7 **31.** 4.13 **33.** 2,000 **35.** 0.0075

37.	2,380	2.38
39.	1	0.01
41.	0.3	0.0003
43.	450,000	0.45
45.	0.709	0.000709
47.	17,000,000	17,000

49. 3,250 m, or 3.25 km **51.** 98,025.6 g, or 98.0256 kg **53.** 840 **55.** 4 **57.** 1.2 **59.** 2 **61.** 3.5 m or 350 cm **63.** 2,500 **65.** a **67.** Approximately 3.3 qt **69.** 6 g **71.** Yes **73.** 3 in. **75.** 2,400 mm **77.** 750 cm **79.** 10.1 kg **81.** 1,350 g **83.** 3.2 L

Review Exercises: Chapter 10, *p. 476–478*

1. 15 **2.** $1\frac{2}{3}$ **3.** 2 **4.** $3\frac{1}{3}$ **5.** 3,000 **6.** 136 **7.** 48 **8.** $2\frac{1}{2}$ **9.** 435 **10.** 4 ft 2 in. **11.** 2 **12.** 125 **13.** 8 hr 10 min **14.** 18 ft 9 in. **15.** 1 gal 3 qt **16.** 7 lb 2 oz **17.** c **18.** a **19.** b **20.** c **21.** b **22.** a **23.** a **24.** c **25.** 0.037 **26.** 4,000 **27.** 800 **28.** 2,100 **29.** 0.6 **30.** 5.1 **31.** 112 **32.** 2 **33.** 20 **34.** 15 **35.** 1.2 hr **36.** 12,800 pt **37.** *Frankenstein* **38.** 0.6 L **39.** 2 g **40.** 0.072 g **41.** 1.6 g/L **42.** 750,000 g **43.** 6 hr **44.** 25,000 mg **45.** 40 ft **46.** 9.104 g **47.** 4 min 30 sec **48.** 10 in. **49.** 24 in. **50.** 5.5 L **51.** 16 wk **52.** 178–208 km/hr

Posttest: Chapter 10, *p. 479*

1. 2 **2.** 21 **3.** 1 hr 45 min **4.** c **5.** c **6.** 4 **7.** 0.5 **8.** 2 km **9.** 22,000 mi **10.** Finback whale

Cumulative Review: Chapter 10, *p. 480*

1. 729 **2.** $5\frac{1}{2}$ **3.** 14 **4.** $\frac{5}{3}$ **5.** \$18 **6.** 8 **7.** 84 **8.** −\$1.47 **9.** 28,000 **10.** \$500 thousand

Chapter 11

Pretest: Chapter 11, *p. 482*

1. a. **b.**

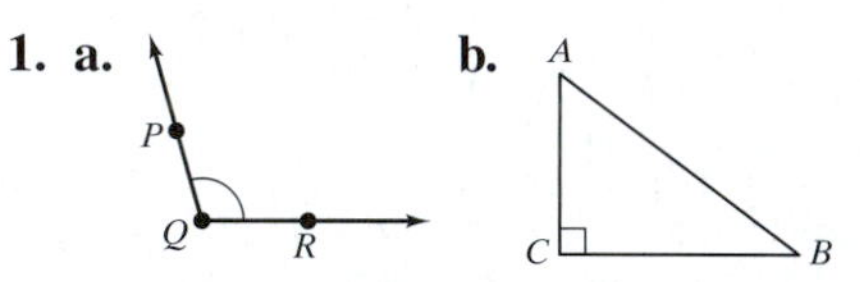

2. a. 6 **b.** 11 **3.** 80° **4.** 54° **5.** 14 in. **6.** 20 ft **7.** Approximately 12.56 in. **8.** 10.4 m **9.** 30 in^2 **10.** 78.5 ft^2 **11.** 125 cm^3 **12.** Approximately 254.34 m^3 **13.** $a = 75°$ **14.** 15 m **15.** $a = b = 90°$; $x = 7$ ft; $y = 5$ ft **16.** $a = 104°$ **17.** 21 cm **18.** 35 sq mi **19.** 6.4 ft **20.** 8,100 cu ft

Practices: Section 11.1, *pp. 487–493*

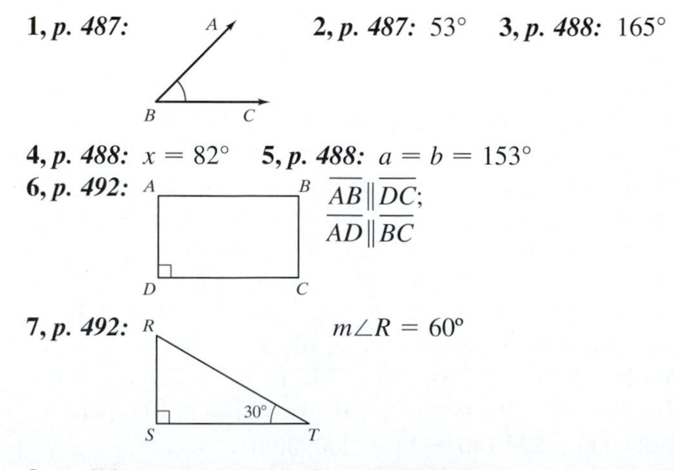

1, *p. 487:* **2, *p. 487:*** 53° **3, *p. 488:*** 165°

4, *p. 488:* $x = 82°$ **5, *p. 488:*** $a = b = 153°$

6, *p. 492:* $\overline{AB} \parallel \overline{DC}$; $\overline{AD} \parallel \overline{BC}$

7, *p. 492:* $m\angle R = 60°$

8, *p. 493:* $m\angle U = 60°$ **9, *p. 493:*** 24 in.

Exercises 11.1, *p. 494*

1. line segment **3.** complementary **5.** perpendicular **7.** parallel **9.** scalene **11.** parallelogram **13.**

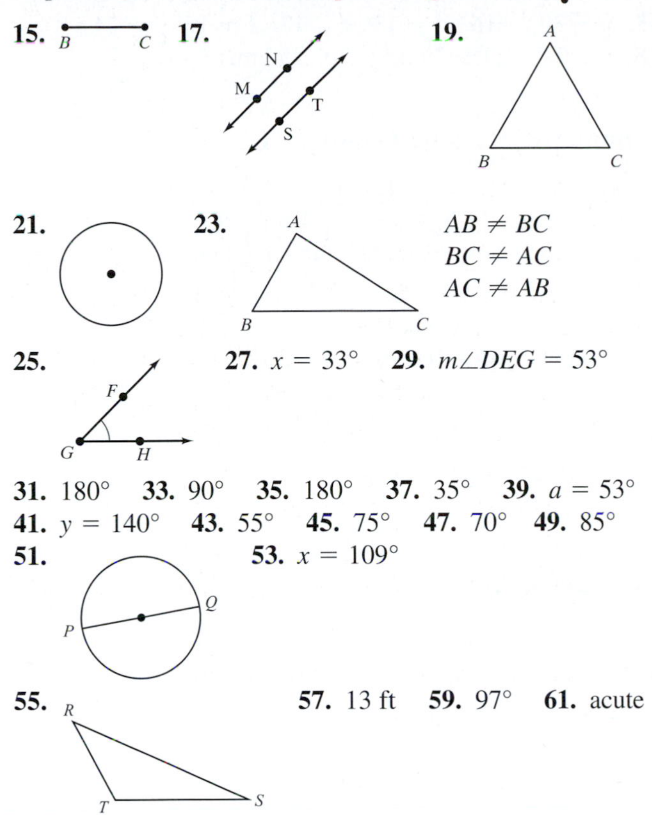

27. $x = 33°$ **29.** $m\angle DEG = 53°$

31. 180° **33.** 90° **35.** 180° **37.** 35° **39.** $a = 53°$ **41.** $y = 140°$ **43.** 55° **45.** 75° **47.** 70° **49.** 85° **53.** $x = 109°$

57. 13 ft **59.** 97° **61.** acute

63. a. 6 in. **b.** 60°

Practices: Section 11.2, *pp. 501–505*

1, *p. 501:* 26 in. **2, *p. 501:*** 3 mi **3, *p. 502:*** \$70 **4, *p. 503:*** Approximately 132 in. **5, *p. 503:*** Approximately 113 ft **6, *p. 504:*** Approximately 164.5 yd **7, *p. 505:*** 82 ft

Exercises 11.2, *p. 506*

1. perimeter. **3.** square **5.** circle **7.** 17 in **9.** 60 m **11.** $10\frac{1}{2}$ yd **13.** Approximately 62.8 m **15.** Approximately 21.98 ft **17.** 42 ft **19.** Approximately 25.12 in. **21.** 40 yd **23.** 21 yd **25.** 18 ft **27.** 19 cm **29.** Approximately 18.84 cm **31.** Approximately 11.12 m **33.** 64 yd **35.** 27 cm **37.** 28 in. **39.** 228 ft **41.** 6 in. **43.** 30 posts **45.** Approximately 50.57 ft **47. a.** Multiplication; **b.** 42,700 km; **c.** 42,000 km

Practices: Section 11.3, *pp. 513–517*

1, *p. 513:* 12 cm^2 **2, *p. 513:*** 12.96 cm^2 **3, *p. 514:*** 7.5 in^2 **4, *p. 514:*** $12\frac{1}{2}$ ft^2 **5, *p. 514:*** $3\frac{1}{2}$ ft^2 **6, *p. 515:*** Approximately 78.5 yd^2 **7, *p. 515:*** No. A customer needs 108 tiles at a total cost of \$538.92. **8, *p. 516:*** 79 mi^2 **9, *p. 516:*** 8.75 m^2 **10, *p. 517:*** 4,421.5 ft^2

Exercises 11.3, *p. 518*

1. area **3.** triangle **5.** circle **7.** 125 m^2 **9.** 30 ft^2 **11.** 290 yd^2 **13.** Approximately 706.5 cm^2 **15.** 32 m^2 **17.** 15.6 m^2 **19.** Approximately 314 in^2 **21.** 6.25 ft^2 **23.** 44.1 yd^2 **25.** 3.64 m^2 **27.** $\frac{1}{16}$ yd^2 **29.** 46 yd^2 **31.** Approximately 64.25 ft^2 **33.** 51.75 ft^2 **35.** Approximately 113.04 cm^2 **37.** 12 mi^2 **39.** 32.5 m^2 **41.** 160 ft^2 **43.** Approximately 0.05 mm^2 **45.** 5,000 ft^2 **47. a.** Multiplication and subtraction; **b.** 100 in^2; **c.** possible estimate: 100 in^2

Practices: Section 11.4, *pp. 524–527*

1, *p. 524:* 72 ft^3 **2, *p. 525:*** 3,375 cm^3 **3, *p. 525:*** Approximately 11 in^3 **4, *p. 525:*** Approximately 113 in^3 **5, *p. 526:*** Approximately 4.19 m^3 **6, *p. 526:*** The ball's volume is about 523 in^3, whereas the box's volume is 1,000 in^3. Therefore the ball occupies more than one-half the box's volume. **7, *p. 527:*** 784 in^3

Exercises 11.4, *p. 528*

1. volume **3.** rectangular solid **5.** cylinder **7.** 216 in^3 **9.** 2,560 m^3 **11.** Approximately 63 ft^3 **13.** Approximately 125.1 ft^3 **15.** Approximately 2,144 in^3 **17.** Approximately 1.95 m^3 **19.** Approximately 92.1 ft^3 **21.** 30,000 ft^3 **23.** Approximately 20.3 m^3 **25.** Approximately 33.5 in^3 **27.** 0.01 kg/cm^3 **29.** 4,187,000,000 mi^3 **31.** 46 in^3 **33.** 15.19 m^3 **35. a.** Multiplication and division; **b.** 2.9 g/cm^3; **c.** possible estimate: 3 g/cm^3

Practices: Section 11.5, *pp. 533–534*

1, *p. 533:* $\overline{AB}$ corresponds to $\overline{GH}$; $\overline{AC}$ corresponds to $\overline{GI}$; $\overline{BC}$ corresponds to $\overline{HI}$. **2, *p. 533:*** $y = 24$ **3, *p. 534:*** $h = 36$ ft

Exercises 11.5, *p. 535*

1. similar **3.** corresponding **5.** $x = 10\frac{2}{3}$ in. **7.** $x = 24$ m **9.** $x = 15$ ft, $y = 20$ ft **11.** $x = 24$ yd, $y = 12$ yd **13.** $x = 3.8125$ m **15.** $x = 12$ m **17.** $y = 1$ km **19.** 576 cm **21.** 200 m **23.** 24 ft

Practices: Section 11.6, *pp. 539–543*

1, *p. 539:* **a.** 7 **b.** 12 **2, *p. 540:*** Between 6 and 7 **3, *p. 540:*** **a.** 7.48 **b.** 3.46 **4, *p. 542:*** 10 in. **5, *p. 542:*** 3.5 ft **6, *p. 543:*** 34 ft

Exercises 11.6, *p. 544*

1. square root **3.** perfect square **5.** hypotenuse **7.** 3 **9.** 4 **11.** 9 **13.** 13 **15.** 20 **17.** 16 **19.** 7 and 8 **21.** 8 and 9 **23.** 6 and 7 **25.** 3 and 4 **27.** 2.2 **29.** 6.1 **31.** 11.8 **33.** 99 **35.** $a = 16$ cm **37.** $c = 3.2$ m **39. b.** 7 m **41. b.** 8 ft **43. c.** 20 m **45. c.** 11.4 cm **47. a.** 8.7 ft **49.** 10 yd **51.** 10 in. **53.** 9.8 km **55.** 16 ft **57.** 240 ft **59.** Approximately 37.4 ft

Review Exercises: Chapter 11, *p. 555*

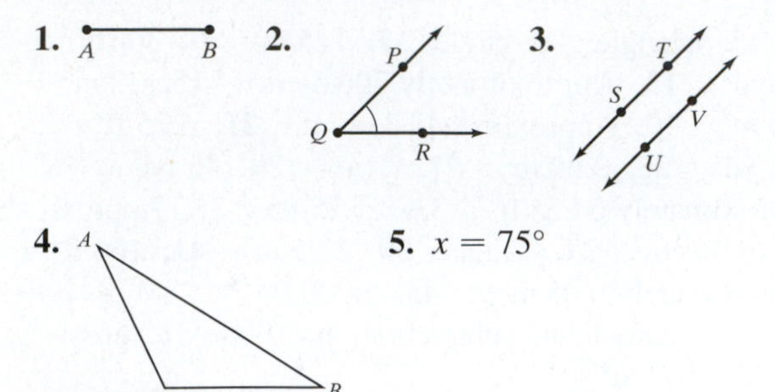

5. $x = 75°$

6. $x = 131°$ **7.** $x = 110°, y = 70°$ **8.** $x = 80°$; $y = 100°$ **9.** 5.4 m **10.** 30 ft **11.** 19 cm **12.** 125.6 in. **13.** 225 yd^2 **14.** Approximately 615 ft^2 **15.** 64 in^2 **16.** 17 m^2 **17.** Approximately 1,318.8 in^3 **18.** 216 ft^3 **19.** Approximately 1.95 m^3 **20.** Approximately 8.18 cm^3 **21.** Approximately 122.82 ft **22.** Approximately 84.78 ft^2 **23.** Approximately 17,850 ft^2 **24.** Approximately 3.81 yd^3 **25.** $x = 8$ ft, $y = 7.875$ ft **26.** $x = 4.5$ m **27.** 3 **28.** 8 **29.** 11 **30.** 30 **31.** 1 and 2 **32.** 9 and 10 **33.** 6 and 7 **34.** 3 and 4 **35.** 2.83 **36.** 35.14 **37.** 13.96 **38.** 5.39 **39. b.** 12 ft **40. a.** 10 in. **41. c.** 9.4 yd **42. c.** 2.8 ft **43.** 28,800 in^2 **44.** 13 mi^2 **45.** Yes, it can; the volume of the room is 2,650 ft^3. **46.** 220 in^2 **47.** 13 mi **48.** 10 ft **49.** $CD = 2\frac{1}{4}$ ft **50.** 7.5 mi **51.** 160 ft **52.** It is not efficient because the perimeter of the work triangle is 25 ft. **53.** 0.2 oz **54.** 1,347 cm^3 of soil **55.** 476.25 in^2 **56.** 2.6 in^3

Posttest: Chapter 11, *p. 560*

1. a. P, Q, R **b.** A, B

2. a. 7 **b.** 15 **3.** 65° **4.** 89° **5.** 14 ft **6.** 4.5 m **7.** Approximately 25 cm **8.** 15 ft **9.** 54 ft^2 **10.** 110 cm^2 **11.** 189 m^3 **12.** Approximately 33 ft^3 **13.** $a = 120°$; $b = 60°$; $y = 5$ m; $x = 10$ m **14.** $a = 69°$ **15.** $x = 80°$ **16.** $y = 13$ yd **17.** $x = 15$; $y = 24$ **18.** 8,100 ft^3 **19.** 79 m^2 **20.** 250 mi

Cumulative Review: Chapter 11, *p. 562*

1. 60 **2.** $\frac{4}{5}$ **3.** 3.1 **4.** 50 **5.** 15 **6.** $y = 8$ **7.** Approximately 707 mi^2 **8.** 1 mi **9.** $\frac{1}{51}$ **10.** Approximately 180,000 more complaints

Appendix

Practices: Appendix, *pp. 563–567*

1, *p. 563:* 25,390,000 **2, *p. 564:*** 8.0×10^{12}, or 8×10^{12} **3, *p. 565:*** 0.0000000043 **4, *p. 565:*** 7.1×10^{-11} **5, *p. 566:*** **a.** 2.464×10^2 **b.** 8.35×10^{11}

Exercises p. *567*

1. 4×10^8 **3.** 3.5×10^{-6} **5.** 3.1×10^{-10} **7.** 317,000,000 **9.** 0.000001 **11.** 0.00004013 **13.** 9×10^7 **15.** 2.075×10^{-4} **17.** 1.25×10^{10} **19.** 4×10^1

Glossary

The numbers in brackets following each glossary term represent the section that term is discussed in.

absolute value [7.1] The absolute value of a number is its distance from zero on the number line. The absolute value of a number is represented by the symbol $| \ |$.

acute angle [11.1] An acute angle is an angle whose measure is less than 90º.

acute triangle [11.1] An acute triangle is a triangle with three acute angles.

addends [1.2] In an addition problem, the numbers being added are called addends.

algebraic expression [4.1] An algebraic expression is an expression that combines variables, constants, and arithmetic operations.

amount (percent) [6.2] The amount is the result of taking the percent of the base.

angle [11.1] An angle consists of two rays that have a common endpoint.

area [11.3] The number of square units that a figure contains is called the area.

associative property of addition [1.2] The associative property of addition states that when adding three numbers, regrouping the addends gives the same sum.

associative property of multiplication [1.3] The associative property of multiplication states that when multiplying three numbers, regrouping the factors gives the same product.

average (or mean) [1.5] An average of a set of numbers is the sum of these numbers, divided by however many numbers are on the list.

bar graph [8.2] A bar graph is a graph in which quantities are represented by thin, parallel rectangles called bars. The length of each bar is proportional to the quantity that it represents.

base (exponent) [1.5] The base is the number that is a repeated factor when written with an exponent.

base (percent) [6.2] The base is the number that we take the percent of. It always follows the word "of" in the statement of a percent problem.

circle [11.1] A circle is a closed plane figure made up of points that are all the same distance from a fixed point called the center.

circle graph [8.2] A circle graph is a graph that resembles a pie, representing the whole amount that has been cut into slices representing the parts of the whole.

circumference [5.1, 11.2] The distance around a circle is called the circumference.

commission [6.3] Salespeople may work on commission instead of receiving a fixed salary. This means that the amount of money that they earn is a specified percent of the total sales for which they are responsible.

commutative property of addition [1.2] The commutative property of addition states that changing the order in which two numbers are added does not affect the sum.

commutative property of multiplication [1.3] The commutative property of multiplication states that changing the order in which two numbers are multiplied does not affect the product.

complementary angles [11.1] Complementary angles are two angles whose sum is 90º.

composite figure [11.2] A composite figure is the combination of two or more basic geometric figures.

composite number [2.1] A composite number is a whole number that has more than two factors.

constant [4.1] A constant is a known number.

corresponding sides [11.5] The sides opposite the equal angles in similar triangles are called the corresponding sides.

cube [11.4] A cube is a solid in which all six faces are squares.

cylinder [11.4] A cylinder is a solid in which the bases are circles and are perpendicular to the height.

decimal [3.1] A decimal is a number written with three parts: a whole number, the decimal point, and a fraction whose denominator is a power of 10.

decimal place [3.1] The decimal places are the places to the right of the decimal point.

denominator [2.2] The number below the fraction line in a fraction is called the denominator. It stands for the number of parts into which the whole is divided.

diameter [5.1, 11.1] A line segment that passes through the center of a circle and has both endpoints on the circle is called the diameter of the circle.

difference [1.2] The result of a subtraction problem is called the difference.

digits [1.1] Digits are the numbers 0, 1, 2, 3, 4, 5, 6, 7, 8, and 9.

discount [6.3] When buying or selling merchandise, the term "discount" refers to a reduction on the merchandise's original price.

distributive property [1.3] The distributive property states that multiplying a factor by the sum of two numbers gives us the same result as multiplying the factor by each of the two numbers and then adding.

dividend [1.4] In a division problem, the number into which another number is being divided is called the dividend.

divisor [1.4] In a division problem, the number that is being used to divide another number is called the divisor.

equation [4.2] An equation is a mathematical statement that two expressions are equal.

equilateral triangle [11.1] An equilateral triangle is a triangle with three sides equal in length.

equivalent fractions [2.2] Equivalent fractions are fractions that represent the same quantity.

evaluate [4.1] To evaluate an algebraic expression, substitute the given value for each variable and carry out the computation.

exponent (or power) [1.5] An exponent (or power) is a number that indicates how many times another number is used as a factor.

exponential form [1.5] Exponential form is a shorthand way of representing a repeated multiplication of the same factor.

factors [1.3] In a multiplication problem, the numbers being multiplied are called the factors.

formula [9.3] A formula is an equation that indicates how a number of variables are related to one another.

fraction [2.2] A fraction is any number that can be written in the form $\frac{a}{b}$, where a and b are whole numbers and b is nonzero.

fraction line [2.2] The fraction line separates the numerator from the denominator, and stands for "out of" or "divided by."

graph [8.2] A graph is a picture or diagram of the data in a table.

hexagon [11.1] A hexagon is a polygon with six sides and six angles.

histogram [8.2] A histogram is a graph of a frequency table.

hypotenuse [11.6] In a right triangle, the side opposite the right angle is called the hypotenuse.

identity property of addition [1.2] The identity property of addition states that the sum of a number and zero is the original number.

identity property of multiplication [1.3] The identity property of multiplication states that the product of any number and 1 is that number.

improper fraction [2.2] An improper fraction is a fraction greater than or equal to 1, that is, a fraction whose numerator is larger than or equal to its denominator.

integers [7.1] The integers are the numbers $\ldots, -4, -3, -2, -1, 0, +1, +2, +3, +4, \ldots$, continuing indefinitely in both directions.

intersecting lines [11.1] Two lines that cross are called intersecting lines.

isosceles triangle [11.1] An isosceles triangle is a triangle with two sides equal in length.

least common denominator (LCD) [2.2] The least common denominator (LCD) for two or more fractions is the least common multiple of their denominators.

least common multiple (LCM) [2.1] The least common multiple (LCM) of two or more whole numbers is the smallest nonzero whole number that is a multiple of each number.

legs [11.6] In a right triangle, the legs are the two sides that form the right angle.

like fractions [2.2] Like fractions are fractions with the same denominator.

like quantities [5.1] Like quantities are quantities that have the same unit.

like terms [9.2] Like terms are terms that have the same variables with the same exponents.

line [11.1] A line is a collection of points along a straight path, that extends endlessly in both directions. A line has only one dimension—its length.

line graph (broken-line graph) [8.2] A line graph (broken-line graph) is a graph in which quantities are represented as points connected by straight line segments. The height of any point on a line is read against the vertical axis.

line segment [11.1] A line segment is part of a line having two endpoints.

magic square [1.2] A magic square is a square array of numbers in which the sum of every row, column, and diagonal is the same number.

markup [6.3] The markup on an item is the difference between the selling price and the cost.

mean (average) [8.1] The mean of a set of numbers is the sum of those numbers divided by however many numbers are on the list.

median [8.1] Given a list of numbers arranged in numerical order, the median is the number in the middle. If there are two numbers in the middle, the median is the mean of the two middle numbers.

minuend [1.2] In a subtraction problem, the number that is being subtracted from is called the minuend.

mixed number [2.2] A mixed number is a number greater than 1 with a whole number part and a fractional part.

mode [8.1] The mode of a set of numbers is the number (or numbers) occurring most frequently in the set.

multiplication property of 0 [1.3] The multiplication property of 0 states that the product of any number and 0 is 0.

negative number [7.1] A negative number is a number less than 0.

numerator [2.2] The number above the fraction line in a fraction is called the numerator. It tells us how many parts of the whole the fraction contains.

obtuse angle [11.1] An obtuse angle is an angle whose measure is more than 90º and less than 180º.

obtuse triangle [11.1] An obtuse triangle is a triangle with one obtuse angle.

opposites [7.1] Two numbers that are the same distance from 0 on the number line but on opposite sides of 0 are called opposites.

parallel lines [11.1] Two lines in the same plane that do not intersect are called parallel lines.

parallelogram [11.1] A quadrilateral with both pairs of opposite sides parallel is called a parallelogram. Also, opposite sides are equal in length, and opposite angles have equal measures.

percent (or rate) [6.1] A percent is a ratio or fraction with denominator 100. A number written with the % sign means "divided by 100."

percent decrease [6.3] In a percent problem, if the quantity is decreasing, it is called a percent decrease.

percent increase [6.3] In a percent problem, if the quantity is increasing, it is called a percent increase.

perfect square [1.5, 11.6] A perfect square is a number that is the square of any whole number.

perimeter [11.2] The perimeter of a polygon is the distance around it.

period [1.1] A period is a group of three digits, which are separated by commas, when writing a large whole number in standard form.

perpendicular lines [11.1] Two lines that intersect to form right angles are called perpendicular lines.

pictograph [8.2] A pictograph is a variation of the bar graph in which images of people, books, coins, etc., are used to represent and to compare quantities.

place value [1.1] Each of the digits in a whole number in standard form has place value.

plane [11.1] A plane is a flat surface that extends endlessly in all directions.

point [11.1] An exact location in space, with no dimension, is called a point.

polygon [11.1] A closed plane figure made up of line segments is called a polygon.

positive number [7.1] A positive number is a number greater than 0.

prime factorization [2.1] Prime factorization is the process of writing a whole number as a product of its prime factors.

prime number [2.1] A prime number is a whole number that has exactly two different factors, itself and 1.

principal [6.3] The principal is the amount of money borrowed.

product [1.3] The result of a multiplication problem is called the product.

proper fraction [2.2] A proper fraction is a fraction less than 1, that is, a fraction whose numerator is smaller than its denominator.

proportion [5.2] A proportion is a statement that two ratios are equal.

Pythagorean theorem [11.6] The Pythagorean theorem states that for every right triangle, the sum of the squares of the two legs equals the square of the hypotenuse.

quadrilateral [11.1] A polygon with four sides is called a quadrilateral.

quotient [1.4] The result of a division problem is called the quotient.

radius [11.1] A line segment with one endpoint on the circle and the other at the center of a circle is called the radius of the circle.

range [8.1] The range of a set of numbers is the difference between the largest number and the smallest number in the set.

rate [5.1] A rate is a ratio of unlike quantities.

ratio [5.1] A ratio is a comparison of two quantities expressed as a quotient.

ray [11.1] A part of a line having only one endpoint is called a ray.

reciprocal [2.4] The reciprocal of the fraction $\frac{a}{b}$ is $\frac{b}{a}$.

rectangle [11.1] A rectangle is a parallelogram with four right angles.

rectangular solid [11.4] A rectangular solid is a solid in which all six faces are rectangles.

reduced to lowest terms (or simplest form) [2.2] A fraction is said to be reduced to lowest terms when the only common factor of its numerator and its denominator is 1.

right angle [11.1] A right angle is an angle whose measure is 90°.

right triangle [11.1] A right triangle is a triangle with one right angle.

rounding [1.1] Rounding is the process of approximating an exact answer by a number that ends in a given number of zeros.

scalene triangle [11.1] A triangle with no sides equal in length is called a scalene triangle.

signed number [7.1] A signed number is a number with a sign that is either positive or negative.

similar triangles [11.5] Similar triangles are triangles that have the same shape but not necessarily the same size.

simplest form (or reduced to lowest terms) [2.2] A fraction is said to be in simplest form when the only common factor of its numerator and its denominator is 1.

solution [9.1] A solution to an equation is a value of the variable that makes the equation a true statement.

solve [9.1] To solve an equation means to find all solutions of the equation.

sphere [11.4] A sphere is a three-dimensional figure made up of all points a given distance from the center.

square [11.1] A square is a rectangle with four sides equal in length.

square root [11.6] The (principal) square root of a number n, written $\sqrt{n}$, is the positive number whose square is n.

statistics [8.1] Statistics is the branch of mathematics that deals with ways of handling large quantities of information.

straight angle [11.1] A straight angle is an angle whose measure is 180°.

subtrahend [1.2] In a subtraction problem, the number that is being subtracted is called the subtrahend.

sum [1.2] The result of an addition problem is called the sum.

supplementary angles [11.1] Supplementary angles are angles whose sum is 180°.

table [8.2] A table is a rectangular display of data consisting of rows and columns.

tessellation [11.1] A tessellation is any repeating pattern of interlocking shapes.

trapezoid [11.1] A trapezoid is a quadrilateral with only one pair of opposite sides parallel.

triangle [11.1] A triangle is a polygon with three sides.

unit fraction [2.3] A fraction with 1 as the numerator is called a unit fraction.

unit price [5.1] The unit price is the price of one item, or one unit.

unit rate [5.1] A unit rate is a rate in which the number in the denominator is 1.

unlike fractions [2.2] Unlike fractions are fractions with different denominators.

unlike quantities [5.1] Unlike quantities are quantities that have different units.

unlike terms [9.2] Unlike terms are terms that do not have the same variables with the same exponents.

variable [4.1] A variable is a letter that represents an unknown number.

vertex [11.1] The common endpoint of an angle is called the vertex.

vertical angles [11.1] Two angles with equal measure formed by intersecting lines are called vertical angles.

volume [11.4] Volume is the number of cubic units required to fill a three-dimensional figure.

weighted average [8.1] A weighted average is a special kind of average (mean) used when some numbers in a set count more heavily than others.

Index

H

I

K

L

M

N

O

P